Contaminated Soils

Volume I

Contaminated Soils

Volume I

Remediation, Analysis
Federal and Regulatory Considerations
Site Assessment, Bioremediation
Risk Assessment
Radionuclides
Environmental Fate
Cleanup Standards

Edward J. Calabrese
Paul T. Kostecki
Marc Bonazountas

ASP

Amherst, Massachusetts

Amherst Scientific Publishers, 150 Fearing Street, Amherst, Massachusetts 01002

ASP recognizes that its business necessarily entails the use of a great deal of paper. We acknowledge a certain irony in working hard to protect the environment through the extensive use of one of the environment's most precious resources. ASP therefore participates in the U.S. Forest Service's Plant-A-Tree Program. This program arranges for the planting of trees in our 156 national forests. Trees are planted for multiple purposes, including recreation area development, wildlife habitat and commercial timber production. A portion of our proceeds is earmarked for this worthy program. Thank you for joining us in our efforts on behalf of the environment.

Printed in the United States of America

ISBN 1-884940-07-2

Preface

"A Retrospective Look at the Last Ten Years"

This year's 10th Annual conference on Contaminated Soils caps off a decade of tremendous growth within the environmental community. Phenomenal changes have been made on all fronts and now after a period characterized by explosive development in knowledge, technology and public awareness, a consolidation in our field and the number of practitioners is about to bring about drastic changes in the industry over the next two to five years.

Much progress has been made in the ten years since the first Contaminated Soils Conference was convened at the University of Massachusetts in 1986. Together, we have created several forums within which we exchange information and share lessons learned through our respective research efforts and applied on-site work.

The environmental industry, too, has evolved. We started out searching for easy solutions to site clean-up, but found no such "silver bullets." Instead, we learned that some sites can and have been cleaned up — those that have contained dissolved constituents of easily attenuated or biodegraded compounds such as petroleum hydrocarbons. Most others, we learned, have not been cleaned up and are not likely to be in our lifetimes. At these sites, the potential for cleanup depends not on the in situ technologies selected for the site, but rather on site conditions. And that's the challenge we face: The more complex the site's hydrogeology and chemistry, the more difficult it will be to meet cleanup standards. Chemical compounds that are strongly sorbed or present as a separate phase (floating or sinking), we have learned, raise the likelihood that cleanup to pristine conditions is not possible.

This new knowledge has led the regulatory community and the public to adopt more realistic attitudes toward site cleanups. Expectations have changed for the better: Interactions between the regulators and industry, the regulators and the public, and the industry and the public that were difficult and usually antagonistic have become much more constructive and win-win on all sides. More realistic approaches to site cleanups are being proposed, more realistic remedial management strategies that include "containment and control" measures rather than cleanup measures are being implemented. Everybody benefits.

Our cleanup issues have become site management issues, with the definition of clean being called into debate. How clean is "clean?" Defining what we mean by such terms as "clean" and "relative risk" and determining the economic impact of cleaning up a site to 10^{-6} risk levels are important issues being addressed on several fronts:

- The National Research Council commissioned a study in 1990 to examine the "how clean is clean" issue and compare the costs of cleaning up groundwater and other media, such as the air we breathe. The study also looked at the effectiveness and technical limitations of pump-and-treat and other methods, this effectiveness must be weighed against available financial and policy considerations. The committee for Groundwater Cleanup Alternatives examined these issues over the next three years, publishing its final report through the National Academy Press in 1994.

- "Better, cheaper, faster" has become the clean-up mantra of the 1990s. The U.S. E.P.A. and several agencies (notably the Department of Energy and the Department of Defense) drew up policies and guidelines that are intended to help streamline the cumbersome cleanup process known as "Superfund." In its Technical Impracticability Guidance of 1992, the U.S. EPA acknowledged that some sites cannot be cleaned up within reasonable timeframes and costs. This guidance outlines the conditions under which a Technical Impracticability Waiver can be issued, as well as the data and site information required. Several waivers have been issued since 1992.

- Several states have enacted legislation to promote more realistic clean-up plans. States are now beginning to adopt remedial management approaches that include soil removal and treatment, if appropriate; groundwater containment and/or control; wellhead treatment where and when needed; and basinwide groundwater resource management. Specific legislation that streamlines the clean-up process and makes allowances for alternative risk-based clean-up goals include the Massachusetts Contingency Plan, the California Draft Containment Zone Policy, and Michigans, Act 451, Part 201.

Several federal, state, and local jurisdictions are even encouraging Brown-fields redevelopment through policies that recognize the economic value and importance of redeveloping industrial or military sites. Not only do these strategies protect areas not yet developed, but they also allow the owners to turn liabilities into assets by returning contaminated property to active and productive use.

The Office of Management and Budget, required the USEPA to perform an economic analysis of what it would cost the nation to implement the draft Resource Conservation and Recovery Act Corrective Action Rule. The Rule is still in DRAFT form, as of this writing.

So what does all this mean, for the future? I believe our challenge over the next decade is to strike a balance — one that protects our environment, while still promoting our future. Sustainable development is indeed possible, and this is the decade during which we can build a solid foundation.

My grateful and sincere thanks to the conference directors for the Contaminated Soils conference held at the University of Massachusetts - Amherst for valuable and insightful interactions, and to the participants at this 10th Annual Conference. The ideas and experiences shared in this publication represent years of remarkable achievement. On to the next ten!

Lynne M. Preslo, R.G.
Senior Vice President, Technical Programs
EARTH TECH
Long Beach, CA

Acknowledgments

We wish to thank all the agencies, organizations and companies that sponsored the conference, without their generosity and assistance the conference and this book would not have been possible.

Benefactors

Association of American Railroads
Gelman Sciences
Massachusetts Department of Environmental Protection
Gas Research Institute
Naval Facilities Engineering Command
Naval Facilities Engineering Service Center

Sponsors

ABB Environmental Services, Inc.
American Petroleum Institute
EMCON
ENSR Consulting and Engineering
Environmental Standards
McLaren/Hart

Supporters

3M
Association for the Environmental Health of Soils (AEHS)
CER Corporation
Earth Tech
Environmental Science Services
New York Department of Environmental Conservation
Millipore Corporation
Northeast Utilities Service Company
Ogden Environmental and Energy Services
Remediation Technologies, Inc.
Roy F. Weston,Inc.
Texaco, Inc.
Tyree Organization, Ltd.
United Retek Corporation

In addition we express our deepest appreciation to the members of the Advisory Committees who volunteered their time to provide guidance and encouragement.

Scientific Advisory Boards

General

Bruce Amos	*Environmental Standards, Inc.*
Katherine Baker	*Environmental Microbiology*
Ralph Baker	*ENSR Consulting and Engineering, Inc.*
Christopher Barkan	*Association of American Railroads*
David Bass	*Groundwater Technology, Inc.*
Michael Battle	*EA Engineering, Science & Technology*

Bruce Bauman	*American Petroleum Institute*
Becky Bozadjian	*ABB Environmental Services, Inc.*
Cliff Bruell	*University of Massachusetts at Lowell*
John Button	*CER Corporation*
Leo Carden	*Roy F. Weston, Inc.*
Mark Chrisos	*Earth Tech*
Rob Clemens	*Ogden Environmental*
Steven Cohen	*Envirogen*
James Colman	*MA Department of Environmental Protection*
Michael Conway	*United Retek*
Richard DeNitto	*ABB Environmental Services, Inc.*
James Dragun	*The Dragun Corporation*
Mohamed Elnabarawy	*3M*
George Furst	*Massachusetts Turnpike Authority*
John Goodin	*Wastewater Technology Centre*
Annette Guiseppi-Elie	*Mobil Oil Corporation*
Robert Hazen	*New Jersey Department of Environmental Protection*
Wallace Hise	*Radian Corporation*
Michael Huesemann	*Battelle*
Paul Jacobsen	*Northeast Utilities*
Jay Johnson	*ABB Environmental Services, Inc.*
John Kubiczki	*EMCON*
Joshua Lieberman	*ENSR*
John Lynch	*RETEC*
Frank Peduto	*New York Department of Environmental Conservation*
Tom Potter	*University of Massachusetts at Amherst*
Lynn Preslo	*Earth Tech*
Thomas Renihan	*Marane Oil Corporation*
Eugene Rutz	*University of Cincinnati*
Steve Simkins	*University of Massachusetts at Amherst*
Kenneth Symms	*Environmental Standards, Inc.*
Patrick Vargo	*Tyree Organization, Ltd.*
Barbara Young	*Millipore Corporation*
Gil Zemansky	*Terracon Environmental, Inc.*

Federal Advisory Board

John Cullinane	*US Army Engineering Waters Experimental Station*
Joseph Graf	*Navy Facilities Engineering Command*
Lynn Kucharski	*McLaren/Hart (Chairperson)*
Katharine Kurtz	*Navy Facilities Health Center*
Paul Rakowski	*Navy Facilities Engineering Command*
Fred Ramsey	*ATSDR*
Debby Tremblay	*US EPA Headquarters*
David Warner	*ABB Federal*
Wade Weisman	*Armstrong Laboratory/OEMH*

Biographies

Edward J. Calabrese is a board certified toxicologist who is professor of toxicology at the University of Massachusetts School of Public Health, Amherst. Dr. Calabrese has researched extensively in the area of host factors affecting susceptibility to pollutants, and is author of more than 300 papers in scholarly journals, as well as 24 books, including: *Principles of Animal Extrapolation; Nutrition and Environmental Health,* Vols. 1 and 2: *Ecogenetic: Safe Drinking Water Act: Amendments, Regulations and Standards; Soils Contaminated by Petroleum: Environmental and Public Health Effects: Petroleum Contaminated Soils,* Vols. 1, 2 and 3; *Ozone Risk Communication and Management; Hydrocarbon Contaminated Soils,* Vols. 1, 2, 3, 4 and 5; *Hydrocarbon Contaminated Soils and Groundwater,* Vols. 1, 2, 3 and 4; *Multiple Chemical Interactions; Air Toxics and Risk Assessment; Alcohol Interactions with Drugs and Chemicals; Regulating Drinking Water Quality; Biological Effects of Low Level Exposures to Chemicals and Radiation; Contaminated Soils; Diesel Fuel Contamination; Risk Assessment and Environmental Fate Methodologies; Principles and Practices for Petroleum Contaminated Soils,* Vols. 1, 2, 3 and 4 and *Performing Ecological Risk Assessments.* He has been a member of the U.S. National Academy of Sciences and NATO Countries Safe Drinking Water Committees, and the Board of Scientific Counselors for the Agency for Toxic Substances and Disease Registry (ATSDR). Dr. Calabrese also serves as Director of the Northest Regional Environmental Public Health Center at the University of Massachusetts and Chairman of the BELLE Advisory Committee.

Paul T. Kostecki, Associate Director, Northeast Regional Environmental Public Health Center, School of Public Health, University of Massachusetts at Amherst, received his Ph.D. from the School of Natural Resources at the University of Michigan in 1980. He has been involved with human and ecological risk assessment and risk management research for the last 12 years. Dr. Kostecki has co-authored and co-edited over 50 articles and 16 books on environmental assessment and cleanup including: *Remedial Technologies for Leaking Underground Storage Tanks; Soils Contaminated by Petroleum Products; Petroleum Contaminated Soils, Vols. 1, 2 and 3; Hydrocarbon Contaminated Soils and Groundwater, Vols. 1 2, 3 & 4; Hydrocarbon Contaminated Soils, Vols. 1 ,2, 3, 4 &5; Principles and Practices for Petroleum Contaminated Soils; Principles and Practices for Diesel Contaminated Soils, Vols. 1, 2, 3, 4 &5; SESOIL in Environmental Fate and Risk Modeling* and *Risk Assessment and Environmental Fate Methodologies.* Dr. Kostecki also serves as Associate Editor for the *Journal of Soil Contamination,* Chairman of the Scientific Advisory Board for *Soil and Groundwater Cleanup Magazine* as well as an editorial board member for the journal of *Human and Ecological Risk Assessment.*

In addition, Dr. Kostecki serves as Executive Director for the Association for the Environmental Health of Soils (AEHS). He is a member of the Navy's National Hydrocarbon Test Site Advisory Board and a member of the Steering Committee for the Total Petroleum Hydrocarbon Criteria Working Group and the Association of American Railroads Environmental Engineering and Operations Subcommittee.

Marc Bonazountas, was born in 1945 in Athens, Greece. He is a civil and environmental scientist, specializing in mathematical modeling of all media and systems: air, soil, water, biota; and global environmental management procedures via advanced technologies as are geographic information systems (GIS) and remote sensing. Dr. Bonazountas, received his engineering degree from the National Technical University of Athens, Greece (1969); his doctorate degree from the National Technical University of Munich, Germany (1973); his diploma in computer sciences from the Date Processing Institute of Athens (1969); and undertook further studies in water resources systems at Massachusetts Institute of Technology (1975), and in public policy at Harvard University (1982). He obtained his Professorship, Habilitation (1984), from the National Technical University of Athens, where he is currently tenured at the professors staff of the Civil Engineering Department, engaged in teaching and research in environmental systems. Prior to joining the National Technical University of Athens, he was for almost ten years with Arthur D. Little, Inc., Cambridge, Massachusetts. He has been involved in numerous U.S. Environmental Protection Agency research and technology projects, has developed mathematical models such as SESOIL, and is currently engaged in environmental modeling and telematics related programs of the European Commission, Brussels. He has authored more than one hundred scientific publications, and is on the editorial board of several scientific journals: *Journal of Soil Contamination, Journal of Mathematical Environmental Modeling, Journal of Coastal Zone Management* and others. He is a member of the council for the Health and Environmental Safety of Soils, and a member of several European Commissions evaluation and planning committees.

Contents

REMEDIATION

CHAPTER 1

A Novel Approach to Minimize Remedial Soil Volumes Using Geostatistical and Linear Optimization Algorithms

K. Hoenke, Chevron Research and Technology, Richmond, California
A. Davis and **M. Keating**, Geomega, Boulder, Colorado
J. Wyatt, Chevron Chemical Company, San Ramon, California

INTRODUCTION

A major problem confronting both regulators and potentially responsible parties at contaminated sites is uncertainty in the volume of soil requiring remediation. This problem is exacerbated by different cleanup scenarios that result in different cleanup standards, and in regional variability in interpretation of the criteria by which a site is judged to have been effectively remediated.

The conventional decision-making based on removal of all soil with concentrations exceeding the target cleanup level generally results in treating excessive volumes of soil, resulting in overremediation and unnecessary expenditures. In general, problems associated with soil volume estimation are (1) inadequate modeling capability, (2) insufficient or inappropriately collected samples, (3) inaccurate or uncertain data analyses, and (4) the mistaken understanding that all soil exceeding the target cleanup level must be treated to satisfy the risk-based land use requirements.

Site soil cleanup goals are usually based on assumptions of future land use (residential, commercial, or industrial) and exposure pathways (dermal, inhalation, or ingestion), plus risk-based calculations that incorporate site-specific contaminant fate and transport, toxicological, and physical properties. Logically, air and groundwater contaminant concentrations at exposure points are the result of the average site condition at some downgradient receptor location (Figure 1). The computer model RISC (Remediation of Industrial Soil Constituents) was developed to better link this type of risk-based cleanup with the remediation process. The RISC model allows treatment of surface and subsurface soils in a similar fashion (i.e., on a site-wide average basis) when assessing compliance with a cleanup standard.

The use of the site-wide average approach, as an alternative to conventional "point-by-point compliance," introduces some flexibility into the remediation

Figure 1. **Conceptual paradigm describing interface between excavation and residual risk**

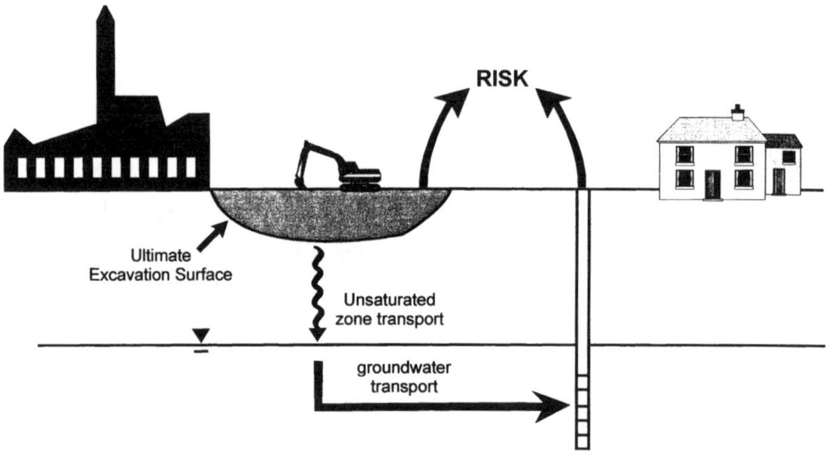

process. After remediation, there will be, by definition, areas on the site that are below the cleanup criteria , and some locations that remain above the criteria (warm spots). However, the net condition results in an average contaminant concentration below which the site must be cleaned to ensure satisfying the risk assumptions and protecting human health and the environment.

To accurately quantify soil volumes requiring remediation, we developed RISC and introduce its application in this chapter.

INTRODUCTION OF RISC

Previous efforts to describe cost minimization have focused primarily on better spatial resolution of compound distribution, using data collected at a sample density of 10 foot centers (Englund and Heravi, 1992). However, to date there has been no published method that interfaces risk assessment with soil remedial volumes from a three-dimensional perspective. Therefore, computer code, written in IDL (Interactive Data Language) was developed to optimize soil removal volumes (Figure 2). The model successively removes soil volumes, beginning with the highest concentrations and progressing to lower and lower concentrations, until the site-wide risk-based remediation (RBR) goal is reached.

In the conventional risk-based approach, cleanup targets are set for each chemical, and the entire site is cleaned up above the criteria set for the site. If the site is defined on a grid basis, then every unit has a concentration, the sum of which divided by the total number of cells generates the site-wide average (SWA) for the site. Typically, the post remediation SWA for a conventional cleanup is far less (sometimes an order of magnitude) than the risk-based

Figure 2. **Decision stucture to optimize soil cleanup**

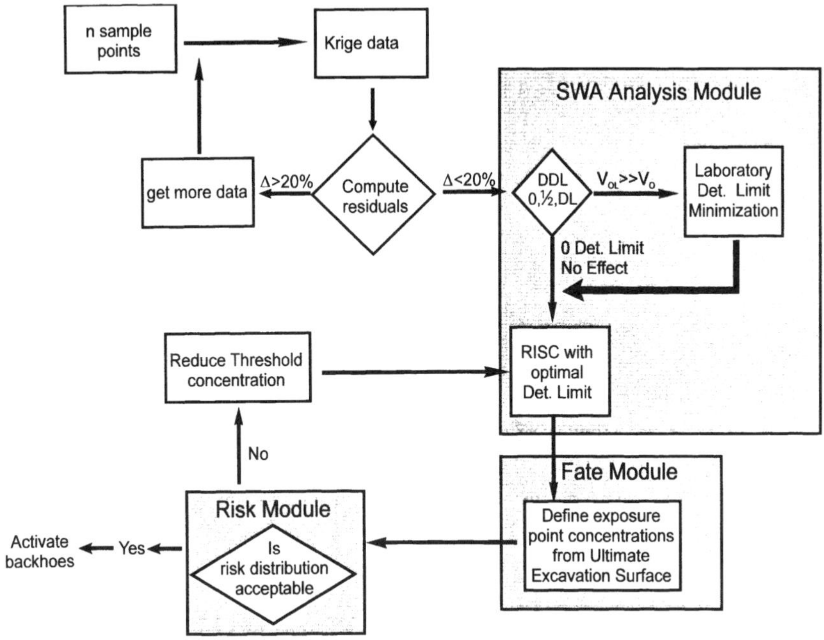

cleanup target concentration. The RISC model addresses this problem by optimizing soil cleanup volumes to reduce overremediation.

The first step in the RISC process is to determine the chemical drivers of the site. The risk-drivers are the few chemicals which present the most risk at the site and whose cleanup will encompass other, less important contaminants. To this end, RISC predicts which cells must be cleaned up for each chemical. Typically, the risk-drivers have a relatively larger proportion of cells requiring cleanup.

Statistical correlation is then used as a 'weight' to emphasize those chemicals which need to be remediated the most. For example, at one site, where θ represents the normalized weight attached to a given chemical, θ_{DDT} =.994, θ_{TOX} =.997 and $\theta_{ATRAZINE}$ =.17, showing that DDT and toxaphene will drive soil cleanup at the site. Since there is also a strong pairwise correlation (r^2>0.9 using 375 kriged data points) between DDT, toxaphene, and atrazine, it is reasonable to assume that remediating one of the driving chemicals will clean the other drivers in the soil.

The RISC program is based on a stack approach which orders the cells from the highest to lowest concentrations. A cell is prioritized for cleanup if one compound substantially exceeds the cleanup criteria, or if it contains many compounds that exceed the criteria by a smaller margin. Each cell is considered in creating the ordered list by evaluating each chemical, and the concentration normalized to prevent chemicals with very large values (that may not exceed the criteria by very much) from dominating over smaller ones. This strategy allows greater emphasis to be apportioned to those chemicals which require the most cleanup, thus allowing statistical delineation, and prioritization of hotspots.

Subsequently, the program peels cells from the stack until the RBR has been reached. RISC also takes into consideration the fact that each chemical as well as each level may have different cleanup criteria.

Delta Reduction

Definition of soil removal volumes is dependent on the adequacy of the available soil data. RISC has also been applied to identify specific locations where additional information will decrease uncertainty in the predicted soil Chemical of Concern (COC) concentrations, and hence increase predictive rigor. The value of RISC in optimizing the number and location of soil bores is to maximize information while simultaneously minimizing sampling and analytical costs.

Any spatial interpolation (i.e., by kriging) of a soil database has inherent error, often due to the paucity, and suboptimal location of sample points. The percent difference between a measured concentration, and the kriged estimate at that location (Δ) determines the model validity at any given point, so that the closer delta becomes to 0%, the more robust is the model variogram used to generate the kriged simulation of soil concentration at any given location.

Figure 3. **Comparison between the original sampled concentrations, and the co-located kriged estimate.**

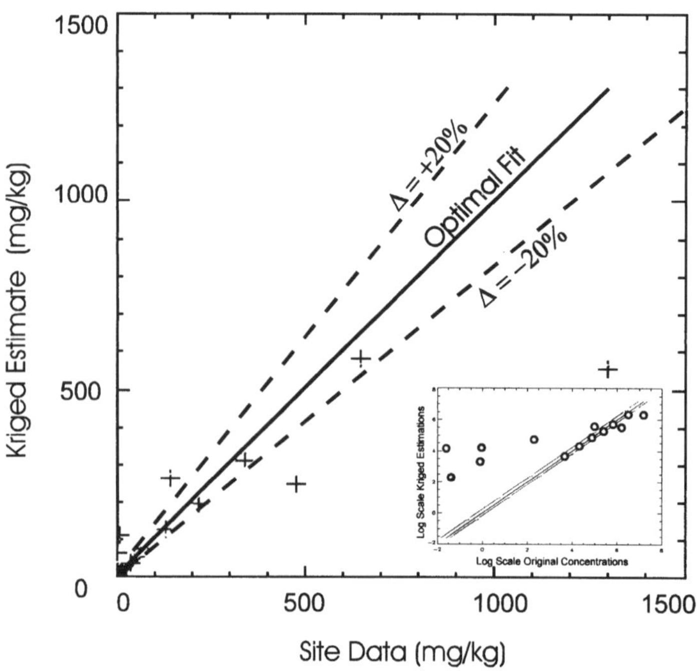

For example, where Δ exceeds a tolerance of +/-20%, additional data will probably result in a more robust interpretation, thus enhancing model accuracy (Figure 3). Those locations with the greatest uncertainty as well as isolated measurements drive the location of new sampling points. As new data are collected, they are incorporated into the database, a new variogram generated, the site soil population re-kriged, and Δ recomputed until an acceptable sitewide tolerance is achieved.

To locate sample points, and avoid oversampling at depth, delta must be computed throughout the soil profile using the approach described above to optimize sample location at depth. Currently, the kriging routine utilizes two-dimensional sets of measured sample points, computing each layer separately. RISC computes Δ isopleths, and then interpolates between equal Δ contours between layers to visually identify the locations of greatest uncertainty in the three-dimensional data set.

Effect of Implementing RBR Approach

Conceptually different remediation strategies result in substantially different soil volumes. For example, the remediation strategy of removing all soil which exceeds the cleanup level (Figure 4a) may be contrasted with the risk-based remediation strategy whereby the soil contaminant concentration attains a site wide average, resulting in a significant reduction in soil remediation volume (Figure 4b).

At one former pesticide formulation site, removing all soil above a residential 10^{-6} cleanup level for toxaphene (0.7 mg/kg) would require remediation of approximately 23,200 yards (Figure 5). Using the RISC model to implement a risk-based remediation to the same 10^{-6} risk level results in a reduction in soil removal of 10,300 yards, requiring remediation of only 12,900 yards (Figure 6). The difference between these remediation volumes is displayed in Figure 7. The RBR approach leaves a few residual "warm" spots, but results in savings of over $1 million compared to conventional "point-by-point compliance."

Minimization of Compound Detection Limits

During verification runs, it became apparent that soil detection limits, from the perspective of the site-wide background concentrations, ultimate excavation surface (UES; Figure 1), and replaced remediated material, have a substantial impact upon the RBR volumes requiring remediation, and therefore an effect upon the individual COC concentrations. Figure 8a depicts the effect of the Detection Limit (DL) on the RBR. In this example, all soil with concentrations above 0.7 mg/kg was removed, and the resulting RBR computed for three cases, i.e., compound DL=0.17 mg/kg, 0.085 mg/kg, and 0 mg/kg. The remediation volume does not change with variable DL, because all soil above the decision criterion (0.7 mg/kg) is remediated. However, if the DL can be decreased from

Figure 4a. **Conceptual decrease in the RBR concentration as the soil is remediated to the 0.7 ppm Clean Up Level.**

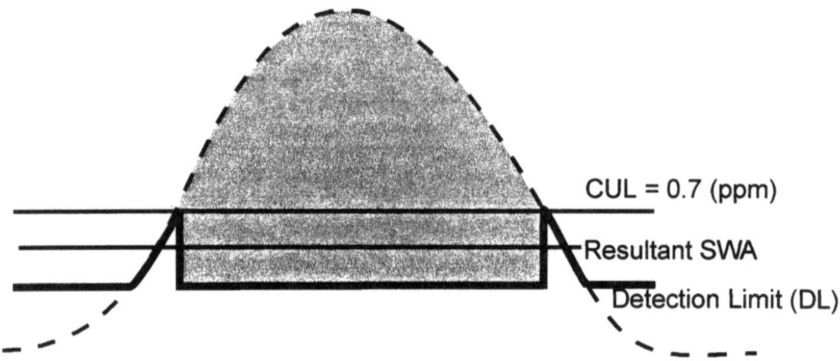

Figure 4b. **Conceptual difference between remediating the site on a point-wise as opposed to a risk-based remediation approach.**

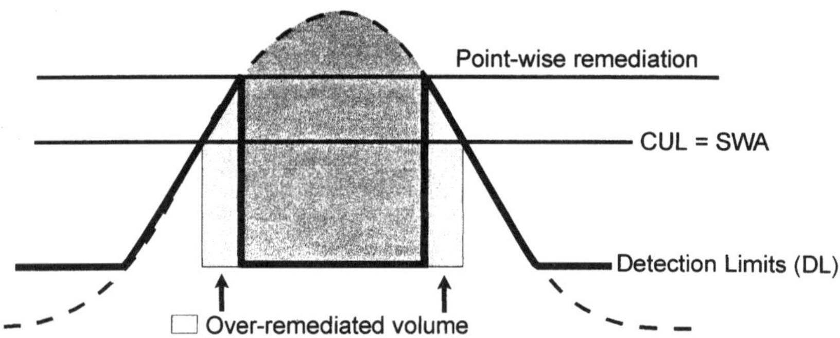

the full DL to ½ DL, the total analyte mass decreases on the site, the residual COC concentrations diminish, and correspondingly, the site-wide risk decreases.

Figure 8b shows that while maintaining a constant RBR goal, decreasing the detection limit results in a reduction in the soil volume to be remediated. This also has the effect of allowing an increase in the maximum concentrations remaining in the soil, while still satisfying the site-wide risk criteria (0.7 mg/kg).

Removing all soil above the cleanup level (0.7 mg/kg) results in remediation of approximately 23,200 yards (Figure 9). For comparison, an RBR-based remediation to the same risk level requires remediation of 12,900 yards at a DL of 0.17mg/kg. Further reductions (i.e., to 12,000 yards) may be achieved by halving the DL (Figure 9). Thus, for a typical remedial scenario for which soil remediation costs are $450/yard, halving the DL saves approximately $400,000.

Figure 5. **Remediation volume necessary assuming all soil above the action level treated.**

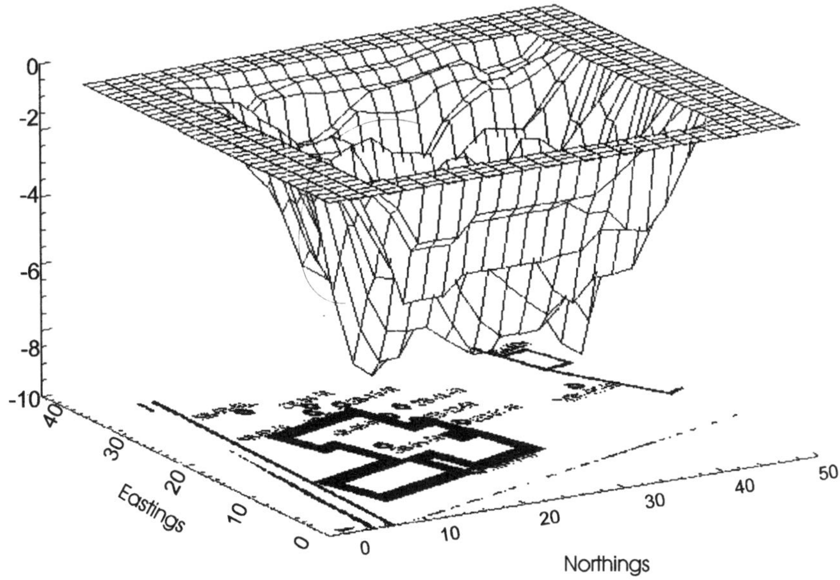

Total Remediation Volume = 23,000 yards

Figure 6. **Remediation volume necessary assuming that the site is remediated to the RBR goal.**

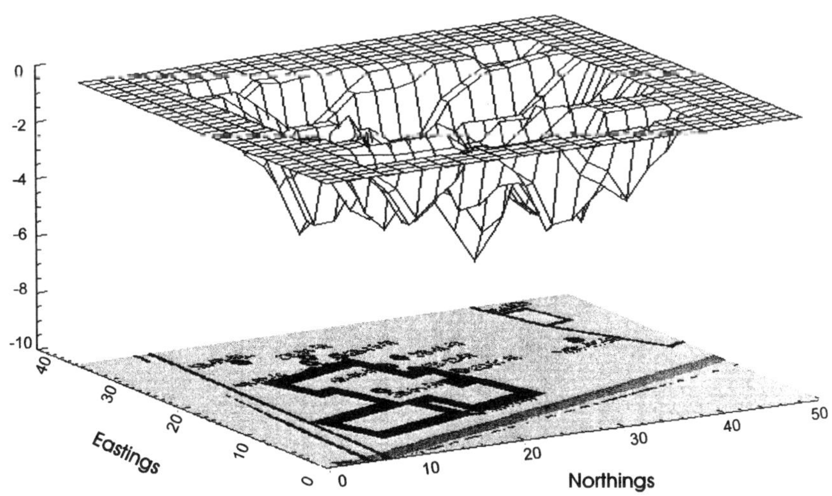

Total Remediation volume = 12,900 yards

Figure 7. **Comparison between risk-based remedial volume (solid projection) and value required assuming a point-wise analysis (hatched).**

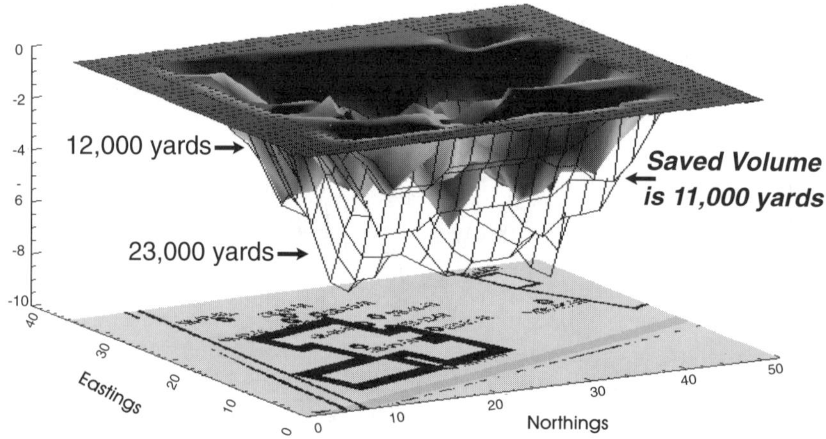

Figure 8a. **Demonstrates the effect of the DL on the RBA when the CUL is fixed at 0.7 ppm.**

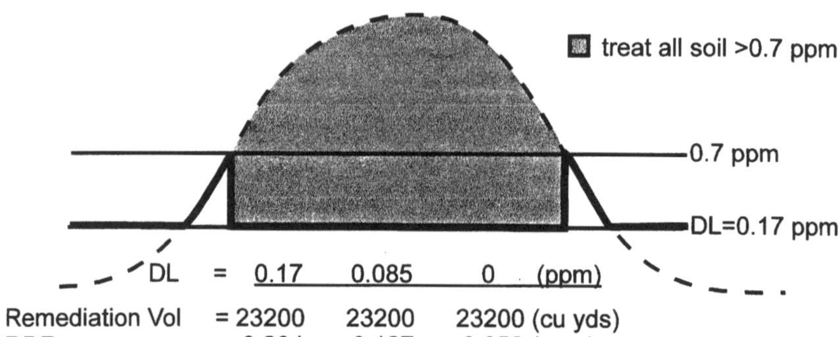

Figure 8b. **Demonstrates the effect of the DL on the risk-based remediation volume when the RBR CUL is fixed at 0.7 ppm.**

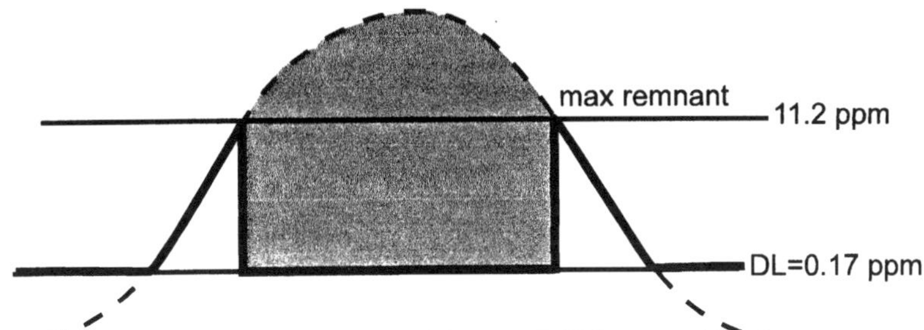

DL	=	0.17	0.085	0	
RBR Remediation Vol	=	12900	12000	11200	(cu yrds)
RBR	=	0.7	0.7	0.7	(ppm)
Max remnant concentration	=	11.2	13.1	13.6	(ppm)

Figure 9. **Graph of RBR volumes vs point-wise remediated volumes. The top line is the point value remediation volume, while lower lines are the RBR remediation volumes at different method detection limits.**

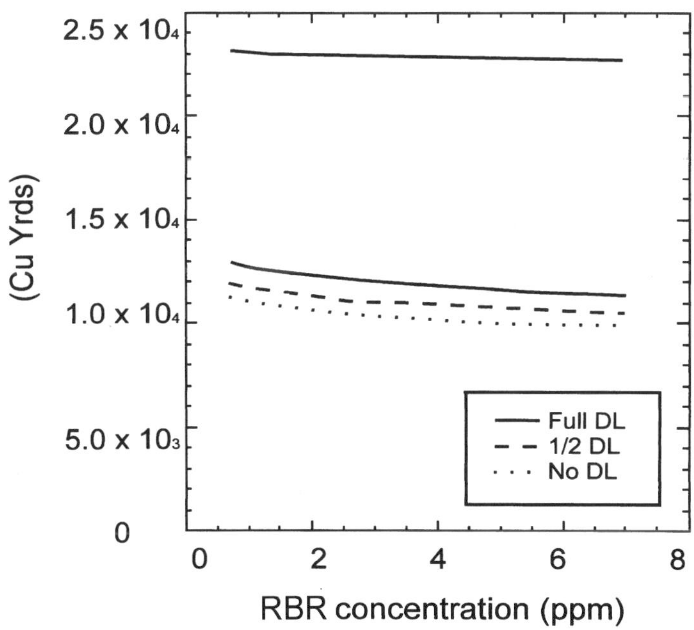

This example shows the importance of fully evaluating method detection limits during remediation and confirmation sampling.

REFERENCES

Englund, E. J. and Heravi, N. 1993. Conditional Simulation: Practical Application for Sampling Design Optimization. Geostatistics Troia '92. A Soares (Ed.) Klower Academic Publishers

CHAPTER 2

Remediation of Chlorinated Hydrocarbons in Vadose Clay by Radio Frequency Heating at the Department of Energy's Savannah River Site

Raymond S. Kasevich, Stephen L. Price, and **David L. Faust,** KAI Technologies, Inc., Portsmouth, New Hampshire

INTRODUCTION

Radio frequency heating (RFH) was successfully used in situ as an enhancement to soil vapor extraction (SVE) of chlorinated hydrocarbons from vadose clay deposits at the Department of Energy's Savannah River Site in Aiken, South Carolina. The chlorinated hydrocarbons consisted primarily of trichloroethylene (TCE) and tetrachloroethylene (PCE). This work was funded by the Office of Technology Management within the Department of Energy's (DOE's) Office of Environmental Management, under the Integrated Demonstration Program for Cleanup of VOCs in Non-Arid Soils. The team assembled for this demonstration included participants from the Westinghouse Savannah River Technology Center, the Westinghouse Science and Technology Center (Pittsburgh, PA), and KAI Technologies, Inc., which provided the RFH technology.

The demonstration featured the integration within a single horizontal well of both RFH and SVE. The horizontal well was screened and passed through a region containing a layer of contaminated clay. Offgases from the well were destroyed with a skid-mounted, thermal-catalytic oxidation system (CATOX) so that they could be vented directly to the atmosphere and be within permissible guidelines established by the state of South Carolina. For round-the-clock operations the RFH system employed a computer monitored 13.56 MHz generator.

During the demonstration, RF energy was coupled to the clay layer from the generator by a specially designed RFH applicator. Over the 25 days of operation the RF heating system delivered a total of 11,000 kilowatt-hours to the subsurface soil. This energy raised the temperature of approximately 27,000 cubic feet of soil above ambient (i.e., 20°C), and caused the nearest 800 cubic feet to exceed 60°C.

An analysis of soil data from this demonstration reveals that the RFH-enhanced SVE system mobilized the chlorinated solvent contaminants. This is exhibited by both a dramatic increase in the concentration of solvents within the treated zone (TCE by 150% to 215%, PCE by 375% to 450%), and by a fundamental change in the way they were distributed after the demonstration was completed. In all, an estimated 116.5 kg of PCE and 6.9 kg of TCE were extracted (based on quantitative analysis of the offgas vapor stream), and a further 20 to 93 kg of PCE and 6 to 13 kg of TCE were moved into the treatment zone. More solvents could have been removed if the extraction system ran for a longer period following the completion of RF heating, or if extraction had not been conducted from the same well, which allowed cool air to be brought from the unheated areas into the treatment zone.

This demonstration of RFH technology provides an understanding of the important components, system design, operating parameters, and site characterization needs necessary for the successful application of RF heating for in situ remediation.

PREPARATION

Prior to the field demonstration, both laboratory and field tests were performed to determine soil treatability. Surrogates of the targeted sediment were analyzed to characterize their dielectric properties as a function of both frequency and water content. These data were used as the input for programs which supported the design of the RF applicator and predicted the heating rates and temperature contours produced throughout the demonstration. This information was used to set the initial operating parameters of the field RFH system. Further testing was conducted onsite to verify the dielectric properties of the sediment and to confirm the design of the RF applicator. These measurements were critical to the control and optimization of the RF heating process, ensuring that the RF applicator would couple energy efficiently into the target contaminated soil under the moisture and temperature conditions present during the field demonstration.

RADIO FREQUENCY HEATING SYSTEM DESIGN

The Savannah River demonstration RF heating system consisted of a 25-kW RF generator operated at 13.56 MHz, 290 ft of transmission line, and an 18 ft, 2-1/4" diameter RF applicator designed to the specific requirements of the site. The RF heating system was controlled by a computer which sensed changes in the way the RF energy was coupled to the environment, determined the status of the RF generator and the soil vapor recovery system, and monitored other environmental sensors such as those which were used to measure the temperature of the borehole liner. The RF heating system was contained in an 8 ft by 8 ft by 20 ft equipment shelter on a 28 ft long mobile trailer. Power was provided by a 100-kW diesel generator.

The computer-controlled RF heating system enabled RFH to be conducted round-the-clock without the presence of onsite personnel. During RFH operations, the control computer sent periodic alphanumeric status messages to pagers that were carried by each of the two system operators. In the event of an alarm condition, this paging system would also be used to immediately inform the operators, so that corrective action could be undertaken. Additionally, a laptop computer, data modem, and cellular phone link allowed remote access to the control computer. An identical computer was used to analyze data collected by the control computer and also served as a backup.

RF energy was delivered to the soil by the RF applicator which received energy from the RF generator through a coaxial transmission line. The applicator was designed to efficiently couple RF energy into this particular soil by determining the best possible design, with the help of an electromagnetic computer model of the subsurface environment. Measured dielectric properties of the clay soil were used as inputs for this model.

A matching network was employed to tune the RF generator to the transmission line. The matching network allowed the RF heating system to be adjusted to the changing electromagnetic environment of the RF applicator caused by the dynamic process of heating the clay soil. In the later stages of heating, tuning became unnecessary, due to the significant volume of dry clay that formed in the immediate vicinity of the applicator.

The RF heating system was in compliance with OSHA safety guidelines for RF exposure and FCC guidelines for interference during the demonstration. This was verified by measurements made with an isotropic field probe and calibrated receiver.

THE RADIO FREQUENCY HEATING DEMONSTRATION

Figure 1 shows the integrated SVE and RFH systems, along with the placements of the piezometers and the fiber optic temperature probes. A clay layer up to 10 ft thick located at 40 ft below grade was the target for RF heating.

This demonstration used a single horizontal borehole to house the RF applicator and for SVE. The horizontal borehole was selected so that the RF energy could be targeted directly into the clay layer which held the contamination. To reach the targeted clay layer during RF heating operations, the applicator was inserted a distance of 250 ft into the borehole.

A fiber optic temperature system was employed to measure temperatures near the RF applicator. The use of fiber optic technology is ideal for RF applications because there are no metal parts. Fiber optic temperature probes were placed on the liner, applicator feed point, and transmission line. Monitoring of the borehole liner temperature during heating was important because the liner was constructed from fiberglass reinforced epoxy, which loses mechanical integrity above 130°C. To prevent this from occurring, the control computer

Figure 1. **Integrated RF Heating and SVE Systems (Reprinted by permission of Westinghouse SRTC)**

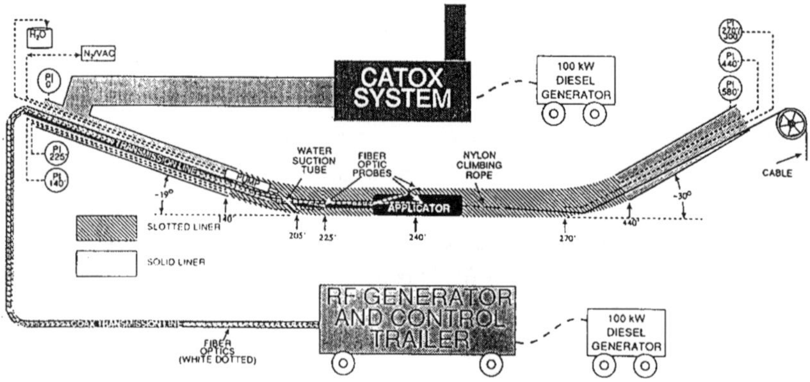

adjusted the duty cycle of the RF generator to maintain temperatures of 100°C to 130°C, resulting in 6 to 15 minutes of downtime each hour.

The vacuum and offgas treatment was performed using a combined vacuum extraction and thermal-catalytic destruction system. The system destroyed the chlorinated hydrocarbons with more than 95% efficiency while producing carbon dioxide and small amounts of hydrochloric acid.

The effectiveness of pretest studies was demonstrated by the measurement of a voltage-standing-wave ratio (VSWR) of 1.75 at the applicator feed point. Further tuning with the matching network presented a 1.05 VSWR to the generator, resulting in an energy transfer efficiency to the soil of more than 85%.

The maximum RF power output ranged from 10 to 25 kW. During the three weeks of this demonstration, a total of 11,675 kWh of RF energy was successfully coupled to the contaminated clay formation. This was calculated by measuring the power supplied by the diesel generator and by measuring all losses in the components of the RF heating system. It was determined that the generator used 21,200 kWh of AC power and converted 65% of it to RF power, 85% of which was delivered to the formation. The overall system efficiency from the diesel generator to the formation was, therefore, 55%. The energy delivered by the RF heating system resulted in the nearest 1,500 cubic feet of the formation surrounding the applicator being heated above 60°C, and with an observed heating rate of 0.06°C/hr at an approximate 6-ft radial distance from the center of the applicator.

Figure 2 shows the cumulative contaminants extracted in kilograms from the initiation of RF heating through cooldown. In all, 116.5 kg of PCE, 6.9 kg of TCE, and 48.1 kg of 1,1,1 trichloroethane (TCA) were extracted from the clay

Figure 2. **Cumulative Contaminates Extracted During RF Heating Demonstration (Reprinted by permission of Westinghouse SRTC)**

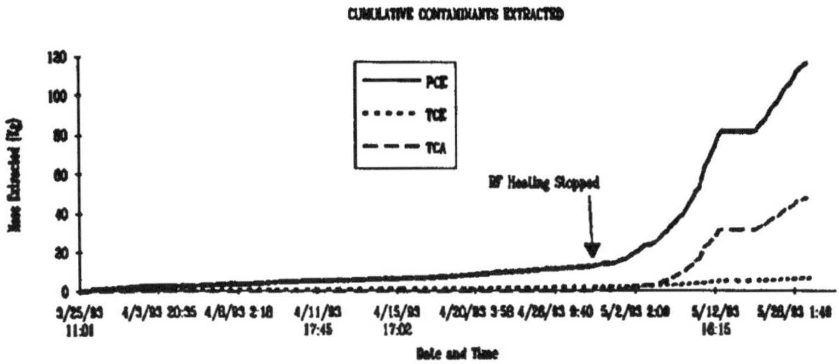

layer from the start of RF heating through the end of the three-week cooldown period.

REFERENCES

Jarosch T.R., Beleski R.J., and Faust D.L. 1993. "Final Report: In Situ Radio Frequency Heating Demonstration," Westinghouse Savannah River Company, Prepared under Department of Energy Contract No. DE-AC09-89SR18035

Kasevich R.S., Holmes R.E., Faust D.L., and Beleski, R.J. 1993. "Remediation of Hydrocarbon Contaminated Soils Using Radio Frequency Heating", Chapter 25, *Hydrocarbon Contaminated Soils*, Vol. III, Chelsea, MI, Lewis Publishers.

CHAPTER 3

Slantwell Sparging

William B. Kerfoot, K-V Associates, Inc., Falmouth, Massachusetts

INTRODUCTION

The introduction of air bubbles into a porous substratum saturated with water, whether surface soils or sand outwash deposits, presents rather novel problems of fluid flow through capillary-like structures. To define the dynamic conditions, the physical nature of soils and the characteristics of gases must first be understood.

Soil Porosity and Permeability

The word "soil" has different definitions, depending upon the discipline. It is generally agreed that soil is the upper portion of the loose material covering the surface of the earth. Soil is a mixture of mineral particles, organic material, air, and water. The common cross section of soil presents a tortuous path to movement of gas. Pore spaces may vary up to 50% of the total volume, but more commonly lie within the range of 10% to 40%. The nature of the maze of interconnecting pores and their moisture content collectively defines the capacity of gas to move (gas conductivity) through the soil.

Table 1. Estimated Porosity of Common Soils

Soil Type (rounded)	Porosity
Gravel	0.40
Coarse sand	0.38
Medium sand	0.33
Fine-medium sand	0.30
Silty sand	0.25

Source: Sevee, 1991.

The ability of gas to move through the soil is a function of soil porosity and permeability. The porosity of a soil is the measure of the interstitial space (void volume). Representative porosities for various substrates are presented in Table 1. Silty sands exhibit porosities between 15% and 20%. Most sands have a porosity between 25% and 38%. Gravels range from 38% to 45%.

Permeability is the measure of the ease of movement of a gas through the soil. The ability of a porous soil to pass any fluid, including gas, depends upon its internal resistance to flow, dictated largely by the forces of attraction, adhesion, cohesion, and viscosity. Because the ratio of surface area to porosity increases as particle size decreases, permeability is often related to particle size (Table 2).

Table 2. **Permeability of Glass Beads Compared with Permeability of Soil Fractions**

Mean Bead Diameter (mm)	Pore Space (microns)	Permeability (Darcy)	Gas Equivalent Conductivity (cm/sec)	Equivalent Soil Classification
2.000	860	1000	1.000	Very coarse sand
1.200	516	250	0.250	Coarse sand
0.655	281	147	0.147	Medium coarse sand
0.327	140	85	0.085	Medium sand
0.167	72	22	0.022	Fine-medium sand
0.083	36	9	0.009	Fine sand
0.041	18	5	0.005	Very fine sand

Source: Modified from Anderson et al., 1987.

An estimate of the permeability of a soil can be obtained by comparing its grain size in millimeters with glass beads of a similar size (see Table 2). This method is generally limited to uniformly graded sands, i.e., sands with a uniformity coefficient of less than 5.0. Permeability (k) is a function only of the soil medium, and is expressed as an area (cm^2).

Interstitial Gas: Velocity and Soil Conductivity

Gas is a fluid that, unlike water, is compressible. Vapor flow rates through porous material, such as soil, are affected by the material's porosity and permeability, as well as the viscosity, density, and pressure gradient of the gas. The movement of gas through soil can be approximated by Darcy's law. A simple formulation of Darcy's law for saturated gas flow in one dimension is:

$$\overline{V} = \frac{V}{n} = \frac{q}{An} = \frac{k\left(dP/dm\right)}{un}$$

where:

$\bar{V}$	=	seepage velocity (cm/sec)
V	=	gas yield $[cm^3)/(cm^2)(sec)]$
q	=	flow rate (cm^3/sec)
k	=	gas permeability (cm^2) (Darcies)
A	=	cross-sectional area (cm^2)
u	=	viscosity [g/(cm)(sec)]
dP/dm	=	pressure gradient $(g/cm)(sec^2)/cm$
n	=	specific porosity (i.e., void nonwetted volume)

The simplified Darcy equation can be used in conjunction with simple vadose-zone well tests to directly relate soil permeability to gas viscosity, flow rate, and pressure gradient. By using direct gas velocity and rearranging the Darcy equation to solve for gas permeability (k), the following equation is derived and compared with its groundwater equivalent (Masserman, 1989):

Gas flow: $$k = \frac{\bar{V}n}{dP/dm}$$

Groundwater equivalent: $$K = \frac{\bar{V}n}{dh/dl}$$

The slope (dh/dl) change in water head with change in distance (dl) is replaced by the pressure gradient (dP/dm) in the soil gas equivalent. The solution for k can be found for a known gradient and porosity. Effective porosity (n) remains unchanged, except that moisture content must be considered with gas movement. The viscosity of air (u) is estimated from Table 3.

Petroleum engineers have defined the Darcy as a unit of permeability. Technically, one Darcy is defined as the permeability that will lead to a specific discharge (V) of 1 cm/sec for a fluid with a viscosity of 1 centipoise under a pressure gradient that makes the term pg/u (dp/dl) equal to 1 atmosphere, where p

Table 3. **Viscosity of Air**

Temperature (°C)	Viscosity [g/(cm)(sec)][a]
0	0.000170
9	0.000176
18	0.000182
29	0.000186
40	0.000190

[a]The units are called poises.
Source: Handbook of Chemistry and Physics, 1972.

is density, u is viscosity, and g is the force of gravity. To convert Darcies to cm^2, multiply by 9.88×10^{-9}. To convert Darcies to gas conductivity in cm/sec, multiply by 9.11×10^{7}. To convert Darcies to cm/sec, divide by 10^3.

The differential equations that govern pressure flow of gas and vapor in soil are nonlinear, since gas density depends upon gas pressure. Masserman (1989) has pointed out, however, that if the maximum pressure difference between any two points in the flow field is less than approximately 0.5 atmospheres, the differential equations developed to model groundwater flow provide good approximations of gas flow. Analytical models used to evaluate ground-water flow can then be designed to estimate gas flow in sandy soils. The particular uniqueness of injection of bubbles into an aquifer (here a saturated sand deposit) is that bubbles rise upward due to the density difference with water.

AQUIFER SPARGING

The term "aquifer-sparging" refers to the injection of air into a saturated zone of an aquifer (water-bearing substratum) for the purpose of removing volatile organic substances. The production of bubbles through water has been used to transfer volatile gases from water to air. Bubbling air through oxygen-demanding organic waste has been used for many years in conventional waste-water treatment processes to provide oxygen for microbial activity and growth. Bubbling through water-saturated soils is a relatively recent process which offers two treatment advantages. When air is bubbled through an aquifer, it simultaneously strips volatile organics out of the groundwater and provides oxygen for microbial activity. The following text explains how to optimize bubble movement through the aquifer using a Spargepoint™ installation.

Comparison of Screen Slot Bubbles with Microporous Filter Bubbles

A standard slot of a 10-slot (.010) PVC monitoring well screen presents an opening of .010 inch by 2.0 inches for bubble production. The cross-sectional width of its bubble would be .010 inch, with a total cross-sectional area of .02 square inches. The interstitial area of the openings presented by medium sand (mean diameter of .4mm or .015 inch) would be $.26d^2 = .26\,(.015)^2 = .000058$ square inches. The bubble size area from the screen is 344 times greater than the mean orifice presented by the sand. The well screen bubble would have a width on its narrow side of .010 inch, compared to the pore triangular opening having a side of about .43d or .43(.015) = .006 inches. The bubble width would be about double the height of the triangular opening (see Figure 1).

It is preferable to match the air bubble size to the size of the soil particle interstitial space. This can be done by using porous plastic cylinders as bubblers instead of slotted well screens. With a 50 micron pore size, the bubbling tubes would create bubbles .05 mm in diameter, or about 2 slot (.002 inch) diameter. This is about 1/3 the cross section of the mean pore size (.006 inch). The area of the bubble would be r^2, or .0000125 square inch, about .22 or 1/5 the area of

Figure 1. **Pore size compared with air bubble size (50 micron) and a 10-slot well opening**

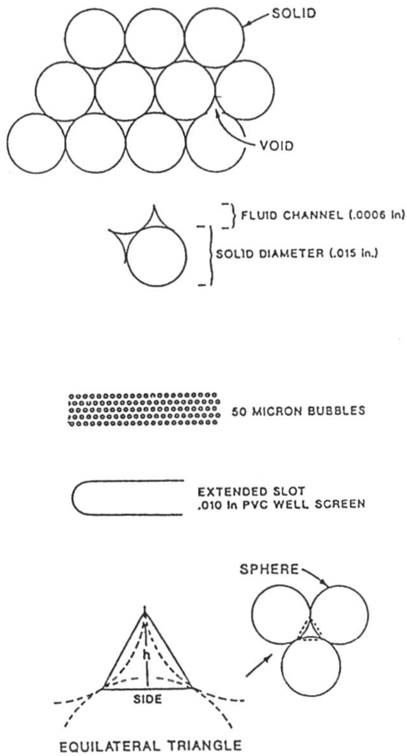

the pore opening. Bubbles this size would become part of the fluid (liquid) flow instead of an obstacle to it.

Choosing the Bubbling Equipment

The design of an in-situ sparge point aeration system should provide uniform coverage of the contaminated saturated soil zone, but not present a barrier to groundwater flow. The bubbling system consists of a prefilter, compressor (20 cfm, 15 psi), pressure relief valve, pressure gauge, flowmeter, manifold, individual point flowmeters, and injectable sparge points. The compressor unit is usually placed aboveground in a protected shed or building, while the main, wellhead flow controllers, tubing, and sparge points are below ground.

One of the simplest ways to avoid blocking groundwater flow is to alternate on/off bubbling periods (12 hours on, 12 hours off). Vadose zone vacuuming normally operates continuously with a soil vacuum system (SVE) set at a flow intake double the air injection rate. In this way, vacuum withdrawal is always greater than injection, eliminating the possibility of offsite gas migration.

Design of Areal Aeration Array (Points) Bubbling Diameter as a Function of Depth and Pressure

The most important design criterion of an air sparge system is the radius of bubbling. Three factors influence the dimensions:

- pressure
- depth of installation below static water
- fluid conductivity of soil

The creation of bubbling occurs when the gas pressure overcomes the hydraulic head (depth of water from static elevations to bottom of bubbler), the line friction, the membrane resistance of the bubbler wall, and the back pressure of the formation. The hydraulic head is converted to psi equivalents by multiplying depth of water by .43. The resistance of a 1/2 inch tube is negligible under ten feet. The membrance resistance of a 3/4 inch sparge point is roughly 2 psi. For a 10-foot installation of a 3/4 inch sparge point, the critical bubbling pressure would be the following:

Hydraulic head:	10 ft. x .43	= 4.3 psi
Line friction:	negligible	= 0.0 psi
Bubbler wall resistance:		= 2.0 psi
Critical bubbling pressure:		= 6.3 psi

The most crucial pressure to overcome is the formation back pressure, which varies with the surface-to-volume relationship of the pore spaces and the extent of their occlusion by fines. For a rough approximation, previous field tests have shown the following ranges:

Gravel	.2 to 3 psi
Coarse sand	.3 to 4 psi
Medium sand	.5 to 6 psi
Fine sand	1.0 to 10 psi
Silty sands	3.0 to 30 psi

There is some disagreement on the relationship between the depth of aeration and the diameter of the zone of aeration. Based upon deicing of waste lagoons by bubbling, Rothmann (1983) found that a ratio of about 1:1 exists between the depth of aeration and the diameter of the zone of aeration. Based on this logic, Repa and Kufs (1985) assumed that an aeration well 10 feet into pure water would give a radius of aeration of about 10 feet.

Angell (1991), on the other hand, found radii of influence of 78 feet for installed sparge points. The actual depth below static water was not presented, although only 10, 15, and 20 psi pressures were used. Assuming that the necessary pressure to bubbles must balance hydrostatic pressure (i.e., 43 psi equals 100 feet of water head), the screen depth was probably 23 feet below static water.

Based on lagoon studies, Angell's reported radius was more than 3.3 times the expected bubbling diameter. The relationship between diameter, pressure, and depth needed further examination.

Field Tests

To compare the diameter of bubble zone with depth, a simple test was devised (Kerfoot, 1994). The bubbler (KVA Spargepoint™) was injected at different depths into medium-coarse sand just below surface water on a sandy beach. The bubble radius could be observed and compared with the measured hydrostatic pressure rise in a set of installed minipiezometers.

An 18-inch long, 3/4-inch diameter porous cylinder with 50 micron pores (KVA Model 14-287) was placed 5 feet below the static water in a medium sand deposit at the edge of a lake (Ashumet Pond, Cape Cod, MA). An air pressure of 25 psi was applied and the flow measured by a gas flow meter (Dwyer Model RMB). A flow of about 2.0 scfm was observed.

The bubble distribution grew from 2.8 feet to 7.5 feet over a period of 20 minutes. At stabilization, a quadrat (90° slice) contained over 40 bubble points roughly 8 inches (.66 ft) apart. The spot of emergence appeared as a small pile of white sand, contrasted against a thin dark lake sediment background.

To compare the water head pressure distribution with the bubbling zone, a series of small point piezometers was placed one foot below static water along the shoreline. Although bubbling was found to achieve only a 9-foot radius, the noticeable rise in water head extended to a 14-foot radius.

At a second location, an 18-inch long by 2-inch diameter porous bubbler (KVA Model 14-278) with 10 micron pores was placed 10 feet below static water in the same medium sand deposit. A compressor system supplied 20 cfm at 15 psi for the test. After 30 minutes, fine bubbles were found at 10-foot distances from the sparge well. After 60 minutes, volatile organic release and fine bubbles were found at a 42-foot radius. A 50-foot radius of bubbling was confirmed after two hours (120 minutes) of continual operation. *Note:* The zone of hydrostatic influence has always been observed to extend further out than the bubble zone.

Depth of Bubbler

The relationship was observed between depth of the bubbler and radius of the bubble zone with air pressure set to only 10 psi above critical bubbling pressure with a 3/4-inch diameter sparge point (Kerfoot, 1993). The diameter observed for bubbling was noticeably less than the measured zone of influence of the displaced water. At 10 feet below static water, a 10-foot pressure radius was observed at the top of the water when operated at critical bubbling. The radius of the observed bubble zone fit closely the relationship predicted by Repa and Kufs (1985).

At a fixed pressure set at 10 psi above critical bubbling, the radius expands linearly (directly proportional) to increasing depth.

Pressure Influence

A second test was conducted on an 18-inch Spargepoint™ located 5 feet below static water. Pressure was increased in increments (5.0 psi) well above critical bubbling pressure (see Figure 2). Although the bubbling zone radius was 5 feet at critical bubbling pressure, it expanded with increased pressure to approximately the square root of the pressure increase:

$$r = \frac{\text{Pressure}}{\text{Critical Pressure}} \wedge 0.5 \ dc$$

where: r = radius of bubbling zone
dc = depth of installation

Pressure Versus Flow

As pressure increases, the gas flow to the Spargepoint™ also increases (Figure 2). For comparison, the gas yield (flow) was measured with the bubbler in air, the main resistance being through the porous sidewalls of the cylinder. The sparge point was also placed in medium sand with less than one foot of water head. The same pressure was applied.

If the critical bubbling pressure is subtracted, the sand and water curve will show the expected flow in medium sand. For example, at a 10-foot depth (critical bubbling = 7.8 psi) and 15 psi pressure, about 1.2 cfm would be expected.

If a fine porous diffuser (10 micron) is used with a highly permeable deposit (medium sand, 100 ft/day hydraulic conductivity), the resistance to flow may be so low that a shallow curve of pressure versus flow occurs. If so, assume that the radius of bubbling will increase by the square root of 2 (i.e., 1.4) times each time the flow volume is doubled.

Degree of Overlap of Bubble Zones

It is important to achieve overlap of the zones of aeration. To achieve 30% overlap, the distance between aeration zones should be set at 3/4 d^b (d^b = the diameter of bubble zone) (see Figure 3).

Installation of an air sparge/SVE system requires that the locations of sparge points take into account the components of the SVE system. Consideration should be given to:

1. The radius of influence of the vacuum extraction system.
2. The radius of bubbling of the air sparging system.
3. The presence of any subsurface barriers (vaporcludes) which may affect air flow.

Figure 2. **Influence of depth and pressure on radius of bubble zone**

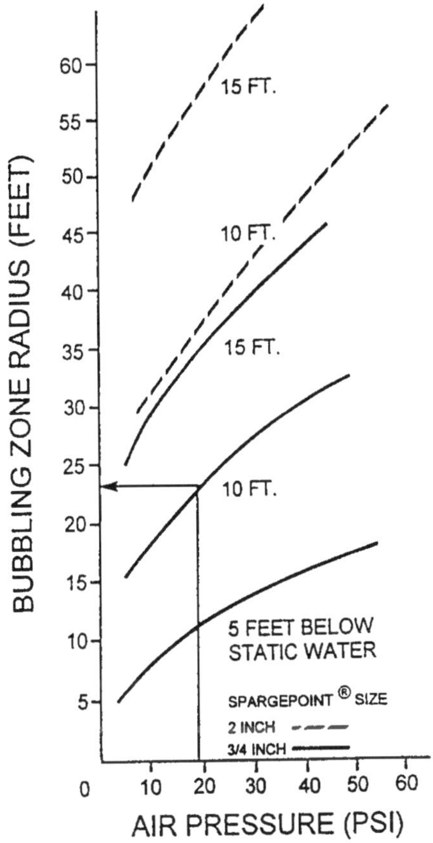

Borings should be done to characterize the soil conditions before placement of a sparge point in a field pilot test. It is important to note that the zone of influence of air injection (bubble zone) should not be confused with the diameter of hydraulic influence. Injected air creates hydrostatic pressure outward, even though bubbles may not be present. The placement of an x-y array (perpendicular arrangement) of point piezometers just below the water table allows detection of hydrostatic pressure rise (head rise) and sampling of air bubbles to confirm diameter.

Table 4 presents a summary of guidelines for predicting the area of influence of the diffuser.

GASEOUS EXCHANGE

If gaseous exchange is proportional to available surface area, with partial pressures and mixtures of volatile gases being held constant, a halving of the radius of bubbles would quadruple (i.e., 4x) the exchange rate. If, in the best case, a

Figure 3. **Overlap of bubble zones to allow even distribution of bubbles**

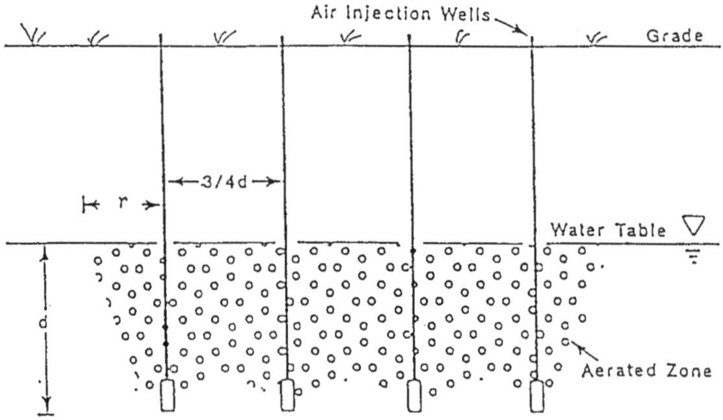

standard well screen creates air bubbles the size of a medium sand porosity, a microporous diffuser of 20-micron size creates a bubble 1/10 the diameter and 10 times the surface/volume ratio.

Diameter	Surface Area	Volume	Surface Area/Volume
(microns)	$4\pi r^2$	$4/3\pi r^3$	
200	125600	4186666	.03
20	1256	4186	.3

Theoretically, the microporous bubble exhibits an exchange rate of 10 times the rate of a comparable bubble from a standard 10-slot well screen.

SLANT ANGLE POSITIONING

Instead of arranging sparge points at a fixed depth, the boring can be angled to allow multiple spacings. This is different from directional horizontal drilling, in that the hole remains straight, but is inclined. First employed to allow sparging under buildings, the concept has far wider application, particularly to adjusting bubble density to contaminant concentration in dispersing plumes, and in the construction of bubble fences.

Concept of Effective Envelope

In order to understand the basis for three-dimensional bubble dispersion, the following must be visualized. Sources with identical flow placed at varying depths will eject bubbles outward in a parabolic manner. As the bubbles rise, they will first disperse, the distance between bubbles gradually broadening. The bubbles

Table 4. **Design Criteria for a Microporous Sparging System**

Air Sparging Design - Using microporous plastic sparge points critical bubbling pressure (pressure to initiate bubbling) is defined as:

Pc (psi) = [0.43 X Depth below Water (ft)] + 3.5 (psi)

The diameter of the bubbling zone produced by the sparge point when supplied with the critical bubbling pressure is equal to the installation depth of the sparge point below the static groundwater surface:

Dc (ft) = Installation Depth below Water (ft)

The increase in the radius of the bubbling zone produced by the sparge point when supplied with greater than the critical bubbling pressure is defined as:

R = [(Pressure / Pc)^0.5] X Dc

Input	Sparge point depth below static groundwater level	10.0 (ft)
	Critical bubbling pressure is calculated as	6.0 (psi)
	Critical bubbling radius is calculated as	20.0 (ft)
Input	Input proposed delivery pressure to sparge points	12 (psi)
	Bubbling zone radius based on input pressure 12.0 and volume (20 scfm)	30 (ft)
	Recommended horizontal spacing between sparge points	22.0 (ft)
	Estimated air flow thru Spargepoint based on input pressure	20 (cfm)

Correction for vertical/horizontal (V/H) permeability if ratio of V/H is:

1:1 multiply R by 1
1:10 multiply R by 1.5
1:100 multiply R by 2.0

To see a graph that relates varying bubbling zone diameters vs. input pressures above and below your selected input pressure for your chosen installation depth, go to the back of the spreadsheet and print the graph BUBBLE-DIAMETER.

will enlarge their diameters as the pressure decreases toward the water table. Once the size of the bubble enlarges to substantially greater than the pore size of the stratum, the bubble no longer acts as a fluid passing through the pores, but tends to form channels. The channeling may allow faster rise of the bubbles to the surface, but interferes with their ability to diffuse through the stratum. In the upper regions of a bubble zone, even though the diameter of the zone has enlarged, the effectiveness of gas transfer has substantially diminished.

For example, a diffuser placed at 100 feet below static water releases 50-micron bubbles into a medium-sand aquifer under a pressure of 60 psi. The diameter of the bubble increases to the cube root of the pressure decrease. By the time the bubbles have ascended to 50 feet, they are 1.2 times larger. At 25 feet, they are 1.4 times larger. At 10 feet, the bubbles have expanded to 2.2 times larger. The optimal diffusion zone ranges from 100 to about 25 feet, assuming no coalescing with other bubbles occurs.

In three dimensions, there exists an effective envelope (or zone) for optimal gas transfer. This region can simply be referred to as the region where the bubble diameter is maintained at less than mean pore size. Beyond this range the vertical transport becomes dominant, transferring the contents of the bubble to the surface, but the gaseous exchange with the soil/water matrix is substantially reduced. Counter to what one might initially assume, placing sources well below a plume region to allow for a greater bubble diameter could be counterproductive if, when the bubbles make contact with the contaminated soil/water region, the bubbles are in their ascension phase rather than diffusive phase.

In the case of a contaminated plume of water which is being depressed with rainwater recharge and whose leading front is dispersing outward, the slant angle drilling can be very efficient. The shallow emplacement produces a concentrated density of bubbles where the contaminant source is strongest. Downgradient, the deeper sparge-point diffusor creates a broader zone to correspond to the broader, but less concentrated, plume region (Figure 4).

Multiple Installations

One of the most intriguing applications of slant angle emplacements is to create deep "bubble fences." By offsetting parallel installations such that the distance from the second to the first is less than the horizontal surface-projected length of the first (i.e., they are overlapped), a bubble zone is created laterally. The degree of overlap vertically can adjust the total bubble density per surface area.

One interesting effect of the overlapping is likely to be the creation of a centralized region of intense stripping of volatile organics (Figure 5). Finer bubbles, as they start to expand, may intersect larger bubbles with higher vertical velocities, aiding in transport to the surface. The vertical location of sparge points can be modified to concentrate the bubbles' effective envelopes in the vicinity of a

Figure 4. **Density changes of bubbles in three-dimensions with a slant-well installation beneath a plume.**

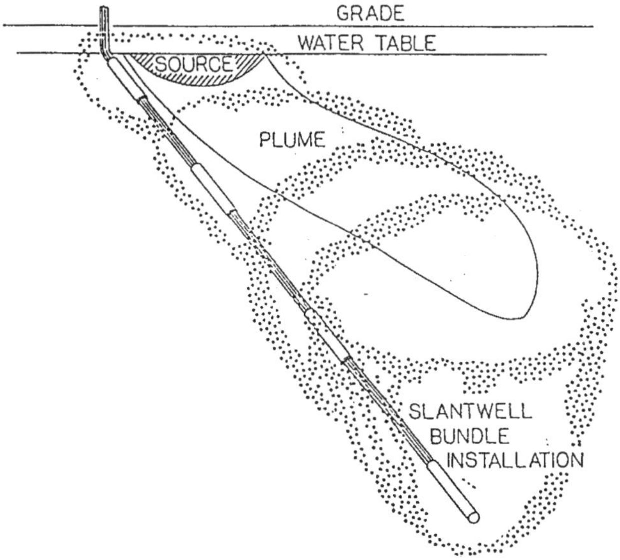

known plume region, instead of an even distribution of sparge points throughout the groundwater column. A logarithmic distribution (toward the static water zone) may be more appropriate for a shallow plume region.

INSTALLATION METHODS

Pass-Through Diffusers

With porous cylinders which have flow-through connections, up to three bubblers can be connected with 1/2- to 1-inch PVC pipe. The sections are lowered into a hollow stem auger similarly to a monitoring well. For 1-inch or less diameter sparge points, a 4-inch hollow stem auger is sufficient. To avoid sand heave, water is usually added to the hollow stem up to slightly above static water. Distilled water may be placed in the sparge point array to avoid reflux of contaminants during flooding. A sand pack equivalent in size to the mean particle size of the formation is desirable. Sand packing to 2 feet above the screens allows a little margin of error for a bentonite pellet or grout seal of about 2 feet in depth above the sand pack. A drop weight may be used to check the sand pack before pellet release.

Figure 5. **Deepslant well installations to create a bubble fence using equal spacing of ten diffusers.**

The expansion of bubble zones, viewed from above is shown in (A). The side view is presented in (B)

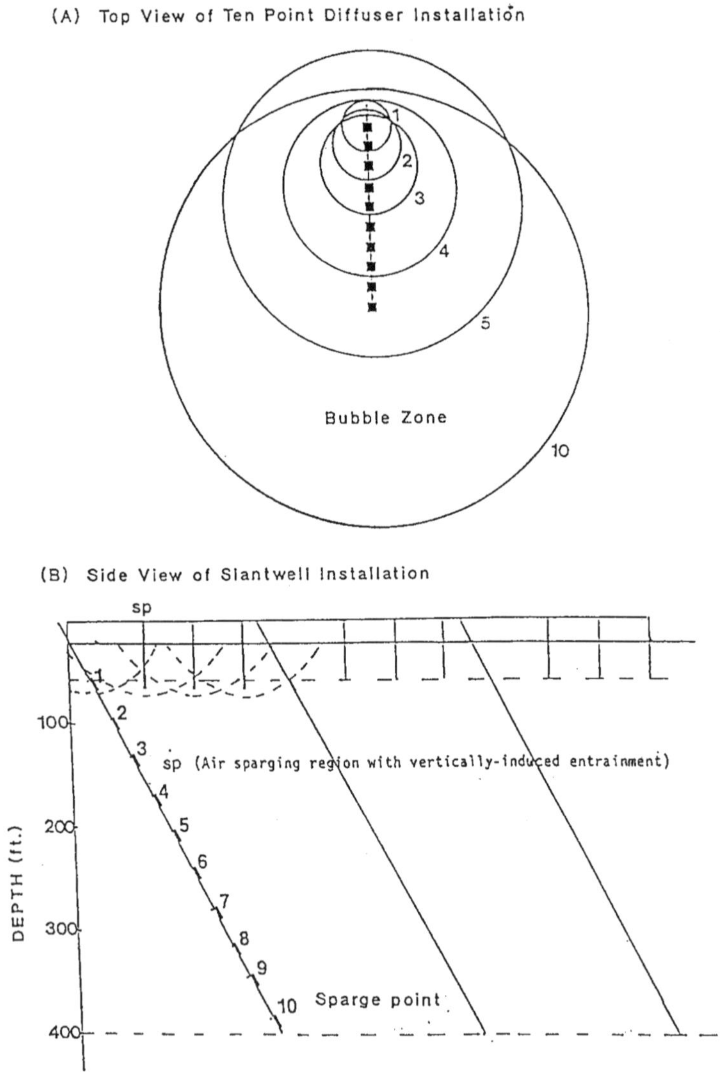

Multiple Bundles

A recent alternative for deep emplacements involves bundling the sparge points with individual connecting tubes. The tubing is run across the outside of the sparge points and radially twirled around the other rising tubes. With the use of a 6-inch hollow stem auger, the farthest point can be attached to steel pipe, and the remaining tubing spirally wrapped around it for support. Each tube leads to a separate wellhead control assembly at the surface completion.

RANDOMIZATION OF BUBBLE PATHWAYS

Entrainment of bubble pathways often occurs with continuous emission of air from diffusors. To improve contact with soil and water, it is desirable to operate in a manner which forces randomizing of pathways. Two methods have been very effective in (1) zone sparging, where the bubblers are turned on and off in cycles, and (2) the use of pumps to agitate the water or counter-circulate in timed sequences alternating with bubbling. In both cases, the bubble pathways are disrupted and the water mass is shifted.

CONCLUSIONS

A descending series of sparge points can be spaced along a slantwell installation to produce unique bubble distributions. The overlapping of the slanted sections allows a bubble barrier to be constructed across the path of a contaminant plume to significant depths. A dense, overlapping stripping zone would be produced without having to bring the end shaft of the drilled hole back up to the surface.

The use of larger porous materials with higher volumes of flow can create stripping zones over 100 feet (30 m) in width. The relationship between depth, pressure, and effective bubble radius is presented with a few simple equations. Multiple installations can be used for volatile organic in-situ stripping, as well as oxygen transfer for biodegration.

REFERENCES

Anderson et al. 1987. National Water Well Association, Dublin, OH.

Angell, K.G. 1991. Air Sparging Case Studies: Proof That an Innovative Technique Works. In: *The Proceedings of the Focus Conference on Eastern Regional Ground Water Issues*, Ground Water Management, Dublin, OH 43017.

Handbook of Chemistry and Physics (R.C. Weast, Ed.). 1972. Cleveland, OH, The Chemical Rubber Co.

Kerfoot, W.B. 1993. Injectable Points for Sandy Soil Sparging Systems In: *Hydrocarbon Contaminated Soils*, Volume III, E.J. Calabrese, P.T. Kostecki, Eds.). Chelsea, MI, Lewis Publishers.

Kerfoot, W.B. 1994. A Simple Test To Determine the Bubbling Radius and Bubble Conductivity of an Aquifer. In: *Proceedings, The Eighth National Outdoor Action Conference*, Minneapolis, MN.

Massermann, J.W. 1989. Applying Groundwater Flowmeter Models in Vapor Extraction Design Systems, *J. Civil Eng.* **115**, 1.

Repa, E. and Kufs, C. 1985. Leachate Plume Management, EPA/540/2-85/004, JRB Associates, Inc. Prepared for the Hazardous Waste Engineering Research Laboratory, Cincinnati, OH 45268.

Rothman, M. 1983. Personal Communication, Schramm, Inc. West Chester, PA.

Sevee, J. 1991. Methods and Procedures for Defining Aquifer Properties. In: *Practical Handbook of Ground-Water Monitoring* (D.M. Neilsen, Ed.) Chelsea, MI, Lewis Publishers.

CHAPTER 4

Restoration of a Heavily Oiled Salt Marsh Using *Spartina alterniflora* Seedlings and Transplants: Effects on Petroleum Hydrocarbon Levels and Soil Microflora

Andrew Bergen, New York City Department of Parks and Recreation, Salt Marsh Restoration Team, Staten Island, New York

Michael Levandowsky and **Thomas Gorrell**, Haskins Laboratories, Pace University, New York, New York

Carl Alderson, New York City Department of Parks and Recreation, Salt Marsh Restoration Team, Staten Island, New York

INTRODUCTION

The NYC Department of Parks and Recreation's Salt Marsh Restoration Team was created in 1991 in order to restore areas of intertidal salt marsh impacted by the January 1, 1990 Exxon spill of 567,000 gallons of No. 2 fuel oil into the Arthur Kill, a tidal strait separating the western shore of Staten Island from New Jersey. The spill impacted 16 linear miles of shoreline, along which are located four salt marshes, each over 100 acres in size. The fuel oil destroyed low marsh vegetation consisting primarily of *Spartina alterniflora* (salt marsh cord grass), a host of invertebrates, winter water fowl, and the nursery grounds for a number of fish in all of these marshes. The effects of fuel oil on salt marshes have been documented since the 1970s (Burns and Teal, 1979; Webb and Alexander, 1991), but no attempt had been made to restore vegetation to areas impacted by oil. The efforts along the Arthur Kill therefore represent an experiment that has implications beyond the scope of the January 1990 spill and subsequent restoration.

Because of the growing interest in bioremediation and its application to oil spills, the Salt Marsh Team decided to assess the Total Petroleum Hydrocarbon (TPH) concentrations of sediments prior to the reintroduction of plants. In order to assess the results reported, the Team returned to the analysis of the same impacted areas conducted immediately after the spill. A consulting firm, Louis Berger and Associates, was hired by the New Jersey Department of Environmental Protection and Energy in 1990 to quantify the damage caused by the spill and to attempt to find a match between oil found in sediments and biological assays with that of the spilled Exxon fuel oil. Table 1 compares the 1990 TPH analysis with samples taken from the same locations by the NYC Dept. of Park's Salt Marsh Restoration Team in 1993. The high standard deviations in both cases indicate the great variability in the TPH concentrations. What was clear was that three years after the spill there was a persistent presence of high levels of TPH in the sediments. In fairness, it should be noted that over two million gallons of petroleum product has been spilled into the Arthur Kill since 1980, according the U.S. Coast Guard. It is probable that test results from both 1990 and 1993 record other spills. However, three years after the spill, areas that had been vegetated were still devoid of vegetation, increasing the chances that the initial damage would be compounded by the effects of erosion on now denuded areas of salt marsh peat. An effort to restore *Spartina alterniflora*, the dominant plant species of the low salt marsh, was considered the best way to reduce the chances of more loss of marsh due to erosion.

Table 1. **Comparision of Total petroleum hydrocarbon concentrations collected and analyzed in 1993 with samples collected and analyzed in 1990 4 months after the January 1990 Exxon oil spill.**

Berger 1990 No. of Samples[a]	Mean + Std. ppm	NYC Parks 1993 No. of Samples	Mean + Std. ppm
40	6441+ 2x104	34	7538+1.3x104

[a]Source: Arthur Kill Oil Discharge Study, Volume 1, Final Report. Submitted to NJDEPE by Louis Berger and Associates, East Orange, NJ.

We were also interested in the relationship between the revegetation of low salt marsh with *Spartina alterniflora* seedlings and the ability of certain heterotrophic bacteria to degrade oil that was trapped in the anaerobic hydric soils. Initially, the project's goal was to restore vegetation to as much as possible of the area heavily impacted by the 1990 spill, and to evaluate success in terms of the survival of the new plants and recolonization of restored areas by ribbed mussels (*Geukensia demissa*). Monitoring parameters suggested by earlier research in restoring salt marsh vegetation have been assessed. These include height, stem density, density of flowers, rhizome spread, basal area of individual plants, aboveground biomass and below-ground biomass, and ribbed mussel population densities in m^2 quadrats (Broome, 1990; Kuenzler, 1961). However, from visual observations it was clear that a significant amount of oil had remained undegraded in the substrate. Thus we thought it might be possible to relate survival of *Spartina alterniflora* seedlings to a reduction of Total Petroleum Hydrocarbon concentrations.

There is a significant body of literature documenting the relationship between bacteria, both anaerobic and aerobic heterotrophs, and reductions of concentrations of a wide spectrum of petroleum hydrocarbons (Aelion and Bradley, 1991; Bauer and Capone, 1988; Hood et al., 1975; Kampfer et al., 1993; Lovely and Lonergan, 1990). From aliphatics (N-paraffins) to polyaromatic hydrocarbons (PAHs), bacteria have been shown to remediate oil in both dry and wet soil conditions. Because *Spartina alterniflora* has the ability to oxygenate its root zone, a significant limiting factor to aerobic bacterial degradation of hydrocarbons may be reduced (Teal and Kanwisher, 1966; Dacy and Howes, 1984; Arenovski and Howes, 1992; Howes and Teal, 1994). The introduction of time release fertilizer (Osmocote 18-6-12 plus Fe, 8-9 month time release) into the substrate with the planting of a seedling or transplant could also supply the bacteria with a nutrient source (Jordan and Payne, 1980). Recent laboratory studies have shown that plants can play an important role as a bacterial substrate in degrading petroleum hydro-carbons (Lee and Banks, 1993; Cunningham and Berti, 1993).

MATERIALS AND METHODS

In this chapter we report on three types of location which were assessed in terms of vegetation, invertebrates, and total petroleum hydrocarbons (TPH). These types are:

L1: Impacted and Restored
L2: Impacted and Unrestored
L3: Unimpacted and Unrestored

Vegetation

Spartina alterniflora plants, both nursery-grown seedlings and transplants from adjacent areas of marsh, were assessed after planting using both nondestructive and destructive sampling parameters. Seedlings and transplants were planted with 18-6-12 + Fe time-release fertilizer, except in a limited number of locations where no fertilizer was used, or a different time-release formula and granular fertilizer were tried (see Table 2 for the analysis of the six fertilizers initially used). 18-6-12 + Fe is approximately 50% nitrate nitrogen and 50% ammonium sulfate. It is likely that the reduced conditions of the salt marsh render the nitrate component less than useful; however, the ammonium sulfate has been a factor in helping establish seedlings and transplants. Iron is added to increase the seedlings and transplants' rate of photosynthesis and respiration. In L1, stem density, density of flowers, and the height of plants were assessed in 43 m^2 quadrats in Year 1 (first-year plants) and 35 m^2 quadrats in Year 2 (second-year plants). Twenty-one aboveground biomass and 18 belowground biomass samples were analyzed for Year 1 plants. Fourteen samples were assessed for Year 2 plants. Aboveground biomass was assessed by cutting the vegetation to the surface, cleaning, oven-drying overnight at 80° C, and weighing the vegetation to the nearest gram. Belowground biomass was dug up from below the same area of vegetation. Live roots (identified by turgor and bright white color) were separated from dead (limp and translucent) and wet-sieved through a 2mm mesh. The remaining live roots were oven-dried overnight at 80° C, and weighed to the nearest gram (0.00).

Table 2. **The Six Fertilizer Applications Used in September 1992 Planting.**

Fertilizer
1. Sierra Blend 17-7-10 +Fe 12-14 month time release
2. Sierra Blend 18-7-9 +Fe 8-9 month time release
3. Osmocote 18-6-12 8-9 month time release
4. Osmocote 18-6-12 +Fe 8-9 month time release
5. Mag Amp 7-40-6 (granular) plus Osmocote 13-13-13 8-9 month time release
6. Osmocote 17-7-12 12-14 month time release

L2 and L3 sites were monitored using the same nondestructive and destructive sampling parameters, but with a different number of sample points. L2 sites had 10 Year 1 nondestructive sample points, 10 Year 1 destructive sample points and 1 Year 2 nondestructive and destructive sample point. L3 sites had 3 nondestructive sample points and 2 destructive sample points for both Years 1 and 2.

Invertebrates

Populations of live and dead ribbed mussels, *Geukensia demissa*, were counted in 71 m^2 quadrats for L1 sites, and 28 m^2 quadrats for L2 and L3 sites. Only surface populations were counted in order to disturb the sites as little as possible.

Table 3. **Culture Media for Bacterial Assays**

A. SMP Medium

	g/100ml distilled H_2O
NaCl	1.8
$MgSO_4.7H_2O$	0.05
$CaCl_2.2H_2O$	0.037
KCl	0.06
NH_4NO_3	0.05
K_2HPO_4	0.005
HEPES	0.1
Na Acetate	0.05
Glutamic Acid	0.05
Glycine	0.05
Sucrose	0.05
Vitamins 8A	1.0 ml/100ml
Soil Extract	10.0 ml/100ml
pH (Adjusted with NaOH)	7.8

Vitamins 8A contains: thiamine, nicotinic acid, putrescine, Ca pantothenate, riboflavin, pyridoxine, pyridoxamine, PABA, biotin, choline, inositol, thymine, orotic acid, folinic acid, folic acid, B12.

B. Medium for oil-degradation assay.

This is the same as SMP medium above, except that the carbon sources (acetate,glutamic acid, glycine, sucrose) are omitted. Made up as slants (1.5% agar), to which 4 drops of fuel oil no. 2 are added.

Total Petroleum Hydrocarbons

Samples were collected from four locations impacted by oil and restored, three locations in the area impacted by oil and denuded of vegetation (unrestored), and two sites outside the impact area with existing vegetation. Samples were collected for Year 1 (prior to restoration) and Year 2 and 3 (after restoration) using a dutch auger. Each sample represents a composite of five cores, 20 cm in depth, collected and split into two 10-cm fractions, 0-10 cm and 10-20 cm depth. It was thought that a difference might be observed in terms of reduction of oil at different depths within the rhizosphere. TPH concentrations were analyzed using EPA method 418.1 (Bossert and Compeau). The use of a 20-cm core was determined based upon the mean depth of the *Spartina alterniflora* rhizosphere.

Bacteria

Aerobic heterotrophic bacteria from these samples were assayed using a dilution culture technique. Replicate soil samples were taken from the top 1 cm and from a depth of 10 cm. Measured quantities of soil were suspended in SMP medium (Table 3b), and replicate dilution series were inoculated, using tenfold dilutions, and cultured at room temperature. The highest dilution tubes showing growth served as source of inoculum for streaking onto agar plates. Thus, estimates of bacterial abundances, and representative isolates of the most abundant culturable forms were obtained. Data reported here are averages of at least two parallel dilution series. In some cases, replicate series gave the same value, and in others they differed by an order of magnitude. No greater differences were seen, and the data should be considered valid to within an order of magnitude. Abundance of hydrocarbon-utilizing bacteria was estimated by suspension and 10-fold dilutions of measured amounts of soil in sterile diluted seawater. The latter served as inocula for agar slants of SM3 medium (Table 3b), to which were added 4 drops of No. 2 fuel oil. In the higher dilutions with substantial growth, no sign of agar digestion was noted.

RESULTS

Vegetation

The initial success of plantings determined approximately 30 days after planting has been 75% survival. This is an estimate from observations of eight planting sites and approximately 200,000 seedlings. The percent success ranged from 99% at three 20,000 plant sites to 20% success at a 15,000-plant site planted in September 1993. The 18-6-12 time release fertilizer was the most effective of the six tried. Table 2 lists the other fertilizers used. Two areas have also been planted with seedlings and no fertilizer. One of these areas has shown very little productivity, while the second was only planted in the summer of 1995.

Tables 4-6 report results for stem density, density of flowers, height and above- and belowground biomass. Table 4 reports on first- and second-year plants in restored areas through 1994. Tables 5 and 6 report on the same parameters in control sites. All nondestructive vegetation parameters increased at restored sites through the second year. There was a slight increase in all nondestructive parameters, except for the density of flowers at impacted unrestored sites, Table 5, type L2. No flowers were reported for Year 2 impacted controls. Unimpacted controls, Table 6, type L3, are still more productive than restored areas, although the gap from Year 1 to Year 2 has been narrowed.

Above- and belowground biomass at restored sites increased by a factor of approximately 2.7, maintaining a ratio between above- and belowground biomass of approximately 1.7. Biomass results for type L2 show an increase in aboveground biomass and a reduction in belowground biomass. Type L3 also showed a reduction in belowground biomass and an increase aboveground biomass.

Table 4. **Site L1: Results (Mean and Standard Deviation) Monitoring Vegetation, Ribbed Mussels (Geukensia demissa) and Total Petroleum Hydrocarbons (TPH) at Restored Areas of Low Salt Marsh, Staten Island, New York.**

Year	Ribbed mussel[a]/m²	AGB[b]/m² (g)	BGB[c]/m² (g)	stems/m²	flowers/m²	height/m² (m)	TPH/m² (ppm)
0	0	0	0	0	0	0	17,534 ±2.3x10⁴
1	nc	313±236	540±282	119±63	7±10	0.81±0.4	6,188 ±6.6x10³
2	6±8	909±443	1395±807	154±70	30±23	1.38±0.4	4,425 ±4.6x10³

Key: [a] = *Geukensia demissus* [b] = above-ground biomass
[c] = below-ground biomass nc = not collected or still to be collected

Table 5. **Site L2: Results (Mean and Standard Deviation) Monitoring Vegetation, Ribbed Mussels (Geukensia demissa) and Total Petroleum Hydrocarbons (TPH) at Unrestored Control Sites in Areas Impacted by the January 1990 Oil Spill.**

Year	Ribbed mussel[a]/m²	ABG[b]/m² (g)	BGB[c]/m² (g)	stems/m²	flowers/m²	height/m² (m)	TPH/m² (ppm)
0	0	nc	nc	nc	nc	nc	6,701 ±4.7x10³
1	nc	183±318	879±1523	14±32	3±5	0.92±0.65	2,651 ±2.8x10³
2	0	225±390	283±490	27±50	0.3±0.9	0.64±0.6	nc

Key: [a] = *Geukensia demissus* [b] = above-ground biomass
[c] = below-qround biomass nc = not collected or still to be collected

Table 6. **Site L3: Results (Mean and Standard Deviation) Monitoring Vegetation, Ribbed Mussels (Geukensia demissa) and Total Petroleum Hydrocarbons (TPH) at Unrestored Control Sites in Areas Unimpacted by the January 1990 Oil Spill.**

Year	Ribbed mussel[a]/m²	ABG[b]/m² (g)	BGB[c]/m² (g)	stems/m²	flowers/m²	height/m² (m)	TPH/m² (ppm)
0	32±32	nc	nc	nc	nc	nc	583±319
1	nc	1072±290	2065±1126	165±31	32±12	1.6±.3	330±106
2	41±24	1680±90	1360±203	137±31	44±10	1.4±0.6	<10

Key: [a] = *Geukensia demissus* [b] = above-ground biomass;
[c] = below-ground biomass nc = not collected or still to be collected

Invertebrates

Table 4, type L1, shows the results for Years 0 and 2. The results show an increase from 0 to a mean of 6 mussels/m^2. There are still no mussels recolonizing areas impacted and unrestored, type L2, while mussel populations at type L3 sites increased by 25% from Year 0 to Year 2.

Total Petroleum Hydrocarbons (TPH)

TPH levels dropped from Year 0 to Year 1 by a factor of 2.8 at L1 and 1.57 at L2. Unimpacted controls at L3 dropped by a factor of 1.76 over the same time period. Levels were 6.6 times higher at L1 than at L2, and 50 times higher than at L3 in Year 1. In Year 2 these difference narrowed between L1 and L2 to 3.6 and 30 times that of L3.

Bacteria

Total culturable aerobic heterotrophic bacteria levels are shown in Table 7. In the winter (before planting), average abundance in surface samples ranged from $7x10^7$ to 10^9 colony-forming units (CFUs) per gram of soil. Abundances in 10-cm depth samples tended to be an order of magnitude lower. In the summer, CFUs remained the same or decreased in a site unimpacted by the oil spill, and in a fertilized planted site. Oil-utilizing organisms were assayed in the summer, after planting, and these data are shown in Table 8. In the fertilized plot, counts of these organisms were as high as those of total heterotrophs, whereas in all other cases they were 1-2 orders of magnitude lower.

DISCUSSION

The success or failure of restoration of *Spartina alterniflora* to areas of low marsh destroyed by the January 1990 oil spill has been determined by a number of factors: the time of year in which the seedlings or transplants were planted; predation by geese; the severity of the winter season, and whether no fertilizer was introduced at the time of planting. There were not major differences in the time-release fertilizers used (Table 2), but we did find that seedlings planted with the one granular fertilizer we used did not do well in a fall planting. It may also be that some substrates were more or less suitable for seedlings and/or transplants, but our study of this is so far incomplete. In no case have we found that oil limits the survival of restored seedlings and transplants. However, recolonization of impacted unrestored sites (L2) has been slow to nonexistent (Table 5). Two sites, one planted with 6,000 seedlings in September 1992 and one planted with 12,000 seedlings in September 1993, did poorly. The 1992 results were a function almost exclusively of the late season, and the area had a survival rate of 60%. In the September 1993 planting there was 20% survival, due to the additional stresses of ice and goose predation. In contrast, seedlings and transplants planted in April and May that were not damaged by geese have been 99% successful.

Table 7. **Aerobic Heterotrophic Bacterial Abundances, as Measured by Dilution Assay. Colony-Forming Units (Cfus) per Gram Soil.**

Type of Site	CFUs/Gram Soil Average of 2 or More Replicates	CFUs/Gram Soil Average of 2 or More Replicates
	Winter	Summer
Unimpacted		
1 cm depth	$7x10^8$	10^8
10 cm depth	10^7	10^6
Impacted		
planted, fertilized		
1 cm	10^9	10^7
10 cm	10^7	10^7
planted, not fertilized		
1 cm	$7x10^7$	$7x10^8$
10 cm	$7x10^6$	10^9
not planted		
1 cm	$7x10^7$	10^{10}
10 cm	10^6	10^9
existing S.alterniflora in impact zone		
1 cm	10^8	10^{10}
10 cm	10^7	$7x10^8$

Table 8. **Comparison of Abundances of Aerobic Heterotrophic and Oil-Utilizing Bacteria in Summer Samples of Impacted Site.**

Type of Plot	Aerobic Heterotrophs Average of 2 or More Replicates	Oil-utilizers CU/gm Average of 2 or More Replicates
Planted, fertilized		
1 cm depth	10^7	$4x10^7$
10 cm depth	10^7	$4x10^6$
Planted, not fertilized		
1 cm	$7x10^8$	$7x10^6$
10 cm	10^9	10^6
Not planted		
1 cm	10^{10}	10^7
10 cm	10^9	$7x10^6$
Existing S. alterniflora in impact zone		
1 cm	10^{10}	$7x10^9$
10 cm	$7x10^8$	$7x10^7$

Results show levels of annual above- and belowground productivity within the ranges cited in by Mitch and Gosselink (1993), although they are somewhat lower than measurements of annual net aerial primary production. Our sampling technique does not account for all the annual production, missing a small percentage of stems which die through the summer, prior to September collection. Most studies tend to separate streamside and inland plants (tall- and short-form *Spartina alterniflora*); we have not done this. Typically inland plants are one-third to one-half as productive as streamside plants. We do not separate them here because we are trying to gauge overall success. Productivity seen in this light compares with studies conducted from North Carolina to New England (Mitsch and Gosselink, 1993). The mean of both tall- and short-form *Spartina alterniflora* aboveground biomass figures cited in Mitch and Gosselink (1993) is 809 g m^{-2} yr^{-1}. Our aboveground figures are 838 g m^{-2} yr^{-1}. Belowground biomass figures are somewhat below the figures cited, 2149 g m^{-2} yr^{-1} as opposed to our 1395 g m^{-2} yr^{-1}. The lower figures could be a sampling error based on the qualitative nature of the live root selection process, or it could be an indicator of slower overall development. The ratio of aboveground biomass to belowground biomass should develop toward 1:2 or 1:3 over time, although the 1.7 for Year 1 and 1.6 for Year 2 do not indicate that the new plantings are under undo stress. It is likely that heavy icing in the winter of 1993-1994 caused the biomass ratio reported for L3 sites to drop in one year from 2 to 0.8. Stems, flowers, and the height of plants at L1 sites have all increased over the two years since their planting, supporting the biomass data.

In the Kuenzler study of ribbed mussels in 1961, conducted near Sapelo Island in Georgia, the mean of 250 quadrats throughout the marsh was 7.8 mussels/m^2. This number is significantly lower than studies which have looked at ribbed mussel populations at streamside or seaward locations, where elevations may be slightly lower, and plant productivity as a function of nutrient and oxygen levels and periods of submergence are all higher (Bertness, 1984; Lin, 1989). In the Bertness study, mussel densities were over 1,000 m^{-2}. Kuenzler found that 46% of the mussels identified occurred within only 6% of the marsh, the streamside tall *Spartina alterniflora.* Our restored areas, type L1, reflect a distribution pattern closer to that of Kuenzler's study, where mussels were counted throughout the marsh, not only at their most productive streamside locations. We should, however, expect these numbers to increase if the results from L3, Table 6 are any indication.

In all cases, the trend of TPH is down. This is likely to be a function of time, but given the large volume of TPH reduced in areas restored (Table 4, type L1) the reductions can also be analyzed in relation to plants and bacteria. The restored area had levels of oil that are thought to be enough to kill salt marsh plants (>10,000 ppm) (Baker et al., 1989). With mean levels of 17,500 ppm, and as high as 57,000 ppm prior to restoration, (those levels reduced to 10,000 ppm in one year), the survival of the restored and fertilized plants, and their ability to oxygenate their rhizosphere, may be of significance. Research in Kuwait points to a similar relationship in dry soil conditions (Radwan, Sorkhoh, and El-Nemr, 1995), and we see the reduction in TPH as, at minimum, a function of soil aeration. An oxygenated rhizosphere should increase the activity of microflora capable of degrading a variety of petroleum hydrocarbons. The reductions in the unrestored locations suggest that other unmeasured factors may be rate-controlling. Observed spring and summer temperatures on denuded areas of marsh are higher than those of sediments under a productive and protective stand of *Spartina alterniflora.* High sediment temperatures may create conditions as favorable to degradation of petroleum hydrocarbons as those of the restored locations.

Data on bacteria are preliminary, but there are some tentative patterns. It appears that, among the various plots, the one which was fertilized favored the oil-degrading microorganisms, which became as abundant as the total aerobic heterotrophs (we assume there is much overlap in these two categories). In all the other impacted sites the general heterotroph population appeared to be higher than the oil-degrading population. It is not yet clear whether, as hypothesized, the *Spartina* rhizosphere favors the oil-degrading microorganisms significantly, as was found with desert plants in Kuwait (Radwan, Sorkhoh, and El-Nemr, 1995). As was noted, *Spartina* apparently does aerate its rhizosphere. In addition, it has been found that *Spartina* roots have nitrogen-fixing symbiont bacteria in their tissues (McClung and Patroquin, 1980), which could perhaps impact nitrogen availability in the soil of the rhizosphere, and this also might enhance the growth of oil-degrading species. Some of these issues are being investigated now, using laboratory microcosms containing *Spartina* plants.

The success of the restored seedlings and transplants to areas impacted by the January 1990 fuel oil spill was not predicted by a review of the current literature. Although it is not yet clear whether the planting influences the microbial reduction of TPH, the keystone species of the salt marsh, *Spartina alterniflora,* is thriving where it has been restored, and with it, organisms dependent upon its impressive annual production of biomass.

ACKNOWLEDGMENTS

The authors would like to thank Marc Matsil, the Director of the Natural Resources Group, NYC Parks; Christopher Aquila and Robbin Bergfors of the Salt Marsh Restoration Team, NYC Parks; and Bruce Litus, Terminal Manager of the GATX Corporation's Staten Island Facility.

REFERENCES

Burns, K.T., and Teal J.M. 1979. The West Falmouth Oil Spill: Hydrocarbons in the Salt Marsh Ecosystem, *Estuarine and Coastal Mar. Sci.* **8**, 349.

Webb, J.W., and Alexander, S.K. 1991. No. 2 Fuel Oil Effects on *Spartina alterniflora* in a Texas Salt Marsh, *Contrib. Mar. Sci.* **32**, 9.

Broome, S.W. 1990. Creation and Restoration of Tidal Wetlands of the Southeastern United States, Chapter 2 in *Wetland Creation and Restoration: The Status of the Science*, (Kusler, J.A. and M.E. Kentula, Eds.) Washington, DC, Island Press.

Kuenzler, E.J. 1961. Structure and Energy Flow of a Mussel Population in a Georgia Salt Marsh. *Limnol. Oceanogr.* **6**, 191.

Aelion, C.M., and Bradley, P.M. 1991. Aerobic Biodegradation Potential of Subsurface Microorganisms from a Jet Fuel Contaminated Aquifer, *Appl. Environ. Microbiol.* **57**, 57.

Bauer, J.E., and Capone, D.G. 1988. Effects of Co-Occurring Aromatic Hydrocarbons on Degradation of Individual Polycyclic Aromatic Hydrocarbons in Marine Sediment Slurries, *Appl. Environ. Microbiol.* **54**, 1649.

Hood, M.A., Bishop, W.S., Jr., Bishop, F.W., Meyers, S.P., and Whelan, T., III. 1975. Microbial Indicators of Oil-Rich Salt Marsh Sediments. *Appl. Microbiol.* **30**, 983.

Kampfer, P.M., Steiof, M., Becker, P.M., and Dott, D. 1993. Characterization of Chemoheterotrophic Bacteria Associated with In-Situ Bioremediation of a Waste-Oil Contaminated Site, *Microbial Ecol.* **26**, 161.

Lovely, D.R., and Lonergan, D.J. 1990. Anaerobic Oxidation of Toluene, Phenol, and p-Cresol by the Dissimilatory Iron-Reducing Organism, GS-15, *Appl. Environ. Microbiol.* **56**, 1858.

Teal, J.M., and Kanwisher, J.W. 1966. Gas Transport in the Salt Marsh Grass, *Spartina alterniflora. J. Experimental Bot.* **17**, 355.

Dacy, J.W., and Howes, B.L. 1984. Water Uptake by Roots Controls Water Table Movement and Sediment Oxidation in Short *Spartina marsh, Sci.* **224**, 487.

Arenovski, A.L., and Howes, B.L. 1992. Lacunal Allocation and Gas Transport Capacity in the Salt Marsh Grass *Spartina alterniflora, Oecologia,* **90**, 316.

Howes, B.L., and Teal, J.M. 1994. Oxygen Loss from *Spartina alterniflora* and Its Relationship to Salt Marsh Oxygen Balance, *Oecologia,* **97**, 431.

Jordan, R.E., and Payne, J.R. 1980. *Fate and Weathering of Petroleum Spills.* Ann Arbor, MI, Ann Arbor Science, 1980, p. 56.

Lee, E., and Banks, M.K. 1993. Bioremediation of Petroleum Contaminated Soil Using Vegetation: A Microbial Study, *J. Environ. Sci. Health,* **A 28**(10), 2187.

Cunningham, S.D., and Berti, W.R. 1993. Remediation of Contaminated Soils with Green Plants: An Overview, *In Vitro Cell. Devel. Biol.* **29P**, 207.

Bossert, I.D., and Compeau, G.C. Cleanup of Petroleum Hydrocarbon Contamination in Soil, Chapter 3 in *Microbial Transformations and Degradation of Toxic Organic Chemicals*, (Young, L.Y. and Ceniglia, C.E., Eds.), New York, Wiley-Liss, Inc.

Mitsch, W.J., and Gosselink, J.G. 1993. *Wetlands*, Second Edition, New York, Van Nostrand Reinhold, p. 249.

Bertness, M., 1984. Ribbed Mussels and *Spartina alterniflora* Production in a New England Salt Marsh, *Ecol.*, **65**, 1794.

Lin, J., 1989. Importance of Location in the Salt Marsh and Clump Size on Growth of Ribbed Mussels, *J. Exp. Mar. Biol. Ecol.* **128**, 75.

Baker, J.M., Bayley, J.A., Howells, S.E., Oldham, J., and Wilson, M. 1989. Oil in Wetlands, in *Ecological Impacts of the Oil Industry*, (Dicks, B., Ed.) Institute of Petroleum and John Wiley and Sons, Ltd., p. 37.

Radwan, S., Sorkhoh, N., and El-Nemr, I. 1995. Oil Biodegradation Around Roots, *Nature*, **376**, 302.

McClung, C.R., and Patroquin, D.G. 1980. Isolation of a Nitrogen-Fixing *Campylobacter* Species from the Roots of *Spartina alterniflora* Loisel, *Can. J. Microbiol.* **26**, 881.

CHAPTER 5

Remediation of Petroleum Impacted Soil and Groundwater at a Site Using Soil Vapor Extraction and Air Sparging

Gregory J. Gromicko and **Robert C. Klingensmith**, Groundwater Technology, Inc., Moon Township, Pennsylvania

Duane K. Simpson, Quaker State Corporation, Oil City, Pennsylvania

INTRODUCTION

Closure activities for three underground storage tanks (USTs) were conducted at Quaker State Corporation's (QSC) former service station in Conneaut, Pennsylvania, as part of a property transfer during July 1991. The facility, formerly owned by QSC, was operated from construction (early 1960s) through the sale of the property (early 1980s). Subsequent to sale of the facility, the property was resold and the building reconfigured several times.

The facility is situated on a corner lot located along state highway Route 322 in the business district of Conneaut, as shown in Figure 1. Across the highway, to the north, is Conneaut Lake. The site is bordered by a residential property to the south and commercial properties on the east and west. A Pennsylvania State Game Commission Game Lands is located approximately 150 feet southeast of the property.

Quaker State again became involved with the property in 1991 when the current owner attempted to sell the property, and the lender for the prospective purchaser identified the presence of USTs. Following the confirmation of the USTs, UST closure activities were initiated. Subsurface investigations were conducted to delineate the extent of potential petroleum impacts, and corrective actions were initiated.

PREVIOUS INVESTIGATIONS

The scope of work associated with the UST closure activities included the closure by removal of UST #1 and UST #2, the closure in place of UST #3, and the determination of the presence of UST #4 and UST #5.

Figure 1. **Site map.**

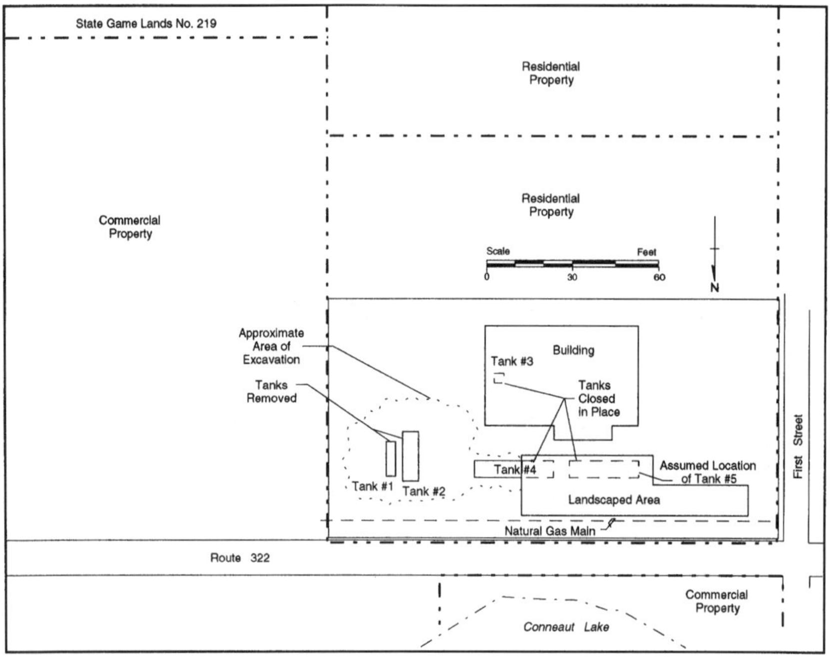

UST's #1, #2, and #3 were known to exist at the site, based upon a site inspection completed by QSC. The presence of UST #4 and UST #5 were reported, but a records search was inconclusive, based upon the dated nature of the initial property sale. Information provided by the current owner of the property indicated that UST #4 and UST #5 were located under the landscaped area, as illustrated in Figure 1.

In accordance with American Petroleum Institute Recommended Practices 1604 and 2015, UST #1 and UST #2 were permanently closed by removal, and UST #3 was permanently closed in place (API, 1987a; API, 1987b). The presence and location of UST #4 and UST #5 were confirmed through the enlargement of the area excavated for closure of UST #1 and UST #2. During the UST closure activities, a series of four test pits were excavated on the western side of the property to determine if any other USTs were present at the site. All test pits were excavated to a depth of five feet, and no evidence of UST systems or petroleum impacts was noted. The locations of all USTs determined to be at the site are illustrated in Figure 2.

Visual and photoionization inspections of the excavations, the resulting excavated soils, and the groundwater that entered the excavations indicated that petroleum hydrocarbons had been released to the surrounding soils and groundwater.

Following UST closure activities, soil samples were collected and analyzed in accordance with the Pennsylvania Department of Environmental Protection (PADEP) site characterization guidelines. The soil samples were analyzed for total petroleum hydrocarbons (TPH) by EPA Method 418.1 and benzene, toluene, ethylbenzene, and xylene (BTEX) by EPA Method 8020. Locations of soil sampling locations are illustrated in Figure 2.

Based upon observations noted during the UST closure and the analytical results of the soil samples collected during the UST closure activities, the presence of petroleum hydrocarbons in the surrounding soils was confirmed (Keystone Environmental Resources, 1991).

Following completion of this investigation, a second phase investigation was conducted to determine the extent and nature of the petroleum impacts to the subsurface soils and groundwater. This investigation consisted of a sampling and analysis program that was completed by advancing three soil borings and constructing monitoring wells (MW-1, MW-2, and MW-3). The soil boring and monitoring well locations are shown in Figure 2.

Soil samples were collected to characterize soil conditions outside of the area of the closed USTs. Two groundwater sampling and analysis events were completed to characterize groundwater quality. The analytical results from the soil boring and groundwater sampling and analyses events indicated that the soils and first waterbearing zone about the closed USTs was impacted with petro-

Figure 2. **Sample Locations**

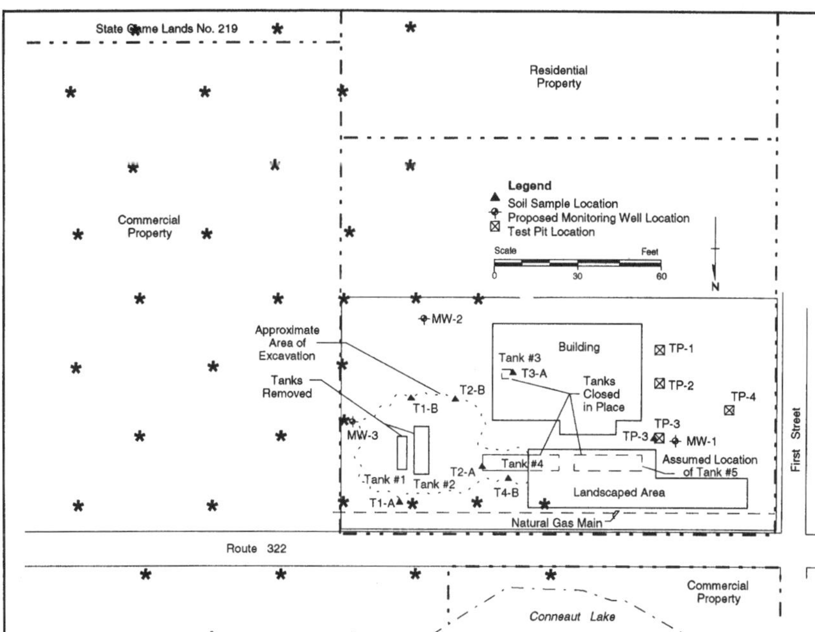

leum products (Keystone Environmental Resources, 1992). The field data also indicated that the groundwater flow direction was in a south to southeastern direction. Following discussions and gaining concurrence with the PADEP, a soil gas survey was conducted to determine the aerial extent of petroleum impacted groundwater.

The soil gas survey was used as a preliminary screening tool to determine the extent of migration of the impacts. By measuring the gaseous phase of the volatile organic compounds migrating to the surface above and the impacted area, the source area and plume extent may be delineated (USEPA, 1987). The advantages of this technique as a screening tool instead of a soil boring and well installation program are that the soil gas sampling method is relatively nondisruptive and less costly, and can provide a thorough evaluation due to the close spacing of the sampling points.

The soil gas survey was completed over a 50 by 50 foot grid pattern, as shown in Figure 2. This grid pattern was surveyed into existing datum points on the site to correlate known site soil and groundwater conditions with the soil gas survey. Of primary concern was evaluating the potential impact of the State Game Land located approximately 150 feet beyond the southeast corner of the site. A total of 22 soil gas and 12 groundwater samples were collected. The analytical results from this survey indicated that the impacted area was confined to the site (USEPA, 1987).

Review of the soil boring logs and groundwater elevation data for the monitoring wells (MW-1, MW-2, MW-3) indicated that migration of the BTEX compounds was limited to the area in the vicinity of the former USTs. This conclusion was developed based upon the low permeability of the fine clayey silt and sandy silt soils, and the low hydraulic gradient within the uppermost waterbearing unit.

CORRECTIVE ACTION

Following the investigative activities, corrective action alternatives were evaluated. Based upon site and constituent characteristics, soil vapor extraction (SVE) integrated with air sparging (AS) was selected as the recommended corrective action alternative.

Soil vapor extraction is the process of removing volatile constituents from soil pore spaces by causing air to flow through the subsurface (Johnson et al, 1990). In SVE, air flow enters through the surface adjacent to the treatment zone and is drawn through the treatment zone, mixing with the soil vapors as it moves. The SVE vapors are withdrawn using a vacuum source through extraction wells or trenches placed within the treatment area. Flow patterns for the air are dependent upon the soil permeability, soil type, soil conditions, and placement of the extraction points. Upon removal, SVE vapors may be treated using an aboveground treatment process, such as activated carbon, prior to discharge of the gas.

Air sparging is the treatment of saturated soil and groundwater by providing air to the subsurface to volatilize organic constituents within the treatment zone (Griffin et al., 1991). Air sparging, using deep well points within the saturated zone, is intended to deliver air to the subsurface in order to transfer the remaining constituents to the vapor phase. Within the unsaturated zone, AS through the shallow well points is intended to supply oxygen for biological activity, and induce air flow to the SVE system. Nutrients may also be added to the AS system to stimulate growth of the indigenous microbial population and promote biodegradation of these constituents (USEPA, 1992). Placement of the injection points is dependent upon site-specific factors such as soil types, permeability, and depth of the water table.

The SVE/AS system installed at this site is an integrated remedial system that has been designed to remove the volatile BTEX constituents from the treatment area. A layout of the SVE/AS system is shown in Figure 3.

The goal of the system is to balance the amount of air that is introduced through AS and removed through SVE with the rate of volatilization and natural biodegradation. Vapor recovery trenches were installed around the perimeter of the site, as shown in the Figure 3. The southern trench is approximately 100 feet in length, and the northern trench extends approximately 170 feet. Construction of these trenches was completed by excavating to a depth of 10

Figure 3. **SVE/AS System Layout**

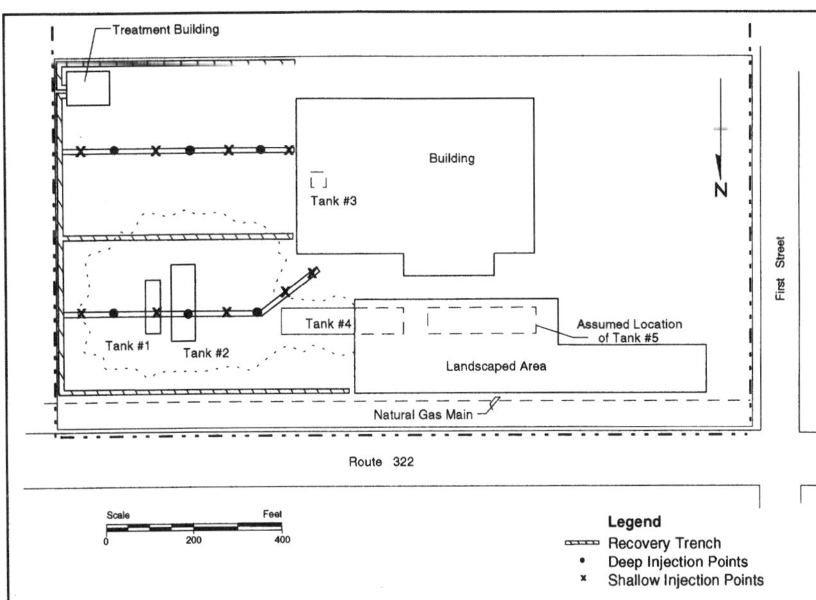

feet below ground surface and backfilling the trench with permeable select fill (2 inch limestone gravel) to a depth of 4 feet below ground surface. On top of the gravel, perforated 4-inch PVC piping was placed, and the trenches were backfilled to grade with less permeable native material. An additional recovery trench was installed in the area of the former UST #1 and UST #2. This third trench, extending approximately 50 feet east to west midway within the treatment zone, was installed to address the petroleum impacts in the shallow soil immediately in the former UST area. This trench was constructed in a similar manner, but only excavated to a depth of 5 feet below ground surface.

Figure 4 shows a schematic of the aboveground components of the system. A positive displacement blower was used to create a vacuum in the recovery trenches. To prevent any moisture collected from the recovery trenches from damaging the blower, a moisture knockout tank was installed in front of the blower. The SVE system is capable of recovering a maximum of 100 scfm at a vacuum of up to 4 inches of mercury. Instrumentation and controls for monitoring the SVE system include vacuum gauges, regulators, flowmeters, and control valves. The extracted vapors are treated prior to discharge to the atmosphere using an air-phase, granular activated carbon (GAC) system. The GAC system utilizes two carbon canisters in series to treat the 100 scfm flow. The system is equipped with sampling ports to monitor the vapors prior to discharge to the atmosphere.

The AS system consists of a positive displacement blower capable of a maximum flow rate of 75 scfm at a pressure of 12 psi. The air is cooled through a heat exchanger prior to movement to the shallow and deep injection points. Instrumentation and controls consist of pressure regulators, pressure gauges, temperature indicators, and control valves.

The injection trenches and injection points are shown in Figure 3. Injection trench #1 is 50 feet in length and injection trench #2 is 60 feet in length. The trenches were excavated to 3 feet below ground surface to permit installation of the injection points and header piping below-grade. Two-inch, schedule 40 PVC well points were utilized with a 2 1/2 foot screened interval. The shallow points were installed to a depth of approximately 6 feet, and the deep injection points were installed to a depth of approximately 18 feet. Sand was placed around the screened intervals, and the remaining riser pipes were grouted to the surface using bentonite and concrete. The injection points are connected to header piping leading to the injection blower.

All trenches were then backfilled, and the entire site surface was paved with asphalt to reduce infiltration of surface run-on and to induce air flow through the treatment area. A small treatment building was erected on the southeast corner of the property to house all aboveground equipment. Separate utilities were installed to allow operation of the system, independent of site operations.

Figure 4. **Schematic of SVE/AS System**

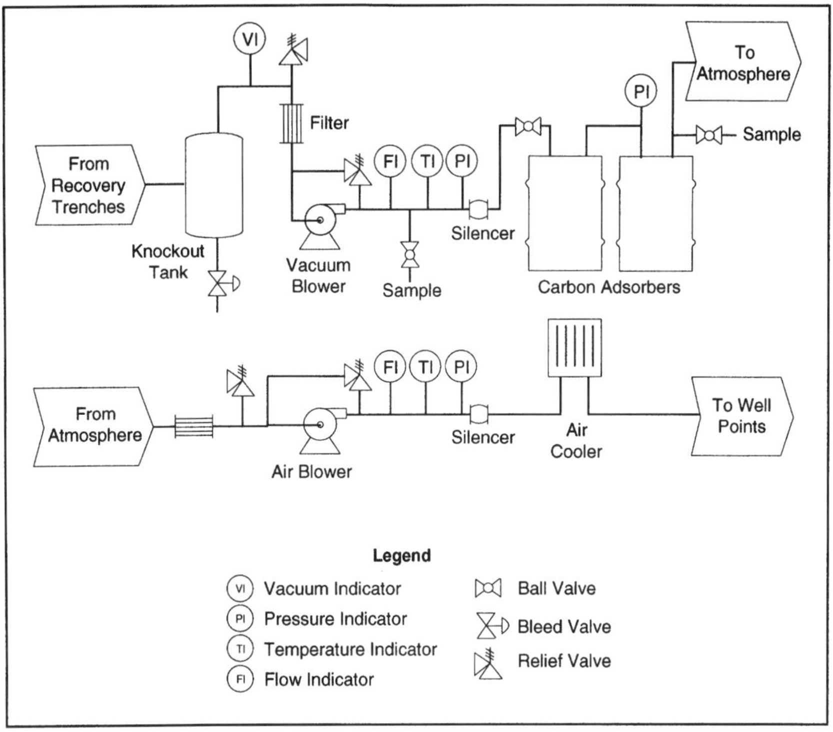

OPERATION

Operation of the SVE/AS system was initiated in August 1993. Startup consisted of setting flowrates and collecting samples of the recovered vapors. The system operates continuously and is monitored monthly. Vacuum and pressure readings are recorded and air flowrates are monitored. Mechanical operation of the equipment is also inspected during the monthly monitoring. System maintenance includes changing the lubricating oil from the blowers, greasing bearings, and cleaning the air intake filter.

Air sampling of the SVE vapors is also performed monthly. Samples are collected prior to the carbon adsorbers to represent recovered vapors, and following the carbon adsorbers as a discharge sample. Samples are collected by inserting a syringe directly into the SVE line via a sampling port. The sample is then extracted and injected into an evacuated vial, and immediately shipped to the laboratory for analyses for BTEX using modified EPA Methods 3810 and 8000.

Groundwater samples are collected quarterly from MW-1, MW-2, and MW-3, and analyzed for BTEX using EPA Method 8020, to determine the progress of the remediation. The first sampling event occurred prior to startup in August 1993, and serves as a baseline.

Baseline soil samples were not collected during startup of the SVE/AS system. However, results from soil sampling conducted in October 1991 have been used as baseline data. Soil samples were collected following one year of operation to determine the effect of the SVE/AS system in addressing the vadose zone soil. Samples were collected near the three initial sampling locations at MW-1, MW-2, and MW-3, and analyzed for BTEX using EPA Method 8020.

RESULTS AND CONCLUSIONS

The results from the SVE air sampling are presented in Table 1. Data show a general reduction in the BTEX recovered in the soil gas. Initial gas samples contained high levels of BTEX, and the concentrations decreased rapidly during the initial months of operation. Additionally, after the decrease, the concentrations generally remained nondetectable. This trend indicates that the easily volatilized constituents were removed initially. As expected, BTEX concentrations in the recovered SVE vapors increased during the summer months because of increased volatilization of the compounds at the higher ambient temperatures.

The results of the groundwater analyses are presented in Table 2. The highlighted values are those that exceed the current PADEP cleanup criteria. As shown, a general reduction in groundwater BTEX concentrations have occurred over time. After one and three-quarters years, MW-1 and MW-3 have met the cleanup criteria. The total BTEX removal percentages in these two wells has approached 100%. The total BTEX removal percentage for MW-2 has exceeded 99%; however, the groundwater benzene concentration has not yet reached the PADEP cleanup levels. The system will be operated until groundwater cleanup criteria have been achieved at all wells for all parameters.

The results of the soil sampling are presented in Table 3. Also shown in the table are the results of the soil sampling performed during the Phase II investigation, which serves as a baseline. As can be seen, the BTEX concentrations in the soil at depths of 5-7 feet and 9-11 feet have decreased during operation of the SVE/AS system. The only increase was observed at location SB-3 at the 7-9 foot depth. This is attributed to the potential presence of a "smear zone" that may contain residual BTEX and may be affected by the groundwater table. Biological degradation of these remaining constituents, enhanced by the oxygen addition from air sparging, may further reduce their concentrations. Future activities will consist of continued operation of the system and, once groundwater BTEX concentrations are below cleanup criteria, the soil will again be sampled to confirm that treatment levels have been achieved. In addition, microbial enumerations and nutrient analyses of the site soils will be performed to confirm the presence of microbial activity.

Table1. **SVE Air Sampling Results** (Results in ppm)

Date	Benzene	Toluene	Ethylbenzene	m&p Xylenes	o Xylenes
9/93	0.45	3.82	1.07	13.95	6.36
9/93	0.38	3.14	0.93	11.98	5.45
9/93	<0.07	0.73	0.22	3.26	2.01
10/93	0.11	0.21	<0.07	0.23	0.16
10/93	<0.07	0.23	<0.07	0.34	0.25
1/94	0.13	0.2	<0.07	<0.07	<0.07
2/94	<0.07	<0.07	<0.07	<0.07	<0.07
3/94	<0.07	<0.07	<0.07	<0.07	<0.07
4/94	<0.07	<0.07	<0.07	<0.07	<0.07
5/94	<0.07	<0.07	<0.07	<0.07	<0.07
6/94	0.7	0.3	<0.07	0.2	<0.07
7/94	0.16	0.13	<0.07	<0.07	<0.07
8/94	0.37	0.33	<0.07	0.13	0.08
9/94	<0.07	<0.07	<0.07	<0.07	<0.07
10/94	<0.07	<0.07	<0.07	<0.07	<0.07
11/94	<0.07	<0.07	<0.07	<0.07	<0.07
1/95	<0.07	<0.07	<0.07	<0.07	<0.07
2/95	<0.07	<0.07	<0.07	<0.07	<0.07
3/95	<0.07	<0.07	<0.07	<0.07	<0.07
5/95	<0.07	<0.07	<0.07	<0.07	<0.07
6/95	<0.07	<0.07	<0.07	<0.07	<0.07
7/95	<0.07	<0.07	<0.07	<0.07	<0.07

Table 2. **Groundwater Analytical Results** (Results in mg/L)

	10/91	7/93	10/93	2/94	5/94	8/94	11/94	2/95	5/95
MW-1									
Benzene	0.227	<0.2	**50**	**17**	1.3	0.4	**8.2**	<0.2	0.6
Toluene	<0.2	<0.2	6	1.1	<0.2	<0.2	<0.2	<0.2	<0.2
Ethylbenzene	<0.2	<0.2	3	1.4	<0.2	<0.2	1.4	<0.2	<0.2
Xylene	<0.3	<0.3	5	6.9	<0.3	<0.3	1.2	<0.2	<0.3
MW-2									
Benzene	**6,510**	**6,080**	**45**	<20	**8.9**	**39**	0.8	**12**	**53**
Toluene	**4,070**	360	15	<20	0.45	<2	<0.2	<0.2	<1
Ethylbenzene	**2,620**	**3,840**	161	24	18	81	4.1	2.3	70
Xylene	**17,800**	**24,800**	2,100	155	78	140	6.4	3.8	32
MW-3									
Benzene	**2,950**	**13,500**	**25**	**19**	1.5	3	<0.2	<0.2	<0.2
Toluene	**2,870**	<200	<220	91	0.52	0.5	<0.2	<0.2	<0.2
Ethylbenzene	**1,530**	**6,770**	35	103	9.6	11	1.3	<0.2	<0.2
Xylene	**13,400**	**41,200**	229	769	78	35	2.1	<0.2	<0.3

Note: Bold values indicate exceedance of PADEP clean-up standards.

Table 3. **Soil Analytical Results** (Results in mg/Kg)

	10/91	8/94
SB-1 (5-7 ft bgs)		
Benzene	37	35
Toluene	40.5	<20
Ethylbenzene	BDL	<20
Xylene	118	<30
SB-1 (9-11 ft bgs)		
Benzene	BDL	<20
Toluene	BDL	<20
Ethylbenzene	BDL	<20
Xylene	BDL	<30
SB-2 (5-7 ft bgs)		
Benzene	BDL	62
Toluene	**11,600**	56
Ethylbenzene	**37,100**	54
Xylene	**297,000**	300
SB-2 (9-11 ft bgs)		
Benzene	**3,570**	**4900**
Toluene	**32,900**	**1000**
Ethylbenzene	**67,100**	**1500**
Xylene	**473,000**	**8000**
SB-3 (5-7 ft bgs)		
Benzene	106	46
Toluene	60.3	<20
Ethylbenzene	249	42
Xylene	**2000**	250
SB-3 (7-9 ft bgs)		
Benzene	BDL	<1
Toluene	462	**6500**
Ethylbenzene	**2,620**	**9.800**
Xylene	**20,100**	**160,000**

Note: Bold values indicate exceedance of PADEP clean-up standards.

Following treatment, a closure report will be prepared and submitted to the PADEP for approval. The site may then be considered "clean," and the site may once again be used commercially.

REFERENCES

(API) American Petroleum Institute. 1987a. Removal and Disposal of Used Underground Storage Tanks, Recommended Practice 1604, Second Edition.

(API) American Petroleum Institute. 1987b. Safe Entry and Cleaning of Petroleum Storage Tanks, Recommended Practice 2015.

Griffin, C.J., Armstrong, J.M., and Douglass, R.H. 1991. Engineering Design Aspects of an In Situ Soil Vapor Remediation System (Sparging), In: *In Situ Bioreclamation Applications and Investigations for Hydrocarbon and Contaminated Site Remediation.* Stoneham, MA, Butterworth-Heinemann.

Johnson, P.C., Stanley, C.C., Kemblowski, M.W., Byers, D.L., and Colthart, J.D. 1990. A Practical Approach to the Design, Operation, and Monitoring of In Situ Soil Venting Systems, *Ground Water Monitoring Rev.*

Keystone Environmental Resources, Inc. August 1991. Phase I Soils Sampling Analytical Results and Proposed Phase II Subsurface investigation Work Plan in the Area or Former Underground Storage Tank Systems, Quaker State Corporation Former Conneaut Lake Facility.

Keystone Environmental Resources, Inc. May 1992. Soil Gas Survey, Quaker State Corporation Former Conneaut Lake Facility.

(PADEP) Pennsylvania Department of Environmental Resources. 1993. Closure Requirement for Underground Storage Tank Systems, December.

(USEPA) U.S. Environmental Protection Agency. 1987. Soil-Gas Measurement for Detection of Subsurface Organic Contamination, EPA/600/S2-87/027, June.

(USEPA) U.S. Environmental Protection Agency. 1992. A Citizen's Guide to Bioventing, EPA/542/F-92/008, March.

CHAPTER 6

Implementation of a Full-Scale Enhanced Recovery System for Coal Tar at a Former MGP Superfund Site

Mark W. Moeller and **Jason A. Gerrish**, Remediation Technologies, Inc., Concord, Massachusetts

James F. Villaume, Pennsylvania Power & Light Company, Allentown, Pennsylvania

Lyle A. Johnson, Western Research Institute, Laramie, Wyoming

INTRODUCTION

This chapter documents the design, construction, and operation of an enhanced recovery system for subsurface coal tar at a former manufactured gas plant (MGP) site. The enhanced recovery technology, Contained Recovery of Oily Wastes (CROW™), was developed by the Western Research Institute (WRI). This technology was implemented by RETEC, and funded by the Gas Research Institute (GRI) and Pennsylvania Power and Light Company (PP&L) at the Brodhead Creek Superfund Site in Stroudsburg, Pennsylvania.

Site Description

The Brodhead Creek site occupies a floodplain area of approximately 12 acres at the confluence of Brodhead Creek and McMichael Creek in Stroudsburg, Pennsylvania. A schematic of the site is shown in Figure 1 (ERMI, 1989). A large man-made earthen levee provides a natural barrier which will deter, but not prohibit, access. To the west, the levee extends out of the site area. To the north, the levee abuts a natural land surface and a concrete flood wall. The flood wall extends northward, is 22 feet high, and is constructed of reinforced concrete. A fence runs between the Stroudsburg Gas Company and the site to the north and west. A Publicly Owned Treatment Works (POTW) bounds the site to the south, and a levee bounds the site to the east. Other site access is inhibited by natural barriers, which include thick vegetation and wetlands.

Figure 1. **Site map.**

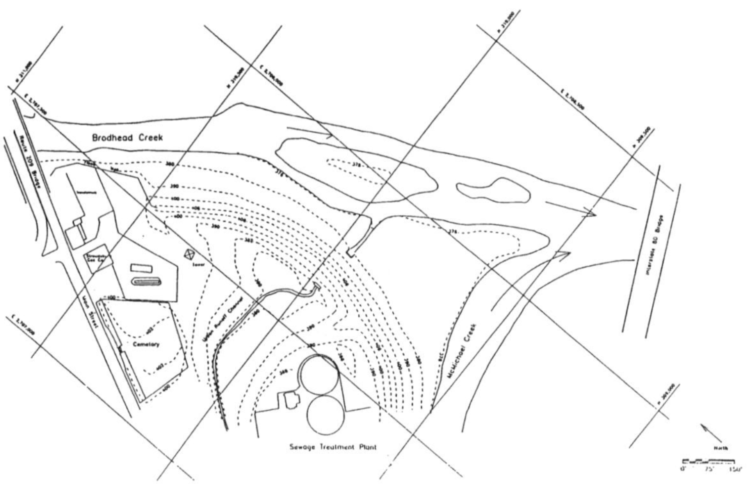

Site History

The Brodhead Creek site is a former manufactured gas plant (MGP) site. The plant was in operation from 1888 to 1944. In 1955, Brodhead Creek overran its banks and flooded the site. Due to this flood, the Army Corps of Engineers designed a flood control levee for the local creeks. In 1980, as the flood control levee on Brodhead Creek was extended, a coal tar discharge was noted. Emergency response measures included installation of a filter fence and underflow dams, and the installation of a coal tar recovery pit. In 1981, a cement and bentonite slurry wall was installed along Brodhead Creek. The slurry wall is 648 feet long and 17 feet deep, and serves to protect Brodhead Creek from the discharge of free coal tar from the site. In 1982-1983, a free coal tar recovery pumping system was installed, and 8,000 gallons of coal tar were recovered from the subsurface. In 1982, the site was placed on the National Priorities List (NPL). A Remedial Investigation (RI) was conducted from 1987 to 1989, and a Feasibility Study (FS) was conducted from 1989 to 1991.

According to the RI and subsequent investigations, free subsurface coal tar existed in two areas at the site: the area near recovery well RCC and the area near monitoring well MW-2. These areas became known as the RCC and MW-2 areas, respectively. Figure 2 (ERMI, 1989) shows the locations of these areas within the site. The coal tar appeared to have permeated the porous stream gravel layer near the surface, and accumulated on a silty sand layer 20 to 30 feet below the surface. Investigations revealed that the silty sand layer effectively contained further vertical migration of the coal tar. The two areas of free coal tar accumulation are stratigraphical depressions within the stream gravel layer. The RCC area is natural depression area, but the MW-2 area is not. Coal tar appears to

Figure 2. **Locations of RCC and MW-2 areas.**

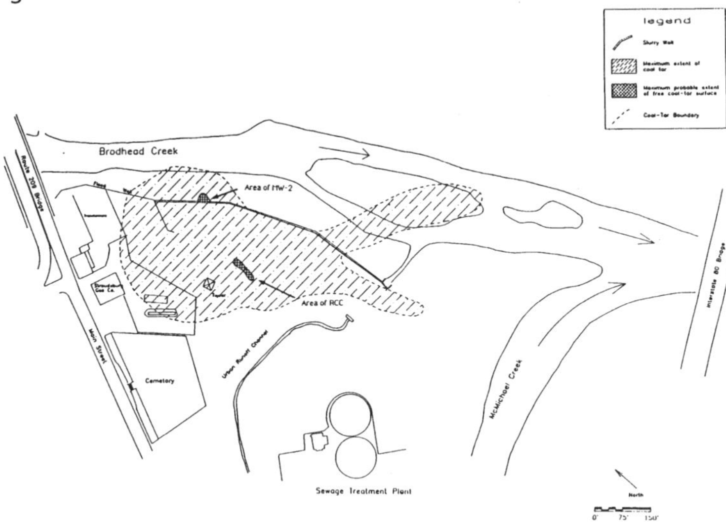

Figure 3. **Vertical distribution of free coal tar.**

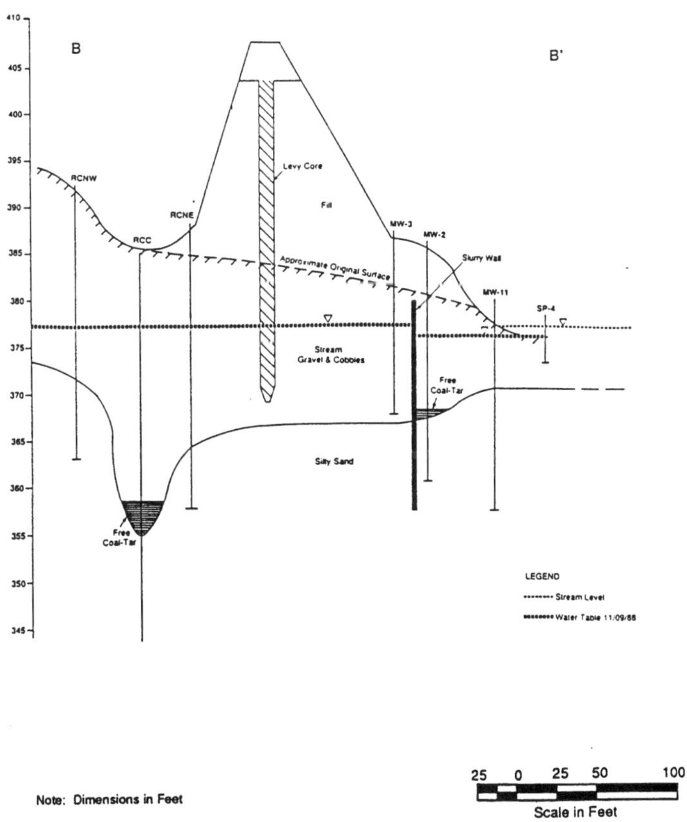

have accumulated in the MW-2 area due to the presence of the slurry wall. Figure 3 (ERMI, 1989) shows that dense coal tar may have accumulated near MW-2 only after the slurry wall was installed. The total volume of free coal tar in the RCC area is estimated at 6,000 gallons, and the total volume of free coal tar in the MW-2 area is estimated to be less than 300 gallons.

Regulatory Framework

In March of 1991, the Record of Decision (ROD) was issued that targeted enhanced product recovery as the selected remedial technology for Operable Unit 1 (OU-1), the free, subsurface coal tar. The Consent Decree was negotiated and signed in December of 1991 by the potentially responsible parties (PRPs). The Consent Decree was entered in the U.S. District Court for the Eastern District of Pennsylvania on September 2, 1992. EPA Region III agreed to allow coal tar removal in the RCC area first, before committing to a removal technology for the MW-2 area. According to the ROD, all recovered coal tar is to be incinerated.

This chapter documents the design, construction, operation, and performance of the CROW™ enhanced recovery system for free coal tar within the RCC area at the Brodhead Creek Superfund Site in Stroudsburg, Pennsylvania.

REMEDIAL DESIGN

Introduction to CROW™

CROW™ is a hot water or steam flushing technology that was first developed by the petroleum industry in the 1970s. When oil prices were relatively high, several technologies were developed that were aimed at increasing domestic oil production. CROW™ was developed to aid in the extraction of oil from oil sands and deep shale deposits. During the 1980s, the concept of hot water flushing was revisited as a potential remedial technology. CROW™, as a remedial technology, was first used at a wood preserving site to recover pentachlorophenol and fuel oil. WRI performed a two-well test which substantially reduced the hydrocarbon plume at this site.

Because of the proven successes associated with CROW™, the utility industry decided to implement the technology at a MGP site. MGP sites often have fairly pure accumulations of dense, viscous, coal tar somewhere below the ground surface. These sites are often close to a surface water body and are in populated urban areas. These factors can make excavation very difficult. A percentage of the free coal tar can generally be removed with standard pumping techniques. However, the groundwater will generally "break through" the tar at some point, limiting the recovery of the tar. This is due to the difference in viscosity between the tar and the water. The theory behind CROW™ is that heat will reduce the viscosity of tar and allow the tar to flow with the water. If an affected area can be hydraulically isolated and a gradient can be induced toward a production well, heated water or steam can be injected such that a bank of tar can be swept toward and captured by a production well. After a sufficient number of pore volume flushes, most of the free tar can be removed, leaving behind a tar residual on the soil matrix. Figure 4 (Leuschner and Johnson, 1991) shows how the absolute viscosity of tar from the Brodhead Creek Site can be reduced with elevated temperatures.

CROW™ has since proved successful in a laboratory-scale treatability study using MGP site residuals and soils from the Brodhead Creek Site. Figure 5 (Leuschner and Johnson, 1991) shows that the oil content of the site soils can

Figure 4. **Simulation of tar distillation**

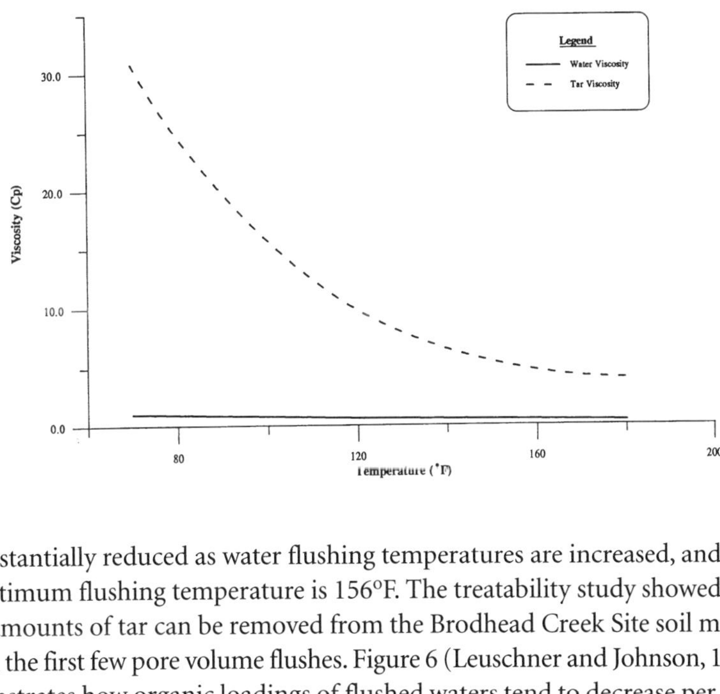

be substantially reduced as water flushing temperatures are increased, and that the optimum flushing temperature is 156°F. The treatability study showed that large amounts of tar can be removed from the Brodhead Creek Site soil matrix within the first few pore volume flushes. Figure 6 (Leuschner and Johnson, 1991) demonstrates how organic loadings of flushed waters tend to decrease per pore volume over time. According to these results, 98% of the removable coal tar will be removed after 19 pore volume flushes of 156°F water, regardless of the initial tar concentration. Other treatability results have indicated that the addition of various chemicals (surfactants) will improve the total removal rates of subsurface tars. However, chemical injection into an actual aquifer may need to be studied further before a regulatory agency will permit it as part of an enhanced recovery operation. The CROW™ system at the Brodhead Creek Site will inject heated water only.

Figure 5. **Reduction in oil content vs. temperature**

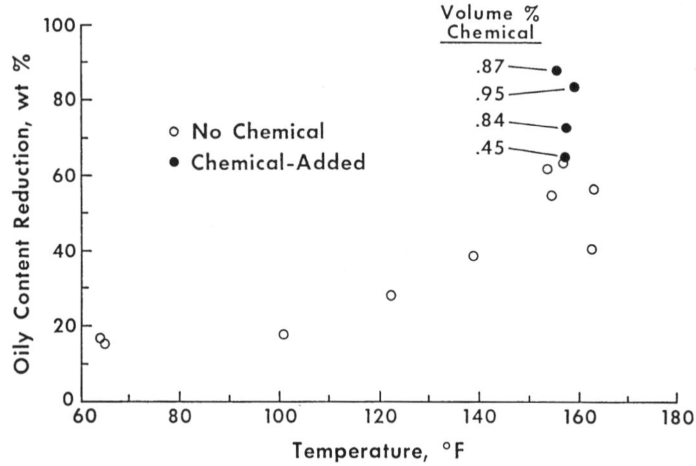

Figure 6. **Organic loadings of produced waters.**

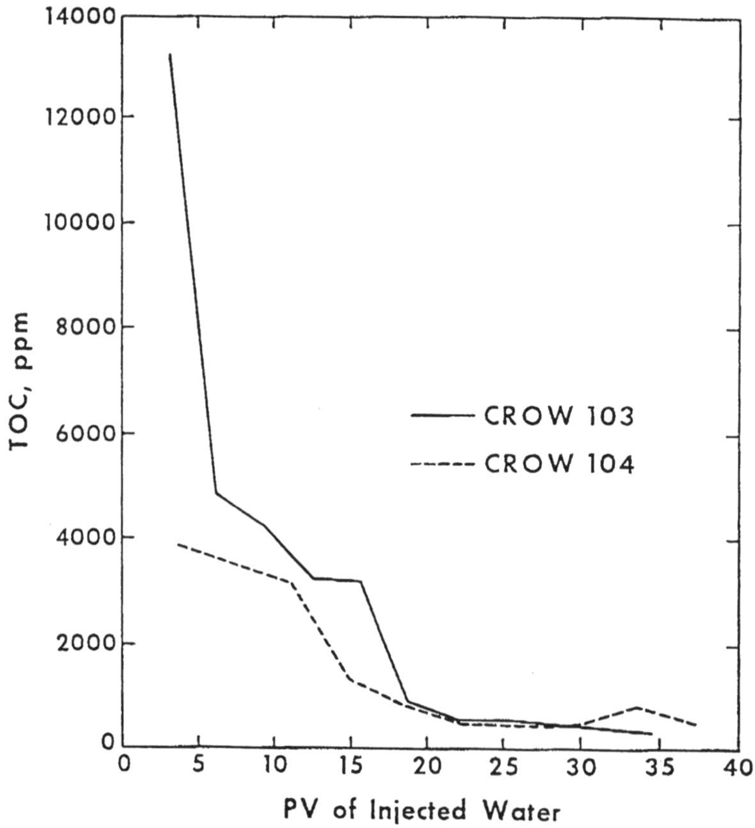

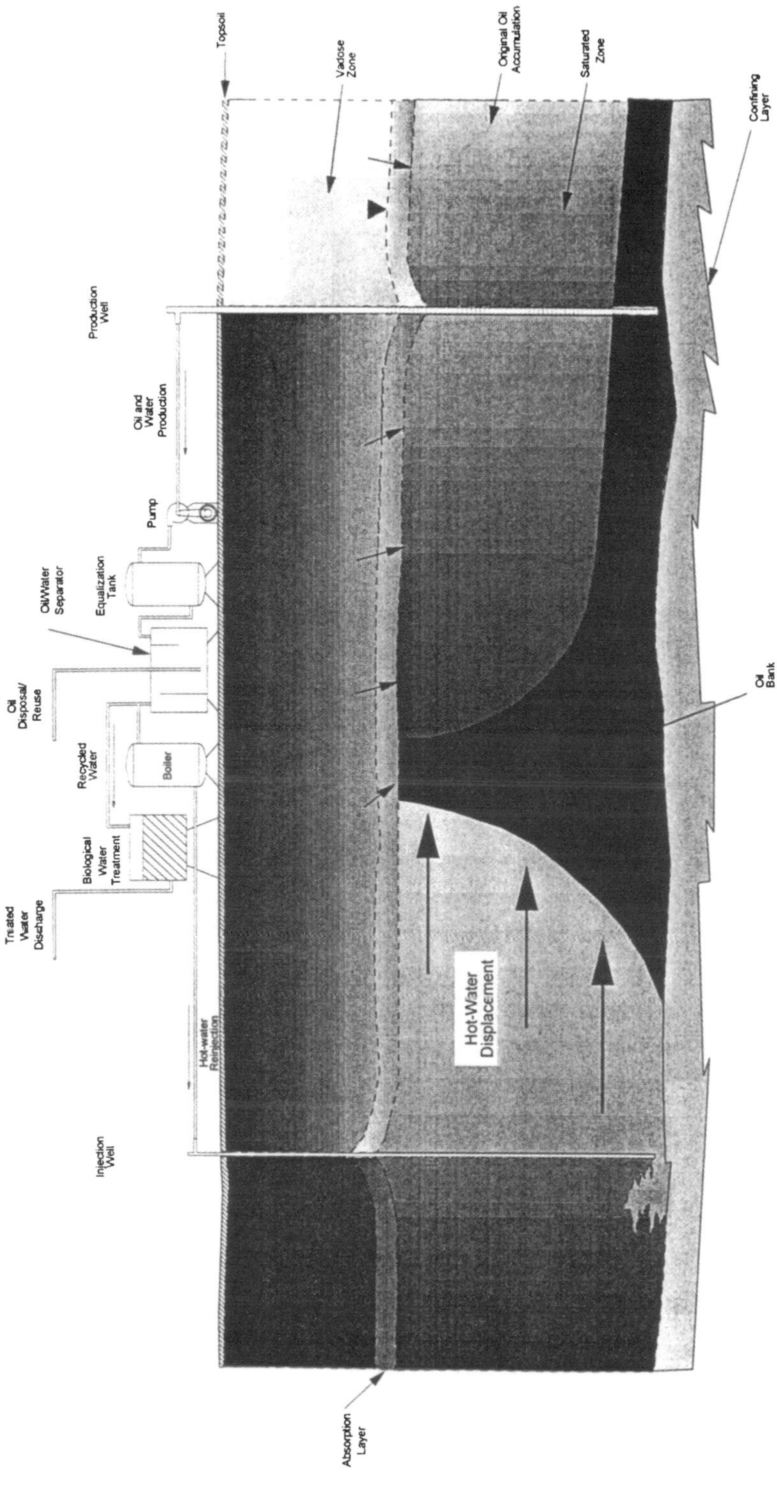

Figure 7. **CROW™ technology for DNAPL recovery.**

Basis of Design

The CROW™ system for the RCC area at the Brodhead Creek Site was designed to recover approximately 6,000 gallons of subsurface coal tar. The CROW™ process relies on injection and recovery wells drilled in such a pattern that groundwater will be laterally contained within the area of concern. Hot water is injected into the aquifer at several points surrounding the tar accumulation in order to heat and mobilize the main accumulation. Heating the tar reduces both the density and the viscosity of the tar to enable the tar to migrate with the groundwater flow.

Hot water injection and product recovery rates will be controlled to sweep accumulated tars through the aquifer sediments. As in conventional petroleum waterfloods and tar sand steamfloods, displacement of oily waste increases organic liquid saturations in the subsurface pore space to form an "oil bank." Within the "oil bank," high saturations of the organic liquid phase increase the relative permeability of the oily wastes, so the injected hot water displaces the "oil bank" to the production wells. Behind the "oil bank," the oil saturation is reduced to a residual saturation in the subsurface pore space. Figure 7 shows a schematic of the subsurface during the operation of a CROW™ system for dense nonaqueous phase liquid (DNAPL) recovery.

Tars are contained vertically during the hot water flush by controlling temperatures during the hot water displacement. Downward migration of tar is reduced by thermal expansion of the tar to a lower density. Balancing the hot water injection and production rates controls the upper boundary of the contaminated area, preventing fluid displacement into the overlying material. Volatilization of the lighter constituents of the tar can be controlled by maintaining a cold water cap over the pattern area. However, at the Brodhead Creek Site, approximately 15 feet of groundwater exists over the area to be swept. This acts as a natural cold water cap. Any volatile organics will subsequently condense once they reach the cold water and migrate back to the sweep area.

Mobilized tar and soluble constituents are contained laterally by groundwater isolation. Based on specific hydraulic measurements from the Brodhead Creek Site, WRI modeled several injection and production scenarios in order to design the appropriate well pattern for the RCC area. The well pattern and pumping rates were specifically designed to contain all the mobilized liquid phases by controlling the potentiometric surface around the affected area.

Recovering subsurface deposits of coal tar to residual saturation levels is expected to improve conditions for either natural or induced microbial degradation. In addition to reducing the volume, mobility, and toxicity of oily wastes, the CROW™ process can also increase the relative permeability of the aquifer by removing tar from pore spaces, thus creating more permeable conditions. Bacterial inoculation, nutrient addition, environmental manipulation, and by-product removal can also be controlled more effectively to accelerate complete remediation of oily waste sites. Groundwater remediation at the Brodhead Creek

Site, however, will take place under a different phase (or Operable Unit) than the current phase, which addresses free coal tar only.

The system will be operated until significant quantities of tar are no longer recoverable from the subsurface. The actual criteria states that when the latest pore volume flush produces less than 0.5% of the total recovered coal tar, then the operation may be terminated. It is at this point that 98.5% of all recoverable coal tar will have been recovered.

CROW™ Process Description

The CROW™ system, as designed, was comprised of the following components:

- two 14-inch diameter production wells;
- six 5-inch diameter injection wells;
- three 20,000 gallon tar separation tanks;
- one 20,000 gallon flow equalization tank;
- one 10,000 gallon oil storage tank;
- one water heater;
- one granular activated carbon - fluidized bed reactor (GAC/FBR);
- four 200 pound carbon adsorption (CA) units; and
- various pipes, pumps, filters, probes, and instrumentation and controls.

A Process Flow Diagram for the Brodhead Creek CROW™ system is shown in Figure 8. The system was designed to recover heated aquifer water at a rate of 115 gpm, remove tar from the water, heat and reinject 100 gpm, and biologically treat 15 gpm which discharges to Brodhead Creek. All separated tars are

Figure 8. **Process flow diagram.**

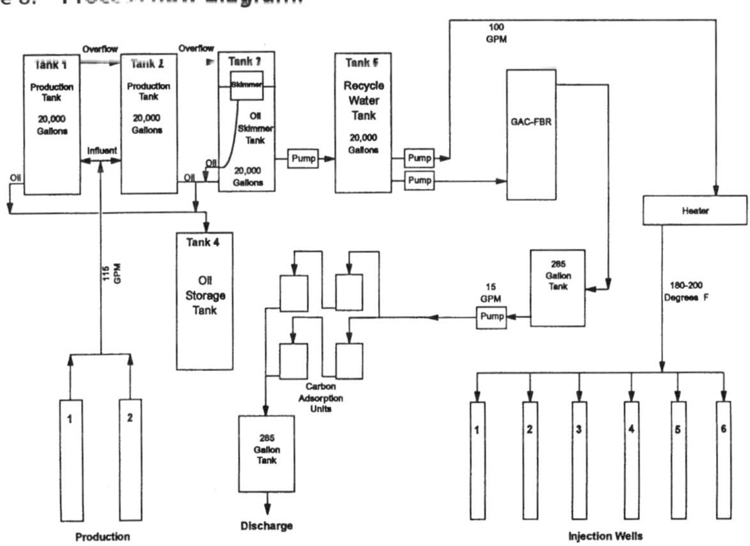

pumped to the oil storage tank, where they can be accessed for disposal in a hazardous waste incinerator or permitted cement kiln.

Tar Recovery System

The primary purpose of the CROW™ system is to recover tar from the subsurface. Tar recovery has two components: bringing the tar to the surface, and separating the tar from the water. The heated aquifer water will alter the tar density and viscosity such that it can be recovered with the water phase. However, the energy from the submersible pumps may create a tar/water emulsion that is pumped into the separation tanks on the surface. The tar at the Brodhead Creek Site has a specific gravity of 1.06 at 70°F. However, when heated to 100°F, the tar has a specific gravity of approximately 1.00 due to thermal expansion. Therefore, in order to separate the tar from the water, any emulsion must be broken and the temperature must be cooled.

During the treatability testing, WRI attempted to break simulated emulsions by adjusting the water chemistry. WRI concluded that reducing the pH to 5.0 would sufficiently break any tar/water emulsion. Therefore, an acid tank was plumbed into the production line so that the pH of the produced waters could be lowered to 5.0 before beginning the separation process. The remainder of the separation process relied on gravity separation. In order to accomplish this, the temperature of the liquid in the tanks would have to be reduced as much as possible. Because this system was designed to operate during the winter, all separation tanks were specified to be 0.25-inch, noninsulated, carbon steel. This would allow the ambient air to sufficiently cool the liquids and promote gravity settling.

Each of the tar separation tanks was designed to provide 2.9 hours of retention time. Because the three separation tanks were connected with gravity lines with no pumps or mixers, the total retention time available was 8.7 hours. As the heavy tars accumulate at the bottom of the separation tanks, a conductivity sensor activates a solenoid valve and product pump that transfers the tar accumulation to the oil storage tank. As the conductivity sensor senses a change from tar to water, the solenoid valve closes and tar transfer is stopped. In the event a floating fraction was encountered, a skimmer pump was installed within the third separation tank. This pump would remove the floating phase and pump it to the same oil storage tank.

The liquid levels in all tanks are monitored by the data acquisition and control system. As the liquid level in the oil storage tank surpasses 5,000 gallons, pump out and disposal arrangements are made.

One hundred (100) gpm of the remaining water is pumped to the water heater, which raises the water temperature to 180°F. The heated water is then reinjected into the ground. The remaining 15 gpm is pumped to a GAC/FBR for further treatment.

Water Treatment System

In addition to recovering tar from the subsurface, the CROW™ system must also contain all heat and mobilized constituents in order to prevent the unnatural spread of contamination outside of the affected area. In order to contain lateral movement of liquids, the injection/production system must be operated in a manner to induce a gradient toward the production wells. The gradient must be significant enough to overcome other natural gradients such that all injected water will be recaptured by the production wells. This is accomplished by producing more water than is injected. This system was designed to continuously produce 115 gpm from the aquifer while injecting only 100 gpm. However, this 15 gpm "slipstream" must be managed.

RETEC and PP&L applied for a National Pollution Discharge Elimination System (NPDES) Permit from the Commonwealth of Pennsylvania. The Brodhead Creek Site is a Superfund Site. Therefore, permits of this nature are not required. However, the Pennsylvania Department of Environmental Resources (PADER) issued a "permit equivalence" with water quality criteria based on GAC/FBR treatability studies on site groundwater. The discharge water quality criteria are presented in Table 1.

The 15 gpm "slipstream" is being treated by an Envirex Model 190 GAC/FBR, followed by four 200-pound CA units (two parallel sets of two). The GAC/FBR provides a continuous reactor feed of 190 gpm. The forward flow from the tar separation system is 15 gpm, and the reactor recycle flow is 175 gpm. The reactor tower is 15 feet high and 4½ feet in diameter. The reactor is filled with water and granular activated carbon, which remains in a fluidized suspension by the upward 190 gpm flow through the reactor. The fluidized carbon provides a tremendous amount of surface area in which to grow bacterial colonies. The Model 190 unit is equipped with nutrient feed and oxygen injection systems which are calibrated to provide optimal bacterial growth conditions. The CA units were installed at the request of the Environmental Protectional Agency (EPA) to provide a polishing step, and to reduce the sampling, analytical, and water storing requirements prior to discharge.

Submitted Documents

The remedial design was performed in four steps:

- the 30% Design;
- the 60% Design;
- the 90% (or Prefinal) Design; and
- the 100% (or Final) Design.

The design process was initiated after the lodging of the Consent Decree in September 1992. The Final Design was approved in May 1994. Included with the

Table 1. **Brodhead Creek Site PADER Effluent Limitations.**

Parameter	Monthly Average µg/1	Daily Maximum µg/1
VOCs		
Benzene	5.00	10.00
Ethylbenzene	10.00	20.00
Toluene	5.00	10.00
Phenol	10.00	20.00
PAHs		
Acenaphthene	10.00	20.00
Acenaphthylene	10.00	20.00
Anthracene	1.00	2.00
Benzo(a)anthracene	10.00	20.00
Benzo(a)pyrene	0. 12	0.24
Benzo(ghi)perylene	0.09	0 18
Chrysene	10.00	20.00
Dibenzo(a,h)anthracene	10.00	20.00
Fluoranthene	1.10	2.20
Fluorene	10.00	20.00
Indeno(l,2,3-cd)pyrene	1.00	2.00
Naphthalene	3.00	6.00
Phenanthrene	1.40	2.80
Pyrene	0.98	1.96
CBOD5	25,000	50,000
Total Suspended Solids	8,000	16,000
pH	6-9 at all times	

design were a total of 12 documents produced by RETEC. These documents included:

- the Predesign Work Plan,
- the Plans and Specifications,
- the Field Sampling Plan,
- the Groundwater Monitoring Plan,
- the Site Security Plan,
- the Permitting Requirements Plan (including NPDES permit application),

- the Health and Safety Plan,
- the Contingency Plan,
- the Quality Assurance Project Plan,
- the Construction Quality Assurance Plan,
- the Operations and Maintenance Plan, and
- the Decontamination Plan.

Each of these documents was submitted and approved by EPA Region III and PADER. The Health and Safety Plan was reviewed, but not officially approved by the agencies, as per their policy.

CONSTRUCTION

The Construction of the CROW™ system took place between June and October of 1994.

Site Preparation

Several site preparation activities were performed prior to initiating construction. Because the Brodhead Creek Site is a small and hilly site, a significant amount of grading of the surface soil was performed. Site preparation activities included:

- installing erosion control fencing along all downgradient boundaries of the work area to prevent site soil erosion and siltation of the wetlands;
- constructing a clean gravel subgrade for the concrete slab which serves as the floor of the process control building;
- constructing a gravel access road into the work areas of the site;
- clearing all work areas of vegetative debris and stockpiling the debris in a non-working portion of the site;
- constructing a subgrade, clay liner, and clay berms for the tank farm (all clay materials contained sufficient quantities of gravel to provide enough shear strength to serve as a foundation for the 20,000 gallon tanks. All clay materials were compacted to 95% of their maximum density.); and
- coordinating, with PP&L, the installation of three new transformers to provide sufficient power to run all site equipment, including a 480-volt, 2.2 megawatt water heater.

The construction contractor finished the tank farm by installing a layer of nonwoven filter fabric over the liner and berms, overlain by a two-foot thick gravel drainage layer.

CROW™ System Wells

The CROW™ well pattern included six external injection wells and two internal production wells. The injection wells were installed by boring with a 10-inch hollow stem auger, placing a 20 to 25-foot segment of 5-inch stainless steel well and well screen within the hole, backfilling the annulus with a 0-variety sand pack, grouting the sand pack to the surface, and sealing the well in place with concrete. One of the wells was installed through the flood control levee. In order to install this well, all well installation plans had to be approved by the Bureau of Flood Protection, and a soil ramp had to be constructed adjacent to the dike in order to access the well installation point. A special track-mounted rig was mobilized to install this injection well.

The production wells were installed by initially advancing a 24-inch borehole to the bottom of the stream gravel layer. However, some of the cobbles within this layer were eight inches in diameter, thus slowing the advancement of the hole. The first attempt utilized a bucket auger rig. After this rig advanced the hole approximately 14 feet, the surrounding earth caved into the hole, backfilling it to only 8 feet. At this point, a 24-inch casing was advanced and the material within the casing was removed with the rig. However, the casing met refusal at 21 feet, four feet short of its final position. Without the ability to drive the casing, a Barber rig was mobilized to the site. Several of the existing monitoring wells were cut back and flush mounted in order to provide the clearance necessary to bring in the large equipment trucks that were mobilized with a Barber rig. The Barber rig was successful at installing the two 25-foot segments of 14-inch galvanized well and well screen. The annulus was backfilled with 0-variety sand, grouted to the surface, and sealed with a concrete cap. All soil cuttings were stored within a 20 cubic yard rolloff box and disposed of at an industrial waste landfill. The well installation activities produced approximately 30 cubic yards of soil waste materials.

Tanks

After the tank farm had been constructed and all the wells were installed, the tanks were erected. The tanks were prefabricated and constructed of carbon steel. Four of the tanks had 20,000 gallon capacities, and one of the tanks had a 10,000 gallon capacity. The 20,000 gallon tanks stand over 30 feet tall and are 10 feet in diameter. The 10,000 gallon tank stands over 20 feet tall and is 8 feet in diameter. Each of the tanks were delivered to the site in one piece on a flatbed trailer. A crane was used to erect each one of the tanks at its proper location within the tank farm. Structural I-beams were used to tie each of the tanks together near their tops. These prevented the tanks from being toppled due to potential wind loadings.

Process Control Building

A process control building was constructed to house the GAC/FBR unit, the water heater, the CA units, all electrical control panels, and to provide a field office which housed the data acquisition and control system.

The first step to construct the process control building was to pour a 32-foot by 32-foot by 6-inch thick, reinforced concrete slab. This slab was poured very early in the construction sequence in order to allow the appropriate curing time. The next step was to erect the GAC/FBR unit on top of the slab. The GAC/FBR unit consisted of a reactor tower and an 8-foot by 18-foot equipment skid that contained all the piping, nutrient tanks, and controls needed to operate the system. After the GAC/FBR was installed, the building was erected on top of the slab and around the GAC/FBR. The building consisted of a 22-foot high, sheet metal Butler building. A separate office was framed out within the process control building. This office was used to house sampling supplies, field logs, design and operation documents, the computer, data logger, and other components of the data acquisition and control system.

Most of the operational parameters were designed to be controlled from within the process control building. All pumps could be turned on and off, solenoid valves opened and closed, and all tank and well levels, temperatures, and pressures were displayed graphically on a computer screen.

Plumbing and Electrical

After each of the major components of the CROW™ system were constructed, the system was linked together with carbon steel piping. All of the piping was supported on pipe stands with cement footings. All of the exposed piping was insulated and the potentially freezable piping was lined with heat tracing. The site was also linked together with electrical conduit and data acquisition conduit. All electrical conduit terminated in the process control building at the main electrical control panel that was linked to the onsite transformers. All data conduit was linked to the data acquisition and control system in the office portion of the process control building.

OPERATION

Startup

The CROW™ system was initially started up in November 1994. The startup sequence was as follows.

1. The GAC/FBR was inoculated with bacterial sludge provided by EFX.
2. Cold water was extracted from the aquifer to fill the water treatment tanks.
3. Cold water was injected into the aquifer after all water treatment tanks were full.
4. Forward flow to the GAC/FBR was initiated after the fluidized bed reached a pre-specified height.
5. All water lines were checked for leaks, and the data acquisition system was checked for evidence of malfunctioning meters.
6. Groundwater levels were monitored to ensure a cone of depression around the production wells.

Once all systems were working and the proper number of operators were trained, the heater was slowly turned on.

It was at this point the first problems arose. The wellhead pressure in the injection wells rose from 0 to 5 psi in a matter of a few hours. Water then started leaking around the well seals. The injection flow was reduced until the leaking stopped. After 24 hours, three of the six injection wells were accepting less than 1 gpm of heated water, instead of 15 gpm as designed.

Problems Encountered

Reduced Capacity Injection Wells

The first course of action was to redevelop the wells. The wells had been sitting idle for several months and, therefore, may have silted in. Redevelopment was successful at first, but after the heat was turned back on, the problem arose again. During early February 1995, a TV camera was lowered into the injection wells to attempt to identify the cause of the constriction. We had expected to see an iron or calcium carbonate scaling on the well screens. Instead, we saw very turbid water with a lot of suspended flake-like material. As the camera neared the bottom of the wells, we saw a large amount of growth on the well screens. The bottom of each well was filled in with 6 to 12 inches of this material. On February 28, we collected a sample of this material to be analyzed for total and dissolved iron, calcium, aluminum, and manganese. The inorganic fraction of the material was comprised almost exclusively of iron.

According to Ground Water Associates, Inc. (GWA), this material was the result of the growth of iron-oxidizing bacteria. Excessive growth of this material can dramatically reduce injection rates. In addition, GWA believed the hollow-stem augering method, by which our wells were installed, caused a disturbance to the well-formation interface that may have reduced the injection capacity of the wells.

In order to maximize the injection rates within these wells, we were forced to reduce the amount of bacteria within the wells. This can be accomplished by removing organics, oxygen, and/or iron from the injection water, redeveloping the wells, and periodically disinfecting the wells. At this point, we decided to focus on iron removal, redevelopment, and periodic disinfecting.

The wells were redeveloped once more using HTH (a chlorine-based detergent) to kill all bacterial mass and to remove suspended particulates from the well and sand pack. We then installed juttering heads on each injection well to allow the wells to be pneumatically agitated at a low pH, followed by a pumpdown that will remove solids. We designed the juttering head system to allow the operator to jutter one well at a time while injecting in the remaining five.

The CA Units

The CA units were designed to be a water polishing step. We expected very low organic loadings to the CA units; therefore, we expected them to last for most of the duration of the project. However, the initial CA units were clogging very quickly and were useless after five weeks. We were seeing a brownish-orange slimy matter coating carbon near the influent portion of the units. We also measured dissolved oxygen (DO) in the influent, effluent, and connection piping between the units. We noticed a DO drop from about 5 mg/L to about 2 mg/L. This was a sign of biological activity. The slimy matter was also a sign of the presence of iron-oxidizing bacteria. We also noticed an increase in turbidity of the effluent from the GAC/FBR and an increased presence of solids in the bag filter prior to the CA units. At this time, the height of the fluidized bed within the GAC/FBR was at its maximum. Therefore, we suspected that wasted biomass was exiting the reactor and clogging the filters and CA units.

According to EFX, this problem could be substantially reduced by removing dissolved iron from the influent water. However, the presence of spent biomass would likely remain throughout the project. Therefore, we installed a sand filter prior to the bag filter before the CA units. The sand filter was capable of removing particles larger than 20 microns and could be backwashed as necessary. The existing bag filter was capable of removing particles larger than 5 microns. The installation of the sand filter appeared to significantly increase the flow through both the bag filter and the CA units.

These problems, as well as the problems with the injection wells, both pointed to iron as the primary factor of concern.

Elevated Constituents in Discharge Water

The NPDES permit equivalence required weekly analytics on the discharge water. The chemical concentrations started to significantly increase during this same time frame. The EPA had asked us to stop discharging water into Brodhead Creek until the discharge water quality could be improved to the limits issued by PADER. At this point, we stopped the forward flow through the GAC/FBR and ran the reactor on 190 gpm of total recycle.

We noticed that the increases in chemical concentrations coincided with increases in total suspended solids. Figure 9 shows the total PAHs in the discharge water in parts per billion, and the total suspended solids in parts per million. This indicated that chemicals were probably breaking through the treatment system by attaching themselves to solid particles. Neither the Envirex unit nor the CA units were capable of removing organic constituents that were in a free phase or in an adsorbed phase. Therefore, the solution to the water quality problem must include removing suspended solids from the discharge stream. This required addressing, once again, the issues of iron, bacteria, and silt. The presence of these materials appeared to be the source of all the major operational problems.

Figure 9. **Discharge water quality data (Broadhead Creek Superfund Site Stroudsburg, Pennsylvania).**

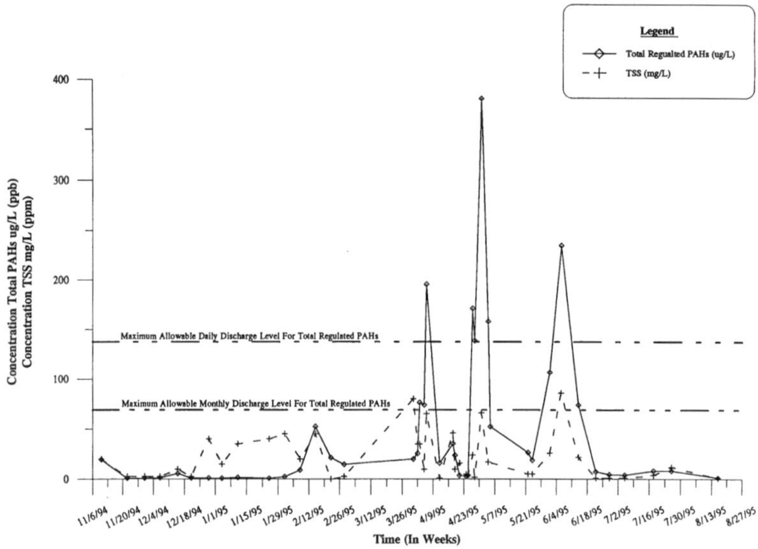

Superheating of Injection Water

An additional problem, that was initially masked by the problems with the injection wells, was the problem with the water heater. Once we were able to filter the injection water such that all injected water was accepted to the aquifer, the heater started shutting itself down at frequent intervals. This heater was designed to heat 100 gpm of water 150°F. As the water was heated, it appeared the iron oxidizing reaction was stimulated. This caused a sudden precipitation of remaining reduced iron that was caught by the effluent bag filter. This bag filter quickly clogged, choked off the injection flow, and caused a superheating of the water within the heater. The element shutdowns within the heater would quickly kick in and shut down the entire heater. Under this scenario, it became very difficult to heat the aquifer to the appropriate temperature. Figure 10 shows how the injection temperatures cycled during the early phases of the project, yet the production temperatures remained fairly steady.

In order to combat this problem, we had to design an iron removal system that totally oxidized and removed all iron in the process water before the water reached the GAC/FBR or the water heater.

Solutions

Iron Removal System

On February 22, we sent several one-gallon water samples to EFX to perform some iron treatability analyses. The proposed treatment solution design was to

Figure 10. **Injection and extraction temperature vs time (PP&L Stroudsburg Crow Project).**

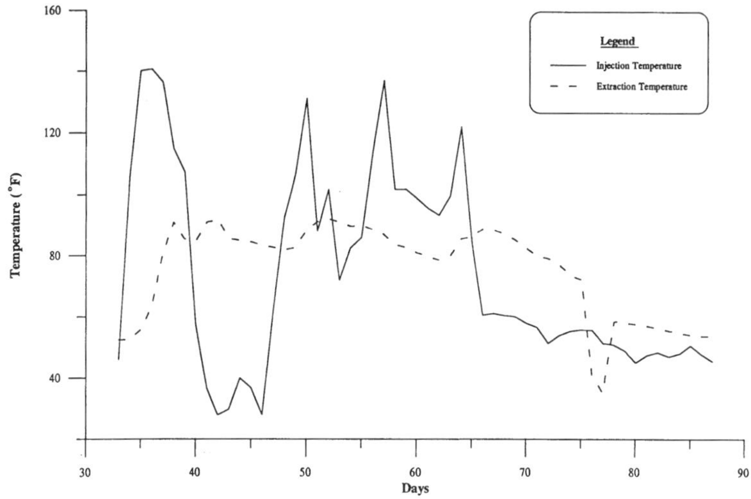

oxidize the dissolved iron within the tank farm. EFX considered several oxidants: potassium permanganate, hydrogen peroxide, oxygen gas, and dissolved oxygen-rich water. Jar tests were performed at varying pHs and dissolved iron concentrations. We selected an appropriate oxidizing agent based on cost, implementability, regulatory considerations, safety considerations, and time considerations.

According to the tests performed by EFX, the hydrogen peroxide was able to oxidize the highest percentage of iron in the shortest period of time. In addition, unlike most precipitation reactions, this reaction was capable of taking place at a neutral pH. Therefore, large amounts of acid or caustic chemicals would not be required. Figure 11 shows that greater than 99% of the reduced iron can be oxidized regardless of temperature or sodium hydroxide addition. An iron concentration of 32 ppm, three to four times the Brodhead Creek Site iron concentration, was used in this study. Figure 12 shows the pHs that resulted from the addition of sodium hydroxide. Combining these figures shows that greater than 99% of the reduced iron can be oxidized by 40 ppm of hydrogen peroxide at a pH of 7.0.

In order to implement this solution, we replumbed the tank farm such that all production water would initially be acidified to a pH of 5.0 (sulfuric acid 93%) and diverted to Tank 1, where the emulsion could be broken and tar could begin to settle. Then, the Tank 1 water would overflow to Tank 2, where the water would be neutralized (sodium hydroxide 35%) and oxidized (hydrogen peroxide 50%). As the iron oxidized, it would settle to the bottom of Tanks 2 and 3.

Figure 11. **Iron removal effectiveness using hydrogen peroxide (40 ppm).**

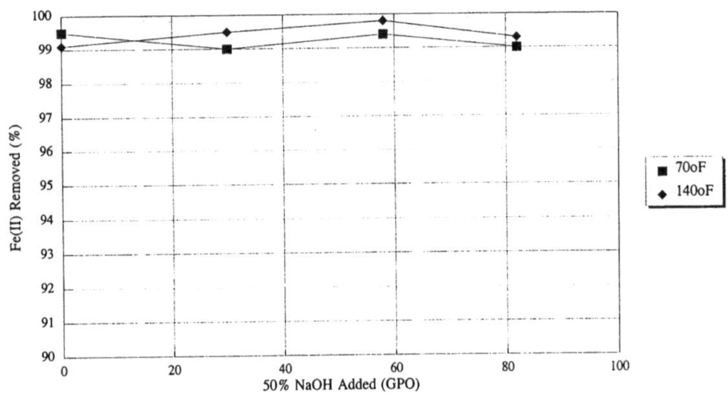

Figure 12. **Resulting pH from caustic soda addition.**

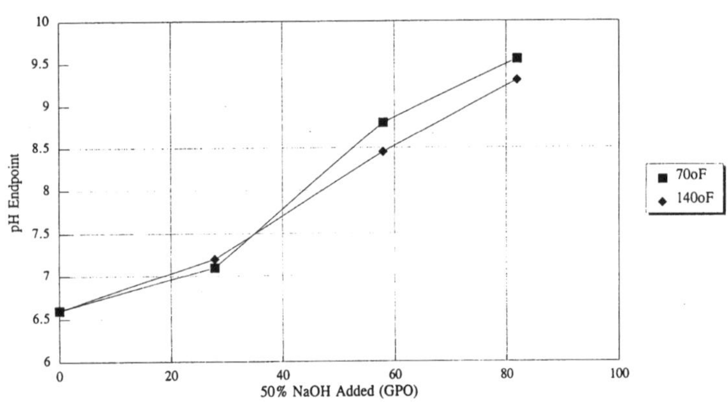

The remaining water would then flow through a series of bag filters to remove any remaining iron floc prior to entering the GAC/FBR or the water heater. Figure 13 shows the revised process flow diagram after the iron removal system was installed. Figure 14 shows how concentrations of total, reduced, and oxidized iron decreased as the water passed through the iron removal portion of the system. From this figure, a 2 ppm concentration of oxidized iron is passed on to the bag filters. In order to preserve the lives of the bag filters, we also decided to add a flocculent to the system. A flocculent should help the iron oxide settle more quickly, thus lowering oxidized iron concentrations in the heated and treated water streams. Based on experience at other sites, RETEC chose polyacrylic acid, a polymer, as the flocculent. Polyacrylic acid was tapped into the system immediately prior to Tank 2. This caused an immediate increase in

Figure 13. **Proccess flow diagram as built.**

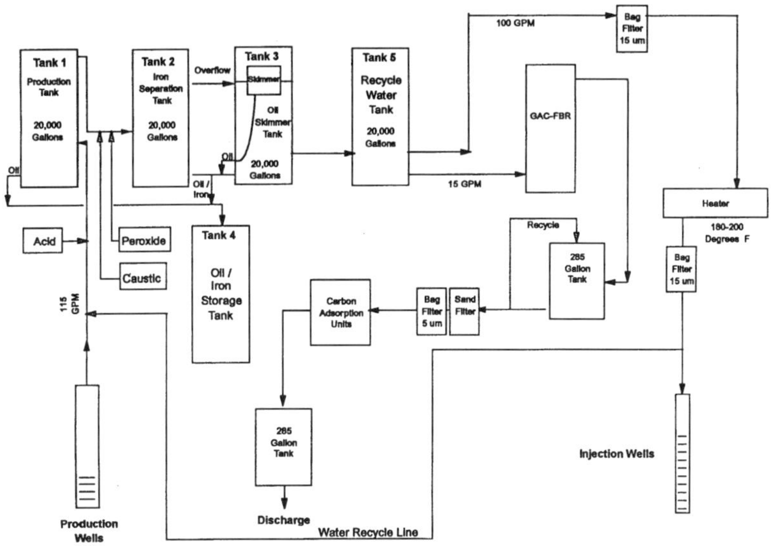

Figure 14. **Management of Iron in groundwater during operation of CROW™ process.**

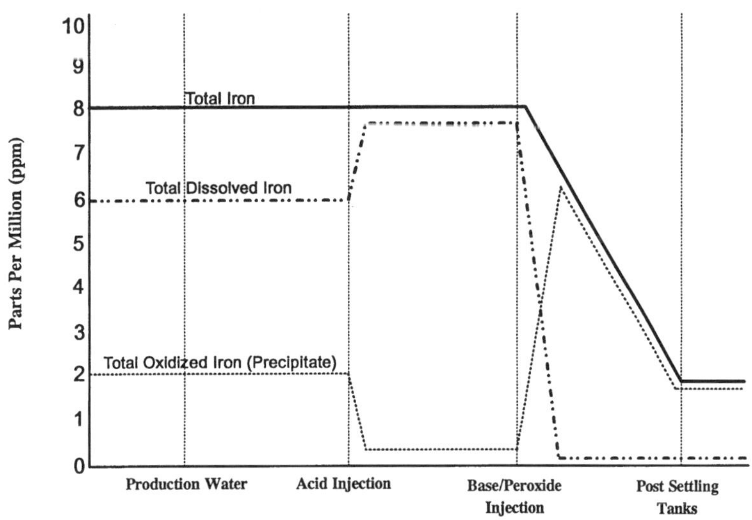

Figure 15. **Pore volumesvs. cumulative recovery.**

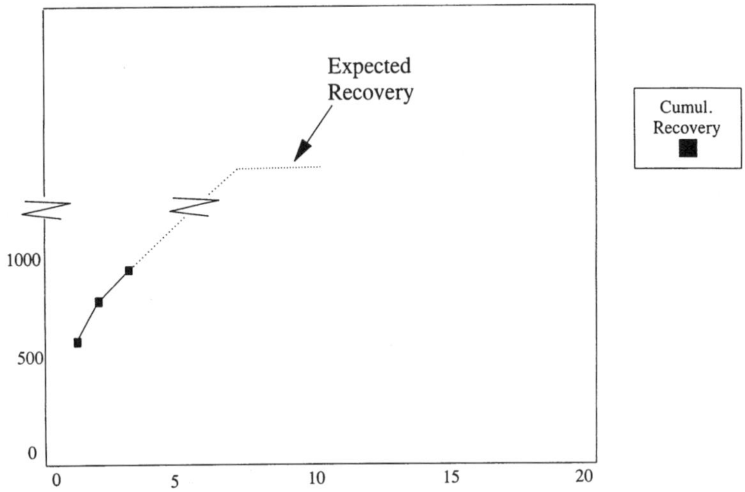

Table 2. Field Data from Envirex Model 190 Granular Activated Carbon - Fluldized Bed Reactor PP&L - Brodhead Creek Site Stroudsburg, Pennsylvania Sampled May 5, 1995

PAH	Influent	Effluent	% Removal
2-Ring	131.00	0.26	99.80%
3-Ring	373.00	2.50	99.33 %
4-Ring	13.80	15.50	-12.32 %
5-Ring[a]	2.50	2.29	8.40%
6-Ring[a]	1.00	0.54	46.00%
Total	521.3	21.09	95.95%

All concentrations in ppb.

All samples were filtered with a 0.45 micron filter prior to analysis.

[a] Concentrations of non detected compounds were taken at one half the reported detection limit.

tar and iron oxide accumulations at the bottoms of Tanks 2 and 3. It also greatly clarified the process water.

Silicon Controlled Rectifier

After the iron removal system was installed, each component of the system began to perform better. The one exception was the water heater. Even though iron was no longer present in large quantities, the system flow rate had been reduced from 100 gpm to approximately 30 gpm. The flow rate reduction was the result of head losses from the bag filters and slow injection within the injection wells.

This reduced flow rate caused continuous overheating within the water heater. This problem was resolved by installing a silicon controlled rectifier within the heater element controls. This device allowed proportional control over the heater, thus allowing a constant temperature at varying flow rates.

Steady State Operations

By the middle of July 1995, the system was operating at a steady state. Approximately 25 gpm of 180°F was being injected into the ground and approximately 40 gpm of 120°F water was being extracted from the ground. Tar was separating from the water in Tank 1, but was settling very slowly due to the hot summer temperatures. Most of the tar was settling along with the flocculated iron in Tanks 2 and 3. Figure 15 shows the amount of tar recovered as of August 23, 1995. This rate of recovery is expected to slowly decrease per pore volume of flushing until approximately 20 pore volumes, when negligible amounts of tar are expected to be removed. Table 2 shows the PAH removal efficiencies within the GAC/FBR unit. The far majority of dissolved PAHs exist as 2- or 3-ring compounds. Most 4-, 5-, and 6-ring compounds do not exist in a dissolved state, but are most likely adsorbed to a <0.5 micron particle that was not filtered.

RETEC and WRI expect to operate this CROW™ system through December 1995.

CONCLUSIONS

Much was learned throughout the construction and operating phases of the CROW™ system at the Brodhead Creek Superfund Site. This project was the first to attempt to utilize this technology for a full-scale remediation effort. Based on these experiences, several changes will be made during the design process of future CROW™ systems. The purpose of this section is to review the lessons learned from the Brodhead Creek Site Project.

1. When installing injection wells, use a cased boring method instead of an augering boring method. This allows injected water to pass through the borehole interface more easily.
2. Do not attempt to inject water into a very shallow aquifer (<5 feet). Unless the well and the aquifer are very accepting of the flow, the injected water is likely to short-circuit around the well and end up on the ground surface.
3. Use a sand pack around the well that is at least four inches thick at all points. The sand should be significantly coarser than that used for monitoring wells.
4. Expect to surge and disinfect the wells on a regular basis. The capacity of the wells will always diminish over time.
5. Model the groundwater flow with a wide range of assumed hydraulic conductivities. Choose a very conservative (low) hydraulic conductivity when sizing and positioning the injection and production wells.

6. Select a water heater or boiler that can maintain precise temperatures at a variety of flow rates.
7. Assume that iron will become a problem in some component of the system and that all iron must be removed.
8. Size all pumps to enable the water streams to be filtered as much as possible. Install many backwashable filters and low micron filters in order to protect the design capacity of the more expensive equipment.

Additional tar recovery results are expected during the fall and winter of 1995. These results will be published in 1996.

REFERENCES

Environmental Resources Management, Inc., *Brodhead Creek Site Draft Remedial Investigation Report*, August 15, 1989.

Leuschner, A.P., and Johnson, L.A., *In Situ Soils Treatment Using the CROW™ Process and Bioremediation*, In Situ and On-Site Bioreclamation Symposium, San Diego, CA, March 19-21, 1991.

CHAPTER 7

Bench- and Pilot-Scale Evaluation of a Carbon-Based Fluidized Bed Biotreatment System for Groundwaters from a Manufactured Gas Plant Site

Gary M. Grey, **O. Karl Scheible**, and **Joy A. Maiello**, HydroQual, Inc., Mahwah, New Jersey

William Guarini, Envirogen, Princeton, New Jersey

Paul Sutton, P.M. Sutton & Associates, Inc., Bethel, Connecticut

INTRODUCTION

Fluidized-bed biological treatment was evaluated as part of a comprehensive bench- and pilot-scale groundwater treatability study for a manufactured gas plant (MGP) site located in Toms River, New Jersey (HydroQual, 1994). The study was sponsored by Jersey Central Power and Light Company (JCP&L), New Jersey Natural Gas (NJNG), and the Electric Power Research Institute (EPRI), and was conducted over the period of April 1993 through August 1994.

The overall project focused on the evaluation of an integrated process sequence for treating contaminated groundwater. At the Toms River site, as well as most MGP sites, the groundwaters contain dense nonaqueous phase liquids (DNAPL), tars, and emulsified oils. Physical/chemical processes were evaluated for removal of these constituents. The primary soluble contaminants are PAHs and VOCs, primarily benzene, toluene, ethylbenzene and xylene (BTEX), and styrene. Biological treatment, air stripping, and carbon adsorption were evaluated for their capacity to remove these constituents. Air stripping and carbon adsorption were found to be effective as alternatives to biological treatment for VOC and PAH removal. This chapter presents the results of the bench- and pilot-scale evaluations of fluidized-bed biological treatment. The results of the comprehensive study are presented in a companion paper (Scheible, Grey, and Maiello, 1995).

EXPERIMENTAL METHODS

The objective of the study was to evaluate candidate treatment technologies through bench-scale testing, followed by an evaluation of appropriate treatment trains through pilot testing. The bench-scale studies were conducted on groundwater samples collected at the site. The pilot studies were conducted onsite with groundwater drawn directly from selected wells to the pilot plant. Both study phases included operation over a range of groundwater characteristics that might be encountered by the full-scale system over its operating life.

The bench studies determined that dissolved air flotation (DAF) was an appropriate pretreatment process producing a subnatant suitable for biological treatment. The bench-scale biological treatment evaluations included activated sludge, sand-based (SFB), and carbon-based (CFB) fluidized-bed processes. The activated sludge unit consisted of a 3-L reactor and an external clarifier. The fluidized-bed units consisted of a 1-L reactor with a 0.5 L fluid-bed volume and an external recycle aeration cell. The activated sludge and sand-based fluidized-bed systems were initially seeded with an activated sludge from an oil refinery to hasten the acclimation period. The bench-scale CFB was started with carbon used in a previous MGP site treatability study. All three bench units were fed DAF subnatant from groundwater taken from a highly contaminated (high strength) source area of the site. After this, only the CFB was operated and fed a DAF subnatant from more moderately contaminated (moderate strength) site groundwaters.

The pilot study consisted of a series of three campaigns which encompassed a wide range of groundwater contamination levels. The onsite pilot-scale biological treatment process was preceded by a DAF system when treating high (Campaign One) and moderate strength (Campaign Two) groundwaters. The DAF process was not included when treating low strength (Campaign Three) groundwater since free oils or tars were not present.

The CFB system was supplied by Envirogen, Inc., Lawrenceville, NJ. The system was a skid-mounted, microprocessor-controlled pilot unit consisting of a 1-ft diameter by 15-ft deep reactor with feed and recycle pumps. In addition, it included an integral high rate clarifier/recycle tank, chemical feed systems for nutrient addition or pH control, data logger, and dissolved oxygen control loop for regulating the pure oxygen supply. The system continuously monitors DO, pH, temperature, bed height, influent flow, and oxygen consumption, and has the ability to respond to off-spec operating conditions (valve or pump malfunctions, pH upsets, etc.) by automatically switching to alternate operating modes or shutdown to protect the bed.

The CFB system was initially charged with coconut-based granular activated carbon, 8 x 30 mesh, and operated at a hydraulic flux of approximately 1350 ft^3/ft^2-d with tap water to wash out fines. Coconut-based carbon is a harder carbon than coal or wood-based carbons, and, as such, suffers less attrition due to abrasion. High strength DAF subnatant was then fed to the unit for approximately 3 weeks to saturate the carbon with organics before seeding. This start-up procedure was used to minimize any subsequent biases to the organics removal that might be ascribed to the adsorptive capacity of the virgin carbon. Once saturated, the bed was seeded with activated sludge from a local oil refinery's treatment plant.

BENCH-SCALE EVALUATION RESULTS

Table 1 presents a summary of the operations and performance data for the bench-scale systems. The organic loadings to the bench-scale systems were relatively low: approximately 0.2 gCOD/g-d for the suspended growth system and 1.8 to 2.0 gCOD/L-d for the fluidized-bed bioreactors. The high strength feed to the units had a relatively low concentration of BTEX compounds, 11 μg/L, which was due to the volatile nature of the BTEX compounds and handling procedures and not the result of pretreatment operations. The influent PAH level was 1224 μg/L, of which 480 μg/L were carcinogenic compounds such as benzo(a)pyrene, benzo(b)fluorantherene, benzo(a)anthracene, and chrysene. No attempt was made to optimize loadings; rather, the bench studies were conducted to assess the biodegradability of the dissolved organics and to determine which process configuration would be most appropriate.

The average mixed liquor suspended solids (MLSS) concentration in the activated sludge system was 2,480 mg/L, with an estimated sludge age of 40 days. Biomass levels were not directly measured in the fluidized-bed units; however, based on their removal efficiencies relative to the activated sludge system, biomass levels on the order of 10,000 mg/L were estimated. In all reactors, the dissolved oxygen concentrations were maintained above 5 mg/L.

The feed to all the biological systems had an average total and filtered COD of 128 and 83 mg/L, respectively. The effluents from the activated sludge, sand-, and carbon-based fluidized-bed units had average soluble COD levels of 48, 59 and 37 mg/L, respectively. When considering performance of the units, it is best to assess removals on the basis of total COD in and soluble COD out, since the particulate COD in the effluent is more a function of downstream filtration processes which are not controlled in the biotreatability evaluation.

All three biological treatment systems were found to provide a substantial reduction of COD and PAHs. The CFB provided the best treatment, with a 71% reduction of COD and effluent PAH concentrations less than 5 μg/L. Based on this performance, the CFB process was selected for continued operation on the moderate strength DAF subnatant. The effluent quality was similar to that produced from treating the high strength subnatant, with an effluent soluble COD of 38 mg/L and effluent PAHs less than 5 μg/L.

Table 1. **Summary of bench-scale biological treatment performance.**

		Activated Sludge	High Strength Fluidized-Bed Sand	High Strength Fluidized-Bed Carbon	Moderate - Strength Carbon-Based Fluidized-Bed
Organic Loading		0.19[(a)]	1.8[(b)]	1.8[(b)]	2[(b)]
Detention Time	(hrs)	6.5	1.7	1.7	0.9
Hydraulic Loading Rate	(gpm/sf)	-	12	12	12
Feed Flow	(L/d)	3.7	6.4	6.5	12.4
Recycle Flow	(L/d)	0.1	1.0	1.0	1.0
Influent					
COD-Total	(mg/L)	128	128	128	74
COD-Filtered	(mg/L)	83	83	83	62
TSS	(mg/L)	53	53	53	15
BTEX	(μg/L)	11	11	11	6
Carcinogenic PAH	(μg/L)	480	480	480	<5
Total PAH	(μg/L)	1224	1224	1224	78
Effluent					
COD-Total	(mg/L)	63	74	57	44
COD-Filtered	(mg/L)	48	59	37	38
TSS	(mg/L)	16	31	16	6.3
BTEX	(μg/L)	<5	<5	<5	<5
Carcinogenic PAH	(μg/L)	18.1	27	<5	<5
Total PAH	(μg/L)	24	45	<5	<5

[(a)]gCOD/gMLSS
[(b)]gCOD/L-d

The bench-scale studies demonstrated that biological treatment was viable and that the CFB was the most appropriate process. Given the behavior of the system with time, it was believed that the mechanism was not simple adsorption. Rather, the carbon was likely near saturation, but with capacity continually renewed by biological degradation. The larger, longer-chained organics have a greater affinity to the carbon, but are less amenable to biodegradation. Holding them onto the carbon for periods of time greater than the hydraulic retention time of the system enhances the ability of the attached biomass to degrade the material.

Other advantages of fluidized-bed technology include smaller space requirements because of the significantly higher biomass levels it can maintain with an equivalent volume; it can be easily configured for offgas capture; conservation of heat; and it has a high oxygen transfer capacity. For these reasons, the CFB process was incorporated into the pilot-scale treatment system at the Toms River MGP site. The studies suggested that there was sufficient nitrogen, but that phosphorus was deficient for effective long-term biological treatment. Phosphorus addition was thus included as standard practice for the pilot studies.

PILOT STUDY RESULTS

The CFB biological treatment system was operated continuously. The feed rate ranged from approximately 1 to 3 gpm, depending on the target loading and the contaminant level in the groundwater. The system was operated at 20 to 30 percent bed expansion, with an expanded bed depth of 10 to 12 feet. Analysis of the fluidized-bed media indicated biomass concentrations of 5,000 to 10,000 mg/L. The recycle pump rate was automatically adjusted to maintain a constant, preset flux through the reactor, typically 7 to 9 gpm/ft^2. Pure oxygen was used to maintain a desired minimum DO of 1.5 mg/L in the reactor; oxygen input and utilization rates were controlled by the microprocessor. Oxygen utilization was also monitored by weighing the oxygen cylinders. The CO_2 produced as a by-product of biochemical oxidation tended to stay in solution and depress the pH, rather than strip out as an offgas. A pH depression was observed in the pilot study to a point where the caustic feed system was activated to maintain pH at approximately 7. Caustic requirements averaged 30 to 40 mg/L as $CaCO_3$.

Figure 1 presents a chronological plot of CFB loadings and performance, denoted by campaign. During the three campaigns, COD loads ranged from 2 to 4.5 gCOD/L-d, progressively increasing during the program. Influent total and filtered COD decreased from one campaign to the next, reflecting the change in groundwater characteristics. Effluent quality, as measured by filtered COD, averaged 30 to 80 mg/L, responding slightly to influent COD concentration. During Campaigns 1 and 2, net sludge production was estimated to be between 0.2 and 0.4 gTSS/gCOD removed. Sand filtration was practiced to remove CFB effluent suspended solids. In a full-scale system, a clarifier or high-rate prefiltration may be more appropriate for effluent solids removal. Subsequent filtration would then be applied if carbon adsorption were to be used for final effluent polishing.

Figure 2 presents COD removal as a function of COD loading for all campaigns. The line shown through the data is a least squares regression, indicating an average COD removal rate of 56%. COD removal dropped to approximately 42% during Campaign 3, although the effluent-filtered COD was the lowest of all campaigns, averaging 37 mg/L. The lower removal rate is an artifact of the low initial COD of the groundwater.

Figure 1. **Chronological plot of COD loading and performance.**

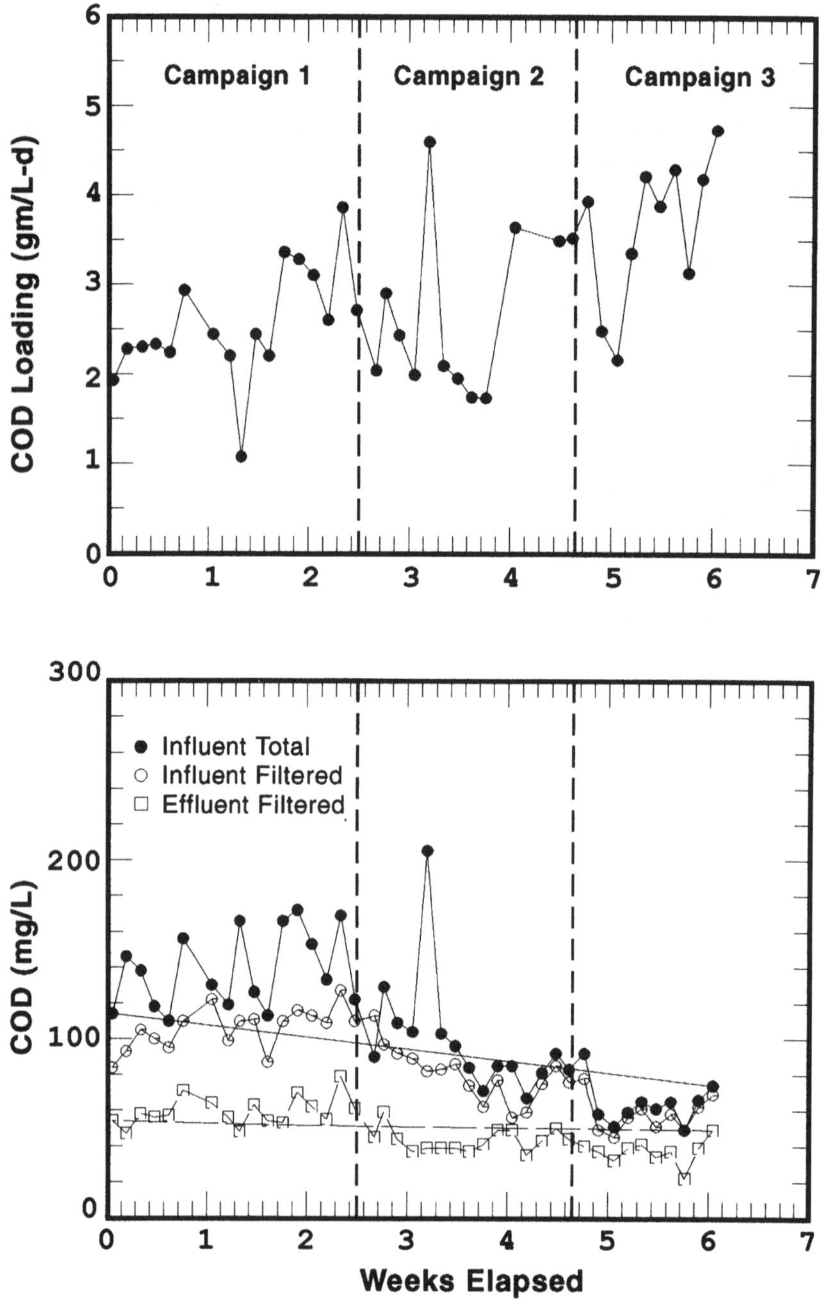

Figure 2. **COD removal as a function of the total COD loading rate.**

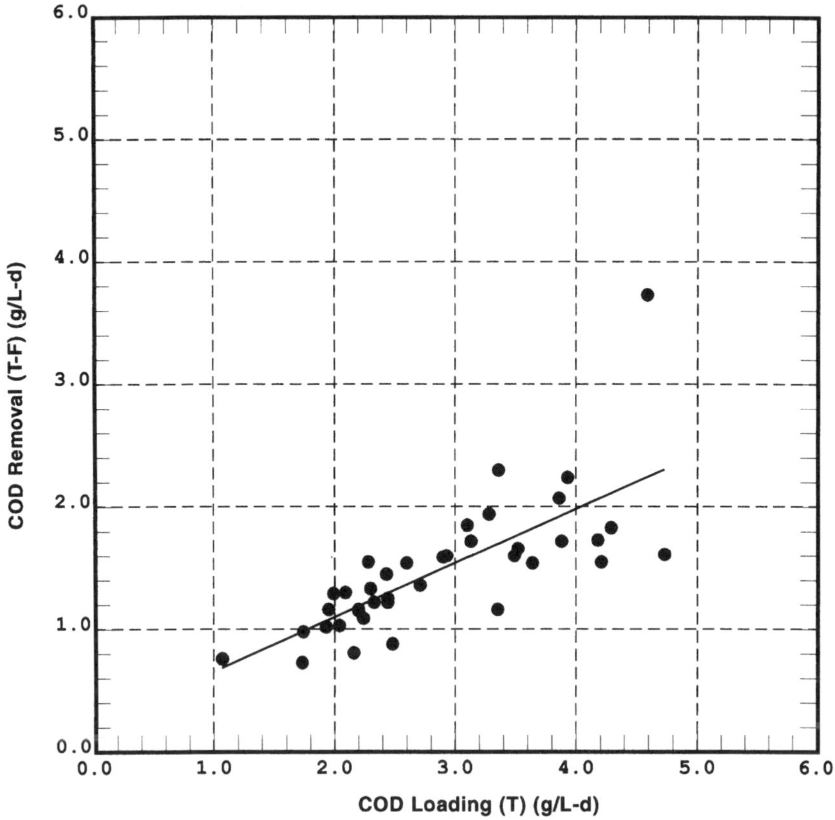

Table 2. **Summary of CFB influent and effluent BTEX and PAH results.**

	Campaign 1		Campaign 2		Campaign 3	
	Influent	Effluent	Influent	Effluent	Influent	Effluent
Volatiles (μg/L)						
Benzene	1900	6.4	960	<2	540	<2
BTEX	5200	66	2500	<2	1600	<2
Total VOCs	17000	68	2500	<2	1600	<2
PAHs (μg/L)						
Carcinogenic PAH	75	43.8	<5	<5	<5	<5
Total PAH	1800	92	210	<5	270	<5

Table 2 presents a summary of influent and effluent quality with respect to the organic constituents for each campaign. During Campaign 1, the CFB process was capable of reducing total volatiles by 99% to a total effluent VOC concentration of 68 μg/L. PAHs were reduced by 94% to 92 μg/L. All of the residual PAH was adsorbed to effluent suspended solids; the PAH concentrations were less than 5 μg/L after filtration. The residual PAHs were mainly those with 4 or 5 fused benzene rings, which are more difficult to biograde. The CFB feeds for Campaigns 2 and 3 were generally similar, with influent total VOCs ranging from 1600 to 2500 μg/L, and PAHs ranging from 210 to 270 μg/L. During both campaigns, essentially complete removal of BTEX and PAH was observed, with effluent concentrations of less than 2 μg/L and less than 5 μg/L, respectively.

A sample of the fluidized-bed clarifier solids was taken at the end of the study and analyzed for volatile and semivolatile organic compounds. TCLP analysis of the biological sludge indicated compliance with all Toxicity Characteristic Leaching Procedure (TCLP) regulatory limits. The sludge solids concentration was 9500 mg/L. All volatile organic compounds analyzed for were below analytical detection levels. On a dry weight basis, several of the PAHs were found in the solids, likely by bioaccumulation or adsorption mechanisms. The levels shown, however, confirm that the primary route of removal was by biodegradation. Total PAHs were measured at 1.31 mg/g in the solids. If 7.3 kg solids were discharged from the system overall (determined from a net solids production calculation), and assuming that the loss of PAHs through the effluent is negligible, less than 5 percent, then a total of approximately 9600 mg of PAHs were also discharged with the solids. In order to obtain this amount of PAHs, the influent to the CFB system had to average approximately 28 μg/L. In fact, the overall weighted average was approximately 820 μg/L, indicating removal by a mechanism other than adsorption.

CFB offgas flows were measured and samples taken to determine the significance of volatilization as a removal mechanism for VOCs. Essentially no offgas flow was detected, and gas samples of the reactor head space contained less than 1 ppm VOCs. The lack of any offgas was attributed to the high oxygen transfer efficiencies of the oxygenator with a pure oxygen supply.

No offgas production is an advantage of this type of biological treatment system over other secondary treatment processes such as air stripping, which is primarily applicable to VOC contaminants. In fact, in a parallel study, air stripping was evaluated as an alternative treatment process to the CFB system. Bench- and pilot-scale testing using high strength groundwater with similar characteristics suggested that air stripping was capable of reducing VOCs to trace or less than detection limit levels. However, only 32 to 44 percent removal of PAHs was observed, indicating that a polishing step may be required (i.e., carbon adsorption) to removal any remaining organics. Air stripping provided similar performance to the biological process for low to moderately contaminated waters.

Figure 3. **Capital and O&M costs for treating high strength groundwater**

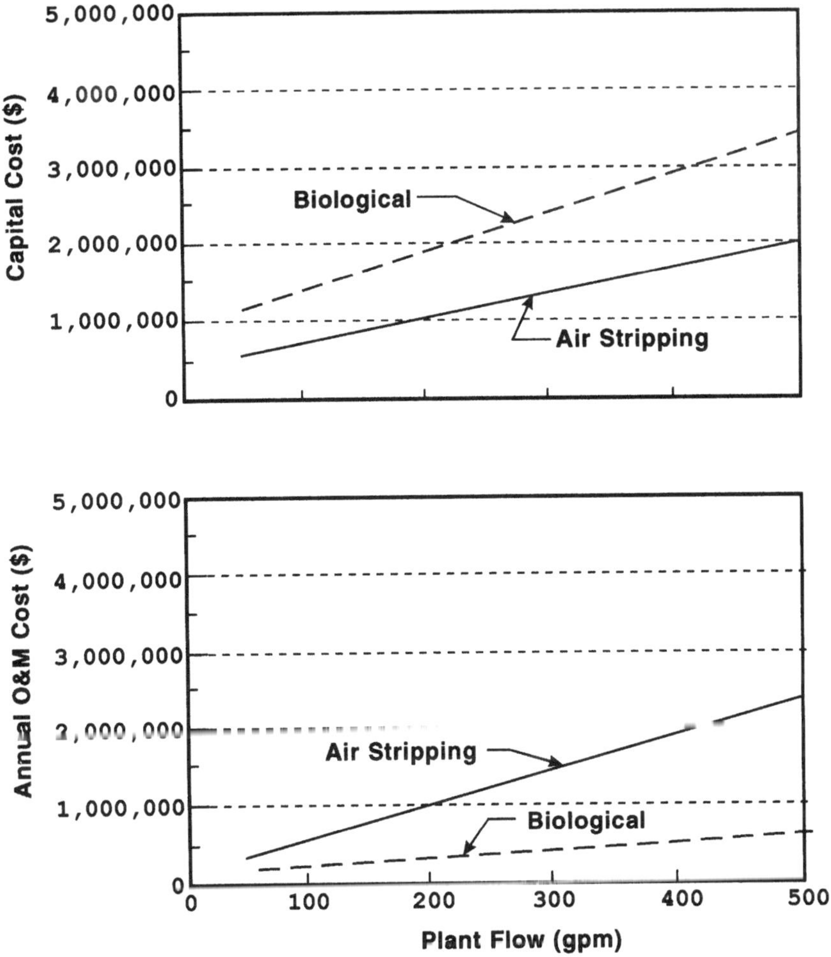

Figure 3 presents the capital (top panel) and operation and maintenance (bottom panel) costs for both fluidized-bed biological and air stripper based treatment trains for high strength groundwater. These treatment trains include DAF as a primary treatment process, and filtration and carbon adsorption as a polishing step. As shown, the biological treatment train is more capital intensive; however, this can be offset rather quickly (3 to 5 years for a 50 gpm system, 1 year for a 250 gpm system) in operation and maintenance savings. The most significant element of the operation and maintenance cost for the air stripper train is liquid phase carbon consumption. This highlights the advantage of biological treatment being a destructive process.

SUMMARY

The carbon-based fluidized-bed biological treatment system was successful in treating DAF subnatant generated from high and moderate strength groundwaters, and for direct application to low strength groundwaters. Substantial COD and complete BTEX and PAH removals were observed. Residual PAHs present in the CFB effluent were found to be adsorbed to effluent suspended solids, and removed by subsequent filtration. Biodegradation is believed to be the primary removal mechanism for BTEX; volatilization was found to be insignificant. Phosphorus addition was necessary to satisfy nutrient requirements. Alkalinity addition was also necessary for effective pH control. Oxygen requirements and sludge production estimates appear to be similar to those of an activated sludge system. The biosolids produced are nonhazardous, based on TCLP criteria.

For treating high strength groundwaters, air stripping, although a very efficient removal process for VOCs (and PAHs to a more limited extent), may require offgas treatment. Performance of biological and air stripping processes under moderate and low strength groundwater conditions were similar; their selection may be dictated by capital, operating, and maintenance life-cycle costs.

REFERENCES

HydroQual, Inc. 1994. "Bench- and Pilot-Scale Treatability Studies for Contaminated Ground Waters from the Toms River Former Manufactured Gas Plant Site." Prepared under subcontract to Foster Wheeler Environmental Corporation, Lyndhurst, NJ, for the Electric Power Research Institute, Jersey Central Power and Light, and New Jersey Natural Gas.

Scheible, O. K, Grey, G.M., and Maiello, J.A. 1995. "Bench- and Pilot-Scale Evaluation of an Integrated Treatment System for Ground Waters from a Manufactured Gas Plant Site," *Proceedings of the Third Annual International Symposium on In-Situ and On-Site Bioreclimation*, San Diego CA, April 1995 (In Press).

CHAPTER 8

A Quantitative Evaluation of TPH and PAH Reduction in Diesel-Contaminated Soils as a Result of Enhanced Biodegradation Through Bioventing

David A. Nickerson and **James N. Baker**, Parsons Engineering Science, Inc., Liverpool, New York

INTRODUCTION

Two remediation systems were recently installed at an underground storage tank (UST) facility in Pittsburgh, Pennsylvania as part of a remedial action program approved by the Pennsylvania Department of Environmental Protection. The remediation systems were designed to work together to reduce residual diesel contamination in soils and prevent further degradation of groundwater quality at the site.

An electric submersible pump was installed inside a four-inch diameter groundwater monitoring well in which moderate levels of dissolved TPH had been previously detected. The pumping well is located in the central area of the contaminated soil plume. Recovered groundwater is treated with granular activated carbon at the surface prior to discharge to a permitted, treated sewer system. The pump is operated by a programmable timer system with the objective of main taining a continuous depression of the water table in the vicinity of the well.

By continuously lowering the water table in the area of greatest soil contamination, the groundwater treatment system exposes contaminated soil to the injected air, further facilitating the biodegradation process at the capillary fringe, and accelerating site cleanup. This dual remediation process has proved very successful at similar diesel-contaminated sites (Nickerson et al., 1994).

During the installation of the groundwater treatment system, a bioventing pilot test was conducted to evaluate the effectiveness of low-flow air injection into the subsurface as a means of enhancing the natural biodegradation of the remaining diesel fuel residuals partitioned to the soil matrix. Results of the pilot study indicated that an average initial biodegradation rate of 2,100 milligrams of TPH per kilogram of soil per year could be achieved under oxygen-enhanced conditions, making bioventing a feasible remedial alternative for cleanup of contaminated soils beneath the facility.

Soil samples were collected in the areas of highest known soil contamination during the installation of soil vapor monitoring points and injection wells. Soil samples representing three depth intervals at each location were analyzed for BTEX, TPH, and PAH concentrations to provide baseline data for comparison with confirmatory samples collected later during the course of the full-scale remediation program, and to estimate cleanup time. During the past several years, bioventing and groundwater aeration systems have been shown to successfully enhance biodegradation of PAH compounds in soil and groundwater, especially lower-molecular-weight compounds such as naphthalene (Alleman et al., n.d.; Symons et al., n.d.). Although the remediation goal for this site was based on a TPH target level of 500 ppm, the regulatory agency has established soil cleanup guidelines for PAH compounds.

The following sections provide a summary of the full-scale bioventing demonstration and results obtained to date.

SITE DESCRIPTION

The project site is located in Allegheny County in the southwestern part of Pennsylvania. This area is characterized by warm, humid summers and cold winters, with an average precipitation of 36 to 40 inches. The area surrounding the downtown Pittsburgh facility is typical of a moderate to heavy commercial setting consisting of lands occupied by business and industry.

Surficial geology near the facility consists primarily of fine sand and gravel with lesser amounts of silt and clay overlying Pennsylvanian age bedrock. The shallow sediments immediately underlying the site consist of construction fill material and light brown and tan, cohesive silt and silty clays. This unit is five to seven feet thick and is underlain by approximately 10 feet of moderately- to well-sorted reddish-brown fine-grained sand. Underlying the fine sand is a mixed silt, sand, and gravel unit from a depth of approximately 17 feet to greater than 40 feet below ground surface.

The principal surface water bodies in the area are the Allegheny, Monongahela, and Ohio Rivers, the confluence of which is located approximately one mile west of the facility. The shallow, unconfined groundwater table occurs at a depth of approximately 35 feet within the silt, sand, and gravel unit, with observed seasonal fluctuations of about two to three feet. Local groundwater flow direction is toward the northeast, with an average hydraulic gradient of 0.001 feet per foot.

In 1993, five underground storage tanks were removed and replaced by two upgraded 12,000 gallon diesel fuel storage tanks. TPH concentrations up to 14,000 mg/kg were detected in soil samples collected during the UST closure project. A subsequent site investigation revealed high TPH concentrations of 1,900 mg/kg and high total BTEX concentrations of 564 μg/kg in soil samples collected from the immediate vicinity of the former diesel fuel USTs. Soil contamination was found to be greatest at a depth interval of 20 to 30 feet below ground surface.

REMEDIAL ACTION PLAN DEVELOPMENT

Although a significant volume of source soil was removed during the UST closure project (700 cubic yards), several areas surrounding and beneath the previously excavated area remained impacted by the former UST operations. Residual diesel fuel-contaminated soil presented a potential long-term source of groundwater contamination. A remedial action plan was developed to address both soil and groundwater contamination with the goal of reducing contaminant concentrations to below the regulatory cleanup objectives of 500 ppm TPH (soil). Due to space restrictions, other logistical constraints, and the depth of contamination (35 feet), only in-situ technologies were evaluated for soil remediation. Bioventing was selected for pilot study evaluation.

The bioventing technology has been widely applied for remediation of fuel-contaminated sites over the past 10 years. It has been successfully applied at locations with widely ranging soil conditions and depths of contamination. Bioventing involves the addition of oxygen to the subsurface to enhance the naturally occurring biodegradation process. In a typical fuel spill scenario, available oxygen in the subsurface is quickly used up by indigenous soil microbes, which use the fuel hydrocarbons as their primary carbon source. In an oxygen-limited environment, the natural aerobic biodegradation of hydrocarbons by indigenous soil bacteria is greatly reduced or eliminated altogether. Once oxygen is supplied by soil venting, natural rates of biodegradation are significantly increased as microbial populations become reestablished in the fuel-rich environment, provided that there is a sufficient supply of essential soil nutrients and favorable soil moisture conditions. At sites where diesel fuel or heavier hydrocarbon contaminants are present, biodegradation plays a much more significant role in reducing the contaminant mass, as compared to volatilization.

A bioventing pilot test was conducted at the facility in December 1994. The objectives of the pilot test were to determine the minimum venting rate required to uniformly distribute air (oxygen) through the unsaturated zone in the areas of highest contamination, and to measure the natural rates of hydrocarbon biodegradation under oxygen-enhanced conditions.

BIOVENTING PILOT TEST

Two vertical air injection wells (IWs) were installed within the source area of the initial release adjacent to the groundwater pump and treatment well MW-3 (Figure 1). Air injection, rather than vapor extraction, was selected as the bioventing test method because diesel fuel contamination in soil is more susceptible to natural biodegradation than volatilization associated with conventional vapor extraction techniques. Another advantage is that low flow-rate air injection did not require costly offgas treatment or air permitting.

The IWs were constructed of four-inch diameter schedule 40 PVC with a screen slot size of 0.06 inches. The total depth of the first IW installed was 35 feet,

Figure 1. **Site location map.**

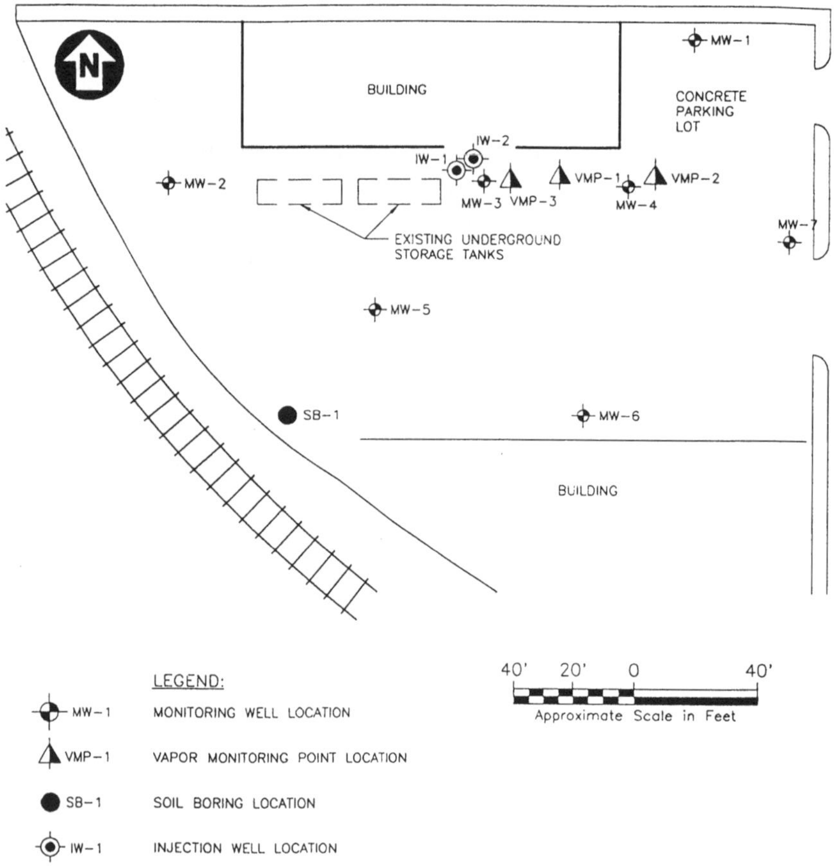

with a screened interval from five to 35 feet below ground surface. Due to concerns that injected air would preferentially flow into the more permeable tank pit backfill material encountered from just below the surface to a depth of 16 feet, a second IW (IW-2) was installed. The screened interval of IW-2 was placed 4 feet below the bottom of the former tank pit and sealed off at the top with bentonite within the natural formation to prevent preferential flow of air into the more permeable, clean tank pit backfill material. IW-2 was subsequently used for air injection during pilot testing.

Three soil vapor monitoring points (VMPs) were installed at distances of 15 to 65 feet from the air injection well (Figure 1) to provide multiple-depth soil gas monitoring at discrete depth intervals during subsequent testing. The VMPs were constructed using 1/2-inch diameter schedule 40 PVC with a screen slot size of 0.02 inches. VMP-1 was installed to a total depth of 32 feet with screened intervals at 24 to 25 feet and 30 to 32 feet below ground surface. VMP-2 was installed to a total depth of 31 feet, with screened intervals of 16 to 17 feet, 22 to 23 feet, and 29 to 31 feet. VMP-3 was also installed to a total depth of 31 feet, with screened intervals at 20 to 23 feet, and 29 to 31 feet. A cross-section illustrating IW and VMP placement is included in Figure 2. Screened sections were sand-packed and sealed in discrete intervals corresponding to the highest levels of contamination, based on field observations and organic vapor concentrations in the soil samples collected. The tops of each of the VMPs were fitted with 1/2-inch quarter-turn ball valves and 3/16-inch hose barbs to allow for connection of monitoring instruments during pilot testing.

Figure 2. **Air injection system and vapor monitoring points cross section.**

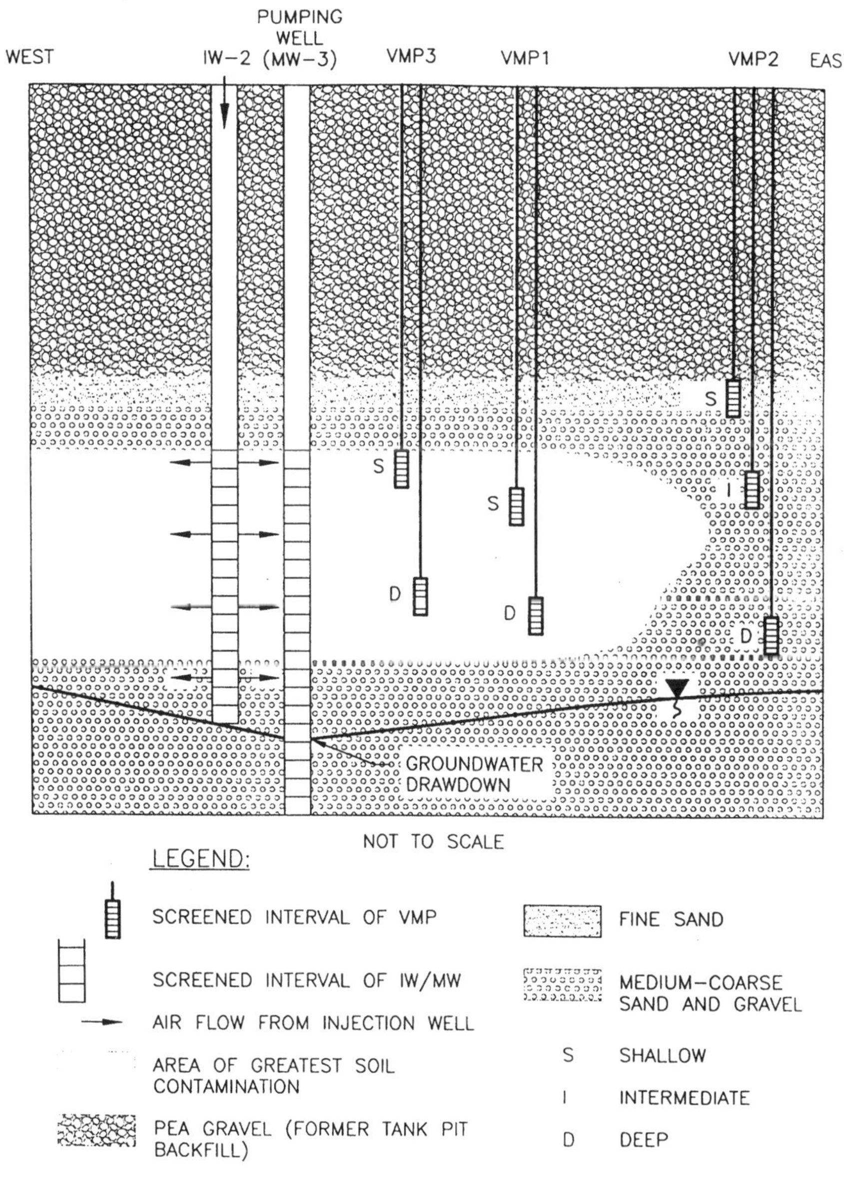

BASELINE SOIL SAMPLING

Soil samples were collected continuously to the total depth of each boring during the installation of the IWs and VMPs. All samples retrieved from the borings were visually inspected for signs of staining, and screened for the presence of hydrocarbon vapors using a photoionization detector (PID). To provide baseline data on the depth and extent of contamination remaining beneath the facility, soil samples collected from the zone of highest contamination and from one to two feet above the water table at each location were analyzed for BTEX by EPA Method 8020, TPH by EPA Method 8015M (diesel range), and PAHs by EPA Method 8100. Analyzed soil samples corresponded with the screened intervals of the VMPs and IWs.

TPH concentrations in the soil samples ranged from nondetect to 2,100 mg/kg. Total BTEX concentrations ranging from 3.0 to 122 µg/kg and total PAH concentrations ranging from nondetect to 26,651 µg/kg were detected in soil samples submitted from IW-1, VMP-1, and VMP-3. Little or no contamination was detected in soil samples collected from VMP-2 located at the eastern boundary of the contaminant plume. Initial soil analytical data are summarized on Tables 1 and 2.

Several soil samples collected during installation of the VMPs were analyzed for parameters which are known to directly affect bioventing/biodegradation potential including pH, alkalinity, total iron, total phosphates, total Kjeldahl nitrogen (TKN), and soil moisture. Test results indicated that the soil in the treatment area contained adequate nutrients and soil moisture to sustain biological activity even if little nutrient recycling takes place.

BIOVENTING SYSTEM COMPONENTS

A 2-horsepower regenerative blower rated for continuous duty and 50 cubic feet per minute (cfm) was installed inside a maintenance building adjacent to the study area. Air injection piping (2-inch diameter PVC) was installed from the blower package to the outside of the building, and then routed below grade and securely fitted to the top of IW-2. Pressure and temperature guages were installed at the intake and outlet end of the blower, and a dilution valve was placed online to control the air flow rate to the subsurface.

INITIAL SOIL GAS CHEMISTRY

Prior to initiating any air injection, soil gas was monitored at each of the VMPs. Total volatile hydrocarbons (TVH), oxygen, and carbon dioxide concentrations were measured. The VMPs were purged using a 1 cfm vacuum pump for approximately 30 seconds to remove stagnant soil gas prior to monitoring. Oxygen and carbon dioxide concentrations were measured as percent O_2 and CO_2 using a Gastechtor Model 32520X portable gas analyzer.

Oxygen concentrations in the VMPs placed within the central area of the contaminant plume (VMP-1 and VMP-2) were low, ranging from five percent

Table 1. **Initial TPH/ BTEX Analytical Data**

Sample Location	Depth (feet below ground surface)	Benzene (ug/kg)	Toluene (ug/kg)	Ethyl-Benzene (ug/kg)	Xylene (ug/kg)	Total BTEX (ug/kg)	TPH (mg/kg)
IWIS	14-16	ND	3.0	5.0	11.0	19.0	330
IWID	22-24	ND	5.0	ND	20.0	25.0	1400
VMPIS	24-25	ND	3.0	15.0	17.0	35.0	1400
VMPID	30-32	ND	2.0	15.0	21.0	38.0	590
VMP3S	20-23	ND	6.0	43.0	73.0	122.0	2100
VMP3D	29-31	ND	ND	1.0	2.0	3.0	120

ND- Not detected above the practical quantitaion limit (PQL).

BTEX anaysis by EPA Method 8020

TPH- Total petroleum hydrocarbons analysis by EPA Method 8015(diesel).

Table 2. **Initial PAH Analytical Data**

PAH Compound	IWIS	IWID	VMPIS	VMPID	VMP3S	VMP3D
Naphthalene	240	530	6100	560	ND	46
Acenaphthylene	ND	2100	ND	ND	2500	ND
Acenaphthene	290	870	5900	630	ND	ND
Fluorene	160	520	3600	600	ND	74
Phenanthrene	96	50	2000	200	ND	ND
Anthracene	48	170	5800	290	ND	ND
Fluoranthene	ND	100	ND	ND	ND	ND
Pyrene	ND	ND	51	ND	ND	ND
Benzo (a) anthracene	ND	ND	ND	ND	ND	ND
Chrysene	ND	ND	ND	ND	ND	ND
Benzo (b) fluoranthene	280	450	200	ND	ND	ND
Benzo (k) fluoranthene	ND	ND	ND	ND	ND	ND
Benzo (a) pyrene	ND	ND	ND	ND	ND	ND
Dibenzo (a,h) anthracene	ND	ND	ND	ND	ND	ND
Benzo (g,h,i) perylene	ND	ND	ND	ND	ND	ND
Indeno (1,2,3-cd) pyrene	ND	ND	ND	ND	ND	ND
Total PAH	1,114	4,970	26,651	2,280	2,500	120

PAHs- Polynuclear aromatic hydrocarbons analysis by EPA Method 8100. Results reported in µg/kg.

ND - Not detected above the analytical practical quantitation limit.

See Table 1 for sample depths.

O_2 in the deepest monitoring zones to 13 percent in the shallower zones. Corresponding CO_2 concentrations were somewhat elevated, ranging from 4.2 to 8.5%. These readings suggested that the indigenous soil microorganisms had been using the residual petroleum hydrocarbons as a primary source of carbon, significantly reducing the natural oxygen supply in the source area of the release, and producing carbon dioxide as a by-product.

RADIUS OF INFLUENCE

After initial soil gas monitoring was completed, the blower system was started and ambient air was injected into the subsurface through the injection well (IW-2) for a period of 23 hours at a rate of 50 CFM and average pressure of 5 inches of water. Pressure gauges with a range from 0 to 5 inches of water were used to measure pressure responses at all depth levels in each of the VMPs. Steady-state pressure readings ranged from 0.1 to 3.4 inches of water within five hours. To estimate the radius of influence of IW-2, a steady-state pressure versus distance from the injection well graph was constructed using pressure response data from VMP-3S, VMP-3D, VMP-2S, and VMP-2D. An extrapolation of the pressure/distance graph indicated that the radius of influence of IW-2 during the air-injection test was approximately 70 feet, a large enough radius to impact the entire volume of contaminated soils remaining beneath the facility.

In addition to pressure response and permeability testing, oxygen and carbon dioxide concentrations were measured at each of the VMPs at all depth levels during the air-injection period. TVH concentrations were also recorded. These data were recorded immediately upon system start-up and at selected time intervals throughout the 23-hour air injection period.

The ability of air injection at IW-2 to promote oxygen flow through the entire contaminated soil column was evidenced by significant O_2 increases and corresponding CO_2 decreases at each of the VMPs. Hydrocarbon vapor concentrations fluctuated widely during the air injection test as stagnant, contaminated soil gas was slowly replaced by fresh oxygenated air. Oxygen concentrations increased by as much as 14% and carbon dioxide concentrations decreased by as much as 8%. Table 3 summarizes initial soil gas concentrations and concentrations measured at the end of the 23-hour air injection test. These results indicated that low flow-rate air injection would be an effective method for increasing soil gas oxygen levels at the site, and that no major modifications to the pilot test system would be required to operate the system on a full-scale basis.

Table 3. **Pilot Study Soil Vapor Concentrations**

Monitoring Point	Depth Interval (feet below groundsurface)	Initial Oxygen (%)	Oxygen After 23 Hours Air Injection (%)	Initial Carbon Dioxide (%)	Carbon Dioxide After 23 Hours Air Injection (%)	Initial Hydrocarbons (ppm)	Hydrocarbons After 23 Hour: Alr Injection (ppm)
VMP1S	24-25	8.0	19.1	8.5	0	222	115
VMP1D	30-32	5.0	19.0	8.0	0	180	97
VMP3S	20-23	13.0	19.2	4.2	0	110	113
VMP3D	29-31	6.0	19.8	7.2	0	200	96
VMP2S	16-17	18.8	19.0	2.2	0.5	96	110
VMP2I	22-23	18.9	17.0	1.5	1.1	62	50
VMP2D	29-31	18.7	19.0	1.5	0	50	0

POTENTIAL AIR EMISSIONS

No detectable hydrocarbon vapors were measured in the breathing zone, at the ground surface, or in any of the buildings or adjacent sewer lines during vapor monitoring conducted during the first four hours of the permeability test using a TVH meter. Because the potential for air emissions is highest during the first hour of air injection and no emissions were detected, the long-term air emission potential for a low flow-rate air injection bioventing system was considered low. Subsequent air monitoring performed to date has supported this conclusion.

In-Situ Respiration Test

As aerobic soil bacteria consume hydrocarbons, they utilize oxygen and produce carbon dioxide. Oxygen consumption provides one way of estimating the hydrocarbon degradation rate. Although production of carbon dioxide may provide an equivalent estimate of fuel biodegradation, the buffering capacity of alkaline soils will react with the CO2 produced by bacterial respiration. This reaction may have the effect of masking CO2 production associated with bacterial respiration.

At the completion of the air injection test, the blower was shut down and an in-situ respiration test was conducted. Oxygen consumption and production of carbon dioxide by indigenous soil bacteria were measured over a period of 22 hours. Data collected from VMP1 and VMP3 were used to establish the average biodegradation rate of the indigenous microorganisms, and to predict the time required to biodegrade the remaining diesel fuel residuals to below regulatory action levels using the bioventing technology.

The readings obtained during in-situ respiration testing are summarized in Table 4. Oxygen utilization rates calculated from O_2 data collected at the vapor monitoring points averaged 0.006% oxygen per minute. This average oxygen utilization rate was used to estimate the average rate at which diesel fuel TPH levels could be biodegraded at this site. Complete biological degradation of fuel hydrocarbons (e.g., n-decane) can be described by the equation:

$$C_{10}H_{22} + 15.5\ O_2 \rightarrow 10\ CO_2 + 11\ H_2O$$

Approximately 3.5 grams of oxygen is required to biodegrade one gram of fuel hydrocarbons. Based on this relationship, the average oxygen utilization rate within the contaminated soils, and the soil's estimated air-filled porosity of 0.105 liters/kilogram; the average rate of fuel biodegradation was estimated to be 2,100 milligrams of fuel per kilogram of soil per year (mg/kg/yr).

FULL-SCALE SYSTEM OPERATION

Based on the results of the pilot study, no changes to the pilot test equipment were recommended, and full-scale bioventing was initiated at the site after receiving authorization from the local air quality regulatory agency. A monthly operation and maintenance program was established to monitor the performance of the bioventing and groundwater treatment systems. Additionally, a 90-day respiration test was conducted to determine if oxygen utilization (biodegradation) rates were remaining constant, increasing, or decreasing with time.

Table 4. **Initial Respiration Test Data**

Location	Screened Interval (feet below grade)	Date	Time	Oxygen (%)	Carbon Dioxide (%)	Hydrocarbons (ppm)
VMP1S	24-25	12/11/94	9:00	19.8	0.0	50
		12/11/94	9:21	20.9	0.0	64
		12/11/94	10:07	20.5	0.0	64
		12/11/94	10:40	20.5	0.0	54
		12/11/94	11:22	20.0	0.0	70
		12/11/94	12:35	20.0	0.0	70
		12/11/94	14:31	19.0	0.0	67
		12/11/94	16:32	19.0	0.0	82
		12/12/94	7:17	15.0	0.3	NA
VMP1D	30-32	12/11/94	8:58	19.8	0.0	110
		12/11/94	9:19	19.9	0.0	113
		12/11/94	10:05	19.0	0.0	114
		12/11/94	10:39	18,5	0.0	113
		12/11/94	11:19	17.7	0.0	140
		12/11/94	12:33	16.8	0.0	160
		12/11/94	14:30	15.0	0.0	150
		12/11/94	16:30	14.0	0.2	160
		12/12/94	7:15	10.0	0.5	210
VMP3S	19.5-23.5	12/11/94	8:55	19.1	0.0	180
		12/11/94	9:15	19.8	0.0	111
		12/11/94	10:02	19.5	0.0	100
		12/11/94	10:34	18.9	0.0	100
		12/11/94	11:15	18.0	0.0	111
		12/11/94	12:27	17.5	0.0	110
		12/11/94	14:22	16.5	0.0	105
		12/11/94	16:25	16.3	0.0	105
		12/12/94	7:10	16,0	0.3	94
VMP3D	28.5-31	12/11/94	8:57	20.0	0.0	68
		12/11/94	9:17	20.5	0.0	110
		12/11/94	10:04	19.2	0.0	111
		12/11/94	10:36	18.4	0.0	111
		12/11/94	11:17	17.3	0.0	125
		12/11/94	12:29	16.0	0.0	140
		12/11/94	14:24	13.9	0.0	140
		12/11/94	16:27	12.5	0.0	150
		12/12/94	7:12	9.0	0.8	200

Results of the 90-day respiration test indicated an average biodegradation of 5,200 milligrams of TPH per kilogram of soil per year, more than twice the pilot study predicted rate of 2,100 mg/kg/yr (Figure 3). Increases in biodegradation rates of this magnitude have been observed at other bioventing sites, and may be attributed to the establishment and growth of the microbial population as oxygen is supplied to the subsurface.

SIX MONTH CONFIRMATORY SOIL SAMPLING

On August 29, 1995, after approximately six months of full-scale bioventing, two soil borings were advanced to the water table (approximately 35 feet) using hollow-stem auger and continuous split spoon sampling techniques in accor-

Figure 3. **Average biodegradation rates.**

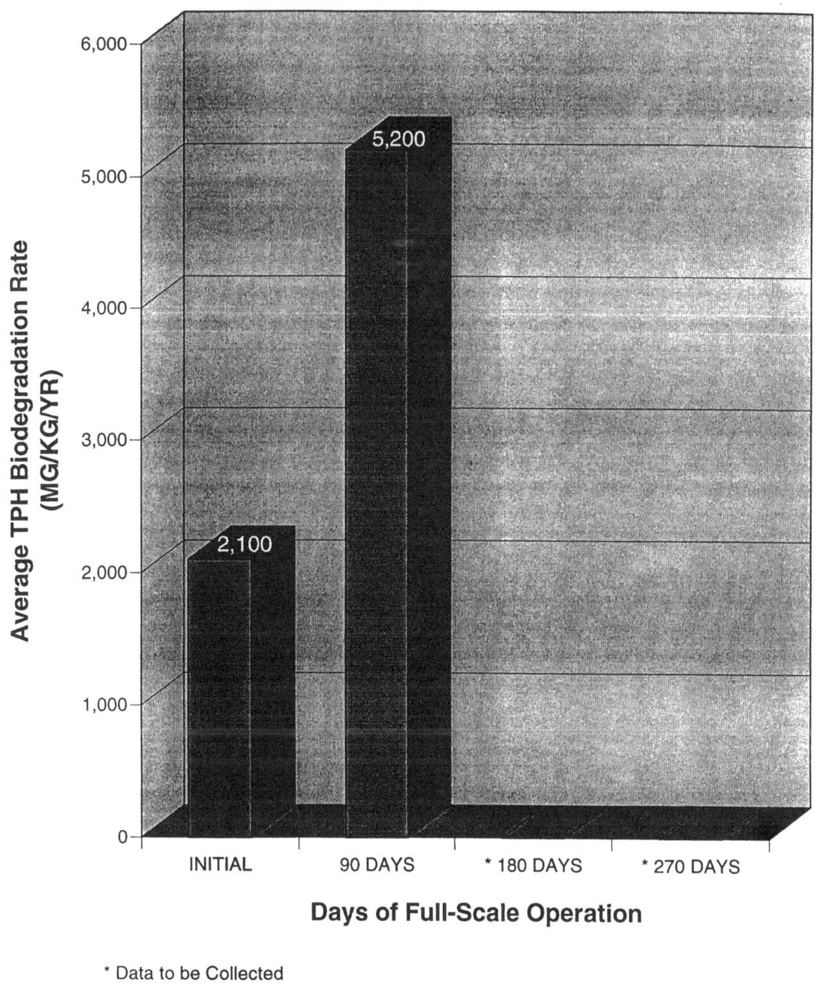

dance with ASTM Standard D1586. Soil boring B-1 was drilled between vapor monitoring points VMP-1 and VMP-3 (within four feet of either VMP), and B-2 was drilled immediately adjacent to injection well IW-1 (Figure 4). The boring locations were selected to confirm the effectiveness of the bioventing system at reducing diesel fuel residuals in the soil.

At each drilling location, all soil samples retrieved from the boring were field-screened for the presence of volatile organic compounds using a PID. Samples selected for laboratory analyses corresponded to the previous sampling depth intervals. Additionally, the sample(s) with the highest field-screened PID reading were also analyzed to account for any vertical redistribution of contaminants due to seasonal groundwater table fluctuations. Soil samples were anlayzed for TPH by EPA Method 8015 (modified-diesel), PAHs by EPA Method 8100, and BTEX by EPA Method 8020 for the purpose of direct comparison with prebioventing analytical data.

Analytical results for the pre- and post-bioventing soil samples are summarized on Table 5. Results indicate that a significant decrease in residual soil contamination has occurred after the first six months of full-scale bioventing. Table 5 summarizes the soil analytical data collected before and after six months of bioventing by grouping the data into three distinct sampling intervals; 15 to 20 feet, 22 to 24 feet, and 30 to 32 feet. Initially, the highest concentrations of TPH and PAH detected within the source area were encountered at the 22 to 24 foot depth interval. High concentrations of TPH (2,100 mg/kg) and PAH (26.65 mg/kg) were detected at that depth interval during the installation of VMP-1 and VMP-3. The soil sample recently analyzed from the corresponding depth interval in soil boring B-1 exhibited a TPH concentration of 1,000 mg/kg and total PAH concentration of 2.97 mg/kg, representing an average TPH reduction of 43% and average total PAH reduction of nearly 80%. Similarly, reductions in total BTEX concentrations of 77 to 100% were observed at all depth intervals. TPH and PAH concentrations in the area of VMP-1 and VMP-3 appear to have increased slightly at the 30 to 32 foot depth interval. This increase may be attributable to redistribution of the contaminants due to the influence of the groundwater pumping operations which were initiated along with the start of full-scale air injection.

Table 5. **Soil Analytical Summary.**

Sample Location	Sample Depth (feet below grade)	TPI-I (mg/kg) Initial	TPI-I (mg/kg) 6 mo.	Total BTEX (ug/kg) Initial	Total BTEX (ug/kg) 6 mo.	Total PAH (ug/kg) Initial	Total PAH (ug/kg) 6 mo.
1W-1	15-20	330	ND	19	ND	1,114	ND
1W-1	30-32	1,400	1,000	25	14	4,970	ND
VMP-I/VMP-3	22-24	1,400/2,100	1,000	35/122	2	26,651/2,500	2,970
VMP-IIVMP-3	30-32	590/120	1,000	38/3	ND	2,280/120	3,008

ND = Not Detected Above the Analytical Detection Limit

DISCUSSION

The average reduction in TPH indicated by the results of the confirmatory soil sampling generally correlates well with the theoretical average biodegradation rates calculated from the pilot study in-situ respiration data. Based on these results, it is anticipated that residual TPH concentrations will be reduced to below regulatory action levels within one year.

Alleman et al. have used the stoichiometric oxidation of phenanthrene to calculate theoretical PAH biodegradation rates from oxygen utilization rates measured during in-situ respiration tests at a bioventing site. Their results indicated that biodegradation of PAH was occurring at a rate of as much as 17.3% of the original concentration within a one-year period under oxygen-enhanced conditions. Initial results of this study indicate that it may be possible to achieve even higher rates of PAH biodegradation using this technology. These observations should stimulate further study of aerobic degradation pathways for PAH compounds, and applications to enhance and accelerate the biodegradation process.

REFERENCES

Nickerson, D.A., Baker, J.N., and Guest, P.R. 1994. Something Old, Something New: Conventional Pump and Treat System Teamed with Full-Scale Bioventing Results in Rapid Cleanup of Diesel-Contaminated Soils, *Proceedings of the Water Environment Federation 67th Annual Conference and Exposition*, Volume II, p. 429, WEF, Alexandria, VA.

Alleman, B.C., Hinchee, R.E., Brenner, R.C., and McCauley, P.T. Bioventing PAH Contamination at the Reilly Tar Site, In: *Proc. Int. In Situ and On-Site Bioreclamantion Symposium*, p. 473, Battelle Press, Columbus, OH.

Symons, B.D., Linkenheil, R., Pritchard, D., Shanke, C.A., and Seep, D. In Situ Groundwater Aeration of Polycyclic Aromatic Hydrocarbons, In: *Proc. Int. In Situ and On-Site Bioreclamation Symposium*, p. 135, Battelle Press, Columbus, OH.

CHAPTER 9

Design of a Full-Scale Soil Vapor Extraction/Aquifer Air Sparging System for Remediating Soil and Groundwater at a Former Petroleum Marketing Terminal: A Case Study

Michael S. Battle, P. G. and **Curtis J. Varner**, EA Engineering, Science, and Technology, Newburgh, New York

INTRODUCTION

In cooperation with the U.S. Environmental Protection Agency (EPA) and New York State Department of Environmental Conservation (NYSDEC), a comprehensive, full-scale field demonstration of the utility of soil vapor extraction (SVE), applied in conjunction with aquifer air sparging (AAS), is underway at an inactive petroleum storage terminal in New York. SVE is an in situ remedial process for the removal of volatile organic compounds (VOCs) from the subsurface by the process of volatilization via induced airflow. AAS is an innovative in situ remediation process designed to aerate the groundwater and remove VOCs from the saturated zone. AAS, in conjunction with SVE, has been demonstrated through pilot testing to remove high concentrations of VOCs within site soil and groundwater. The effectiveness of SVE/AAS at the subject site, which is underlain by 11-16 ft of heterogeneous fill material impacted by elevated concentrations of various VOCs and semivolatile organic compounds (SVOCs), is being monitored through an extensive groundwater, soil, and soil vapor sampling and analysis program.

In response to the presence of residual light non-aqueous phase liquid (LNAPL), sorbed, vaporous, and dissolved-phase petroleum hydrocarbons in site soil and groundwater, corrective actions utilizing SVE/AAS were recently undertaken. SVE/AAS was selected as an alternative to traditional groundwater pump-and-treat based on several factors: portions of the site are subjected to significant groundwater fluctuations due to the proximity of the site to a tidally influenced creek, high groundwater extraction rates would be required to maintain adequate hydraulic control of the shallow water table aquifer, extensive time would be required to remove residual hydrocarbons trapped below the water table interface, and supplemental vadose zone remediation would be required.

Invasive SVE/AAS remedial system construction was performed over a 3-month period at the beginning of 1995. Components of the system include 17 ambient air injection (sparge) wells designed to cause the partitioning of dissolved-phase and sorbed hydrocarbons trapped beneath the water table into the sparged air, and to promote bioremediation. Additionally, 18 lateral SVE trenches have been installed to withdraw VOCs from the vadose zone and recover sparged air. The design of the air sparging system provides for precise control of operating conditions through the use of pressure and flow regulators coupled with an air-drying system. The SVE system provides for maximum recovery of VOC-laden sparged air and hydrocarbons within the vadose zone, while simultaneously promoting biodegradation by drawing ambient air through the soil pores to increase the oxygen content. Offgas treatment is accomplished with a thermal/catalytic oxidation system.

SOIL VAPOR EXTRACTION/AQUIFER AIR SPARGING TECHNOLOGY AND APPLICATION

SVE/AAS combines in situ aeration of groundwater with vacuum-enhanced recovery of both the sparged air and hydrocarbon vapors which may be present in the unsaturated soil. AAS is an emerging in situ technology used as a method of saturated zone remediation for sites impacted by dissolved-phase and adsorbed VOCs. SVE is typically operated in conjunction with AAS to ensure retrieval of sparged air and prevent the uncontrolled pressure-induced transport of VOCs in the vadose zone.

The major components of SVE/AAS consist of specially-designed wells used to inject (sparge) ambient air through impacted groundwater, and vapor extraction points (installed either horizontally or vertically) which are used to withdraw sparged air while inducing air flow through the application of a negative pressure (vacuum) in the vadose zone. Dissolved-phase and sorbed VOCs partitioned into the sparged air migrate through soil pore channels to the surface of the water table, at which point an active SVE system recovers the hydrocarbon-laden vapor. The aeration of the groundwater functions both to volatilize dissolved and sorbed VOCs and to increase the dissolved oxygen concentration. VOCs present in the vadose zone are withdrawn by the SVE system through volatilization and advection. Enhancement of oxygen concentration (both in groundwater through sparging and the vadose zone through venting) in turn promotes in situ bioremediation by stimulating the growth of the indigenous aerobic microorganism population. The dual action of sparging and soil venting, under optimal conditions, can achieve simultaneous remediation of both the groundwater and overlying unsaturated soil. A schematic diagram of these processes is shown in Figure 1.

Figure 1. **Conceptual schematic of SVE/AAS remediation system.**

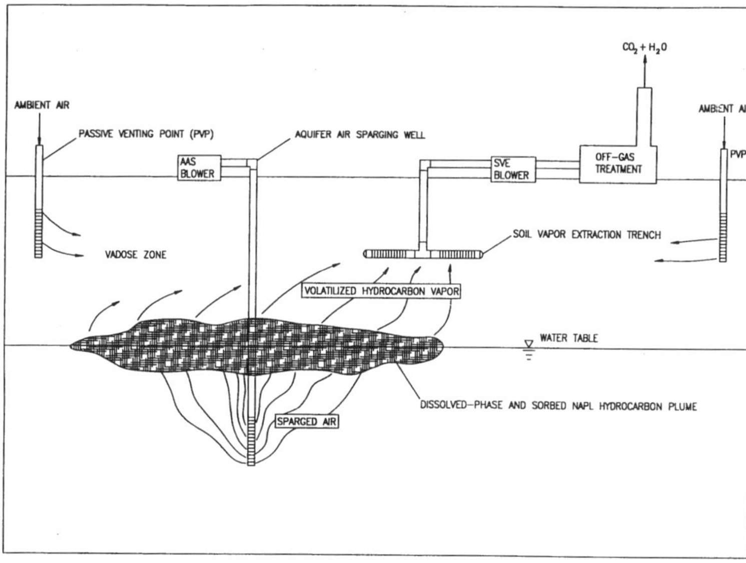

The effectiveness of SVE/AAS is highly dependent upon the volatility of the compounds to be remediated. A compound's volatility is directly related to the degree to which it will partition into the vapor phase. The parameter which best describes a compound's volatility is its vapor pressure. It has been suggested that compounds with vapor pressures greater than 1 mm Hg at 20°C are amenable to removal via SVE (USEPA, 1991; Angell, 1992). The compounds to be mitigated at the subject site, including benzene, toluene, ethylbenzene, and total xylenes (BTEX) and methyl tertiary-butyl ether (MTBE), all possess vapor pressures in excess of 1 mm Hg.

Partitioning of dissolved-phase hydrocarbons into the sparged air is governed by the equilibrium relationships between the hydrocarbons in the gaseous and liquid phases. For dilute concentrations near atmospheric pressure, this relationship is governed by Henry's law. Of the constituents of concern, the aromatic compounds (i.e., BTEX) can be readily removed by air sparging because they possess sufficiently large Henry's law constants (e.g., $>10^{-5}$ atm-m^3/mole (Angell, 1992; Brown et al., 1991)).

Partitioning of hydrocarbons sorbed onto soil below the water table is also a function of the soil adsorption coefficient. The soil adsorption coefficient is dependent upon soil composition, the chemistry of the hydrocarbons, and the geochemistry of the groundwater. Generally, as the soil adsorption coefficient decreases, the effectiveness of AAS increases. In cases where the soil adsorption coefficient is sufficiently low, the hydrocarbon concentration in the sparged groundwater will be a function of the coupled processes of volatilization from the groundwater and desorption from the soil in the saturated zone.

Favorable soil conditions are a prerequisite for successful SVE/AAS. Saturated zone soil that is not highly heterogeneous, and with an intrinsic permeability (hydraulic conductivity) in the range of 10^{-3} to 10^{-5} cm/sec or greater, is required to permit sufficient air transmission through the saturated zone (USEPA, 1991; Brown et al., 1991; Middleton and Hiller, 1990). During completion of the remedial investigation and pilot tests at the site, undisturbed soil samples were obtained for geotechnical analysis, including intrinsic permeability. The results of these analyses suggest that favorable soil conditions exist at the site for SVE/AAS; however, the heterogeneity of the fill material at isolated areas throughout the site may produce localized restrictions to the AAS process.

During AAS operations, the concentration of dissolved oxygen in the groundwater will be increased, promoting aerobic biodegradation of dissolved phase hydrocarbons. The maximum solubility of oxygen in groundwater is a function of pressure and temperature. For the SVE/AAS system, the net pressure will be the resultant pressure from air sparging and vapor extraction. There may be a slight increase in groundwater temperature near the AAS injection wells due the temperature of the sparged air (approximately 160°F). Enhancement of dissolved oxygen, however, will not independently lead to enhanced microbial activity since hydrocarbon degrading bacteria also require optimal ranges of hydrocarbon concentrations (as a carbon source), pH, and inorganic nutrients.

SITE DESCRIPTION

Site History

The site is located in a heavily industrialized area. Figure 2 provides the general site features prior to facility decommissioning. The layout of the SVE/AAS system designed and installed at the subject site is shown in Figure 3. The portion of the site undergoing remediation is approximately 2.5 acres.

The site had been operated as a petroleum bulk storage facility for approximately 60-70 years prior to decommissioning in 1991-1992. Petroleum products stored at the site included kerosene, gasoline, diesel fuel, and Nos. 2, 4, and 6 fuel oils. Petroleum storage was accomplished primarily by a series of aboveground and mounded (installed partially below grade) bulk storage tanks. Additionally, five underground storage tanks were located at the terminal. All aboveground, mounded, and underground storage tanks, in addition to the loading racks, vapor recovery unit, foam house, and oil/water separator systems, were taken out of service and removed during 1991-1992.

Figure 2. **Petroleum marketing terminal prior to decommissioning.**

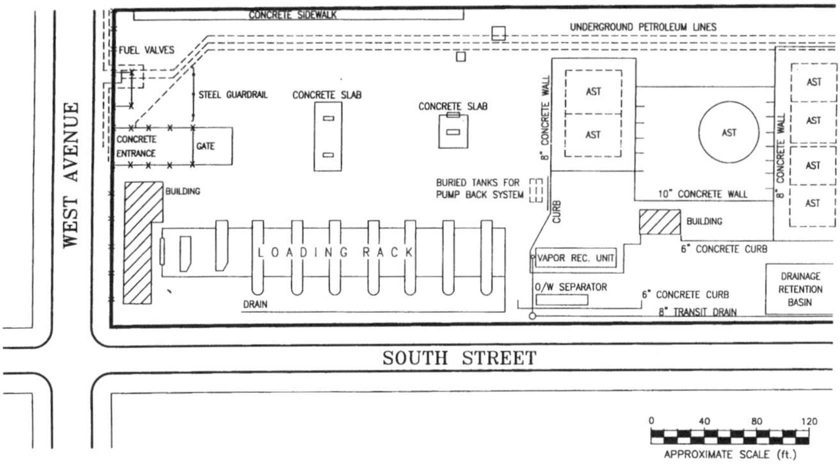

Figure 3. **Layout of SVE/AAS remediation system.**

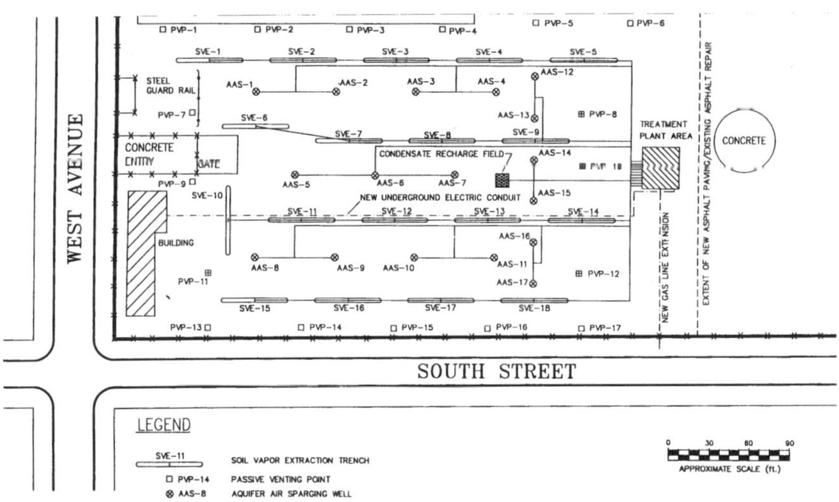

SITE GEOLOGY

Based upon geologic logs of the test borings and monitoring wells, the site-specific shallow stratigraphy is comprised of surficial fill which is underlain by an extensive silty clay deposit. The silty clay is considered to be the natural formation upon which much of the surrounding area has been built. The surficial fill material is comprised primarily of fine to coarse sand with varying amounts of silt, fine to coarse gravel, concrete, wood, brick, and rubble. The thickness of the fill ranges from 11 to 15 ft across the site. Depth to groundwater at the site ranges from approximately 7 to 10 ft below grade.

Baseline Environmental Investigation

EA performed a baseline environmental investigation at the petroleum terminal in 1990 to assess the soil and groundwater quality and hydrogeology of the site. A total of 106 soil vapor monitoring points were sampled across the site for VOCs (e.g., BTEX) and total volatile hydrocarbons (TVH). The highest concentrations (>10,000 ppm) of TVH in soil vapor were observed in the vicinity of the former loading rack.

A series of soil borings/groundwater monitoring wells were installed across the site at depths ranging from 6.5 to 18 ft. Soil sampling during the drilling program confirmed the presence of VOCs (e.g., BTEX, MTBE) at 10 of the 12 borings at concentrations ranging from 15 to 7,610 μg/kg. SVOCs (e.g., polycyclic aromatic hydrocarbons [PAHs]) were reported at 11 of the 12 borings at concentrations ranging from 6,600 to 1,924,800 μg/kg. The elevated concentrations of PAHs are believed to have occurred based on historical site use and/or the composition of the fill material (containing ash and coal).

Groundwater sampling at site wells confirmed the presence of VOCs at concentrations ranging from 87 to 26,680 μg/L. SVOCs were also reported, though at lower concentrations than the VOCs.

SOIL VAPOR EXTRACTION PILOT TESTING

Prior to design and construction of a full-scale SVE/AAS system, pilot testing was performed to assess the applicability of SVE and AAS for remediating elevated concentrations of petroleum hydrocarbons at the subject site. SVE pilot testing was conducted using an existing monitoring well (EAW-12) and a lateral vent trench. EAW-12 is located along the eastern portion of the site. The lateral vent trench was installed just south of the center of the former loading rack area. The locations of EAW-12 and the vent trench are shown in Figure 4.

EAW-12 was constructed with 10 ft of screen (0.01 in.) from 2 ft below grade to the bottom of the well (12 ft). Depth to groundwater was measured at 3.75 ft; thus, 1.75 ft of screen was positioned in the vadose zone, and depths between 2 and 3.75 ft were most actively vented. The lateral vent well trench was constructed with 20 ft of screen (also 0.01 in.) and installed at an approximate depth

Figure 4. **Locations of SVE/AAS pilot tests.**

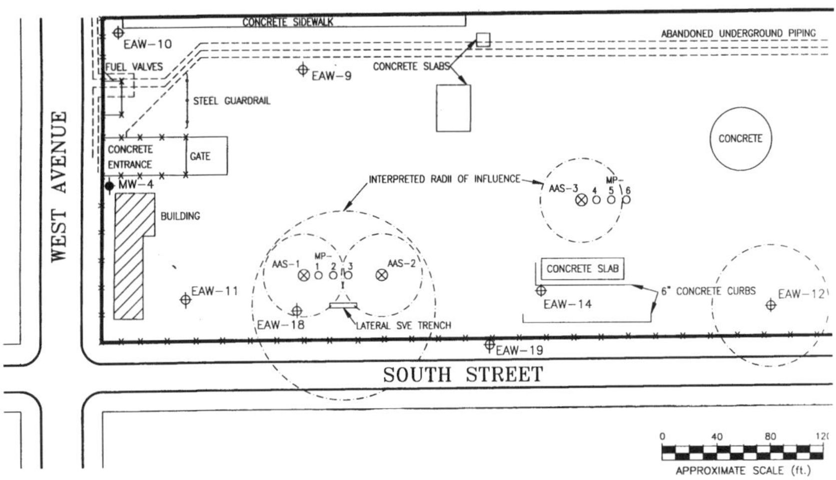

of 5 ft. The lateral vent trench was backfilled with pea gravel to facilitate air flow in the vicinity of the vent screen. During pilot testing activities, vacuum was applied to the well/vent manifolds using a field portable 1 HP regenerative blower system. The blower exhaust rate (i.e., SVE flow rate) averaged 76 standard cubic feet per minute (scfm) during SVE pilot testing at both EAW-12 and the lateral vent trench.

The SVE radius of influence was assessed at both pilot test locations using soil vacuum probes driven into the subsurface at radial locations around the pilot test area. The soil vacuum probes, constructed of 0.75-in. steel pipe, each with a 12-in. screened section, were used to monitor vacuum response by attaching differential pressure gauges to hose barb fittings. The radial zones of SVE influence were calculated at approximately 44 ft and 68 ft, adjacent to EAW-12 and the lateral vent trench, respectively (Figure 4).

The average concentrations of benzene and TVH withdrawn from the soil during SVE pilot testing at EAW-12 were 243.6 ppm_v and 6,587.5 ppm_v, respectively. During SVE pilot testing at the vent trench, the average concentrations were 13.4 ppm_v benzene and 2,312.5 ppm_v TVH. Based on the blower exhaust rate of 76 scfm, the above concentrations correspond to an initial extraction rate of 4.6 lb/day benzene and 147.8 lb/day TVH at EAW-12 and 0.3 lb/day benzene and 51.9 lb/day TVH at the vent trench.

AQUIFER AIR SPARGING AND COMBINED SOIL VAPOR EXTRACTION/AQUIFER AIR SPARGING PILOT TESTING

AAS pilot testing was undertaken to assess the effectiveness of saturated zone aeration in removing VOC from the groundwater. AAS pilot testing was conducted using two pilot AAS wells installed in the vicinity of the former loading racks (AAS-1 and AAS-2), and 1 pilot AAS well installed north of the former vapor recovery unit (AAS-3). The locations of the sparge test wells are shown in Figure 4.

The test parameters used to assess treatability of the saturated zone by active sparging included: collection and analysis of soil samples for intrinsic permeability (hydraulic conductivity), radial pressure influence in the vicinity of each sparge well, flow rate and pressure of ambient air injection, volatilization rate (air:water ratio), and delivery of atmospheric oxygen to the indigenous microbial population. The remainder of this section discusses the findings of the testing conducted to assess the above parameters.

Geotechnical Analyses of Soil

During the installation of the pilot test sparge wells, undisturbed soil samples were taken from both the fill material and the underlying silty clay to perform geotechnical analyses of the physical properties which are known to affect the utility of SVE/AAS application, including intrinsic soil permeability and grain-size distribution. The results of the geotechnical analyses performed on the three undisturbed samples are provided in Table 1.

As indicated in Table 1, the intrinsic permeability (also referred to as the hydraulic conductivity) of the fill material is 1.5×10^{-3} cm/sec. Grain-size distribution analyses show the fill material to be comprised of greater than 93% sand and gravel. Permeabilities greater than approximately 10^{-4} cm/sec are generally an indication of favorable soil conditions for implementing SVE/AAS.

In contrast to the relatively high permeability of the fill, the permeability of the underlying silty clay averaged 8.75×10^{-8} cm/sec. Grain-size distribution analyses of the silty clay samples confirmed their composition at greater than 94% silt and clay. These results confirmed that the silty clay serves as a natural aquiclude restricting the vertical migration of petroleum-related hydrocarbons. Based on the geotechnical analyses, and chemical analyses, discussed below, remediation at the site is restricted to the fill material.

Table 1. **Summary Of Geotechnical Analyses Performed On Soil Samples Collected During The Air Sparging Well Installation Program**

Sample Parameter	AAS-1 (18-20 ft)	AAS-2 (10-12 ft)	AAS-3 (16-18 ft)
Soil Description	MH, Elastic Silt	Fill	MH, Elastic Silt
Liquid Limit	60	Non-Viscous	58
Plastic Limit	33	Non-Plastic	33
Plasticity Index	27	None	25
Specific Gravity	2.7	2.27	2.7
Moisture Conductivity (%)	60.5	34.5	78.6
Intrinsic Permeability (cm/sec)	9.5E-08	1.5E-03	8.0E-08

Chemical Analyses of Soil

As a means of evaluating the chemical composition of the fill material and uppermost portion of the silty clay unit, standard penetration test samples were taken at the three pilot air sparging wells, including two from the uppermost portion of the silty clay unit (Well AAS-1, 16-18 ft; and Well AAS-3, 14-16 ft). Each of the samples was retained for analysis for BTEX and MTBE by EPA Method 602, total petroleum hydrocarbon (TPH) screening by modified EPA Method 8015, total recoverable hydrocarbons by EPA Method 418.1, and total organic carbon by EPA Method 415.1. The results of these analyses are summarized in Table 2.

The analytical results indicate that the concentrations of BTEX and TPH decrease significantly from the more permeable fill material to the top of the less permeable silty clay. At both sampling locations (i.e., Wells AAS-1 and AAS-3), BTEX and TPH concentrations are an order of magnitude higher in the fill material than in the silty clay immediately underlying the fill. The total hydrocarbon screening performed by modified EPA Method 8015 suggests the hydrocarbons encountered onsite to be either degraded gasoline or diesel. Total organic carbon in the five soil samples ranged from 18,000 mg/kg (1.8%) to 118,000 mg/kg (11.8%); however, four of the five samples reported total organic carbon levels ranging from 1.8 to 2.1%. Both soil samples collected in the silty clay reported total organic carbon levels of 1.8%.

Results of Air Sparging Pilot Tests

The effective radial influence of each of the sparge tests conducted at Wells AAS-1, AAS-2, and AAS-3 was evaluated using four data sets: differential pressures, hydrocarbon loading to the vadose zone, variations in hydrocarbon concentrations in shallow groundwater, and dissolved oxygen measurements versus distance in the saturated zone. A fifth data set, water table elevation fluctuation, was not evaluated due to difficulties in differentiating between the effects of pilot

Table 2. **Summary Of Chemical Analyses Performed On Soil Samples Collected During Air Sparging Well Installation Program**

Parameter	Units	AAS-1 (5-7 ft)	AAS-1 (16-18 ft)	AAS-2 (12-14 ft)	AAS-3 (5-7 ft)	AAS-3 (14-16 ft)
Soil Classification	N/A	SP/GM	MH	SP/GM	SP/GM	MH
Total BTEX	mg/kg	4.77	0.28	36.9	17.87	0.32
MTBE	mg/kg	(U<1.0)	(U<1.0)	8.6	(U<1.0)	(U<1.0)
TPH as Gasoline	mg/kg	(U<120)	(U<17)	(U<140)	74	(U<16)
TPH as Diesel	mg/kg	1500	360	400	(U<12)	48
Total Recoverable Hydrocarbons	mg/kg	1800	61	220	810	(U<52)
Total Organic Carbon	Percent	2.1	1.8	11.8	1.9	1.8

NOTE: (U<__) indicates not detected above limit shown.

BTEX = Benzene, toluene, ethylbenzene, and total xylenes; MH = Elastic clayey silts; MTBE = Methyl tertiary-butyl ether; N/A = Not applicable; SP/GM = Poorly graded gravel-sand-silt mixtures (fill material); TPH = Total petroleum hydrocarbons.

testing versus tidal influence. Due to the limited duration of the tests, gradient (parallel to groundwater flow) versus cross-gradient (normal to flow) sparging influence was not evaluated.

Differential pressure versus distance is typically used as a primary data set in assessing AAS areal influence. The differential pressure corresponding to the radial zone of AAS influence is site-specific and has been reported to range from 0.01 to 1 in. H_2O (Leonard and Brown, 1992; Brown et al., 1991). For the tests completed at the subject site, the differential pressure threshold for sparging was assumed at 0.10 in. H_2O.

To assess the data, the logarithms of differential pressures were regressed against linear distance from the sparge well for each step of the test. Differential pressure data were evaluated as two separate data sets: the first for monitoring points in the vadose zone only, and the second for monitoring points installed partially or entirely in the saturated zone, with the latter data set considered more representative of actual sparge zone conditions.

Monitoring of dissolved oxygen versus distance from the sparge well was also used to infer AAS radial influence, particularly at locations where the background dissolved oxygen concentrations in groundwater were negligible (i.e., less than 0.5 mg/L). For purposes of evaluating radial influence of the sparge tests in this study, net increases in dissolved oxygen concentrations in groundwater greater in magnitude than 1 mg/L were considered the limit of effective response.

It should be noted that determinations of radial influence based on differential pressure or dissolved oxygen concentrations are not known to have been scientifically correlated to radial zone of remediation influence, other than by limited data sets consisting of groundwater sample analysis. The collection and analysis of soil samples (in both the vadose and saturated zones) prior to and following active sparging, and the determination of diffusion-driven sparge velocity thresholds, may serve to validate such correlations.

Pilot Testing Results

AAS pilot testing was performed at Wells AAS-1 and AAS-2. Combined SVE/AAS pilot testing was performed at Well AAS-3. AAS pilot testing was conducted in two steps at Well AAS-1. The first step was run at 2.5 pounds per square inch gauge (psig) pressure and an injection rate of 6.5 cfm. The second step was operated at a pressure of 3.5-4 psig, with corresponding injection rates of 13-26 cfm. AAS pilot testing at Well AAS-2 was conducted in three steps. The first test step was operated at 2.5-3 psig and a sparge rate of 5.75 cfm. The second step was performed at 3-4 psig with a corresponding sparge rate of approximately 12 cfm. The third and final step was performed at a pressure of 4 psig and a sparge rate of approximately 24 cfm.

At Well AAS-3, combined SVE/AAS pilot testing was conducted in three steps including SVE-only operations and combined SVE/AAS. Prior to the first combined SVE/AAS test step, SVE-only was performed at a vacuum of 44 in. H_2O and extraction rate of 57 cfm. The first step of SVE/AAS testing was performed at 2.5-3 psig, with sparge rates between 7.5 and 16.5 cfm. The second step was performed at 3 psig with sparge rates ranging from 16.5 to 25 cfm. During each step, SVE operating parameters remained at 44 in. H_2O vacuum and 57 cfm vapor extraction flow rate.

Differential Pressure

Differential pressures observed in the saturated zone monitoring points during the pilot tests were used to evaluate AAS radial influence. Regression analysis on differential pressures measured in the saturated zone for Well AAS-1 tests suggests the radial influence at 34 ft and 46 ft for test Steps 1 (2.5 psig) and 2 (3.5 psig), respectively (Figure 5).

Regression analysis of differential pressure measurements recorded in the saturated zone during testing at Well AAS-2 yielded calculated radial influences of 36 ft, 52 ft, and 53 ft for test Step 1 (2.5-3 psig), Step 2 (3-4 psig), and Step 3 (4 psig), respectively (Figure 6). However, a close linear relationship was not observed based on the linear regression performed (average $R^2 = 0.61$). Differential pressures measured at the outlying distances contributed to the deviation from linearity.

The operation of the pilot SVE and AAS systems concurrently had no impact on the radius of influence of the pilot sparge system. Based upon concurrent operation of SVE and AAS pilot systems, regression analyses were only performed on the differential pressure data collected from saturated zone monitoring points during AAS-3 step tests. Regression analysis suggests a moderate to poor linear correlation ($R^2<0.6$) for both test steps. However, review of the data shows that, with the exception of one monitoring point, differential pressure measurements at set zone monitoring points were in excess of 0.10 in. H_2O. Thus, the AAS radial influence during each of the test steps was likely in excess of 40 ft.

Dissolved Oxygen

Dissolved oxygen measurements were recorded during AAS pilot testing to delineate AAS radial zone of influence. The six locations monitored during pilot testing at AAS-1 showed net increases in dissolved oxygen, ranging from 0.32 mg/L to 6.84 mg/L. Using a net increase in dissolved oxygen concentration of 1 mg/L to define effective AAS, radial influences of 20 ft and 31 ft were observed during test Steps 1 and 2, respectively (Figure 7).

Dissolved oxygen measurements at AAS-2 (five locations total) reported net increases, though wide variability in the data existed. Net changes in dissolved

Figure 5. **Sparge pressure versus distance at Test Well AAS-1.**

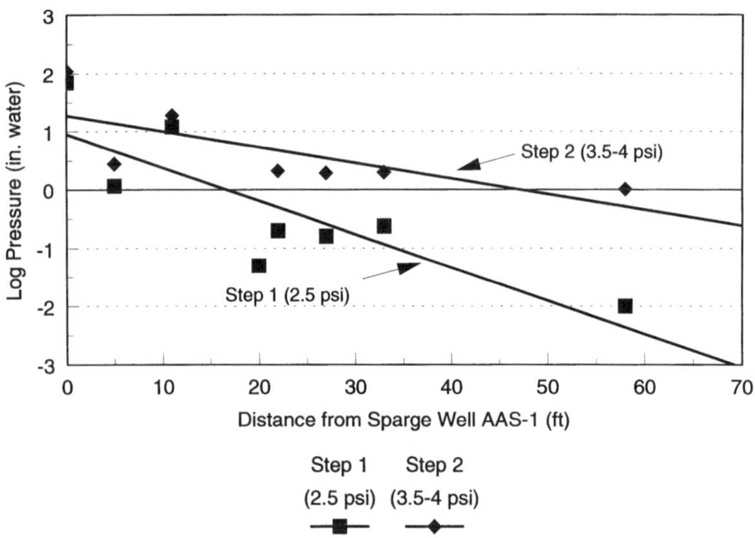

Figure 6. **Sparge pressure versus distance at Test Well AAS-2.**

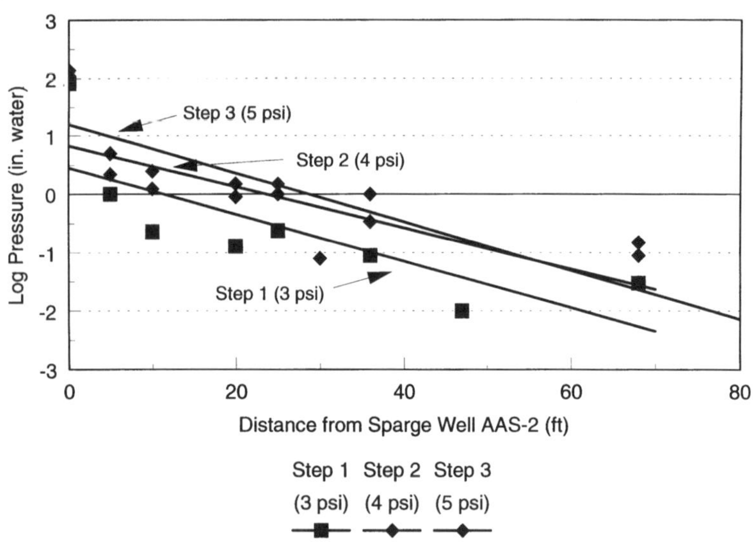

oxygen ranged from -0.04 mg/L to +1.53 mg/L. Using 1 mg/L of dissolved oxygen to define effective AAS, radial influences of 24 ft, 24 ft, and 32 ft were observed during test Steps 1, 2, and 3, respectively (Figure 8).

Dissolved oxygen measurements were recorded on a limited number of monitoring points prior to and during AAS testing at Well AAS-3. The four locations monitored during the test showed net increases in dissolved oxygen, with

Figure 7. **Dissolved oxygen versus distance at Pilot Test Well AAS-1.**

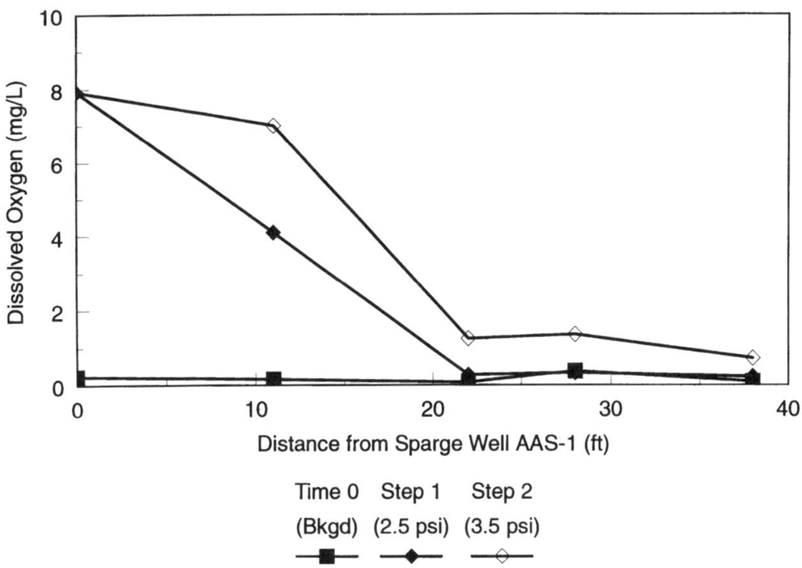

Figure 8. **Dissolved oxygen versus distance at Test Well AAS-2.**

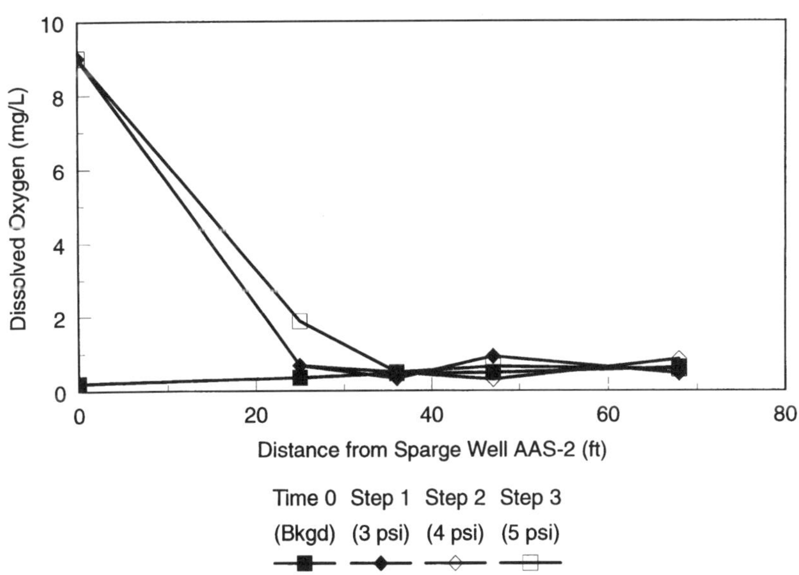

little variation in data at radial distances of 10 ft, 20 ft, and 30 ft. As a result of pilot testing, net increases in dissolved oxygen ranged from 1.80 mg/L to 2.69 mg/L at the monitoring points; dissolved oxygen increased 6.27 mg/L at sparge

Well AAS-3. Due to the limited spatial distribution of the monitoring points in the saturated zone, and the trend of sustained elevated dissolved oxygen levels versus distance, sparging radial influence at Well AAS-3 was not quantified.

Total Volatile Hydrocarbons

TVH concentrations were measured in soil vapor samples using a mobile laboratory equipped with a gas chromatograph/flame ionization detector prior to, during, and after AAS pilot testing. At each of the AAS pilot test locations, TVH concentrations primarily increased, but with no apparent correlation to distance from the sparge wells. TVH concentrations measured in soil vapor samples collected immediately following pilot testing, in contrast to background TVH concentrations, ranged from -85 to +6,158 percent. The average increase in TVH concentration during pilot testing activities ranged from 55 ppm_v (+253 percent) at Well AAS-3 to 2,883 ppm_v (+737 percent) at Well AAS-2.

Groundwater Sampling

Groundwater samples were collected prior to and following AAS pilot testing to assess resultant concentration changes (Tables 3, 4, and 5). Significant changes in BTEX and TVH concentrations are an indication of effective AAS activity. Analytical results for groundwater samples collected prior to and following AAS and SVE/AAS pilot testing demonstrated significant changes in BTEX and TVH concentrations. In general, the concentrations of BTEX and TVH were observed to increase during pilot testing, although some sample locations reported decreases in BTEX and TVH concentrations. The significant changes in BTEX and TVH concentrations reported following only several hours of pilot testing provided an indication that active air sparging dramatically affects the equilibrium of dissolved-phase VOCs in the saturated zone.

Based upon known site conditions, significant variations in TVH concentrations in groundwater samples were considered the result of the tendency toward phase equilibrium between the dissolved-phase, sorbed-phase, and vaporous phases within the saturated zone. This feature will typically be more pronounced during the initial stages of active AAS, and is a function of the distribution of hydrocarbons within the saturated zone and the composition of site soil.

Hydrocarbon Removal Rates

Vapor sampling was conducted throughout pilot testing to obtain information on the variability of concentrations to be anticipated during full-scale SVE/AAS operations. Seven SVE and SVE/AAS influent vapor samples were collected for analysis during pilot test activities, including SVE-only between step tests at Well AAS-1, SVE-only between step tests at Well AAS-2, SVE-only prior to AAS testing at Well AAS-3, and during SVE/AAS testing at Well AAS-3. A summary of

Table 3. **Analytical Results For Ground-Water Samples Collected During Pilot Testing At AAS-1**

Testing Stage	Total BTEX (mg/L)	TVH (mg/L)
	Well EAW-18	
Prior to sparging	5.9	12.7
Following sparging	19.3	45.1
	Sparge Well AAS-1	
Prior to sparging	0.5	2
Following sparging	ND	1.6
	Piezometer MP-1	
Prior to sparging	4.9	24.8
Following sparging	5.1	29.9
	Piezometer MP-2	
Prior to sparging	9.9	42.1
Following sparging	19.9	106
	Piezometer MP-3	
Prior to sparging	8.8	63.7
Following sparging	25.7	220

Table 4. **Analytical Results For Ground-Water Samples Collected during Pilot Testing At AAS-2**

Testing Stage	Total BTEX (mg/L)	TVH (mg/L)
	Well EAW-18	
Prior to sparging	19.3	45.1
Following sparging	1	22.8
	Sparge Well AAS-1	
Prior to sparging	ND	1.6
Following sparging	ND	3
	Piezometer MP-1	
Prior to sparging	5.1	29.9
Following sparging	10.0	62.3
	Piezometer MP-2	
Prior to sparging	19.9	106.0
Following sparging	30.7	177.0
	Piezometer MP-3	
Prior to sparging	25.7	220.0
Following sparging	29.9	148.0

Table 5. **Analytical Results For Ground-Water Samples Collected-During Pilot Testing At AAS-3**

Testing Stage	Total BTEX (mg/L)	TVH (mg/L)
	Sparge Well AAS-3	
Prior to sparging	ND	9.6
Following sparging	ND	2.3
	Piezometer MP-4	
Prior to sparging	0.5	7.8
Following sparging	0.9	5.5
	Piezometer MP-5	
Prior to sparging	0.2	4.8
Following sparging	0.3	3.3

the benzene, toluene, and TVH concentrations and corresponding removal rates is provided in Table 6. Hydrocarbon removal rates varied considerably between test locations. TVH removal rates were as follows: approximately 1,032 lb/day in the vicinity of Well AAS-1; 141.5 lb/day in the vicinity of Well AAS-2, and approximately 1.1 lb/day during active sparging at Well AAS-3. Hydrocarbon removal rates are anticipated to display similar variability as a result of the employment of additional extraction wells/trenches.

Aquifer Air Sparging Pilot Test Conclusions

Pilot test results suggest the applicability of AAS technology at the subject site. Though not quantified via long-term operations, mass removal efficiency via air sparging was favorable during pilot testing. Significant changes in TVH concentrations were observed in shallow groundwater, an indication of physical and chemical activity resulting from active AAS operations. Hydrocarbon loading into the vadose zone during active AAS resulted in an average increase in TVH concentration of greater than 540% (1,700 ppm_v), in comparison to background TVH concentrations . Areal influence of individual sparge wells, based on measurements of differential pressure in the saturated zone during testing, was calculated at greater than 30 ft. Soil characteristics, including permeability and composition, indicate site soil is amenable to in situ aeration technologies.

As with many in situ technologies, the effectiveness of SVE/AAS may also be potentially limited at localized portions of the site due to site characteristics (e.g., the heterogeneity of the fill material comprising the shallow stratigraphy). Despite the short duration of pilot testing, heterogeneity in the fill material was identified, indicating the potential for limited removal efficiencies or air channeling locally within some areas. Differential pressures and dissolved oxygen levels were observed to vary considerably with respect to distance from sparge

Table 6. **Hydrocarbon Removal Rates By Soil Vapor Extraction During Aquifer Air Sparging Pilot Testing**

Test Location	Soil Vapor Extraction Rate (cfm)	Total Volatile Hydrocarbon (ppmv)	Removal Rate (lb/day)
AAS-1	63	47000	1032.5
AAS-2	63	6440	141.5
AAS-3	57	56	1.1

wells, and followed a poor to moderate linear relationship. These potential limiting factors were considered in the design of the full-scale SVE/AAS remedial system.

DESIGN CRITERIA

Operating Parameters

During the AAS pilot test conducted at Well AAS-1, AAS injection pressures ranged from 2.5 to 4 psig. Corresponding flow rates ranged from 6.5 to 26 cfm. Injection pressures greater than 3.5 psig were difficult to maintain due to the ability of the formation to accept large air injection rates. The sparge injection rate increased significantly when the sparge pressure was increased from 3.5 to 4 psig, indicating the potential for air channeling within the shallow water table. AAS injection pressures ranging from 3 to 5 psig and flow rates ranging from 5.75 to 24 cfm were observed during AAS pilot testing at Well AAS-2. As occurred at Well AAS-1, sparge rates increased rapidly with correspondingly minor increases in injection pressure.

During the pilot test conducted at Well AAS-3, AAS injection pressures ranged from 2.5 to 3 psig. Corresponding flow rates ranged from 7.5 to 16.5 cfm. Injection pressures greater than 3 psig could not be sustained due to the difficulty in the sparging equipment to maintain an injection rate the formation was able to accept. Similar to tests at Wells AAS-1 and AAS-2, the injection rate increased significantly when the sparge pressure was increased from 2.5 to 3 psig. Correspondingly, control of air injection rates and pressures was a primary consideration in the full-scale AAS system design.

The sparge pressures tested were nominally above the observed hydrostatic head pressures (1.8 to 2.2 psig), suggesting little resistance by the formation to sparging of ambient air. The observed flow rates in all test areas were primarily within the range of AAS flow rates reported by the EPA (USEPA, 1992) (i.e., 2-16 cfm), with the exception of instances when the hydrostatic head pressure was exceeded generally by more than 1.5 psig.

Based on test results, sparge pressures should not exceed hydrostatic head pressure by more than approximately 1.5 psig. Corresponding flow rates should be maintained at or below approximately 15 cfm at each sparge well. The intent of operations should be to maximize the partitioning of hydrocarbons into the vapor phase and, therefore, higher flow rates are not necessarily more efficient. Pressure and flow are parameters designed to be adjusted in the field during operations. SVE flow rates in both test areas were several times greater than the corresponding sparge flow rate, indicating the minimum injection rate to removal rate ratio of approximately 1:3 (necessary to control vapor migration) will be achieved.

Preliminary Estimate of Area of Influence

From the pilot tests, it was evident that a pervasive, low pressure injection of ambient air through saturated zone soil yields a favorable hydrocarbon removal rate. In addition, site conditions such as depth to water and soil permeability, and the physicochemical properties of the hydrocarbons present onsite, are favorable for implementation of an AAS system.

The radial influence of each AAS well was estimated based on regression analyses of differential pressures within the saturated zone and direct measurement of dissolved oxygen concentrations. Observations of total hydrocarbon concentrations in soil vapor and groundwater were not utilized to assess radial influence. The preliminary estimates described below take into account these measurements and the impact observed on total hydrocarbon concentrations within the vadose and saturated zones.

For the AAS system design, an AAS well radial influence of 30 ft was employed. Assessments of effective radial influences assume injection pressures nominally above the hydrostatic pressure created by the water table (as were applied in the first test step at each of the sparge wells). At sparge wells AAS-1 and AAS-2, the interpreted effective radial influence of AAS was approximately 30 ft. At sparge well AAS-3, the interpreted effective radial influence of AAS was 40 ft.

The intent of the design of the sparge wells is to maximize the vertical area of influence such that sparged air is distributed through the entire zone of saturation. At this site, the saturated zone subject to AAS extends to the top of the underlying silty clay, which is considered an aquiclude. Air sparged beneath the water table will enter the saturated zone formation from the screened interval through the path of least resistance, which is the sand pack surrounding the screen.

Mass Removal

TVH removal during the three pilot studies ranged from 1.1 lb/day to 1,032 lb/day, indicative of much variability in hydrocarbon distribution within the subsurface at the site. Based on the potential to achieve elevated hydrocarbon removal rates during the initial stages of active AAS, and the observance of similarly variable hydrocarbon removal rates during prior SVE pilot studies, advanced vapor abatement technology (i.e., thermal/catalytic oxidation) was engineered into the design of the full-scale SVE/AAS remediation system.

SYSTEM DESIGN AND OPERATION

The SVE/AAS remedial system consists of three primary components: the AAS injection wells and compressor, the vapor extraction trench network and associated vacuum blowers, and the offgas treatment system. The AAS component operates by injecting air into the saturated zone to facilitate the transport of volotilized hydrocarbons into the vadose zone. The SVE component serves to capture all air introduced via the sparging process, and to induce air flow through the vadose zone to remove hydrocarbons resident above the water table interface. The offgas treatment system oxidizes the hydrocarbon vapors received from the SVE system (converting incoming VOCs into carbon dioxide and water vapor) via thermal/catalytic oxidation prior to atmospheric discharge.

At the commencement of remedial system operations, only the SVE and offgas treatment system will be active. In this manner, reduction of hydrocarbon contamination within the vadose zone can be quantified without interference resulting from hydrocarbons extracted from the groundwater during air sparging. The thermal/catalytic oxidation system will initially be run in thermal mode to minimize offgas treatment costs until the concentration of incoming VOCs decreases sufficiently to warrant catalytic operation. Once vadose zone remediation is complete (i.e., asymptotic recovery is reached), the AAS system will be activated to initiate groundwater treatment. The SVE system will remain active to recover sparged air and continue vadose zone "restoration." It is likely that activation of the AAS system will introduce sufficient hydrocarbon concentrations into the vadose zone to warrant renewed thermal operation of the oxidation system. Once incoming VOCs are sufficiently diminished (through source reduction), the oxidation system will be returned to catalytic operation. The SVE/AAS system will continue to run until hydrocarbon concentrations in groundwater and soil are reduced to levels acceptable for site closure.

Aquifer Air Sparging System Components

The AAS system consists of 17 air-injection wells located throughout the treatment area. The locations of the AAS wells were specified based on a 30-ft radius of influence (determined during pilot testing) to maximize groundwater aeration within the treatment area. To prevent offsite migration of dissolved phase hydrocarbons, the radius of influence for each AAS well was restricted to the boundaries of the treatment area, and was entirely contained within the SVE capture zone. Additionally, six of the AAS wells were installed as a "cut-off" wall at the eastern portion of the treatment area (oriented so that groundwater flow must pass through the "cut-off" wall area prior to migrating downgradient of the treatment zone), to prevent offsite migration of dissolved-phase hydrocarbons. The layout of the AAS wells is shown in Figure 3.

The AAS wells were constructed of 2-in. Schedule 80 PVC with 2 ft of screen (0.010 in.) installed 6 in. above the top of the silty clay unit. The borehole annulus surrounding the well screens was filled with silica sand filter pack to facilitate air flow into the saturated zone. To prevent "short-circuiting" of sparged air, each AAS well was sealed immediately above the filter pack and below the groundwater interface with bentonite.

During pilot testing, it was determined that control of AAS injection pressure and flow rate was essential to prevent over-pressurization of the aquifer and short-circuiting of sparged air to AAS wells with limited hydrostatic resistance. Therefore, pressure reducing regulators and globe valves were installed at the head of each AAS well to modulate injection pressure and flow rate, respectively. The globe valves were installed upstream of the regulators to serve as shutoff devices when necessary (i.e., for monitoring activities), allowing continued operation of the AAS system during individual line shutdown. The pressure regulators were specified to reduce incoming line pressure from approximately 15 psi to a low range of 0.0 to 6 psi (sufficient to overcome hydrostatic head throughout the site). Air flow and pressure indication were provided by in-line, direct-reading, rugged-vane flow gauges and pressure gauges mounted at each AAS well head. A schematic of a typical sparge well and well head equipment configuration is shown in Figure 9.

Sparge air is provided by a 20 HP rotary lobe compressor rated for 200 cfm at 15 psig. The compressor is skid-mounted and encased in a noise-reduction housing, and is fitted with a filter/silencer to maintain decibel levels below 85 dCB at 3 ft. A process and instrumentation diagram depicting the operational elements of the AAS compressor skid is provided in Figure 10. After compression, moisture is removed from the air by an air-drying system. The air-drying system is comprised of three primary components: an aftercooler with condensate trap, a 120-gal air reservoir tank, and a single tower desiccant air-dryer. The aftercooler reduces the approach temperature of the incoming air by advection with atmospheric air. This temperature reduction induces condensation prior to the air-drying tower, thereby reducing desiccant consumption. Condensate arising from the aftercooling process is trapped and drained automatically. Air flow then enters a 120-gal air reservoir tank which serves to minimize pulsations in line pressure and maintain air pressure during fluctuations in compressor performance. The air reservoir tank is also fitted with an automatic condensate purge valve. After the reservoir tank, air flows through the air-drying tower where residual moisture is absorbed by the desiccant (a nontoxic hygroscopic compound compacted into tablets to maximize surface area). An afterfilter removes any airborne particulate material, including spent desiccant, to protect downstream equipment (i.e., AAS pressure regulators and flowmeters). Clean,

Figure 9. **Construction diagram of typical AAS well.**

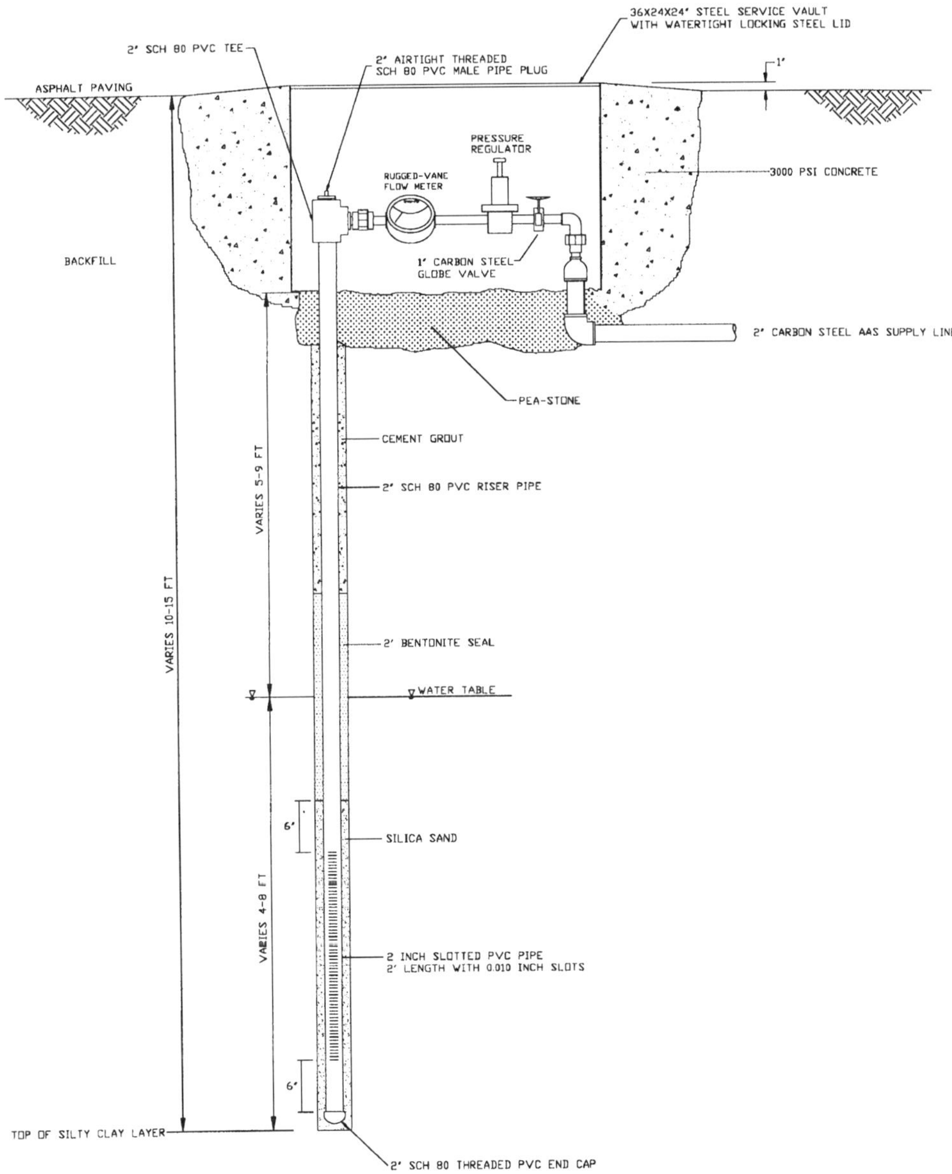

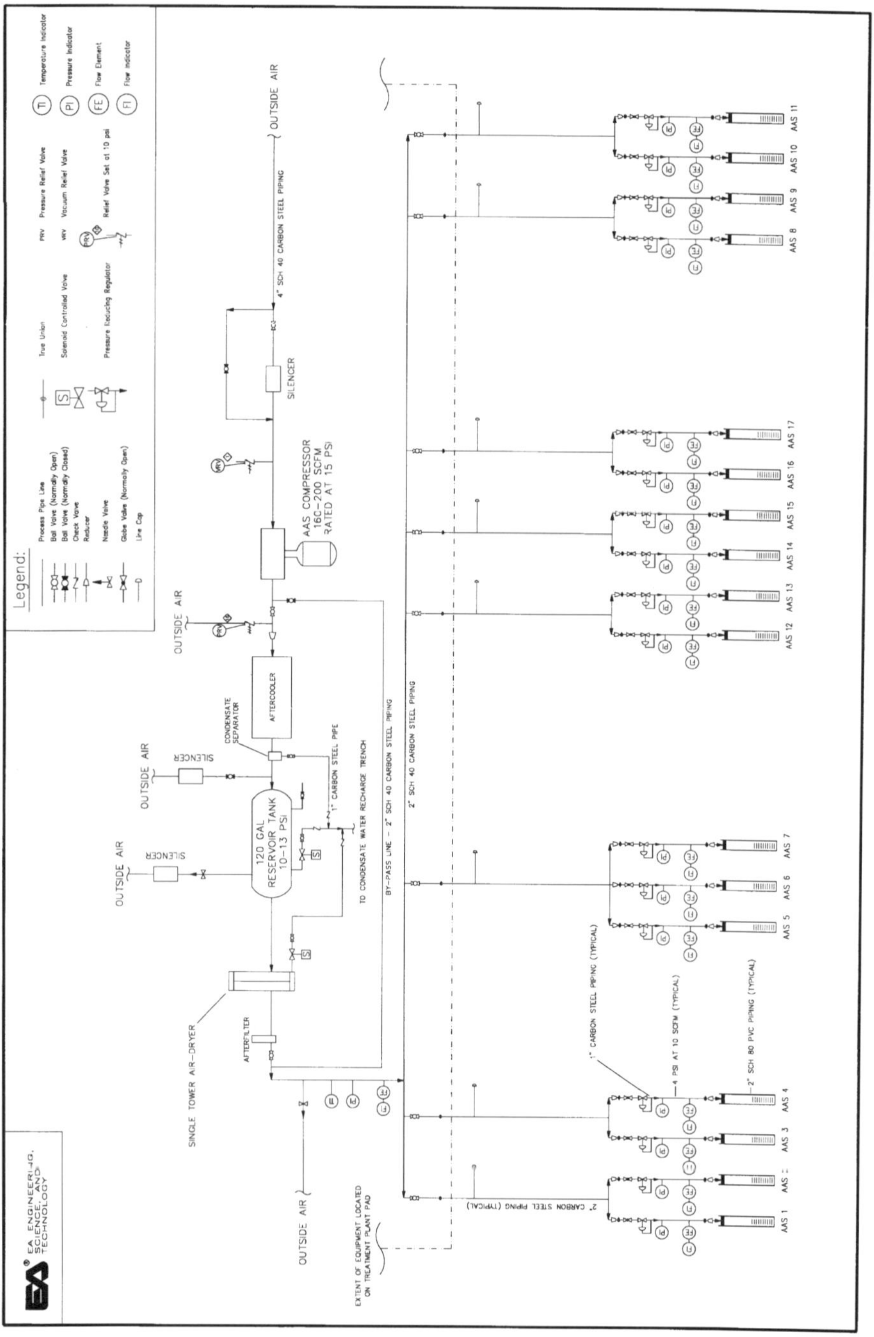

Figure 10. Aquifer air sparging system process and instrumentation diagram.

dry air is then distributed to the sparge wells via the central sparge manifold. A bypass line is provided to allow continued AAS system operations during air-dryer maintenance, or periods when air-drying is not required.

Soil Vapor Extraction System Components

The SVE system was designed to remove volatile hydrocarbons by providing vacuum-induced air flow in the vadose zone throughout the treatment area, and to capture sparged air. The layout of the SVE system is shown in Figure 3. Figure 11 provides a process and instrumentation diagram for the SVE blower skids and oxidation unit.

The invasive portion of the SVE system consists of 2 sets of 9 vapor extraction trenches (18 total) and 17 passive venting points. The passive venting points are constructed of 2-in. Schedule 40 PVC with 3 ft of 0.010 in. well screen installed 1 ft above the seasonally high-water table elevation. The function of the passive venting points is to promote subsurface air flow from the perimeter of the treatment area to the SVE trenches. Figures 12 and 13 provide schematics of a typical SVE trench and a typical passive venting point, respectively.

Vacuum to each set of SVE trenches is supplied by two positive displacement blowers rated for flow between 270 and 495 cfm at 60-90 in. H_20 vacuum. Soil vapor withdrawn from the subsurface is directed to a catalytic oxidation unit, rated for a maximum of 1,000 cfm, for oxidation of VOCs. The catalytic oxidation unit achieves in excess of 99% destruction efficiency of incoming VOCs. Each SVE trench has a design flow rate of 30-55 cfm. This flow rate was selected based on pilot test results to provide an SVE radius of influence sufficient to cover the remediation zone and capture sparged air. SVE flow rate and vacuum settings will be normalized upon system start-up. Normalization of flow is achieved using butterfly valves at the head of each SVE pipe run (located at the intake manifold). Flow rate is quantified by employing pitot tubes and factory-calibrated differential pressure gauges accurate to within 4% of the instrument range. Vacuum in each intake line and at each vent trench riser is indicated with a differential pressure gauge.

Moisture separators and particulate filters are provided upstream of the blowers to remove water and airborne particles, respectively. If left unattended, such contaminants would significantly reduce blower life and overall system performance. Each moisture separator is evacuated by a ⅙ HP centrifugal draw-off pump activated by a high level switch, and shut down by a low level switch. Upon activation of either of the centrifugal draw-off pumps, condensate from the respective moisture separator is pumped through a common dual cell carbon treatment unit (connected in series) to remove dissolved-phase hydrocarbons. The treated water is then gravity-fed to a groundwater recharge trench located within the treatment zone. In the event of centrifugal pump failure, or other system error resulting in high water level in a moisture separator, a high level alarm is activated, the entire SVE/AAS system is shut down, and the system alerts the operator through an automated paging system.

The SVE blowers are designed to shut down in the event of high running temperature and/or pressure, or upon indication of any catalytic oxidizer or moisture separator alarm. Failure of either SVE blower automatically shuts down all SVE/AAS system components and notifies the operator through an automated paging system. Additionally, should insufficient line pressure be indicated downstream of the blowers, the same shutdown procedure is followed.

Vapor sampling ports are provided at the head of each SVE manifold riser (18 total) and at the combined influent location for each SVE skid to monitor incoming VOC concentration. A vapor sample port is also provided on the oxidizer stack to monitor offgas treatment efficiency. Liquid sampling ports are provided upstream and downstream of the carbon adsorption units to confirm treatment and assess the need for replacement of the lead granular activated carbon unit (when necessary).

Thermal/Catalytic Oxidation System

The thermal/catalytic oxidizer provides offgas treatment for the SVE/AAS system. The oxidizer operates in two modes to achieve greater than 99% VOC destruction efficiency (e.g., oxidation of hydrocarbons to CO_2 and H_2O). During periods of high incoming combustible vapor concentrations (i.e., greater than 25% of the lower explosive limit), the oxidizer is operated in thermal mode. When in thermal mode, oxidation is achieved by subjecting the hydrocarbons to a temperature of 1,400°F for 1 second. A turbulence generator is included to ensure complete mixing during the oxidation process. When the incoming VOC concentration falls below 25% of the lower explosive limit, the oxidizer is switched to catalytic mode to minimize natural gas use. Natural gas supplements the hydrocarbon-laden soil vapor as a fuel source to ensure continual burner temperature. During catalytic operation, the oxidizer is operated between 600°F and 1,250°F.

The oxidation unit is self-regulating with respect to air-flow and natural gas fuel supplementation. However, in the event of insufficient VOC removal efficiency, or other oxidation system error, the entire SVE/AAS system is shut down, and the operator is notified through an automated paging system.

Layout of Soil Vapor Extraction/Aquifer Air Sparging and Oxidation Equipment

The layout of the SVE/AAS treatment buildings and skid-mounted equipment is shown in Figure 14. The dual SVE skid-mounted blower units, together with the 18 SVE manifold risers, are housed within a 16 ft × 24 ft explosion-proof (Group D, Class 1) steel building. The SVE system condensate removal components (two centrifugal draw-off pumps and two 55-gal liquid-phase granular-activated carbon vessels) are also located in the SVE building. As required by

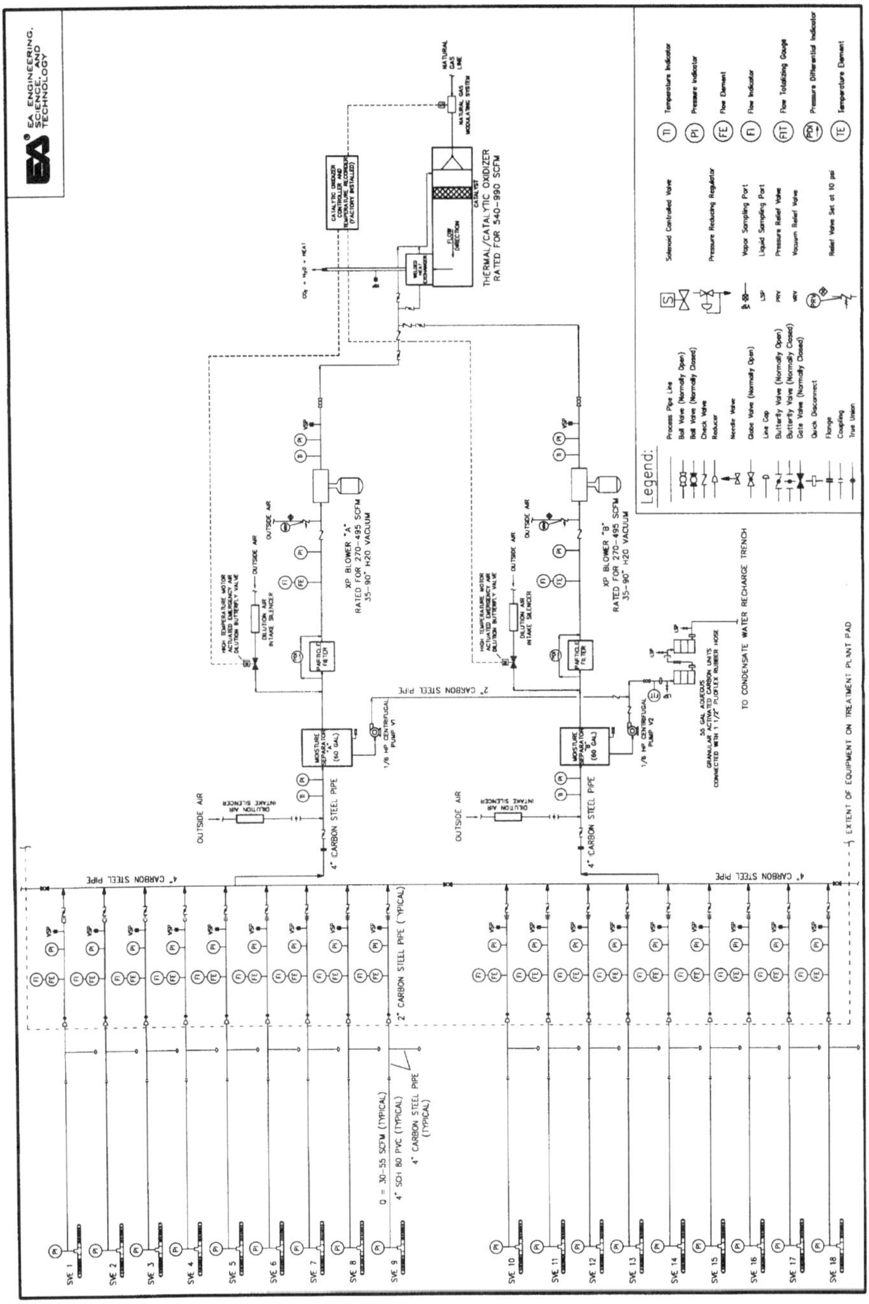

Figure 11. Soil vapor extraction system P&ID.

Figure 12. Construction diagram of typical SVE trench.

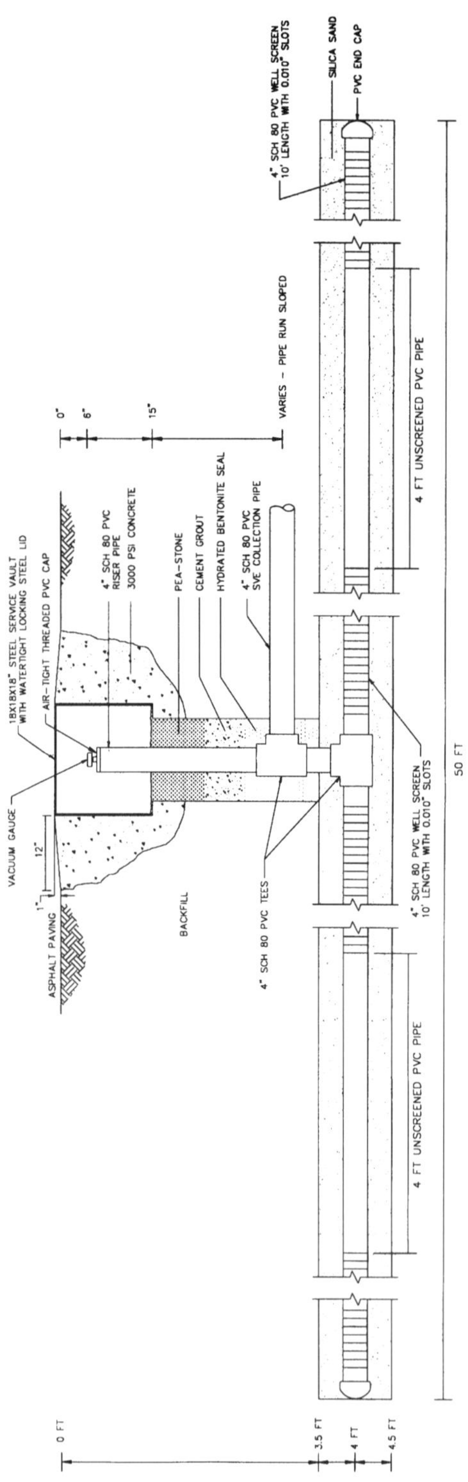

Figure 13. **Construction diagram of typical passive venting point.**

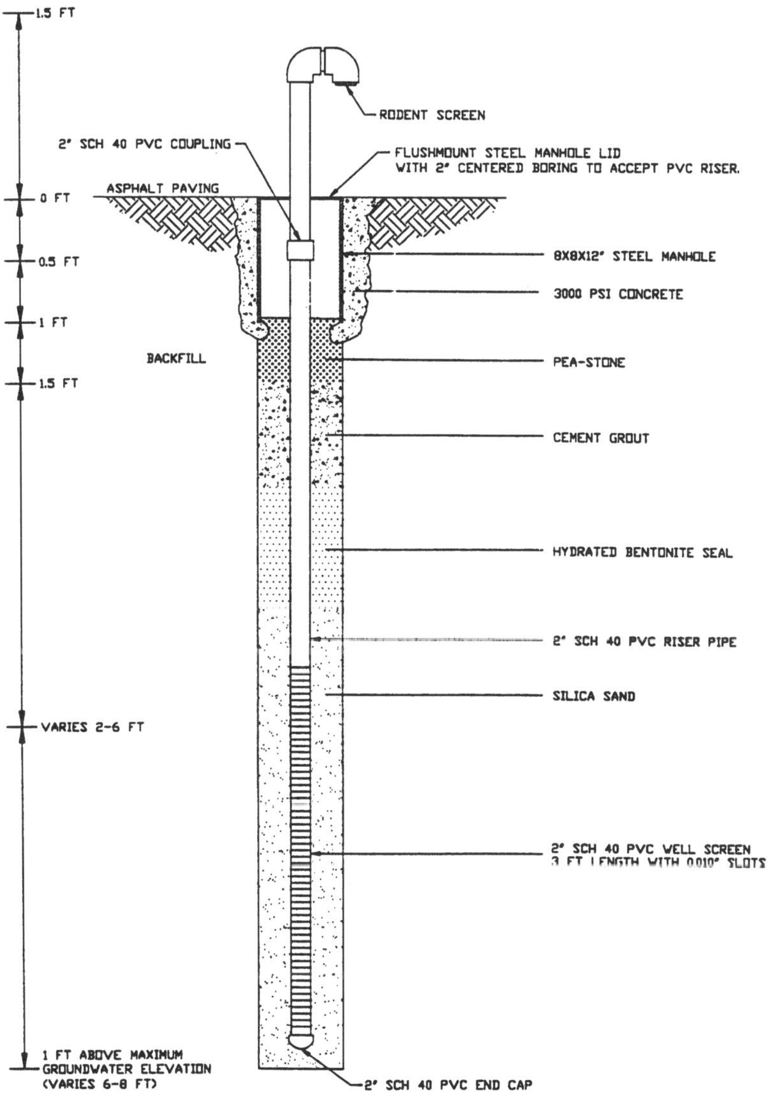

Figure 14. **Plan view of SVE/AAS treatment plant.**

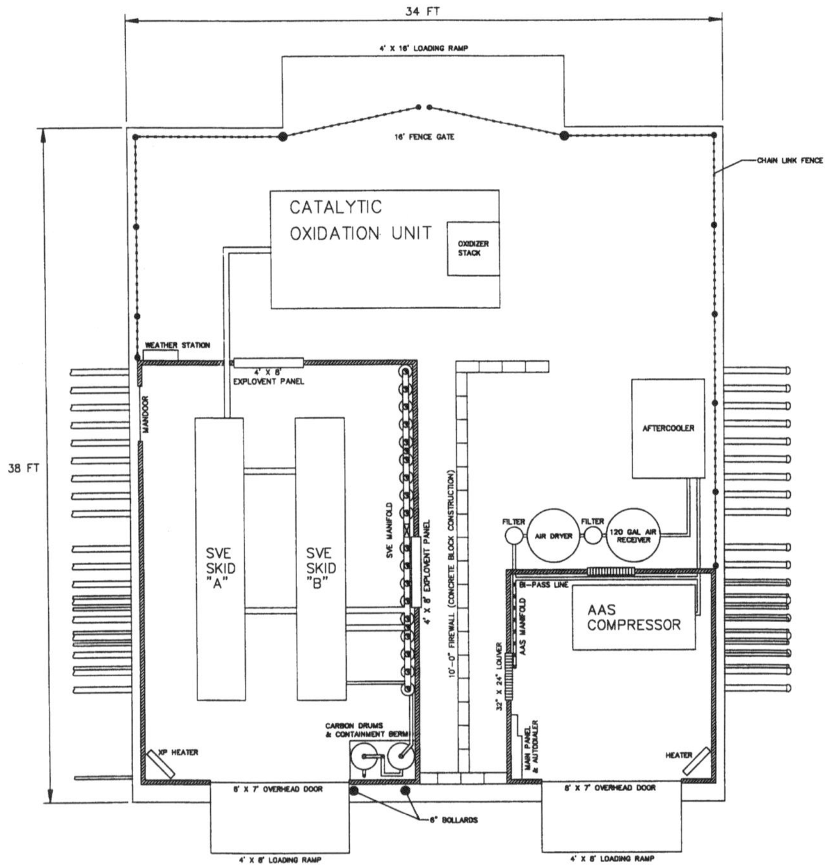

local fire department code, the building is mechanically ventilated (during all hours of SVE operation) at 425 cfm to provide internal air volume turnover at a rate of six exchanges per hour. Also per fire department code, two 4 ft × 8 ft explosion-relief blow-out panels are provided to release building overpressurization in the event of an explosion during operations. Access to the SVE building is provided by one 3-ft mandoor and one 7 ft × 8 ft overhead door.

The AAS compressor unit and sparge manifold are housed within a 12 ft × 12 ft steel, nonclassified, building. The main electrical panel and emergency stop button are also located in the AAS building to avoid the explosion-proof requirement of the SVE building.

Both buildings, the oxidation unit, and the air-drying system are housed on a 30 ft × 34 ft concrete foundation. A concrete block fire wall, providing an additional two-hour fire wall rating, is constructed between the explosion-proof and non-explosion-proof buildings. The entire treatment plant perimeter is enclosed by a 10-ft chain-link security fence.

ACKNOWLEDGMENTS

The authors wish to acknowledge Ms. Gloria McCleary and Ms. Michele Predmore of EA Engineering, Science, and Technology for their technical assistance and review of the manuscript.

REFERENCES

Angell, K.G. 1992. In Situ Remedial Methods: Air Sparging. *The National Environmental Journal*, January-February, pp. 20-23.

Brown, R.A., Henry, E., Herman, C., and Leonard, W. 1991. The Use of Aeration in Environmental Clean-Ups. *Proceedings of the Petroleum Hydrocarbons and Organic Chemicals in Ground Water: Prevention, Detection, and Restoration. A Conference and Exposition.* Houston, TX, November.

Brown, R., Herman, C., and Henry, E. 1991. The Use of Aeration in Environmental Clean-Ups. HazTech International Pittsburgh Waste Conference. 14-16 May. Pittsburgh, PA.

Leonard, W.C. and Brown, R.A. 1992. Air Sparging: An Optimal Solution. National Water Well Association/American Petroleum Institute Conference on Petroleum Hydrocarbons and Organic Chemicals in Groundwater: Prevention, Detection and Restoration. 4-6 November. Houston, TX.

Middleton, A.C. and Hiller, D.H. 1990. In Situ Aeration of Ground Water: A Technology Review. A Conference on Prevention and Treatment of Soil and Groundwater Contamination in the Petroleum Refining and Distribution Industry. 16-17 October. Montreal, Quebec.

U.S. Environmental Protection Agency (USEPA) . 1991. Soil Vapor Extraction Technology Reference Handbook, U.S. Environmental Protection Agency, Office of Research and Development, Washington, DC. Document EPA/540/2-91/003.

USEPA. 1992. A Technology Assessment of Soil Vapor Extraction and Air Sparging. Document EPA/600/R-92/173. September.

CHAPTER 10

A Model for Predicting Air Sparging Performance

James A. O'Neill, II and **Brian D. Symons,** Remediation Technologies, Inc., Fort Collins, Colorado

This chapter describes a conceptual model for air sparging and incorporates this model into a simple computer spreadsheet for predicting air sparging performance. The conceptual model describes the air sparging process in a mass balance equation. This chapter briefly describes each term in the equation (initial and final contaminant mass, contaminant mass flow in groundwater, volatilization, and biodegradation). The conceptual model has been incorporated into a computer spreadsheet to predict air sparging performance. The spreadsheet provides the model equations, a graphical illustration of the change in contaminant concentration in groundwater with time, and a mass balance of the contaminant in the system. Data from application of air sparging at different sites were used to calibrate the model and estimate typical volatilization and air sparging efficiency factors.

Although air sparging has been conducted at numerous sites (Johnson et al., 1993; Middleton and Hiller, 1990; USEPA, 1992), the technology has generally been applied using rule-of-thumb design parameters, pilot testing, and experience at other sites. An air sparging model is needed to better predict system performance. Model-based predictions will apply to a wide range of site-specific conditions without the need for conducting lengthy pilot studies.

Several models have been developed that describe air sparging. These models have generally focused on predicting air flow and related physical effects (water level increase and pressure distribution) and on microscale oxygen transfer and volatilization (Mohr, 1995; Gvirtzman and Gorelick, 1992; Lundegard and Anderson, 1993). However, these models have limited utility in describing the fate of a contaminant in a full-scale, complex system. This suggests the need for a model that can evaluate air sparging on a site-wide basis.

CONCEPTUAL MODEL FOR AIR SPARGING

The overall objective of subsurface remediation is to reduce contaminant concentrations to below regulatory or risk-based levels. The primary focus of air sparging is to inject air into the saturated zone to increase oxygen levels and enhance biodegradation and/or to purge volatile organics and enhance volatilization. Air entering the system is essentially clean, but air exiting the system may contain volatile organic compounds, and biodegradation may have caused some oxygen to have been consumed. Figure 1 shows a conceptual diagram of an air sparging system using a horizontal injection well.

Figure 1. **Control volume of contaminated subsurface.**

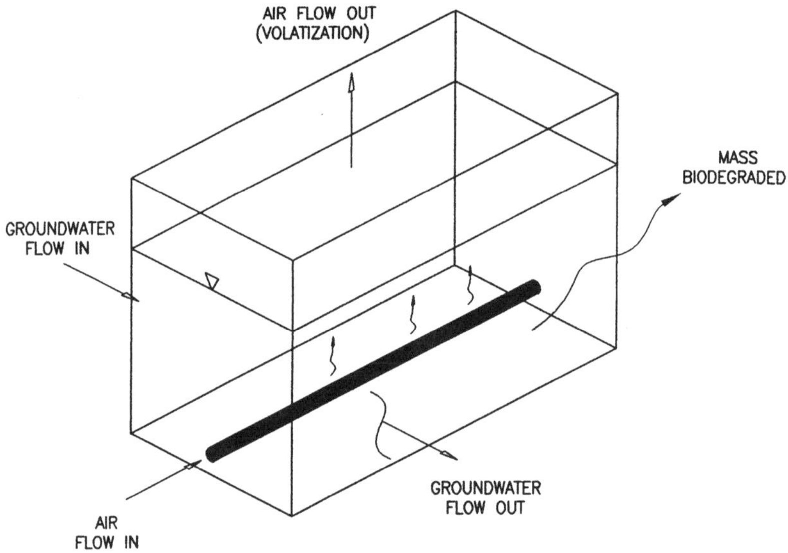

Simple Mass Balance Equation

The combination of initial contaminant concentrations in the soil, water, air, and oil in the saturated zone and the effects of air moving through this zone can be evaluated using a mass balance approach with the following equation:

Final Mass - Initial Mass = Mass Flow In - Mass Flow Out - Mass Volatilized - Mass Biodegraded (1)

The dimensions of the system are somewhat arbitrary because they may include the total horizontal and vertical extent of contamination at a site, the area of influence of a single air sparging system within the contaminated area, or a unit volume (e.g., one cubic foot).

Initial and Final Mass in System

The initial and final contaminant masses indicate the total mass stored in the system initially and at a later time. Cumulatively, the initial contaminant mass (M_T) at a site is comprised of contamination in groundwater, soil, air, and oil. This relationship is shown in the following equation:

$$M_T = M_w + M_s + M_a + M_o \tag{2}$$

where:

M_w = the containment mass in the groundwater
M_s = the containment mass in the soil
M_a = the containment mass in soil-pore air
M_o = the containment mass in the oil phase

Groundwater Flow and Chemical Transport

The masses of a contaminant in groundwater flowing in and out of the system are defined by the groundwater flow rate and contaminant concentration in groundwater. The flow of groundwater is calculated using Darcy's law:

$$Q = kiA \tag{3}$$

where:

Q = groundwater flow rate (ft^3/day)
k = hydraulic conductivity (ft/day)
i = groundwater hydraulic gradient (ft/ft)
A = cross-sectional area through which groundwater is flowing (ft^2)

The mass of a contaminant in groundwater flow is equal to the flow rate of groundwater (ft^3/day) multiplied by the contaminant concentration (lb/ft^3).

Volatilization

Some mass flows out of the system in air through volatilization. This is accounted for in the volatilization term. The mass of a contaminant removed in the exiting air by air sparging is a function of the air flow rate out of the system (ft^3/min) and the concentration of the contaminant in the air (lb/ft^3). In an air sparging system, the maximum concentration of a contaminant in the air phase can be estimated using the Henry's law constant for that contaminant:

$$C_a = K_h z_v C_w \tag{4}$$

where

C_a = air concentration (mg/m^3)
K_h = dimensionless Henry's law constant

z_v = volatilization efficiency factor
C_w = water concentration (mg/m^3)

The air flow rate out of the system equals the air flow rate into the system.

Biodegradation

Biodegradation is chemical changes caused by microorganisms. For the purposes of this chapter, biodegradation has been more narrowly defined as the aerobic transformation of an organic compound into water and carbon dioxide. Additional cell mass may be temporarily generated in this process. A generalized equation for the biodegradation is:

$$C_6H_{14} + 9.5O_2 = 7H_2O + 6CO_2 \qquad (5)$$

Equation 5 indicates that oxygen is the key element for aerobic biodegradation of an organic compound. Based on this equation and the gram molecular weights for hexane, or $C_6\ H_{14}$ (86), and oxygen (32), approximately 3.5 pounds of oxygen are required to completely biodegrade 1 pound of hexane. This ratio of oxygen mass to degraded organic compound mass is similar for other hydrocarbons as well. When dissolved oxygen concentrations in the system are maintained above 1 or 2 milligrams per liter (mg/L), oxygen is not considered to be limiting, and biodegradation rates follow first order decay kinetics; this is the conceptual model assumption. However, under some conditions, biodegradation is controlled by the rate of oxygen transfer. Therefore, the rate of contaminant biodegradation is directly proportional to the rate of oxygen transfer or the reaction is zero order. The degradation rate can therefore be defined by the zero order decay rate (K) as described in the following equation:

$$K = Q_a d z_b / 3.5 \qquad (6)$$

where:

Q_a = air flow rate (ft^3/day)
d = oxygen density in air (lb O_2/ft^3); and
z_b = oxygen transfer efficiency factor.

SPREADSHEET DESCRIPTION

Figure 2 presents a computer spreadsheet that incorporates the conceptual air sparging model. The computer spreadsheet is presented in three parts: the model equations and assumptions, a graphical illustration of the change in contaminant concentration in groundwater with time, and a mass balance of the contaminant in the system.

The spreadsheet equations further define the conceptual model by treating the air sparging system as a continuously stirred batch tank reactor (Metcalf and

Figure 2. **Example air sparging model (benzene).**

Assumptions

Aquifer Conditions

		Units	Conditions	Stands For
Qi=	182	ft³/day		Groundwater flow rate
Vt=	240,000	ft³		Total Volume
Xa=	0.03			Fraction of Air
Ps=	109	lb/ft³	Dry basis	Density of Soil
Xw=	0.31			Fraction Water
Xs=	0.66			Fraction Soil
Xo=	0			Fraction Oil
z=	0.0175			Efficiency factor
Qa=	4320	ft³/day		Air Flow Rate

Contaminant

Benzene		Units	Stands For
Ks=	0.00196	ft³/lb	Soil/Water Partition Coefficient
Ko=	89		Oil/Water Partition Coefficient
k=	0.043	/day	First Order Decay Rate
Kh=	0.2279		Air/Water Partition Coefficient

Equations

Cw(t)=((YCw(0)+B)*exp(Yt)-B)/Y

Ca(t)= Kh z Cw(t)

Cw(0)=	0.021817	lb/ft³
Cw(0)=	350	mg/L
Cw(in)=	0.021817	lb/ft³
Cw(in)=	350	mg/L

Y= -0.04483

B= 3.16E-05

$$Y = \frac{-[Q/Vt + k(Xw + Ps\ Ks) + Qa/Vt\ Kh\ z]}{(Xw + Ps\ Ks + Xa\ Kh\ z + Xo\ Ko)}$$

$$B = \frac{Q/Vt\ Cw(0)}{(Xw + Ps\ Ks + Xa\ Kh\ z + Xo\ Ko)}$$

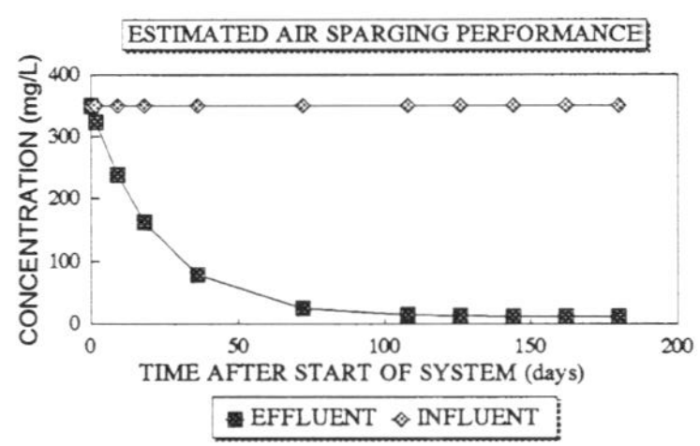

Time (days)	Cw(t) (mg/l)	Ca(t) (mg/m³)
0	350	1,396
1.8	324	1,291
9	238	947
18	162	648
36	79	314
72	25	99
108	14	56
126	12	50
144	12	47
162	12	46
180	11	46
365	11	45

MASS BALANCE (lbs)

SYSTEM(a) MASS	INFLUENT(b)	BIO.(c)	VENTED(d)	EFFLUENT(e)
2,741.8	7.1	205.6	0.7	6.6
2,536.2	28.6	682.2	2.2	19.4
1,861.0	35.7	605.6	1.9	16.6
1,272.6	71.5	708.7	2.3	16.1
617.0	142.9	554.0	2.0	10.1
193.8	142.9	220.8	0.7	5.7
109.5	71.5	80.3	0.3	2.6
97.9	71.5	74.0	0.2	2.4
92.7	71.5	71.2	0.2	2.4
90.4	71.5	70.0	0.2	2.3
89.4	734.6	798.0	2	23.7

Note:

(a) Trench Mass = GW mass + Soil Mass = Cw(t) * V(t) * (Xw + (Ps * Ks))

(b) Influent = Cw(t) * Qi * (Time)

(c) Bio = Initial trench mass - Final trench mass − Influent - Vented - Effluent

(d) Vented = Ca(t) * Qa * (Time)

(e) Effluent = Cw(t+1) * Qi * (Time)

Eddy, 1979). This approach has been applied to groundwater pump-and-treat systems (RETEC, 1991; Brubaker, 1991; Gelhar and Wilson, 1974; Zheng et al., 1991), but also applies to air sparging systems. The reactor depth is the depth of the aquifer under consideration. The area of the reactor depends on the geometry of the site.

The total mass of a contaminant in specific phases is calculated by substituting the following equations into Equation 2:

$$M_w = V_T \Theta_w C_w \tag{7}$$

$$M_s = V_T \rho_s C_s \tag{8}$$

$$M_a = V_T \Theta_a C_a \tag{9}$$

$$M_o = V_T \Theta_o C_o \tag{10}$$

$$C_s = K_s C_w \tag{11}$$

$$C_o = K_o C_w \tag{12}$$

$$M_T = C_w V_T (\Theta_w + \rho_s K_s + \Theta_a K_h z + \Theta_o K_o) \tag{13}$$

This yields the following mass equation for the change in contaminant mass as a function of time:

$$\frac{d[C_w]V_T (\Theta_w + \rho_s K_s + \Theta_a K_h z_v + \Theta_o K_o)}{dt} = Q_w C_{w(IN)} - Q_w C_{w(t)} - K(V_T \Theta_w C_{w(t)} + V_T \rho_s K_s C_{w(t)}) \tag{14}$$

where:

V_T = total system volume (ft^3)
Θ_w = fraction water in the system
Θ_a = fraction air in the system
Θ_o = fraction oil in the system
K_s = soil/water partition coefficient (ft^3/lb)
K_o = oil/water partition coefficient
ρ_s = dry soil density (lb/ft^3)

Equation 14 can be rearranged to describe the change in contaminant concentration in groundwater as a function of time:

$$\frac{dC_w}{dt} = \frac{\frac{Q_w}{V_T} C_{w(IN)} - [\frac{Q_w}{V_T} + K(\Theta_w + \rho_s K_s) + \frac{Q_a}{V_T} K z_v] C_{w(t)}}{(\Theta_w + \rho_s K_s + \Theta_a K_h z_v + \Theta_o K_o)} \tag{15}$$

The solution to Equation 15 can be simplified by defining two variables (α and β) as follows and by substituting the values into Equation 15:

$$\alpha = \frac{[\frac{Q_w}{V_T} + K(\ominus_w + \rho_s K_s) + \frac{Q_a}{V_T} K_h z_v]}{(\ominus_w + \rho_s K_s + \rho_a K_h z_v + \ominus_o K_o)} \tag{16}$$

$$\beta = \frac{\frac{Q_w}{V_T} C_{w(IN)}}{(\ominus_w + \rho_s K_s + \ominus_a K_h z_v + \ominus_o K_o)} \tag{17}$$

$$C_W(t) = \frac{((\alpha C_w(o) + \beta)e^{\alpha t} - \beta)}{\alpha} \tag{18}$$

MODEL CALIBRATION

The model has been calibrated based on actual performance data from several of RETEC's air sparging projects. Relevant data describing these projects are summarized in Table 1. Figures 3 through 8 illustrate several air sparging configurations that have been used to calibrate the model. These configurations include drilled vertical and horizontal or vertical wells and horizontal wells in a gravel-filled trench. In addition, air sparging has been used to treat a number of contaminants including benzene, toluene, ethyl benzene, xylenes, polycyclic aromatic hydrocarbons, trichloroethylene, and ammonia. A number of soil types were encountered in the air sparging applications ranging from fine sand to clay.

The model calibration demonstrated that the model could predict system performance after site-specific characteristics had been determined. Furthermore, values for decay rates and partition coefficients determined from the reported literature were sufficient as model inputs to adequately predict system performance.

Model calibration yielded an appropriate range of volatilization efficiency factors. The volatilization efficiency factor (measured concentration in air/equilibrium concentration in air) was approximately 0.02 for the sites evaluated.

Table 1. **Summary of Site Characteristics and Volatilation Effeciency Factors**

DESCRIPTION	Site A Wood Treating Site	Site B Petroleum UST	Site C Petroleum Refinery	Site D Natural Gas Process	Site E Spill Site
Remediation Goals/Objectives	0.050 ppm naphthtalene	0.005 ppm benzene	0.005 ppm benzene 1 ppm ammonia	0.05 ppm benzene	0.005 ppm carbon tetrachloride
Contaminants	Semi-volatiles (naphthalene, creosote)	BTEX (gasoline)	BTEX, NH_4 (gasoline)	BTEX	Carbon Tetrachloride
Site/Area Lithology	9 acre plume fine sand	0.13 acre plume fine sand	742 acre refinery silt and sand	80 acre plume layered sand, silt, clay	1 acre plume salty clay/clayey silt
System Description					
Number/depth of injection wells	9 vertical/47.5 ft deep	2 vertical/43 ft deep	11 vertical/18 ft deep	3 vertical/40 ft deep	14 vertical in trench/20 ft deep
Number/depth of extraction wells	no extraction	1 vertical/20 ft deep	no extraction	no extraction	2 horizontal/6 ft deep
Length of system	260 ft		200 ft	200 ft	250 ft
Well spacing	25 ft		20 ft	100 ft and 50 ft vertical	20 ft
Air injection equipment	30 HP rotary screw	5 HP liquid ring	20 HP rotary screw	20 HP rotary screw	positive displacement
Total Volume, ft^3	55,000	55,000	76,200	240,000	5,000
Total Groundwater Flow, ft^3/day	731	240	1,217	182	16.3
Total Air Flow, ft^3/day	38,880	8,640	47,520	19,440	198,700
Volatilization Efficiency Factor	0.0175	0.017 - 0.02	0.0175	0.0175	0.0215

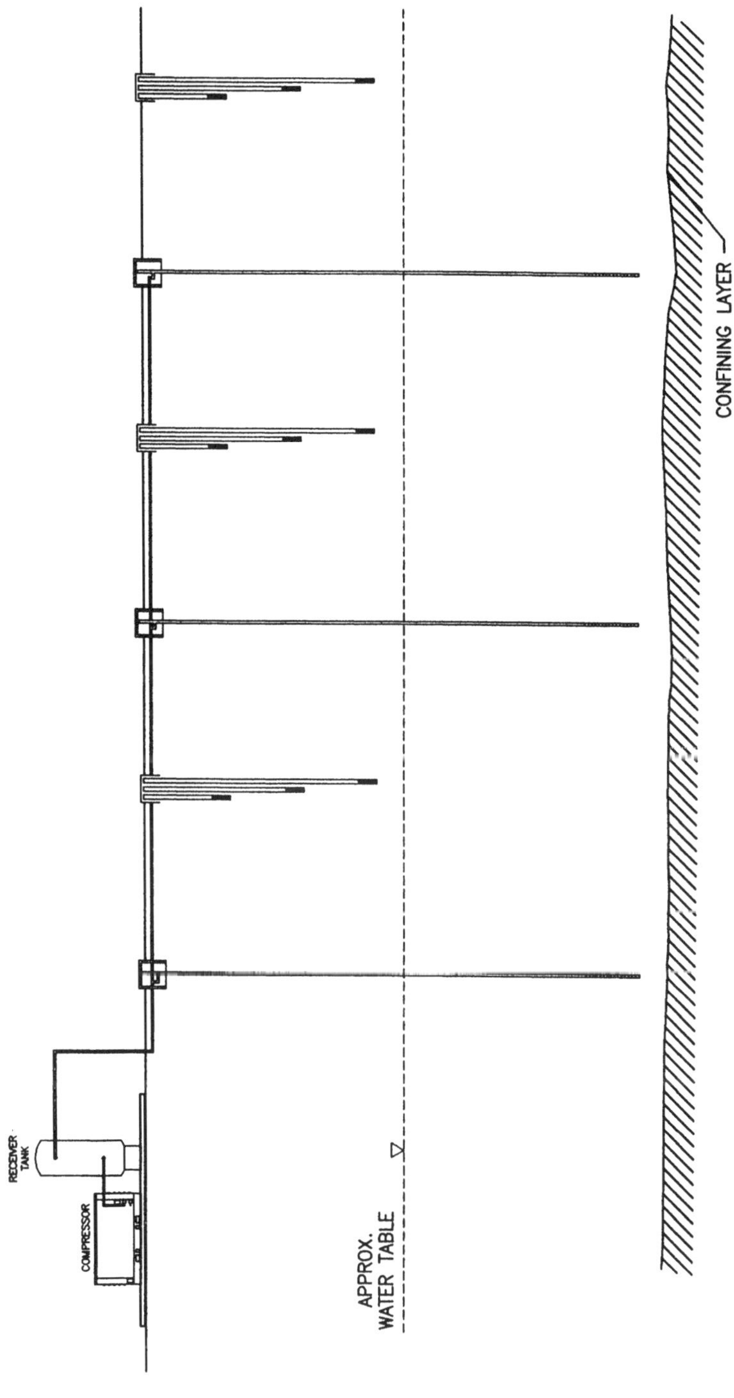

Figure 3. **Vertical air sparging wells - no vapor recovery.**

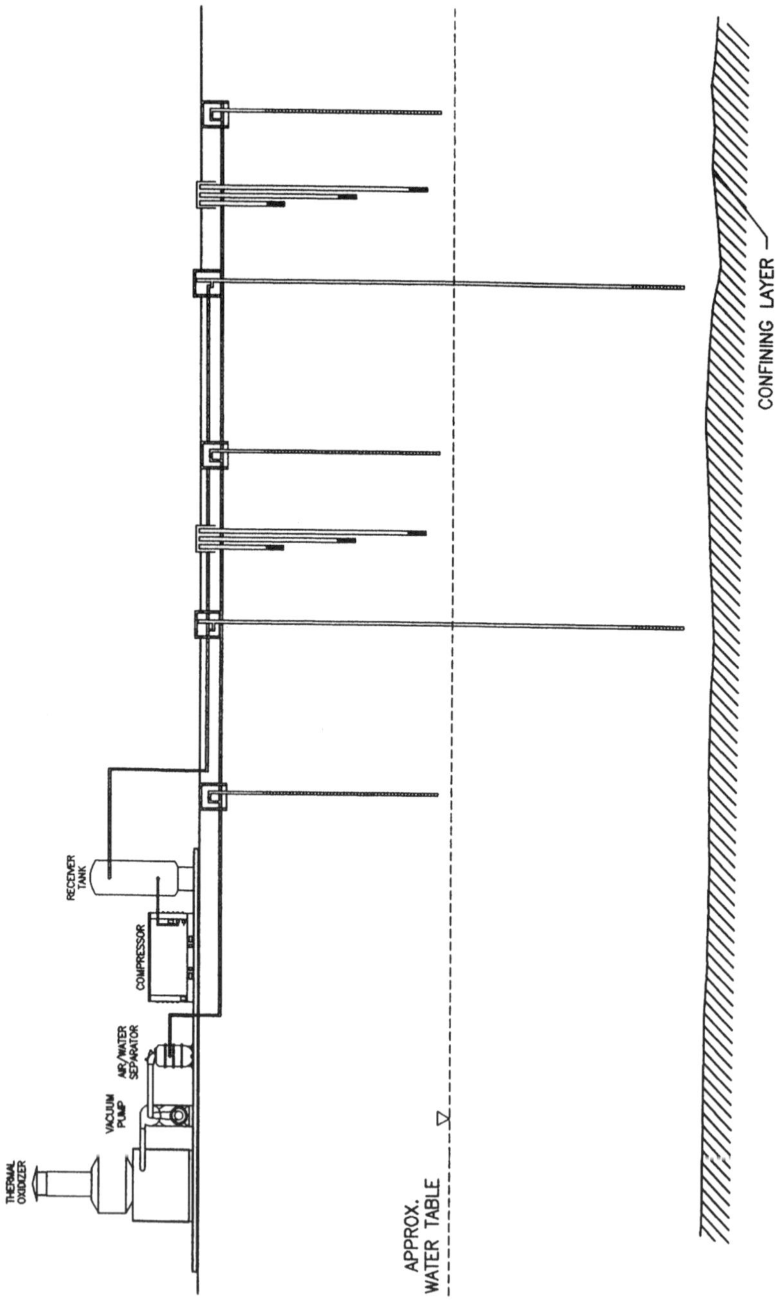

Figure 4. **Vertical air sparging and vertical soil venting wells.**

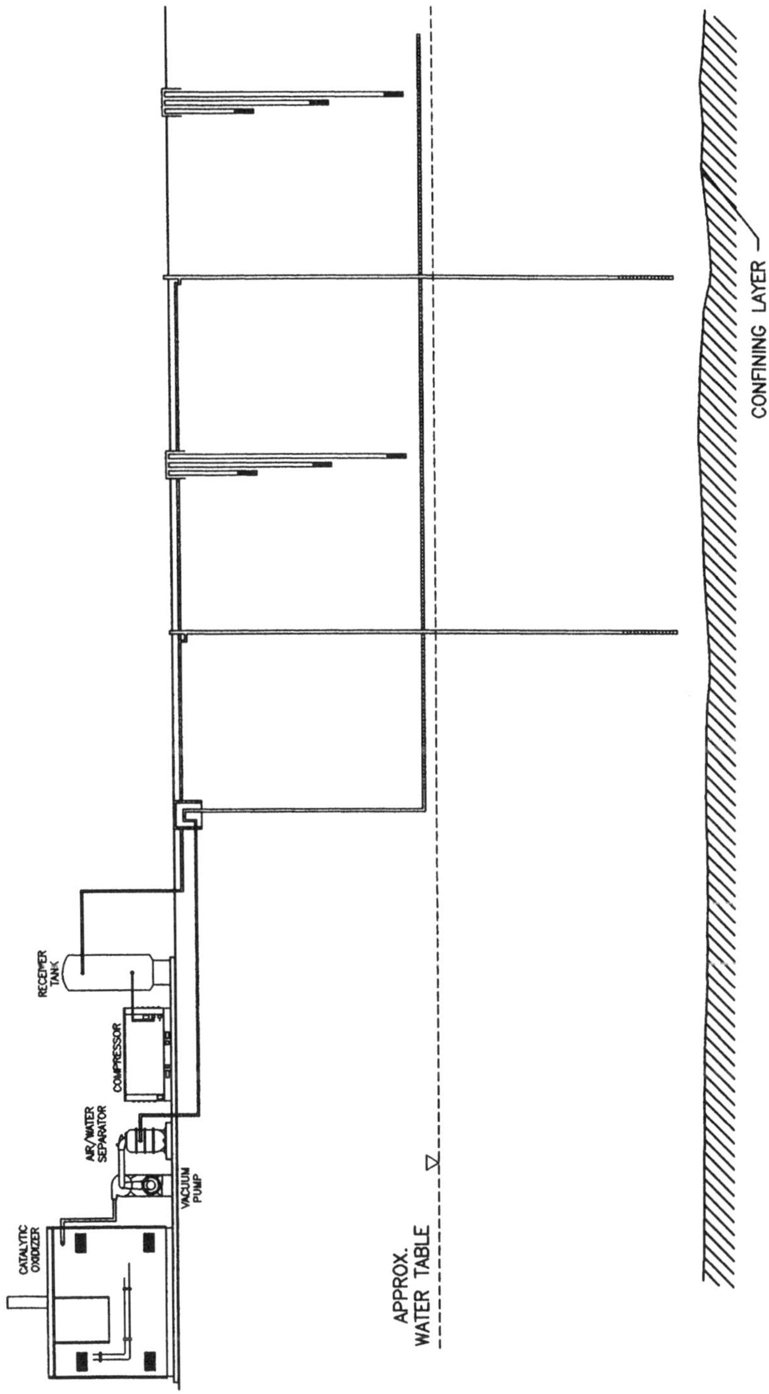

Figure 5. **Vertical air sparging and horizontal trenched soil venting wells.**

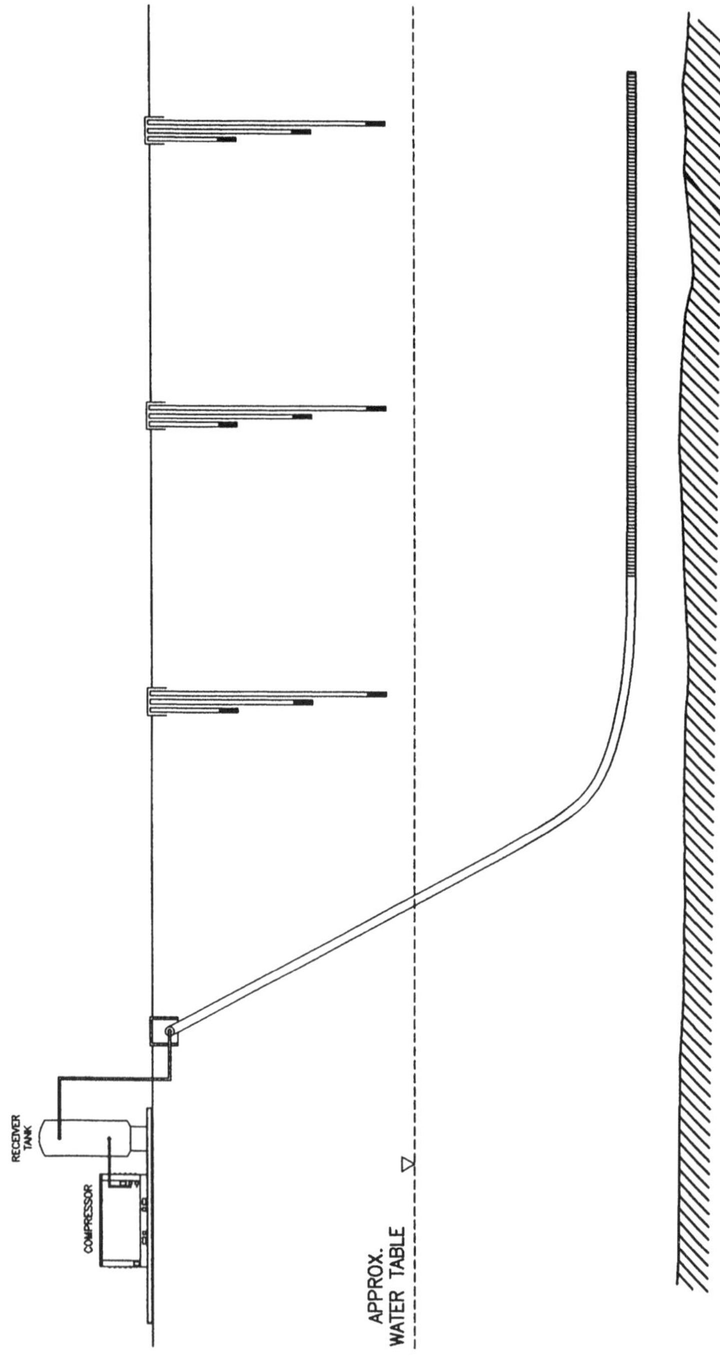

Figure 6. **Drilled horizontal air sparging well - no vapor recovery.**

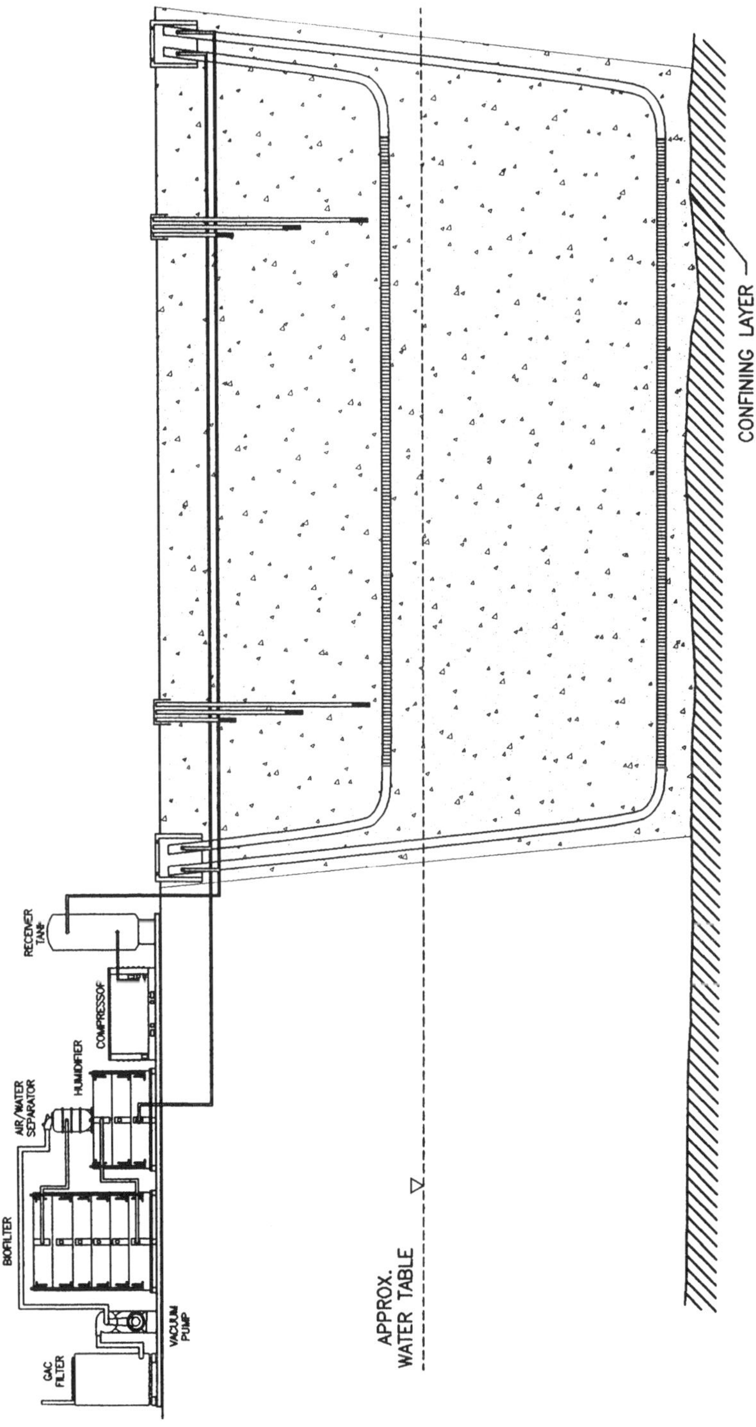

Figure 7. **Continuous trenched horizontal air sparging and soil venting wells.**

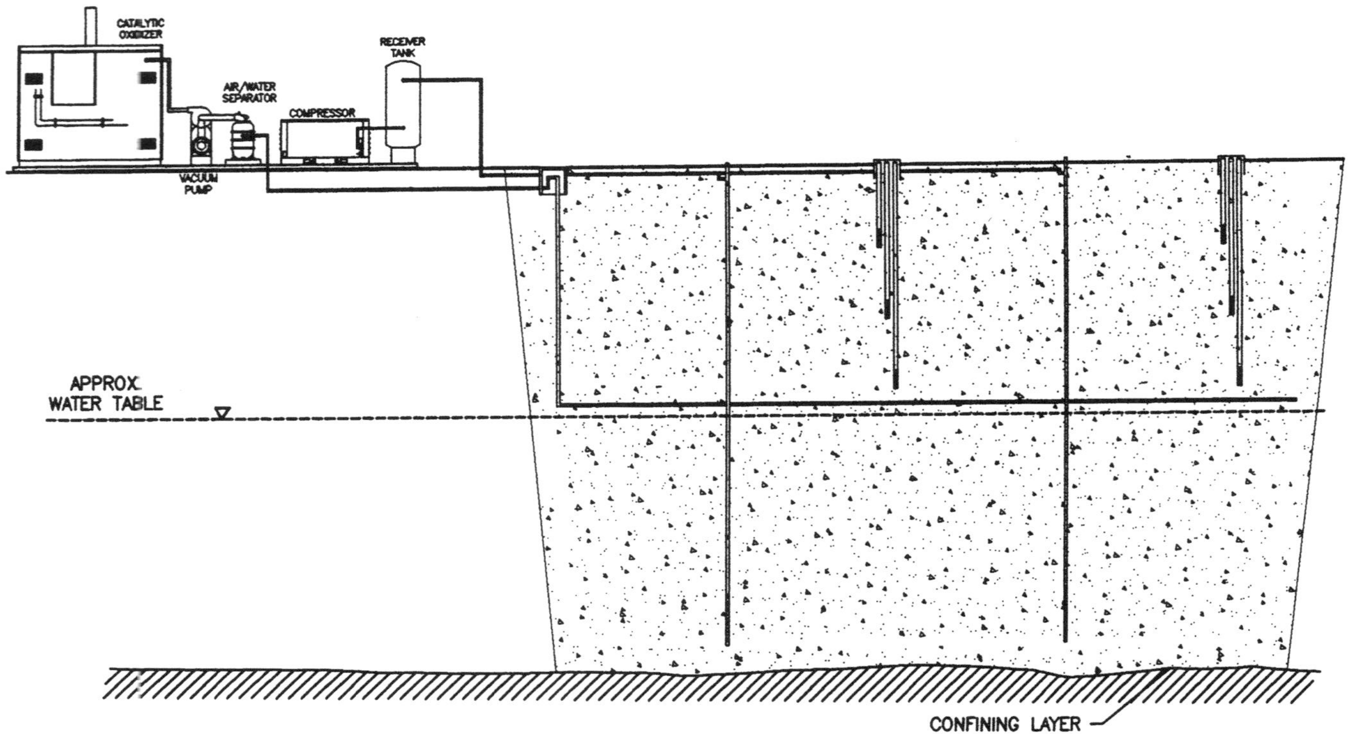

Figure 8. **Vertical air sparging and horizontal soil venting wells in gravel trench.**

REFERENCES

Brubaker, G.R. 1991. In-Situ Bioremediation of PAH-Contaminated Aquifers. In: *NWWA Conference of Petroleum Hydrocarbons and Organic Chemicals*, 1991.

Gelhar, L.W. and Wilson, J.L. 1974. Groundwater Quality Modeling, *Groundwater* **12**, 399.

Gvirtzman, H. and Gorelick, S.M. 1992. The Concept of In-Situ Vapor Stripping for Removing VOCs from Groundwater, *Transport in Porous Media*, **8**, 71.

Johnson, R.L., Johnson, P.C. McWhorter, D.B. Hinchee, R.E., and Goodman, I. 1993. An Overview of In-Situ Air Sparging, *Groundwater Monitoring Rev.* Fall, p. 127.

Lundegard, P.D. and Anderson, G. 1993. Numerical Simulation of Air Sparging Performance. In: *Proceedings of the 1993 Petroleum Hydrocarbons and Organic Chemicals in Groundwater: Prevention, Detection, and Restoration*, National Groundwater Association, Houston, TX, 1993.

Metcalf and Eddy. 1979. *Wastewater Engineering*, p. 166. New York, McGraw-Hill, Inc.

Middleton and Hiller, In-Situ Aeration of Groundwater, a Technology Overview. 1990. In: *A Conference on Prevention and Treatment of Soil and Groundwater Contamination in the Petroleum Refining and Distribution Industry*, Quebec, 1990.

Mohr, D.H. 1995. Mass Transfer Concepts Applied to In-Situ Air Sparging. In: *Proceedings of the 1993 Petroleum Hydrocarbons and Organic Chemicals in Groundwater:Prevention, Detection, and Restoration*, Battelle, San Diego, CA, 1995.

RETEC. 1991. *Phase III Feasibility Study Report*, Remediation Technologies, Inc., Fort Collins, CO.

(USEPA) U.S. Environmental Protection Agency. 1992. A Technology Assessment of Soil Vapor Extraction and Air Sparging, EPA/600/R-92/173, Office of Research and Development, Risk Reduction Engineering Laboratory, Cincinnati, OH.

Zheng, C., Bennett, G.D., and Andrews, C.B. 1991. Analysis of Groundwater Remedial Alternatives at a Superfund Site, *Groundwater* **29**, 838.

CHAPTER 11

The Effect of Water-Table Elevation on the Efficiency of Soil Vapor Extraction

Scott C. Swanson, Timothy J. Peck, and **Ian D. MacFarlane,** EA Engineering, Science, and Technology, Inc., Sparks, Maryland

INTRODUCTION

Soil vapor extraction (SVE) or venting, commonly in conjunction with groundwater pumping, has been successfully used to remove volatile organic chemicals (VOCs) such as hydrocarbons from the vadose zone (Guest et al., 1993; Marley, 1993; Johnson et al., 1990; Hutzler et al., 1989; Hinchee et al., 1987). VOCs are generated when hydrocarbons volatilize from residual contamination in the subsurface. Vented gases are treated typically using carbon adsorption, or a catalytic oxidizer or incinerator unit before atmospheric discharge, if required under regulatory control or to meet other emissions limits.

The generation of residual hydrocarbon contamination from a release of nonaqueous phase liquid (NAPL) has been described in detail in several previous studies (Huling and Weaver, 1991; Abdul, 1988; Tyler et al., 1987). The zone of residual contamination not only exists within the vadose zone below the release location, but also along the capillary fringe and within the upper portion of the unconfined aquifer. The vertical limit of residual contamination increases with fluctuations of the water table.

Because SVE technology is designed to work on unsaturated soil containing residual hydrocarbons to allow the free flow of air through the contaminated zone, the maintenance of unsaturated soil conditions including the dewatering of the upper portion of the aquifer is essential for successful, and efficient, venting and removal of the source of contamination. Also, the design of SVE equipment is based on given subsurface conditions including expected VOC concentrations, usually conditions found before the onset of venting. However, as subsurface conditions change due to removal of excess VOCs and fluctuations in the water table, the efficiency of SVE equipment operation will vary with these conditions.

CASE DESCRIPTION

Subsurface Conditions

An example of a typical SVE application is used as a case study, which is a gasoline service station located in the central Maryland Piedmont. The site is located on a topographically elevated area underlain by a highly fractured schist bedrock that is mantled by an overburden of weathered rock about 50 ft thick. Soil developed on the weathered rock is a deep, well-drained loam with a permeability ranging from 0.63-2 in./hr (Matthews, 1969). Groundwater occurs in fractures and joints within the bedrock, and within the overburden, which locally provides recharge to the bedrock aquifer. Groundwater flow within the bedrock is influenced by the orientation of the fractures, and occurs under semiconfined conditions. Within the overburden, the aquifer is unconfined. Average transmissivity and storativity values of the overburden-shallow bedrock aquifer system are 400 ft^2/day and 0.03, respectively, based on onsite aquifer tests.

Remediation History

Gasoline Release and NAPL Recovery

A leak of over 5,000 gallons of gasoline from underground tanks at a station in the early 1980s resulted in a threat to nearby private and municipal water-supply wells screened within bedrock (about 90-300 ft deep). Based on gauging records of monitoring wells, the full extent of free-phase NAPL occurred over approximately 1 acre (Figure 1). From January 1984 to the spring of 1988, about 1,500 gal of free-phase NAPL were removed using jack pumps in one offsite private supply well and four onsite, shallow, monitoring wells. Because of the continued potential threat to municipal wells (private wells were abandoned) and high dissolved hydrocarbon concentrations in monitoring wells, a groundwater remediation system was installed.

SVE Design and Operation

To evaluate the potential effectiveness of SVE to remove residual NAPL, vent tests were conducted on five, existing onsite monitoring wells. Each well had about 5 ft of screen extending above the water table. Each test was conducted by applying a vacuum at one well head using a 1-HP explosion-proof, regenerative blower (EG&G Rotron Model DR404A) and monitoring vadose zone pressure at remote points using soil probes. Blower exhaust rates ranged from 15-23 cubic feet per minute (cfm) at a vacuum of about 38 in. of water. The zone of influence radius, as defined by a pressure differential of 0.05 in. water, ranged from 9 to 25 ft. Total VOC concentration in the vented soil gas ranged from 780 to 7,700 ppm, as measured by direct injection of soil gas samples into an onsite gas chromatograph with flame ionization detection (GC). These values corresponded to a relatively high mass-loading of approximately 6-60 lb/day of total VOC at the time of the vent test.

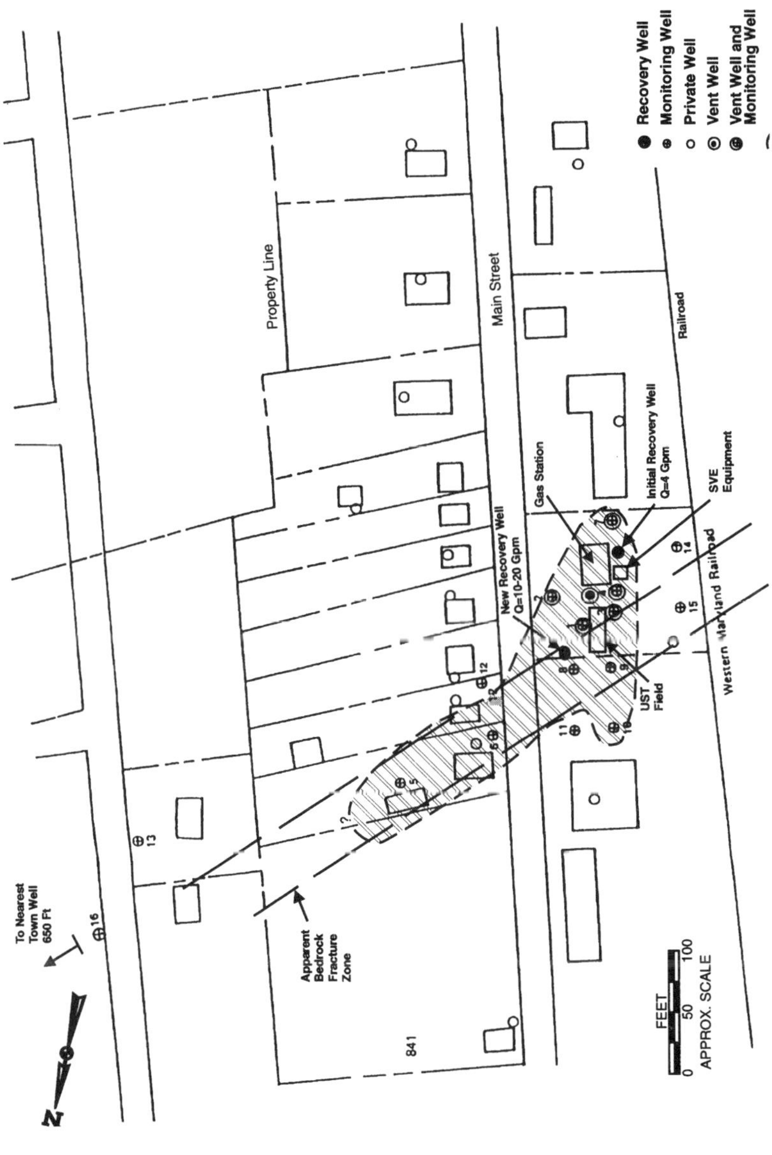

Figure 1. **Plan of project area.**

Figure 2. **Schematic of soil venting system layout.**

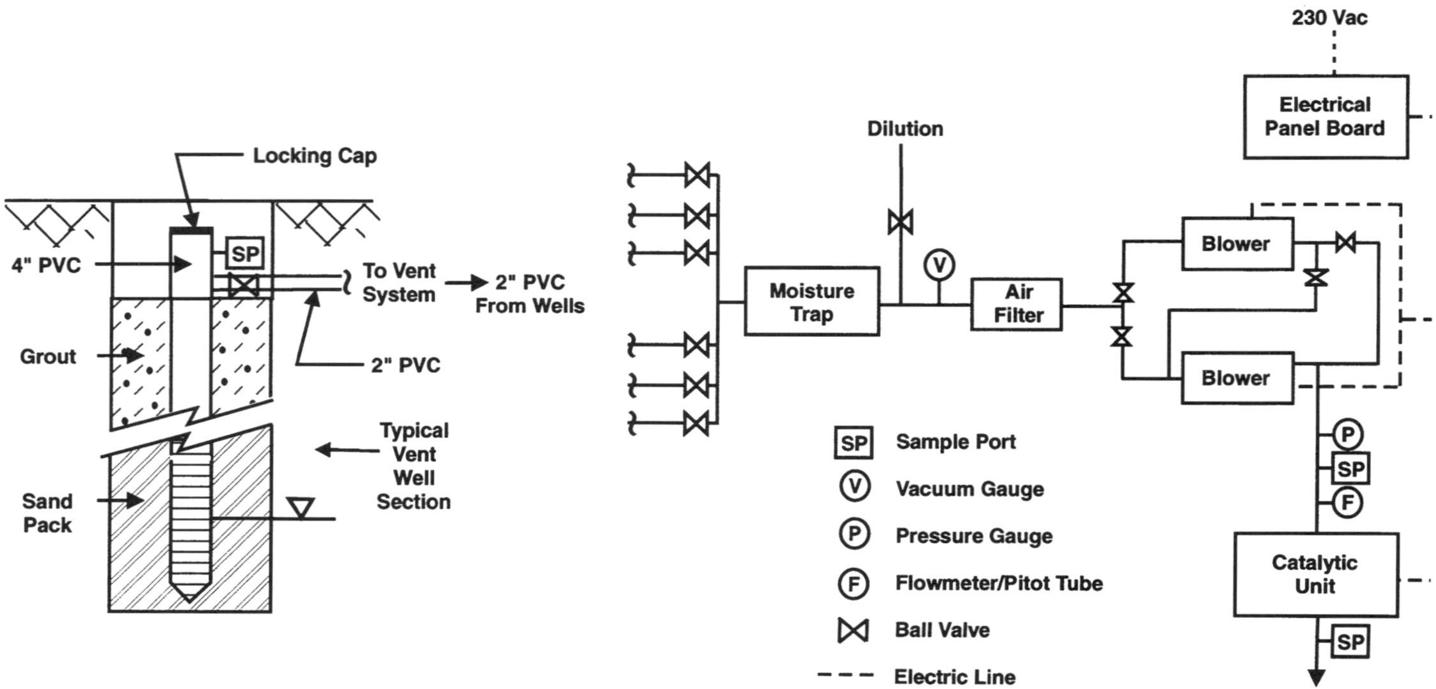

The SVE system design depicted in Figure 2 included two EG&G Rotron Model DR505A 2-HP regenerative blowers. The blowers were installed in parallel to achieve up to 25 cfm at each well, and 100 cfm total system design flow. Two individual blowers were used to allow more flexibility in future venting operations. Each vent well head consisted of a tee and ball valve to individually control flow. A moisture separator installed in-line before the blowers was used to trap excessive soil vapor moisture before entering the blowers.

The influent air was routed into a 100-cfm rated, Falmouth Inc. catalytic oxidizer (CO) and thermally treated to destroy VOCs to ensure emissions were less than the state regulatory limits of 20 lb/day total VOCs and less than 0.02 lb/hr benzene. The CO is designed to maintain a minimum catalyst temperature (about 330° C) for VOC destruction, utilizing the heat generated from combustion of the influent vapors, and supplemented with an internal heating element. Sampling and flow measurement ports were installed at each well head, before the treatment system, and after the CO. Flow and vacuum gauges were installed to monitor system operating parameters. VOC concentrations were measured initially by onsite GC and later with a Foxboro Century 128 organic vapor analyzer.

Initial Phase of Remediation

In early 1990, the initial phase of remediation was expeditiously implemented to prevent a delay in cleanup: pumping an existing, onsite well at 4 gpm and venting six onsite shallow wells. Groundwater pumping lowered the water table about 1-3 ft across the site. This lowering enhanced the process of hydrocarbon vapor extraction by SVE through the enlargement of the unsaturated zone of residual contamination. Based on remediation system monitoring data, SVE removed about 98% of the total recovered hydrocarbon and, thus, was mitigating the source of the dissolved plume.

Second Phase of Remediation

Because the initial remediation effort proved to be successful (i.e., about 1,900 lb of hydrocarbons were removed in the first four months of operation), a new groundwater recovery well was installed onsite and within the bedrock fracture zone. The goal of additional pumping was to further depress the water table within the residual NAPL area to enhance the efficiency of the existing SVE system and further mitigate plume migration. The new well was designed to pump at 10-20 gpm to induce a drawdown of up to 6 ft across the site to ensure that the residual NAPL zone was dewatered. A 6-ft drawdown was critical because, based on gauging records over 10 years, the water table locally fluctuated over this distance. Pumping of this new well, in addition to the existing recovery well, was started in February 1995.

SVE EFFICIENCY

History

The importance of maintaining a depressed water table to maximize SVE efficiency, including the VOC removal rate and efficient equipment operation, is illustrated by the venting history at the site. The total hydrocarbon concentrations in the extracted soil vapor over the 5 and one-half year system operation are plotted in Figure 3. The fluctuation of water table at the site over the same period is shown in Figure 4. At the start of venting in March 1990, influent concentration was at 2,500 ppm with a removal rate of about 150 lb/day. VOC concentrations from individual wells initially varied from 55 to 5,400 ppm. After two weeks of operation as excess VOCs were removed, the influent concentration decreased to approximately 1,000 ppm and individual well concentrations ranged from 5 to 1000 ppm.

In the following year, influent concentration and removal rate decreased to less than 100 ppm and less than 5 lb/day, respectively, as the water table rose about 2 ft, and excess VOCs were vented from the soil. After state approval, the CO was deactivated. In the subsequent eight months, when a drought caused an approximately 5-ft drop in the water table, influent concentration and average removal rate, respectively, jumped to 1,800 ppm and 35 lb/day in November 1991. To ensure regulatory emission limits were not exceeded, the CO was

Figure 3. **Total hydrocarbon concentration in extracted soil vapor during venting operation.**

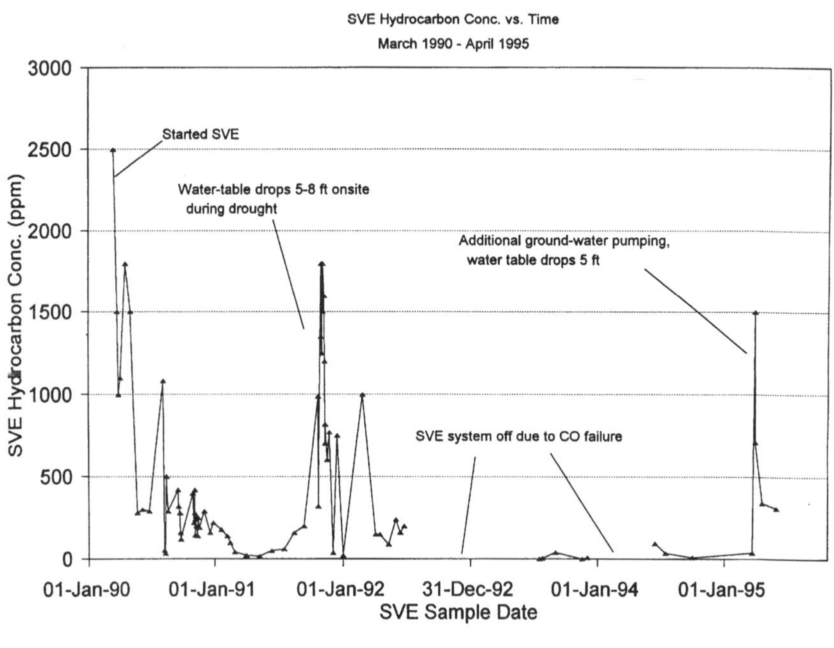

restarted. Within one month, influent concentration dropped back to less than 1,000 ppm as the excess VOCs in the soil were quickly depleted.

In the next three years, as the water table rebounded up to 8 ft, SVE efficiency declined, with influent concentration remaining below 100 ppm and removal rates below 5 lb/day. Overall, the soil venting system and catalytic unit operated as designed at concentrations greater than about 200 ppm, when sufficient catalyst temperature was maintained. However, at influent concentrations below 200 ppm, the overuse of the internal heating element frequently caused failures of the CO and shutdowns of the venting system. State regulatory control limited the flexibility of utilizing the CO only when needed. A 14-day air emission monitoring test was required to verify emission levels were below permitted levels before removal of a control device could be granted. This test was intended as a last resort, due to its cost and the possibility of the need for later restart (as in November 1991).

When the new groundwater recovery well was brought online in February 1995, within one month the water table at the site was depressed more than 5 ft, exposing a larger area of residual contamination. Vented total hydrocarbon concentration increased from 15 to about 1,000 ppm. In the following three months, influent concentration was about 300 ppm and the average removal rate was about 16 lb/day. This success in enhancing the venting of VOCs from the subsurface with additional groundwater pumping indicated that after 5 years of remediation at the site, a substantial amount of residual contamination remained in the shallow subsurface. This remaining contamination was not attacked until the water table was further depressed.

Figure 4. **Changes in water-table elevation over duration of venting.**

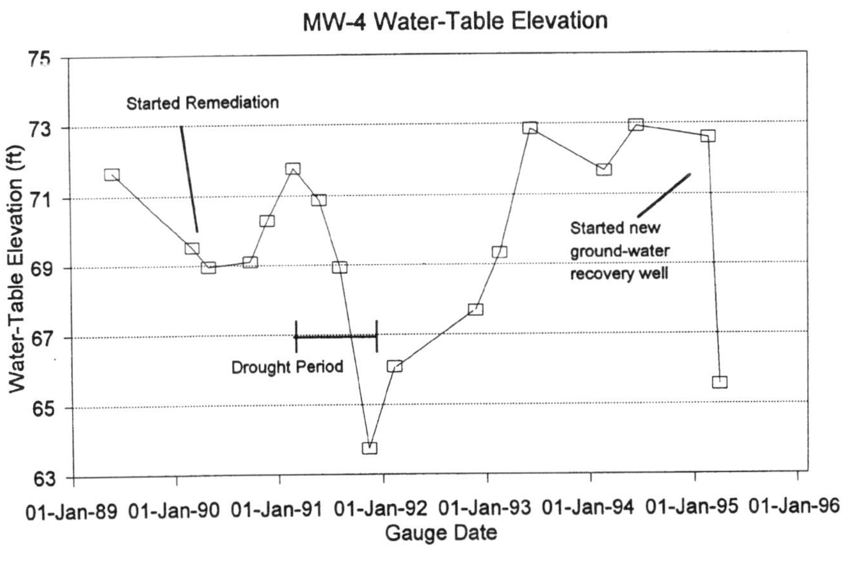

Statistical Analysis

Over the course of the five years of venting, higher VOC removal rates generally occurred when the water table was lower. The statistical significance of water-table elevation with SVE efficiency can be observed in a linear regression of the logarithm of influent VOC concentration versus elevation. The linear regression is plotted in Figure 5. The goodness-of-fit (R^2) of this regression is 0.67, indicating the importance of maintaining a depressed water table to obtain higher VOC concentration. The R^2 value for untransformed VOC data is 0.49. The higher significance for the transformed VOC data apparently indicates that as the water table drops, VOC concentration increases exponentially (i.e., a larger area of residual hydrocarbon occurs near the water table, where NAPL previously spread across the capillary fringe).

Figure 5. **Best-fit linear regression of soil vapor VOC concentration vs. water-table elevation.**

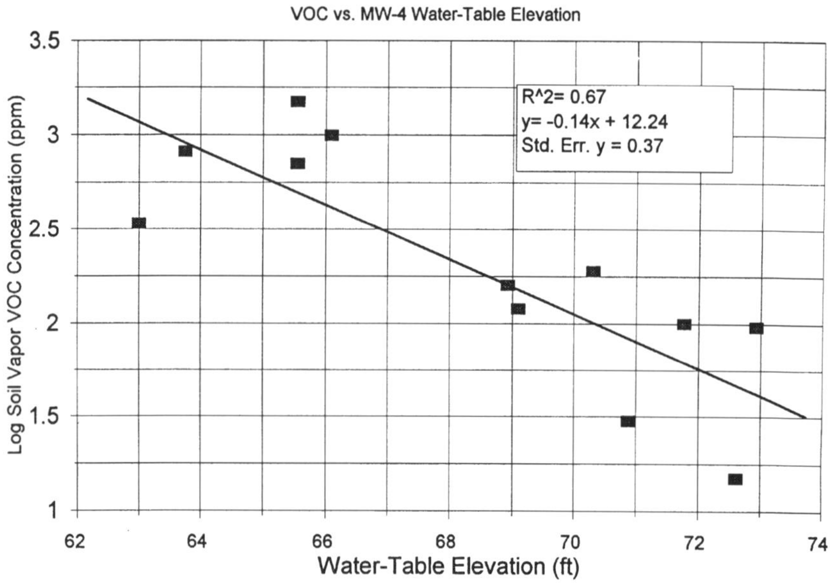

Achieving Remediation Goal

After five years of operation of the remediation system, a total of about 9,100 lb or 1,500 gal of residual hydrocarbons was removed. Soil venting removed about 97% of the residual removed or about 1,500 gal, which illustrates the efficiency of SVE to remediate a residual source. Assuming an initial gasoline release of 5,000 gal, about 2,000 gal remain in the subsurface in some form. The proportions of the fate of the spill (i.e., pumped NAPL, remaining in subsurface, and residual removed) are illustrated in Figure 6. The decrease of dissolved petroleum concentrations in the subsurface over time is illustrated in Figure 7, which shows dissolved total benzene, toluene, ethylbenzene, and xylenes (BTEX) in the initial groundwater recovery well. BTEX levels decreased asymptotically from about 2,000 to 40 µg/L. Dissolved total BTEX in onsite monitoring well MW-1 decreased from about 20,000 to 500 µg/L. With the depressed water table, SVE will continue to remove the residual source, thus helping to mitigate the generation of the dissolved plume.

Figure 6. **Estimated proportions of recovered hydrocarbon by different subsurface remediation techniques.**

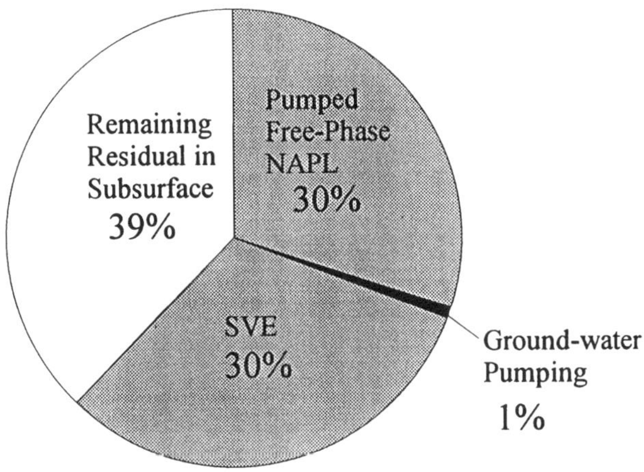

Figure 7. **Changes in total dissolved BTEX in pumping well during remediation.**

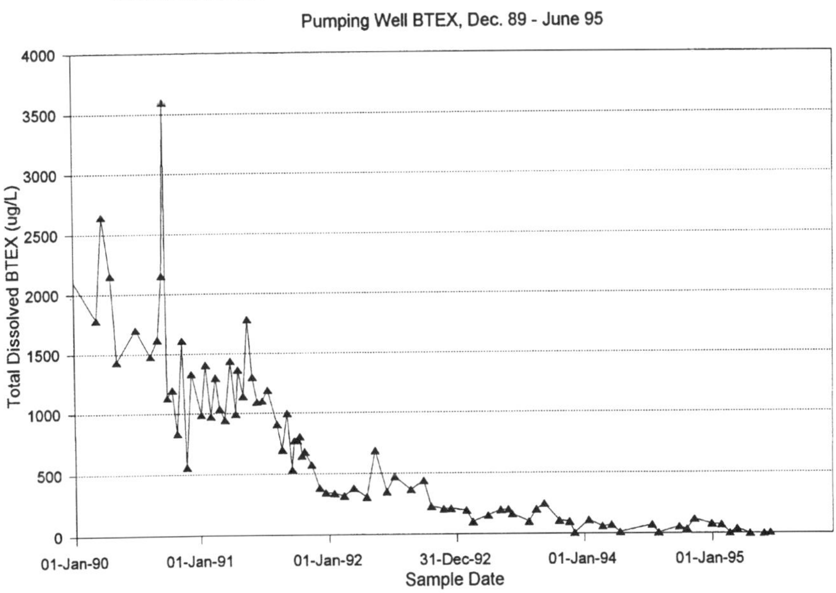

CONCLUSIONS

SVE can be highly effective, for appropriate subsurface conditions, in removing residual hydrocarbon from the unsaturated zone. However, even after several years of groundwater recovery and soil venting, a significant residual source may continue to linger within the upper portion of the unconfined aquifer. The maintenance of a depressed water table to completely expose the residual contamination is critical for effective source removal and efficient operation of SVE equipment. The statistical significance of VOC concentration being dependent on water-table elevation further substantiates this goal. Thus, in most SVE applications, groundwater pumping is an integral part of the remediation system.

REFERENCES

Abdul, A.S. 1988. Migration of Petroleum Products Through Sandy Hydrogeologic Systems, *Ground Water Monitoring Rev.* **8**(4), 73.

Guest, P.R., Friesen, K.A., and Blicker, B.R. 1993. An Integrated Approach to Remediating a No. 2 Fuel Oil Release in Soil and Groundwater. In: *Hydrocarbon Contaminated Soils,* Volume III, Chapter 22 (Calabrese, E.J. and Kostecki, P.T. Eds.), Chelsea, MI, Lewis Publishers.

Hinchee, R.E., Downey, D.C., and Coleman, E.J. 1987. Enhanced Bioreclamation, Soil Venting and Ground-Water Extraction: A Cost-Effectiveness and Feasibility Comparison. In: *Proceedings of the NWWA/API Conference on Petroleum Hydrocarbons and Organic Chemicals in Ground Water - Prevention, Detection and Restoration*, p. 147, Houston, TX.

Huling, S.G. and Weaver, J.W. 1991. Ground Water Issue: Dense Nonaqueous Phase Liquids, EPA/540/4-91, Office of Solid Waste and Emergency Response, U.S. EPA, Washington, DC.

Hutzler, N.J., Murphy, B.E., and Gierke, J.S. 1989. State of Technology Review Soil Vapor Extraction Systems, EPA/600/2-89/024, Risk Reduction Engineering Laboratory, Office of Research and Development, U.S. EPA, Washington, DC.

Johnson, P.C., Stanley, C.C., Kemblowski, M.W., Byers, D.L., and Colthart, J.D. 1990. A Practical Approach to the Design, Operation, and Monitoring of in Situ Soil-Venting Systems, *Ground Water Monitoring Rev.* **10** (2), 159.

Marley, M.C. 1993. Evaluation of Vadose Zone Air Flow Pathways Utilizing Tracer Gases and the Subsequent Implications on Vapor Extraction System Design. In: *Hydrocarbon Contaminated Soils,* Volume III, Chapter 23 (Calabrese, E.J. and Kostecki, P.T. Eds.), Chelsea, MI, Lewis Publishers.

Matthews, E.D. 1969. Soil Survey of Carroll County, Maryland, U.S. Department of Agriculture, Soil Conservation Service, Washington, DC.

Tyler, S.W., Whitbeck, M.R., Kirk, M.W., Hess, J.W., Everett, L.G., Kreamer, D.K., and Wilson, B.H. 1987. Processes Affecting Subsurface Transport of Leaking Underground Tank Fluids, EPA/600/6-87/005, Environmental Monitoring Systems Laboratory, Office of Research and Development, U.S. EPA, Las Vegas, NV.

CHAPTER 12

Landfill Stabilization Focus Area: A National Technology Development Program

Jaffer Mohiuddin, Office of Technology Development, U.S. Department of Energy, Washington, D.C.

David Mark, Office of Waste Management, U.S. Department of Energy, Washington, D.C.

George Schneider, Advanced Technology and Planning, U.S. Department of Energy, Idaho Falls, Idaho

Paul Zielinski, Office of Environmental Restoration, U.S. Department of Energy, Washington, D.C.

James Brown, Office of Technology Development, Savannah River Operations, U.S. Department of Energy, Aiken, South Carolina

Andrew Bouie, BDM, Federal, Inc., Gaithersburg, Maryland

THE NEW APPROACH

The U.S. Department of Energy (DOE) Office of Environmental Management has developed a new approach toward using science and technology resources to solve its environmental problems. This process has been outlined in "A New Approach to Environmental Research and Technology Development at the U.S. Department of Energy Action Plan" (U.S. DOE, 1993). According to this action plan, the Office of Environmental Management is faced with five major high-priority problems. Therefore, the Office of Environmental Management established five focus areas, as follows: Mixed Waste Characterization, Treatment, and Disposal; High-Level Waste Tank Remediation; Landfill Stabilization; Contaminant Plume Containment and Remediation; and Facility Transitioning, Decommissioning, and Final Disposition. More focus areas may be established on an as-needed basis. The Office of Environmental Management, in cooperation with other DOE research organizations, will use the best science and technology available to solve the most challenging set of environmental problems in the world. Integral to this new approach is an up-front awareness of program needs obtained from customers, users, regulators, and stakeholders. These needs will be matched to the suppliers of technological solutions. The goal of this new approach is to conduct a research and technology development program that is focused on overcoming the major obstacles and takes advantage of the best talent in DOE, other national science organizations, industry, and universities to advance the cleanup of the DOE sites.

To manage and implement the "New Approach" to environmental research and technology development, the DOE has assembled an integrated team. This structure is depicted in Figure 1. Elements of this structure are briefly discussed below:

Figure 1. **Integrated team organizational structure.**

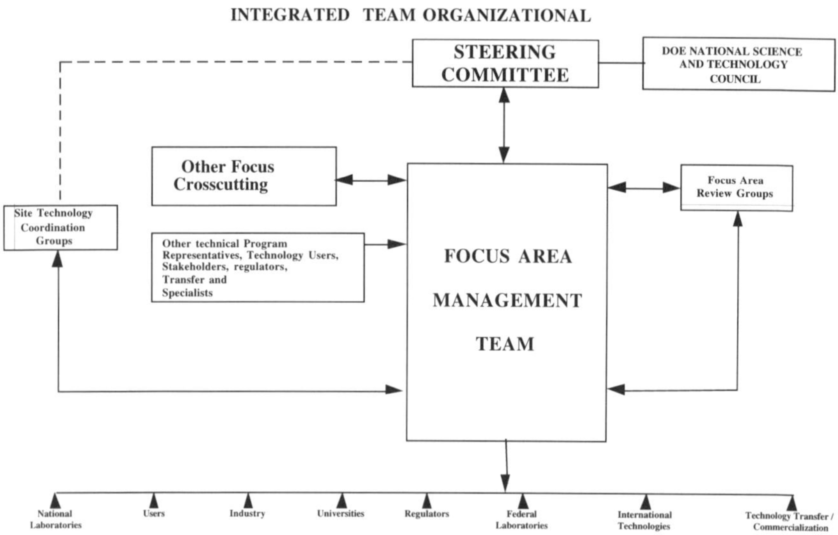

Steering Committee

The Steering Committee provides top-level oversight and ensures that the technology development program continues to be responsive to the Office of Environmental Management's priorities. The Committee reviews actions of each Management Team, establishes overall priorities and policy, provides programmatic direction, allocates financial resources for the focus areas, and addresses procedural and coordination issues. Membership includes senior-level representatives from other offices within the DOE and other government agencies such as the U.S. Environmental Protection Agency, National Institutes of Standards and Technology, and the Department of Defense.

The DOE National Environmental Science and Technology Council

The DOE National Environmental Science and Technology Council reports to the Assistant Secretary for Environmental Management, as an ax-officio member of the Steering Committee, and advises the Steering Committee. Scientists and engineers with national reputations in their fields of expertise comprise this Council. The DOE National Environmental Science and Technology Council serves as technical outreach to nationally recognized talents and resources outside the DOE system. As such, the Council will provide an avenue for key contacts to the scientific community to ensure the best possible scientific and engineering approaches are pursued.

Management Team (one for each focus area)

The Management Team translates the Steering Committee's policy guidance and direction, validates and prioritizes needs within the focus area, issues program guidance for technology development, commits resources, coordinates, and integrates with the other focus areas, solicits and receives customer needs from the Site Coordination Groups and appropriate Headquarters elements. In addition, it manages and is accountable for project cost and schedule, provides system integration, including technical, cost and schedule components within the focus area, develops a life-cycle approach to solve problems and satisfy customer needs, maintains awareness of available and developing technologies, integrates national technology transfer and commercialization efforts through private sector involvement. The Management Team consists of program managers from various DOE Headquarters and Field Offices that have significant landfill problems and need for technologies to address those problems.

Focus Area Review Groups (one for each focus area)

The Focus Area Review Groups conduct independent reviews to recommend priorities and evaluate the technical merit of proposed and ongoing programs and technologies.

Site Technology Coordination Groups (one for each DOE site)

The Site Technology Coordination Groups (STCG) were established at each DOE site to provide prioritized site needs and required implementation dates. The STCGs consists of DOE officials, regulators, and other stakeholders. Their involvement at the local level will provide for early stakeholder input, which will be coupled with identified technology needs to better shape the technology selection and development process.

Crosscutting Technology Development Areas

The Crosscutting Technology Development Areas are the Office of Technology Development programs that develop technology that is common among all or several of the focus areas to avoid duplication and to further coordinate development activities. These technology areas are characterization (site characterization, waste assay, and monitoring technologies), robotic technology (remote controlled, telerobotic, and automated technologies), and efficient separations technologies (technologies for the separation of heterogeneous wastes and for selective separation of a waste component from a mixture pre- or post-treatment).

The overall objectives of the this "New Approach" are to: (1) accelerate development of graded risk-based technology solutions to meet the Office of Environmental Management's highest priority requirements; (2) ensure applicable technologies are identified and/or developed and delivered ready for use; (3) facilitate programmatic decision-making, reduce uncertainty, and minimize/reduce net remediation costs through a better understanding of site problems; (4) ensure that basic research necessary to develop remediation technologies is addressed; and (5) maintain a program that is under managerial and financial control.

LANDFILL STABILIZATION FOCUS AREA

Mission

The mission for the Landfill Stabilization Focus Area is: to use the best science and technology available to develop, demonstrate, and deliver safer, more cost-effective technologies to address the most challenging landfill remediation problems; develop new technologies for the design and operation of new landfills; expeditiously effect the implementation of developed technologies.

The Problem

Waste Volume and Description

Numerous DOE landfills pose significant remediation challenges. The size of the problem is large, and the current estimate of the total amount of buried waste is 3.1 million cubic meters. This figure does not include an estimate of the associated contaminated soil. The waste types also present the DOE with a unique problem. These wastes contain hazardous materials and radioactive materials as well as radioactive and hazardous contaminants, mixed together.

The Landfill Stabilization Focus Area will develop technologies to address all disposal sites where hazardous and/or radioactive waste has contaminated soil and/or been buried. This includes all buried waste and contaminated soils (buried and near surface), whether on pads or in trenches, sumps, ponds, pits, sanitary landfills, cribs, heaps and piles, auger holes, and caissons. Although the problem defined here focuses on DOE contaminated soil and buried waste, the problems and associated technological solutions are not unique to the DOE. However, the presence of mixed waste is unique to the DOE. Most land disposal

Table 1. **DOE Complex Landfill Waste**[a]

Facility	Low Level Waste Volume (m3)	Low Level Mixed Waste Volume (m3)	Buried Transuranic Waste Volume (Pre-1970) (m3)	Stored Transuranic Waste Volume (Post-1970) (m3)	Total Volumes (m3)
INEL	145,371	27,000	57,100	64,838	294,309
Hanford	582,839	10,800	109,000	15,566	718,205
Savannah River	636,645	9,000	4,534	9,701	659,880
Los Alamos	215,687	300	14,000	8,036	238,023
Oak Ridge	207,372	1,300	5,947	2,007	216,620
Rocky Flats		8,500		934	9,434
Others	1,028,428	49,900		1,075	1,079,403
DOE Site-Wide Total	2,816,342	106,800	190,584	102,151	3,215,874

[a] Source: Oak Ridge National Laboratory, 1992.

records are poor to nonexistent. A wide variety of radioactive, hazardous, industrial, and sanitary wastes have been land disposed. Often, different types of wastes have been disposed of in the same disposal unit. Therefore, dual regulatory requirements often apply to the mixture of waste material. These wastes exist as solids, liquids, or sludge, and have a diverse range of chemical constituents including radionuclides, polychlorinated biphenyl, heavy metals, organic solvents, and reactive compounds. These wastes were disposed of directly or in containers such as steel drums, wooden boxes, and cardboard cartons. The direct disposal or the degradation of the containers, coupled with the absence of release barriers, has resulted in the contamination of surrounding soils and, in some cases, subsurface soils and groundwater, exacerbating the nature and extent of the problem.

The waste is predominantly located at the Hanford Reservation, the Savannah River Site, the Idaho National Engineering Laboratory, the Los Alamos National Laboratory, the Oak Ridge Reservation, and the Rocky Flats Plant. Table 1 presents a listing of DOE facilities with associated buried waste volumes.

Technological Problem

An extensive set of characterization and remediation technologies are needed to investigate, evaluate, and remediate the DOE's landfills. These needed technologies fall into five technical areas. Each of these technical areas have limitations with respect to the applicability to varied site conditions, safety of remediation workers, scale-up for large volumes of buried waste, and the associated extreme costs.

Assessment technologies have traditionally relied on environmental media sampling and laboratory analysis for contaminant detection or for determining soil properties. The major weaknesses in assessment methods are generally found in their inability to resolve spatial values in heterogeneous soil for contaminant detection and location. Monitoring refers to the need to monitor the progress of, or final success of, a remedial action or closure method. These methods have generally been an extension of the environmental media sampling and laboratory analysis methods for characterization. These methods also have limitations relative to near real-time feedback to remediation operations and assessments of long-term post-closure verification of stabilized or contained buried waste.

Waste stabilization technologies can provide a temporary or permanent reduction of landfill contaminant mobility or toxicity. This is often accomplished through in situ treatment or encapsulation technologies. In situ stabilization technologies can avoid removal costs or reduce removal risks by controlling the contaminant mobility during waste handling operations. Existing stabilization technologies are primarily cement-based grout injection into the waste or soil. Cement-type grouts do not have the necessary long-term durability nor widespread applicability across the DOE complex.

Containment technologies are based on the installation of surface or subsurface barriers to isolate the waste or soil contamination. This type of approach can eliminate near-term threats to human health and the environment. However, natural geologic forces could eventually allow some wastes to escape from the containment system. Containment technologies for in-surface covers currently use clay-based hydraulic barriers to limit precipitation infiltration into buried waste zones. Current trends have shown that in arid environments, traditional clay-based covers tend to desiccate, crack, and fail. Subsurface barriers have been limited to groundwater containment or diversion with little application to unsaturated soil barrier applications.

Removal of contaminated soils and/or buried waste can be achieved today, but at a high cost and significant risk to onsite workers. For radioactive or mixed wastes, costs are driven by contamination control issues, waste segregation and characterization, the limited capacity for offsite disposal, and liability transfer issues. Unless the restoration site is very small and has a high demand for future use, excavation as a remedial measure is often far too costly, and is not selected in favor of containment strategies.

Treatment (ex situ) technologies need to be versatile in their ability to handle many waste forms and waste mixtures, and provide a long-lived (greater than 100 years) final waste form. Incineration has been the technology of choice for the treatment of a large number of hazardous wastes. However, incineration has come under a great deal of scrutiny in the last few years due to questions about stack emissions. Nonthermal processes generally have only been capable of handling limited waste forms and mixtures.

Strategy

A prime focus of the Landfill Stabilization Focus Area is the consolidation and integration of all research and developmental activities pertaining to this focus area currently conducted by all elements of the Office of Environmental Management. This consolidation and integration will provide the basis for a rigorous, systematic, and effective management of the technology development process. The Landfill Stabilization Focus Area will work closely with the Site Technology Coordination Groups to identify specific environmental problems as they relate to technology needs for the remediation of landfills - this includes the remediation schedules and the formulation of associated plans. It is this coordination and communication that will provide focus to the technology development effort. These activities will also allow for an evaluation of existing and newly developed technologies to determine if effective and acceptable solutions currently exist.

The Landfill Stabilization Focus Area will establish the technical and programmatic goals of the program by the determination of specific technology gaps. This determination will be based on detailed global analyses of the available technology and the requirements of the Office of Environmental Management. The focus area will also coordinate collaborative efforts with other federal agencies.

The Landfill Stabilization Focus Area will establish a directed research, development, demonstration, testing, and evaluation program that integrates closely all activities from basic research to implementation. The guiding principle for the Landfill Stabilization Focus Area is "product delivery." The common thread binding all the activities under this focus area is enhanced communication, cooperation, and collaboration between technology developers, the customers (problem holders), the stakeholder, and the regulators.

The Landfill Stabilization Focus Area is committed to using the best minds and technology available to address the technological challenges facing the DOE concerning the remediation of its landfills. Therefore, the focus area has continued to try to increase the participation of universities, the private sector, and regulators in the technology development process. In particular, the emphasis is on getting all parties (the customers, universities, private sector, and regula-

Figure 2. **Landfill stabilization focus area configuration options.**

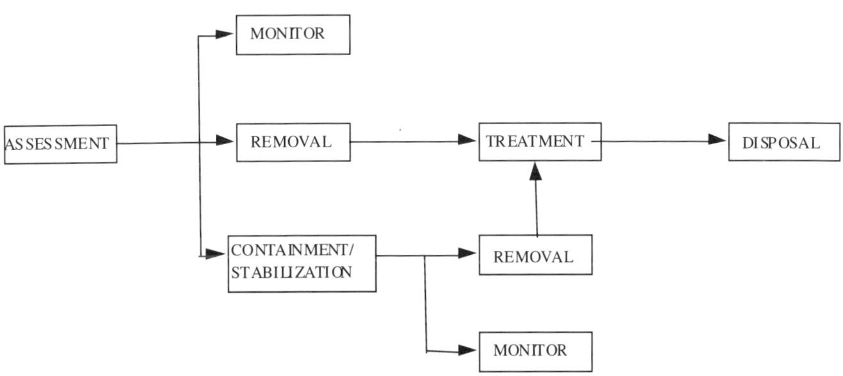

tors) involved with technology development efforts as early in the developmental process as possible. This early involvement will likely increase the potential for a given technology to be demonstrated, implemented, and commercialized. The goal for fiscal year 1996 is to have 30% of the fund for the focus area go in support of universities and the private sector.

Systems Approach

The Landfill Stabilization Focus Area is using a systems approach to the development of its program. This systems approach is taking a look at the entire remediation solution, from cradle to grave. The intent of the Landfill Stabilization Focus Area is to enhance all or part of a system for the remediation of a landfill site. The Landfill Stabilization Focus Area is therefore using a configuration option scheme to illustrate the technology development options available and necessary. For the purposes of this chapter, the four basic configuration options for the development of landfill remediation technologies will be discussed (see Figure 2). The options all begin with site assessment, and branch into other configurations as follows: (1) Retrieve the waste; (2) Treat the waste in situ (stabilization); (3) Contain the waste; and (4) Monitor the waste. The remaining branches on Figure 2 show the options available after the primary remediation option is selected. The resultant technology groupings are: assessment technologies for site and waste assessment; containment technologies (which we define as a process that confines the waste to an area); stabilization technologies (which we define as a process which does something to the waste); ex situ treatment technologies; and disposal technologies.

In addition, the Landfill Stabilization Focus Area has been divided into four product lines responsible for taking the technologies developed and demonstrated in the focus area, and delivering products for implementation and commercial-

Figure 3. **Landfill stabilization focus area crosscutting product lines.**

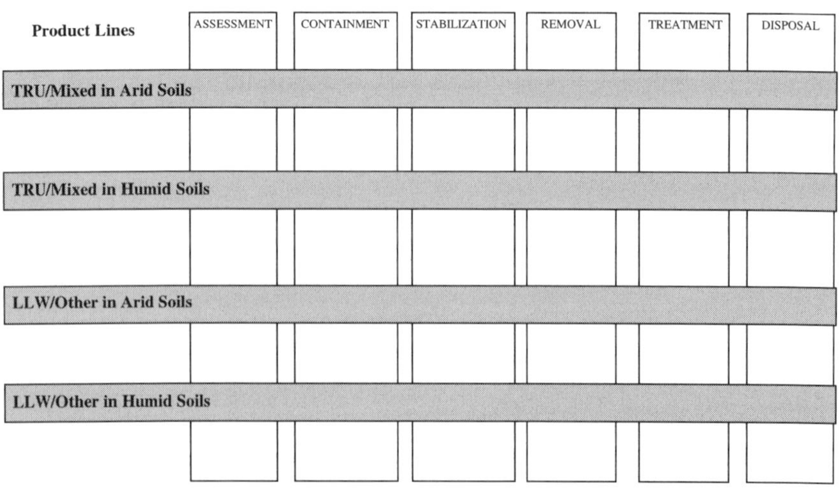

ization. The products lines were developed based on the general waste type and the region in which the wastes are found in DOE landfills. The product lines are: transuranic/mixed waste in arid soils (TRU/Mixed Arid); transuranic/mixed waste in humid soils (TRU/Mixed Humid); low-level/other waste in arid soils (LLW/Other Arid); and low-level/other waste in humid soils (LLW/Other Humid). The product lines also cut across the technology development areas. User needs can be addressed in each of the technical areas necessary to complete the enhancements in the proposed system, or replace it completely (see Figure 3).

Near Term Programmatic Focus

The Assistant Secretary of the Office of Environmental Management has challenged this focus area to demonstrate the DOE's ability to contain source terms that show the potential for migration by January of FY1997 (U.S. DOE, 1994). Therefore, one of the major thrusts of the Landfill Stabilization Focus Area will be to investigate, develop and demonstrate the necessary technology(ies) to enable the focus to meet the January 1997 deadline.

Management Structure

The management of the Landfill Stabilization Focus Area has been structured to accommodate the technical areas and the need to have the customer as part of the developmental process to allow technologies to be more quickly implemented and commercialized. The Savannah River Operations Office was selected as the lead organization, and the Idaho National Engineering Laboratory as its primary partner. The Management Team of the Landfill Stabilization Focus Area

Figure 4. **Landfill stabilization focus area virtual organization.**

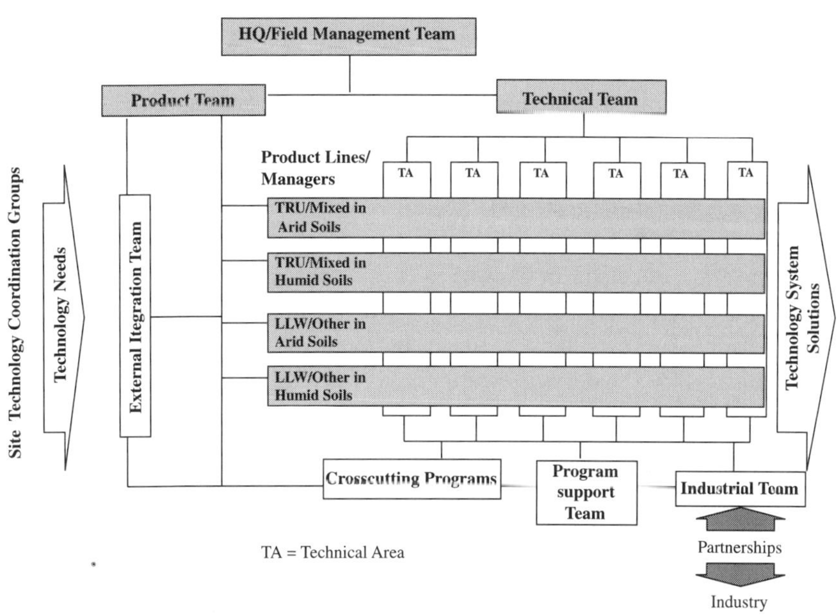

has members from the DOE headquarters' Office of Technology Development, the Office of Environmental Restoration, and the Office of Waste Management, as well as field partners from the Office of Technology Development, Savannah River Operations, Idaho National Engineering Laboratory, and Sandia National Laboratory.

The Management Team also has the participation/partnership of other DOE laboratories, DOE offices, and other federal agencies. The federal agencies include the Environmental Protection Agency, the Department of Defense, various services and agencies such as the U.S. Army Corp of Engineers, United States Geological Society, and the United States Bureau of Mines.

The Landfill Stabilization Focus Area has coordinated closely with the Crosscutting Programs (Robotics, Characterization, and Efficient Separations), the other focus areas, private industry and universities, and other federal agencies. The Landfill Stabilization Focus Area has an External Integration Team that is responsible for coordinating across the boundaries of the focus area. The flow of information into the focus area (the technology needs) and the flow of solutions out of the focus area are depicted in Figure 4.

Activities of the Program

The focus area, in late FY 1995, completed two major events. The first is the Third Annual Landfill Stabilization Focus Area Technology Exhibition. The exhibition has been a success for the Landfill Stabilization Focus Area in helping to bridge several gaps between our technology developers and the customer, the regulators, and the private sector. These exhibitions have launched several licensing and cooperative agreements between the focus area and our customers EM-30 and EM-40, other government agencies, universities, and the private sector. The second event is the International Containment Technology Workshop. The purpose of this meeting was to identify for the DOE, definitively, the technology development efforts for containment technologies that are necessary to continue. The product from this meeting will provide the basis for the selection of appropriate technology(ies) to invest in, to assist the Landfill Stabilization Focus Area in meeting its January 1997 deadline.

The Landfill Stabilization Focus Area is completing a complex-wide survey with the Site Coordination Groups to ensure that the technologies and priorities in the program continue to be reflective of the expressed needs. The product from this survey will assist the focus area in further defining the needs it must address for the development/enhancement of DOE-complex landfill-related remediation technologies. The Landfill Stabilization Focus Area will periodically solicit needs and information from the Site Technology Coordination Groups so that the development program may be adjusted appropriately.

The Landfill Stabilization Focus Area has been fairly successful in its attempts to demonstrate, implement, and/or transfer technologies to industry or the Office of Waste Management or to the Office of Environmental Restoration. In

Figure 5. **Landfill stabilization focus area FY 1995.**

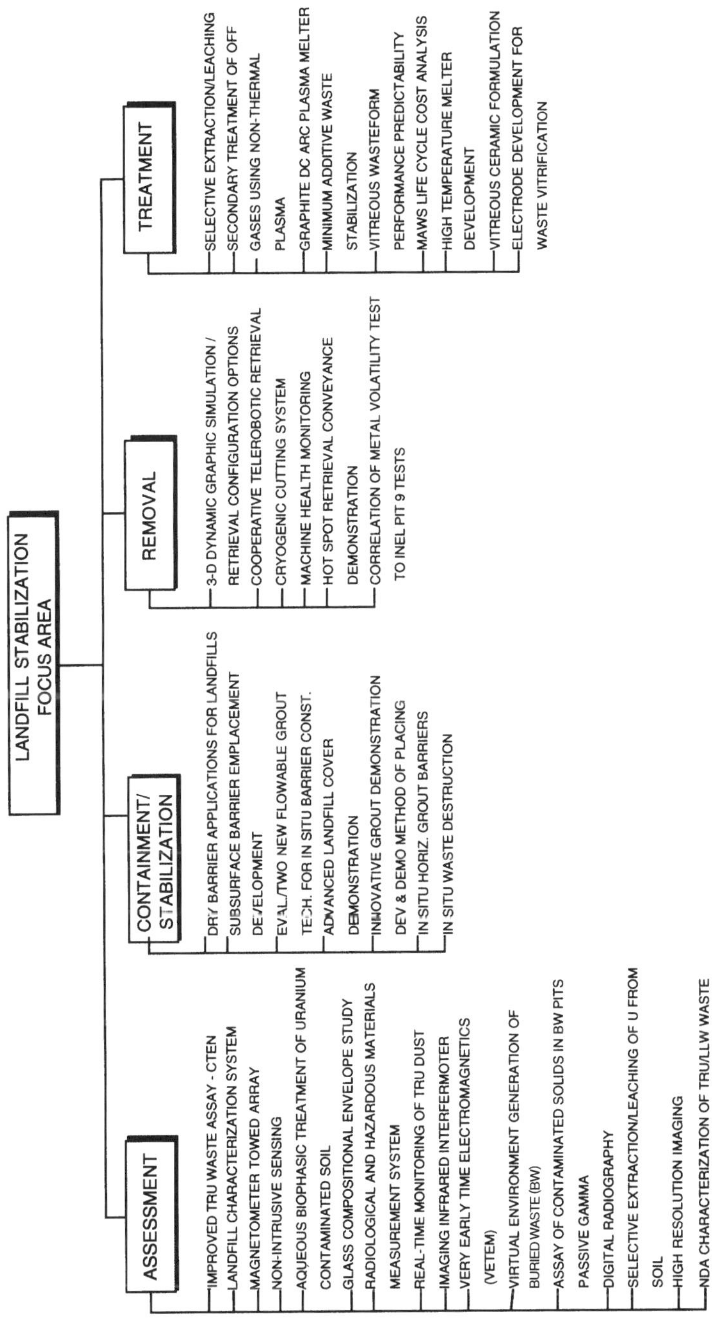

fiscal year 1995, the program has demonstrated, at some scale, 25 technologies. In addition, the focus area has implemented, transferred, or had included as part of the solution for a remediation, six technologies at several sites inside and outside the DOE complex.

In fiscal year 1995, the Landfill Stabilization Focus Area organized itself by the technology areas. Figure 5 shows a list of the technology development activities for fiscal year 1995. The following are descriptions of select technologies developed, demonstrated, and/or evaluated in fiscal year 1995. Details of these and other technologies can be found in the office of Environmental Technology Development, Land Stabilization Focus Area, Technology Summary, June 1995, DOE/EM-0251

Assessment

Combined Thermal and Epithermal Neutron Assay

The Combined Thermal and Epithermal Neutron Assay (CTEN) system analyzes retrieved buried waste drums prior to interim storage or treatment and also provides post treatment assay of the final packaged waste. The CTEN interrogates the sample drum with thermal and epithermal neutrons to determine the amount of fissile isotopes contained. The resulting fast-fissile neutrons are differentiated by detectors that are insensitive to the interrogating neutrons. The CTEN system provides more accurate and sensitive fissile and transuranic assays of TRU (transuranic) drums of waste by reducing inaccuracies caused by self-shielding. Self-shielding occurs when lumps of fissile material are present, and can result in assay errors of several hundred percent.

- Application:
  - Assay of 55 and 83-gallon waste containers
  - Evaluate drums for certification to meet DOT shipping limitations and receiving facility Waste Acceptance Criteria
- Benefits:
  - Evaluates waste containers using both thermal and epithermal neutrons
  - Minimizes the effects of self-shielding by lumps of fissile material to provide more accurate and repeatable transuranic assay of waste drums
- Demonstration:
  - FY1995, LANL
- Participation:
  - LANL, DOE-Idaho, LITCO, Canberra

Very Early-Time Electromagnetic System

The very early-time electromagnetic (VETEM) system will collect and interpret data from the shallow subsurface. This instrument will operate in the region of the EM spectrum between low frequency EM induction and ground penetrating radar (GPR) frequencies.

- Applications:
  - High-resolution electromagnetic (EM) imaging system for shallow environmental problems (less than 10 meters) such as DOE buried waste sites
  - For sites where traditional EM equipment and interpretation techniques do not produce adequate results
- Benefits:
  - Enhanced resolution of the shallow subsurface
  - Operates at sites where the physical properties of the soil make high resolution GPR difficult
- Demonstration:
  - Summer 1995, INEL Cold Test Pit
- Participation:
  - USGS; EPA; Others

Landfill Assessment and Monitoring System

The Landfill Assessment and Monitoring System (LAMS) is a method to characterize metal and mixed waste contaminants, their sources, and their migration beneath landfills. The LAMS emphasizes minimally intrusive technologies and downhole sensors. The LAMS consists of five separate subsystems: screening and sampling optimization techniques, innovative drilling technologies, onsite analysis and in situ sensors, subsurface monitoring technologies, and data evaluation and risk analysis techniques.

- Applications:
  - Metal and mixed-waste contaminants beneath landfill sites
- Benefits:
  - Systematic implementation of technologies for site-specific conditions and needs
  - Ensures optimal approach to characterization, monitoring, and remediation
- Demonstration:
  - FY1995 Sandia
- Implemented:
  - Target Verification and Calibration Task at SNL's Technical Area II by SNL-ER, 1994
- Transferred:
  - Several individual technologies patented or have licensing agreements

- Six technologies are commercially available

Containment/Stabilization

Dry Barrier Applications for Landfills

The dry barrier concept was developed for application as landfill cover or liner systems in arid environments. The dry barrier is a coarse layer of geologic material that is dried by air flow. The moving air accumulates and removes water or leachate from the landfill as it passes through the dry barrier layer to limit infiltration and contaminant migration.

- Applications:
  - Component of a landfill cover or liner system
  - Method for stripping gas-phase contaminants
- Benefits:
  - Reduces the risk of contaminant migration from landfills
  - Extends longevity of cover and barrier systems
- Demonstration:
  - SNL Chemical Waste Landfill, Summer 1995
- Implementation
  - Incorporated into LANL EM-40 plan for landfill liner system, 1995
- Participation:
  - D.B. Stephens & Associates, University of New Mexico

Advanced Landfill Cover Demonstration

The Advanced Landfill Cover Demonstration (ALCD) is developing technology to improve upon current landfill cover systems. The project will provide alternatives to the EPA's landfill cover designs that will work more effectively in arid climates. The ALCD is coordinating input from four projects to improve evaluation of landfill barrier selection, materials, and monitoring technologies. These four areas are: (1) migration barriers, (2) dry barriers, (3) post-closure monitoring, and (4) decision support systems.

- Applications:
  - Landfills containing mixed, hazardous, and metal waste
- Benefits:
  - Landfill covers that are superior to current EPA cover designs, particularly in arid environments
  - Planned demonstrations should result in greater regulatory acceptance and interstate reciprocity
- Demonstrations:

  - Planned for summer 1995 at SNL

- Participation:
  - Demonstrations planned in collaboration with the WGA DOIT Committee, U.S. EPA, and other state environmental agencies

Flowable Grout Barrier Materials

The grout materials are a mineral wax/bentonite emulsion developed in Germany (montan wax), and a glyoxal-modified sodium silicate grout developed in France. The focus of this investigation is to determine the feasibility of using conventional grout permeation methods for placement of a barrier, and to evaluate the resistance of these materials to waste components.

- Applications:
  - Injectable barrier system for use beneath leaking landfills and underground storage tanks

- Benefits:
  - Injectable from the surface eliminating the need to excavate trenches and construct barriers
  - Reduces risk of worker exposure to hazardous materials and reduces the generation of secondary waste

- Demonstration:
  - FY1995, Pilot-scale single well injection in sandy soils at SNL's Mixed Waste Landfill site

- Participation:
  - Golder Associates and the European licensees MIBRAG and Societe Hoechst Francaise

Subsurface Barrier Emplacement

The Subsurface Barrier Emplacement project is adapting equipment and technologies currently used in civil engineering applications to emplace impermeable horizontal barriers beneath a waste site. The approach integrates several different technologies including subsurface access techniques such as directional drilling, barrier emplacement technologies, barrier materials, barrier verification, and post emplacement monitoring.

- Applications:
  - Leaking lined or unlined landfills and waste sites

- Benefits:
  - Wastes are contained, allowing additional time to develop remedial treatment

- Demonstrations:

  - SNL Geotechnical Test Range in FY1995

- Participation:
  - Applied Geotechnical Engineering and Construction, Inc.

Retrieval

Dig Face Characterization

The Dig Face Characterization system consists of multiple sensors that are integrated into the retrieval process. During progressive soil and waste removal, an array of geophysical, chemical, and radiological sensors are continuously deployed by a robotic system to characterize the dig face. Characterization data are interpreted on-line in real-time to identify hazards and to support the retrieval process. Specific sensors deployed on the system include: gamma-neutron sensor, spectrometer, magnetometer, volatile organic sensor, and 3-D holographic radar.

- Applications:
  - Any hazardous waste site undergoing retrieval
- Benefits:
  - Reduces environmental, health, and safety risks by providing continuous real-time monitoring and analysis of hazards
  - Adaptable to meet specific site characterization requirements
- Demonstration:
  - INEL, Summer 1995
- Implementation:
  - Mound/EM-40, Summer 1995, written commitment
  - BNL/EM-40, FY96, work plan currently under development
- Participation:
  - Geonics Ltd., Ecology Int., Colorado School of Mines, and USGS

Innovative Subsurface Stabilization

The innovative grouting/retrieval technology involves three operations: grouting, fracturing, and retrieval. The first step is to grout the waste and surrounding soil. Next, the monolithic grouted block is fractured, using a demolition grout. Finally, the debris is excavated in a relatively dust-free environment with remotely controlled equipment.

- Applications:
  - Subsurface waste storage sites
  - Isolation of point sources or hot shots
- Benefits:

  - Confinement of waste prior to retrieval and treatment with less contamination spread
  - Eliminates need for elaborate contamination control strategies during retrieval and handling, thereby reducing costs without jeopardizing worker safety
  - Cost estimates for retrieval of a one-acre size pit applying this technology is \$30M, compared with \$200M for conventional retrieval

- Demonstration:
  - September 1995 demonstration of grout injection technology with chosen grout material

- Participation:
  - INEL, Hanford, BNL

Real-Time TRU Contaminated Dust Monitor

A real-time dust monitor (RTDM) was developed to monitor TRU contaminated dust based on laser-induced breakdown spectroscopy (LIBS) technology. Continuous Air Monitors (CAM) are incorporated into the system to provide an improved radioactive assay. Monitoring of the potentially contaminated dust particles is essential for personnel protection, due to the extremely low levels of human uptake allowed.

- Applications:
  - Personnel protection and contamination control during buried waste retrieval operations
  - Construction or decommissioning sites where contaminated dust poses a potential health or environmental risk
  - Applied in accident mitigation

- Benefits:
  - Real-time field deployable instrument
  - Characterization of samples to pCi/g levels
  - RTDM uses CAM measurements for accurate alpha determination as a function of dust density
  - RTDM uses LIBS for detection of many hazardous heavy metal emissions in the visible and ultraviolet regions

- Demonstration:
  - Summer 1995 INEL Cold Test Pit

- Implementation:
  - INEL Pit 9 in FY96, EM-40

Treatment

Nonthermal Plasma Offgas Treatment

The Nonthermal Plasma Offgas Treatment is an alternative nonincineration technology for volatile organic vapor destruction and metal oxidation. Electrical energy directed into the process offgas stream creates highly reactive free radicals. The free radicals can oxidize/reduce pollutants, fragment pollutants directly, or promote excited-state chemistry for the destruction of the VOCs.

- Applications:
  - Treating gaseous-based wastes such as VOCs in stack gas and as stand-alone plasma devices
  - Treating offgas streams of incompletely destroyed waste from primary stage units, incinerators, and furnaces
- Benefits:
  - Nonincinerator treatment for VOCs destruction and metal oxidation
  - Treats secondary waste gases from commonly used treatment processes for mixed waste such as vitrification, incineration, and thermal stripping
  - Onsite treatment with high destruction and removal efficiencies.
  - Versatile operation to process a wide range of substances and concentrations
- Demonstration:
  - FY95 lab scale demonstration on arc melter offgases
- Participation:
  - Los Alamos National Lab, INEL, Electric Power Research Institute, University of Illinois

Minimum Additive Waste Stabilization

The Minimum Additive Waste Stabilization (MAWS) concept combines different waste streams as resources for vitrification treatment to achieve both volume reduction and durability of the final waste form. Implementation of MAWS optimally integrates multiple technologies such as vitrification, thermal destruction, soil washing, offgas treatment, and/or ion exchange wastewater treatment into a single system to minimize final and secondary waste generation, and overall costs.

- Applications:
  - Largely inorganic waste streams consisting of soils, sludge, ashes, asbestos, concrete debris, and ion exchange resins
  - Most DOE sites
- Benefits:
  - Produces durable glass final waste forms

- Projected cost savings are greater than 50% over conventional cementation treatment
- Volume reduction of at least 25%

- Implemented:
  - GTS Duratek awarded contract to vitrify 700,000 gal. M-area sludge at Savannah River
  - GTS Duratek awarded contract to design and build melter to process radioactive K-65 silo wastes at Fernald.

Real-Time Furnace Diagnostic Tools

Two innovative diagnostic real-time tools have been developed in conjunction with the Graphite DC arc furnace that should be universally applicable in any furnace/melter environment. The first tool is the milliwave radiometer (MR), which is an improvement over conventional pyrometers to continuously measure temperatures inside the furnace. The second tool is the microwave plasma analyzer (MPA), which is capable of continuously and simultaneously monitoring concentrations of the 11 RCRA metals in the offgas from the furnace.

- Milliwave Radiometer Benefits:
  - Improved accuracy due to continuous calibration for emissivity changes
  - Ability to measure melt temperatures below cold cap in certain instances
  - Wide temperature range from -200¡C to 20,000¡C

- Microwave Plasma Analyzer Benefits
  - True in-situ and real-time monitoring of trace metals
  - Robust to particulate filled and optically obscured offgas flow streams
  - Demonstrated Hg sensitivity in the 0.5 to 3.0 ppb range

- Demonstration:
  - FY95 radioactive demonstration of MPA and MR

- Participation:
  - PNL, MIT, T&R Associates, and Electro-Pyrolysis, Inc.

In fiscal year 1996, the Landfill Stabilization Focus Area has been fully developed and the program has been adjusted to reflect the refinements. Figure 6 shows the technology development activities for fiscal year 1996, organized by product line.

Implementations/Commercialization

The implementation/commercialization activities of the focus area have made very good progress; for example:

Rapid Geophysical Surveyor was licensed to Sage Earth Sciences. The technology is a high speed, high density magnetic data collector. The Rapid Geophysical Surveyor has been implemented at several DOE and DOD sites. The

Figure 6. **Landfill stabilization focus area FY 1996.**

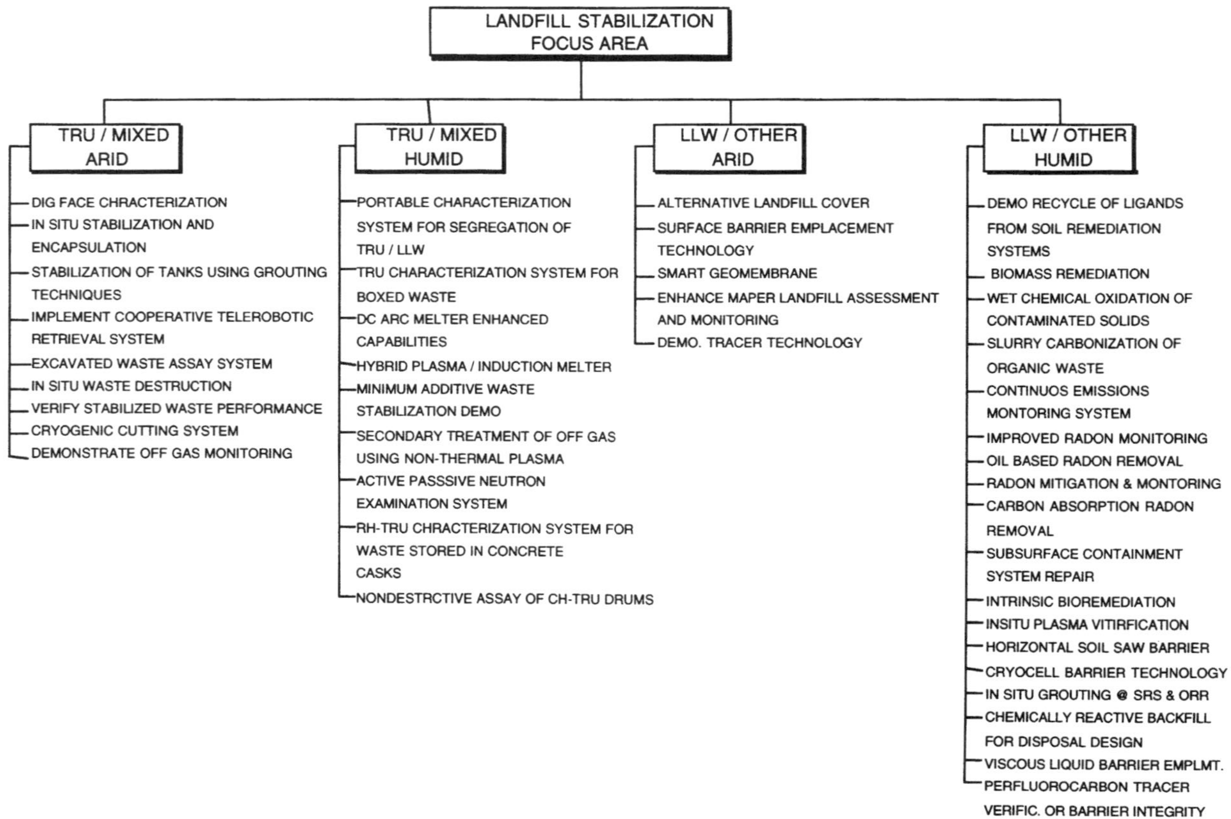

technology was implemented at Mound, and the National Institutes of Justice has selected it for application in courtroom security. The system located an unknown buried tank, and saved Mound $200K in characterization costs.

The Multi-Sensor Analysis System is a multi-data-set integration tool that produces an optimal estimate of the distribution of buried material, and presents the results in a graphic form for ease of reading. The system was implemented at Sandia National Laboratory by Environmental Restoration, and reduced the cost by $500K.

The Cryogenic Cutting System is a waste retrieval tool for sizing of waste materials. The system uses liquid nitrogen and carbon dioxide pellets for cutting and abrading. The system eliminates the secondary waste associated with other cutting systems. A Cooperative Research and Development Agreement was also completed with three industry partners.

The Combined Thermal and Epithermal Neutron, described above, completed a Cooperative Research and Development Agreement with Canberra for development of the software for a production scale system.

The Contamination Control Unit is a mobile unit that provides dust suppression, using soil fixants and misting curtains to control spread of contamination during landfill retrieval operations. The system has been implemented at the Idaho National Engineering Laboratory and the Hanford Reservation. The technology was transferred to Idaho National Engineering Laboratory Waste Management Operations for implementation at the Idaho National Engineering Laboratory TSA-Retrieval Enclosure Facility.

The Rapid Transuranic Monitoring Laboratory is a field deployable laboratory. The system can monitor the airborne transuranic concentration, and rapidly process soil, smear, and sir filter samples for transuranic contamination. The Rapid Transuranic Monitoring Laboratory was implemented during the fiscal year 1994 Idaho National Engineering Laboratory site-wide soil survey. The technology has been licensed to Thermal Analytical and selected for implementation at Oak Ridge National Laboratory and the Savannah River Site.

In fiscal year 1996, the Landfill Stabilization Focus Area intents to build on the successes of fiscal year 1995. The focus area expects to have 10 technologies available for transfer and implementation. Here are a few examples of the expected outcomes for the focus area in fiscal year 1996:

The Cooperative Telerobotic Retrieval System provides the capability to robotically and surgically retrieve buried waste from landfills; the Dig Face Characterization System (see above description); the Holographic Impulse Radar generates real-time, three-dimensional images of buried waste in low conductivity soils; and the Subsurface Barrier Emplacement System provides the means to place underneath and around a landfill to limit horizontal and vertical contaminant migration. These technologies have been scheduled for demonstration and, if successful, implementation at Brookhaven National Laboratory for the remediation of approximately 30 pits.

The Excavated Waste Assay System provides real-time, nonintrusive characterization of excavated waste for support in waste retrieval and processing. The Real Time Monitor for TRU Dirt provides real-time monitoring of transuranic contaminated dirt and dusts to support landfill remediation activities. These technologies are scheduled to be demonstrated/implemented as part of the Remediation of Pit 9 at the Idaho National Engineering Laboratory.

CONCLUSION

The Landfill Stabilization Focus Area will develop technologies for the Office of Waste Management, the Office of Environmental Restoration, and the Office of Facility Transition and Management, for other elements of the DOE such as the Office of Environmental Safety and Health and the Office of Energy Research; other federal agencies such as the Environmental Protection Agency, and the National Institute of Standards and Technology. In the final analysis, using this New Approach and the Landfill Stabilization Focus Area will select, develop, and demonstrate technologies that have the potential for implementation across the DOE complex as well as in the private sector - technologies that will provide for safer, more cost-effective and more permanent solutions to our DOE landfill problems.

REFERENCES

Oak Ridge National Laboratory. 1992. U.S. Spent Fuels and Radioactive Waste Inventories, Projections, and Characteristics Integrated Data Base for 1992, October.

U.S. DOE (U.S. Department of Energy). 1993. A New Approach to Environmental Research and Technology Development at the U. S Department of Energy, Action Plan, December 14, 1993.

U.S. DOE (U.S. Department of Energy). 1994. Environmental Management Program Strategic Plan, March, p. 24.

CHAPTER 13

Evaluation of Biological Degradation of a Dissolved Gasoline Plume by Air Sparging

Edward P. Van Doren, Maureen A. Dooley, and **Dennis G. Tuttle,** ABB Environmental Services, Inc., Wakefield, Massachusetts

Robert J. Shaughnessy, Commonwealth of Massachusetts, Department of Environmental Protection, Bureau of Waste Site Cleanup, Boston, Massachusetts

INTRODUCTION

The site is located on Cape Cod, Massachusetts on a north-south trending peninsula bounded by ponds to the west and east, and Vineyard Sound to the south (Figure 1). The middle portion of the site is approximately two miles from Vineyard Sound. Topography at the site is fairly flat, ranging in elevation from 30 feet National Geodetic Vertical Datum (NGVD) at the source area, to 20 feet NGVD in the midsection of the site, to sea level at the surface of Great Pond (USGS, 1979)

A release of gasoline resulted in a plume of free product and contaminated groundwater; the dissolved-phase plume has migrated toward and is discharging into the Great Pond, located approximately 2,000 feet southwest of the release location. The periodic loss of product from the source area to the underlying, sandy aquifer went undetected for several years, and was discovered indirectly when residential basement air became contaminated with volatile constituents in January 1988. When initially discovered, the maximum thickness of measured product was over 16 inches in a monitoring well located approximately 450 feet downgradient of the release.

The distribution of dissolved-phase constituents in groundwater is illustrated in Figure 2, as of 1990. In 1992 and 1993, no measurable free product was detected by ABB-ES in monitoring wells or recovery wells.

SITE HYDROGEOLOGY

The surficial geology of Cape Cod is characterized by glacial deposits of late-Wisconsin age, and the site is located in the southern portion of the Mashpee

Figure 1. **Site location.**

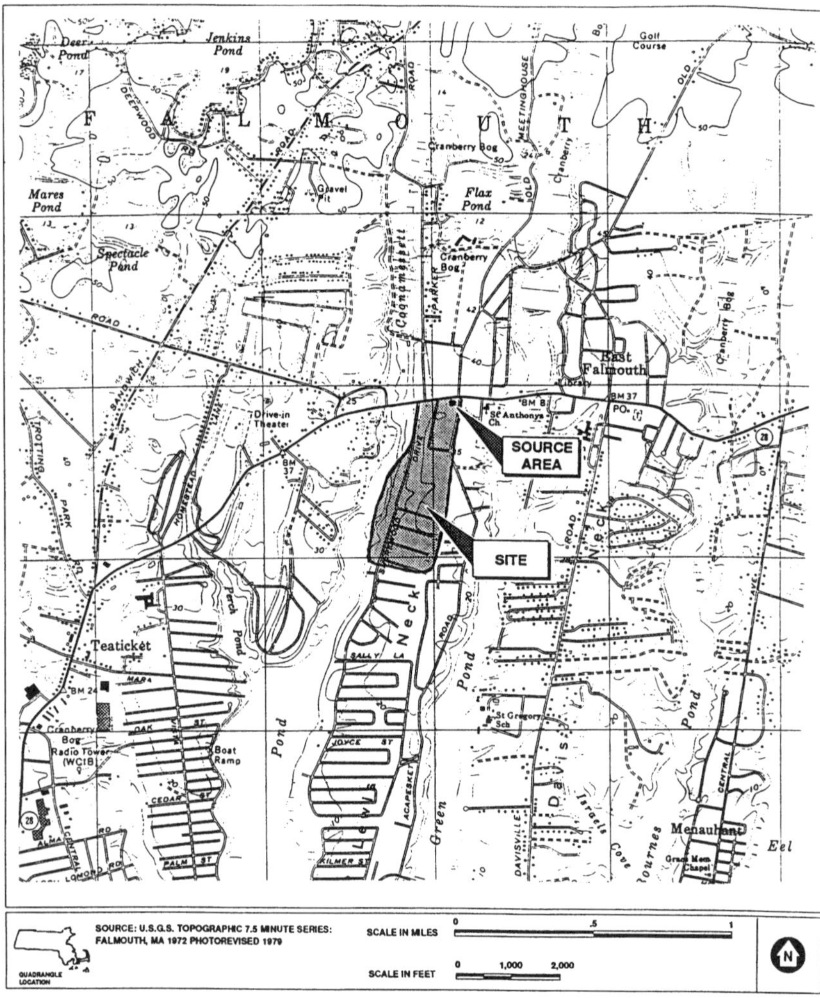

pitted outwash plain (Strahler, 1966). Outwash deposits in this vicinity generally consist of fine to medium sand and gravel. In many areas the outwash is underlain by lacustrine deposits of silt, clay, and very fine sand, as well as basal till deposits. Borings at the site have not extended more than 80 feet below grade to determine site-specific depth to bedrock; however, soils encountered throughout the area to this depth were consistently a fine to coarse sand, with occasional gravel and silt. The depth to groundwater ranges from 30 feet at the source to zero feet where the groundwater discharges to the pond estuary.

Generally, groundwater flows south, then flows either to the east or west in response to the groundwater divide located below the peninsula. The gasoline

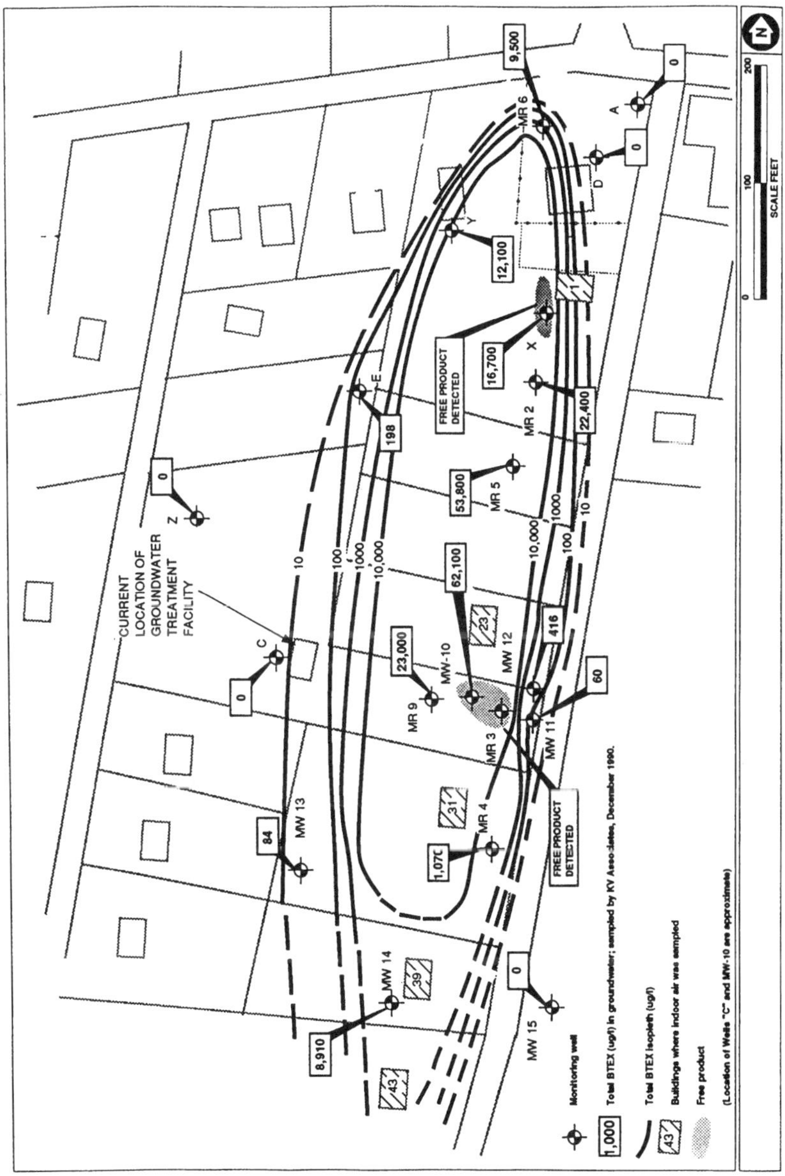

Figure 2. Distribution of BTEX, 1990.

release(s) appear to have occurred close to this divide because dissolved constituents migrate essentially due south for several hundred feet before the trace of contaminant migration bends to the west. The hydraulic conductivity from site-specific measurements ranges from 250 to 312 ft/day, with an average of 288 ft/day. Combined with a gradient at the water table of 0.0025 and an assumed porosity of 0.30, the calculated seepage velocity of groundwater is approximately 2.4 ft/day.

GROUNDWATER EXTRACTION AND TREATMENT SYSTEM

The existing groundwater extraction and treatment system consists of: two groundwater recovery wells screened in overburden material; four aqueous phase granular activated carbon (GAC) vessels (2,000 lb of carbon each) arranged in a series/parallel mode; and reinjection of treated groundwater into two injection wells screened in the overburden. The location of the treatment building, extraction wells, and injection wells is shown in Figure 3. Currently the combined flow rate from the two extraction wells ranges between approximately 80 to 120 gallons per minute (gpm). The flow rates from the individual recovery wells average 70 gpm and 17 gpm for RW-1 and RW-2, respectively. This system began operation in December of 1991.

The average benzene, toluene, ethylbenzene, and xylene (BTEX) concentration in the influent groundwater, prior to GAC treatment, is 2,500 μg/L. In addition to BTEX, methyl-t-butyl ether (MTBE) has also been detected in the influent groundwater to the treatment system at concentrations ranging from 0 to 140 μg/L. The concentrations of BTEX at RW-2, which is closer to the source, are typically much greater than those detected at RW-1. For 1993, the combined influent concentration of BTEX to the treatment system ranged from 340 to 4,900 μg/L. The average mass of contaminant removed by each recovery well is estimated to be 0.19 lb/day and 0.44 lb/day for RW-1 and RW-2, respectively.

CONTAMINANT DISTRIBUTION AND PLUME CONTROL

As of 1992, the majority of free product had been removed from the aquifer, leaving an unknown quantity of residual gasoline at the water table and overlying vadose-zone soils, and a well-defined dissolved phase plume that is relatively narrow (<300 feet) compared to its length of 2,000 feet (see Figure 4). The lateral extent of plume migration is well defined by very low to nondetectable concentrations of BTEX in groundwater from several shallow monitoring wells. The vertical extent of dissolved contaminants is also very limited, as depicted in Figure 5. The centerline of the plume appears to reach a maximum depth below the water table of only 30 feet, more than 1,500 feet from the release.

Since December of 1992, the influent concentration of BTEX has remained persistent, with frequently reported concentrations of over 3,000 μg/L. This occurrence is likely the result of gasoline constituents persisting in the environ-

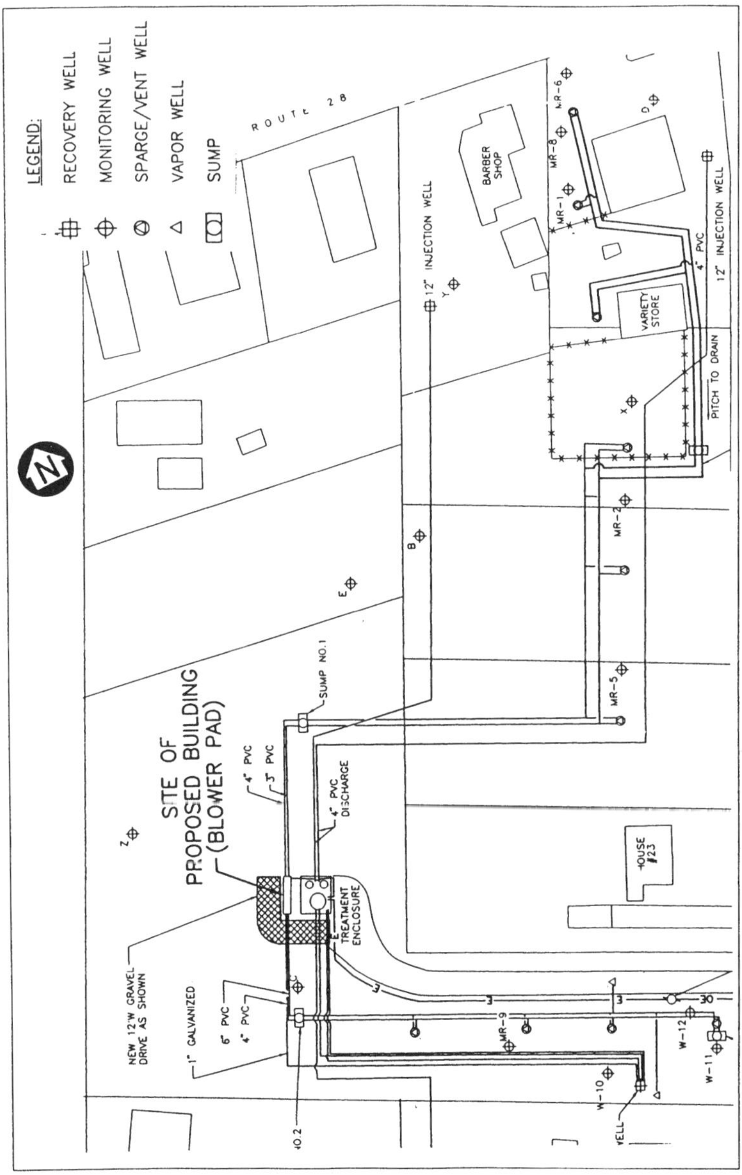

Figure 3. **Venting system layout.**

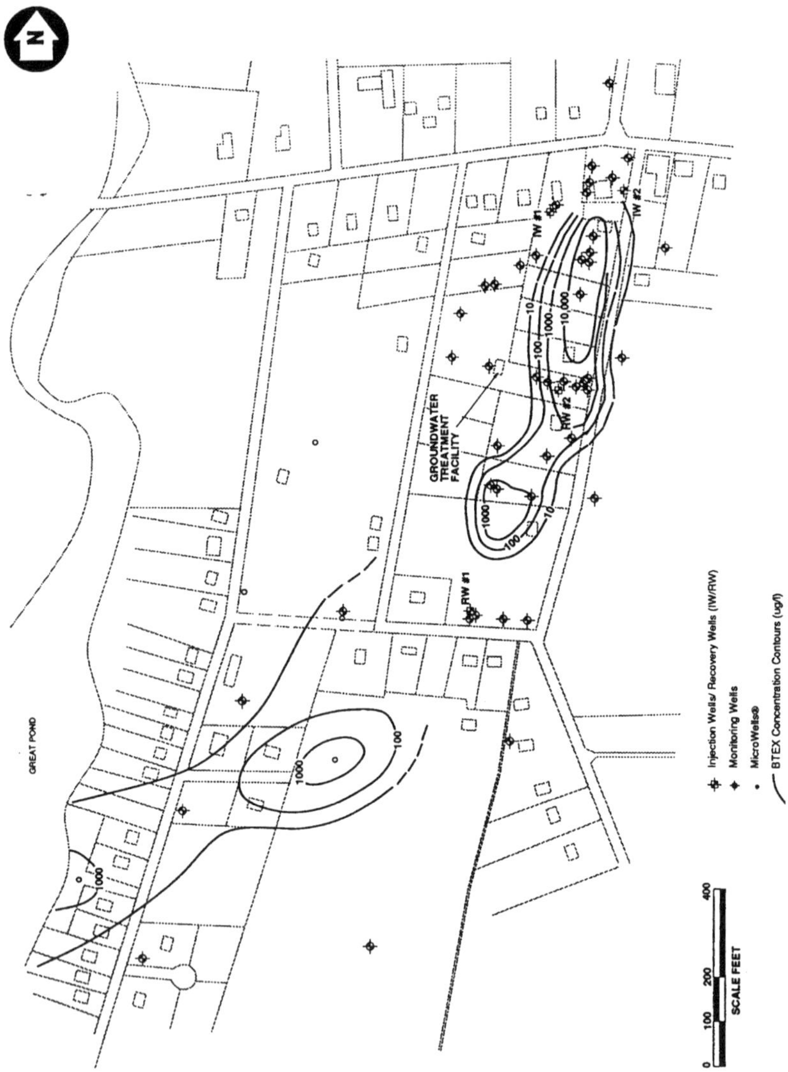

Figure 4. **Horizontal distribution of BTEX, 1992.**

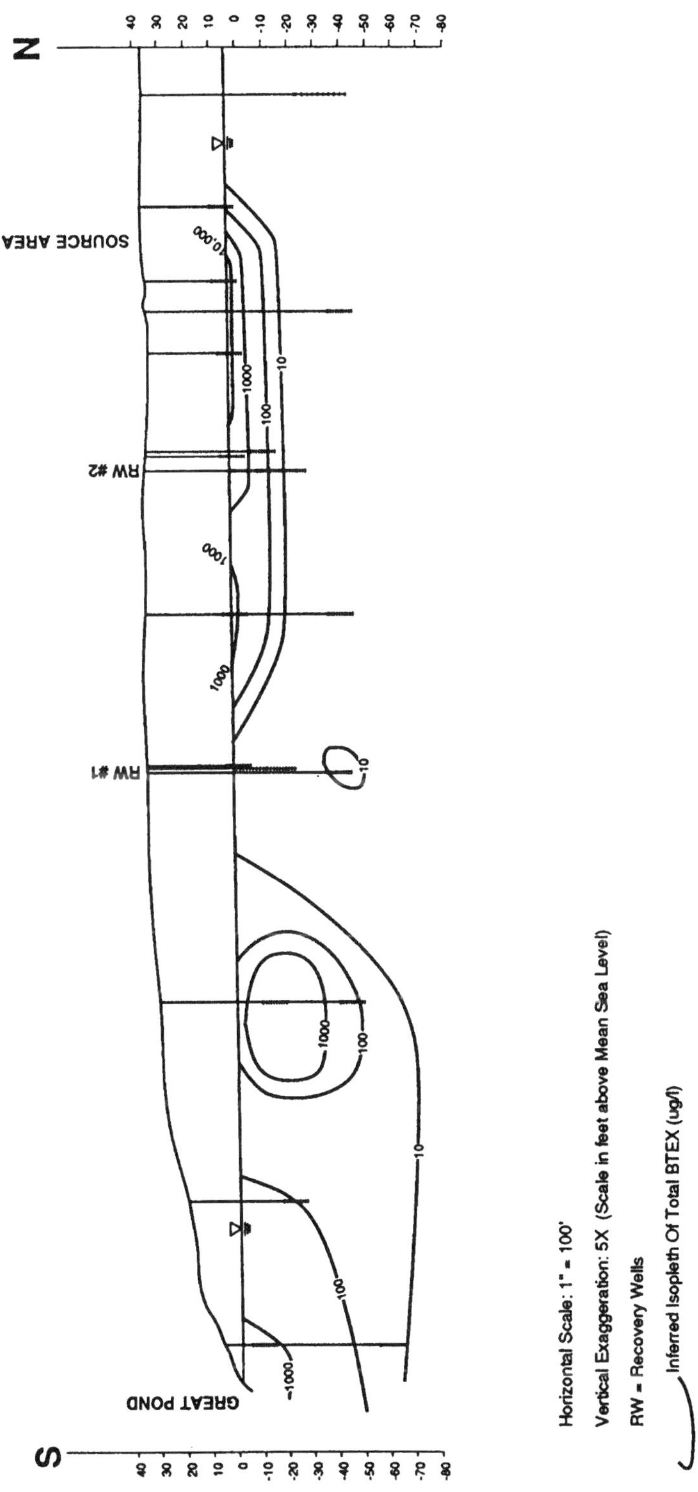

Figure 5. **Vertical distribution of BTEX, 1992.**

ment upgradient of the recovery wells, which slowly release contaminants to the groundwater. As free product migrated along the water table it became depleted as residual amounts of gasoline became trapped in pore spaces in both the vadose and saturated zones. The amount of residual gasoline along the path of free product migration could be substantial. Studies have shown that gasoline removed as a free phase from a recovery well may be less than 30% of the volume released (Parker et al., 1994). The amount of material trapped in pore spaces depends on soil characteristics, the distance from the point of release to the water table, the amount of lateral spreading, and the degree of water table fluctuation. These observations suggest that a substantial amount of soil in the vicinity of the water table contains residual product.

SPARGING AND SOIL VAPOR EXTRACTION SYSTEM

To address these sources of residual contamination, ABB-ES designed an air sparging and soil vapor extraction system in an attempt to accelerate the remediation of the gasoline plume. The full-scale system design includes operation of 10 sparging wells and 12 venting wells (Figure 3). Based on a pilot test, the venting blower was sized to handle a flow rate of 400 to 500 scfm at a well head vacuum of approximately 5 to 7 inches of water column (in. w.c.). The sparging blower was sized to provide an air injection rate of 40 to 50 scfm at a well head pressure of approximately 4 to 5.5 pounds per square inch gauge (psig). The estimated radius of influence of the venting wells, assuming a minimum of one pore volume of air volume per 24 hours, is about 50 feet. The offgas control system consists of a catalytic incineration unit operated using natural gas as a fuel supplement. A system flow diagram is shown in Figure 6. The installation of the system was completed in the spring of 1995, and the system has been operating continuously since the beginning of June 1995.

SPARGING AND SOIL VAPOR EXTRACTION OPERATION AND MONITORING

Since start-up of the system, the venting blower has been operating at an average flow rate of approximately 400 scfm, with the flow from each venting well being approximately equal (35 scfm each). The initial vapor concentration from the combined influent to the venting blower was approximately 280 ppm as measured with a PID. After two months of operation, the combined vapor concentrations had decreased to approximately 90 ppm. The sparging blower has been operating at a rate of approximately 40 scfm, with the flow rate from each sparging well being approximately equal (4 scfm each).

BIOLOGICAL CHARACTERIZATION

A treatability test was conducted to evaluate the feasibility of stimulating hydrocarbon biodegradation during air sparging, and if changes should be made

Figure 6. **Piping and instrumentation diagram.**

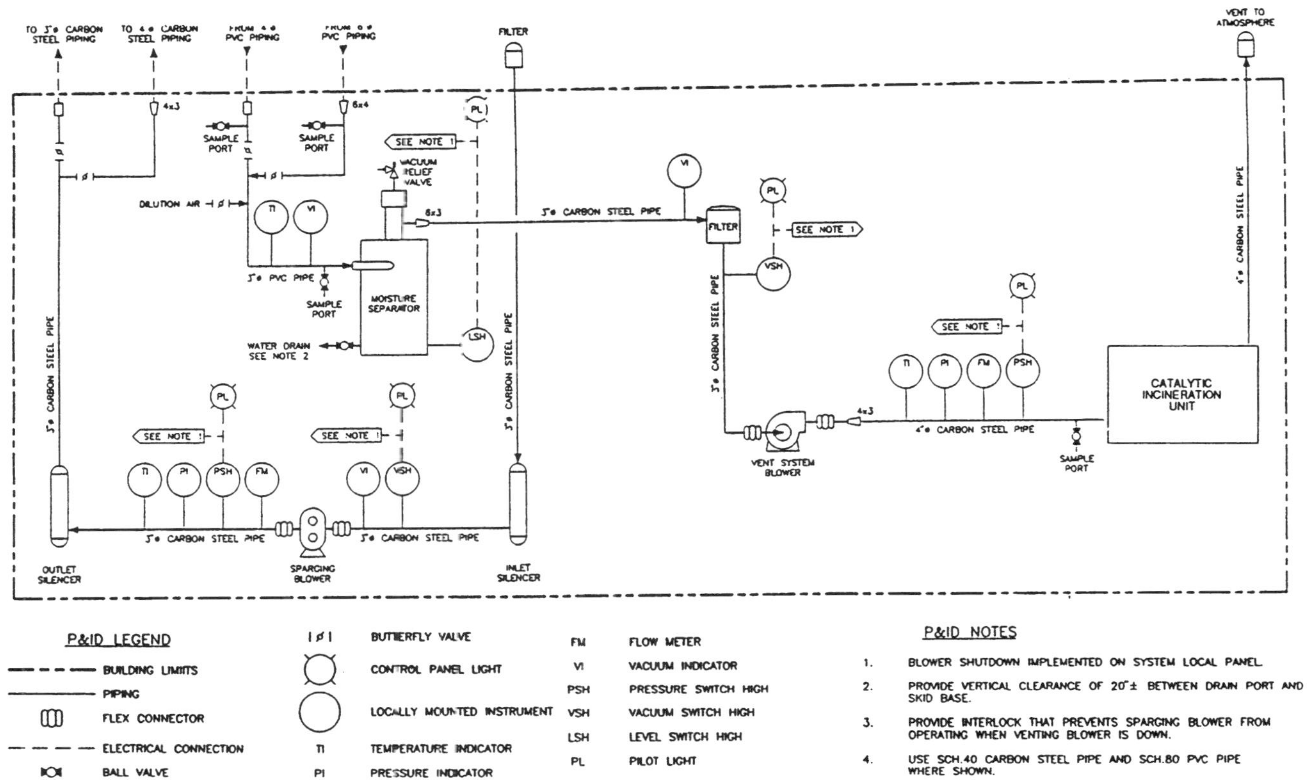

to the system design to enhance the biodegradation process. Initial characterization analyses and biodegradation testing were performed as part of a treatability testing process. The specific objectives of the study were to:

- measure chemical parameters that influence biodegradation
- determine if indigenous hydrocarbon bacteria are present
- characterize petroleum
- evaluate nutrient requirements
- demonstrate petroleum biodegradation

Groundwater was collected from well locations MR-3A, MR-5, MR-9 and MW-119 in February, 1995. The venting and air sparging system had not started operation yet at this time. Samples MR-3A, MR-5, and MR-9 are adjacent to sparging wells. Monitoring well MW-119 is located upgradient of the plume.

Results from TPH analysis, using modified EPA Method 8100, indicated that high levels of petroleum were detected in MR-3A and MR-5 groundwater (> 50mg/L), and lower levels in MR-9 (3mg/L). A significant proportion of the petroleum is made up of volatile compounds based on results of VOA analysis (Table 1) and from GC fingerprint analysis (Modified EPA Method 3550/8100). The petroleum present in the samples exhibited a chromatographic pattern similar to gasoline.

Table 1. **Volatile Organic Analyses (EPA Method 8020)**

Compound	MR-3A (ug/L)	MR-5 (ug/L)	MR-5Dup (ug/L)	MR-9 (ug/L)	MW-119 (ug/L)
Benzene	402	<500	<500	80J	<5
Toluene	6335	9588	9588	5192	1.7J
Ethyl benzene	2316	2839	2564	1894	<5
m,p-xylene	5839	7292	7231	5088	<5
o-xylene	5184	5596	5397	4195	1.1J

J = Estimated Value

Low levels of nitrate nitrogen were detected only in groundwater from MR-5 (3mg/L). No ammonia nitrogen or phosphate was detected in any of the groundwater tested, and the pH was acidic ranging between 5 and 5.5.

Microbial characterization analyses were performed to determine if hydrocarbon degrading bacteria were present, estimate biomass, and evaluate nutritional status of indigenous bacteria. Total heterotrophic bacteria and hydrocarbon degrading bacteria were measured using spread plate-counting procedures in all four groundwater samples, and only MR-5 and MR-9 were analyzed by Microbial Insights, Inc. (Knoxville, TN) for Phospholipid Fatty Acids (PLFA/PHA).

Hydrocarbon bacteria were detected in all groundwater; however, the concentration of both total and hydrocarbon degrading bacteria was much lower (approximately 10-100 times) in MR-9 groundwater (Table 2). PLFA results were also used to estimate biomass, and those data also showed that there was a relatively higher concentration of biomass in MR-5 than there was in MR-9. The biomass value generated from PLFA analysis was higher than the value obtained by plate count methods. This result is not unexpected because the spread plate method is designed to only measure total aerobic bacteria where the PLFA procedure derives the biomass value from phospholipids extracted from both aerobic and anaerobic bacteria.

Table 2. **Bacteria Analysis Results**

Sample ID	Total Bacteria[1] (cells/ml)	Total Bacteria[2] (CFU/ml)	Hydrocarbon Bacteria[2] (CFU/ml)
MW-119	NA	2.4×10^4	2.3×10^3
MR-3A	NA	1.1×10^4	2.3×10^3
MR-5	1.1×10^6	1.3×10^5	5.0×10^3
MR-9	4.7×10^4	2.3×10^3	67J

J = Estimated Value (Below detection Limit)
NA = Not Analyzed
[1] Bacteria numbers estimated based on results from Phospholipid Fatty Acid Analysis.
[2] Results were obtained using spread plate method.

The microbial community appears to be made up of gram negative type bacteria in groundwater from both MR-5 and MR-9, based on results from PLFA analysis. Generally, gram negative bacteria can utilize a wide variety of carbon sources, which can include petroleum. There also was evidence that gram positive bacteria and anaerobic sulfate or iron reducing bacteria were present in lower quantities (5%). The bacteria were also shown to be slow growing, and in a "stationary" growth phase. This is in contrast to bacteria that are very active that would be in a "logarithmic" growth phase. Other results from the PLFA testing showed that bacteria in MR-9 showed signs of environmental stress from starvation or toxicity. The environmental stress is likely due to reduced availability of carbon and nitrogen in groundwater at MR-9, as compared to MR-5.

BIODEGRADATION TESTING

Biodegradation testing was conducted to demonstrate the feasibility of enhancing TPH biodegradation in groundwater. The effect of nutrient availability was also evaluated as part of the testing.

Test microcosms were prepared using groundwater (1:1 mixture MR-5 and MR-3A) from the site. Test vessels were sealed and mineral nutrients, pH adjustment, and oxygen were provided. Two sets of microcosms were prepared with different nutrient formulations; one was a mineral salts media that includes nitrogen, phosphate, and trace nutrients. The second test condition was prepared using ABB-ES' nutrient formulation ACT™, which contains 1:1 ammonia chloride:potassium tripolyphosphate and is used primarily for groundwater bioremediation.

A parallel set of control microcosms were prepared; one without any added mineral nutrients or pH adjustment, and the other was amended with the biocide mercuric chloride. The control vessels were used to measure abiotic losses such as volatilization. The specific conditions prepared for testing include:

- ammonia media
- ACT-50™ mg/L
- no added nutrients
- killed control

TPH was measured twice; when the microcosms were prepared, and after 15 days, by GC/FID (Modified EPA Method 3550/8100). The entire contents of duplicate microcosms were extracted and analyzed for TPH at each time point. Biological activity was monitored over the test period by measuring the concentration of volatile organics in the headspace (direct headspace injection/GC/FID detection), oxygen, and carbon dioxide (direct headspace injection/GC/TCD detection).

Results from testing showed that greater than 99% of the gasoline (TPH) was degraded after 15 days in the two nutrient-amended microcosms, and 95% in the microcosms that were only made aerobic (Table 3). A 10% reduction in TPH was observed in the killed control.

Review of GC chromatograms shows a significant reduction in the chromatographable hydrocarbon in the all the live microcosms after 15 days. Results from both TPH and headspace analysis are presented in Figures 7 and 8.

Organic vapors (including BTEX) in the microcosm headspace were measured several times during the testing period to monitor biodegradation (Table 4). Results showed that TPH levels were reduced by 98%, 96%, and 94% in the ammonia media, ACT™, and no nutrient amended microcosms, respectively. Complete removal of BTEX was observed in the ammonia media, and reductions of greater than 97% were observed in the ACT™ and no nutrient microcosms (Table 5). Volatile TPH biodegradation was more rapid in the nutrient-

Table 3. **Total Petroleum Hydrocarbon Analysis, Treatability Tests**

TIME (day)	Ammonia Media (mg/L)		50 ppm ACT (mg/L)		No Added Nutrients (mg/L)		Killed Control (mg/L))	
0	12.74	12.41	13.53	12.92	14.44	14.75	14.26	14.36
15	0.03	0.04	0.03	0.03	0.36	0.85	12.98	12.98

Figure 7A. **Petroleum fingerprints - total TPH.**

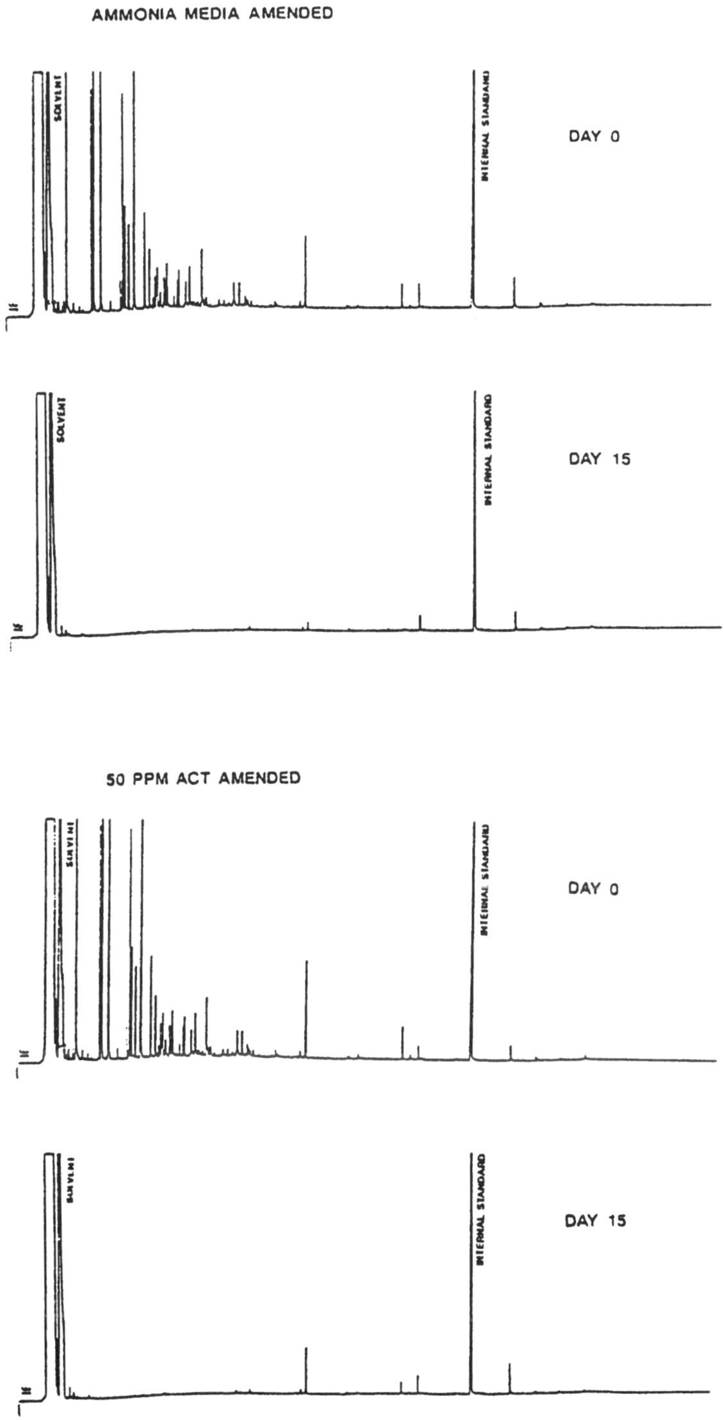

Figure 7B. **Petroleum fingerprints - total TPH.**

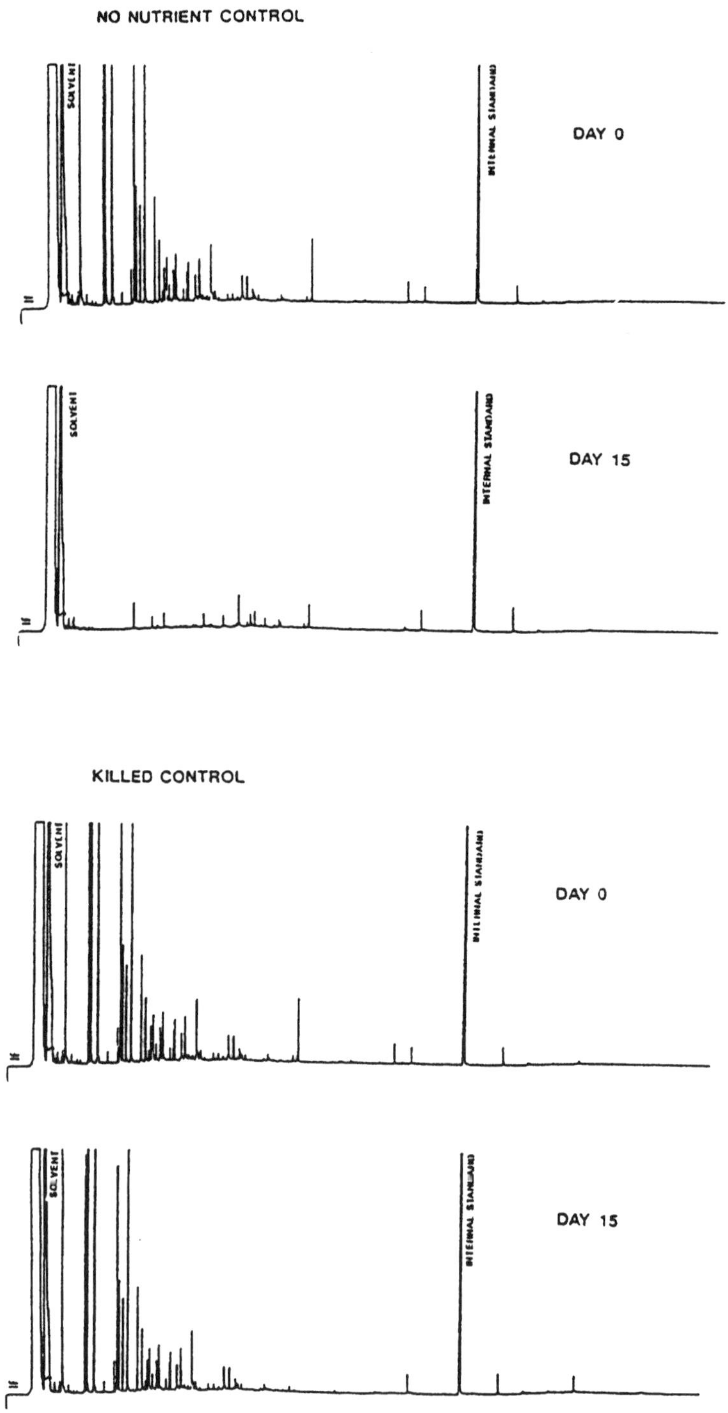

Figure 8A. **Petroleum fingerprints - volatile TPH.**

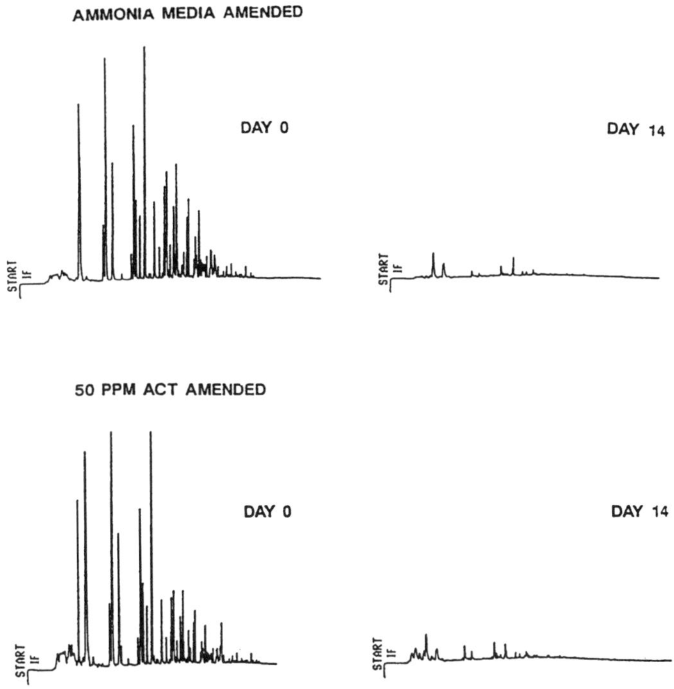

Figure 8B. **Petroleum fingerprints - volatile TPH.**

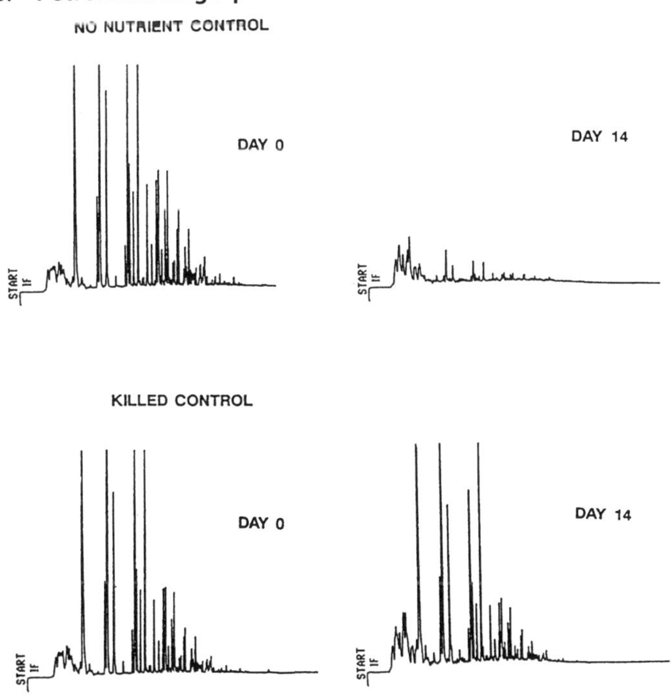

Table 4. **Total Volatile Petroleum Hydrocarbon - Headspace Analysis, Treatability Tests**

TIME (day)	Ammonia Media (mg/L)[1]	50ppm ACT (mg/L)[1]	No Added Nutrients (mg/L)[1]	Killed Control (mg/L)[1]
0	2.81	2.71	3.84	3.11
4	0.11	0.23	0.92	2.46
8	0.15	0.14	0.57	2.20
14	0.06	0.11	0.22	1.71

[1]Units are mg of volatile TPH, as measured by GC/FID per liter of headspace.

amended groundwater; however, extensive TPH removal was observed in groundwater that was not amended with nutrients. There was a 45% loss of volatile TPH measured in the killed control, which is likely due to volatilization. The TPH biodegradation rate was much greater than the volatilization rate, since the greatest extent of biodegradation occurred within the first four days (76-96% removal) versus a 21% reduction of TPH that was observed in the killed control. Carbon dioxide was also detected in microcosms, which confirms the presence of biological activity.

Table 5. **Total BTEX - Headspace Analysis, Treatability Tests**

TIME (day)	Ammonia Media (mg/L)[1]	50 ppm ACT (mg/L)[1]	No Added Nutrients (mg/L)[1]	Killed Control (mg/L)[1]
0	1.0	1.1	1.5	1.4
15	<0.01	0.02	0.04	0.82

[1]Units are mg of volatile BTEX, as measured by GC/FID, per liter of headspace.

Results from testing demonstrated that gasoline biodegradation can be enhanced in the groundwater. The original objectives of the test were met in that:

- gasoline degrading bacteria were detected in groundwater
- the indigenous microbial population is slow-growing and in one case (MR-9) considered to be stressed
- extensive biodegradation of gasoline was observed in groundwater microcosms
- nutrients (primarily nitrogen and phosphate) enhanced the rate and extent of gasoline biodegradation; however, extensive biodegradation was observed in microcosms that were only provided oxygen

These data show that extensive TPH degradation would be expected when oxygen is introduced into the subsurface as part of the air sparging process. Although the addition of nutrients increased the rate of biodegradation, significant reductions should occur when only oxygen is provided to groundwater. Therefore, ABB-ES does not recommend any system modifications that would include the addition of a nutrient delivery system at this time.

REFERENCES

Parker, J.C., Zhu, J.L, Johnson, T.G., Kremesec, V.J., and Hockman, E.L. 1994. Modeling Free Product Migration and Recovery at Hydrocarbon Spill Sites, *Ground Water* **32**, (1), January-February.

Strahler, A.N. 1966. *A Geologist's View of Cape Cod*, p. 115. Garden City, The Natural History Press.

(USGS) United States Geological Survey, Topographic 7.5 Minutes Series, Falmouth, Massachusetts Quadrangle, Scale: 1:24,000, 1972 (Photorevised 1979).

ANALYSIS

CHAPTER 14

Enhanced Mobilization of Heavy Metals Using Sequential Extraction

Elaine Hammick, Environmental Risk Limited, Bloomfield, Connecticut
Nikolaos Nikolaidis, University of Connecticut, Storrs, Connecticut
Gordon Brookman, Environmental Risk Limited, Bloomfield, Connecticut

INTRODUCTION

Sequential extraction is a procedure consisting of a series of chemical extractions performed on the same soil sample. The extractants used in the first few steps of a sequential extraction tend to be very selective in targeting specific components of the soil structure. The extractants used in the last step of the procedure are less specific, and more rigorous and destructive (Asikainen and Nikolaidis, 1993).

A sequential extraction technique was developed by Asikainen and Nikolaidis (1994) to identify the mechanisms which bind metals to soil. The procedure was designed to separate exchangeable and occluded fractions of metals (specifically chromium) bound to a contaminated soil. In addition, the occluded fraction was further broken down into organic-bound, Fe/Mn oxide-bound, and residual portions. The metals bound to the soils by an exchangeable mechanism tend to be easy to remove because the bonds are easily broken. The occluded fraction tends to be harder to remove because the metals are absorbed or complexed to the soil. The metals bound to the residual portion are impossible to remove, except by total acid digestion of the soil (Asikainen and Nikolaidis, 1994). This is because these metals have become incorporated into the permanent soil matrix.

The research of Asikainen and Nikolaidis indicates that, on average, 66% of chromium is extractable. Of this amount, 25% is exchangeable, 11% is bound to organic matter, and 30% is bound to iron and manganese oxide surfaces. The remaining 34% of chromium is tightly bound to the soil matrix (Asikainen and Nikolaidis, 1994).

The sequential extraction procedure was developed on a laboratory scale. This chapter describes a study performed to evaluate the process further through use of column scale tests. If the sequential extraction procedure (or a modification of the procedure) can be shown to be successful in mobilizing chromium and other metals such as lead, zinc, copper, and nickel from a column of soil, it may be practical to apply the procedure in the field as an in-situ soil flushing procedure to enhance remediation. The results of these tests can then be analyzed to assess whether the procedure is practical for application in the field on a pilot scale.

The goal of this experiment was to reduce the concentration of metals remaining in the soil, and to enhance mobilization of leachable metals from the soil.

COLLECTION OF CONTAMINATED SOIL

In many other examinations of solutions which have been used for soil washing or soil flushing, soil samples spiked with known concentrations of heavy metals have been used for experimentation (Espisito et al., 1988). In this experiment, actual field-contaminated soil samples were used for experimentation.

The work of Nikolaidis et al.(1994) and Asikainen and Nikolaidis (1993; 1994) suggest that both fast and slow (kinetically controlled) retardation processes are significant. The fast processes are associated with the surface of the sediments, and include the sorption of chromate on iron oxide cations and the binding of chromium by soil organic matter. The slow processes include the diffusion of chromium into the soil matrix and its subsequent precipitation and coprecipitation in the mineral structure, and the intraparticle diffusion in the secondary porosity. The slow processes are primarily a function of time, which indicates that the longer the soil is exposed to contamination, the higher the total accumulation of metals in the soil matrix. The reversibility and rates of leaching of chromium and other metals that have been diffused in the soil matrix have not been determined yet. However, empirical observations and back of the envelope calculations indicate that the rates are slow and the reactions are in part irreversible. This "aging" process cannot be reproduced in laboratory time scales of experimentation. Field-contaminated soil samples were used in this study, so the effect of aging was taken into consideration. It is believed that this yielded more realistic results when compared to other bench-scale leaching experiments.

The soil used in the experiments described in this chapter was collected from a metal plating facility located in Putnam, Connecticut. Current operations at the facility include chrome and nickel plating. Past operations also included cadmium, copper, and zinc plating. The samples were collected from a crawl space located beneath the former manufacturing building, where plating tanks had been installed above bare ground many years ago.

The collected soil was returned to the laboratory, where it was sifted using a 7mm screen. The soil was wrapped in plastic and kept in a plastic tote to keep it at a field moisture level. The soil was not dried for use during the experiment.

Two samples of soil were analyzed for total metals using EPA Method 3050, and for leachable metals using the Toxicity Characteristic Leaching Procedure (TCLP) method. The extractants of both methods were analyzed for chromium, copper, lead, nickel, and zinc. The averaged results of this analysis on the untreated soil are summarized in Tables 1 and 2.

Table 1. **Summary Averaged Results, Total Metals Analysis of Soil**

Parameter	Untreated	Control	Exp. #1	Exp. #2	Exp. #3	Exp. #4	Exp.#5
Chromium mg/kg	32,000	36,350	28,910	29,850	26,650	32,850	28,730
Lead mg/kg	1,455	1,765	1,615	1,455	1,170	1,825	945
Copper mg/kg	381	420	290	279	203	384	189
Nickel mg/kg	422	480	312	373	195	373	204
Zinc mg/kg	492	427	277	142	181	560	182

Table 2. **Summary Averaged Results TCLP Metals Analysis of Soil**

Parameter	Untreated	Control	Exp.#1	Exp. #2	Exp. #3	Exp.#4	Exp.#5
Chromium mg/L	175.5	50.3	7.14	5.01	30.6	17.2	8.03
Lead mg/L	ND<0.05	ND<0.05	ND<0.05	0.10	0.06	ND<0.05	0.06
Copper mg/L	1.73	0.67	0.22	0.21	0.17	0.11	0.10
Nickel mg/L	2.80	0.29	0.76	0.33	0.17	0.18	0.54
Zinc mg/L	3.35	0.70	0.52	1.27	0.39	0.31	0.30

An analysis of the metal concentrations in background soils at the Putnam site has been conducted by Scherer (1994). Table 3 lists the average metal concentrations found in an uncontaminated area of the Putnam site for chromium, lead, copper, and nickel. These results were determined by EPA Method 3050 analysis of soil from an uncontaminated portion of the site. Comparing these results to the results in Table 1 of the total metals content of the contaminated

Table 3. **Average Total Metal Concentration in Background Soil at Putnam Site**

Parameters	Average Concentration (mg/kg)
Cadmium	4
Chromium	32
Lead	19
Copper	14
Nickel	25
Zinc	25

Source: Sherer, 1994.

soil, it is obvious that the soil is contaminated with chromium, lead, copper, nickel, and zinc from the industrial activities taking place at the site.

The results of the total metals analysis of the contaminated soil and the TCLP analysis were normalized so that the two analyses could be compared. The concentrations were normalized to micrograms of contaminant per gram of soil. The normalized results are presented in Table 4.

Table 4. **Averaged, Normalized TCLP and Total Metals Data for Untreated Soil**

Constituent	Total Metals (μg/g)	TCLP Metals (μg/g)	TCLP/Total %
Chromium	32,000	3510	11%
Lead	1455	1	0.1%
Copper	382	34.5	9%
Nickel	422	56	13.3%
Zinc	492	67	13.6%

The normalized results indicate that the TCLP analysis only detects 11% of the chromium, less than 1% of the lead, 9% of the copper, 13% of the nickel, and 14% of the zinc. The TCLP test measures only a small fraction of the metals which are actually present in the sample. The TCLP is designed to indicate the constituents which would leach out of the waste after disposal in a typical landfill. A weak organic acid solution (acetic acid) is used for the TCLP test.

LABORATORY-SCALE SEQUENTIAL EXTRACTION

Laboratory-scale sequential extraction experiments were performed to assess how much of each contaminant can potentially be removed in each step of the extraction. The results can also be used to assess the completeness of each step of the extraction in the soil columns. The original sequential extraction sequence designed by Nikolaos Nikolaidis and Jay Asikainen (1993; 1994) was used. The experiment is as follows:

1. Measure 2 grams of soil (dry weight) in a 40 mL serum bottle.
2. Remove exchangeable metals:
   a. Add 40 mL 0.1 M KH_2PO_4/K_2HPO_4 (1/1) solution (phosphate buffer solution).
   b. Agitate for 16 hours.
   c. Centrifuge at 2000 to 3000 rpm for 30 minutes to 1 hour.
   d. Withdraw supernatant with pipet and store for analysis.
   e. Wash residue with 10 mL distilled, deionized water.
   f. Repeat centrifuge.
   g. Remove supernatant and discard.

3. Remove organic bound metals:
    a. Add 40 mL 0.1 M $Na_4P_2O_7$ (sodium pyrophosphate) at pH 10.
    b. Agitate for 16 hours.
    c. Repeat steps 2c-2g.
4. Remove iron/manganese oxide bound metals.
    a. Add 40 mL 0.1 M $N_2HOH{\bullet}HCl$ (hydroxylamine hydrochloride) in 25% acetic acid (vol/vol).
    b. Agitate 6 hr at 95°C 3±C in a hot water bath.
    c. Repeat steps 2c-2g.
5. Analyze residual using EPA Method 3050.

The results of the laboratory-scale sequential extraction were normalized so that the concentrations in the liquid extraction and residual phase could be compared. The results were normalized to micrograms of constituent per gram of soil. Table 5 presents the averaged, normalized results and indicates the percent of the metals bound in each phase.

Table 5. **Normalized, Averaged Sequential Extraction Data - Laboratory Scale Experiment**

	Chromium µg/g	Lead µg/g	Copper µg/g	Nickel µg/g	Zinc µg/g
Exchangeable	9650	10	10	45	56
Organic Bound	1104	91	85	81	93
Oxide Bound	1716	319	111	140	135
Residual	17,645	241	100	103	111
Total	30,115	661	306	369	395
% Exchangeable	32%	1.5%	3.3%	12.2%	14.2%
% Organic Bound	3.7%	13.8%	27.7%	21.9%	23.6%
% Oxide Bound	5.7%	48.2%	36.3%	37.9%	34.2%
% Residual	58.6%	36.5%	32.7%	28.0%	28.0%
Total	100%	100%	100%	100%	100%

The normalized results indicate that most of the chromium in the soil is either exchangeable or is held in the residual phase. A small amount (approximately 10%) of the chromium in the soil is bound to the organics or oxides in the soil. The majority of the lead is bound to the oxide phase. Very little of the lead is exchangeable. This may be the reason why little lead is extracted through the TCLP test. Copper is fairly evenly bound to the organics, oxides, or to the residual in the soil. Nickel and zinc have very similar profiles and are fairly evenly spread between all the phases, with slightly less held in the exchangeable phase.

A comparison was made between the total amount of metals extracted in the laboratory-scale sequential extraction and the total metals analysis. This comparison is presented in Table 6. The comparison indicates that with the ex-

Table 6. **Normalized, Averaged Total Metals Analysis vs. Laboratory Scale Sequential Extraction**

Constituent	Total Metal Analysis (μg/g)	Total Removed by Sequential Extraction (μg/g)	% Sequential Extraction/ Total Metals
Chromium	32,000	30,115	94.1%
Lead	1,455	661	45.4%
Copper	382	306	80.1%
Nickel	422	369	87.4%
Zinc	492	395	80.3%

ception of lead, the sequential extraction was efficient in extracting all of the metals in the sample. The sequential extraction only accounted for 50% of the lead present in the sample. This may indicate that the sequential extraction technique tested is not as efficient in removing lead as the other metals.

COLUMN CONSTRUCTION

The soil columns were constructed using 27.5-inch-long, 2.75-inch diameter glass tubes designed for use in chromatography. The columns were equipped with a porous glass frit at the base which provided support for the soil column. The bottom of the column was equipped with a Teflon stopcock to allow control of flow from the column. The columns were cleaned using a laboratory-grade soap solution, followed by a hydrochloric acid rinse and a distilled water rinse.

The filtering medium chosen for the base of the column was a 2-inch layer of packed glass wool, followed by a 1-inch layer of 3-mm diameter glass beads. A stainless steel screen was used to separate the wool from the beads and the beads from the soil. This helped to make the packing of the glass beads more uniform and helped to prevent short-circuiting of the leachate through the beads to the glass wool.

The soil was packed on top of the last stainless steel screen. The soil was packed in 1-inch lifts to an approximate field density of 100 lb per cubic foot. (It should be noted that this is not necessarily the field density at the Putnam site, but a typical field density for a sandy loamy soil based on information from the U.S. Department of Agriculture, Soil Conservation Service. The actual, in-place density at the Putnam site was not measured.) Each 1-inch lift contained 0.68 lb or 154.2 grams of soil. Twelve inches of soil was placed into each column. Total soil treated in the column was 1.854 kilograms per experiment.

EXPERIMENTAL PROCEDURES

Five column study experiments and one control experiment were performed to assess the effectiveness of the sequential extraction procedure on the columns of soil. In each experiment performed, chemicals were applied to the top of the column through a 1/4-inch diameter tube connected to a 20-liter container. The water was applied using a gravity feed, at a rate as close to the maximum rate of outflow from the column as possible. Each experiment was performed in duplicate. Only the average of the results of the two experiments are presented in this chapter.

In each case, after leaching was complete, a sample of the soil in the column was collected for analysis by EPA Method 3050 and TCLP for chromium, lead, copper, nickel, and zinc. The leachate was analyzed only for chromium as an indicator parameter.

Control Experiment

The first experiment conducted was the control experiment, which was essentially a leaching experiment using distilled deionized water. Distilled, deionized water was applied to the column for 44 hours. Six liters of water passed through Column A, and seven liters of water passed through Column B. The columns were allowed to drain for several hours prior to sample collection. A summary of the experimental procedure is included in Table 7.

Samples were collected of both the soil and the leachate. Average leachate generation rates and the chromium concentration in the leachate are summarized in Table 8. The average results of the total metals analysis in the soil and the TCLP analysis of the soil are summarized in Table 1 and 2.

Experiment 1 - Column Scale Sequential Extraction

Experiment 1 consisted of a version of the laboratory-scale sequential extraction study which was modified for use in the columns. The experiment was performed as follows: 0.1 M KH_2PO_4/K_2HPO_4 (1/1) (phosphate buffer) solution was applied to the column for 16 hours. A sample of leachate was collected from the collection bucket after 16 hours. The column was allowed to sit for 8 hours, then it was rinsed with distilled water for 2 hours. In the second step of the experiment, 0.1 M $Na_4P_2O_7$ (sodium pyrophosphate) solution was applied to the column for 16 hours. A sample of leachate was collected from the collection bucket after 16 hours. The column was allowed to sit for 8 hours, then it was rinsed with distilled water for 2 hours. 0.1 M $N_2HOH{\cdot}HCL$ (hydroxylamine hydrochloride) in 25% acetic acid (vol/vol) was applied for 6 hours. A sample of leachate was collected from the collection bucket after 6 hours. The column was allowed to sit for 18 hours, then it was rinsed with distilled water for 2 hours. A summary of the experimental procedure is provided in Table 7.

This method was developed as a modification to the laboratory scale experiment. The column was allowed to sit for 8-18 hours to mimic the centrifuge used in the laboratory scale experiments. The duration of the sit time was chosen to allow the changes from one solution to the next to take place at con-

Table 7. **Experimental Summary Duration of Application (Hours)**

Chemical	Control	Experiment 1	Experiment 2	Experiment 3	Experiment 4	Experiment 5
0.1M KH_2PO_4/K_2HPO_4		16	54	54	54	54
Drainage Time		8	16	16	16	16
Distilled Water	44	2	2	2	2	2
0.1MNaP_2O_7		16	54	54	54	
Drainage Time		8	16	16	16	
Distilled Water		2	2	2	2	
0.1M $N_2HOH{\cdot}HCL$ in 25% Acetic Acid		6	22			
Drainage Time		18	24			
Distilled Water		2	2			
Artificial Acid Rain					54	
0.1M NaP_2O_7						14
5% Acetic Acid Solution						10
0.1M NaP_2O_7						14
5% Acetic Acid Solution						10
0.1M NaP_2O_7						14
5% Acetic Acid Solution						10
0.1M NaP_2O_7						14
5% Acetic Acid Solution						10
Distilled Water						16

venient times, preferably during working hours. The 2 hours of flushing with distilled water typically represented one pore volume of water to flush out the previous solution before beginning the next.

After the experiment was complete, the column was disassembled and samples were collected of both the soil and the leachate. Average leachate generation rates and the chromium concentration in the leachate are summarized in Table 8. The average results of the total metals analysis in the soil and the TCLP analysis of the soil are summarized in Table 1 and 2.

Table 8. **Leachate Generation (Average) Experiment 1-4**

Chemical	Duration	Leachate Generated	Chromium Concentration (mg/L)
Control			
Distilled, Deionized Water	44	6.5	1,470
Experiment 1			
0.1M KH_2PO_4/K_2HPO_4	16	2.37	10,560
0.1M NaP_2O_7	16	1.75	1,745
0.1M $N_2HOH{\cdot}HCL$ in 25% Acetic	6	1.37	1,160
Experiment 2			
0.1M KH_2PO_4/K_2HPO_4	54	3.5	5,605
0.1M NaP_2O_7	54	3	808
0.1M $N_2HOH{\cdot}HCL$ in 25% Acetic	22	1.5	700
Experiment 3			
0.1M KH_2PO_4/K_2HPO_4	54	2.17	6,160
0.1M NaP_2O_7	54	1.15	1,124
Experiment 4			
0.1M KH_2PO_4/K_2HPO_4	54	3.12	3,335
0.1M NaP_2O_7	54	2.4	531
Artificial Acid Rain	54	1	627

Experiment 2 - Long-Term Sequential Extraction

Experiment 2 consisted of a modified version of the sequential extraction that was applied in Experiment 1. In the laboratory-scale sequential extraction, the solution to soil ratio was high at 20 mL of solution to 1 gram of soil. In Experiment 1, the solution to soil ratio was approximately 1 mL of solution to 1 gram of soil. Because of the large difference between the laboratory-scale experiment and Experiment 1, an attempt was made to increase the solution to soil ratio by increasing the duration of the experiment. It was felt that increasing the duration of the experiment would also increase the contact time between the solution and the soil, and would increase removal efficiencies. Experiment 2 was conducted as follows: 0.1 M KH_2PO_4/K_2HPO_4 (1/1) solution was applied to the column for 54 hours. A sample of leachate was collected from the collection bucket after 54 hours. The column was allowed to sit for 16 hours, then it was rinsed with distilled water for 2 hours. 0.1 M $Na_4P_2O_7$ solution was applied to the column for 54 hours. A sample of leachate was collected from the collection bucket after 54 hours. The column was allowed to sit for 16 hours, then it was rinsed with distilled water for 2 hours. 0.1 M $N_2HOH{\bullet}HCL$ in 25% acetic acid (vol/vol) was applied to the column for 22 hours. A sample of leachate was collected from the collection bucket after 22 hours. The column was allowed to sit for 24 hours, then it was rinsed with distilled water for 2 hours.

After the experiment was complete, the column was disassembled and samples were collected of both the soil and the leachate. Average leachate generation rates and the chromium concentration in the leachate are summarized in Table 8. The average results of the total metals analysis in the soil and the TCLP analysis of the soil are summarized in Tables 1 and 2.

Experiment 3 - Modified Sequential Extraction

In the sequential extraction procedure, the last step uses the most hazardous chemicals. Both hydroxylamine hydrochloride and acetic acid are moderately hazardous chemicals. Hydroxylamine hydrochloride can cause irritation to the respiratory tract, and is harmful if swallowed. It is also a skin irritant. Acetic acid is corrosive and can cause damage to the respiratory tract if vapors are inhaled. At high concentrations, it is harmful if swallowed, and can cause burns to the skin.

A review of the results of the laboratory-scale sequential extraction experiment indicates 30 to 35% of the chromium can be easily removed from the soil by the phosphate buffers and the sodium pyrophosphate in the first two steps of the experiment. The oxide bond is a much stronger bond between the metal and the soil. It was theorized that if the exchangeable and organic portion of the metals were removed, the oxide-bound and residual metals may be allowed to remain on the soil because of the strength of those bonds. This would allow application of only the first two leaching solutions in a field situation. This would make the application safer, as these solutions are not very hazardous.

In the modified sequential extraction, only the first two extraction steps described above in Experiment 2 were applied to the column. First, 0.1 M KH_2PO_4/K_2HPO_4 (1/1) solution was applied to the column for 54 hours. A sample of leachate was collected from the collection bucket after 54 hours. The column was allowed to sit for 16 hours, then it was rinsed with distilled water for 2 hours. 0.1 M $Na_4P_2O_7$ solution was applied to the column for 54 hours. A sample of leachate was collected from the collection bucket after 54 hours. The column was allowed to sit for 16 hours, then it was rinsed with distilled water for 2 hours.

After the experiment was complete, the column was disassembled and samples were collected of both the soil and the leachate. Average leachate generation rates and the chromium concentration in the leachate are summarized in Table 8. The average results of the total metals analysis in the soil and the TCLP analysis of the soil are summarized in Tables 1 and 2.

Experiment 4 - Modified Sequential Extraction

In the fourth experiment, the sequential extraction was again modified. The purpose of the third experiment was to see if use of the first two sequential extraction chemicals would be sufficient to remove the metals which were not tightly bound to the soil. The results of the small-scale sequential extraction indicated that 30 to 35% of the metals were removed from the soil using the first two extraction steps. The fourth experiment was then designed to assess whether the metals still remaining on the soils after the first two extraction steps would still be mobile, or be primarily immobile and fixed to the soil. To perform this experiment, the third leaching step was replaced with an acid rain wash. The acid rain was developed by adding sulfuric and nitric acid, at a ratio of 60% sulfuric acid to 40% nitric acid, to distilled, deionized water to a pH of 4. The purpose of the acid rain wash is not to remove metals from the soil, but to determine if the metals remaining on the soil after the first two steps are immobile, or are leachable by acid rain.

The experiment was performed as follows: As in Experiment 3, 0.1 M KH_2PO_4/K_2HPO_4 (1/1) solution was applied to the column for 54 hours. A sample of leachate was collected from the collection bucket after 54 hours. The column was allowed to sit for 16 hours, then it was rinsed with distilled water for 2 hours. 0.1 M $Na_4P_2O_7$ solution was applied to the column for 54 hours. A sample of leachate was collected from the collection bucket after 54 hours. The column was allowed to sit for 16 hours, then it was rinsed with distilled water for 2 hours. A solution of artificial acid rain was then applied for 54 hours.

A drop in flow was observed during the application of the artificial rain solution. It was theorized that metals being mobilized by the sodium pyrophosphate were precipitating out of solution and clogging the pores in the soil column. In the first two experiments, it was theorized that the acetic acid in the final step dissolved the precipitate and removed it from the column. Without the acetic acid step, precipitated chromium remained in the column in Experiments 3 and 4. It is believed that this affected the flow rate and the concentration of chromium in the leachate. It also caused more chromium than was theorized to remain in the soil after leaching.

After the experiment was complete, the column was disassembled and samples were collected of both the soil and the leachate. Average leachate generation rates and the chromium concentration in the leachate are summarized in Table 8. The average results of the total metals analysis in the soil and the TCLP analysis of the soil are summarized in Tables 1 and 2.

Experiment 5 - Modified Sequential Extraction

Experiment 5 was designed to overcome the problem with the precipitation of metals during the application of the sodium pyrophosphate. The first step of the experiment remained the application of the phosphate buffer solution for 54 hours. A longer distilled water rinse was added to wash out any metals that had bonds that were weakened by the phosphate buffer. The waiting time was eliminated to minimize precipitate formation.

To minimize precipitation during the application of the sodium pyrophosphate and to dissolve any precipitate that did occur, a two-step process was devised. The sodium pyrophosphate step was broken from a single long application into four shorter applications. Each sodium pyrophosphate step was followed by the application of a 5% acetic acid solution. A 5% acetic acid solution was selected because it is approximately the same acetic acid content as vinegar, which is not a particularly hazardous substance. The purpose of the acetic acid solution was to dissolve the precipitate that occurred during the pyrophosphate application. The acetic acid application was also for a short duration. The four rounds of sodium pyrophosphate/acetic acid application were then followed by a longer distilled water rinse. The wait time was again eliminated to minimize precipitation formation.

The experiment was performed as follows: 0.1 M KH_2PO_4/K_2HPO_4 (1/1) solution was applied to the column for 54 hours. Distilled, deionized water was applied for 16 hours. A sample was collected of the leachate for analysis at this point. Sodium pyrophosphate was then applied for 14 hours. Acetic acid solution was applied for 10 hours. This was repeated three additional times, for a total of four applications. At the completion of this step, distilled, deionized water was applied for 16 hours. A sample of the combined leachate from the four applications and the distilled water application was collected at this point. The leachate generated and the concentrations of constituents detected in the leachate is summarized for both experimental columns in Table 9.

After the experiment was complete, the column was disassembled and samples were collected of the soil. Average leachate generation rates and the chromium concentration in the leachate are summarized in Table 9. The average results of the total metals analysis in the soil and the TCLP analysis of the soil are summarized in Table 10. After the leaching was completed, the samples of the leachate and the treated soil were collected.

Table 9. **Leachate Generation and Constituent Concentration, Experiment 5**

Experiment	Phase	Duration of Application (hrs)	Leachate Generated (liters)	Chromium Concentration (mg/L)	Lead Concentration (mg/L)	Copper Concentration (mg/L)	Nickel Concentration (mg/L)	Zinc Concentration (mg/L)
Experiment 5A	0.1 M KH_2PO_4/K_2HPO_4	54	3.1					
	Distilled Water	16	0.4					
	Combined Leachate	60	3.5	2700	ND<0.05	14.1	31.6	28.0
	0.1 M NaP_2O_7	14	0.6					
	5% Acetic Acid	10	0.25					
	0.1 M NaP_2O_7	14	0.23					
	5% Acetic Acid	10	0.31					
	0.1 M NaP_2O_7	14	0.8					
	5% Acetic Acid	10	0.35					
	0.1 M NaP_2O_7	14	0.8					
	5% Acetic Acid	10	0.55					
	Distilled Water	16	0.9					
	Combined Leachate	112	4.79	470	0.48	23.7	1.58	2.07
Experiment 5B	0.1 M KH_2PO_4/K_2HPO_4	54	3.3					
	Distilled Water	16	0.65					
	Combined Leachate	60	3.95	2650	ND<0.05	6.61	17.7	16.9
	0.1 M NaP_2O_7	14	1					
	5% Acetic Acid	10	0.35					
	0.1 M NaP_2O_7	14	0.55					
	5% Acetic Acid	10	0.35					
	0.1 M NaP_2O_7	14	0.40					
	5% Acetic Acid	10	0.35					
	0.1 M NaP_2O_7	14	0.25					
	5% Acetic Acid	10	0.25					
	Distilled Water	16	1.7					
	Combined Leachate	112	5.20	434	1.22	9.05	0.85	1.87

Table 10. **Leachate Generation Normalized Results, Experiment 5**

	PO_4^{-3} Buffer Step (Exchangeable)		NaP_2O_7/AA Step (Organic)	
	5A	5B	5A	5B
Chromium, µg/g	5107	5657	1217	1220
Lead, µg/g	0.09	0.1	1.2	3.4
Copper, µg/g	27	14.1	6.1	25.4
Nickel, µg/g	60	37.8	4.1	2.4
Zinc, µg/g	53	36.1	5.4	5.1

RESULTS AND DISCUSSION

A summary table which presents the averaged results for total metals analysis of the soil is presented in Table 1. Overall, the total metals results appear to be a poor indicator of the amount of metals that were removed from the soil. The total metals analysis showed an increase in metal content after treatment with distilled water in the control experiment. Based on the results of the leachate for the indicator parameter chromium, it is known that metals were removed during the control experiment. The total metals analysis does not reflect that removal. This is due to the soil variability and inherent analytical errors in the EPA Method 3050 analysis.

The leaching experiments showed some reduction in metal content through total metals analysis; however, the results did not follow expected patterns. The total metals content showed the lowest concentration after Experiment 3. This experiment was the same as Experiment 2, except the hydroxylamine hydrochloride/acetic acid leaching step was eliminated. It would be expected that Experiment 3 would have higher levels of total metals than Experiment 2. Experiment 4 had elevated levels of metals compared to Experiment 2, and while Experiment 4 would be expected to have slightly elevated levels compared to Experiment 2, the levels in Experiment 4 are higher than predicted.

These results show some of the problems with the use of total metals analysis. Total metals analysis uses only two grams of soil, which are digested in strong acid solution to extract the metals. The small amount of soil which is used for the sample causes the results to vary, depending on the spatial variability of the sample. Although every effort was made to homogenize the soil prior to packing in the column, and to mix the soil from the column after treatment prior to sample collection, it is still believed that spatial variability had an impact on the total metals results.

Another potential problem which may have impacted the total metals analysis are the dilutions required for the total metals analysis. Because of the high levels of metals found in the soils, the commercial laboratory that performed the analysis had to dilute the samples after extraction to perform the analysis. This also adds error into the results. Because the initial sample is only two grams, any error introduced through dilution would be magnified.

It is therefore believed that the total metals analyses are not a good indicator of the results of these experiments. The results do not consistently show that metals were removed, while it is known that metals were removed because chromium (as an indicator parameter) was detected at high levels in the leachate. The averaged results of the TCLP analysis are included in Table 2. The TCLP analysis seems to be a better indicator of the results of the experiments. The TCLP uses 100 grams of soil which are leached using a weak organic acid solution. The use of 100 grams causes some averaging of the results, which removes some of the spatial variability. A problem with the TCLP analysis is that the lead present in the soil does not appear to be leachable by the TCLP extraction. The levels of lead detected are either below detection limits, or in a few cases, just above detection limits. Therefore, the effectiveness of the leaching experiments on the lead cannot be judged by the TCLP results.

The untreated samples, as expected, had the highest levels of metals based on the TCLP analysis. The control experiment where distilled, deionized water was applied to the column caused a 72% reduction in chromium, 60% reduction in copper, 88% reduction in nickel, and 78% reduction in zinc. Further reduction was noted after chemical treatment in Experiment 1. The chromium, copper, and zinc content all were reduced from the control and untreated sampled. In Experiment 1, nickel showed a slight increase when compared to the control experiment. The cause of this slight increase is not known; however, the reduction from the untreated sample value for nickel content is clear.

In Experiment 2, the duration of the experiment was increased over that of Experiment 1. In Experiment 2, chromium concentration was reduced even more when compared to Experiment 1. Experiment 2B contained a level of chromium which was below the level that would cause the soil to be considered characteristically hazardous. Lead was detected for the first time, but at very low levels. It is possible that application of the treatment chemicals weakened the lead bonds sufficiently for the lead to be detected during the TCLP analysis. The lead was still detected by the TCLP at very low levels. Copper showed a reduction similar to that detected in Experiment 1. Nickel showed a level which was approximately the same as that detected through the control experiment. Zinc showed a slight increase. This increase was consistent for both samples. Like the lead, this may be the result of bonds which were weakened but not broken during the application of the leaching chemicals. The bonds were then broken during the TCLP analysis, and the zinc level appeared higher than in the control experiment and Experiment 1, where chemical application durations were shorter.

In Experiment 3, the longer duration of chemical application used in Experiment 2 was applied; however, the third step, application of the hydrolyamine hydrochloride/acetic acid step, was eliminated. Experiment 3 showed elevated levels of chromium. The chromium is believed to be elevated because the hydroxylamine hydrochloride/acetic acid step was dropped in this experiment. It is believed that the second step, the application of sodium pyrophosphate, caused chromium to be removed from the soil; however, the chromium formed a precipitate which blocked the pores in the soil. It is believed that this precipitation caused the reduced flow rates detected in step 2 throughout all of the experiments. In Experiments 1 and 2, the hydroxylamine hydrochloride/acetic acid mixture seemed to dissolve the precipitate, improving flow and removing the precipitated chromium. In Experiment 3, the third step of the process was eliminated, so the precipitate remained in the soil. This caused chromium levels to be elevated after this experiment. The precipitate did not seem to impact the other metals. Levels of these metals remained reduced.

In the fourth experiment, the third step of the experiment was the application of an acid rain solution. The purpose was to see if contaminant removal during the first two steps was sufficient so that metals would not continue to leach from the soil after treatment. It was this experiment that indicated that it may be precipitates forming during step 2 that could be causing a problem with the experiments. Because the acid rain solution is such a weak acid solution, it was not acidic enough to dissolve the precipitate that formed during the application of the sodium pyrophosphate. This caused a reduction in flow rate from step 2 (pyrophosphate) to step 3 (acid rain). Because the acid rain did not dissolve all of the chromium precipitate, the level of chromium remaining in the soil was elevated. It was reduced from Experiment 3, however. The levels of lead, copper, nickel, and zinc all remained low—very similar to those detected in Experiment 3.

In the fifth experiment, the use of the alternating cycles of sodium pyrophosphate and dilute acetic acid solution seemed to work to dissolve the precipitate formed by the sodium pyrophosphate. Total metals levels showed an overall downward trend for all of the metals. Very low levels were achieved using the TCLP results, and although a 95% reduction in the amount of chrome detected by the TCLP test was achieved, the final level was still above the level by which the sample would be determined to be characteristically hazardous. It should be noted that the initial chromium concentrations found in the soil were extremely high. If the initial concentration were slightly lower, a 95% reduction would have easily brought the TCLP value for chromium below 5 mg/L.

A separate analysis was performed on the leachate from Experiment 5. Table 10 indicates the results of the analysis of the leachate generated, which has been normalized to represent micrograms of metals mobilized per gram of soil.

Table 11 compares the results of the normalized leachate concentrations from Experiment 5 to the total metal concentrations. Table 12 compares the percent of metals removed in the exchangeable and the organic step of Experiment 5 to the results of the laboratory-scale sequential experiment results.

Table 12 indicates that in the exchangeable step of Experiment 5, 50% of the chromium was removed in the exchangeable step that was available for removal. None of the lead was removed, and all or nearly all of the copper, nickel, and zinc was removed.

In the organic-bound step, all of the chromium available for removal was removed. Almost none of the lead, nickel, or zinc was removed. Half of the organically bound copper was removed. This indicates that the alternating sodium

Table 11. **Experiment 5 Average, Normalized Extractable Metal Concentration Compared to Total Metal Concentration**

	Exp. 5 PO_4^{-3} Buffer (µg/g)	% of Total	Exp. 5 NaP_2O_7/AA (µg/g)	% of Total
Chromium	5382	16.8%	1218.5	3.8%
Lead	0.1	0	2.3	0.2%
Copper	20.6	5.4%	43.2	11.3%
Nickel	48.9	11.6%	3.3	0.8%
Zinc	44.6	9.1%	5.3	1.0%

Table 12. **Percent Extractable Metal Concentration Compared to Sequential Analysis, Experiment 5**

	Exchangeable		Organic Bound	
	Exp. 5	SEA	Exp. 5	SEA
Chromium	16.8%	32%	3.8%	3.7%
Lead	0%	1.5%	0.2%	13.8%
Copper	5.4%	3.3%	11.3%	27.8%
Nickel	11.6%	12.2%	0.8%	21.9%
Zinc	10.1%	14.2%	1.2%	23.6%

pyrophosphate/acetic acid step designed for the removal of the organic fraction is extremely effective in the removal of the chromium. It is not effective in the removal of lead, nickel, or zinc.

Overall, the following conclusions can made based on the experiments that were performed. The total metals analysis is strongly affected by spatial variability in the sample. Although every effort was made to homogenize the soil both prior to packing into the column and after removal from the column, the analytical results showed wide variations. Consequently, total metals analysis should be used cautiously when used to represent even a previously homogenized soil, and with extreme caution when used to represent a large amount of soil. It is recommended that multiple samples be collected and the results averaged.

The method developed on the laboratory scale by Asikainen and Nikolaidis (1993) can be modified for use to mobilize metals on the bench scale. Comparison of the Control experiment to the results of Experiment 1 indicate that the method definitely increases the mobilization of chromium from the soil. Examination of the results from Experiment 5 also confirms that with the exception of lead, metals are being mobilized.

One drawback of the method was determined in Experiments 3 and 4 when it was discovered that the sodium pyrophosphate in Step 2 was causing the reprecipitation of metals in the column which were causing the pores of the soil to be blocked with precipitate. This was successfully overcome in Experiment 5, where relatively nonhazardous chemicals were used to achieve good results. The cyclical application of the sodium pyrophosphate and the dilute acetic acid solution in Experiment 5 seemed to reduce or eliminate the formation of the precipitate.

The experimental method that is recommended for further study in terms of mobilization of heavy metals is the one developed in Experiment 5. Experiment 5 achieved an 95% reduction in chromium, 94% reduction in copper, 78% reduction in nickel, and 90% reduction in zinc, based on the TCLP analysis of the soil. The results of the analysis of the leachate from this experiment indicated that chromium, nickel, copper, and zinc were mobilized. Lead was not mobilized through this procedure. As previously discussed, the total metal analysis was less conclusive; however, the results also showed a general downward trend. These results were achieved using two relatively nonhazardous phosphate solutions and a dilute acetic acid solution. The acetic acid solution was diluted to approximately the same acid concentration as vinegar. This method achieved positive results while using relatively nonhazardous chemicals. The chemicals used also do not include chelating agents, so that the leachate can be easily treated in a standard industrial wastewater treatment system.

Another area for future experimentation could include use of the sodium pyrophosphate as an agent to reduce permeability in the aquifer through precipitation of certain metals. The significant lowering of the permeability of the soil could be effective in creating a barrier to contaminant transport.

ACKNOWLEDGMENTS

This research was supported by a Department of Energy (DOE) Small Business Innovative Research Grant No. DEFG0594ER81666. Such support does not constitute an endorsement by the DOE of the views expressed in this chapter.

The authors would like to thank Kim Walsh, the laboratory technician who performed most of the experiments; Whitby Ellsworth of National Chromium, Inc., Putnam, Connecticut; and Maurice Hamel of the Connecticut Department of Environmental Protection for their assistance and support.

REFERENCES

Asikainen, J.M., and Nikolaidis, N.P. 1993. "Redox Capacity Analysis and Heavy Metal Sequential Extraction for Site Assessment of Subsurface Contamination." Technical Report ERI-93.05 Environmental Research Institute, University of Connecticut, May 1993.

Asikainen, J., and Nikolaidis, N.P. 1994. "Sequential Extraction of Chromium from Contaminated Aquifer Sediments." *Ground Water Monitoring and Remediation*, Spring, 1994.

Espisito, M.P., Loche, B.B., Greber, J., and Travis, R.P. 1988. "Superfund Standard Analytical Reference Matrix Preparation and Results of Physical Soils Washing Experiments." PEI Associates.

Nikolaidis, N.P., G.A. Robbins, M. Sherer, B. McAninch, G. Binkhorst, J. M. Asikainen, and S. Suib. 1994. "Vertical Distribution and Partitioning of Chromium in a Glacio-fluvial Aquifer." *Ground Water Monitoring and Remediation*, Summer, 1994.

Scherer, M.M., 1994. " Equilibrium Studies of Chromate Adsorption on Glacio Fluvial Aquifer Sediments," M.S. Thesis, University of Connecticut.

CHAPTER 15

Chromium Speciation Analysis in Soils/Sediments - Zero Percent Matrix Spike Recoveries May Not Equal Unreliable Data

Rock J. Vitale, George R. Mussoline, and **Kelly A. Rinehimer**, Environmental Standards, Inc., Valley Forge, Pennsylvania

BACKGROUND

Under current environmental investigations, total chromium is among the heavy metals on the U.S. Environmental Protection Agency's (USEPA's) target analyte list examined in soil/sediment samples. However, because of the significant difference in toxicity between trivalent (III) and hexavalent (VI) chromium, it is essential to evaluate these individual species during risk assessment and subsequent remediation. Recent advances in soil/sediment sample preparation and analytical techniques have enabled investigators to successfully differentiate between these two chromium species (Vitale et al., 1993). While the preparation and analysis of predigestion matrix spikes has traditionally provided analytical chemists with an indication of data quality for total metal analyses, this quality control (QC) technique has limited utility for chromium speciation analysis, specifically for the analysis of Cr(VI).

The use of SW-846 Method 3060A, an alkaline digestion procedure for the determination of Cr(VI) in soils, recently proposed in the latest update of SW-846, has been documented to be an effective technique for the analysis for soluble and insoluble forms of Cr(VI) (James et al., 1995; Vitale et al., 1995a). In certain soils/sediments that have highly reducing properties, chromium exists exclusively in the Cr(III) species. The matrix spiking of such soils/sediments with Cr(VI) is predicted to yield a zero percent recovery. Traditionally, such recoveries would have been interpreted as an indication that the resultant data are unreliable. Hence, when such recoveries are observed, it is critical that the redox characterization of the sample be determined so that the quality control information can be correctly interpreted.

SW-846 Method 3060A includes algorithms to assist the data user with interpretation of QC analysis results. The topic of this chapter is the discussion and interpretation of actual field sample Cr(VI) quality control data through the use of redox indicators as applied to SW-846 Method 3060A.

INTRODUCTION

SW-846 Proposed Method 3060A (Vitale et al., 1995) is an alkaline digestion method for extracting hexavalent chromium from soils, sediments, and solid wastes. This method is of significance because of the differences in toxicity between Cr(III) and Cr(VI). Cr(VI) is a human carcinogen (via inhalation) and Cr(III) is an essential dietary element for humans (Anderson, 1989; IRIS, 1993; Paustenbach et al, 1991). Method 3060A utilizes a hot alkaline solution containing 0.28 M Na_2CO_3/0.5 M NaOH 0.28 M Na_2CO_3 to solubilize both sparingly-soluble and water-soluble Cr(VI) compounds in solid samples (James et al, 1995; Vitale et al., 1995a; Vitale et al., 1995b). Once Cr(VI) is solubilized, the digest is analyzed by adding a diphenylcarbazide (DPC) solution in acetone, and adjusting the solution to pH 2 using sulfuric acid. The Cr(VI) reacts with the DPC to produce a red-violet complex, and its absorbance is measured spectrophotometrically at 540 nm. This analytical method is designated as SW-846 Method 7196A (USEPA, 1992). An additional analytical technique (SW-846 Method 7199) using ion chromatography with a post-column reaction provides an alternative to the manual method referenced.

In order to determine the reliability of the extraction procedure, predigestion matrix spike recovery data is one quality control measure that is evaluated. Traditional interpretation of predigestion matrix spike recoveries was such that if the matrix spike recovered well (i.e., within 80-120%), then this was an indication that the analytical method was performing well. Conversely, (in the traditional interpretation of matrix spike recoveries), if the spike recovery was not within 80-120%, then this was an indication of an analytical bias. One of the major points of this chapter is to demonstrate that poor predigestion matrix spike recoveries for the analysis for Cr(VI) in soils/sediments may not necessarily be indicative of a deficiency in the analytical method.

Certain reducing sample types (e.g., anoxic sediments) cannot support chromium in the hexavalent state, regardless of whether the hexavalent chromium was added in the natural field conditions or during the chemical analysis. A major portion of SW-846 Proposed Method 3060A is the assessment of ancillary parameters in order to determine whether or not a sample type exhibits the ability to maintain chromium in the hexavalent valance state. The following ancillary parameters aid in this characterization of a sample: pH, oxidation reduction potential (ORP), total sulfides, total organic carbon (TOC), biological oxygen demand (BOD), and chemical oxygen demand (COD). The analysis of the aforementioned parameters establishes a better understanding of the tendency of Cr(VI) to exist in the unspiked sample, and assists in the interpretation of quality control data for predigestion matrix spike recoveries.

RESULTS AND DISCUSSION

The analysis of ancillary parameters in Method 3060A is designed to help interpret variable recoveries of Cr(VI) predigestion matrix spikes. Collectively, the following parameters play an important role in determining whether or not the soil is reducing in nature and thus, if the sample can sustain Cr(VI) in the hexavalent oxidation state.

pH

In general, soil samples need to be alkaline in nature (pH>7.0) in order to sustain chromium in the hexavalent state. It appears that the higher the pH of the soil, the less likely the Cr(VI) will be reduced to Cr(III). Based on the data evaluated, those spike samples with pH values < 7.0 tend to exhibit lower spike recoveries.

ORP

Generally, the larger the negative oxidation reduction potential (ORP) value, the stronger the reducing environment. In addition, the more positive the ORP, the greater the tendency Cr(VI) can be sustained in the soils; thus, the better the possibility of obtaining "acceptable" predigestion matrix spike recoveries. Similarly, negative ORP readings exhibit like degrees of reducing capacity, comparable to their positive value counterparts. As shown in the Eh/pH diagram (Figure 1), a reducing environment may exist at a high pH (10) if the ORP value is between -600 mV and 0 mV. (NOTE: A value of 199 mV was subtracted from the Eh values on Figure 1 in order to convert the measurement to an ORP reading). In addition, a reducing environment can occur at higher ORP values (~+300 mV), provided the pH of the soil is acidic (~4.0).

TOC

Chromium may exist in a number of oxidation states; however, they are not all of the same stability. As shown in the reduction potential diagram for chromium (Figure 2), the reduced form is favored when positive values are observed for the standard electrode potential (E°), whereas the oxidized form is relatively stable when negative values are obtained for the standard electrode potential. Since considerable energy would be required to convert Cr(III) to lower or higher oxidation states, Cr(III) is the most stable form of chromium in solution at acidic pHs. Although CrO_4^{2-} (a hexavalent form of chromium) is relatively stable, its high positive reduction potential denotes that it is strongly oxidizing and is unstable in an acid solution, as well as in the presence of organic molecules with oxidizable groups (alkanes, alkenes, alcohols, aldehydes, ketones, carboxylic acids, etc.). Therefore, the greater the levels of total organic carbon (TOC) in the soil, the more

Figure 1. **Eh/pH Phase Diagram**

The dashed lines define Eh – pH boundaries commonly encountered in soils and sediments.

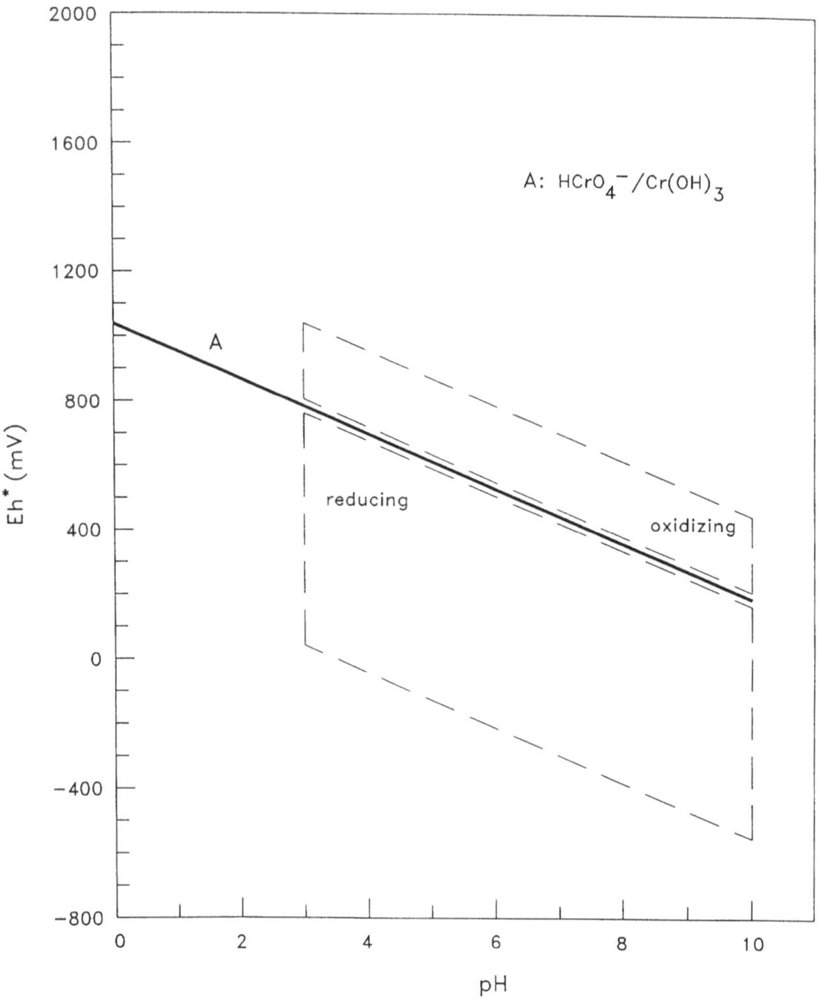

* Note the Eh values plotted on this diagram are corrected for the reference electrode voltage: 244 mV units must be added to the measured value when a separate calomel electrode is used, or 199 mV units must be added if a combination platinum electrode is used.

Figure 2. **Reduction potential diagram for Chromium.**

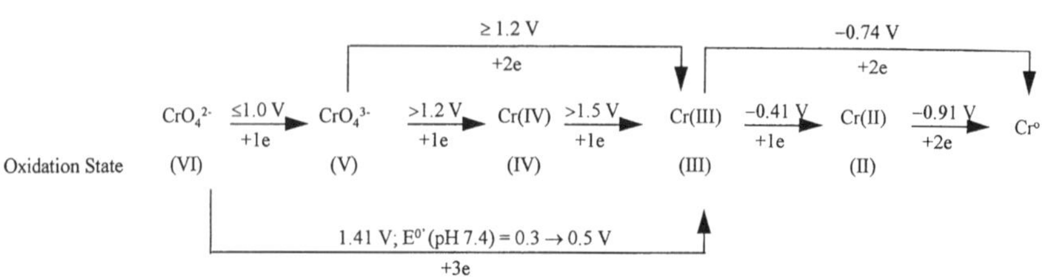

likely the hexavalent chromium will be reduced to trivalent chromium (*Chromium in the Natural and Human Environments*, 1988a).

Sulfides

The presence of sulfides in soils is a strong indicator that the soil is reducing in nature. A water-soluble reagent such as sodium sulfide can reduce hexavalent chromium and precipitate it to chromium hydroxide. This reduction seems to occur in nature when a secondary metal is present, such as iron. One method that is used by wastewater treatment plants for removing heavy metals (i.e., chromium) from water samples, is through sulfide precipitation (*Chromium in the Natural and Human Environments, 1988b*).

BOD

Biochemical oxygen demand (BOD) is an empirical analytical procedure which measures the dissolved oxygen consumed by microbial life while absorbing and oxidizing the organic matter present in a sample. It is expressed as the quantity of dissolved oxygen required during stabilization of the decomposable organic matter by aerobic biochemical action (Lewis, 1993). Therefore, the greater the levels of BOD, the greater the levels of organic matter, and the more likely Cr(VI) can be reduced.

COD

Chemical oxygen demand (COD) is designed to determine the amount of oxygen required to oxidize the organic matter in a waste sample, under specific conditions of oxidizing agent, temperature, and time. Because the test utilizes a specific chemical oxidation, the result has no definite relationship to the BOD or TOC of the sample. The greater the levels of COD, the higher the levels of organic matter, and the less likely Cr(VI) will remain in the +6 oxidation state.

If numerous indicators are observed for a reducing environment for a given sample, poor Cr(VI) predigestion matrix spike recoveries are predicted to be the result of soil reduction, and not representative of a method-induced reduction or technical error. The characterization of the above parameters is essential in establishing the sample's oxidation/reduction environment.

Discussion of Eh/pH

The following discussion will refer to the data presented in Table 1. Plotting Eh/pH ancillary parameter data for samples S-01-1, S-01-2, and S-01-3 demonstrated how additional field data aid in interpreting Cr(VI) pre-digestion matrix spike recoveries. When the Eh value for sample S-01-1 is plotted against the pH value for this sample, the result falls in the reducing area on Figure 1. Low recoveries were observed for both the low-level predigestion matrix spike (0%) and high-level pre-digestion matrix spike (15%) samples analyzed for Cr(VI) for sample S-01-1. When the Eh value for sample S-01-2 is plotted against the pH value for this sample, the result again falls in the reducing area on Figure 1. This result (alone) would indicate the presence of a reducing soil. However, the recoveries for both the low-level predigestion matrix spike (97%) and high-level predigestion matrix spike (98%) that were observed for hexavalent chromium in sample S-01-2 would indicate that Cr(VI) was not reduced to Cr(III) (these recoveries were within the traditional acceptance range of 80 - 120 %). When the Eh value for sample S-01-3 is plotted against the pH value for this sample, the result falls in the oxidizing area on Figure 1. This would indicate that reducing characteristics do not appear to be present in this sample. Again, the low-level and high-level predigestion matrix spike recoveries (97%, 97%) observed for the Cr(VI) analysis of sample S-01-3 were within the traditional acceptance range of 80%-120% and confirmed the nonreducing soil type.

The Eh/pH diagram (Figure 1) indicates that sample S-01-2 would be a reducing sample matrix; however, "traditionally acceptable" predigestion matrix spike recoveries were observed for this sample. The plot of Eh/pH is a good first indicator of the oxidation/reduction potential of a sample; however, the additional ancillary parameters must be evaluated to further clarify the sample's potential for oxidation/reduction of Cr(VI). The use of any one of the ancillary parameters alone may not be sufficient to characterize the oxidation/reduction potential of a sample type.

Discussion of Additional Data

Based on the data evaluated, recoveries within the traditional acceptance ranges were observed for the low-level predigestion, high-level predigestion and postdigestion matrix spikes for samples S-01-2 and S-01-3 analyzed for hexavalent chromium. The ancillary data collected were successful in the prediction that "acceptable" recoveries should have been obtained, even though sample S-01-2 indicated a reducing sample matrix and sample S-01-3 indicated an oxidizing sample in their respective Eh/pH plots on Figure 1. Low levels of TOC and COD indicate that these samples are low in organic matter. In addition, BOD was not detected in the samples. These data indicate that Cr(VI) can be stable under these conditions, and that one could expect to observe "acceptable" recoveries of Cr(VI) added to a sample matrix. The plot of sample S-01-2 on Figure 1 was close to the cross-over threshold for an oxidizing/reducing environment. Because of this proximity to the cross-over threshold, the additional ancillary parameter data was instrumental in further characterization of sample S-01-2 as an oxidizing sample matrix.

In the traditional interpretation of matrix spike data, poor recoveries were observed for the low-level and high-level matrix spikes for sample S-01-1 analyzed for hexavalent chromium. As seen in Table 1, high levels of TOC, COD, and BOD were observed in the sample, indicating high levels of organic matter present in the sample. In addition, the ORP and pH values measured for sample S-01-1 indicate that the sample is reducing in nature. These data are indicative of reducing tendencies of the soil type, and therefore Cr(VI) may not be sustainable. These additional parameters were utilized to predict that native Cr(VI) cannot be present in sample S-01-1. Unlike samples S-01-2 and S-01-3, sample S-01-1 had high levels of organic matter and an ORP that would indicate that the sample would reduce Cr(VI) to Cr(III) regardless of whether the source of the Cr(VI) was from an environmental source, or if the Cr(VI) was added to the sample matrix in the laboratory. Figure 3 graphically interprets the differences observed in the oxidation/reduction capacities, as well as the organic content of samples S-01-1, S-01-2, and S-01-3.

Example of How Ancillary Data Is Interpreted

Since Cr(VI) does not readily exist in reducing samples in the field or during chemical analysis, it is clear that the ancillary parameter data noted herein will aid in the interpretation of the predigestion matrix spike recovery data. For example, if the predigestion matrix spike recovery is less than the traditional acceptable limits of 80%-120%, then the ancillary data need to be assessed. The predigestion matrix spike recovery data for Cr(VI), in conjunction with the ancillary data of the soil, will aid data users in the understanding of the soil chemistry, and aid in the development of an explanation for any reduction of Cr(VI) under field conditions and during chemical analysis.

In assessing the characteristics of samples, it is necessary to determine whether the soil type has reducing propensities: the pH and ORP values should be within the reducing dashed lines as seen in Figure 1, high levels of TOC, COD, and BOD may be observed, and the soil may contain some level of sulfides. It should be noted that each of these parameters by themselves cannot definitively determine whether or not the sample is reducing (as demonstrated by sample S-01-2). The ancillary parameters must be evaluated collectively to give a better understanding of the chemical properties of the soil.

Table 1. **Soil Sample Characterization**

SDG	Sample ID	Percent Moisture	TOC (mg/kg)	COD (mg/kg)	BOD (mg/kg)	pH	Eh (mV)	Cr(VI) (mg/kg)	Cr(VI) QC Results
SDG 1	S-01-1	79.7	20000	377000	2040	6.28	380	N.D.	-
	S-01-1DUP	76.7	30000	381000	2320	6.13	370	N.D.	-
	S-01-1LMS	76.7	-	-	-	-	-	N.D.	0.0%
	S-01-1HMS	76.7	-	-	-	-	-	250	15.0%
	S-01-1PDS	76.7	-	-	-	-	-	1570	91.0%
SDG 2	S-01-2	19.2	200	1600	N.D.	6.85	370	N.D.	-
	S-01-2DUP	19.2	160	1600	N.D.	7.05	390	N.D.	-
	S-01-2LMS	19.2	-	-	-	-	-	48	97.0%
	S-01-2HMS	19.2	-	-	-	-	-	486	98.0%
	S-01-2PDS	19.2	-	-	-	-	-	198	100.0%
SDG 3	S-01-3	21.3	550	1800	N.D.	7.61	510	N.D.	-
	S-01-3DUP	21.3	420	2300	N.D.	7.58	530	N.D.	-
	S-01-3LMS	21.3	-	-	-	-	-	49	97.0%
	S-01-3HMS	21.3	-	-	-	-	-	492	97.0%
	S-01-3PDS	21.3	-	-	-	-	-	200	98.0%

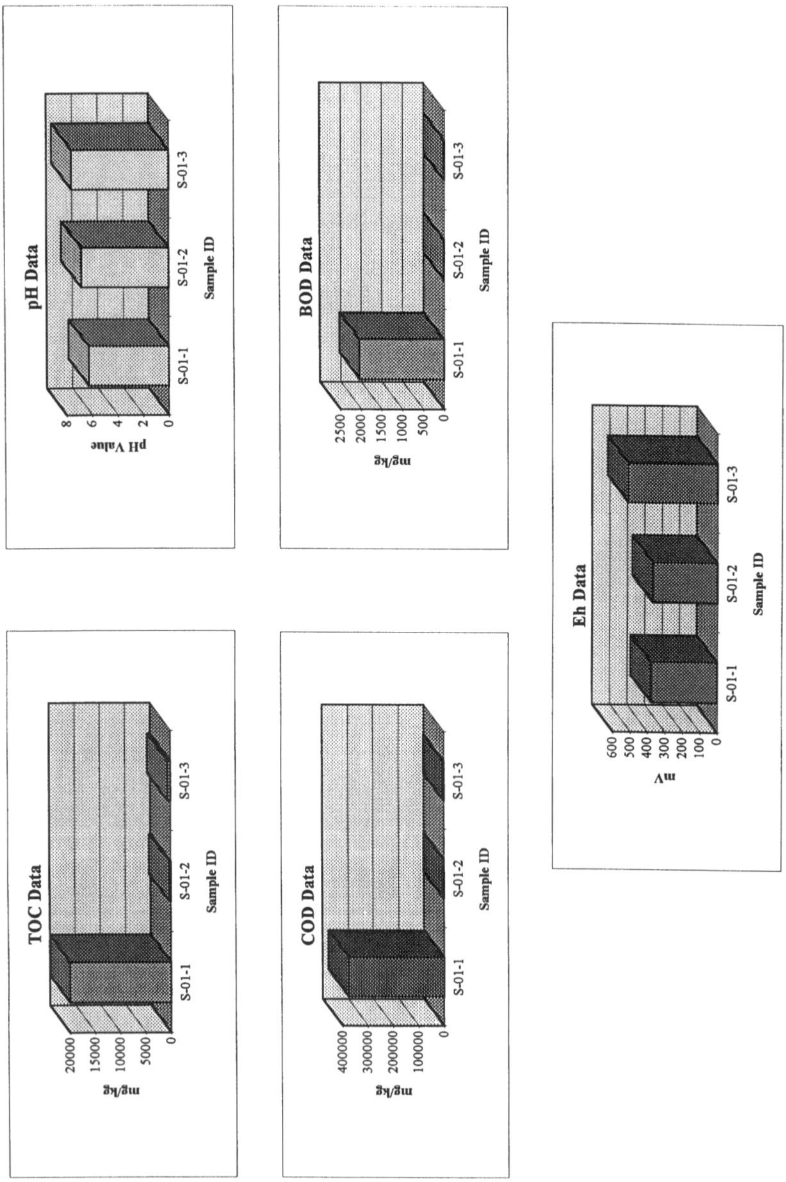

Figure 3. **Ancillary Parameter Data.**

CONCLUSION

It is to be concluded that zero or low predigestion matrix spike recoveries do not necessarily mean the data obtained are unreliable. Ancillary parameter data need to be evaluated to determine if the sample type is reducing in nature and cannot support Cr(VI), either in the natural environment or after being spiked into a sample by a laboratory. If reducing conditions are observed (high levels of TOC, BOD, COD and/or sulfides, acidic pH, negative ORP values), and other quality control samples such as a laboratory control samples (*viz.*, blank spikes) are acceptable, then the data should be viewed as acceptable for use.

REFERENCES

Anderson, R.A. 1989. Essentiality of Chromium in Humans, *Sci. Total Environ.* **86**, 75.

Chromium in the Natural and Human Environments, p. 25. 1988. New York, John Wiley & Sons, A Wiley-Interscience Publication.

Chromium in the Natural and Human Environments, p. 237. 1988. New York, John Wiley & Sons, A Wiley-Interscience Publication.

IRIS (Integrated Risk Information System) 1993. A continuously updated electronic database maintained by the U.S. Environmental Protection Agency, National Library of Medicine, Bethesda, Maryland.

James, B.R., Vitale, R.J., Mussoline, G.R., and Petura, J.C. 1995. Hexavalent Chromium Extraction from Soils: Evaluation of an Alkaline Digestion Method, *Environ. Sci. Technol.*, Volume 29, Number 9, pgs. 2377-2381.

Lewis, R.J., Sr. *Hawley's Condensed Chemical Dictionary*, Twelfth Edition. 1993. New York, Van Nostrand Reinhold.

Paustenbach, D.J., W.E. Rinehart, and P.J. Sheehan. 1991. The Health Hazards Posed by Chromium-Contaminated Soils in Residential and Industrial Areas: Conclusions of an Expert Panel, *Regulatory Toxicol. Pharmacol.* **13**, 193.

USEPA (United States Environmental Protection Agency). 1992. Test Methods For Evaluating Solid Waste, Physical/Chemical Methods, SW-846, 3rd Edition, Revision 1, July.

Vitale, R.J., Mussoline, G.R., Petura, J.C., and James, B.R. 1993. A Method Evaluation Study of an Alkaline Digestion (Modified Method 3060) Followed by Colorimetric Determination (Method 7196A) for the Analysis of Hexavalent Chromium in Solid Matrices, presented to the United States Environmental Protection Agency, June 15.

Vitale, R.J., Mussoline, G.R., Petura, J.C., and James, B.R. 1995a. SW-846 Method 3060A, Third Update to the Third Edition of SW-846, July.

Vitale, R.J., Mussoline, G.R., Petura, J.C., and James, B.R. 1995b. Hexavalent Chromium Quantification in Soils: An Effective and Reliable Procedure, *Am. Environ. Lab.*, 1.

CHAPTER 16

Delineation and Closure of a PCB Transformer Leak at an Active Railroad Yard Utilizing Enzyme Immunoassay Screening Analysis

Matthew A. Cousino, ENSR Consulting and Engineering, Pittsburgh, Pennsylvania

Sabina Strautman, Consolidated Rail Corporation, Philadelphia, Pennsylvania

INTRODUCTION

This chapter presents the results of an investigation conducted during the week of July 25, 1994 at the Consolidated Rail Corporation (Conrail) Frontier Yard transformer spill area (Spill No. 9207920) in Buffalo, New York (see Figure 1). The investigation was conducted according to the Soil Sampling and Remedial Work Plan (ENSR Document No. 1880-008-200) submitted to the New York State Department of Environmental Conservation (NYSDEC) in May 1994.

The transformer station was located in an active part of the railyard, and supplied power to Conrail buildings and operations. The remediation subcontractor's proposed remedy would have resulted in the loss of the transformer station, costly excavation and disposal expenses, and the disruption of railyard activities for 2 to 3 weeks.

As opposed to removing the transformer station, ENSR proposed a traditional investigative approach and onsite screening technology to confirm the extent of potential contamination and document that polychlorinated biphenyls (PCBs) were not a problem. ENSR's approach would have closed the spill in a timely (2 days) and cost-effective manner.

The objective of the soil sampling program was to obtain site closure. This was accomplished by the collection and analysis of soil samples in the area of the spill. The NYSDEC has indicated that the spill area will be closed when sampling results show PCB levels in soil are less than 10 micrograms per gram (μg/g).

Twenty-three soil samples were analyzed in the field using enzyme immunoassay (EIA) screening analysis. Eleven of these soil samples were also sent to

Figure 1. **Site location.**

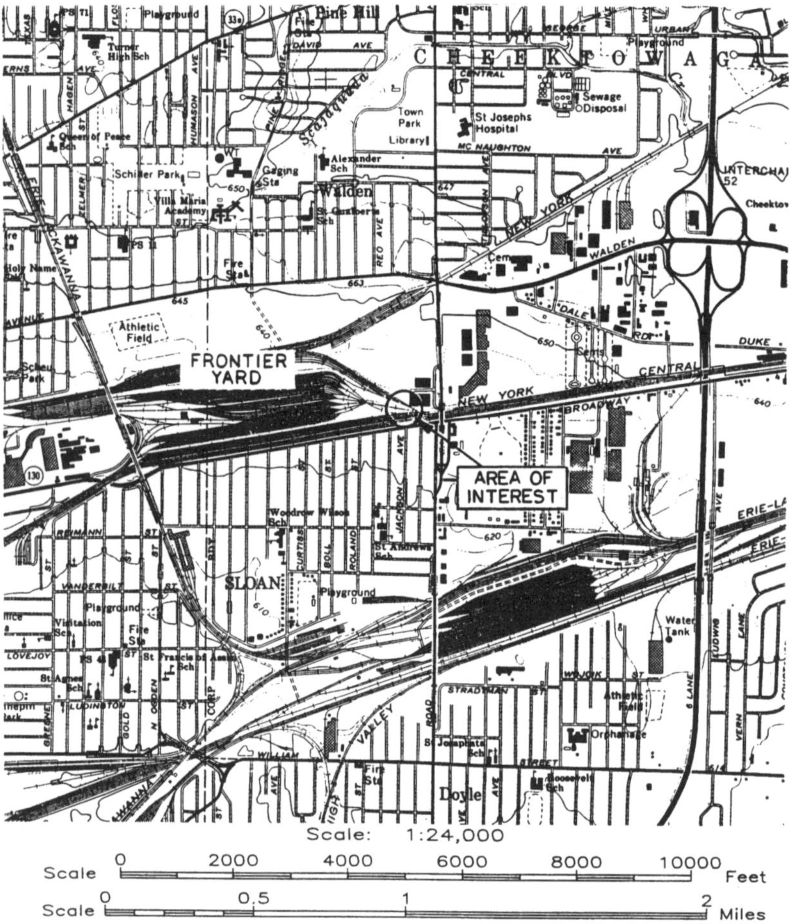

an analytical laboratory for verification using EPA Method 3540 (Extraction) and 8080 (Analysis).

Both field and laboratory results show PCB concentrations of less than 10 µg/g for all soil samples tested. An excellent correlation was also obtained between the EIA results and the laboratory verification results.

Based on the site analytical results, no further action was recommended to the NYSDEC and the spill site was closed.

BACKGROUND

Site Description and Location

The Conrail Frontier Yard is located in the City of Buffalo and the Town of Cheektowaga in Erie County, New York. The location of the transformer pad where the spill occurred is near the eastern end of the property, northeast of the hump yard office and west of the Harlem Road overpass. The transformer pad is approximately 23 feet by 17 feet in size, and at the time of the spill contained three new transformers and three older transformers (see Figure 2). A 6-foot high chain link barbed wire fence surrounds the transformer pad. The area around the transformer pad is topographically flat.

An asphalt roadway borders the southern side of the transformer pad, and an asphalt parking area borders the northern side of the transformer pad. The area immediately east and west of the transformer pad is not paved and is covered with gravel.

Former Site Activities

The PCB spill occurred on October 7, 1992. Conrail discovered one transformer leaking from its base onto the transformer pad and adjacent soil along the southern edge to the gravel roadstone base (see Figure 2). PCB concentrations in the transformer oil ranged from 110 to 154 mg/kg of Arochlor 1260. Conrail responded to the release by notifying the NYSDEC and engaging an emergency spill subcontractor. Initial response activities included the use of absorbent pads and clay material to capture oil remaining on the transformer pad. The concrete pad was subsequently cleaned with a No. 2 fuel oil wash. Visibly stained gravel and stone adjacent to the transformer pad was hand-shoveled into drums.

On October 9, 1992, Conrail discovered a second transformer that was starting to leak on the transformer pad. The oil remaining in the transformer was

Figure 2. **Site plan.**

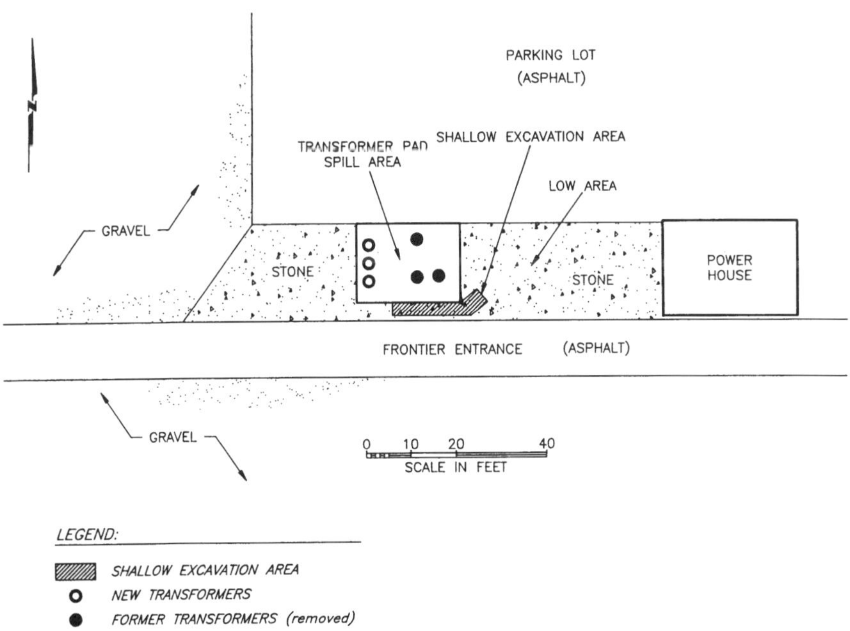

transferred to a drum, and the pad was recleaned. The three older transformers, in addition to three drums of PCB transformer oil, were removed from the concrete pad and disposed of at an approved facility on July 23, 1993.

On October 13, 1992, the remaining gravel and stone fill showing visual evidence of oil impacts was excavated and drummed. The excavation along the south side of the concrete pad was approximately 6 inches deep at its western end, and graded to 1 foot deep at the eastern end. The excavation was approximately 2 to 3 feet wide, and was covered with plastic at completion.

METHOD OF INVESTIGATION

The investigation approach included determination of the spill boundaries and the presence of PCBs beneath the concrete transformer pad, and along the bottom and sidewalls of the shallow excavation south of the concrete pad. The presence or absence of PCBs in soil or fill material was determined through the use of EIA screening analysis. A subset of these samples were split in the field and sent for verification sampling to a laboratory in Buffalo, New York.

Sampling was performed beneath the concrete transformer pad and in the shallow excavation south of the concrete pad. Core holes were drilled through the concrete pad in order to collect soil samples beneath the pad. Surface soil samples were collected from the shallow excavation after the plastic covering was removed. The soil boring and sampling locations are shown in Figure 3. Sample locations were biased toward the southern side of the concrete pad where oil was observed during the response activities.

Figure 3. **Soil boring and sampling locations.**

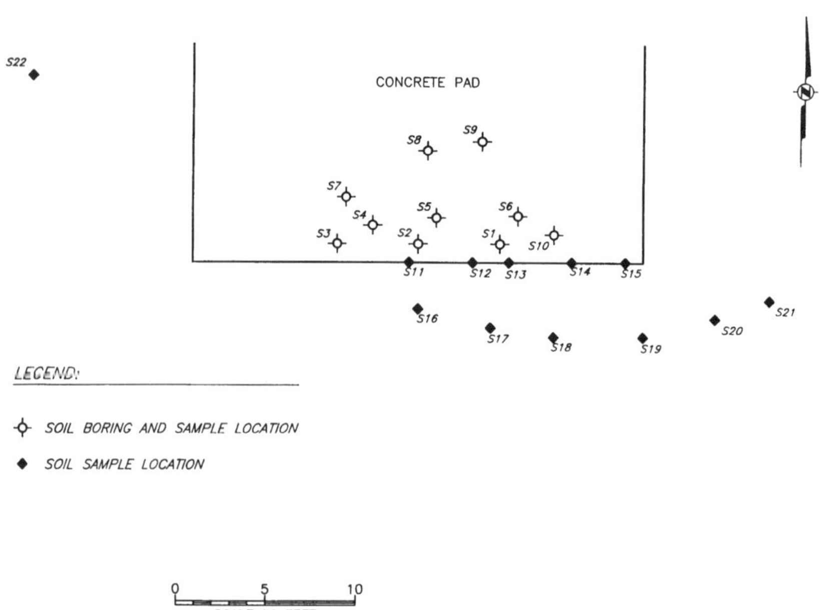

In order to evaluate the presence of PCBs beneath the concrete, a total of 10 core holes were drilled through the concrete pad (boring locations S1 to S10). The pad was approximately 7 inches thick and reinforced with 1/4 inch steel fencing. At three locations (S8, S9, and S10) two samples were collected for vertical delineation.

Soil samples were collected from the 10 coreholes from a 6-inch interval beginning approximately 2 inches beneath the base of the concrete pad and the underlying crushed limestone base. Soil samples were collected from a 1-inch diameter core sample with a 6-inch vertical catch. At locations S8, S9, and S10, a second 6-inch soil sample was collected (S8A, S9A, and S10A) to further evaluate the distribution of PCBs with depth. The second soil sample was collected approximately 8 to 14 inches beneath the base of the concrete pad and the crushed limestone base.

In addition to the core hole sampling, 11 soil samples (S-11 to S-21) were collected from the sidewalls and bottom of the shallow excavation south of the transformer pad. These samples were collected to provide information on the potential extent of PCBs in the excavation and at the base of the transformer pad where it meets the excavation.

Soil samples S-11 to S-15 were collected approximately 2 inches immediately beneath the edge of the concrete pad. These samples were collected from a 6-inch interval using disposable PVC sampling scoops. Soil samples S-16 to S-21 were collected at the base of the shallow excavation next to the sidewall. These samples were also collected from a 6-inch interval beginning approximately 6 inches below ground surface.

Each soil sample was placed into a glass jar and labeled. Soil samples collected for EIA analysis were transferred to the EIA field work station for sample preparation and analysis. Split soil samples collected for laboratory verification were transferred to a cooler for temporary storage until the laboratory picked up the samples within 24 hours. A completed chain-of-custody form accompanied the split samples to the laboratory.

RESULTS

The collected soil samples generally consisted of silt, sand, and clay fill in a crushed stone matrix. The material beneath and adjacent to the transformer concrete pad was generally black to brown in color. At several locations, 2 to 3 inches thick, a black organic muck layer was observed. This layer was underlain by a dark grey clay layer. This layer appeared to be relatively impermeable. The fill above this layer was generally moist to wet.

Figure 4 shows the field EIA results, and Figure 5 shows the laboratory PCB analytical results. The field and laboratory analytical results did not detect PCB above the NYSDEC closure criteria of 10 μg/g.

Figure 4. **PCB immunoassay field results.**

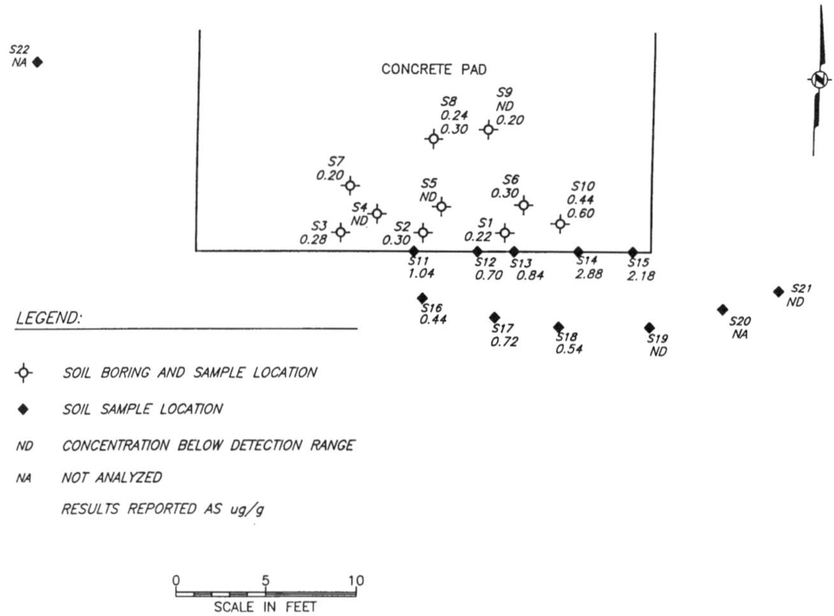

Figure 5. **Laboratory PCB analytical results.**

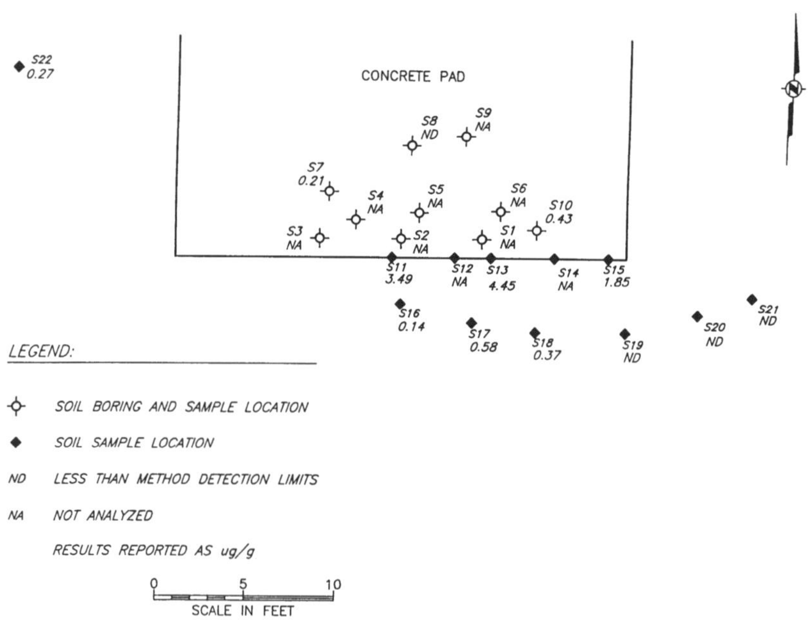

Table1. **Summary of PCB EIA and Laboratory Analytical Results.**

Sample Number	Sample Depth	EIA Results (ug/g)	Laboratory Results (ug/g)
S-1	2-8 in. below concrete	0.22	-
S-2	2-8 in. below concrete	0.30	-
S-3	2-8 in. below concrete	0.28	-
S-4	2-8 in. below concrete	ND[1]	-
S-5	2-8 in. below concrete	<0.14	-
S-6	2-8 in. below concrete	0.30	-
S-7	2-8 in. below concrete	0.20	0.21
S-8	2-8 in. below concrete	0.24	<0.11
S-8A	8-14 in. below concrete	0.30	-
S-9	2-8 in. below concrete	<0.14	-
S-9A	8-14 in. below concrete	0.20	-
S-10	2-8 in. below concrete	0.44	0.43
S-10A	8-14 in. below concrete	0.60	-
S-11	2-8 in.	1.04	3.49
S-12	2-8 in.	0.70	-
S-13	2-8 in.	0.84	4.45
S-14	2-8 in.	2.88	-
S-15	2-8 in.	2.18	1.85
S-16	6-12 in.	0.44	0.14
S-17	6-12 in.	0.72	0.58
S-18	6-12 in.	0.54	0.37
S-19	6-12 in.	<0.10	<0.09
S-21	6-12 in.	<0.08	<0.08
S-22	4 to 10 in.	-	0.27
S-20	6-12 in.	-	<0.9

- = Not Analyzed
ND[1] = Concentration below calculated Detection Range

Table 1 presents the field and laboratory analytical results. Twenty-three enzyme immunoassay field tests were performed. Eleven of these samples were split and sent to the analytical laboratory for verification.

As shown in Figure 6, EIA and laboratory split sample results are closely correlated. The differences in the EIA and laboratory analytical results for samples S-11 and S-13 may be due to variations in sampling consistency, which is a common problem with soil samples.

The EIA field tests detected PCB at concentrations ranging from 0.20 µg/g to 2.88 µg/g. Five EIA tests were nondetect for PCBs (S-4, S-5, S-19, S-21, and S-9). The laboratory verification samples (S-7, S-10, S-11, S-13, S-15 to S-18)

Figure 6. **Comparison of EIA and laboratory results.**

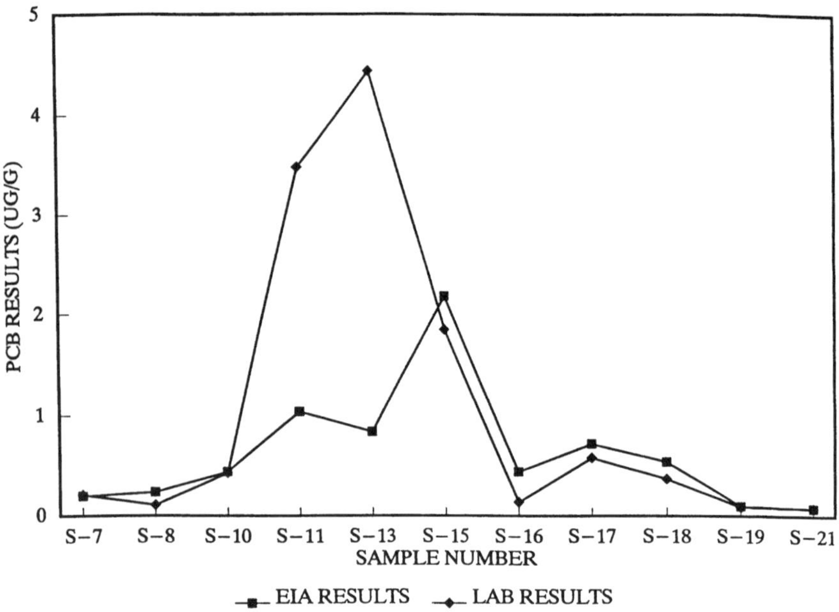

contained PCBs at concentrations ranging from 0.14 μg/g to 4.45 μg/g. Three laboratory verification samples (S-19, S-21 and S-8) were nondetect for PCBs (method detection limits ranged from 0.08 to 0.11 μg/g).

Sample S-22 was collected west of the transformer concrete pad in an area believed to be undisturbed by the activities associated with the transformer spills. This sample contained PCBs at a concentration of 0.27 μg/g.

The laboratory tested the verification samples for seven different Arochlor compounds and only detected Arochlor 1260. This is the same Arochlor compound detected during the initial spill response activities.

QUALITY CONTROL

Quality assurance in the field was maintained through the use of disposable polyvinyl chloride (PVC) sample scoops and vinyl gloves, clean working conditions, and decontamination procedures for the metal drive sampler before and between sample locations. Collected soil samples were immediately placed in the sample collection jar, labeled, and placed in a cooler at 4°C. EIA samples were analyzed in the field each day (i.e., two runs were conducted). Laboratory verification samples were picked up in the field by the laboratory courier and transported to the laboratory for analysis.

Quality control in the laboratory included duplicate analysis of soil sample S-21 (both results showed PCB concentrations at less than 0.08 μg/g), in addition to standard quality control procedures associated with EPA Method 8080.

Field EIA quality assurance and quality control consisted of PCB standards (at concentrations of 0.25, 1, and 5 μg/kg), a PCB control sample (at a concentration of 2 μg/kg) and three PCB proficiency samples. The PCB standards are used to establish the curve for absorbance interpolation. The control sample is analyzed in each run and treated in the same manner as the unknown extracted soil samples. Proficiency samples (P-A, P-B, and P-C) were included in the beginning, middle, and end of each run. Proficiency samples are additional quality assurance checks used to verify the performance of the field analytical technique.

EIA field data reduction was performed using linear regression (the ETA testing equipment is preprogrammed to perform the assay method) in which immunoassay optical readings are automatically converted into sample concentrations. Calibration data consists of a blank and PCB standards which are used to describe a best fit or linear regression straight line. The correlation coefficient [Corr (R)] represents the "goodness of fit" of the blank and standard data to the best fit line. The square of this value represents the proportion of variance that is explained by the linear regression (this is a measure of how well the sample preparation, extraction, and dilution procedures were performed). Correlation coefficients of 1.000 and 0.9984 were obtained. EIA assay results should have a Corr(R) value of 0.990 or greater.

Another measure of the quality of the data is the coefficient of variation (% CV) between the PCB standard sample duplicates. The % CV values should be 10% or less to be acceptable; below 5% is excellent. Run 1 achieved % CVs of 0.3 to 4.2, and Run 2 achieved % CVs of 0.2 to 2.8. The concentration values obtained on the control and proficiency samples were within the specified ranges for each run.

SUMMARY AND CONCLUSIONS

Twenty-three grab soil samples were collected beneath and adjacent to the transformer concrete pad, and analyzed in the field for PCBs using field screening analysis. Eleven of these samples were split and confirmed in the laboratory using EPA-approved methodology.

- Based on the results, no PCBs were detected above the NYSDEC action level of 10 μg/g, and it was recommended that the spill be closed. The EIA testing results show PCBs were not present above NYSDEC closure criteria of 10 μg/g.
- The verification sample results closely correlated with the EIA results.
- Using a traditional investigative approach and EIA screening technology, Conrail:

Was not required to remove the transformer station;

Sustained no disruption of train traffic;

Reduced field work from an estimated 2 to 3 weeks to 2 days; and

Reduced estimated closure costs by two-thirds (not including the potential costs to replace the concrete pad and transformers).

FEDERAL

CHAPTER 17

Using Reflectance Fourier Transform Infrared Spectroscopy (FTIR) for Rapid Analysis of Contaminated Soils

Jane W. Adams and **Hollis (Jay) H. Bennett**, U.S. Army Engineer Waterways Experiment Station, Vicksburg, Mississippi

INTRODUCTION

Spectroscopic methods that can identify chemical compounds are being examined as possible new technologies for in-situ soil probes. With the advent of IR-transmitting fibers, Fourier transform infrared (FTIR) spectroscopy is a strong candidate. This chapter reviews work done to investigate FTIR in a field-simulated situation in which we used direct reflectance spectroscopy to examine a variety of contaminated soils. Contaminants of interest were explosives, volatile organic compounds (VOC) consisting mainly of solvents, and petroleum, oils, and lubricants (POLs).

A primary goal of the Department of Defense (DOD) Environmental Cleanup effort is to develop technologies for in-situ soil probes that expedite the identification and position of a variety of liquid and solid contaminants in soils. The Army Corps of Engineers has been tasked to develop onsite sensors to identify contamination at locations which are known to contain waste products and which require remediation. Site characterization methods which give inexpensive, rapid, and accurate analyses using minimal sample preparation are of continuing interest for a number of reasons. These include ease of use, broad applicability in diverse environments, and broad applicability across all governmental regulatory levels. The Tri-Service Site Characterization and Analysis Penetrometer System (SCAPS) is one such tool used to identify and 3-dimensionally map subsurface contamination plumes and soil stratification (Lee et al., 1993). The purpose of this work was to investigate FTIR spectroscopy to identify and quantify contamination with a view toward its inclusion in SCAPS.

Background

FTIR spectroscopy is a technique which can characterize single or mixed chemical compounds in all states of matter. It is quick, simple, and versatile. FTIR technology uses collimated, broadband infrared radiation to excite a sample, whereupon chemical compounds are differentiated on the basis of the vibrations of their constituent atomic bonds. The peaks in an FTIR spectrum are generated by a collection of motions by atomic bonds, such as bending or stretching, that fingerprint a specific chemical compound. By contrast to chromatographic or mass spectrographic analyzers, infrared analyzers are fast, accurate, and call for extremely simple sample preparation procedures. In addition, FTIR operators require little advanced analytical training to operate the equipment. FTIR spectroscopy has been broadly used as a workhorse analytical method for research and manufacturing processes for the past 50 years and is now gaining use for contaminant studies (AWMA, 1994).

Due to the wide applicability of FTIR, sampling techniques exist to evaluate nearly any form of matter. The preferred method is to examine a spectrum resulting from IR radiation transmitted through the sample. Naturally, for a remote sensor application where one wishes to look at soil below the surface of the ground, this is not possible. The diffuse reflectance technique enables non-destructive handling of solid samples with rough surfaces without contacting the sample surface.

Spectra collected in the diffuse reflectance mode represent a combination of transmission and reflection characteristics, although the spectrum itself resembles a transmission spectrum. However, due to the fact that the reflected IR radiation is scattered throughout a hemisphere, the collected sum of indeterminate reflected pathlengths leads to intensity errors in a diffuse reflectance spectrum. Emission spectra may be corrected to remove the effects of diffuse scattering in order to give a spectrum more analogous to an absorbance plot of a pure transmission spectrum. The conversion of reflectance spectra is necessary for comparison against common spectral libraries.

With respect to soil analyses by FTIR, spectra of single mineral compounds have been collected by the U.S. Geological Survey, professional mineralogical societies, and others (Coates, 1995; Farmer, 1974; McMillan and Hofmeister, 1988). The majority of these data are transmission spectra. Not much work has been done on complex soil mixtures. King and Olhoeft (1989) performed the first FTIR analyses of a soil mineral to examine the spectral response of toluene physically absorbed onto the exchangeable cations of montmorillonite clays. Their work shows that it is possible to interpret FTIR spectra when the material being examined is a single chemical of a single phase. The interpretation of complex mixed states of matter, as soil is, becomes much more difficult.

Objectives

The primary objective of this work was to collect diffuse reflectance FTIR spectral data for a variety of soils and contaminants of interest to the Army. A further emphasis was to determine the applicability of FTIR technology for rapid analysis of materials at field sites, either as a mobile field FTIR laboratory or integrated in some manner with the SCAPS application (Lee et al., 1993). We also evaluated quantitative analysis procedures using the partial least squares linear regression technique.

EXPERIMENTAL PROCEDURES

Test Soils

The test soils represent a range of physical properties. Included were neat soils used for establishing baseline spectrographic signatures, as well as soils taken from several DOD sites contaminated with explosive compounds. The baseline soils were collected, transported to the laboratory, cleaned of organic matter, and sieved to 4 mm to remove large aggregates. The soils were air-dried at 100°C in a vented oven for 24 h, then stored in sealed containers until further treatment, as discussed below. The soils taken from explosive-contaminated sites were tested in the as-received state.

Contaminants

The contaminant categories were explosives, petroleum/oils/lubricants (POL), and volatile organic compounds (VOC). The contaminants we examined were 2,4,6-trinitrotoluene (TNT); JP4 jet and #2 diesel fuels; and toluene and trichloroethylene. These chemicals were added to dry, sieved soils on a dry weight percent basis. Another additive we evaluated was water. Water is a naturally occurring compound in soils. Water gives varying spectral signatures depending on soil chemistry, so we examined each of the clean and contaminated soils at six different moisture levels. Deionized water was added on a weight percent basis in the amounts of 0, 2, 7, 12, 20, and 50 wt.%. TNT and liquids with low vapor pressures were thoroughly mixed with soil and then stored in glass vials with Teflon septa. VOCs were added to soil by injection through the septum of a sacrificial cap, which was replaced immediately after the contaminant was mixed into the soil. Samples were stored 24 h before spectra were taken.

FTIR Spectroscopy

All measurements were made in air at room temperature. FTIR spectra of the soils were obtained over the range from 4000 cm^{-1} to 400 cm^{-1} (2.5 to 25 μm) at 4 cm^{-1} resolution using a Nicolet 740 FT-IR spectrometer. Initial experiments were made using 0.5 to 4 cm^{-1} resolution and from 32 to 128 scans to determine the proper spectral resolution. It was found that higher resolution and increased scans did not significantly improve the signal-to-noise ratio beyond that found using an experimental setup of 32 scans collected at a resolution of 4 cm^{-1}. Spectra were taken in the reflectance mode using a Spectra-Tech Collector* biconical diffuse reflectance attachment, as shown in Figure 1. This de-

* Spectra-Tech, Inc., Stamford, CT.

vice makes it possible to mimic the way that data may be collected in actual environmental conditions in the field. The collector uses two 90° off-axis ellipsoidal mirrors that act as condensers to the beam of diffuse, reflected IR radiation. The specular component of this reflectance is not useful. The unit is fitted with a blocker device to eliminate the problem of a superimposed specular reflected spectrum (Messerschmidt, 1985).

Figure 1. **Schematic diagram of the biconical relectance apparatus. The ellipsoidal mirrors of the collector focus incoming infrared radiation (A) onto the sample surface and the diffuse components (B) radiated from the sample surface into the spectrometer.**

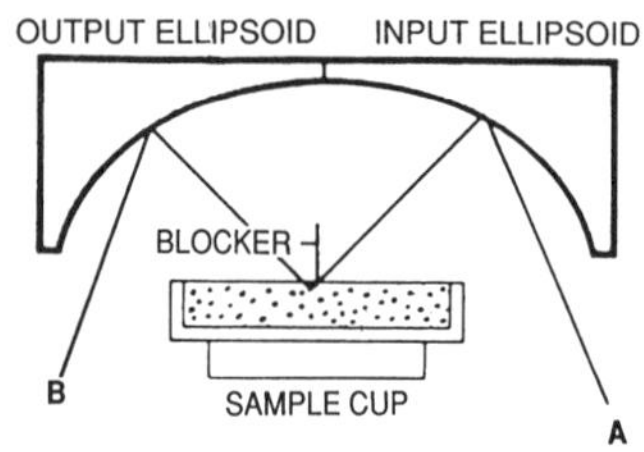

RESULTS

Spectral Data

Table 1 presents the matrix elements of our experiments. Tables 2 through 4 give physical characteristics of the soils.

Spectra of pure contaminant compounds and water presented in Figure 2 show the diagnostic regions of the IR spectra, primarily the absorption due to C-H stretching. Spectra of the five neat, baseline soil types are presented in Figure 3. Representative spectra in Figure 4 show the results of increasing watercontent, and Figure 5 the result of water and fuel additions to soils. Figure 6 presents the FTIR spectra of soils taken from explosives-contaminated sites.

Table 1. **Test matrix elements: contaminants and amounts added to soils[a]**

Contaminants Compound	Mol.wt., $g mol^{-1}$	Density, $g cm^{-1}$	Contaminant concentration levels, ppm	Soil moisture levels, wt.%
• Water	18.02	1.00	0	0
• Toluene, $C_6H_5CH_3$	92.14	0.87	1000	2
• Trichlorethylene $ClCH\text{-}CCl_2$, (TCE)	131.39	1.46	1500	7
			3000	12
• #2 Diesel fuel	--	0.87-0.90	5000	20
• JP4 Jet fuel	--	0.75-0.81	10,000	50
• 2,4,6-Trinitrotoluene, $C_7H_5N_3O_6$ (TNT)	227.13	1.65	15,000	

[a] Weast, 1987.

Table 2. **Textural Analysis of Neat, Control Soils[a]**

Soil type, identifier	Sand, %	Silt, %	Clay, %
Cay	2.0	20.6	77.4
Sand	97.7	2.3	0
Silty sand	77.5	13.6	8.9
Cayey sand	0.4	92.9	6.7
Washed sea sand[b]	100		

[a] Curtis et al.,1995.

[a] Product no. S25-10, Fischer, Fair Lawn, NJ

Table 3. **Mineral Phases of Neat Soils[a]**

	% Mineral phase								
Soil type, identifier	Quartz	Smectite	Calcite	K-feldspar	Na-plagioclase	Mica	Kaolinite	Dolomite	Cristobalite
Washed sea sand[b]	100								
Clay	3	80	trace	1	1	0	0	0	15
Sand	71	0	0	23	6	trace	0	0	0
Silty sand	89	0	trace	6	0	0	5	0	0
Clayey sand	50	38	0	5	3	trace	2	2	0

[a] Curtis et al.,1995.

[b] Product no. S25-10, Fischer, Fair Lawn, NJ.

Table 4. **Characteristics of Explosives Site Soils[a]**

Contaminated Site	Texture			Explosive Content, ppm		
	Sand	Silt	Clay	TNT	RDX	HMX
Hastings	18	54	28	12,637	--	--
				12,800	--	--
				18,572	19.3	--
McAlester AAP	42.5	42.5	15	17,605	3,375	--
Weldon Springs	27	54	19	41,100	--	--
				41,800	--	--
Louisiana AAP	35	47.5	17.5	377	2,470	471
Sub Base Bangor	77.5	22.5	0	9.5	--	--
Lone Star AAP	67.5	20	2.5	3	0.6	2.4
Iowa AAP	62.5	25	12.5	1.6	24.7	2.9
Crane Sifter	19	58	23	1,495	11,200	1,250
Umatilla	92.5	2.5	5	34,600	8,590	2,140

[a] Adams and Bennett, 1995; Pennington et al.,1994 Pennington et al., 1995; Gunniston et al., 1993.

Figure 2. **Baseline FTIR reflectance spectra of volatile organic contaminants and water.**

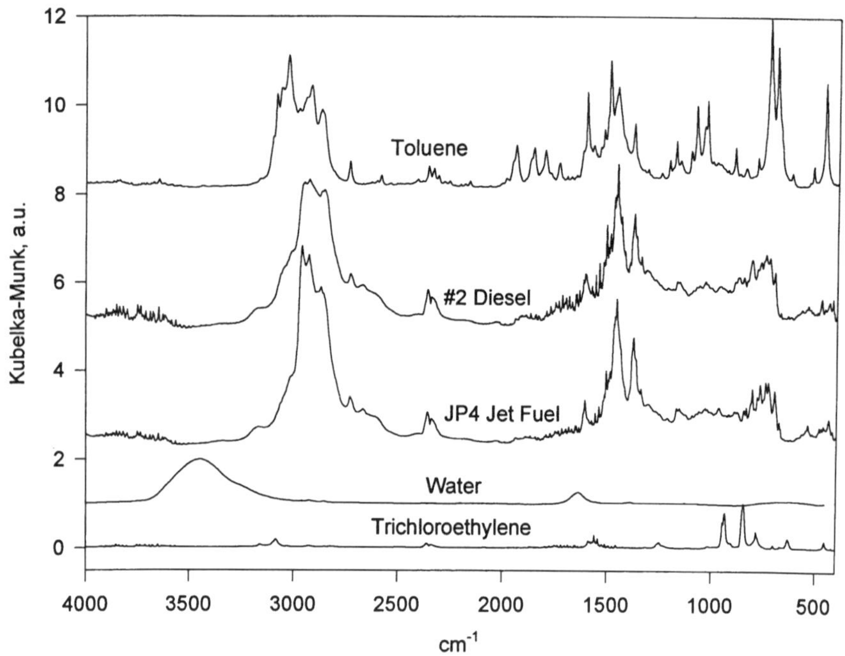

Figure 3. **FTIR reflectance spectra of baseline, uncontaminated soils.**

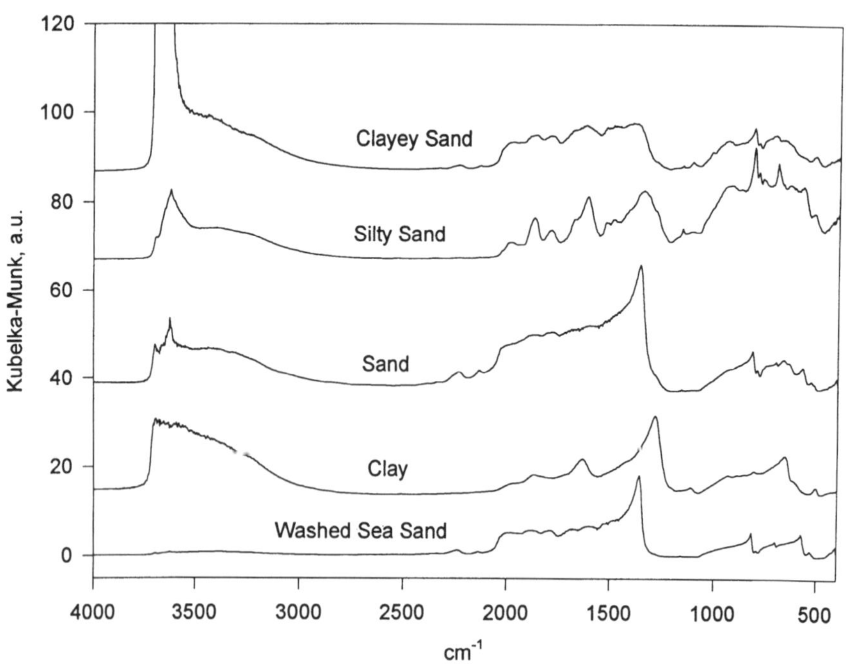

Figure 4. **FTIR reflectance spectra of two soils, clay and sand, showing the effect of increased water content on the spectral response:**

(a) a natural clay soil

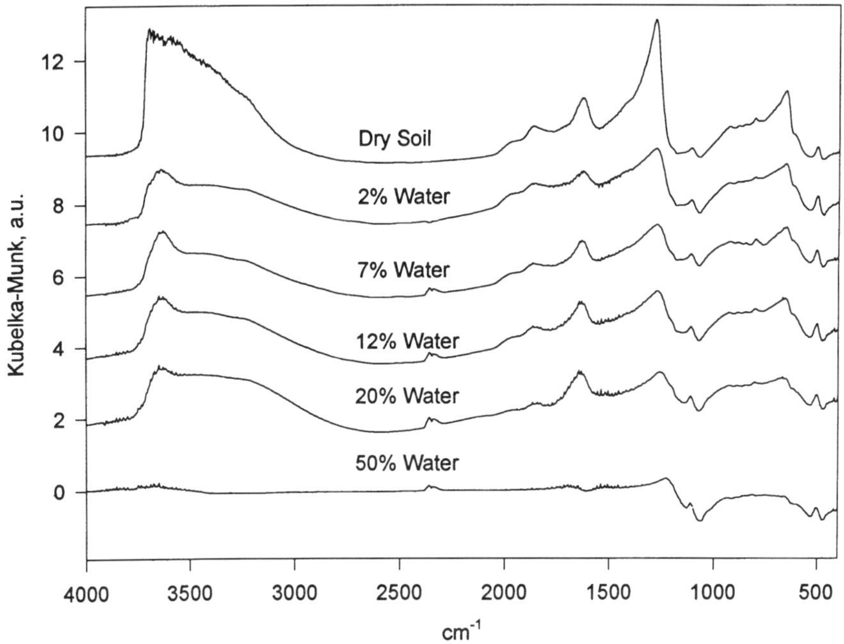

(b) sand.

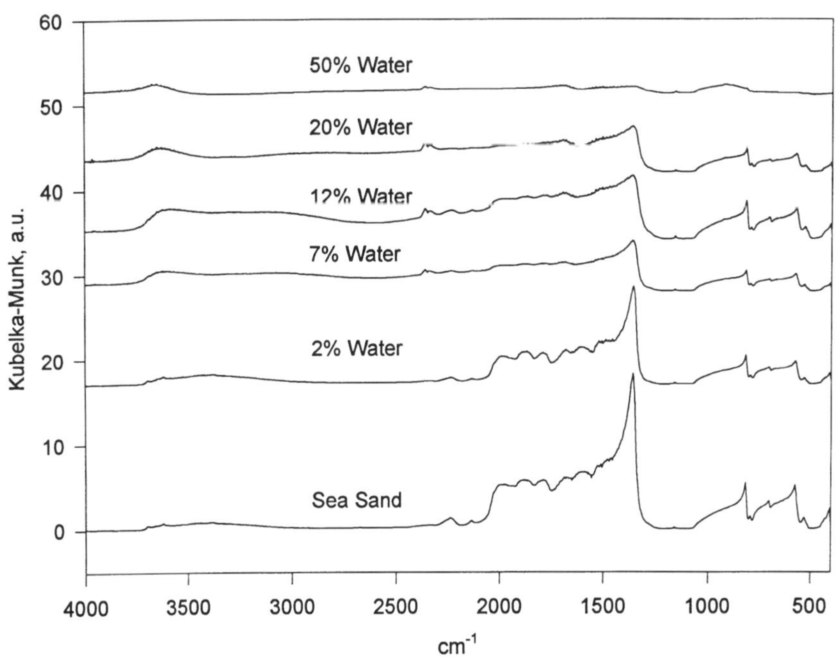

Figure 5. **FTIR reflectance spectra of a sandy soil showing the addition of JP4 jet fuel and water.**

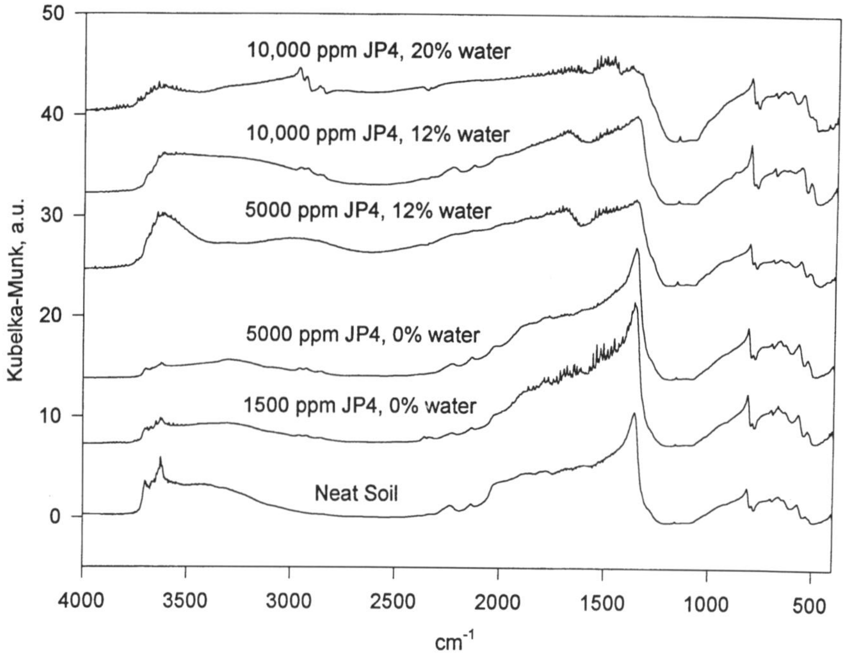

DISCUSSION

Spectral Interpretation

Soil texture is a critical factor to the ability to detect the FTIR signatures of contaminant agents in soils. The texture is determined by the amounts of sand, silt, and clay. These terms define particle size, and loosely define the physical and chemical characteristics of soil. The amounts of these components determine the permeability of air and water and also the capacity a given soil has to absorb contamination. Particle size also affects the incident infrared light by scattering it. Thus, the texture of a platy clay or sandy soil can modify the FTIR spectrum that can be collected from it. In general, very sandy soils gave results that were easy to interpret at low moisture levels and more difficult to interpret with increasing moisture. Spectra of clayey soils were the most difficult to interpret.

Many common soil chemical compounds have functional groups of atoms that are the same as those which are diagnostic of contaminating chemicals soil, and thus can complicate the interpretation of a soil FTIR signature. The most important is water, which exhibits nearly 100% absorbance at ~3500 cm^{-1} in a broad peak extending nearly 700 wavenumbers. Another effect of water is to enhance the spectral diffuse component in an FTIR reflectance spectrum because it can act as a mirror surface on wet particles. This is shown in spectra of water in a clay and sandy soil in Figures 4a and b. Some other compounds which

Figure 6. **FTIR spectra of soils taken from explosives-contaminated sites.**

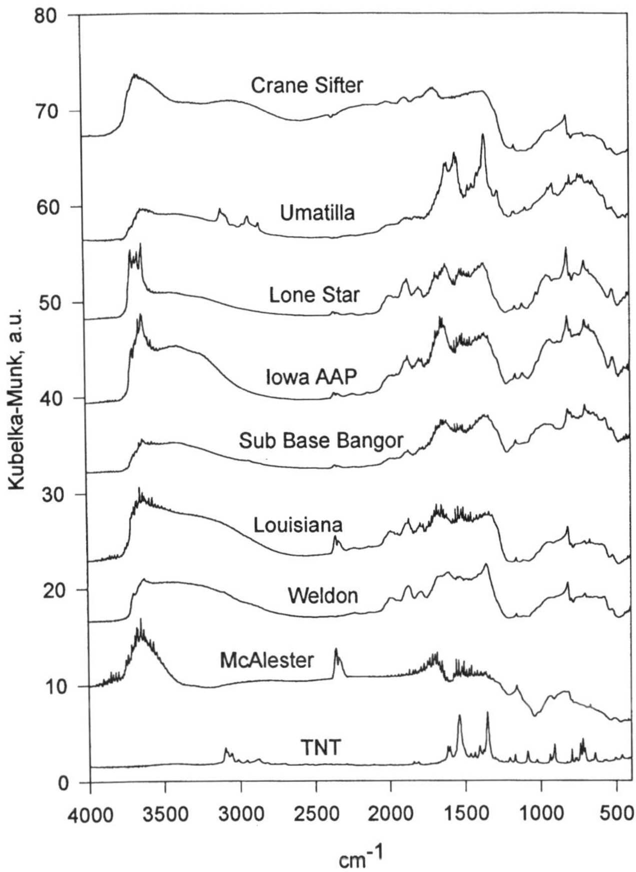

occur naturally in soils are nitrate groups, carboxylics and hydroxyls, phenols and other aromatic groups. Humic acids give a strong signal if present in amounts over about 0.5%.

The effects of the soil matrix were strongly seen in spectra of soils taken directly from contaminated ammunition plant sites. For the Umatilla soil shown in Figure 6, which is nearly 100% sand, the explosive compound is easily detected. In the same figure, a shoulder on the spectrum of the Weldon Springs soil at ~3100 cm^{-1} is interpreted to be from TNT because the soil was taken directly from an explosive-contaminated location, and the subsequent laboratory analysis indicated nearly 4 wt.% TNT. The other spectra are not diagnostic. Either they were from relatively clean areas, as was the case for three of the soils, or the spectral interactions of soil matrix compounds made interpretation inconclusive. The spectra of soils in Figure 6 from the McAlester and Crane Sifter sites distinctly illustrate this problem.

Quantitative Analysis

Semiquantitative indication of contaminant concentrations is possible with state-of-the-art spectrometers and software, providing that some basic feature of the site soil and probable contaminants are known. The partial least squares linear regression method estimated contaminant concentrations within an accuracy of about 0.5 wt.%, or ~5000 ppm, when there are distinctive spectral features. This limit may be improved upon with further analysis of the specific site soils and signal processing. Soil classification is readily determined with the SCAPS probe, and thus is not a problem. DOD sites generally have historical data available which tells what compounds are likely contaminants. The established regulatory cleanup standards levels required by each state for soil and groundwater hydrocarbon cleanup vary widely, ranging from lows of "any amount" and 0.02 ppm for toluene, to as high as 2000 ppm for diesel fuels (Oliver, 1994). In a dry sandy soil it should be possible to detect 2000 ppm of fuel, but this amount is probably not detectable with a few percent of water present. A point to be taken from this comment is that FTIR spectroscopy of soil will not be able to specify soil contamination to the same detection levels that have been established for soil by benchtop analytical techniques. Furthermore, difficulty in interpretation arises for mixtures of soil and waste compounds due to the fact that some diagnostic peaks of naturally occurring soil organic compounds and contaminant chemical functional groups may overlap. On a positive note, pattern recognition algorithms have been shown to be able to alleviate this problem (Klawun and Wilkins, 1995).

FTIR for Site Characterization

Until recently, the lack of reliable infrared-transmitting waveguides has been the main limitation to performing subsurface FTIR soil analysis. Commercial fibers are beginning to be available for FTIR applications where the length of the remote fiber optic is <10 m (AWMA, 1994; Ge et al., 1995). The Naval Research Laboratory has developed a research-grade, low-loss infrared-transmitting fiber optic which is physically robust and long enough to be used in a cone penetrometer sensor system (Sanghera and Aggarwal, 1995; Ewing et al., 1995). These low-loss chalcogenide glass, IR-transmitting fibers permit use up to 10 μm in the mid-IR region. The Naval Research Laboratory and WES are currently developing an FTIR sensor based on this new fiber for use in a SCAPS probe.

CONCLUSIONS

A primary goal of the DOD Environmental Cleanup effort is to develop technologies for in-situ soil probes that expedite the identification and location of a variety of liquid and solid contaminants in soils. This study examined soils from DOD facilities selected for remediation. Contaminant concentrations ranged from as low as 2 ppm to ~40,000 ppm by mass, and moisture ranged from 0 to 50%. Soil particle size, chemistry and mineralogy completed the analysis. FTIR readily detected most contaminants directly in many kinds of dry soils at 1% mass concentration. For some specific contaminant and soil combinations lower limits are possible.

Despite detection and quantification limitations, FTIR is a method which may directly detect the presence of some important soil contaminants. It can be ideally used as a rapid site analyzer to constrain the region where lab samples should be taken, and thus substantially reduce the need for laboratory analytical analyses of contaminated site samples. The detection of a specific contaminant compound depends on the soil type, moisture level, and contaminant concentration. The following lists the positive and negative factors of using FTIR to directly evaluate soil contamination.

Pro	Con
• no sample prep	• very sensitive to moisture
• extremely simple operation	• IR-fibers are developmental
• rapid, real-time data collection	• complex spectra, competing absorption bands may require extensive signal processing
• inexpensive	
• can evaluate all states of matter	
• chemical specificity—most compounds of interest are IR active	• soil matrix may substantially reduce reflectance energy, e.g., clays, water

ACKNOWLEDGMENTS

This work was jointly funded by the U.S. Army Environmental Quality Technology program, the U.S. Army Environmental Center under the sponsorship of Mr. George Robitaille, and the WES Laboratory Discretionary Research and Development Program which is managed by Dr. Robert Whalin, Director, WES.

REFERENCES

Lee, L.T., Chrestman, A.M., Douglas, D.H., Powell, J.F. and Malone, P.G. 1993. Site Investigations with the Site Characterization and Analysis Penetrometer System at Fort Dix, New Jersey, Technical Report GL-93-17, U.S. Army Engineer Waterways Experiment Station, Vicksburg, MS, p. 11.

AWMA Proceedings. 1994. Optical Sensing for Environmental Monitoring, SP-89, Air and Waste Management Assoc., Pittsburgh, PA.

Coates, J.P. 1995. Resources and References for Spectral Interpretation, *Spectroscopy* **10**, 14.

Farmer, V.C., ed. 1974. *The Infrared Spectra of Minerals, Mineral. Soc. Monograph 4*, London, Mineralogical Society.

McMillan, P.F. and Hofmeister, A.M. 1988. IR and Raman Spectroscopy, *Spectroscopic Methods in Mineralogy and Geology, Reviews in Mineralogy 18*, p. 99 (Hawthorne, F.C., Ed.), Washington, DC, Mineralogical Society of America.

King, T.V.V. and Olhoeft, G.R. 1989. Mapping Organic Contamination by Detection of Clay-Organic Processes, Proceedings Petroleum Hydrocarbons and Organic Chemicals in Ground Water, National Water Well Association, Dublin, OH, p. 627.

Messerschmidt, R.G. 1985. Complete Emission of Specular Reflectance in Infrared Diffuse Reflectance Measurements, *Appl. Spect.* **39**, 737.

Weast, R.C., Ed. 1987. *CRC Handbook of Chemistry and Physics*, 68th ed., Boca Raton, FL, CRC Press, Inc.

Curtis, J.O., Weiss, C. and Everett, J. 1995. Effect of Soil Composition on Complex Dielectric Properties, Technical Report EL-95-Draft, U.S. Army Engineer Waterways Experiment Station, Vicksburg, MS.

Adams, J.W. and Bennett, H.H., Jr. 1995. Rapid Identification and Quantitative Analysis of Selected Contaminant Materials Using Fourier Transform Infrared (FTIR) Spectroscopy Techniques, Technical Report EL-95-Draft, U.S. Army Engineer Waterways Experiment Station, Vicksburg, MS.

Pennington, J.C. et al. 1994. Interactions of Liquid Propellant/LP XM46 with Soils, Technical Report EL-94-10, U.S. Army Engineer Waterways Experiment Station, Vicksburg, MS.

Pennington, J.C., Myers, T.E., Davis, W.M., Olin, T.J., McDonald, T.A., Hayes, C.A., and Townsend, D.M. 1995. Impacts of Sorption on in situ Bioremediation of Explosives Contaminated Soils, Technical Report IRRP-95-1, U.S. Army Engineer Waterways Experiment Station, Vicksburg, MS.

Gunnison, D., Pennington, J.C., Price, C.B., and Myrick, G.B. 1993. Screening Test and Isolation Procedure for TNT-Degrading Microorganisms, Technical Report IRRP-93-2, U.S. Army Engineer Waterways Experiment Station, Vicksburg, MS.

Oliver, T. 1994. State Cleanup Standards for Hydrocarbon Contaminated Soil and Groundwater, *Soils* **12**, 14.

Klawun, C. and Wilkins, C.L. 1995. Neural Network Assisted Rapid Screening of Large Infrared Spectral Databases, *Anal. Chem.* **65**, 374.

Ge, Z., Brown, C.W. and Alberts, J.J. 1995. Infrared Fiber Optic Sensor for Petroleum," *Environ. Sci Technol.* **29**, 878.

Sanghera, J.S. and Aggarwal, I.D. 1995. Development of Low-Loss IR Optical Fibers for Optical Sensors. Optical Sensors for Environmental and Chemical Process Monitoring, p. 99 (Aggarwal, I.D., Farquharson S., and Koglin, E., Ed.), Proceedings of the SPIE 2367, SPIE-The International Society of Photo-Optical Instrumentation Engineers, Bellingham, WA.

Ewing, K.J., Bilodeau, T., Nau, G., Aggarwal, I.D., King, T., Clark, R., and Robitaille, G. 1995. Fiber Optic Infrared Reflectance Probe for Detection of Hydrocarbon Fuels in Soil. Optical Sensors for Environmental and Chemical Process Monitoring, p. 17 (Aggarwal, I.D. Farquharson S., and Koglin, E., Ed.), Proceedings of the SPIE 2367, SPIE-The International Society of Photo-Optical Instrumentation Engineers, Bellingham, WA.

CHAPTER 18

Soil Washing/Soil Leaching: An Innovative Soil Treatment for Metals Contamination

Matthew K. Bowers and **Norman C. Wenck**, Wenck Associates, Inc., Maple Plain, Minnesota

Martin R. McCleery, Department of the Army, Twin Cities Army Ammunition Plant, Arden Hills, Minnesota

Peter Rissell, U.S. Army Environmental Center, Aberdeen Proving Ground, Maryland

Darryl L. Terho and **Bridgette A. Manderfeld**, Federal Cartridge Company (Army Contractor), Twin Cities Army Ammunition Plant, Arden Hills, Minnesota

INTRODUCTION

Lead and other metals can pollute industrial sites like foundries, platers, smelters, battery reclaimers, and ordnance plants. For example, dumps and unlined stockpiles with metallic materials can contaminate the soil, especially posing a threat when contaminated dust can be inhaled or water supplies are affected. The U.S. Environmental Protection Agency (EPA) has named lead as their top-priority pollutant, followed by other substances like chromium, cadmium, and mercury.

For years, metals-contaminated soil could not be treated. Now an innovative treatment method has been found, thanks to an ambitious project at an Army ammunition plant in Minnesota. This project, which combined *soil washing* and *soil leaching*, was selected for an EPA-sponsored demonstration under their Superfund Innovative Technologies Evaluation (SITE) Program in 1994. It earned a Grand Award for outstanding achievement from the Minnesota Consulting Engineers Council, and has been featured at Army R&D conferences and shown on the Discovery Channel. Onsite tours of this project were conducted weekly. Completed in Summer 1995, this is the country's *first* full-scale project to use soil washing/soil leaching to remove metals from soil.

THE PROBLEM

The Twin Cities Army Ammunition Plant (TCAAP) lies among densely populated Minneapolis-St. Paul suburbs. The sprawling 2,400 acres include many contaminated disposal areas, making this Minnesota's top-ranked Superfund site.

One of the most contaminated areas was a former burning and burial ground for scrap ammunition and powder. Beginning in the 1950s, the ash and residue were dumped on the surface, with cartridge casings buried in trenches. These commonly accepted disposal practices left high levels of hazardous metals in the shallow soil and the disposal trenches. Lead was the worst, averaging 750 parts per million (ppm) and peaking at 120,000 ppm. In addition, tons of old bullets and cartridge casings remained at the site. Metals migrate slowly through soil, so they did not reach the major drinking water aquifer 120 feet below the site.

TCAAP stopped making ammunition in the 1970s, and is directing hundreds of millions of dollars toward pollution cleanup. The burning ground in particular had to be remediated under Superfund, as well as TCAAP's RCRA permit governing hazardous waste activities.

Undesirable Alternatives

The central dilemma was that no methods had been established to *remove* metals from soil. Instead, there were three traditional options:

1. Cap the site so rainwater and snowmelt do not wash contaminants further down;
2. Immobilize contaminants in place, by injecting cement or other material into the soil; or
3. Excavate the soil and then immobilize it for disposal in a landfill.

Capping and in-place immobilization were undesirable because they would leave contaminants onsite. As a result, options for site reuse would be sharply limited and require long-term monitoring, interfering with the Army's plans to eventually transfer or redevelop the land.

Excavating the soil would leave the site clean, but the contaminants would not be reclaimed; instead, they would simply be shifted to another location, sustaining the Army's liability. Furthermore, landfilling large amounts of contaminated soil is increasingly discouraged by regulators.

Incineration was not an option because the metals remain in the ash or become airborne. Methods to *treat* metals-contaminated soil were unestablished at the time. Nevertheless, the Army and regulators wanted a permanent solution, and their goal was "clean closure" of the site without transferring long-term risk to a landfill.

New Treatment Solution

Based on a 1993 agreement, the Army and Minnesota Pollution Control Agency decided to explore an untried treatment technology called *soil washing/soil leaching.*

Soil washing is an old sorting technique from the mining industry which has been used as a remediation technique in the Netherlands, Germany, and to a limited extent in the U.S. Physical processes break apart, separate, and rinse the soil, generating a clean soil fraction and a contaminated soil fraction. As a volume reduction technique, soil washing reduces how much contaminated material may have to go to a landfill.

However, metals still remain after soil washing, especially smaller metallic particles and ionic metals bound to the finer soils, along with discrete metallic fragments in the sand fraction. Therefore, a critical new addition is *soil leaching,* which chemically dissolves and recovers nearly all of the remaining metals. The recovered metals are recycled, eliminating the Army's liability. To our knowledge, there has been only one demonstration test elsewhere and no full-scale project, making this the first soil washing/soil leaching project in the United States.

Besides the challenge of implementing an untried technology, TCAAP faced strict cleanup goals to meet RCRA requirements. These goals were to reduce nearly all the metals to negotiated background levels (see list), rather than higher levels based on health risk. For lead, which was the highest-priority pollutant, the Minnesota Pollution Control Agency initially pursued a background level below 40 ppm, but agreed to a "soil exposure guideline" based on health risk set by the Minnesota Department of Health. This was 300 ppm, which is still challenging, considering that EPA's typical health-based cleanup levels for lead range from 500 to 1000 ppm.

Soil Cleanup Goals (ppm)	
Lead	300 ppm
Chromium	100 ppm
Copper	80 ppm
Nickel	45 ppm
Silver	5 ppm
Antimony	4 ppm
Cadmium	4 ppm
Mercury	0.3 ppm

Bench-scale studies for this TCAAP site reached these goals. They also confirmed that the site soil (mainly silty sand) would be amenable to treatment, and determined the optimum combination of soil washing, density separation, leachant type, and leachant contact times. Based on these convincing results, the Minnesota Pollution Control Agency agreed to bypass pilot testing, and required only an acceptance period of the first six days of actual treatment. During this period, all cleanup goals were met, including the reduction of lead to below 60 ppm in each run.

The soil washing process used at TCAAP was developed by Bescorp of Fairbanks, Alaska, who performed an EPA SITE demonstration project in Alaska. The crucial new soil leaching process (TerraMet™) was created by COGNIS of Santa Rosa, California. Wenck Associates, Inc. of Maple Plain, Minnesota, was the environmental engineering consultant who managed the project and coordinated with the Army; their operating contractor Federal Cartridge Company, Arden Hills, Minnesota; the Minnesota Pollution Control Agency; and U.S. EPA Region V.

THE PROJECT

Beginning during the summer of 1993, the site was cleared by burning vegetation and removing trees, small buildings, and underground utilities. A concrete pad had already been built nearby for an unrelated soil incineration project, so the treatment equipment was set up there. Six flatbed trucks brought the equipment onsite, where the soil washing and leaching processes were integrated into a single operation covering about half an acre. For safety, the site was restricted and monitored, workers wore protective clothing, and everything was decontaminated before leaving the site.

Over a three-acre area, soil was excavated in 6-inch layers down to 1 to 2 feet, with excavation down to about 12 feet at the disposal trenches. The extent of excavation was directed in the field using x-ray fluorescence (XRF), with confirmation in the laboratory. XRF is a relatively new and cost-effective technique that allowed greater precision during excavation. Because the site contained extensive amounts of ordnance fragments, specialists were contracted to oversee the excavation and removal of ordnance from the soil. The excavated material was transported the short distance to the treatment pad in covered trucks and staged in covered stockpiles.

One complication was that much more contaminated soil was found than had been expected. Earlier studies by others had found one disposal trench, but 19 more were discovered during the excavation, which increased the quantity of soil from 7,500 tons to nearly 20,000 tons. Furthermore, contaminant concentrations were higher than expected; for example, lead levels were 10-12% higher, with peak concentrations rising from 86,000 ppm to about 120,000 ppm. These higher levels required longer leachant contact times, further slowing the project. These complications were in addition to those inherent in designing a full-scale treatment system based on bench-scale studies alone.

As a result, the actual treatment rate was typically 6-12 tons of soil per hour instead of the design rate of 20, with 60-120 tons of soil treated daily, six days a week. Due to difficulties operating an aqueous process during Minnesota winters, treatment occurred September-November 1993, resumed for May-October .1994, and finished up during April-July 1995, with a total treatment period of 13 months.

Treatment began by feeding the excavated soil into a hopper and then a trommel that broke up soil clumps and rinsed it with high-pressure water. A logwasher from the mining industry was added in 1995 to better break up the soil. Material over 1/4-inch in diameter was screened out and sent to a conveyor, where inspectors hand-picked any remaining ordnance, while the rinsed stones and gravel were tested and returned to the site as clean.

The smaller rinsed material (below 1/4-inch in diameter) was sent on to a chamber that separated it into sand and fines (silt/clay), because they require different equipment for mixing and conveyance. Next, the sand went through density separation equipment (a spiral classifier and a mining industry jig) that removed larger metallic fragments for recycling. After soil washing and density separation, the sand and fines were sent on to the leaching process to remove the remaining metals (most of which were in the fines fraction). In separate, parallel circuits for sand and fines, the soil underwent an acid leach that dissolved fine metallic fragments and ionic metals into solution. For sand, this took place in a paddle washer, while for fines, this occurred in a series of mixers and clarifiers. In both leaching circuits, the leachant flowed against the incoming soil current, with the leachant getting progressively cleaner as the soil proceeded through the equipment series.

The treated soil was dewatered, neutralized, and laboratory tested, with clean material returned to the site. Due to the moisture content of the soil, especially the fines, dewatering was a problem that was solved during the project by adding a centrifuge. The metals-laden leachant was sent to recovery units, where it flowed through an electrochemical reduction system so that the metals precipitated out into a cake. The leachant was recycled internally.

At the conclusion of the project, equipment was decontaminated and removed, the remaining process water and leachant sent offsite for disposal, and the site graded, covered with topsoil, and revegetated with native prairie grasses.

RESULTS

The project was highly successful for lead removal, which was the primary contaminant of concern at the site. Approximately 98% of the soil passed the enforceable lead standard of 300 ppm.

The process was less successful lowering antimony, copper, and mercury to background levels because they were more difficult to leach and more difficult to recover from the leachant. An important finding, however, is that nearly all the metals were reduced to levels below typical values based on health risk. For example, the soil that failed the antimony, copper, or mercury background-based goals was still below typical health-risk levels, which are 10 to 100 times higher.

Therefore, the COGNIS/Bescorp process is a strong choice for sites with cleanup goals based on health risk. Since disposing of soil that poses little or no health risk is a questionable use of economic resources and limited landfill space, TCAAP is meeting with the state to renegotiate the antimony, copper, and mercury cleanup goals. If this negotiation is successful, virtually all of the soil can be returned to the site as clean; otherwise, about 13,000 tons of soil will be deemed clean, with 7,000 tons held back for further treatment or disposal.

The treatment process recovered 246 tons of metals for recycling. The metallic fragments were recovered from sand using density separation filled 234 55-gallon drums, while the metal cake precipitated out of the leachant-filled 392 drums. They were trucked via a licensed hazardous waste transporter to a lead smelter in Minnesota. The project also recovered approximately 9 tons of live ammunition, 44 tons of bullet projectiles, and 83 tons of cartridge casings. All ammunition was sent to an offsite incineration facility, where the explosive portions are destroyed and the metallic portions are recycled.

The schedule was lengthened because more soil was found, and with higher contaminant concentrations. Assuming 7,500 tons of soil treated at 20 tons/hour, the project would have taken 7 weeks. However, 20,000 tons of soil had to be treated, and at a slower rate of 6-12 tons/hour to account for greater contamination, requiring 12 months of treatment.

Originally, the estimated cost was approximately $3.5 million for 7,500 tons of soil. The actual cost increased to approximately $11.5 million for 20,000 tons of soil. Although some of the conditions worsened, the increased quantity of soil allowed some of the fixed costs to spread out, and the overall unit cost remained the same. The cost for soil washing/soil leaching is comparable to the traditional solutions of capping, stabilization, or landfilling, especially considering that treating the soil dramatically reduced the Army's liability and eliminated the need to monitor the site. The cost for treating each ton of soil to the strict cleanup goals was $240, with an overall unit cost of $575 per ton, which includes excavation, ordnance removal, health/safety measures, engineering oversight, laboratory analysis, shipping, and so on.

As with any new technology, it is inevitable that there will be improvements to the process in the years ahead, but we believe great progress has been made thus far. It is a big accomplishment to *treat* metals-contaminated soil rather than leaving it onsite or in a landfill, and to then return the hazardous metals to beneficial reuse. With metals impacting sites around the country, this project is a vital demonstration of the emerging treatment technology of soil washing/soil leaching.

CHAPTER 19

Decontamination of PCB in Soil Using the Base-Catalyzed Decomposition Process (BCDP) at U.S. Navy PWC in Guam

D.B. Chan and **S. Laura Yeh**, Naval Facilities Engineering Service Center (NFESC), Port Hueneme, California

INTRODUCTION

The Navy, along with other Department of Defense (DOD) Tri-Service, Department of Energy (DOE), and Environmental Protection Agency (EPA), have identified thousands of polychlorinated biphenyl (PCB) contaminated sites in the United States and overseas that require remediation. This requirement results from the implementation of the Comprehensive Environmental Response, Compensation, and Liability Act (CERCLA), and the Superfund Amendments and Reauthorization Act (SARA), the Toxic Substance Control Act (TSCA), and the Resource Conservation and Recovery Act (RCRA), for the cleanup of hazardous substances that have been released to the environment in the past (before 1984). PCBs have significant potential effect on the human health, general welfare, and the surrounding environment's ecosystem and natural resources.The Navy identified a PCB contaminated site at the vicinity of a former transformer repair/rework facility (Building 3009) located within the Public Works Center (PWC) complex near Apra Harbor, Guam. At this facility, electrical transformers were cleaned and repaired, and transformer oils were purified, stored, and recycled. The transformer oil storage and recycling systems were removed from service in 1977, when the EPA started to regulate the handling of PCB-containing materials. Based on such a regulatory requirement, the Navy initiated an investigation at the site, and subsequently identified a PCB oil being released to the environment. The PCB oil leakage contaminated soil in an area that covers a nearby surface runoff zone and a storm drainage ditch, and that requires a remediation.

Studies to characterize and estimate the extent of PCB contamination at the PCB oil release site were conducted from 1983 to 1990. It was estimated that approximately 4,000 cubic yards or 5,500 tons of soil were contaminated with PCBs (Aroclor 1260) having an average of 2,500 parts per million (ppm) concentration, ranging from 25 to 45,860 ppm.

The Navy, in search of an applicable, cost-effective, onsite remedial technology for PCB removal in soil in the Guam environment, initiated a collaborative effort with EPA/RREL in Cincinnati, Ohio, to develop an innovative alternative technology that culminated in the Base-Catalyzed Decomposition Process (BCDP). Both incineration and landfilling were determined to be very expensive and impractical, due to Guam's remote location. Biological treatment has been determined to be ineffective for the PCB soil contamination in Guam, specifically for the Aroclor 1260 PCB, and for the high concentrations; an average of 2,500 ppm. Therefore, a physicochemical process like BCDP became an economically attractive alternative to be developed for use in Guam.

The BCDP, patented by EPA/RREL, is an efficient, but relatively inexpensive and low temperature treatment process for PCBs. It is also effective in treating other halogenated compounds, such as insecticides, herbicides, pentachlorophenol (PCP), lindane, and chlorinated dibenzodioxins and furans.

BCDP TECHNOLOGY DESCRIPTION

Catalytic Transfer Hydrogenation

Hydrogenation of an unsaturated carbon structure using hydrogen gas and a metal catalyst is a reaction well known to the chemical industry. The possibility of achieving reduction with the aid of an organic molecule that functions as the hydrogen donor in the presence of a catalyst is a process known as catalytic transfer hydrogenation.

In the catalytic transfer hydrogenation reaction, the chlorine atoms on chlorinated molecules are removed and replaced by hydrogen atoms. It is illustrated in an equation as shown below:

$$\underset{\text{(chlorinated)}}{\text{Donor-H} + \text{Acceptor}} \xrightarrow[>300^{\circ}\,\text{C}]{\text{Base Catalyst}} \text{Donor} + \underset{\text{(dechlorinated)}}{\text{Acceptor-H}} + \text{Salt (chloride)}$$

In principle, the donor compounds can be any organic compound whose oxidation potential is sufficiently low so that hydrogen transfer can occur under relatively mild conditions. This principle is successfully applied to dehalogenate many toxic environmental contaminants, such as PCBs, dioxins, and furans.

The Rotary Kiln Reactor

The rotary kiln reactor (see Figure 1) provides a first stage treatment of PCB contaminated soil/solids. The processed (effluent) product meets the EPA's TSCA permit standard of 2 ppm per PCB congener. The cleaned soil, exiting from the rotary kiln reactor, can be returned to the site from which it was excavated.

In the processing, the contaminated soil is first screened, processed with a rock crusher, and stockpiled (see Figure 2). This is followed by feeding it to a pugmill, where a measured quantity of sodium bicarbonate ($NaHCO_3$) is uniformly mixed with the stockpiled soil (in the amount of 10% by weight) prior to its being fed to the rotary kiln reactor, in a continuous mode of operation. The chemical-laden soil is heated to 332°C (630°F) for an hour inside the rotary kiln reactor. By this treatment, the PCBs in the soil are either decomposed or being thermally desorbed. As a result, the soil contains less than 2 ppm per PCB congener. This treatment meets the EPA TSCA permit standard, and the cleaned soil may be returned to where it was excavated.

Generation and Treatment of the Residues

Residues generated from the rotary kiln reactor consist of dust (particulate), water vapor, and gases. These residues contain residual PCBs that must be captured and treated to prevent release to the atmosphere, and to the receiving sewer.

As shown in Figure 2, the offgases pass through a cyclone and a baghouse, where all the particulates (larger than one micron in size) are removed. The residual vapor, gases, and fines are processed through a venturi-(wet) scrubber, then are demisted and passed through carbon adsorbers. The cleaned gases, which are free of PCBs, are then discharged into the atmosphere. The scrubber liquid is allowed to settle, and the supernatant is recycled for additional scrubbing. The residual liquid (which is not used for recycling) is mixed with activated carbon for removing any remaining organics, including PCBs, and then passed through a filter press, where all carbon loaded with organics are separated. The filtrate from the filter press is pumped through an activated carbon adsorber to remove any residual PCBs. The carbon adsorber effluent contains less than one part per billion (ppb) of PCBs, meeting sewer discharge limits.

The Stirred Tank Reactor (STR)

The STR is used to treat all residues collected from the first stage treatment, and acts as the second stage or the final stage of treatment (see also Figure 2). The residues include dust collected from the cyclone and baghouse; precipitate from scrubber liquid settling, and the "spent" carbon from exhaust gas treatment, filter press (cake), filtrate treatment, and condensate resulting from gas/vapor emission control.

The STR, operated in batch mode, is charged with all the residues described above, and the following chemical reagents: sodium hydroxide (NaOH, in the amount of 2.5 times the PCB load by weight); a high boiling point oil such as Bunker oil #6 (in the amount of 1 to 1 of total residues by volume); a hydrogen donor (in the amount of 1 to 1 of PCB load by weight); and a proprietary catalyst (in the amount of 1% of residue weight). The mixture is continuously stirred

Figure 1. **Rotary kiln reactor.**

Figure 2. **BCDP flow diagram.**

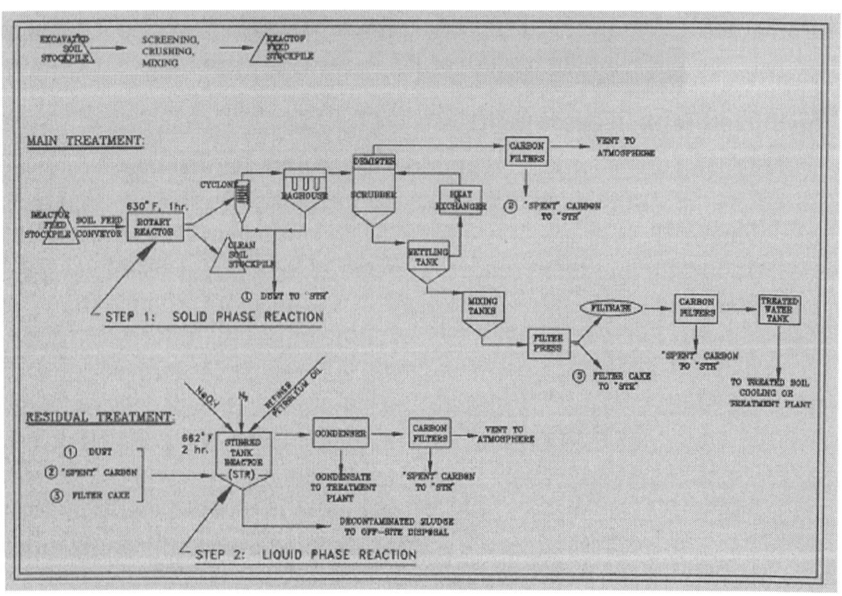

while it is being heated to a temperature of 350°C or 662°F, and is held at that temperature for a period of 3 hours.

Nitrogen gas is continuously pumped into the STR during the treatment to avoid any potential fire hazard. All PCBs in the residues are destroyed in this STR treatment. The Bunker oil is reclaimed for reuse in the STR operation. The accumulated oily waste from the STR operation can be used as boiler feed when its PCB content is less than 2 ppm. The gases emitted from the STR are purified with an activated carbon adsorber prior to being discharged to the air, while the vapor is condensed, and the liquid is passed through a set of activated carbon adsorbers for the removal of any residual PCBs. The treated vapor condensate contains less than 1 ppb PCBs, which is acceptable for discharge to the sewer. The final sludge may be solidified for conventional landfill disposal, or it may be fed to the rotary kiln reactor for reprocessing.

BCDP TECHNOLOGY DEMONSTRATION

The BCDP System

The BCDP system consists of the following major components: a grizzly and crusher; a pugmill; a sodium bicarbonate (baking soda) feeder; a rotary kiln reactor (with 1-2 ton per hour processing capacity); a cyclone; a baghouse; a venturi (wet)- scrubber; a demister; a carbon mixing tank; a filter press; a STR (with 400 gallon capacity); a condenser; a group of activated carbon absorbers; and a set of belt conveyors to transport soil from one component to the others. All unit components used in the system are off-the-shelf in nature, and therefore allow for easy operation and maintenance.

The Technology Demonstration Test Runs

Two series of test runs were conducted during the technology demonstration period, in February and November of 1994, respectively. The results for each run are summarized and presented in Tables 1 and 2. Table 1 shows that the first series BCDP treatment on 15 tons of PCB-containing soil with the rotary kiln reactor did achieve the removal of PCB in the soil to meet the EPA standard of 2 ppm per congener. It also shows that the desorbed PCB was deposited and concentrated in the cyclone and baghouse dust/particulates. In a rough material balance, shown in the table, there were about 29% uncounted PCBs presumably being decomposed inside the reactor. Figures in the water treatment columns indicate the effectiveness of carbon treatment, i.e., the effluent is acceptable to the sewer system. Sample analysis indicated that there were very trace amount of dioxins and furans in the reactor's feed, but none in the treated soil. However, as noted in the table, elevated levels were found in the cyclone and baghouse dust/particulates, as with PCB concentration.

The STR was not tested during this first series of test runs; therefore, no data exist for its performance in the field. Table 2 shows that the second series

BCDP rotary kiln reactor tested on about 45 tons of PCB soil, performed consistently to produce clean soil, even with much higher PCB concentrations (1,360 ppm average in Run 3, compared to 630 ppm in Run 1) in the feed. Again, it indicates the presence of PCBs in the cyclone and baghouse dust/particulates. It is worth noting that the baghouse particles contained extremely high concentrations of PCBs. This was primarily due to condensation of oily droplets (rich in PCBs) within the baghouse.

The second stage carbon treatment appeared to experience carbon breakthrough in that the effluent did not meet 1 ppb standard. However, this problem can easily be corrected. The observed concentrations or ratios of PCBs, dioxins, and furans in the feed, product, and the dust/particulates from cyclone and baghouse during the second series test run were comparable to February's first series run. The STR was operated during the second series test run. Results indicate that it was capable of completely treating the PCBs, dioxins, and furans in the feed. The STR operation was interrupted several times by a steam-induced foaming problem that caused a slowdown in heating. This problem was mitigated later by using sodium hydroxide flake, rather than its solution.

A further test run of the STR was conducted in May 1995, after some modifications were made to the STR operation, and a new hydrogen donor, one of the reaction ingredients, was introduced.

As described previously, the STR is used for treating residues resulting from operating the first stage treatment, a rotary kiln reactor. Particulates, spent carbons, and liquid condensate are collected and charged to the STR along with the chemicals, sodium hydroxide flakes, a high boiling point oil, a proprietary catalyst, and a hydrogen donor. A mixture of the residue and chemicals in the STR is heated to a temperature of 670°F and held for 3 hours while stirring.

Samples are drawn for analysis when the STR has cooled down to an ambient temperature. Three typical test runs were conducted between May 15 and 25, 1995. As the data show below, it was a total success.

STR Operation Results (batch mode operation, May 15-25, 1995, ~ 1,200 lb)
PCB Concentration (in ppm)

Run No	Before Treatment	After Treatment
1	850	nondetect *
2	25,000	nondetect *
3	42,000	nondetect *

Note : * PCB detection limit set at 0.5 ppm for each congener

Table 1. **BCDP Treatment Results for First Test Series (STR was Not Tested) Run 1 & 2 (performed respectively on February 5, 10 tons/10 hours & February 10, 5.7 tons/5.25 hours)**

A. Soil Treatment		B. Water Treatment		C. Rough Material Balances		
SAMPLE POINT	PCBs CONCENTRATION, ppm [a]	SAMPLE	PCBs CONCENTRATION	STREAM	TOTAL STREAM wt (lb)	PCBs Conc./Mass, ppm/lb
Feed	Run #1, 630 +/- 100, (10% $NaHCO_3$) Run #2, 1,300 +/- 100, (5% $NaHCO_3$)	Before treatment	1,054 ppm	Feed	20,000	630/13
Product	< 0.5, (0.0049 - 0.3)	After filter press	129 ppb	Cyclone	65	20,000/1.3
Cyclone	20,017	After first stage carbon	4.64 ppb	Baghouse	22	18,000/0.4
Baghouse	17,913	After second stage carbon	< 0.5 ppb	Scrubber water	6,000	1,100/6.6
Offgas	ND [b]			Carbon filter press	100	8,900/0.89
				Uncounted PCBs		/3.81 or 29 %

[a] Used Method 680; [b] ND = No Detectable PCBs.

Notes: (1) Sample analysis was done by the Southwest Research Institute (SWRI).
(2) Dioxin, very trace amount in feed; none in product; and elevated levels in cyclone and baghouse dust/particles.

Table 2. **BCDP Test Results for Second Test Series (STR Tested) Run 3 (November 19 - 21, 44.411 tons/36 hours)**

A. Soil Treatment		B. Water Treatment	
Sample Point	PCBs Concentration, ppm [a]	Sample	PCBs Concentration
Feed	1,360 (Avg.)	Before treatment	12 ppm
Product	0.2 (Avg.)	After filter press	2.6 ppm
Cyclone	15,742	After 1st stage carbon	3 ppb
Baghouse	109,331	After 2nd stage carbon	1.75 ppb
Offgas	ND [b]	Filter cake	48,000 ppm

[a] Used Method 680.; [b] ND = Undetermined

Note : Sample analysis was done by the Southwest Research Institute (SWRI)

CONCLUSIONS

1. The BCDP can cost-effectively (cost is estimated at about $400 per ton) treat PCBs in soil onsite to meet the EPA TSCA permit standard of 2 ppm per congener. Thus, the treated soil can be returned to the excavated area.
2. The BCDP is a simple process for detoxifying highly toxic chlorinated compounds, such as PCBs, dioxins, and furans, and uses inexpensive and nontoxic chemicals. It also requires minimally trained personnel to operate and maintain, since all unit components used in the system are off-the-shelf in nature.
3. The BCDP system is an automated and mobile system. It can easily be transported to project site, is quick to set up and get ready for operation.

CHAPTER 20

Update on Navy Technology Transfer Projects for Soil Contamination

John P. Collins, Naval Facilities Engineering Service Center, Port Hueneme, California

INTRODUCTION

The Naval Facilities Engineering Service Center (NFESC) is the Navy's Center for specialized facilities engineering and technology. In partnership with our customers, we deliver world-quality products and services in:

- shore, ocean, and waterfront facilities
- environment
- amphibious and expeditionary operations
- energy and utilities

As a member of the Naval Facilities Engineering Command (NAVFAC) team, we provide worldwide support to:

- NAVFAC HQ, EFDs/EFAs and PWCs
- Fleet and Shore Activities
- Marine Corps
- SYSCOMS/Claimants
- Other DOD Agencies

The Center provides solutions to problems through engineering, design, construction, consultation, test and evaluation, technology implementation, and management support. We leverage technology to enhance the effectiveness and efficiency of our customers. We use existing technology where we can, and identify and adapt breakthrough technology when appropriate. The Center performs research and development, when required, to meet Navy and Marine Corps needs.

This chapter will be limited to the Environmental Program. The Program includes, but is not limited to, Installation Restoration (IR or Cleanup), Pollution Prevention (PP), and Compliance. In particular, for the technology transfer issues, I will present only the programs associated with IR, PP, and Compliance.

The Secretary of Defense is charged by the Superfund Amendments and Reauthorization Act (SARA) of 1986 to carry out a program of environmental restoration at facilities under the jurisdiction of the Secretary. This program is known as the Defense Environmental Restoration Program (DERP). Traditionally, the program was centrally managed for the cleanup of past contamination from toxic and hazardous substances, low-level radioactive materials, and petroleum, oil, and lubricants (POL) at Department of Defense (DOD) sites. The Deputy Under Secretary of Defense for Environmental Security [DUSD(ES)] centrally manages the account, develops and defends the budget, and allocates funds among the Army, Navy, Air Force, and the Defense Agencies. The program is "devolving," or transferring management of funds to the services level. Base Realignment and Closure (BRAC) facilities follow procedures similar to DERP, except funding is through a separate BRAC account.

Facilities of the United States Government must comply not only with all federal laws and regulations, but also with all state and local requirements pursuant to federal environmental laws, such as CAA, CWA, RCRA, FIFRA, TSCA, CERCLA/SARA, and SDWA. Federal agencies need to recognize their obligations to comply with state and local environmental requirements, and incorporate such requirements into their environmental compliance and auditing programs. In many cases, state and local regulations are more stringent than Environmental Protection Agency (EPA) requirements, and federal facilities within those jurisdictions are required to comply with them. Executive Order #12088 also requires that federal agencies cooperate and consult with the state and local agencies in the prevention, control, and abatement of environmental pollution. This is particularly important, since the EPA has delegated or approved pollution control program administration authority in qualified states for many programs. States with delegated or authorized programs have the primary responsibility for ensuring federal facility compliance and enforcement of federal environmental requirements, as well as any additional applicable state requirements.

PROBLEM SUMMARY

The Navy, along with other government organizations and the public sector, is faced with a monumental task: mitigating the damage, or risk of damage, to health and the environment posed by hazardous materials released into the environment. The subject materials are released into the environment because of fuel and chemical spills, leaking storage tanks and transporters, disposal of wastes and off-spec materials, and past industrial operating practices.

The Navy's effort to address the problem began in 1981 with the implementation of a four-phased NACIP (Navy Assessment and Control of Installation Pollutants) program, which evolved into DERP in 1986. The Navy has identified five major waste/site problems at the Navy activities:

a. Organics (including fuel oil) contamination of groundwater and soil
b. Combined waste in landfills resulting in soil and groundwater contamination
c. Polychlorinated biphenyls (PCBs) and pesticides contaminating groundwater and soil
d. Ordnance-related compounds contaminating groundwater and soil
e. Heavy metal contamination of groundwater and soil/sediment

We are developing and transferring the technologies which are/will be cheaper, faster, and better in cleanup, pollution prevention, and compliance that are associated with these contaminants identified above.

TECHNOLOGIES

NAVFAC assigns roles and responsibilities to its commands in support of and consistent with the environmental requirements and programs of OPNAVINST 5090.1B. The NFESC has a primary responsibility to support the needs of the Fleet through engineering expertise. The NFESC will develop, transition, and implement environmental technologies and products in support of our customers' (Navy/Marine Corps activities) needs. The transition and implementation of new environmental technologies to shore facilities requires a sequence of steps that is similar to the acquisition procedures outlined in DODINST 5000.1. Each step involves data gathering, defining customer requirements, determination of deficiencies, address deficiencies, and final implementation of the technology. In addition, for any technology to be successfully implemented, life cycle costs need to be identified and parameters including labor, training, and infrastructure requirements need to be addressed.

Technologies can be acquired through in-house or joint development, transfer of proven technology from government or private industry, or modification of existing technologies. Government-developed technologies should be shared and be transferable to private industry where common needs arise.

Technologies the NFESC is currently engaged in demonstrating, transferring, and/or implementing are listed below (Note: many technologies in the list may also be associated with compliance problems).

REMEDIATION TECHNOLOGIES

SCAPS (Site Characterization Analysis Penetrometer System)

- three units now in the field as model for technology implementation
- Tom Flor discussed at September 1995 conference at Amherst, MA

BCDP (Base-Catalyzed Decomposition Process) for PCB Decontamination

- completed technology demonstration, Navy contractor is using the BCDP system at PWC Guam for cleanup of 5,500 tons of PCB-contaminated soil
- details will be presented later by Dr. Chan from NFESC

Biopile Bioremediation (resembling landfarming)

- demonstrate and implement the technology at 16 sites

In-Situ Bioslurping and Bioventing

- transferring technology for cleanup at two sites for bioslurping, and eight sites for bioventing
- also demonstrated at two sites for intrinsic bioremediation

Landfill Capping

- technology developed by DOE has been demonstrated at MCAS Kaneohe, HI
- five more sites have been identified for the technology in Hawaii, where heavy precipitation required by implementing the technology is readily available

Red Hill Leak Detection

- developing fuel tank monitoring system for temperature and pressure changes due to leakage

Small Arms Range Remediation

- test at two sites for lead and copper removal

Anaerobic Bioremediation

- gasoline removal at Seal Beach, CA, moving toward technology demonstration phase

SIVE (Steam Injection and Vacuum Extraction)

- demonstrated at Lemoore NAS for in-situ removal and recovery of JP-5
- 100,000 gallons of JP-5 (free product) was recovered from an estimated 200,000 gal spill site

NTS (National Test Site)

- biotechnology and physicochemical processes application for gasoline, diesel fuels, solvents, and TNT/RDX removal

Stabilization/Solidification

- chemical fixation and asphalt/cement solidification of heavy metals
- prepared a specification for technology transfer and implementation

POLLUTION PREVENTION TECHNOLOGIES

SCWO (Supercritical Water Oxidation)

- cooperation with DOE to treat mixed waste (including radioactive material) and highly toxic waste, such as chemical warfare agents

Plating Waste Reduction/Electrolytic Recovery

- prepared a User Data Package (UDP) for technology implementation (for rinse water only)

Hydroblast Wastewater Treatment

- have 53 units in operation in the field, managed by the NAVSSES

Bilge Oily WW Treatment

- three operational systems tested at three locations, processed 1.5 million gallons of bilge oily wastewater

Recycling/Purification of Plating Baths

- successfully tested at two sites for reduced use of chemicals in bath and membrane concentration technique. Technology is in transfer stage

IWTP Zero Discharge

- developed membrane technology for recovery and reuse of metals

Sodium Nitrite Treatment

- perfected nitrite reduction to nitrogen gas process. 45 units are in use in the field

TECHNOLOGY TRANSFER PROCESS

Navy-wide environmental technology transfer is a major function of the Naval Facilities Engineering Service Center (NFESC). To meet our customer's needs and requirements, we must search and develop the most cost-effective technology alternatives, and transition them to the field in a timely manner. To accelerate the Navy-wide transition, the NFESC has formed many Technology Application Teams/Process Action Teams (TATs/PATs). The TATs/PATs review all remediation/pollution prevention technologies currently being developed in-house and out, and provide recommendations for fast tracking the developmental and Navy-wide implementation of each technology.

The transition and implementation of new environmental technologies to shore facilities requires a sequence of steps, as shown in Figure 1, that is similar to the acquisition procedures outlined in DoDINST 5000.1. Each involves data gathering, defining customer requirements, determination of deficiencies, addressing deficiencies, and final implementation of the technology. In addition, for any technology to be successfully implemented, life cycle costs need to be identified as well as parameters, including: manpower, training, and infrastructure requirements.

When implementing environmental technologies additional issues need to be considered, including: commercial sources of innovative technologies, site-specific parameters, local regulatory requirements, changes in customers' needs, and differences in priorities of customers experiencing similar needs.

Technologies can be acquired through in-house joint development, transfer of proven technology from government or private industry, or modification of existing technologies. Government-developed technologies should be shared and be transferable to private industry where common needs arise. The strategy outlined in Figure 1 is used, independent of how the technology is acquired by the NFESC, as long as it provides a solution to the customer's need.

It is important to note that a given technology does not have to follow every step in the process as shown in Figure 1. The NFESC will emphasize available technologies where required before any RDT&E. This is necessary to provide fast track technologies with minimal developmental costs. These decisions are made on a case-by-case basis.

KEYS TO SUCCESSFUL TECHNOLOGY TRANSFER

A successful technology transfer process is extremely difficult to achieve and requires a broad range of skills. Past failures in technology transfer have often been due to the lack of appreciation for the interdisciplinary skills required for success, and a lack of recognition regarding the complexity and difficulty of bringing technologies from development to practical day-to-day field use. There are three significant functions or roles which, when properly fulfilled, can contribute significantly to a successful technology transfer process. These are: Gatekeeper, Entrepreneur, and Technology Partnerships. The functions of each position, as well as the characteristics and skills required to fulfill each role, are discussed below.

GATEKEEPERS

The function of a Gatekeeper as identified on Figure 1 is to facilitate go/no-go decisions throughout the Technology Development Process. This is not a popular

Figure 1. **RDT & E flow chart.**

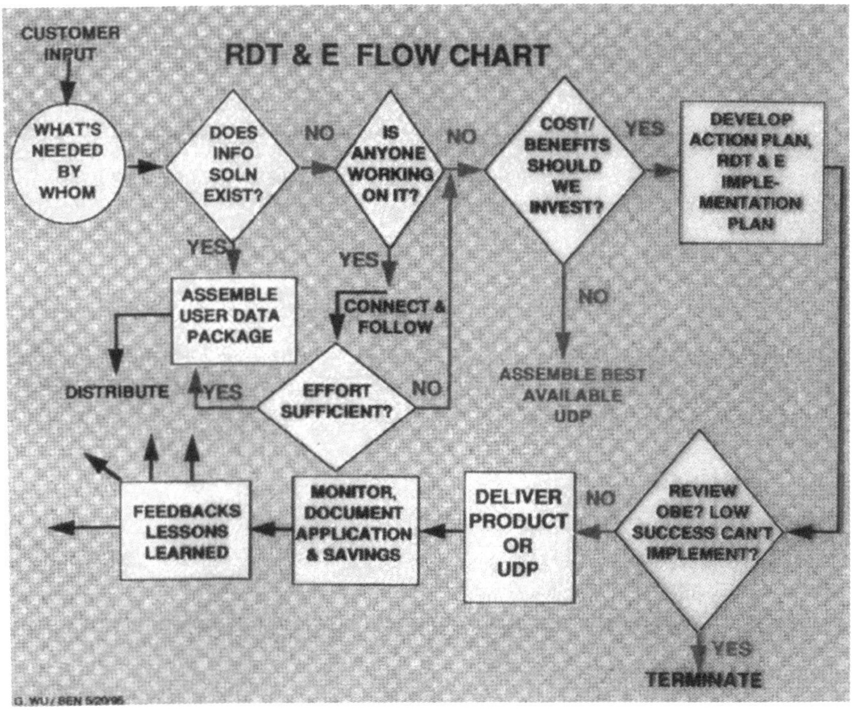

role, and has often not been fulfilled. Technology Gatekeeper "failures" are often due to lack of clear lines of authority and responsibility for go/no-go decisions. Gatekeepers also are often chosen from members of Technology Development Team, and therefore can have an in-built bias to continue projects to the next stage.

The characteristics of Gatekeepers are they must be independent, have decision authority or direct access to decision-makers, and should not be part of the Technology Development Team. They must not be part of the "R&D Club," and should feel no collegial loyalty to the development scientists and engineers. Gatekeepers must be "reality driven," with a laser-like focus on future needs and changing circumstances. They must be able to recognize when a developmental project is achieving its technology objective, is on schedule, on budget, but should be stopped because either the need has changed or been satisfied by an alternate technology from a different source.

ENTREPRENEURS

Perhaps the biggest mistake often made in the Technology Transfer Process is assuming the individual developing the technology is the best individual to transfer the technology to field use. In fact, in most cases, this is the worst person to be transitioning technology to field use. It must be recognized that the skills and talents of transitioning technologies, procuring equipment, constructing facilities, programming operation and maintenance funds, training personnel, and bringing the technology into full field use are entirely different skills than those scientific skills required in the development phase. Effective technology transfer individuals are entrepreneurs in every aspect. They are results-driven. They have a focus that success is only measured when a technology is in use and effectively showing results at the field level.

PARTNERSHIPS

Partnerships can significantly facilitate the Technology Transfer Process. For many projects, partnerships are essential where a single agency does not have the resources, the regulatory authority, or the comprehensive skills necessary to complete the Technology Transfer Process. Having a technology in place in the field, but lacking a permit to operate, is no more valuable than having the technology back in the laboratory. The most effective partnerships are those that bring together the Technology Transfer entrepreneurs; activity level personnel; local, state, or federal regulators; the public; and other interested parties. The technology transfer of the SCAPS truck to field use on the East and West Coast was greatly facilitated through a public/private partnership which involved academia, state and federal regulators and the Western Governor's Association.

SUMMARY AND CONCLUSION

The Naval Facilities Engineering Service Center (NFESC) is the Navy's Center for specialized facilities engineering and technology. In partnership with our customers, we deliver world-quality products and services.

In response to recent national and international emphasis on environmental issues in cleanup, pollution prevention, and compliance to the public laws and regulations, the NFESC has been tasked with significant roles in installation restoration, pollution prevention, and compliance by DoD/Navy. The NFESC is developing, transferring, and implementing remediation and pollution prevention technologies that will result in accomplishing site remediation and pollution prevention efforts faster, cheaper, and more effectively.

The technology transfer process is complex and must be approached by a team with individuals fulfilling the correct roles and assessing the requisite skills and expertise. Three roles which can facilitate the process are those of gatekeeper, entrepreneur, and partner. Properly selecting individuals with the skills to fulfill these roles can make technology transfer a reality.

CHAPTER 21

Army Overview of Environmental Remediation Technology

Michael J. Dette, Technology Demonstration and Transfer Branch,
U.S. Army Environmental Center, Aberdeen Proving Ground, Maryland

BACKGROUND

Contaminated soil presents a large problem to the Army, as it does many other federal and private sites across the United States. The Army has many of the same soil contamination problems as others do: petroleum products, solvents, and metals, but the Army has some that are not common to the rest, such as explosives, propellants, chemical agents, and depleted uranium. Also, the Army has the problem of having all of the above, combined, at some sites.

HISTORY

As a result, the Army has been developing technologies to deal with these problems. The Army Environmental Center has been developing, demonstrating, and implementing environmental technologies since 1980. Starting as part of the Program Manager for Chemical Demilitarization, we focused on explosive demilitarization technologies. This expanded into restoration technologies as the Army entered into the Installation Restoration Program. Along with restoration, the U.S. Army Toxic and Hazardous Materials Agency, as the Center was known in 1980, also developed pollution abatement technologies. In 1988, compliance technologies were added to the function of the organization. Today the U.S. Army Environmental Center no longer researches and develops technologies as its primary mission, but demonstrates and implements technologies that will benefit the environment in restoration, pollution prevention, compliance, and conservation.

PARTNERS

The Army environmental center is not alone in this endeavor. We rely on the work of many laboratories to research and develop tomorrow's technologies and provide the basic principles for full-scale implementable answers to the Army's environmental problems. Paramount to our success in demonstrating technologies is the work of organizations such as the Waterways Experiment Station,

Construction Engineering Research Laboratory, Cold Regions Research Engineering Laboratory, Topographic Engineering Center, Army Research Laboratories, and private sector companies. These organizations provide the bench-scale development and basic technology expertise to give the Army Environmental Center the ability to scale up good ideas, and make fully implementable technologies.

The rest of this discussion will focus on soil contamination technologies - those that have been proven to work on Army problems, and those that will hopefully be the answers soon.

ARMY SOIL PROBLEMS

The two largest problems the Army faces in soil contamination are explosives and heavy metals; the explosives being primarily TNT, RDX, HMX, Nitro Cellulose, and their breakdown products. The metals are lead, chromium, cadmium, and depleted uranium. At present, the most excepted and proven technologies are incineration for explosives, and monolithic stabilization for metals.

EXPLOSIVE CONTAMINATION

It was the Army Environmental Center that pioneered the use of incineration for explosive contaminated soil. It has been used to clean up soil contamination at Louisiana Army Ammunition Plant, Cornhusker Army Ammunition Plant, Savanna Depot Activity, and Alabama Army Ammunition Plant.

In recent years, new technologies have been developed to deal with explosives. These have moved into the realm of biological degradation. The first to be demonstrated is composting. At Umatilla Army Depot Activity, soils from a washout pond were composted with local manures in static piles and by windrow. TNT was shown to be degraded, and the resultant mixture was nontoxic. Based on this demonstration, windrow composting was chosen to be the remedial action in the record of decision.

Upcoming technologies for explosives include aerobic and anaerobic bioslurries and phytoremediation. An aerobic bioslurry process has been under demonstration at Joliet Army Ammunition Plant. The process uses a semibatch reactor process with a consortium of microbes and a cometabolite of molasses. The greatest advantage of the bioslurry process is the process control available. By using reactors process, variables such as moisture, mixing rate, nutrient loading, etc., can be monitored and adjusted to ensure an efficient process. This demonstration has just been completed, but a variation on this process using a surfactant is now being demonstrated to determine if the surfactant can speed the degradation of explosives. In fiscal year '96, a demonstration is scheduled for an anaerobic process at Iowa Army Ammunition Plant to treat explosive contaminated soil.

The use of plants to degrade explosives is presently getting greater attention. At present, the most favorable technologies to remediate explosive-contaminated soil are all ex-situ. No truly in-situ process has yet been found, but phytoremediation could be that process. Many plants have been found that have a nitroreductase enzyme. These plants can degrade explosive contaminants in the laboratory. This is not just a process of plant uptake, but one of actual degradation. The Army Environmental Center is presently working on a demonstration of phytoremediation in a wetland environment, for explosive contaminated groundwater, but this could be transferred directly to soils if proven successful.

HEAVY METAL CONTAMINATION

Heavy metal soil contamination has also been a long-standing problem. The two most used remedial actions at present are stabilization, and dig and haul. While both technologies take the immediate problem away, they do not necessarily offer a permanent solution or an end to liability. So now, the use of soilwashing and leaching and vitrification are being demonstrated. At the Twin Cities Army Ammunition Plant this process was used to remediate a lead contaminated site. The site was cleaned up and the lead was removed and recycled, thus ending the chain of liability. While this process is effective for many metals, it still has some disadvantages. The process is ex-situ, requires a large amount of equipment, and can be troubled by some metals such as mercury. A vitrification process is being demonstrated on explosive and heavy metal contaminated soil using plasma arc torches to supply the energy. While viable and in-situ, the process is also expensive, and will probably only be used on the more difficult wastes such as chemical agents.

Upcoming technologies for heavy metal soil contamination include phytoremediation and electrokinetics. Plants have also been shown to uptake metals; therefore, certain plant species can be grown in contaminated soil, and after the growing season be cultivated, ashed, and the ash smelted. Electrokinetics is a process of placing an anode and a cathode in the soil, applying an electric current across the electrodes, resulting in a current flow. Basically, by sending a current through the soil, metal ions will migrate according to charge to the appropriate electrode, and either be plated out onto the electrode, or can be removed with the electrolyte solution produced. This has been pilot-tested at Fort Polk where, with the electrodes spaced six feet apart, 95% of the lead in the soil was removed. If demonstrated successfully on a large scale, this process has the ability to be the in-situ answer to heavy metal contamination in many cases.

OTHER SOIL CONTAMINATION

While other soil contamination problems exist on Army sites, the Army is focusing on those that are more unique to the Army. Explosives and heavy metals are the Army's responsibility under a system called Project Reliance, which split contaminant responsibilities between the military services. The Air Force and the Navy are responsible for contaminants such as solvents and petroleum products. Therefore, the Army will implement environmental technologies for these other soil contaminants. Their technology development is not actively being pursued by the Army, but in the past, the Army has developed and demonstrated soil remediation technologies for volatile organic compound contamination. At Twin Cities Army Ammunition Plant, an In-Situ Vapor Extraction system was installed at a landfill that contained Trichloroethylene (TCE) from past operations and disposal. This system, combined with another installed later, has removed over 90 tons of TCE contamination from these sites. Also, at Letterkenny Army Depot the Army demonstrated the use of Low Temperature Thermal Stripping to remove TCE and other contaminants from an industrial lagoon and landfill. Both of these technologies are now in widespead use.

One other area of concern for the Army is Unexploded Ordnance (UXO). While not a soil contaminant in itself, we do consider it a soil remediation problem, especially at installations selected for closure. The Army has a very active program for unexploded ordnance detection and remediation. The Army Environmental Center has set up a seeded test bed at Jefferson Proving Ground to allow primarily private sector companies to come in and demonstrate their capabilities in over-the-land and airborne unexploded ordnance detection.

In addition, the Center is also working on the development of better unexploded ordnance detection and remediation technologies such as the Subsurface Ordnance Characterization System (SOCS). SOCS is a test-bed system that simultaneously acquires data from a variety of mounted UXO detection sensors. The system's flexible, open architecture enables it to readily use different sensor technologies for UXO detection, identification, and localization. Currently, magnetometer and ground penetrating radar (GPR) sensors are incorporated on SOCS. The SOCS design minimizes interferences among the GPR, magnetometers, and platform. These interferences were minimized by selecting and replacing materials, shielding GPR antennas, and isolating structural members that could form current loops in the platform and produce secondary magnetic fields.

In April 1995, SOCS was successfully demonstrated at Tyndall AFB. The demo included a survey of a 1.5-acre test site that contained metal storage drums and buried inert ordnance. The results of the demo showed that the SOCS subsystems (e.g., navigation, autonomous vehicle control, sensors and data acquisition) could be successfully integrated and operated to characterize a site containing UXO. The sensor suite operated as designed, and the data acquisition subsystem successfully collected position and time-tagged information from the magnetometer array and the GPR. The sensor data was post-processed, and target maps were generated and evaluated. Further analysis indicated that the sensor data was collected with a position accuracy within 7 cm.

CHAPTER 22

SCAPS as a Model for Implementing Technologies

Thomas H. Flor, Stephen E. Eikenberry, and **Vern Novstrup**, Naval Facilities Engineering Service Center, Port Hueneme, CA

TECHNOLOGY IMPLEMENTATION OVERVIEW

Terms

Terms such as innovative technology and technology implementation have been used extensively in recent years in technical and management discussions. Although these terms would appear to be clearly defined, various people and organizations commonly use the terms in very different ways. Therefore, we feel any meaningful discussion of innovative technology implementation must first define key terms. For purposes of this chapter, we will define the terms as follows:

Innovative Technology

Any technology which is significantly better, cheaper, or faster than existing technologies that is not broadly applied, due to limited knowledge or established standards within the engineering community.

Technology Implementation

A process that includes expansion of stakeholder[1] awareness, information dissemination, development of standard procedures and protocols, development of cost and performance reports, and other actions that encourage technology users to evaluate, procure, install, and operate previously unused or underused technologies.

Technology Transfer

Transfer of a government-developed technology to private and/or commercial applications for the purpose of stimulating commercial activity and/or national technology leadership.

---

[1]Stakeholders include regulators, community representatives, project designers, project implementers, and technology users.

Certified Technology

A technology that has undergone a rigorous evaluation by an independent organization or agency which issues a formal approval or certification.

Validated Technology

A technology that has undergone a rigorous test and evaluation program designed to measure specific operational parameters to determine the overall suitability of the technology for a specific purpose.

Demonstrated Technology

A technology that has been tested at one or more locations for the purpose of providing data for the specific purpose of encouraging stakeholders to gain a better understanding of its capabilities.

Accepted Technology

A technology that has received general acceptance as safe, cost-effective, and reliable within the stakeholder community for its intended purpose.

Historical View of Technology Implementation for Site Cleanup

Many key factors impacting on site cleanup are not clearly understood or well defined. In particular, investigation, cleanup, and permitting standards are highly variable. As a result, a technology that is well suited for a particular site may be totally unacceptable at a similar site in a different jurisdiction. Because of these inconsistencies it is very difficult, if not impossible, to select design parameters for a new technology that will meet regulatory investigation or remediation requirements for a specific class of sites.

Technology designers have tended to develop technologies without adequate input from the stakeholder community, either because the user community does not know what will be required, or because the system designers simply fail to obtain input at the appropriate time. In addition, many developers have not fully appreciated the social and business aspects of implementing a new technology. As a result, many emerging environmental technologies have failed because they did not meet the users' needs.

Technology developers tend to focus on the part of the project that is unique. As a result, other key parts of the overall project are neglected. Some examples of this problem that we have observed include failure of system designers to fully appreciate (1) safety standards, (2) spill prevention and control standards, and (3) equipment productivity optimization.

Changing the Implementation Paradigm

Although most technology implementation and technology transfer projects are managed by scientists and engineers, we must remember that the majority of the effort relates various aspects of the social and business sciences. The technical aspects of the process are very much secondary to a successful effort of either transferring or implementing a new technology. In order to be accepted, a new technology must not only work, it must also establish credibility with stakeholders. Credibility requires transfer of technical information, as well as the more abstract "warm and fuzzy feeling" that the technology is going to operate as expected, and produce a return on investment for the stakeholders.

Therefore, technology implementation and technology transfer efforts should include validation, certification, and demonstration efforts to achieve final acceptance. Validation efforts allow technology developers to fully define the operational parameters and limitations of the technology. Certification programs independently evaluate the accuracy of developer claims. Demonstrations allow the stakeholder community to preview the technology for the purpose of evaluating how the technology could be used to resolve real world problems. Validation, certification, and demonstration efforts do not necessarily imply acceptance. However, they greatly improve the probability of success.

SCAPS Background

The Navy's Site Characterization and Analysis Penetrometer System (SCAPS) is an innovative technology developed to rapidly characterize subsurface conditions at petroleum-contaminated sites. The technology obtains data by hydraulically pushing a small diameter, instrumented probe into the earth. Fiber optic sensors in the probe provide real-time semiquantitative information about the distribution of subsurface petroleum, and qualitative data about fuel product type, using Laser Induced Fluorescence (LIF) methods. Other sensors measure point penetration resistance, sleeve resistance, and electrical resistivity of the formation; these measurements are used to classify the soil. SCAPS is capable of obtaining a nearly continuous real-time data log of subsurface conditions, which is critical to determining the best configuration for monitoring well placement and selection of remediation technologies.

SCAPS MODEL FOR TECHNOLOGY IMPLEMENTATION

In the late 1980s, the three Services (Navy, Air Force, and Army) started development of a cone penetrometer mounted petroleum sensor that would allow for continuous vertical measurement of contaminant concentrations at spill sites. The system was designed to reduce the cost and time required to characterize petroleum spill sites. The initial system development was funded under the Strategic Environmental Research and Development Program.

In early 1993, the Naval Facilities Engineering Command decided to procure three SCAPS units. One unit was designated to support ongoing research and development efforts related to testing new cone penetrometer sensors. Two units were procured to provide support to Defense Environmental Restoration Account (DERA) and Base Realignment and Closure (BRAC) site investigations.

The cone penetrometer systems were procured from a commercial vendor of cone penetrometer systems. Naval Command Control and Ocean Surveillance Center Research and Development Division (NRAD) was selected to design and install the LIF systems on all three trucks.

In June 1993, Naval Facilities Engineering Service Center (NFESC) was tasked with the initial operation of the two production units and transfer of the units to an unspecified end user. The tasking called for NFESC to become responsible for oversight of LIF implementation in the Navy. Detailed discussions resulted in NRAD retaining responsibility for continued technology development, validation, certification, and technology transfer. NFESC was given responsibility for identification and clarification of potential uses of the technology, preparation of operating procedures, short-term operation of the production units, and training crews to operate the systems.

In late 1994, the Navy Public Works Centers (PWCs) in San Diego and Jacksonville were selected as the Navy's long-term operator of the SCAPS systems. Both commands immediately began to prepare for their role by selecting highly qualified individuals from their existing staff, providing the required training to the team members, and participating in field operations.

In November 1994, NFESC took delivery of the first of the two production units. The second unit was delivered in January 1995. Since delivery, both units have been operated in a production mode which has allowed NFESC and the PWCs to optimize operations and modify equipment where needed. Both production units are currently scheduled to be formally transferred to the PWCs for long-term operation in late 1995. Following transfer, NFESC and NRAD will continue to provide oversight and support needed to assure successful operations.

Validation, Certification, Demonstration, and Acceptance Activities

In an effort to achieve acceptance of the SCAPS technology, the Navy pursued a program that includes validation, certification, and demonstrations.

The validation effort sought to collect both cost and performance data needed to respond to questions from stakeholders. The performance validation effort is managed by NRAD, and focuses on collecting data that compare SCAPS field results to standard EPA test methods, including EPA 418.1 and EPA 8015, for a large number of soil types and petroleum products. Validation data are collected using highly standardized protocols. The data are being used to confirm the overall agreement of SCAPS data to existing test methods. Cost validation is managed by NFESC and is initially focused on developing an economic analysis to support the decision to implement the technology. As further operational data are collected, the analysis is being refined into a rate structure that will be used by the PWCs for long-term operations.

The NRAD is working to obtain "certification" through the California Environmental Technology Certification program, the Environmental Protection Consortium for Site Characterization Technology, and the Western Governor's Association (WGA) DOIT Committee. Although neither the EPA or WGA has authority to issue a certification, both organizations plan to produce a formal report documenting results of their evaluations.

The NRAD and NFESC have conducted a number of formal field demonstrations to allow stakeholders to view the technology and critically evaluate performance characteristics. Some of these include projects at the Hydrocarbon National Test Site in Port Hueneme, Navy Environmental Leadership Program in San Diego, and a static display at the Tenth Annual Conference on Contaminated Soils. In addition, the Navy commonly invites stakeholders to visit production sites to view ongoing operations All of these efforts have allowed interested stakeholders to see the equipment firsthand and to discuss capabilities of the system with the developers and operators.

SCAPS Model for Technology Transfer

In addition to owning and operating the SCAPS technology to support internal site cleanup needs, the Navy is seeking to fully develop commercial sources for the technology to further supplement in-house capabilities. To accomplish this goal, the NRAD has entered into a Cooperative Research and Development Agreement (CRADA) with Loral Corporation and Hogentogler and Company Incorporated for the purpose of encouraging use of the technology. The CRADA further expands on a technology licensing agreement issued by Army Waterways Experiment Station to several commercial vendors and technology development efforts that are sponsored by the Air Force. These efforts have resulted in full commercialization for the technology by Loral Corporation.

The CRADA has allowed for significant sharing of information that streamlines the validation, certification, and demonstration programs that are needed to support commercialization. Where feasible, Loral Corporation has been invited to participate in validation, certification, and demonstration projects.

CONCLUSIONS

The Navy has worked aggressively to achieve stakeholder acceptance of the SCAPS technology by focusing on validation, certification, and demonstration of the technology. Although these efforts have required a substantial investment of time and money, they have proved essential to obtaining regulatory acceptance and encouraging use of the technology.

The Department of Defense has been able to successfully transfer the SCAPS-LIF technology to the private sector by using a CRADA to reduce the risk to the private sector, and by investing in technology development, validation, certification, and demonstration programs.

CHAPTER 23

Expedite Everything: Base Closure at Naval Air Station Cecil Field, Jacksonville, Florida

Andrew J. Lonergan, ABB Environmental Services, Inc., Tallahassee, Florida

Mark E. Davidson,[1] Southern Division, Naval Facilities Engineering Command, North Charleston, South Carolina

Lisa M. Routhier, ABB Environmental Services, Inc., Orange Park, Florida

INTRODUCTION

Naval Air Station (NAS) Cecil Field was placed on the National Priorities List in 1989. At that time, 19 sites had been identified. A Federal Facilities Agreement (FFA) was signed in 1990 by the Navy, the State of Florida, and the U.S. Environmental Protection Agency. The FFA stipulated that a Site Management Plan (SMP) be completed and approved every fall. The SMP summarized the work which would be completed in that year and established dates for the submission of documents for regulatory review. Disagreements concerning the execution or completion of work under the FFA would be resolved with a formal dispute resolution process. In the summer and fall of 1992, disagreements arose and NAS Cecil Field went into dispute resolution. All site work at the base came to a virtual standstill for almost a year until the dispute was resolved in the summer of 1993. Resolution of the dispute coincided with the listing of Cecil Field on the 1993 Base Closure list.

In the fall of 1993, as the reality of closure by 1998 set in, the situation looked pretty grim. The agencies weren't happy with the Navy's program, the Navy wasn't happy with the agencies' requirements, and nobody was happy with the consultant. The parties weren't communicating, and trust was at a minimum. To make matters worse, the dispute resolution stipulated that a SMP had to be agreed upon by all parties and that the SMP must include the completion of four Interim Records of Decision (IRODs) and the submission of two Remedial Investigation/Feasibility Study reports in fiscal year 1994. Needless to say,

[1] Views expressed in this paper are those of the author and not necessarily the Department of the Navy.

the conditions were not favorable for "expediting everything," or for even accomplishing those items listed in the dispute resolution, and all of the work to meet the 1998 closure date had to begin immediately.

Now, in the fall of 1995, we are "expediting everything," and the progress made at NAS Cecil Field is a true success story. The people working at Cecil Field consider themselves to be a team, a team which consists of the Navy, the U.S. Environmental Protection Agency (USEPA), the Florida Department of Environmental Protection (DEP), and two consulting firms. Everyone is working together to meet the same goals as quickly as possible, and it is working.

The following paragraphs summarize how progress has been made, and how the process can be expedited. In order to "expedite everything," three key elements are required. First, trust must be established and maintained between all parties concerned in the effort. Trust is probably the most difficult hurdle to cross. The second key is cooperation; everyone has to work together, and very little will be accomplished without cooperation. Finally, all parties must persevere. Mistakes are made and setbacks occur; the impact of these must be minimized, and they cannot be allowed to derail the process.

PROGRAMS

Base closure requires the execution of work on several fronts. To meet the demands of closing a large federal facility, the parties involved have established several programs to complete the work. These programs have been divided into three broad categories for the purposes of this discussion. They are project management, standard hazardous waste site activities, and activities associated base closure. A listing of Base Realignment and Closure (BRAC) program components for NAS Cecil Field is provided below, in the order in which they are discussed.

- Project Management
- Installation Restoration Program
- Underground Storage Tank Program
- Unexploded Ordnance Program
- Remedial Actions
- Community Relations Program
- Environmental Baseline Survey
- Grey Sites Program
- Lead-Based Paint, Asbestos, and Water/Waste Water Management Plans
- Finding of Suitability for Transfer or Lease
- Reuse Commission

Each of the programs listed is discussed below. The discussion consists of a brief summary of the program and an evaluation of what worked well, what didn't work well, and how the process was expedited, if it could be.

PROJECT MANAGEMENT

As with projects of any kind, project management is a crucial part of the base closure activities for NAS Cecil Field. Project management, more than any of the other programs discussed, is at the heart of expediting everything. Base closure activities for NAS Cecil Field are being managed in a fashion which exemplifies how trust, cooperation, and perseverance, can combine to achieve success.

In the fall of 1993, the Navy, USEPA, and the Florida DEP began partnering. The Navy's contingent to the process included representatives of Naval Facilities Engineering Command Southern Division (SDIV), NAS Cecil Field, and ABB Environmental Services, Inc. The remedial project managers (RPMs) for Florida and the USEPA made up the remainder of the Tier I partnering team. Partnering had already been established between management level personnel at the Navy and the agencies (Tier II) prior to the beginning of partnering at the Tier I level. The goal of the partnering initiative was to expedite the completion of work by empowering the Tier I team to do whatever was necessary to make things happen, while remaining within the law, with the knowledge Tier II support was available whenever needed. A detailed discussion of how partnering works is not the focus of this chapter, but the partnering initiative provided the crucial first step in the road to success.

During the training sessions for partnering, the Navy, FDEP, USEPA, and ABB managers came to know each other, work together as a team, and began to trust each other while not working on environmental issues for NAS Cecil Field. These training sessions and subsequent partnering maintenance activities provided an environment of cooperation which led to trust. Trust and cooperation became vital when addressing the environmental and management challenges base closure presents for NAS Cecil Field.

In addition to the partnering, a BRAC Cleanup Team (BCT) was established. The BCT consists of three persons, the BRAC Environmental Coordinator (BEC) and the RPMs from the FDEP and USEPA. These three managers, functioning as a team within or above the partnering team, are responsible for making base closure work. Trust, cooperation, and perseverance within the BCT are essential, because they must see that the base is closed, the environment is protected, the laws are observed, the funds are available, the resources are properly allocated, the costs are conserved, the problems are resolved, and the various programs are properly staffed. This is a formidable list of items to accomplish, and the last item, staffing, is very important to expediting everything.

Proper staffing of the various programs is a key management element to success and to expediting any process. The managers need to find and use specialists in every area, and ensure that those specialist are given the tools to accomplish their tasks. A specialist, given the proper support and latitude, is the person best suited to safely and efficiently expedite a process. The consultant plays an important role in providing the needed specialists or subcontracting for their services as needed. The NAS Cecil Field BCT has also successfully used technical resources from within the SDIV, the FDEP, and the USEPA. If the team does not know an answer or the best way to proceed, then someone who does must be located.

Trust, cooperation, and perseverance are facilitated by communication. The partnering initiative has served to greatly increase communication among all parties concerned with the closure of NAS Cecil Field. When the partnering team meets, ranks and titles are left at the door, and communication, while structured, is free and open. This open communication has led to trust, and trust has led the consultant to be viewed as a true consultant, rather than a subcontractor. Often, and sometimes rightfully so, the consultant is perceived as someone interested only in profit. This perception can lead to genuinely good ideas not being fully explored. If everything is to be expedited, all ideas, regardless of their source, need to be fully evaluated.

The project managers have to be prepared for the unexpected. The unexpected always occurs, and often more than once a day, when trying to accomplish the closing of a large federal facility. Several unexpected items are discussed below for the various programs, but it is important to note under project management that having a plan for crisis management has helped the closure effort to succeed even when the unexpected occurs. When the unexpected occurs for NAS Cecil Field, the team first endeavors not to overreact or to act singly. The BCT or the partnering team will get together, either in person or on the phone, discuss the unexpected item, draw on the technical resources available to them, and then make a decision that is agreeable to everyone by reaching consensus (rather than voting). Over and over again, this process, which will not work without trust, cooperation, and perseverance, has served to smooth the bumps and accelerate the process. Additionally, the team has grown together by overcoming obstacles as a team.

This type of hands-on project management requires a great deal of dedication from the Navy and agency RPMs. The BCT has to be available at all times to keep everything running smoothly. In order to increase availability, the members of the BCT have developed a means of moving forward when someone is missing. Everyone has personal matters to manage or needs an occasional vacation, but at the same time, the work must go on if the expedited pace is to be met. BCT members routinely pass their proxy to another member of the BCT when one can't be present. Imagine, the USEPA giving the Navy its proxy. This is unprecedented in our experience, and trust in its highest manifestation. This type of hands-on management and flexibility has been extremely helpful in keeping the work going, and managing the unexpected.

Finally, uncertainty must be managed if processes are to be expedited. The management of uncertainty is probably the most difficult aspect of base closure. At the typical Superfund site, uncertainty, and the failure to aggressively manage it, result in tremendous delays in the time it takes to actually do something positive for the environment. Often it can take up to a year to get a work plan approved because of a failure to recognize and manage uncertainty, when that very uncertainty [is unacceptable contamination present?] is the entire reason for performing an investigation. Delays of this type have been successfully and aggressively managed at NAS Cecil Field (discussed below). The management of uncertainty is an ongoing battle.

INSTALLATION RESTORATION PROGRAM

The Installation Restoration (IR) Program at NAS Cecil Field is focused on the characterization and remediation of Comprehensive Environmental Response, Compensation and Liability Act (CERCLA) and Resource Conservation Recovery Act (RCRA). The NAS Cecil Field facility has been designated a CERCLA site and consists of 19 known sites. Eighteen of the sites are CERCLA sites, one is an underground storage tank site, and only part of one site has been addressed under RCRA.

As discussed above in the Introduction, work on the IR sites is completed under a Site Management Plan (SMP). The SMP was originally included in the Federal Facilities Agreement for NAS Cecil Field as an enforcement tool for the agencies. The SMP is still used, but its role is now more of a tracking tool. The failure to negotiate a site management plan in 1992 stopped work on the base for almost a year. Negotiation isn't really necessary any more, as the work is moving so quickly that the dates fall where they may. There is so much going on that team members routinely check the SMP to see where we are and what is coming next.

Several successes have been achieved in the IR program. The Operable Unit 1 (OU 1) Remedial Investigation/Feasibility Study (RI/FS) is a good example. OU 1 consists of two abandoned landfills. Monitoring well installation began in August of 1993, the RI/FS was submitted in March of 1994, finalized in December of 1994, and the final Record of Decision (ROD) was signed in August of 1995, a total of 24 months from monitoring well installation to final ROD for two landfills.

As the Cecil Field Team (Navy, agencies, and consultants) has grown together, success comes more easily and frequently. Learning, up front, what the agencies require in an investigation and then delivering it the first time has permitted a large amount of work to be completed in a short amount of time. Between August 1994 and September 1995, draft RI/FS reports were submitted for

OUs 2, 7, and 8, and final RI/FS reports were submitted for OUs 1, 2, 7, and 8 (a total of six separate sites). Additionally, as stipulated in the dispute resolution, four Interim RODs, were signed during fiscal year 1994 and two final RODs (OUs 1 and 2) will be signed in FY 1995.

The RI/FS process is an area particularly suited for expediting. Several process improvements have led to better and shorter documents which need less agency review. Agency support throughout the RI/FS process has been the key element to expediting everything.

One of the improvements made in the process concerns the preparation and finalization of work plans. Work plans for NAS Cecil Field CERCLA sites state that the site will be screened until the nature and extent of contamination can be determined, and then a confirmation sampling program will be completed. This does not seem that out of the ordinary, but it actually is. There is no discussion of where or how many wells will be installed or soil samples collected until the contamination is located and defined via an aggressive and comprehensive screening program.

Following completion of the screening program, a meeting is held and the confirmation sampling locations and numbers are agreed upon. Following agreement on the confirmation program, the meeting minutes (with figures showing the locations) are submitted as the final step in a dynamic work plan which has the approval of the agencies prior to its submission. This process has worked so efficiently, especially when media are treated sequentially, that the field team does not even need to be demobilized between the completion of the screening program and the beginning of the confirmation program. The alternative (summarize the screening results in a document and propose a sampling program, submit for agency review, receive and address comments, then finalize the document) pales in comparison. Trust and cooperation have led to greatly reduced project planning time.

Several steps have been taken to expedite the RI/FS process. These steps include meetings with the Navy, the agencies, and the consultant following document finalization to determine what worked well and what didn't and, more importantly, what wasn't needed and what could be shortened. Because of increased communication and the trust and cooperation developed over time, entire sections have been deleted from the standard RI/FS. This has made the document smaller and less costly to produce and reproduce. Additionally, meetings are held to discuss the findings of the investigation prior to submission of the RI/FS. These meetings have led to RI/FS reports that can (almost) be approved the first time through. Following the receipt of comments on an RI/FS (one FS did not receive any comments), a meeting is held to formulate a response which is agreeable to all parties concerned. This discussion and resolution of issues has led to a process in which only a draft and a final RI/FS are submitted; the draft final has been eliminated.

Staffing of the IR program has helped to expedite the process. A technical lead is assigned to an OU when the field program starts. That technical lead stays with the OU from the field program to the completion of the remedial design document. The dedication of personnel helps to expedite the preparation of project deliverables and ensure that the memory of the investigation is maintained (questions such as, "Why didn't you do this..." can really disrupt the schedule, especially if no one has an answer). Finally, a dedicated field operations leader for all IR field work has helped to ensure consistency in the acquisition of field data.

UST PROGRAM

Underground storage tank (UST) concerns at NAS Cecil Field are currently being addressed by the same management structure responsible for the other environmental programs. This, however, has not always been the case. There are over 100 USTs on the facility which service individual buildings or operations, and approximately 50 USTs which formerly provided fuel oil to private residences. In addition to these tanks, there are two large tank farms on the base service flight operations with JP-5. The two tank farms are connected to two separate fuel distribution systems which bring fuel to the two runways. The wide variety of uses for USTs has resulted in different concerns having responsibility for their monitoring and maintenance.

The BRAC Cleanup Team, in conjunction with representatives from NAS Cecil Field, has decided to consolidate the various efforts in progress to address UST concerns for the base. This effort is still in the planning phase, and a management plan for the UST program is currently being developed.

UXO PROGRAM

Because NAS Cecil Field's operation included the use, storage, and maintenance of aircraft ammunition and ordnance, an Unexploded Ordnance (UXO) location and removal program was initiated in order to ensure the turnover of property free from the risks of buried explosives. The task, at first, seemed relatively straightforward, with a limited scope - essentially locate, excavate, haul off, and properly dispose of the buried ordnance. But, as with many new projects, things were not quite that easy.

Prior to being placed on the closure list, Cecil Field had already identified and evaluated ordnance and ammunition disposal/detonation areas (Sites 14, 15, 18, and AOI34). Further interviews and record searches following placement on the closure list revealed one additional site, as well as the need for a survey of a large area that was formerly used as a naval gunnery school.

A UXO contractor was procured, and a geophysical survey was initiated. Once the survey was completed, the anomalous areas were then excavated by hand, and the ordnance properly disposed of. Inert material was generally disposed of as scrap metal, while ordnance that potentially was live was either transported off base for detonation and disposal or, in some cases, detonated in place.

Performing an in-depth record search was a very useful tool in helping to locate previously existing ordnance disposal and detonation areas. The record search techniques included a review of all applicable records, interviews of key personnel, a review of aerial photographs, and a search of all drawings in the public works department. These techniques, while used during the initial development of the Environmental Baseline Survey (EBS), were also performed to a limited extent following the completion of the EBS. This follow-on record search led to the discovery of not only one additional site, but it also led to the realization that another one of the sites, the former naval gunnery training area, may have been larger than expected. The area, originally slated for only a 50-acre survey, should have been more like 500 acres, with a maximum downrange area up 5000 acres.

While our budget allowed us to increase our survey of this area to only 460 acres, an underlying question remained; how do you survey an 5000 acre parcel that may have ammunition randomly dispersed over the entire area and come away with a high confidence level that you found everything? The project team decided that instead of doing all 460 acres as one lump parcel and, thus, leaving over 4500 acres unsurveyed, we would take more of a screening approach to the whole area. This was accomplished by subdividing the 460 acres under the existing contract into several 20- to 25-acre plots whose locations were logically chosen, based on aerial photographs and drawings.

As results from portions of the subplots became available, the areas for survey have been continuously divided to expand horizontal coverage. Regular meetings are held to determine trends in the distribution of the anomalies, and plan the division of remaining plots to maximize coverage. This flexibility in rearranging the areas to be surveyed has enabled us to raise our confidence level in the survey we have undertaken; perseverance and cooperation have enabled us to keep the program moving despite what appeared, originally, to be a very unfavorable situation.

The geophysical survey performed by the contractor was accomplished with the DANS system, an acronym for Data Evaluation and Navigational System. DANS is a multicomponent geophysical mapping and surveying system. The configuration we used consisted of the Schonstedt GA-52B Magnetometer as the sensor and the Ultrasonic Navigation System (UNS) to continuously record precise horizontal position of the sensor. The UNS reports location information to a onsite computer, which tracks the location of the sensor in real time and generates a track map. A controller, watching the tracks develop onscreen, directs the technician with the sensor to ensure adequate coverage of the area. Magnetometer readings are also continuously relayed to the computer, and recorded digitally for plotting on the track maps.

The DANS system provides several advantages over conventional "mag and flag" survey methods. The use of DANS enables real-time display and logging of data in a Geographic Information System (GIS) format that is used to produce maps with an accuracy of 6 inches. This geophysical technology gave us both a more accurate and more economical advantage over previous survey techniques. It also allows us to determine, even years later, the location of all anomalies. Mag and flag type surveys typically do not afford you this opportunity.

A cost-saving measure employed for the UXO program was use of the Navy's Explosive Ordnance Division (EOD). EOD is onboard for the excavation of all anomalies located during the DANS survey of the gunnery school. These highly trained individuals are available to Naval facilities to help in the surveying and excavation of UXO. While, in some cases, requiring lead time, they make up for this disadvantage with the relatively inexpensive costs for their services. In the case of Cecil Field, the only cost incurred for UXO excavation was their per diem and supplies.

One of the issues that surfaced midstream in the program was that during the excavation of the anomalies by the contractor (at site 18 and AOI 34), it was determined that EOD was required to be onsite during this excavation. This oversight caused excavation to come to a halt until EOD arrived on site. Although the project team had informed EOD, through phone conversations, that our contractor would be excavating, it was never followed up with any letter or official correspondence. This issue led to a certain amount of fingerpointing, which may have been avoided if a letter to EOD had been prepared, outlining our intentions.

Another issue was how to handle inert ordnance and potentially live ordnance following its excavation. A discussion of the policies, laws, and RCRA issues concerning the handling and disposal of excavated UXO are beyond the scope of this chapter, but the project team learned a good lesson. Before starting any UXO work, a plan, acceptable to all parties, must be in place for the disposal of excavated material. A small amount of planning and communication can really go a long way toward making a UXO excavation effort proceed smoothly.

REMEDIAL ACTIONS

Remedial Action, actually cleaning something up, is the express goal of all work completed for a NPL site such as Cecil Field. However, most actions require that a Record of Decision (ROD) or an Interim ROD (IROD) be signed prior to beginning the cleanup.

As discussed in the Introduction, the 1993 dispute resolution stipulated that an IROD had to be completed for four sites in FY 1994. Completing four IRODs in less than 12 months was difficult because, at the time of the dispute resolution, only one candidate site had been identified. The race was on to locate three other sites at which interim actions were appropriate, and complete the evaluation necessary to support the signing of an IROD. And, as discussed above, trust between all parties concerned with remedial response activities at NAS Cecil Field was at a minimum, the base had just been slated for closure, and partnering had not begun.

Four sites were quickly identified. In all cases, the medium chosen for remediation was soil, but the evaluation (i.e., paperwork) still had to be completed before the IRODs could be signed. Several document had to be completed prior to signature of an IROD. These are:

- a Focused Feasibility Study to identify remedial action objectives and acceptable remedial alternatives;
- a Proposed Plan to select an alternative and solicit input from the public;
- a Public Meeting to explain the proposed activities to the public;
- a Responsiveness Summary to respond to all comments received during a 45-day public comment period; and
- the IROD itself.

The successful completion of these documents in less that 12 months for a single site can be a difficult challenge, when one considers that all of them, with the possible exception of the Responsiveness Summary, need to be prepared as a draft, reviewed by the client, redrafted, reviewed by the agencies, redrafted, and then finalized. But completing four of them, when differing conditions for the chosen sites required separate documents for each site, in less than 12 months, is nearly a miracle. It was done successfully (i.e., on time, within budget, and to the general satisfaction of the client and agencies). All four IRODs were signed in FY 1994, and one of the remedial actions, complete with a 100% design package was also completed in FY 1994.

The completion of these four IRODs required a tremendous amount of cooperation between all parties concerned, and an even larger amount of perseverance. By October of 1994, following the completion of this challenge and the realization that the partnering was actually working, the team was working as a team, and the all-important element of trust was growing.

This area of the remedial response process, post RI/FS activities to remedial action, is ripe for expediting. Acceleration of the process was accomplished largely by free and open communications between all parties (clients, consultants, and agencies) and the commitment that we were going to do it. Focused Feasibility Studies, which evaluated a reduced number of alternatives for only the medium of concern for that action, were used rather than waiting for a completed RI and a FS which addressed all contaminated media.

The meetings, in addition to facilitating communication which helped to build a team feeling and eventually trust, also greatly assisted in speeding up the process. At these meetings, which were difficult at times, all of the persons necessary to make decisions were present, and the meetings lasted until decisions were reached. Going back to one's respective home office, talking to one's people, and responding by letter can really slow the process down. The NAS Cecil Field team, whenever possible, held the meetings after the analysis was completed, but before the documents were written. This is such a simple idea, have everyone agree on what's to be done **before** the documents are written, but it is rarely done. Reaching agreement on a selected remedy or on which alternatives will be evaluated before the document is prepared speeds everyone's review of the document, and can literally shave months off review times.

The road to these four IRODs was not always smooth, as can be imagined, but they were completed. When a document had to be prepared in only a few days, it was. When a document had to be reviewed in only a few days, it was. When flights had to be missed, they were. The team must cooperate and persevere in order to truly expedite a complicated process.

At the time of this writing, August of 1995, the pace has not slackened. Remedial actions are in progress at the three remaining IROD sites, and final RODs will be signed for two Operable Units before the end of FY 1995. Additionally, to expedite future remedial actions and reduce costs, the NAS Cecil Field team has allowed for long-term treatment facilities to be incorporated into the remedial actions for two of the IROD sites. At Site 5, a cell was built for the biological treatment of petroleum contaminated soils. This cell will eventually receive contaminated soils from several sites, and it is planned to remain active until no longer needed. For the cleanup of contaminated soils at Site 17, a thermal desorption unit was assembled onsite. Like the biological treatment cell at Site 5, the thermal desorption unit will also treat soils from other sites and is planned to remain in operation until it is no longer needed. Tremendous cost savings (in transportation alone) may be realized by having these two effective technologies available and operating onsite. Additionally, the pace of site evaluation and ROD or IROD completion will have to remain high to keep these two units operating at capacity. The NAS Cecil Field team is looking forward to the challenge!

COMMUNITY RELATIONS PROGRAM

The community relations program at NAS Cecil Field is a program which is highly successful; however, this success is often masked by low attendance at public meetings and a somewhat disinterested public. Although the team has put in a great amount of time and effort toward this program, the outcome is always the same: very little interest from the community. A public meeting for a proposed remediation alternative will typically have a public turnout of less than five people. But, does that necessarily make the program a failure? Or, on the other hand, does that make it a huge success, since we have had virtually no opposition from the public and, obviously, they trust us? While the truth of the matter probably lies somewhere between these two extremes, the real challenge of the program has been how do you get the general public from a rural area to take an interest in the environmental issues of a base in their own backyard. The answer to this question has yet to be solved.

While the lack of community interest at Cecil Field is probably the exception rather than the rule, the protocol the team has taken has certainly been the norm. Public meetings have always been held for major cleanup milestones. Availability sessions are held. Additionally, fact sheets are continually being developed, and an information repository has been created and maintained. Advertisement of public meetings usually consists of notices being published in the local newspapers, radio spots, and fliers distributed in surrounding neighborhoods.

In the spring of 1994, the Department of Defense (DOD) issued a directive for all DOD installations involved in environmental cleanup to develop an advisory board consisting of local citizens whose charter was to provide input to the cleanup team during the decision-making process, and to disseminate these cleanup resolutions back into the community. This board became known as the Restoration Advisory Board (RAB), and is composed of approximately 20 community members, each representing a surrounding community or a local civic group. While getting 20 community members from a generally uninterested public was a challenge in itself, the board was eventually chosen from the 30 or so applicants to represent a diversified cross section of the local community. The makeup of the community RAB members includes an environmental lawyer, an environmental engineer, an environmental activist, and a person who lives near an adjacent Superfund site. With the exception of those four individuals, the remaining community members have little exposure to the remedial response process.

The RAB started meeting monthly, with the first few meetings generally serving to establish subcommittees, charter, mission statement, as well as to choose the community chairperson whose responsibility it was to run the meetings. The next couple of meetings were devoted to training in the remedial response process. Finally, the last couple of meetings have been presentations of site-specific programs. These site-specific meetings have not only been educational to the RAB, but have also served as a sounding board for the cleanup team regarding the acceptance of the preferred cleanup alternatives by the general public. In lieu of a dry run of a public meeting, it has been the cleanup team's policy to give the upcoming public meeting presentation to the RAB beforehand. This gives the presenter a chance to practice before an audience, keeps the RAB aware of the progress of the program, and acts as a community barometer for the chosen cleanup alternative. The RAB has provided very useful insight and constructive criticism regarding the remedial alternatives chosen, and the format for public meetings.

It is important to have the RAB be a part of the decision-making process. If you can obtain the RAB's buy-in while making use of their input, you will then be laying the foundation for the all-important issue of trust, and trust is rewarded with dividends. In the case of Cecil Field, the development of trust between the RAB and the cleanup team has helped pave the way for the use of innovative technologies and the all important final risk-management decisions.

ENVIRONMENTAL BASELINE SURVEY

An environmental baseline survey (EBS) must be conducted for all property that will undergo transfer. The EBS is based on all existing environmental information related to storage, release, treatment, or disposal of hazardous substance or petroleum products on the property, and attempts to determine the presence or likely presence of a release or threatened release of any hazardous substance or petroleum products. An initial EBS was completed for the facility as a first step in the base closure process. A final will be completed for each individual parcel prior to its transfer.

An EBS was conducted for all property at the Main Base, Yellow Water Weapons Area, Outlying Landing Field (OLF) Whitehouse, Palatka Radar Site, and the JP5 pipeline which extends from NAS Jacksonville to NASCF. A total of approximately 650 parcels and 25,000 acres were physically surveyed.

The EBS categorized the parcels into one of seven categories according to their readiness for property transfer. Each category has a corresponding color for mapping purposes. The categories are summarized below.

- 1/White - Areas where no storage, release, or disposal of hazardous substances or petroleum products has occurred (including no migration of these substances from adjacent areas).
- 2/Blue - Areas where only storage of hazardous substances or petroleum products has occurred (but no release, disposal, or migration from adjacent areas has occurred).
- 3/Light Green - Areas where storage, release, and/or migration of hazardous substances or petroleum products has occurred, but at concentrations that do not require a removal or a remedial action.
- 4/Dark Green - Areas where storage, release, and/or migration of hazardous substances or petroleum products has occurred, but all remedial actions necessary to protect human health and the environment have been taken.
- 5/Yellow - Areas where storage, release, and/or migration of hazardous substances or petroleum products has occurred, removal and/or remedial actions are underway, but all required remedial actions have not been taken.

- 6/Red - Areas where storage, release, disposal, and/or migration of hazardous substances or petroleum products has occurred, but required response actions have not been implemented.
- 7/Grey - Areas that have not been evaluated or require additional evaluation.

The basewide EBS for NASCF was completed in six months. A total of 15 individuals were mobilized to obtain the required information. Every square foot of dry land was visually surveyed, and file searches were conducted at state, local, and Navy activities. Local records and aerial photographs were reviewed to determine past uses of the property. Environmental databases were reviewed to determine if spills or releases had been reported on Navy or adjacent property. A report of the EBS, both in paper and digital format (on CD-ROM) was produced.

Several elements of the initial EBS could have been improved upon, when considered a year later. First of all, the depth to which record searches and attempts to located historical data went, could have been increased. This element of an initial EBS can be time-consuming and costly, but the benefits can be tremendous. The need to verify whether all of the relevant data have been collected continues to cost money and take time.

A second element of the initial EBS which could have been improved upon was the grouping of areas. In most cases, individual parcels, buildings, or structures were evaluated as separate areas. In the cases where we did "lump" areas together (mainly open tracts of land with some rubble or debris), it now appears that we would have been better off splitting out the rubble and debris areas from the larger unaffected tract. The lesson learned here is that during the early evaluation stages, it is better to "split" rather than "lump." Grouping of separate areas together for a block transfer is easier and faster than holding up a large area for an isolated small-scale problem.

GREY SITES, EBS CATEGORY 7

In 1993, NAS Cecil Field was placed on the Department of Defense Base Realignment and Closure (BRAC) list. Along with this decision came DOD's commitment to expedite the environmental cleanup of the base and turn the property over to the private sector for the creation of new jobs. The task of cleaning up the base in this new realm of BRAC closure meant, essentially, that a reevaluation of the entire base had to be performed, and required cleanup actions initiated as appropriate. This reevaluation was accomplished by performing a fence-to-fence walkover and survey of all open parcels of land, and an inspection of every building.

The results of this walkover were documented by building number in the Environmental Baseline Survey (EBS), along with an assigned color designator to classify the building as being transferable or not. Additionally, all open land parcels were documented as to their environmental condition, and color coded for purposes of property transfer. Those buildings and parcels that were environmentally questionable were assigned the color designator grey and, therefore, would require further investigation to properly determine if transferable or not. These buildings and parcels are ineligible for transfer until an environmental assessment and, if necessary, cleanup action is completed and the property reclassified as transferable. The results of the Cecil Field EBS generated over 250 grey sites, which essentially enlarged the environmental cleanup program of NAS Cecil Field from approximately 25 sites to a staggering 275 sites. A means unique to the normal CERCLA RI/FS process had to be developed in order to expedite the cleanup and reclassification of this many sites.

The process established to investigate the grey sites was developed by the project team and essentially consisted of a mini, stand-alone sampling and analysis work plan written for each site. These individual work plans, called a Sampling and Analysis Outline (SAO), were prepared after the project team walked the site again to ensure that the environmental problem that had originally driven the grey classification was still present, and that it, indeed, required a sampling event to positively determine its environmental condition. The SAO consists of a brief description of the building and its location, what type of operations had gone on in the building, a description of the environmental problem that had driven the grey classification, a map showing the location of the questionable problem, and a description of the recommended sampling program and analysis to be performed on the various media. All this was presented in a four- to five-page document for each site. This process has worked very well for the team in light of the fast-track time frame invoked on the cleanup process. In most cases, an average SAO would require only one day to prepare, including the site visit.

The approach taken by the team to write the SAOs and perform the field work was to divide the base up into zones which were established, based on areas of similar function and proximity to each other. Each zone would consist of approximately 20 or 30 buildings. For instance, one group of 20 to 30 SAOs was written for all the industrial buildings along the north-south flightline. Thus, if the reuse commission indicates to the team that there is a potential buyer for a certain area of the base, the team can expedite the SAOs in that corresponding zone to accommodate the reuse. This flexibility has created a good working relation between the project team and the reuse commission. Additionally, doing a small group of SAOs (including field work) at a time has allowed the project team to continually incorporate lessons learned and has, thus, initiated a continuous improvement process into the program. This preparation and execution of small, manageable groups of SAOs has also streamlined the contractual aspects of the program. As actual costs are realized for the earlier groups, these costs can then be used to extrapolate future costs, therefore eliminating much of the guess work that is typically associated with a contractual award.

Another success of the grouping approach with regard to time and, therefore, costs has been the trust factor that has evolved, along with team review of the SAOs. BCT review of the SAOs has been done in a meeting format, with consensus driving all decisions.

While much of the effort to date at Cecil Field has been the development of SAOs, it is important to realize that the SAO is only the investigative portion of the process. In addition to the SAO workplan, the field work results must be evaluated and a decision made as to whether the results indicate a risk to human health or the environment. If they do indicate a risk, the cleanup method must be determined. With 250 grey sites, the project team has developed innovative means and processes in order to keep out of the CERCLA RI/FS world. Otherwise, the costs and added time frame would derail the fast-track concept.

The handling of open areas of land called for a little different approach than that for buildings. During the original development of the EBS, large open tracts of land, which Cecil Field has many of, were grouped into large parcels consisting of hundreds of acres. If a single drum was found in the middle of this parcel, the whole parcel was color-coded grey and categorized nontransferable. This oversight unnecessarily tied up hundreds of acres of land that were actually ready for transfer. Subsequent EBS developed for some of the noncontiguous properties of Cecil Field deviated from this approach by categorizing only a very small portion of the area around a single drum as grey. The area surrounding the grey site was color-coded as transferable and was not tied up because of a single drum. This method has allowed us to categorize more land as being transferable. It is thus very important to develop the EBS in a thorough manner so as to free up as much property in the beginning stages as possible.

MANAGEMENT PLANS

As part of base realignment and closure activities, management plans were developed to address lead-based paint (LBP), asbestos-containing material (ACM), and water/wastewater issues. These management plans provide, on a per parcel basis, a summary of applicable regulatory requirements and Navy policy, an evaluation of permit status and requirements, an assessment of existing conditions (separately for LBP, ACM, and water/wastewater issues, where applicable), recommendations for corrective actions, and an estimate of corrective actions costs.

The original concept of a "management plan" for various compliance issues (e.g., LBP, ACM, Water/Wastewater, and PCBs) was proposed by the BRAC

Environmental Coordinator (BEC, the Navy's lead RPM for activities at a BRAC base) in the very early stages of the base closure process. The original BEC was replaced, and the idea of what the reports would contain and how they would look was not fully developed. The Cecil BRAC Cleanup Team, however, went forward with the management plan concept and, because of the team's perseverance and cooperation to meet an end, the plans have evolved into useful management tools.

To collect the necessary information for the plans, teams were organized to perform evaluations for each particular compliance issue. The first step for each team was the compilation of existing information, and an assessment of what was needed to complete the their plan. A considerable amount of information had been developed by the base regarding several of these issues, but the information was not collected with base closure in mind. For instance, with ACM, three surveys had been completed and several remediations had already been accomplished; many buildings had been surveyed, but still required sampling for verification. Field teams were mobilized to the facility to gather the necessary information and assess the situation.

It became clear that a complete inventory for both LBP and ACM did not exist; the facility already had the PCB issue under control with their own program; and the water/wastewater plan development could proceed. To fill the LBP and ACM gaps, cost estimates and a contract modification, which can sometimes cause significant schedule delays, would have to be completed. The field teams proceeded with inventories of facilities that would not require sampling, while a contract was let to allow for the sampling of the facilities in question. This tactic resulted in virtually no schedule interruption. By the time the teams were done with the first part of the inventory, the contracting was in place to complete the sampling portion. The transition from the nonsampling facilities to those that required sampling was smooth, and produced a complete inventory for NASCF on schedule. Acceleration of the contracting for this "new" step was made possible by trust and open communication between all parties. Perseverance also played a large role; the process was not derailed by an unforeseen condition.

Prior to the field program, team leaders responsible for each of the management plans coordinated closely with NASCF facility contacts to ensure easy access to all buildings/structures in question. The importance of this step and the cooperation of facility personnel cannot be overemphasized. Access barriers to information must be overcome if aggressive schedules are to be met and knowledgeable, cooperative facility personnel are essential.

One of the major successes of the management plan development was the continual input from the BCT and project team. Navy personnel with expertise in lead-based paint, asbestos-containing material, and water and wastewater management, were included early in the development of these plans in order to provide guidance. These experts provided constant dialogue with the consultant while producing these plans. This dialogue served to greatly reduce cycle time on the submission and review of draft documents.

The management plans serve as important tools for the closure of the base. They list, on a per building basis, all of the relevant information concerning current conditions for that parcel. The inclusion of cost estimates for the corrective actions needed, whether it be manage-in-place or removal, is helpful for providing cost to complete information to DOD and prioritize the allocation of available resources. Additionally, the data in the management plans and the manner in which it is organized will serve to expedite property transfer, the goal of base closure.

FINDING OF SUITABILITY FOR TRANSFER OR LEASE

Following the completion of a Final EBS for a parcel, one of two documents will be prepared and signed by the property owner, either a Finding of Suitability for Transfer (FOST) or a Finding of Suitability for Lease (FOSL). The FOST indicates that DOD (in conjunction with the agencies) has determined the property can be sold to a new owner. The FOSL indicates that the property is not currently suitable for resale, but can be transferred.

The decision to complete a FOST or a FOSL for a given parcel of land or building can have several layers of complexity, depending on the environmental condition of the property. A detailed treatment of the criteria which must be evaluated prior to making the lease or transfer decision is beyond the scope of this chapter. Some guidance, however, on the completion of FOSTs and FOSLs has been developed and is available.

Only one property transfer has been completed at NAS Cecil Field. Approximately 100 acres of undeveloped wooded area on the base perimeter were sold for a road construction project. It is too early in the closure process for us to provide a discussion of how the FOST/FOSL process has worked at NAS Cecil Field, and what experience has shown. However, as part of planning for base closure, some steps have been taken to expedite the process when it begins in earnest.

In order to expedite the completion of FOSTs and FOSLs, a system has been developed to manage as much data as possible on a per parcel basis. Management of environmental information and relevant historical information on a per parcel basis has, so far, been a fairly straightforward process, although some types of information, mainly relating to the environmental condition of infrastructure (e.g., sanitary sewer system), cannot be easily assigned to a parcel of land. And the early evaluation and development of a consistent numbering systems can be important; having IR Site 1, UST Site 1, Building 1, and Grey Site 1 actually representing four separate parcels of land can cause data management complications.

All of the property comprising NAS Cecil Field was divided into separate parcels during the initial EBS. During this survey, and all subsequent investigations, the team has endeavored to split rather than lump. Since the initial EBS, all information gathered for a parcel is tracked and stored as being associated with that parcel. The "storing" has been accomplished with databases, scanning all information onto CD-ROM, and physically placing original copies of all relevant information into per-parcel notebooks. The "tracking" has been accomplished with the Site Management Plan, the linking of databases, and vigilance.

The management of information on a per-parcel basis will make the evaluation of data prior to transfer or lease easier to accomplish. If a purchaser wants a single building, the data are already grouped that way. If a purchase wants a group of buildings or parcels, the information can be easily combined. The NAS Cecil Field team believes this "split" situation is better than a "lumped" situation, because of the time and effort which would be required to extract information for "just that building" from the group within which it was evaluated. Additionally, the preparation of FOSTs and FOSLs was put under contract early in the program so that procurement would not delay the transfer or lease of property.

REUSE COMMISSION

With approximately 8,000 civilian jobs slated to be lost due to the closure of Cecil Field, a Reuse Commission has been formed by the City of Jacksonville, whose goal is to develop a strategic plan that will result in creating new opportunities for employment, recreation, and residential development for the nearby community. In order to reach this goal, the commission has hired a consultant to prepare a reuse plan that will identify reasonable reuse options based on the existing physical and economic conditions at Cecil Field and the surrounding communities. Once the plan is completed, the City of Jacksonville and the Navy must approve the plan. Subsequent to the completion of the reuse plan an Environmental Impact Statement (EIS) must be prepared to evaluate the potential environmental effects that disposal and redevelopment of the property will have. Current completion date for the Reuse Plan is the summer of 1995, and for the EIS, the spring of 1996.

With the current trend in Congress to reduce funding for all branches of the armed services, the expeditious completion of a reuse plan cannot be overemphasized. With funds available for environmental cleanup programs shrinking, the Navy was forced to prioritize, based on planned reuse. Unfortunately for Cecil Field, the reuse plan has not been finalized. The Navy wants to see property going back to the community, and is funding those bases first. The predicament we face at Cecil Field is how to get in the position to transfer property if we have no funds to investigate and clean up our sites. The fact that environmental cleanup funding is now keyed to the reuse plan needs to be reinforced early in the reuse plan development process.

With over 250 grey sites identified during the development of the Environmental Baseline Survey, a means to prioritize these sites needed to be established. Since the majority of the grey sites are buildings, and their cleanup impacts property transfer, the project team had to establish a good working relation with the reuse committee so as to stay abreast of areas with high reuse potential. By establishing good communication channels with the committee, the project team has been able to prioritize the grey site investigations in order to expedite the investigation and property transfer of these high-interest areas and, therefore, accommodate the new owner.

Potential reuse alternatives, as established by the reuse plan, need to drive the cleanup standards and levels to which sites are remediated. For instance, if the reuse plan has proposed an area to be used for industrial redevelopment, then the cleanup levels and risk characterization should also be based on industrial cleanup standards and not on a more conservative set of standards, such as residential. Unfortunately, as in the case of Cecil Field, many of the sites have already undergone investigation, and cleanup goals have been established, typically to residential standards. Again, the quicker you establish a reuse plan, the better off your cleanup program will be.

Many state, city, and local government agencies and special interests groups are involved in the review of reuse alternatives in order to protect the interests of the groups they represent. While the development of the reuse plan has flowed relatively smoothly for Cecil Field, there is great potential for disagreement among these groups. The need for early involvement and buy-in of all these groups is essential in order to prevent last minute disagreements which may delay or require portions of the plan to be redone.

As part of the development of the EIS, a historical and cultural survey was conducted at Cecil Field. This survey identified areas of the base that may be archaeologically sensitive. In the case of Cecil Field, these areas were primarily areas of potential former Indian habitats. Again, by the time this survey was completed, a number of sites had already undergone remediation, including one in an area of high sensitivity. It is therefore suggested that field crews be briefed to be aware of and on the lookout for such cultural artifacts, and inform them of the proper procedures to follow when these items are discovered.

CONCLUSIONS

The process of base closure requires that everything be expedited. This is a formidable task and requires a large amount of hard work. The key elements of success in our efforts have proved to be the development and maintenance of trust, a willingness to cooperate to meet common objectives, and the ability to persevere when difficult situations arise.

The partnering program, undertaken by the Navy, the State of Florida, and Region IV US Environmental Protection Agency, has been the most significant stride taken toward closure of NAS Cecil Field. Partnering has allowed trust, cooperation, and perseverance to become integral elements of the closure program. Having the Navy, the state, and the EPA work together to manage the program, via the BRAC Cleanup Team, means that all parties have a stake in decisions made and therefore, everyone wants to see the programs succeed. This seemingly simple statement creates an environment where everything else can fall into place.

CHAPTER 24

Quality Assurance Audit of CERCLA Risk Assessment Programs: A U.S. Air Force Experience

William D. Sweet, **Judy A. Strickland**, and **Stanley O. Hewins**, U.S. Air Force Center for Environmental Excellence, Brooks Air Force Base, Texas

INTRODUCTION

The U.S. Air Force Center for Environmental Excellence (AFCEE) is a service center supporting the US Air Force Installation Restoration Program (IRP). The objectives of the IRP are to characterize former hazardous waste disposal and spill sites, and to implement response actions consistent with the Comprehensive Environmental Response, Compensation and Liability Act and the National Contingency Plan. AFCEE's Environmental Restoration Directorate (AFCEE/ER) implements IRP investigations at Air Force installations using private companies under contract. The AFCEE/ER project managers monitor contractor activities to ensure fiscal accountability; adherence with legal requirements; accurate, valid analytical data; scientifically sound data interpretation and site-specific risk assessments; and consistency in program operation. They are assisted in their oversight function by AFCEE legal, accountant, contracting, and scientific/engineering staff. The in-house scientists and engineers function as a consulting group, and are supplemented by contracted consultants.

Before AFCEE established the risk assessment audit program, contractors were offered only minimal written guidance communicating risk assessment requirements. AFCEE accomplished contractor oversight in the risk assessment arena largely by technical review and critique of risk assessment deliverables. AFCEE established the risk assessment audit program to ensure quality risk assessments by directly, personally communicating to the contractor the risk assessment needs of the IRP. To provide maximum benefit, the audit program emphasizes AFCEE/ER's risk assessment requirements early in an IRP project, before, rather than after, deliverables are produced. The audit does not result in certification or approval. Rather, it is an open identification of strengths and weaknesses in relation to AFCEE expectations. This identification may help the contractor in defining goals to improve the process. The evaluation team uses written standards, and professional judgment and experience to review the contractor's risk assessment capabilities to identify factors that may affect the quality of risk assessment deliverables.

WHY DOES AFCEE PERFORM RISK ASSESSMENT AUDITS?

There are multiple reasons to establish a risk assessment audit program. The major areas of consideration are the federal contracting and procurement process; the dearth of education programs and professional certifications, and no standardized curriculum from a professional or accrediting entity, for environmental risk assessors; and the difficulty of adequately communicating by written guidance. Whereas the first consideration is peculiar to federal agencies, the latter two apply across the environmental restoration industry.

Although AFCEE contractors undergo a rigorous source selection process, the current experience indicates that this process does not consistently result in high quality risk assessments. There are at least three reasons for this. First, the selection process evaluates many competencies in addition to risk assessment. A relatively weak risk assessment portion may be outweighed by other considerations. Next, those writing contract proposals are not necessarily the same as those producing risk assessments for the contract. Therefore, the proposal may not accurately reflect the quality of risk assessments produced under the contract. Third, the duration of many AFCEE contracts is five years, and during this period of time, a contractor's risk assessment and management staff may change several times, thereby changing risk assessment quality. A structured evaluation of risk assessment capabilities can impart "the AFCEE way" to the contractor's staff and reassure the USAF that good quality risk assessments will be performed.

Compared to other environmental disciplines such as engineering and geology, environmental toxicology and risk assessment are new fields of study which are largely without educational programs and standards, or professional organizations and certifying boards. Thus, risk assessment quality is difficult to judge on the basis of the achievement of professional standards. Because of the lack of formally trained risk assessors, many contractors have underqualified staff and/or staff who are qualified in another discipline, but, with some minimal training, function as "risk assessors." Environmental risk assessment requires multidisciplinary effort. Human health toxicology/risk assessment requires education and experience in human anatomy, biochemistry, physiology, epidemiology, and toxicology, at least. Ecological toxicology/risk assessment requires knowledge in wildlife biology, ecology, and toxicology. In addition, risk assessors should be conversant with geology, hydrology, chemistry, the fate and transport of chemicals in the environment, meteorology, physics, and engineering.

Another reason to perform risk assessment audits is that communicating risk assessment needs and requirements is somewhat inefficient through written materials (*e.g.*, guidance documents, contracts). Without personal communication/consultation, it is difficult to know how to interpret written guidance or to know what is expected beyond what is written. The risk assessment audit allows face-to-face communication and establishes a point of contact for future consultation. In the current jargon, this is a "partnering" process.

GENERAL PHILOSOPHY

AFCEE performs risk assessment audits to develop, among AFCEE contractors, a more consistent approach to risk assessment. The audit process attempts to better standardize how risk assessments are scoped and produced, and define the standards, objectives, and criteria for risk assessments acceptable to AFCEE and regulators. Standardizing the approach applies to different projects by a company, across different offices of a company, and among the several contractors. The format of the risk assessment and associated documents is much less important than content. A format is prescribed in US EPA guidance (USEPA 1989a; 1989b; 1991a; 1991b), however.

The overall approach to the risk assessment audit is that it serves primarily as a team-building exercise. The audit provides a forum for information exchange between AFCEE and its contractor. The risk assessment audit can even be viewed as an outside management consultation, for free! It is not intended or used as a certification process, and does not result in a pass or fail determination. The prospect of inspection by the contracting party, most especially by an agency of the federal government, has sometimes resulted in fears that the audit will be used for contractor manipulation, intimidation, or prosecution. This reaction is perhaps understandable, but considering that these motives are mostly unethical, and certainly fatal to the main objective of partnering to produce a better risk assessment, the AFCEE strives to avoid the fact or appearance of these motives. In other settings, but not with AFCEE, the risk assessment audit might be legitimately used for contractor evaluation for contract renewal, or future contracted work.

THE PROCESS

At AFCEE/ER, the risk assessment audit process begins with a request for the in-house risk assessment consultants to perform a risk assessment audit. The request originates from the project managers who are also within AFCEE/ER, and who are ultimately responsible for contractor oversight. An in-house consultant may suggest an audit because a risk assessment was found to be seriously deficient. In addition, the request for a risk assessment audit may be solicited by a consultant as the result of a policy decision that AFCEE wants to audit every contractor, preferably early in the multi-year contract.

With the formal tasking in hand, the audit team, which usually consists of two persons, is named. The in-house consultant telephones the contractor and explains the purpose and procedure of the audit and schedules a firm date for the onsite visit. The date of the audit is set by mutual agreement, largely at the convenience of the company. The contractor's staff members who produce risk assessments, who manage the risk assessment effort, and who participate in the internal, (semi-)autonomous quality assurance group must be available during the onsite visit. The date should be 2-4 weeks from the first contact. Since the contractor's staff already know how they execute risk assessments, a longer period is unnecessary for the little preparation required. Preparation of a formal briefing by the company is always unhelpful, and may be actually obstructive to the process.

During the initial telephone contact, the in-house consultant also requests that the contractor forward written materials for examination by the evaluators prior to the visit. The written materials requested include an organizational chart emphasizing the risk assessment unit's place in the company, brief professional biographies of the entire risk assessment staff, and a recently performed Baseline Risk Assessment for a Superfund site.

Based on these written materials, the evaluation team forms provisional hypotheses of the strengths and weaknesses of the risk assessment group. The following elements are considered:

- staff size and composition. Is the staff size adequate for the range of projects called for in the contract? Composition refers to the field of expertise and to the presence of highly educated analysts and evaluators, and more routine production people in reasonable proportion.
- staff qualifications. What formal training and education do the staff have? What type and duration of work experience do they have?
- the position in the organizational structure of the risk assessment group within the company. Is it well-integrated into the whole Superfund environmental investigation and remediation process, or is it an appendage? In some companies the risk assessment group is left out of the project scoping and planning stages, and is not involved in the project until all the environmental data has been collected. This is contrary to USEPA and AFCEE guidance.
- the overall organization of the company with respect to the quality assurance/quality control (QA/QC) group. Many companies have risk assessment and QA/QC staff distributed among several geographical locations. What is the relationship between the risk assessment staff and the QA/QC staff?
- the company's approach to risk assessment. Does the company have an identifiable policy or philosophy on risk assessment, and is it concordant with those of regulatory agencies and AFCEE? If not, what is the significance of that divergence?

The submitted materials are not sufficient to answer these questions, but the materials do allow the evaluators to become familiar with the structure of the company and the individuals producing risk assessments. The evaluators' estimate of the probable strong and weak areas, and areas that can't be evaluated from the available material, guides the onsite visit, and makes the visit more efficient and productive for both parties.

The onsite visit continues with the evaluation of the same factors for which the written material is evaluated, with the addition of the physical resources. This is done in a seven-step agenda which consists of an introduction, presentation of the company's approach to risk assessment, facilities tour, staff interviews, discussion of selected areas, quality control/quality assurance, and an outbriefing. The agenda usually follows this sequence, but it may be rearranged to accommodate individual circumstances. The onsite visit is often accomplished in one day, but may require one and a half days.

The introduction by the audit team is brief. They introduce themselves and explain the objectives and procedures of the day's activities. Clarifications are given as needed.

The company then describes its organization, including lines of authority, organization, and geographical locations, with emphasis on their risk assessment team. The contractor continues with a description of their risk assessment philosophy and process for doing risk assessments within the overall environmental investigation process. Questions the evaluators are trying to answer from this presentation include the following: At what stage of the investigation are the risk assessors involved? (In accordance with USEPA guidance, AFCEE wants risk assessors involved at the earliest stages of project scoping.) What specific tasks do the risk assessors perform? When is the conceptual site model done, and what part do the risk assessors have in its development? How is the conceptual site model used to guide the overall project, and, specifically, the risk assessment. (Sometimes the conceptual site model is viewed as a deliverable at or near the end of the investigation.) What is the risk assessor's involvement with Data Quality Objectives (DQOs)? (Some contractors feel that DQOs are only for the analytical laboratory, and the risk assessors are uninvolved.) How are decisions about the risk assessment documented and archived? How are information and processes validated? What QA/QC measures are incorporated in the risk assessment process, and when? (This area is evaluated more specifically during the QA/QC agenda item.) How and to what extent do the risk assessors interact with the client and the regulators?

After the discussion of the contractor's approach to risk assessment, the evaluators tour the facility. A walk-through tour shows the audit team the work environment, workspace, equipment, and auxiliary facilities. The adequacy of the computational, statistical, graphical and text processing support for AFCEE projects is judged. In addition, access to governmental regulations and guidance material, and to previous work on the subject facility from the same and other contractors, is examined. Also, access to other information about the Air Force Base (e.g., topographical maps, geological information, demographics, ATSDR Public Health Assessments) is investigated.

The next item on the agenda is staff interviews. Based largely on the material submitted before the site visit, but often modified by the just-concluded presentation, a list of individuals who have been selected for individual interviews is announced. Interviewee selection is solely at the discretion of the evaluators, although contractor suggestions are considered. Management and supervisors are asked about staffing levels and composition, work product (document) review and QC measures, workload management, corporate philosophy of integrating the many fields into a CERCLA investigation, and corporate programs for professional development of the risk assessment staff. The technical staffers are interviewed on some of the preceding items, but the emphasis is on the individual's education, training and experience, and on their roles in accomplishing the risk assessment product.

The preceding segments usually are accomplished before lunch. After lunch, the evaluators select topics for further exploration. The discussion of selected areas provides an impromptu discussion of areas otherwise insufficiently explored. It is not necessarily about problem or deficient areas, but it may be. For example, it may be that the contractor has an innovative management structure, making it difficult to grasp the implications in the few minutes previously allotted. That may be a praiseworthy strength, not a limitation. Ordinarily, the topics are chosen by the evaluators, but the company may suggest topics that it thinks need to be show-cased, e.g., an innovative software producing higher-quality, lower-cost risk assessment calculations. Or the company may perceive a misunderstanding held by the visitors that it wants to correct. The choice of topics remains with the evaluators, however.

Although briefly presented earlier, QA/QC receives special focus in the next segment, which is a presentation by the QA/QC staff. The degree of autonomy and authority of the QA/QC staff is of primary concern, with standards, and procedures and methods of evaluation running a very close second. Despite considerable thought, AFCEE risk assessors have found it difficult to define a checklist of elements and standards. This is another area where professional judgment plays a big part. Two identifiable areas are: (1) mechanisms to verify input values used for risk assessment calculations, and the correctness of the output; and (2) the documentation of the QC review. Documenting the review means, among other things, that the QC review has an auditable trail.

The concluding event of the onsite visit is an outbrief. The outbrief presents the evaluators' comprehensive view of the company's risk assessment effort. Highlighting strengths is as important as discussing areas of concern and possible general approaches to improving these areas. However, AFCEE explicitly recognizes and defers to the management's prerogative and responsibility to manage their company.

The evaluators promptly write a report of their findings. Any substantive issues, especially weaknesses, that were not included in the outbrief will be discussed with the company by telephone before the written report is submitted. Blindsiding the contractor with a previously unmentioned weakness or deficiency will destroy the team-building that is the very reason for the audit. For reasons internal to the AFCEE structure, the evaluators' written evaluation is given to the project manager, who has the option of passing it on to the audited company. This technicality has not been a problem, because every company evaluated to date has received a copy of the written evaluation.

This concludes the evaluation process as it stands now. Two changes are planned for the near term. The next audit will add another step. The company will be invited to respond in writing to the evaluator's comments. This response should include actions taken or being implemented, alternative actions being evaluated, and areas where the company disagrees with the evaluators' perception of needed improvement.

The AFCEE guidance document on risk assessment audits (1995) will be finalized. As experience is gathered, the document will be revised, probably on a one- to two-year cycle.

AFCEE VIEW OF A GOOD RISK ASSESSMENT

The attributes or behaviors that AFCEE looks for during this risk assessment audit process are those that will result in a product consistent with USEPA guidance. Stated another way, the desired attributes will deliver a scientifically and legally defensible risk assessment that meets project needs. Defining what is required to produce a defensible risk assessment is not easy. A risk assessment does contain rather simple "plug and chug" mathematical formulas, but it is most emphatically more than that. It is a craft product - professionally, intellectually crafted to fit the specific situation and circumstances. The difficulty in defining a list of desirable attributes for an intellectual craft product is similar to that faced by Tom Wolfe in writing about the Project Mercury astronauts. He was reduced to calling the astronauts' qualities "The Right Stuff."

So what is "the right stuff" in a risk assessment effort? The right stuff that we're trying to find, evaluate, encourage, and develop is those attributes required to meet the project needs of protecting the health and well-being of current and future humans, protecting the natural environment for the benefit of present and future generations of species human and nonhuman, and being responsible stewards of the taxpayers' money. The attributes identified to date are discussed below.

A risk assessment needs to be more than raw data and bare risk numbers to do this. It needs an analysis and evaluation of what those numbers mean, or may mean in certain circumstances. AFCEE contractors must, of course, leave the risk management decision to the risk manager, the U.S. Air Force, as they must for every legally responsible party. However, it is not the job of the risk manager to analyze the data and arrive at alternatives, and then to chose among the alternatives. For one thing, analysis by the manager may result in unintentional, unconscious bias in the development of alternatives. The risk assessor is in some ways in more of a position to be objective. Further, most managers, at least in the U.S. Air Force, are not equipped by training or experience, or given the time, to analyze the data and develop alternatives. It is our guess that most private companies who contract for an environmental evaluation with risk assessment don't have that expertise either. If they did, they would do it themselves. Our advice is to expect, contract for, and demand sufficient evaluation and analysis in risk assessment.

The risk assessment evaluation and analysis should be realistic and site-specific. For example, although it is frequently appropriate to use default values in exposure parameters, site-specific information should be used if possible. By way of illustration, in one risk assessment done for AFCEE, the residents living in on-base housing were the potential receptors of a toxic substance. The default value of the reasonable maximum residence time for the US population as a whole, 30 years, was used in the risk calculations. However, only active duty military members and their families live in this housing; they typically move every three years. An inquiry made to the base housing office revealed that the longest anyone had ever resided in this housing was 14 years, and stays over seven years were exceptional. Just using the site-specific maximum of 14 years would have cut the exposure estimate, and the calculated risk, by just over half. It wouldn't have cost very much to place one of the contractor's junior employees in the housing office for a day or so, to collect the residence duration over the past 20 years, then calculate an appropriate statistically-derived value for the exposure to have a truly site-specific exposure parameter. In an example of the failure to use realistic scenarios, swimming in a lake on an Air Force base located a few miles from Canada was the route of exposure to toxics. The event (swimming) duration chosen for the risk calculations was 2.6 hours. In this location, the water temperature is unlikely to ever get much above 65°F, the highest monthly average temperature. Such an exposure would reduce core body temperature to about 95°F. This core body temperature is compatible with life, but scarcely tolerated on a voluntary basis. In addition, children were proposed to be the primary swimmers, and because of their greater surface area to volume ratio, they would be even more hypothermic than adults. This unrealistically excessive exposure increased the calculated risk. These experiences illustrate the requirement for thoughtful analysis to reach realistic, site-specific conclusions, but also the need for solid general knowledge across many areas of science to get to a defensible, site-specific, realistic risk assessment. Risk assessment staff with a broad range of knowledge must be available, and used.

Risk assessment must be fact-based. It is never possible to have all the facts, so there will always be uncertainty. However, all facts, which are necessary to make a decision with sufficient confidence and which may be reasonably obtained, should be. For example, every exposure pathway should have data (e.g., analytical results from environmental sampling) to support assertions of the exposure pathway being complete or not. If some data do not exist, good reasons must be given why these data are unobtainable or unneeded. If the missing data are needed to characterize the pathway, the implications of this gap must be discussed. A sufficient number of appropriately located samples must be taken to support statistically-based exposure point concentrations.

The risk assessment must intelligibly present a believable rationale. The most rigorous logic is largely wasted unless it is written into the risk assessment document in an understandable fashion. The risk assessment must record the data, the analysis, the rationale upon which the conclusions are based, and the proposed alternatives. The risk assessment document is the primary vehicle for presenting the site and the remediation decision to the public and the regulators; therefore, it must communicate risk well. The document is also the basis for dispute resolution, and for court proceedings, should this come to pass. Thus, the document's presentation of the site, the analysis and rationale, and the conclusions, must be able to withstand scrutiny for reasonableness of possibility and probability.

Legal requirements must be met. These include use of toxicity values from USEPA's Integrated Risk Information System (IRIS), and where applicable, state-promulgated values. Analytical methods must, if possible, be capable of determining concentrations compatible with using these toxicity values.

How does one get such a risk assessment? A staff with a broad background is most useful. But they must be able, willing, and permitted to bring this knowledge to a risk assessment. The managers of the entire project must understand the place of risk assessment in the CERCLA remediation process, respond to the risk assessor's requirements for a good risk assessment in each situation, and price the effort accordingly (but not deliver a Lamborghini when a tricycle is all that is needed!). The risk assessors must be involved early in the process, during project scoping and development of the work plan, and sampling and analysis plan. They should, if at all possible, visit the site. Generally, the risk assessors should be involved from the first meetings with regulators and customers. This involvement should continue throughout the project. This does not necessarily mean a consistent or high level of effort. The effort will wax and wane with the stage of the project and the complexity of the site.

CONCLUSION

The risk assessment audit is a tool toward a quality risk assessment. The process outlined here is specifically designed for multi-year, multi-project contracts within the U.S. Air Force. Probably, in private industry, single project contracts are more typical. The described process is therefore not strictly applicable, but this audit process may form the basis for evaluating potential contractors and the final product.

The audit is only one tool, even for multi-year, multi-project contracts. An audit is not a panacea for risk assessment woes. An audit only provides some assurance that a quality risk assessment is possible because the required elements are present. It cannot ensure that the capabilities will be adequately employed on any single project or even on any project. Each project may, and should, be monitored and evaluated throughout, from contractor selection, through project scoping, to final report.

ACKNOWLEDGMENTS

The AFCEE risk assessment audit program has had many participants in addition to the authors over its five-year existence. Principal among these individuals are: Ivan J. Boyer, John M. DeSesso II, Stephen M. Hsia, Raymond S. Kutzman, Richard D. Mavis, Fred T. Price, and Thomas F. Wolfinger of The MITRE Corporation; and Robert D. Taylor of Booz-Allen & Hamilton. The authors thank Barbara Smith-Townsend, Darrel R. Cornell, and Daniel L. Welch of AFCEE for thoughtful discussions and critically reading the manuscript.

REFERENCES

Risk Assessment Guidance for Superfund. Volume I, Human Health Evaluation Manual, Part A, Baseline Risk Assessment, USEPA, 1989a.

Risk Assessment Guidance for Superfund. Volume II, Environmental Evaluation Manual, USEPA, 1989b.

Risk Assessment Guidance for Superfund. Volume I, Human Health Evaluation Manual, Part B, Development of Risk-based Preliminary Remediation Goals, USEPA, 1991a.

Risk Assessment Guidance for Superfund. Volume I, Human Health Evaluation Manual, Part C, Risk Evaluation of Remedial Alternatives, USEPA, 1991b.

Guidance for AFCEE Quality Assurance/Quality Control Audits of Installation Restoration Program Contractors Performing Chemical Risk Assessments, draft, February 1995.

CHAPTER 25

Bench- and Pilot-Scale Thermal Desorption of Pesticide-Contaminated Soils from Rocky Mountain Arsenal

Carl P. Swanstrom and **Mark J. Gowdy**, Argonne National Laboratory, Argonne, Illinois

Mark Besmer, Rocky Mountain Arsenal, Commerce City, Colorado

INTRODUCTION

Argonne National Laboratory performed a pilot-scale, direct-fired thermal desorption (DFTD) treatability study on pesticide-contaminated soil from the U.S. Army Rocky Mountain Arsenal (RMA) near Denver, Colorado. The study was performed to provide RMA with performance and cost data in support of its onsite feasibility study. Specific objectives of the study included: (1) measuring the removal efficiencies for organic soil contaminants, including chlorinated pesticides, at different operating conditions; (2) determining the effect of DFTD treatment on heavy metal soil contaminants; and (3) monitoring for the formation of undesirable organic by-products such as polychlorinated dibenzo-p-dioxins (PCDDs) and polychlorinated dibenzofurans (PCDFs). Operating conditions for this pilot-scale DFTD testing were based on the results of previous bench-scale tests that were performed using batch, indirect-heated thermal desorption (IHTD) equipment. This chapter will examine pesticide removal at various operating conditions, the fate of heavy metals in the solids, and the production of PCDDs and PCDFs.

Rocky Mountain Arsenal

The U.S. Army used the Rocky Mountain Arsenal for the manufacturing, assembling, testing, storage, and destruction of chemical warfare agents and incendiary and explosive munitions from 1942 through 1970. In 1947, the Army leased excess RMA facilities to a private company for the manufacture of pesticides. In 1952, a second company assumed the lease and continued to manufacture pesticides and herbicides until 1982. Another company leased facilities at RMA from 1946 to 1948 for the production of chlorobenzene and dichlorodiphenyltrichloroethane (DDT). There are no current production activities at RMA. Production of military agents, munitions, and agricultural chemicals occurred at the South Plants manufacturing facilities. Liquid wastes generated by the manufacturing processes were discharged into Basin A (unlined), Basin F (asphalt-lined), and other unlined basins at the site. Spills, leaks, and other releases from the past production and disposal activities noted above have resulted in the soil and groundwater contamination of various RMA areas.

The Army began the remedial-investigation/feasibility-study (RI/FS) process at RMA in 1984 to define the nature and extent of the contamination, and to evaluate strategies for cleanup. Numerous soil-medium groups were defined at RMA that were contaminated with a wide variety of organic and metallic contaminants of concern (COC). These contaminants include: volatile halogenated organics, volatile hydrocarbons, volatile aromatic organics, organosulfur compounds, organophosphorus compounds, dibromochloropropane, polynuclear aromatic hydrocarbons, semivolatile halogenated organics, and organochlorine pesticides. RMA soils have also been found to contain heavy metals, including arsenic, mercury, cadmium, chromium, copper, lead, and zinc. As part of the RI/FS process, RMA established preliminary remediation goals (PRGs) for 27 of these organic and inorganic contaminants, the values for which are listed in Table 1. At the time the pilot-scale testing was performed, only two of the soil-medium groups at RMA were being considered for thermal treatment: the Former Basin F (Basin F) and the South Plants Central Processing Area (South Plants), totaling nearly 500,000 cubic yards.

Direct-Fired vs. Indirect-Heated Thermal Desorption

Thermal desorbers come in two basic types: direct-fired and indirect-heated. The majority of full-scale thermal desorption systems are continuously fed, direct-fired rotary kilns. For most applications, DFTD systems thermally destroy the offgas in an afterburner. A DFTD is very similar in design and operation to an incinerator. Temperatures in the kiln are the major operating difference. The offgas from an incinerator is typically 1300–2000°F and the ash, 900–1400°F. A DFTD operated in the parallel-flow mode typically has exit gas temperatures of 500–1200°F, with the treated soil exiting at 600–1200°F. Since the gas temperature in a DFTD is considerably lower than that in an incinerator, the refractory lining of the kiln is not used in most DFTDs. The absence of the refractory lining allows the kiln to be much lighter for transport, and require less setup time and maintenance. The burner nozzle may protrude into the kiln or be backed out so that only a small portion (or none) of the flame is in the kiln. The carrier (sweep) gas in a DFTD is air-mixed with the products of combustion from the burner. The amount of oxygen in the carrier gas can be adjusted over only a limited range, since there must be sufficient oxygen for proper burner operation. A DFTD with an afterburner will most likely have feed limitations, based on the metals concentration of the feed soil and the type of air pollution con-

Table 1. **Preliminary Remediation Goals for RMA Soils**

Contaminant of Concern	Goal (ppm)[a]
Aldrin	0.72
Benzene	10
Carbon Tetrachloride	2.3
Chlordane	3.7
Chloroacetic Acid	77
Chlorobenzene	850
Chloroform	48
Dichlorodiphenylethane (DDE)	13
Dichlorodiphenyltricloroethane (DDT)	14
Dibromochloropropane (DBCP)	0.2
1,2-Dichloroethane	3.2
1,1-Dichloroethylene	0.52
Dicyclopentadiene (DCPD)	3,700
Dieldrin	0.41
Endrin	230
Hexachlorochlorocyclopentadiene (HCCPD)	1,100
Isodrin	52
1,1,2,2-Tetrachloroethane	1.5
Methylene Chloride	35
Tetrachloroethylene	5.4
Toluene	7,700
Trichloroethylene	28
Arsenic	4.2
Cadmium	50
Chromium	7.5
Lead	2,200
Mercury	570

[a] PRGs are based on a biological worker's reasonable maximum exposure at a 10^{-6} risk level.[1]

trol equipment used. These limitations may or may not restrict the maximum feed rate to the DFTD. Significant levels of arsenic, cadmium, and chromium, such as were found in several of the soil-media areas at RMA, could restrict feed rates, depending on system design. Mercury could also be a problem, if it is present at high enough levels.

An IHTD has a heat source that is separated from the contaminated medium by a metal wall. For laboratory-scale testing, electric resistance heating is frequently used. At the bench- and pilot-scales, electric resistance heaters or fossil-fueled burners are the most common heat sources. Fossil-fueled burners, electric resistance heating, microwave heating, and heat-transfer fluids have been used at the pilot- and full-scales. The solids exit temperature for a full-scale IHTD is similar to that of a DFTD, but rarely exceeds 1000°F. Additionally, the temperature profile of the solids can be more closely controlled in a multiple-burner IHTD. The exit-gas temperature in an IHTD will typically be 200–800°F, which is considerably lower than that of a DFTD. The exit-gas temperature is important if the offgas is to be chilled to condense the volatilized organics and water. The composition of the sweep gas can be controlled in an IHTD. Many systems use nitrogen and can maintain oxygen levels below 1%. The sweep gas used in the bench-scale tests was a mixture of air and nitrogen, resulting in 5% oxygen in the sweep gas. This concentration was chosen to approximate the oxygen level in a DFTD. An IHTD using an offgas condensing system typically does not have any waste-feed metals limitations, with the possible exception of mercury.

BENCH-SCALE, INDIRECT-HEATED TESTING

Equipment and Test Conditions

Laboratory-scale treatability studies using static-tray and heated-tube testing apparatus showed that organochlorine pesticides present in the RMA soils at total concentration levels of several hundred to several thousand parts per million (ppm) were reduced to less than 50 parts per billion (ppb) at 480–1200°F (Farmayan et al., 1989; International Technology Corp., 1990). On the basis of these initial laboratory data and the thermal desorption data available in the literature, Roy F. Weston, Inc. coordinated a series of bench-scale thermal desorption tests, described below, using indirect-heated equipment. Two objectives of the Weston bench-scale testing were to examine the removal of all RMA COCs expected to be above the PRGs, and determine the effect of thermal treatment on the leachability of metals (Roy F. Weston, 1992a).

The indirect-heated, bench-scale equipment used for the Weston study consisted of a 5-in.-diameter by 12-in.-long Incoloy cylinder rotated at 5 rpm and externally heated by a custom-made electric furnace. Test soil was charged into the cylinder in the batch mode, and the heat-up rates of the soil bed where controlled automatically by thermocouples in the soil. The sweep gas was prepared by mixing nitrogen and air in a ratio of 3:1, which resulted in an oxygen concentration of approximately 5% by weight. Ten bench-scale tests were performed on Mixed Basins (MB) soils consisting of soils collected from Basins A, D, and F, and nine tests were performed on the Mixed South Plants (MSP) soils consisting of soils collected from two central processing locations and the salt storage pads. Temperature, time at temperature, and sweep gas rates varied between tests, but were held constant during each test after the system equilibrated at 390–750°F for 5 to 20 min, with sweep gas flows ranging from 1 to 3 L/min.

Bench-Scale Test Results

The bench-scale test results from the final 1992 Weston report for Mixed South Plants and Basins soils are shown in Tables 2 and 3, respectively (Roy F. Weston, 1992b). Only data from tests performed at a 10-min residence time are shown, as this residence time and temperature are most like those experienced during the pilot-scale DFTD tests. Furthermore, it was difficult to see any significant removal-rate increase at 20 min over that at 10 min for the same temperature. The bench-scale temperatures selected for comparison were the low, high, and midpoint for each series.

From Table 2, we see that at 750°F all pesticides in the Mixed South Plants soil were reduced to below their respective detection levels. At both 570°F and 660°F, some DDE remained in the treated soil just below the PRG. These results suggested that 660°F was a good baseline value for the pilot-scale DFTD testing. The only other organic in the feed above the PRG value was DBCP, which was effectively removed at 570°F. Both benzene and toluene appeared to increase

Table 2. Bench-Scale Results for Mixed South Plants Soil

Test Name	Feed Soil	MSP-1	MSP-3	MSP-6	
Soil Temp. (°F)	NA	570	660	750	
Time at Temp. (min)	NA	10	10	10	
Gas Flow Rate (L/min)	NA	1	1	1	
			Analysis		PRGs
pH	9.00	7.50	7.52	7.86	NA
PESTICIDES (mg/kg)					
Aldrin	665	(0.08)*	0.098	(0.08)	0.72
Chlordane	11.05	(0.08)	(0.08)	(0.08)	3.7
4,4-DDE	3.35	0.96	0.13	(0.16)	13
4,4-DDT	36	(0.16)	(0.16)	(0.16)	14
Dieldrin	41	(0.16)	(0.16)	(0.16)	0.41
Endrin	(305)	(0.16)	(0.16)	(0.16)	230
Isodrin	34.5	(0.08)	(0.08)	(0.08)	52
METALS (mg/kg)					
Arsenic	920	787	532	1,070	4.2
Cadmium	16	15.7	9.0	22.1	50
Chromium	56	45.3	43.1	67.8	7.5
Lead	508	443	394	602	2,200
Mercury	1.4	0.003	0.02	0.02	570
PCDD/PCDF (μg/kg)					
TEQ	9.8	2.3	1.6	0.3	NA

Each value in parentheses is the minimum detection limit. Concentration was below this limit.

Table 3. **Bench-Scale Results for Mixed Basins Soil**

Test Name	Feed Soil	MB-2	MB-6	MB-10	
Soil Temp. (°F)	NA	480	570	660	
Time at Temp. (min)	NA	10	10	10	
Gas Flow Rate (L/min)	NA	1	1	1	
			Analysis		**PRGs**
pH	11.76	11.07	11.02	10.96	NA
PESTICIDES (mg/kg)					
Aldrin	160	0.34	(0.08)*	(0.08)	0.72
Chlordane	36.5	(0.08)	(0.08)	(0.08)	3.7
4,4-DDE	4.25	0.36	0.15	(0.16)	13
4,4-DDT	2.2	(0.16)	(0.16)	(0.16)	14
Dieldrin	130	(0.16)	(0.16)	(0.16)	0.41
Endrin	(72)	(0.16)	(0.16)	(0.16)	230
Isodrin	140	(0.08)	(0.08)	(0.08)	52
METALS (mg/kg)					
Arsenic	139	116	141	127	4.2
Cadmium	3.2	2.7	4.7	3.9	50
Chromium	23	20.4	31.2	33.8	7.5
Lead	29	29.2	33	29.9	2,200
Mercury	164	70	3.8	0.89	570
PCDD/PCDF (μg/kg)					
TEQ	1.7	0.7	NA	0.2	NA

Each value in parentheses is the minimum detection limit. Concentration was below this limit.

in concentration after treatment, but the levels of these organics were considerably below their respective PRG values. As expected, there was no evidence of significant metals removal at any temperature, except for mercury. Mercury was removed even at the lowest temperature. The slightly alkaline feed soil was reduced in pH to near neutrality at all temperatures.

Similar results were obtained with the Mixed Basins soils at somewhat lower temperatures, as shown in Table 3. At 480°F, there was still detectable Aldrin and DDE, with the Aldrin level being only slightly below the PRG. At 570°F, the Aldrin level was nondetectable, and the DDE level was almost one order of magnitude below the PRG. At 660°F, all pesticides were below their respective detection limits. For pesticide removal, it appears that 570°F would be sufficient. Again, the benzene and toluene levels appeared to increase in concentration, but they were still considerably below their respective PRG values. The metals behaved as expected, with no evidence of volatilization, except for mercury. Increasing the temperature continued to improve the removal efficiency for mercury. The Toxic Characteristic Leachate Procedure (TCLP) limit for mercury is

0.2 mg/L, which allows a total constituent value of approximately 4.0 mg/kg because of the 20-to-1 dilution in the TCLP test. This would require a temperature of 570°F on the basis of this test. The pH of the alkaline soil was slightly reduced at all temperatures.

PILOT-SCALE, DIRECT-FIRED TESTING

Equipment and Test Conditions

Pilot-scale, direct-fired testing was performed at the EPA's Incineration Research Facility (IRF) in Jefferson, Arkansas, which is operated by Acurex Environmental Corp. The rotary kiln system (RKS) at the IRF consists of a 3-ft-diameter by 7-ft-long refractory-lined rotary kiln with a 2.0-million-Btu/h burner mounted at the feed end of the kiln. Offgas is treated sequentially by a 1.5-million-Btu/h afterburner; a primary air pollution control system consisting of a quench, an ionizing wet scrubber, and a fabric-filter baghouse; and a secondary air pollution control system consisting of an activated-carbon bed and a HEPA filter. For this pilot-scale study, the RKS was equipped with a 2-1/2 in. screw auger and hopper system capable of continuously feeding RMA soils at rates up to 900 lb/h. The rotary kiln is also equipped with seven thermocouples to measure soil-bed temperatures at approximately 1-ft intervals along the axis of the kiln. A schematic of the RKS at the IRF appears in Figure 1.

The planned DFTD test conditions are given in Table 4. During the baseline test (Test A), the soil was to be heated to 675°F, which is just above the minimum value determined for the Mixed South Plants soil by bench-scale testing.

Figure 1. **Schematic of the IRF Rotary Kiln DFTD.**

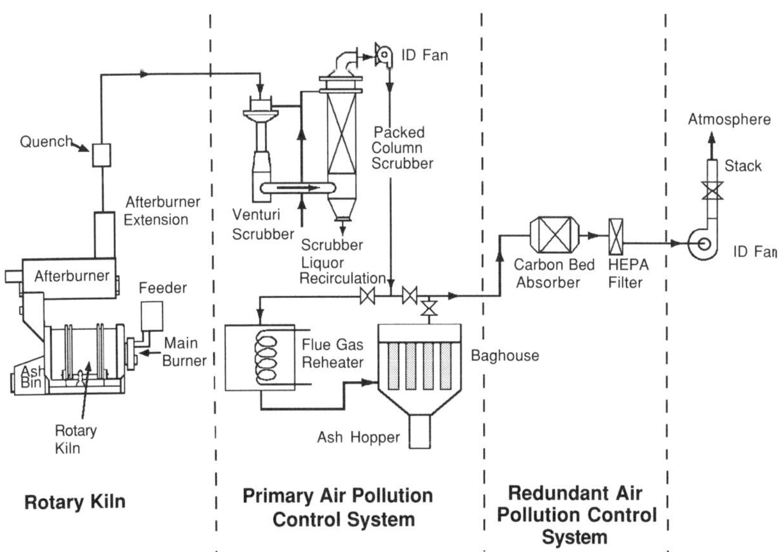

Table 4. **Planned Direct-Fired Test Conditions**

Test ID	Test Name	Feed Rate (lb/h)	Residence Time (min)	Exit Soil Temp. (°F)
A	Baseline	500	35	675
B	High Temperature	500	35	825
C	Low Residence Time/High Feed Rate	750	25	675

Heating to 675°F would put the soil at or above 660°F for 5–15 min. The design of the IRF kiln limited the fill volume to approximately 6%. It was decided to keep the fill volume approximately constant for all tests to eliminate one variable. The potential variables then were solids temperature, residence time, and feed rate. Temperature is of primary concern, and the tests were planned at the baseline value of 675°F and at 825°F (Test B). At 825°F, all of the pesticide contaminants should be reduced to below their detection levels. Since fill volume is being held constant, the increased feed rate test (Test C) should also have a shortened residence time.

The actual test conditions are presented in Tables 5a and 5b. The desired baseline temperature for both soils was increased from 675°F to 700°F, based on preliminary results from a shakedown test not reported in this chapter. The target temperature was held to ±25°F for all six tests. The feed rate for South Plants Test C was considerably below the target of 750 lb/h due to the excessive abrasion of the feeder screw auger. Lack of feed material and other project constraints prevented a repeat of the test. For all other tests, the feed rate was maintained at the desired level ±50 lb/h. The residence time is a calculated value and was held within ±10% for all tests.

Table 5a. **Actual Test Conditions at South Plants**

Test ID	Test Name	Feed Rate (lb/h)	Residence Time (min)	Exit Soil Temp. (°F)
A	Baseline	530	35	700
B	High Temperature	480	35	835
C	Low Residence Time/High Feed Rate	565	25	840

Table 5b. **Actual Test Conditions at Basin F**

Test ID	Test Name	Feed Rate (lb/h)	Residence Time (min)	Exit Soil Temp. (°F)
A	Baseline	460	35	710
B	High Temperature	510	35	830
C	Low Residence Time/High Feed Rate	740	25	695

Test Results

The primary goal of this treatability study was to determine if all of the organic COCs can be reduced to below their respective PRGs at reasonable temperatures and residence times, as predicted from the earlier bench-scale study. Since the PRGs for metals are based on their total concentration and not their leachable value (TCLP), thermal desorption was not expected to achieve the PRGs for metals. However, metals recovery in the treated solids and changes in leachability are discussed. Since there is always a concern about dioxin and furan formation in thermal processes treating chlorinated organics such as pesticides, this issue is also addressed.

Solids Mass Balance

It is necessary to perform a mass balance to assure that all process streams are accounted for and that accurate test data are recorded. In thermal processes, material balances are often difficult to close with respect to both dry-solids and ash content. Dry solids represent the solids mass remaining after overnight oven drying at 221°F (105°C). When conducting thermal desorption studies at temperatures 700–825°F, the mass balance will not close, because the treated solids mass will be reduced by an amount equal to the mass that volatilizes between 221°F and the temperature of the test. Conversely, if the ash value at 1022°F (550°C) is used, it is possible to recover over 100%, since the treated solids never reach that temperature. Table 6 contains mass balances calculated on a dry-solids and an ash basis. Since the soil temperatures reached during the tests were between the temperatures used for the dry-solids and ash measurement procedures, a mass balance based on the average of the two values was calculated. Mass closure for this average ranged from 96.8% to 99.1%. These data indicate that very little material was not accounted for.

Pesticides

There are seven pesticide COCs as shown in Table 1. Of the seven, five were found in the South Plants soil and four at Basin F. Dieldrin, which has the lowest PRG at 0.41 mg/kg, had the highest soil concentration at about 3,000 mg/kg in the South Plants samples. This combination allows for a severe test of the efficacy of thermal desorption.

Table 6. **Thermal Desorption System Mass Balance**

	Percent Solids Recovery					
	South Plants			Basin F		
Basis for Calculation	Test A	Test B	Test C	Test A	Test B	Test C
Based on Dry Solids @ 221°F (105°C)	94.9	97.1	97.6	92.1	92.0	94.0
Based on Ash Solids @ 1022°F (550°C)	98.8	98.7	98.3	105.6	102.2	104.1
Average of 2 Methods	96.8	97.9	97.9	98.9	97.1	99.1

Table 7. **Feed Soils Pesticides and PRGs**
(Average of 9 samples as mg/kg on a wet basis)

Pesticide	South Plants	Basin F	PRGs
Aldrin	1,030	663	0.72
Chlordane	66.6	<5.69	3.7
4,4-DDE	<19.8	<7.92	13
4,4-DDT	<24.0	<9.59	14
Dieldrin	2,542	341	0.41
Endrin	600	275	230
Isodrin	74.2	35.0	52

Pesticide concentrations in the feed soils, reported on an as-received (wet) basis, are presented in Table 7, and can be compared with those from Tables 2 and 3 for the bench-scale studies. For South Plants, Aldrin, Chlordane, and Isodrin are at the same order of magnitude. DDT and DDE were below 50 mg/kg in both sets. Dieldrin and Endrin were much higher in the pilot-scale samples. The pilot-scale Basin F soil samples were similar for Aldrin, Dieldrin, and Isodrin. No Chlordane (<5.69 mg/kg) was detected in the Basin F samples, while the Mixed Basins soils had 36.5 mg/kg. Endrin was much higher in the pilot-scale samples. Since contaminated soil samples frequently differ in contaminant concentrations by several orders of magnitude, the two sample sets can be considered suitably similar for the testing being performed.

To calculate the percent removal of the specific pesticides, the concentrations have been recalculated on a dry-weight basis. The percent removal was calculated by dividing the reduction in contaminant concentration by the initial concentration. Table 8a presents the results for the South Plants tests.

The baseline South Plants Test A conducted at 700°F failed to meet the PRG for Dieldrin by 0.60 mg/kg, and the Chlordane residual of 1.13 mg/kg was close to the PRG of 3.7 mg/kg. The results from Test B at 800°F were significantly better. All pesticides except Dieldrin were reduced to below their detection levels. Dieldrin in the treated soil at 0.045 mg/kg was well below the PRG of 0.41 mg/kg. Test C was originally planned to be at the baseline temperature with a 50% increase in the feed rate and a reduction in residence time from 35 to 25 minutes. Since the preliminary analytical data from Test A indicated residual pesticides in the treated soil, the temperature was elevated to 825°F. Due to mechanical problems with the soil feed system, the feed rate during sampling was 564 lb/h instead of the desired 750 lb/h. After Test C, a new feed screw had to be constructed due to excessive wear from the South Plants and shakedown tests. For comparison purposes, Test C is similar to Test B with a 10% increase in feed rate and about a 25% decrease in residence time. One would expect somewhat higher pesticide residuals from Test C compared to Test B. Indeed, that is what

Table 8a. **South Plants Pesticides Before and After Thermal Desorption**
(All concentrations in mg/kg on a dry-weight basis)

	Test A Baseline			Test B High Temperature			Test C Short Residence Time and High Temp.			
Pesticide	Feed	Product	% Removal	Feed	Product	% Removal	Feed	Product	% Removal	PRGs
Aldrin	993	0.211	99.979	1,310	<0.005	99.9996	1,220	0.041	99.9966	0.72
Chlordane	75.2	1.13	98.499	74.6	<0.006	99.992	78.1	<0.006	>99.993	3.7
4,4-DDE	<22.5	<0.008	NA	<22.7	<0.008	NA	<22.6	<0.008	NA	13
4,4-DDT	<27.3	<0.010	NA	<27.5	<0.010	NA	<27.3	<0.010	NA	14
Dieldrin	2,280	1.01	99.956	3,160	0.045	99.9986	3,260	0.157	99.9952	0.41
Endrin	631	0.047	99.993	749	<0.027	99.996	674	<0.027	>99.996	230
Isodrin	80.7	<0.01	99.992	86.6	<0.006	99.993	86.3	<0.006	>99.993	52

Table 8b. **Basin F Pesticides Before and After Thermal Desorption**
(All concentrations in mg/kg on a dry-weight basis)

	Test A Baseline			Test B High Temperature			Test C Short Residence Time and Baseline Temp.			
Pesticide	Feed	Product	% Removal	Feed	Product	% Removal	Feed	Product	% Removal	PRGs
Aldrin	798	<0.005	>99.999	645	<0.005	>99.999	730	0.005	99.999	0.72
Chlordane	<6.47	<0.006	NA	<5.69	<0.006	NA	<6.48	<0.006	NA	3.7
4,4-DDE	<9.00	<0.008	NA	<7.92	<0.008	NA	<9.02	<0.008	NA	13
4,4-DDT	<10.9	<0.010	NA	<9.59	<0.010	NA	<10.9	<0.010	NA	14
Dieldrin	423	<0.006	>99.999	340	<0.006	>99.998	355	<0.006	>99.998	0.41
Endrin	312	<0.027	>99.991	300	<0.027	>99.991	284	<0.027	>99.991	230
Isodrin	51.4	<0.006	>99.988	26.6	<0.006	>99.976	37.9	<0.006	>99.983	52

happened. The Test C conditions produced measured pesticide residuals for Aldrin and Dieldrin, but both were below their respective PRGs.

The pesticide removal efficiencies presented in Table 8a are excellent. In Test A, the poorest percent removal was 98.499% for Chlordane. For Tests B and C, the percent removal was greater than 99.99% for all pesticides that had a measurable concentration in the feed.

The pesticide removal efficiencies for the Basin F tests are shown in Table 8b. The Basin F soils required ambient air drying from about 14% moisture to 6%, and were tested after the South Plants soils. The drying was required to accommodate the screw feeder system being used, and did not result in any measurable loss of pesticides. The baseline Test A reduced all of the pesticides to below their detection levels. Test B was executed as planned, and also reduced all pesticides to below their detection levels. Preliminary analytical results indicated that there were no measurable residual pesticides in the soil from Test A or B, so the original planned operating conditions were implemented for Test C. The 700°F baseline temperature was used for Test C, and the feed rate was increased 50% to 750 lb/h, while the residence time was decreased from 35 min to 25 min. It was hoped that the Test C conditions would result in detectable levels of one or more pesticides in the treated soil. The data in Table 8b show a measurable amount of Aldrin, but this can be misleading. The product results

Table 9a. **South Plants Total Metals**
(All concentrations in mg/kg on a dry-weight basis)

	Test A			Test B			Test C			Average(3)
Metal	Feed	Product	%Change	Feed	Product	%Change	Feed	Product	%Change	% Change
Arsenic	249	324	30	193	247	28	162	195	21	26
Cadmium	0.85	1.4	64	1.26	0.89	-29	0.85	0.63	-26	3
Chromium	13.5	29.7	120	14.3	46.7	227	13.1	39.1	199	182
Lead	40	45.5	13	52	50.3	-3	39	41.3	5	5
Mercury	3.52	1.4	-60	4.23	0.33	-92	3.76	0.25	-93	-82

Table 9b. **Basin F Total Metals**
(All concentrations in mg/kg on a dry-weight basis)

	Test A			Test B			Test C			Average(3)
Metal	Feed	Product	%Change	Feed	Product	%Change	Feed	Product	%Change	% Change
Arsenic	<16.0	23.1	>45	<15.8	24.6	>56	4.4	6.3	44	48
Cadmium	<0.53	<0.50	NA	<0.53	<0.50	NA	<0.53	<0.50	NA	NA
Chromium	20.7	28.0	36	20.4	19.4	-5	21.1	24.7	17	16
Lead	27.7	24.8	-10	20.8	25.1	21	25.2	28.8	14	8
Mercury	0.564	<0.078	->86	0.664	<0.079	->88	0.717	<0.080	->89	->88

are an average of three samples, and the Aldrin analysis produced two results below the detection level of 0.046 mg/kg and one at 0.067 mg/kg. It is difficult to draw any valid conclusions with this data set. In any case, it indicates that the baseline temperature with a reduced residence time and a high feed rate will still produce treated soil that meets the pesticide PRGs.

Metals

Thermal desorption is not generally capable of removing heavy metals from contaminated media such as soil, with the exception of mercury. There has been some concern that arsenic may volatilize, especially if it exists as an organic compound. Although extensive intermediate gas-stream and stack sampling were performed for metals, this chapter will only address the effect thermal desorption had on the metals in the treated soil. The five heavy-metal COCs at RMA (Table 1) include arsenic and mercury.

Arsenic appears to have increased in all three South Plants and Basin F tests (Tables 9a and 9b). Since the data are reported on a dry-weight basis, varying moisture content does not explain the apparent arsenic increase, which could well be within experimental error. In any case, the data do not indicate that arsenic was volatilized at the temperatures tested. Chromium in the Basin F soils behaved similarly. The increase in chromium in the South Plants samples may be attributed to the excessive wearing of the chrome-plated feed screw. Since the chromium increase from the Basin F tests, where the new feed screw had minimal wear, is much lower, this seems to be a reasonable explanation. For both

Table 10a. **South Plants TCLP Metals**
(All concentrations in mg/L)

	Test A			Test B			Test C			Average(3)
Metal	Feed	Product	% Change	Feed	Product	% Change	Feed	Product	% Change	% Change
Arsenic	7.42	3.88	-48	5.71	3.28	-43	7.00	2.35	-66	-52
Cadmium	<0.0190	<0.0167	NA	<0.0191	0.0183	NA	0.0219	0.0074	-66	NA
Chromium	<0.0316	<0.0278	NA	<0.0318	<0.0278	NA	0.0075	<0.0056	->26	NA
Lead	<0.0302	<0.0222	NA	<0.0302	0.0382	NA	<0.0379	<0.0333	NA	NA
Mercury	0.0008	0.0004	-55	0.0009	<0.0002	->78	0.0009	<0.0002	->78	-70

Table 10b. **Basin F TCLP Metals**
(All concentrations in mg/L)

	Test A			Test B			Test C			Average(3)
Metal	Feed	Product	% Change	Feed	Product	% Change	Feed	Product	% Change	% Change
Arsenic	0.1363	0.0834	-39	0.1496	0.0690	-54	0.119	0.002	-98	-64
Cadmium	0.0249	0.0300	20	0.0327	0.0349	7	0.033	0.028	-16	4
Chromium	0.0207	<0.0056	->73	0.0148	<0.0056	->62	0.021	0.006	-72	-69
Lead	<0.0592	<0.0556	NA	<0.0590	<0.0556	NA	<0.060	<0.056	NA	NA
Mercury	0.0014	<0.0002	->86	0.0015	<0.0002	->86	0.0019	<0.0002	->90	-87

soil types, cadmium and lead showed no significant change on the average, while the mercury was reduced by 60–93%, as expected.

The TCLP metals values for the treated soils are presented in Tables 10a and 10b. Of particular interest is arsenic, which decreased for both the South Plants and the Basin F soils by 39–98%. It is worth noting that the arsenic concentration in the South Plants soil is at least 10 times greater than that in the Basin F soil, yet they both showed similar percent reductions in their leachability after thermal treatment. There is no apparent difference between the soils when heated to 700–825°F. As discussed previously, the PRGs for the metals are based on total concentration, not the TCLP value; however, it is interesting to compare the treated soils to the EPA's regulatory toxicity characteristic level for arsenic, which is 5.0 mg/L (40 CFR 261.24 Table 1). The untreated South Plants samples would be classified as toxicity characteristic waste D004, but the treated samples would not.

It appears that the thermal treatment had no effect on leachable cadmium. The leachable chromium behaved in a similar manner to arsenic. Even though there was a large increase in the total chromium in the South Plants soil tests, the leachable amount was below the detection level in two samples and less than the feed in the third. Leachable lead was found near the detection level in only one of the six treated samples, and in none of the untreated samples. The TCLP value for mercury was reduced by approximately the same percentage as the total mercury concentration.

Table 11a. **South Plants Feed and Product PCDD and PCDF Analysis**
(All concentrations in ppb)

Analyte	Feed Avg.(3)	Treated Soil Test A T=700°F	Treated Soil Test B T=835°F	Treated Soil Test C T=840°F
Total TCDD	0.0019	0.404	0.658	0.322
2,3,7,8-TCDD	0.0005	0.0068	0.0055	0.0032
Total PCDD	0.287	0.955	1.686	0.822
Total TCDF	0.529	20.53	6.87	6.31
2,3,7,8-TCDF	0.148	16.39	4.25	3.96
Total PCDF	4.127	32.086	14.9	13.18
Total PCDD+PCDF	4.41	33.04	16.59	14.00
PCDD/PCDF Ratio	0.07	0.03	0.11	0.06
TEQ	0.074	1.953	0.638	0.554

Table 11b. **Basin F Feed and Product PCDD and PCDF Analysis**
(All concentrations in ppb)

Analyte	Feed Avg.(3)	Treated Soil Test A T=710°F	Treated Soil Test B T=830°F	Treated Soil Test C T=695°F
Total TCDD	0.0122	0.015	<0.008	0.169
2,3,7,8-TCDD	0.0029	0.0016	<0.008	0.0127
Total PCDD	10.359	0.132	0.439	1.194
Total TCDF	0.226	2.44	2.05	0.203
2,3,7,8-TCDF	0.0274	0.422	0.416	0.0343
Total PCDF	2.406	10.471	31.38	0.542
Total PCDD+PCDF	12.77	10.60	31.82	1.736
PCDD/PCDF Ratio	4.31	0.01	0.01	2.20
TEQ	0.063	0.294	0.902	0.058

PCDD/PCDFs

There is always a concern about the formation of PCDDs or PCDFs when using thermal treatment methods on materials containing chlorinated organics such as chlorinated pesticides, certain wood-treating compounds, and PCBs. In addition to the feed and treated solids, the scrubber liquid, scrubber solids, baghouse ash, and afterburner and baghouse exit flue gases were analyzed for PCDDs and PCDFs. This chapter will address only the dioxin and furan concentrations in the feed and treated soils.

Tables 11a and 11b summarize the dioxin/furan data for the South Plants and Basin F soils, respectively. Intensive discussion of the data in these two tables could easily constitute another chapter. However, there are a few results that must be mentioned. For the South Plants soil, the 2,3,7,8-TCDD increased in the treated soil over the feed soil by approximately one order of magnitude at both temperatures. Total PCDD+PCDF increased for all tests, but by a lesser value at the higher temperature. The toxicity equivalency (TEQ) also increased by approximately one-order magnitude, but with the largest increase occurring at the lowest temperature. The ratio of PCDDs to PCDFs was fairly consistent for all three tests.

The data from the Basin F tests shown in Table 11b are not as consistent. Tests A and B, representing baseline and high temperatures, are similar. Test C, which was similar in temperature to Test A, is markedly different in all reported parameters from Tests A or B. The system operating data sheets, data tracking sheets, and laboratory reports are being reviewed to see if a cause can be found. At this time, no explanation is available. If one looks only at the TEQ value, Test A increased the concentration by a factor of 4, Test B resulted in a 14-fold increase, and Test C produced essentially no change. At the conditions tested, the direct-fired thermal treatment of Basin F soils produced a change ranging from none to greater than a ten-fold increase in the TEQ.

CONCLUSIONS

The pilot-scale DFTD testing clearly demonstrated that all the pesticides listed as COCs can be reduced to well below their current PRGs for either soil type. The batch-mode, bench-scale IHTD testing predicted somewhat lower treatment temperatures than were required by the DFTD. Besides differences in the testing procedures and temperatures, the soil samples used for the two testing programs were not from exactly the same source. With all these differences between the two treatability studies, the slight difference in the required temperatures is not surprising. It does, however, support the value from bench-scale treatability studies. Bench-scale testing predicted that thermal desorption could reduce all of the pesticides to below their respective PRGs at temperatures in the 500–700°F range with residence times of 25–40 min. Pilot-scale DFTD testing required the somewhat higher temperatures 700–825°F with residence times of 25–35 min.

There were no surprises with regard to the total metals. With the exception of mercury, there were no measurable changes in the total metals concentration for either the bench- or pilot-scale tests. The mercury concentrations in the Basin soils for the two studies were significantly different. The bench-scale Mixed Basins soil contained 164 mg/kg of mercury, while the pilot-scale Basin F soil contained only 0.56 mg/kg. Bench testing showed a direct relationship between temperature and residual mercury, with only 0.89 mg/kg remaining in the Mixed Basins soil after treatment at 660ºF; the pilot testing results were all below the detection level. The Mixed South Plants and South Plants soil samples had similar mercury contamination levels (Tables 2 and 9a). The bench-scale tests were inconclusive, since the mercury was reduced to a lower level at 570°F than at 660°F or 750°F. The pilot-scale tests produced a 60% reduction at 700°F, 92% at 835°F, and 93% at 840°F.

The leachability of the metals before and after thermal treatment, as defined by the TCLP test, proved interesting. There were no TCLP metals data available from the bench-scale tests. The apparent increase in chromium for the South Plants tests was discussed previously. The data in Table 9a clearly show a reduction in TCLP concentration for arsenic in all of the pilot-scale tests for both soils. A similar phenomenon was seen for the chromium in the Basin F soils. It appears that thermal treatment had no effect on the leachability of cadmium, since two of the tests showed a slight increase and one test produced a similar decrease. Since the metals PRGs are based on total concentration at RMA, the reduction in the TCLP value does not apply. At other sites requiring soil remediation, thermal desorption should not be ruled out if the metals treatment standards are based on the TCLP values. Caution must be exercised, however, since it is possible that the TCLP value of some metals could increase for different soil types and different operating conditions.

The increase of PCDDs and PCDFs in the treated soil, while not surprising, is somewhat troubling, due to the rather large increase in several of the tests. It is generally accepted that the potential for PCDD or PCDF production exists in thermal processes when PCDD or PCDF precursors are present. Total PCDD+PCDF increases of up to almost 650% (South Plants Test A) with a TEQ of 1.953 μg/kg in the treated soil are cause for concern. The process gas streams and stack emissions data are still in review, but stack emissions were well within all proposed air regulations. All scrubber water samples contained less than 50 parts per quintrillion as TEQ, and all scrubber solids were less than 10 parts per trillion.

Bench-scale testing with the IHTD produced markedly different results. Tables 2 and 3 present the TEQ values from the bench-scale study. The TEQs were calculated assuming that all PCDDs and PCDFs reported were 2,3,7,8-substituted. This produced high values for the TEQs. The bench-scale IHTD data produced lower TEQ values in all of the treated soils compared to the feed soils.

The data also show a trend toward lower TEQ values as the soil temperature increased.

This discrepancy regarding PCDD/PCDF formation can most likely be attributed to the difference in the type of thermal desorber used. The indirectly heated system operated in the batch mode and consistently produced reductions in the TEQ of up to 97%, while the continuously fed, direct-fired system produced increases ranging from 0% to 650%. It is not within the scope of this chapter to postulate why one system increases the PCDD/PCDFs and the other reduces them. This difference in results can be used to stress the need for pilot-scale testing for remedy selection. The pilot system used must be of the same type as the system being considered at the full scale. The most important parameter is the heating mechanism. Indirect heating and direct firing cannot be interchanged. Bench-scale testing was more than adequate for remedy screening and predicted the required times and temperatures well enough to perform budgetary economic estimates. It did not, however, warn of the possibility of PCDD/PCDF formation.

The next logical step to complete the testing of thermal desorption for RMA soils would be to perform a pilot-scale continuous IHTD treatability study on the Basin F and South Plants soils. Such a study would determine (1) whether similar temperatures and times would produce similar pesticide COC removals, (2) whether the difference in heating would affect the metals leachability, and, most importantly, (3) whether the increase in the TEQ was due to the direct heating. No such testing is currently planned.

ACKNOWLEDGMENT

This work was supported under a military interdepartmental purchase request from the U.S. Department of Defense, Program Manager for Rocky Mountain Arsenal, through U.S. Department of Energy Contract W-31-109-Eng-38.

REFERENCES

EBASCO Services, Inc. 1994. Proposed Detailed Analysis of Alternatives Report, Version 3.0-Soils DAA, DAAA-05-92-D-0002, EBASCO Services, Inc., December.

Farmayan, W.F., Dicks, L.W.R., and Rhodes, I.A. 1989. Laboratory Thermal Desorption Studies on RMA Basin Sediments and Contaminated Soils, WRC Technical Progress Report RMA 2-89, August.

International Technology Corp. 1990. Final Report for MK Environmental Services (Thermal Desorption Studies), International Technology Corp., August.

Roy F. Weston, Inc. 1992a. Treatability Test Plan for Bench-Scale Thermal Desorption Studies for Rocky Mountain Arsenal Soils, Final, DAAA 15-88-D-0023, Roy F. Weston, Inc., July.

Roy F. Weston, Inc. 1992b. Final Treatability Test Report Bench-Scale Thermal Desorption Studies for Rocky Mountain Arsenal Soils, Volume I, DAAA 15-88-D-0023, Roy F. Weston, Inc., November.

CHAPTER 26

Loring Air Force Base: CERCLA/SARA Interim Remedial Measure Case Study

Jeffrey J. Tasca and **Keri J. Ford**, Ogden Environmental & Energy Services Co., Inc., Somerset, New Jersey
Paul J. Gretsky, IT Corporation, Las Vegas, Nevada

INTRODUCTION

The Air Force Center for Environmental Excellence (AFCEE) has assigned Ogden Environmental and Energy Services Company, Inc. (Ogden) a series of Delivery Orders (DO), for Environmental Remedial Action/Soil and Tank Removal at various Air Force installations nationwide. This chapter focuses on the Low Level Radioactive Waste (LLRW) and Underground Storage Tank (UST) Removal conducted at Loring Air Force Base (LAFB), Limestone, Maine. AFCEE issued the Notice to Proceed (NTP) to Ogden on 07 September 1994 to commence the remedial operation. Ogden personnel were mobilized on 12 September and commenced site operations on 19 September 1994.

The LAFB DO encompassed remedial work in the Weapons Storage Area (WSA) which provided for the excavation of soil and LLRW debris from two trenches (designated as Trench C and E), and the removal of five USTs previously thought to have been used for wash/waste water operations. This work was performed under U.S. Environmental Protection Agency (USEPA) Region I guidance as a Time Critical Removal Action. Related remedial activities included: the excavation, packaging, and transportation of LLRW; post-excavation sampling and analysis to demonstrate clean closure; backfilling of the excavations, and documentation of the activities, findings, and conclusions.

Field activities and laboratory analyses were conducted in compliance with applicable federal, state, regional, and county laws and regulations pertaining to safety and environmental quality, with certain exceptions where AFCEE policy and/or requirements differ from either promulgated regulations or nonpromulgated agency guidelines. Quality Project Plans (QPPs) were prepared, including a site-specific Health and Safety Plan (HSP), Construction Quality Plan (CQP), and Sampling and Analysis Plan (SAP). An initial site characterization of the WSA was performed in accordance with the approved QPPs.

The Remedial Investigation (RI) Removal Action (RA) field activities concluded on 22 November 1994. On 16 February 1995 Ogden submitted the Draft RI/RA Report for Underground Storage Tanks and Low Level Radioactive Waste Trenches at Loring Air Force Base (Ogden, 1995) to AFCEE, LAFB, and the involved regulatory agencies. The RI/RA Report documented the Interim Remedial Measure (IRM) by summarizing the technical performance and the analytical data generated during field activities. The report also provided the analytical data and corresponding calculations which demonstrated compliance with the Intended IRM Cleanup Level Objectives for the two trenches and five USTs. Upon receipt of regulatory, Department of Defense (DoD), and Radioisotopic Committee (RIC) reviews, the Final RI/RA Report was prepared for submittal to the Maine Department of Environmental Protection (MEDEP) and USEPA - Region I as part of the Remedial Investigation/Feasibility Study (RI/FS) process. An AFCEE Final Report required under the AFCEE/Ogden contract was also prepared to serve as a comprehensive record of the DO's execution. This report documented the performance of the delivery order-specific tasks, with a focus on the contractual aspects of the project including: technical performance, schedule performance, and major issues and resolutions.

Ogden received notice that the Record of Decision (ROD) for the Clean Closure of the LAFB WSA Operable Unit 1 RI was signed August 1995. The LAFB remediation effort reflects the United States Air Force (USAF) commitment to both environmental compliance and environmental and economic stewardship

IRM OVERVIEW

LAFB, officially closed 30 September 1994, is part of the DoD Installation Restoration Program (IRP) that investigates suspected past releases of toxic or hazardous materials at DoD facilities. The IRP at LAFB includes Site Inspections (SIs), Remedial Investigations/Feasibility Studies, and IRMs at Areas of Concern (AOCs) on the Base. These environmental activities are governed by a Federal Facilities Agreement (FFA) among USEPA, MEDEP and the USAF which was signed 30 January 1991 and amended December 1993.

The AOCs have been organized into Operable Units (OUs) based on geographical location, relations to surface drainage and man-made structures, and possible hydrogeologic connections. Based on data collected from the RI/FS conducted at OU 1 - the LLRW Disposal Sites, located in the WSA - it was determined that an IRM was necessary. The IRM was justified as a Time Critical Removal Action and was performed in 1994, with final offsite disposal of the LLRW being completed in January 1995. All activities conducted as part of the IRM were performed in accordance with the approved, site-specific CQP, HSP, and SAP. Direct oversight of IRM activities was performed by AFCEE, LAFB, MEDEP, USEPA, and the USAF RIC.

Oversight Responsibilities

The MEDEP was responsible for the technical oversight of the project to ensure that the RA was conducted in accordance with the FFA and state requirements. The University of Maine provided technical support to MEDEP concerning the LLRW. The USEPA was responsible for the technical oversight of the RA to ensure that federal regulations and the criteria of the FFA were met. The USAF RIC was responsible for directly overseeing all aspects of the project related to the investigation and remediation of radiological materials. Under the direction of the DoD, the RIC has been designated as the authorized representative of the USAF for reviewing and approving all plans and project execution on USAF properties that involve radiological materials. Authorization from the RIC was required prior to the start of intrusive activities, for backfilling the sites, and for final site release. Armstrong Laboratories was designated by the RIC for all laboratory support at LAFB, as well as being responsible for reviewing the site-specific QPPs and for monitoring the project's execution. Final approval for intrusive activities was given, in writing, by AFCEE after approval from the USAF RIC was received. The DoD designated and permitted Treatment, Storage, and Disposal Facility (TSDF) that received the LLRW excavated during LAFB DO activities was located in Clive, Utah.

A number of regulatory and compliance activities were completed by Ogden prior to and during the execution of this work. Compliance related activities included the development of QPPs; submittal of a Time Critical Removal Action Memorandum; characterization of the trenches and USTs; trench waste characterization and profiling for the waste disposal facility; establishment of the background and cleanup levels for the contaminants of concern; waste manifesting and transportation; and approval of clean closure for the USTs and trenches located within the WSA. Throughout the IRM process, the public was provided with comment periods for the Action Memorandum, and public meetings were conducted to provide detailed information during each phase of the IRM. The IRM was broken into two elements: Site Characterization and Site Remediation.

IRM Activities

IRM Element I - Site Characterization was performed from 20 June 1994 through 01 July 1994. The objectives of the site characterization were to: (1) identify the contaminants of concern; (2) delineate the locations, extent, and volume of the LLRW trenches; (3) characterize the contents of the five USTs and underlying soils associated with the WSA maintenance structures; (4) refine the Element II QPP and SAP; and (5) provide documentation for the preparation of the USEPA Time Critical Removal Action Memorandum. Element I activities included: a topographic survey, geophysical surveys, cone penetrometry investigation, and test pit excavation. Sampling and analysis of LLRW encountered in the test pits and of the contents of the five USTs for the presence of tritium were conducted as the final phase of Element I. The data collected during the site characterization element were presented in the Final Radioactive Waste Sites Operable Unit 1, Site Characterization Report for Soil and Underground Storage Tank Low Level Radioactive Waste Removal (Ogden, 1994).

The USAF RIC approved the execution of IRM Element II - Site Remediation, with concurrence from USEPA and MEDEP, based on the analytical results generated from the site characterization. The objectives of Element II were to remove and dispose of the five USTs, remediate the two LLRW trenches, and receive regulatory-approved Clean Closure.

IRM Element II - Site Remediation commenced on 12 September 1994 with mobilization to LAFB. The removal of the five USTs, which were determined to be free of tritium contamination through the efforts of Element I, was initiated on 21 September 1994 and completed on 23 September 1994. UST Clean Closure samples were collected on 23, 24, and 26 September 1994. Based on these sample results, UST Clean Closure was approved by the RIC, and the UST sites were restored to preconstruction condition on 16, 17 and 18 November 1994.

The removal action for Trench C commenced on 21 September 1994 with the establishment of work zones, site preparation activities, and the excavation/stockpiling/sampling of clean overburden. The removal action at Trench C was temporarily delayed from 27 September to 17 October 1994 while rehabilitation of the commercial rail line to be used for shipment of the LLRW was completed. Excavation at Trench C recommenced on 07 November 1994 and was completed on 15 November 1994. A total of 739.10 tons of LLRW was excavated from Trench C and loaded into nine lined gondola rail cars. A gridded radiation walkover survey utilizing Ludlum Model 12™ meters with alpha scintillator and beta/gamma probes and Micro R™ meters was performed on 16 November 1994, and Clean Closure sampling was performed on 17 November 1994.

Excavation at Trench E commenced on 20 October 1994 and was completed on 06 November 1994. A total of 721.60 tons of LLRW was excavated from Trench E and loaded into 10 lined gondola rail cars. A radiation walkover survey was performed on 14 November 1994 and Clean Closure Sampling was performed on 15 November 1994.

LLRW-contaminated personal protective equipment (PPE) and site materials (liners, stabilized decontamination water, wood, etc.) derived during the remedial action were loaded into the final rail car in accordance with the specifications required by the TSDF. The transportation of 20 gondola rail cars loaded with LLRW occurred in two separate shipments. The first shipment of 12 cars was made on 11 November 1994, and the final shipment of eight cars was made on 22 November 1994. All 20 gondola cars had been accepted at the TSDF by 25 January 1995.

Radionuclides of Concern

Based on technical discussions with the USAF, involved regulatory agencies, and a review of existing historical and analytical data, a list of radionuclides of potential concern for the LAFB WSA was developed. Radiological material found in weapons stored and maintained at LAFB consisted primarily of plutonium-239 (Pu-239) and uranium-235 (U-235), enriched from natural uranium. No incidents involving the release of plutonium were ever reported in connection with operations at the LAFB WSA. Clean Closure samples were analyzed for the following radionuclides: uranium-234 (U-234); uranium-235 (U-235); plutonium-238 (Pu-238); plutonium-239 (Pu-239); americium-241 (Am-241); radium-226 (Ra-226); radium-228 (Ra-228); thorium-227, 228, 230, 231, 232 and 234 (Th-227, -228, -230, -231, -232 and -234); protactinium 234 (Pa-234); and neptunium-237 (Np-237).

For the purposes of ensuring Clean Closure, the following isotopes were determined by all involved regulatory groups and the USAF to be the radionuclides of concern: U-234, U-235, Pu-239, Am-241, and Ra-226. Radium-228, Th-227, 228, 230, 231, 232, and 234 and Np-237 were not considered to be radionuclides of concern because of their natural occurrence in the environment or relatively short half-lives (Th-227 18.2 days, Th-231 25.5 hours, Th-234 24.1 days).

While not evaluated in this removal action, samples analyzed for Pu-238; Np-237; Ra-228; Pa-234; and Th-227, 228, 230, 231, 232, and 234 were collected in order to provide LAFB with the information necessary to complete ongoing RI/FS and CERCLA risk assessment in compliance with the FFA.

In addition to the radiological contaminants of concern, select closure samples at each site were analyzed for volatile organics, semivolatile organics, pesticides/PCBs, and metals. These analyses were completed in order to confirm that mixed waste was not present at, or being removed and disposed of from the LAFB WSA site as LLRW.

Cleanup Levels

It was necessary to determine background activities for the radionuclides of concern since the guidelines used to calculate dose rates for residual radioactivity or acceptable cleanup levels are based on radiation levels measured above background. The Clean Closure level for each radionuclide of concern was defined for this project by the USAF and involved regulatory agencies as the average activity (i.e., background) detected, plus two standard deviations. If activities for the radionuclides of concern in the closure samples were below the criteria, the site was considered clean.

Based on previous investigations, historical records, and data, and the site characterization study of the USTs and trenches conducted in July 1994, hazardous waste was not expected to be encountered in the disposal trenches. The analytical results of the chemical analyses were compared with Preliminary

Remediation Goals (PRGs) for all TCL/TAL constituents. The PRGs were established for RI activities conducted at LAFB OUs 8 through 11 by LAFB and MEDEP representatives.

Background values for the radionuclides of concern were obtained from samples collected during August of 1994 by the LAFB RI/FS contractor. Uranium-238 levels were measured so that the ratios of the uranium isotopes could be determined through percent-weight calculations. Isotopic ratios of elemental uranium were then compared to determine whether the levels present were indicative of naturally occurring, enriched or depleted uranium. In addition, background values for gross alpha and beta were obtained from studies conducted during the LAFB OU1 RI/FS. Measured gross alpha and gross beta levels were utilized as a screening tool during the site remediation to identify material that was "clean" (indistinguishable from background) or potentially contaminated.

UST REMOVAL ACTION ACTIVITIES

Initially, the locations of the USTs, denoted as areas A, B, C, D, and F of the WSA, were delineated utilizing historical data. Based on Site Characterization data which determined that the USTs did not contain hazardous waste or LLRW, no exclusion zones (EZ) or contaminant reduction zones (CRZ) were established around the excavations. However, 50-foot safety zones were posted around the work zones to ensure the general health and safety of both workers and the public during excavation activities. All removal activities, with the exception of Clean Closure sampling, were conducted in standard OSHA Level D PPE.

The overburden at Tanks A, B, D, and F was removed utilizing a CAT 330L Excavator. A CAT 416B rubber tire backhoe was used at Tank C. Soils were stockpiled adjacent to the excavations on 10 mil polyethylene sheeting. All stockpiled soils were covered with 10 mil polyethylene sheeting, bermed, and staked with hay bales to prevent removal by wind and/or surface water runoff or runon. Soils determined to be clean were held for eventual backfill into the excavations.

The maximum volume of excavation for the USTs was determined during the Site Characterization study. These areas extended, at a minimum, 1 foot below the tank bottom (including anchors and/or saddles) and three feet beyond the tank sides and ends. Care was exercised during excavation so as not to damage or crush the USTs, pipelines, or utility lines present. Entry into the excavations for either tank removal or sampling was not required; therefore, the excavations were not shored or sloped.

UST Purging and Removal

Based on historical use of the USTs for water/wastewater storage, the USTs were not believed to contain combustible vapors. However, all standard precautions regarding flammable and hazardous vapors were observed. As a safety precaution, the air within the tanks was analyzed with a combustible gas indicator at several stages during excavation and removal activities. Prior to excavation, vapors were purged by adding solid carbon dioxide (dry ice) in the amount of 1.5 pounds per 100 gallons of tank capacity. The dry ice was crushed and distributed evenly over the greatest possible area to secure rapid evaporation. The tanks were then hoisted from the excavations and staged on 10 mil polyethylene sheeting. Once closure samples were collected, the tanks were rendered unusable by cutting large holes in both tank heads and interior, and exterior surfaces were shovel cleaned. The tanks were labeled prior to removal from the site with the phrase *Not suitable for human consumption* and the date of removal.

The excavations were barricaded with a combination of yellow and black caution tape and orange fencing, and maintained until the results of the Clean Closure sampling were received. Each UST with the exception of Tank A was wrapped in 10 mil poly sheeting to retard against water infiltration. Hay bales instead of poly sheeting were stacked against the open ends of UST A.

Analytical Data

Each UST was sampled for three matrices, soil from beneath either tank end, scrapes from the interior of the tank and from the region of the delivery pipes, and sediment from the floor of the tank. Two soils, two scrapes, and one sediment sample were obtained from each tank, with the exception of UST F. There was not adequate volume of sediment in the bottom of this tank to collect a sediment sample.

Soils

The soil samples from each tank were collected from two small stockpiles that had been brought up from the excavation by the backhoe or excavator operator. The stockpiles were collected from the floor of the excavation near the ends of the former UST sites and within two feet of the base where the UST had been previously positioned. The stockpiles were placed on plastic sheeting and covered with the sheeting until the sampling occurred. The stockpiles were equal roughly in volume to one-half cubic yard. Stainless steel trowels and mixing bowls were used to obtain individual scoops that made up composite samples. One composite was taken per pile for a total of two samples. None of the stockpiles brought up from the floor of the tank excavations contained stained or discolored soil or gross contamination. Field screening of the excavated soils with the photoionization detector (PID) and the hand-held radiation monitors demonstrated further that the soils were free from volatile organic and/or gross radiological material.

Soil samples were analyzed for both radiological and chemical contaminants. Radiological parameters included gross alpha and beta activity, and alpha and gamma spectrometry. Chemical constituent analyses included: TAL metals; TCL volatiles; TCL semivolatiles; and SW-846 pesticides and PCBs.

Scrapes

The scrape samples were collected from the inside of the tank once it had been rendered inert and determined not to be a confined space. Personnel did not have to actually enter the tanks to obtain scrape samples from either ends of the tanks because the holes cut in the ends allowed easy access to the interior walls and delivery pipes. Two composites were collected for radiological analyses, one from the inner walls and one from around the delivery pipe and inside the pipe if adequate volume was available. Stainless steel trowels and mixing bowls were used to collect the samples. All of the USTs, except UST A, were lined with black latex-like material that made up the majority of the scrape samples from these tanks. The scrapes from UST A consisted mainly of scale, rust, and soil.

Scrape samples were analyzed for gross alpha and beta activity, alpha and gamma spectrometry. In addition, sediments samples, if present, were collected and analyzed for the same parameters plus TAL metals. Wipe samples were counted onsite on the Ludlum Model 29™ Dual Scaler.

Sediments

One sediment sample was collected from each tank, except for UST F, as there was inadequate volume of sediments in the tank to collect a sample. The samples were composites of the sediment on the floor of the tank, which was comprised of mud, gravel, latex-like liner, and rust/scale. Sampling personnel did not have to enter the tanks to obtain adequate sample volume. Stainless steel implements were used to obtain the sediments.

Data Validation

Data validation of closure samples collected from the USTs in Areas A, B, C, D, and F, within the WSA was completed on 4 November 1994. Validation of radiological data was performed in accordance with the USEPA Guidance for Data Useability in Risk Assessment (USEPA, 1990). Validation of organic and inorganic data was performed in accordance with the most recent versions of the USEPA Contract Laboratory Program (CLP) National Functional Guidelines for Organic Data Review, and the USEPA CLP Laboratory Data Validation Functional Guidelines for Evaluating Inorganic Analyses (USEPA, 1988; USEPA, 1989).

Site Restoration

Upon MEDEP acceptance and approval by the USAF RIC of clean closure, the UST excavations were backfilled with the previously excavated uncontaminated soils and certified clean fill. Imported material was procured from a local quarry and was of equal or less permeability than the soil removed. Fill material was free of biodegradable material such as wood products, sheet rock, or tree remnants. The material was also free of any metals, extraneous debris, solid waste, and had no stones larger than 3 inches in diameter.

Backfill was placed in layers that, when compacted, did not exceed 12 inches. Each layer was spread evenly and thoroughly mixed during the spreading to obtain uniformity of material to the greatest extent practical. After each of the top lifts had been placed, mixed, and spread evenly, they were bucket-compacted to the specified density. A minimum compaction of 85% of the maximum density was maintained in all grassed areas. Final grade of the backfill areas was the same as existing grade prior to excavation. Once the site was regraded to existing contours, the area was reseeded in accordance with the State of Maine DOT specifications for seed and mulch.

Summary of UST Removal

The removal of the UST in Area A involved excavation, removal of a concrete ballast slab, dewatering, removal and disposal of the 5,000-gallon UST and vent piping, backfill, and site restoration. As groundwater was encountered in the excavation, a sump with 4" trash pump was utilized to dewater the excavation. Based on historical laboratory data, the groundwater at the WSA was determined not to be contaminated. Therefore, groundwater was not containerized during dewatering activities. The discharge from the excavation was pumped into a lined surge pond/settling basin created to control erosion, and then allowed to overflow onto the ground. No water was discharge directly into any surface water body. Erosion control measures employed at UST A included the lined surge pond/settling basin, hay bales, and/or a silt fence, as required.

A concrete ballast slab measuring approximately 16' x 20' x 2' was encountered at 14 feet below grade. The slab was poured in place over the tank and reinforced with steel bar. This slab was demolished with a 1000 pound *headache* ball, excavated, and stockpiled alongside the excavation. Once the material was determined to be clean through testing, the concrete was then used as backfill material in the lower limits of the original excavation.

Clean Closure Sampling was conducted on 23 September 1994. Two soil samples were taken from the excavation floor, two sediment, and two scrape samples from the walls of the UST were also collected and analyzed. All sample results for the radionuclides of concern were below the established cleanup criteria. Upon receipt of the laboratory data, a determination and approval of Clean Closure was made by LAFB, RIC and MEDEP. The open excavation was backfilled with the stockpiled soil, demolished concrete ballast slab, and imported fill material. Restoration to preconstruction conditions was completed on 16 November, and the tank was stenciled and transported to an offbase scrap facility on 17 November 1994.

The removal of USTs B, C, D, and F was performed in the same sequence as Tank A. However, these tanks, as opposed to Tank A, did not have concrete ballast slabs nor did they require dewatering. The sizes of these USTs were: 1,500-gallon Tank D, and 1,000-gallon Tanks B, C, and F. The removal of these tanks involved excavation, inerting, cleaning, removal and disposal of the tanks and vent piping; soil sampling and analysis; and backfill and site restoration as described in the previous sections. Removal activities commenced on 24 September 1994.

Clean Closure sampling was conducted on 24 and 26 September 1994. Soil samples were taken from the excavation floors. Sediment, scrape, and wipe samples from the walls of each UST were also collected and analyzed. All sample results for the radionuclides of concern were below the established cleanup criteria. Upon receipt of the laboratory data, a determination and approval of Clean Closure was made by LAFB, RIC, and MEDEP. The open excavations were backfilled with the stockpiled soils and imported fill material. Restoration to preconstruction conditions was completed on 15 November for Tanks B and F, and 16 November for Tanks C and D. Tanks B and F were stenciled and transported to an offbase scrap facility on 16 November, Tank D on 17 November, and Tank C on 18 November 1994.

LLRW TRENCH REMOVAL ACTION ACTIVITIES

The remediation of the two LLRW trenches (Trench C and E) was performed in sequence, utilizing a consistent operational approach, which included the following activities: (1) excavation and stockpiling of clean overburden soil; (2) excavation and containerization of LLRW and contaminated soil; (3) transportation of the LLRW by railroad to the TDSF; and (4) post-excavation sampling and laboratory analysis for Clean Closure confirmation.

Site Preparation

Site preparation included the installation of erosion control measures (silt fencing and silt barriers) to control the migration of sediment, and implementation of surface water diversion measures to protect the excavations. In addition, the chain-linked fence adjacent to Trench C was relocated to allow room for overburden soil stockpiles and layback of the excavation. Radiation walkover surveys were performed at the loading dock area and operations areas to document the pre-existing activity at each location. These surveys documented the existing condition and served as the baseline for project closeout.

Prior to the commencement of remedial efforts, the limits of the excavations were laid out with wooden stakes and flagging tape. From these initial limits, the EZ and CRZ were then established for the two trenches. The EZ lines were delineated with 3/8" nylon rope (yellow and magenta radiation rope) supported by wooden stakes. The CRZ lines were delineated with yellow and black caution tape supported by wooden stakes. The design and setup of the work zones complied with the approved HSP and standards set forth in 29 CFR 1910.120 (OSHA, 1993). Also, areas within the work zones where overburden soils could be stockpiled were identified and marked.

Following initial site preparation activities, access ways around the trenches were prepared. To stabilize the soils around the entrance to the decontamination pads, a CAT D3 bulldozer was used to compound bevel the decontamination pad area and then level the access route onto the pad. A hydraulic roller was then used to compact the base soils. The CAT D3 bulldozer was used to create a bermed surface water runoff diversion ditch on the upgradient side of Trench E to facilitate the movement of surface water runoff away from the excavation. Geotextile fabric was laid out over the area and covered with approximately 6 inches of 3/4" minus gravel. The gravel roadways provided a sound base and traction for rolloff tractor operation.

A decontamination pad was constructed in each of the EZs at the two trench areas. The decontamination pad areas were graded from each end with slopes angling toward the center and back of the pad to facilitate drainage of the decontamination rinsate to a common collection point (sump), in the pad. The boundaries of the pads were constructed utilizing spruce ties (6" x 6" x 8') to form a berm around the outer perimeter of the pads. A 60-mil, impermeable, geomembrane was placed on the floor of the pads and extended over the ties to serve as a barrier. Plywood sheeting (5/8" x 4' x 8') was placed on top of the liner and then rough-cut planking placed as a final layer to serve as the working surface for operations occurring on the pad.

Excavate and Stockpile Overburden

Information from the Site Characterization was used to define the limits of the excavations and determine the quantity of clean overburden covering the waste material. To minimize the amount of material containerized for disposal, clean overburden excavated from the trenches was stockpiled on 10 mil polyethylene sheeting alongside each of the excavations. Stockpiled material was sampled and analyzed, and when certified clean, utilized for backfill in the trenches. Excavation and stockpiling of the overburden was performed from the perimeter of the trenches using a CAT 330L excavator. The excavations proceeded in approximately 1-foot lifts, until waste materials were encountered. Each bucket of material from a lift was radiologically surveyed using Ludlum Model 12™ meters with alpha scintillator and beta/gamma probes to assure that only clean soil was being stockpiled. In addition to the field screening, composite samples for gross alpha and beta analyses were collected every ~5 cubic yards. One composite sample for alpha spectrometry and one for gamma spectrometry were collected for every ~50 cubic yards of stockpiled soil. A final determination as to the suitability and reuse of each stockpile as backfill was then made following receipt of analytical results.

Excavate and Load Waste to Rolloffs

Once LLRW was encountered in a trench, all material within that level of the excavation was presumed to be contaminated. Potentially contaminated waste was excavated with the CAT 330 L excavator and loaded into lined rolloff boxes staged on the decon pad. Each bucket of waste material was field-screened for radiological and volatile organics constituents prior to being loaded into the rolloff box.

All rolloff boxes were staged and lined on the decon pad prior to loading. Rolloff boxes were lined so that excess material generated during loading operations would not adhere to the external portions of the rolloff box, and potentially contaminated material was not in direct contact with the box. Once a box was filled to the desired capacity, the liner was folded closed over the waste material and the exterior surfaces visually inspected and cleaned as needed. If cleaning was required, the rolloff box was brushed down and/or wiped on the decon pad to remove any fixed material resulting from the loading operation. Each rolloff box was tarped, once loaded for transport to the LAFB coal yard. Wipe samples (smear patches) were collected from each rolloff box for onsite screening prior to the release of the box from the EZ/decon pad. The wipe samples were counted using the onsite Ludlum Model 29™ Dual Scaler. The purpose of this procedure was to prevent the spread of contamination onto the access way and the rolloff tractors which shuttled the rolloff boxes to and from the excavation. Following counting of the wipe samples and release from the EZ, the loaded rolloff boxes were transported to the loading dock at the coal yard for offloading into a gondola car.

Each trench excavation proceeded to the discernable limits of the waste. At this point, approximately one foot of material beyond the original limits of the waste was excavated to ensure the removal of all LLRW. The bottom and sides of the trenches were then radiologically surveyed to determine if hot spots (i.e., anomalies) remained. Spot excavation, as necessary, was performed to ensure that the trench sides and bottom were clean. After which, a sample grid was laid out within the excavation to establish clean closure sampling locations.

Sampling Strategy

Sampling conducted during the removal action activities was broken into two elements: (1) samples collected to determine the suitability of stockpiled overburden as backfill material; and (2) samples collected to determine whether or not the conditions of clean closure were met. Sample grid methodologies were derived from the Manual for Conducting Radiological Surveys in Support of License Termination (NUREG/CR-5849 ORAU-92/C57) (U.S. NRC, 1992). Confirmatory survey/sampling was conducted during the months of October and November to confirm the adequacy and accuracy of soil excavation in removing the LLRW material from the WSA.

Excavation/Removal

Based on the Site Characterization investigation, it was anticipated that approximately 325 cubic yards of soil would be removed during the excavation phase of Trench C, and 1,180 cubic yards would be removed from Trench E, of which 230 cubic yards from Trench C and 622 cubic yards from Trench E were expected to be "clean" soil. This would leave a total of 653 cubic yards of LLRW material to dispose of, and 850 cubic yards of overburden to stockpile for Trench C and E.

Clean soil (overburden) excavated during the removal phase was staged in ~5 to 10 cubic yard stockpiles. Each of these piles was sampled and analyzed for gross alpha and beta activity. In addition, every ~50 cubic yards of material, was analyzed for a full list of radiological isotopes (gamma and alpha spectrometry) to determine the levels of radionuclides present in the overburden soils as a composite sample of the individual stockpiles that added up to 50 cubic yards. This information was necessary in order to determine if the material could be used as clean backfill. These samples were only to be analyzed for radiological constituents since the soil were screened with a PID for the presence of volatile organic constituents. At the completion of the excavation/removal, a total of 40 gross alpha/beta and seven alpha/gamma spectrometry samples were analyzed from stockpiles of "clean" overburden created at Trenches C and E.

Post-Excavation Clean Closure

The Post-Excavation sampling was designed to ensure that the activities of radionuclides from Trenches C and E remaining after excavation and removal operations did not exceed the average background activities plus two standard deviations for these radionuclides. To make this determination, gross alpha, gross beta, gamma spectrometry, and alpha spectrometry were performed on the closure samples.

Additionally, in order to meet OU1 site cleanup requirements concerning other hazardous wastes at Trench C and E, samples were collected for TAL metals, TCL volatiles, TCL semivolatiles, and SW-846 pesticides/PCBs. The number of samples collected for each analysis and the reference grid systems utilized are provided below.

Establishment of Reference Grid Systems

Grid systems were established at the trenches within the WSA to: (1) Facilitate systematic selection of measuring/sampling locations; (2) Provide a mechanism for referencing a measurement/sample back to a specific location so that the same survey point could be relocated; and (3) Provide a convenient means for determining average activity levels.

The grid used consisted of a system of intersecting lines, referenced to a fixed point. The grid lines were arranged in a perpendicular pattern, dividing the survey locations into squares or blocks of equal area. The objective of the grid system was to ensure the collection of quality radiological data from sufficient representative locations (points), so that a statistically sound conclusion regarding the status of the entire trench area could be developed. The reference grid point spacings for each trench were proportional to the final dimensions of the excavated areas.

For Trench C, the reference grid pattern was composed of 8 systematic sampling locations (i.e., grid points numbers 6, 7, 8, 9, 15, 16, 17, and 18) and 15 additional sampling locations to provide the closed-spaced triangular grid patterns (i.e., grid points numbers 1, 2, 3, 4, 5, 10, 11, 12, 13, 14, 19, 20, 21, 22, and 23). In all, 23 locations were identified through the grid system for clean closure sampling. These points were approximately 7.5 ft/2.29 m apart on diagonal or staggered 6 ft/1.83 m on centers. Figure 1 illustrates the grid pattern and sampling locations established for Trench C.

For Trench E, the reference grid pattern was composed of 8 systematic sampling locations (i.e., grid points numbers 6, 7, 8, 9, 15, 16, 17, and 18) and 15 additional sampling locations to provide the closed-spaced triangular grid patterns (i.e., grid points numbers 1, 2, 3, 4, 5, 10, 11, 12, 13, 14, 19, 20, 21, 22, and 23). In all, 23 locations were identified through the grid system for clean closure sampling. These points were approximately 14.1 ft/4.31 m apart on diagonal or staggered 10 ft/3.05 m on centers. Figure 2 illustrates the grid pattern and sampling locations established for Trench E.

Pre-Sampling Survey

Prior to the collection of samples for clean closure, radiological field surveys were performed along the grid in the floor and side walls of the excavations to identify if locations of elevated activity existed using the Ludlum micro R™ meter. If areas above the established action levels were detected, the soils were spot remediated with shovels until "clean" soils were uncovered. If no areas of elevated activity were encountered, the clean closure samples were collected and shipped offsite for analysis.

Samples

Samples obtained for clean closure analyses were collected from the top 6"/ 15 cm of soil at the locations established along the reference grid. Sufficient quantities of soil were collected and homogenized in stainless steel mixing bowls so that samples could be split with the RIC and MEDEP.

Split samples for the MEDEP were randomly drawn from the eight systematic sampling locations. Four split samples, one from each quadrant, were collected from Trench C and E and forwarded to the University of Maine for a total of eight splits. Split samples for the RIC were collected from all locations; both, systematic and additional. In total, 46 samples were forwarded to Armstrong Laboratory Brooks Air Force Base, Texas.

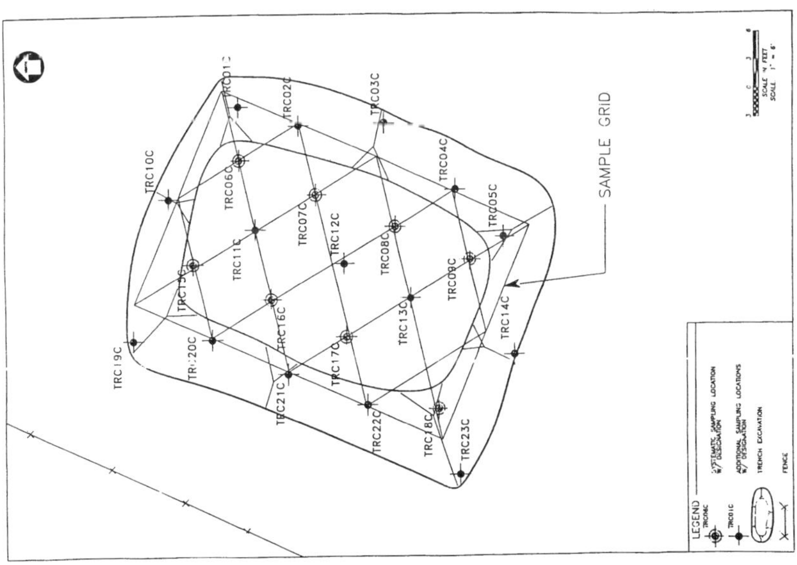

Figure 1. **Trench C - sampling grid, weapon storage area, Loring Air Force Base, Caribou, Maine.**

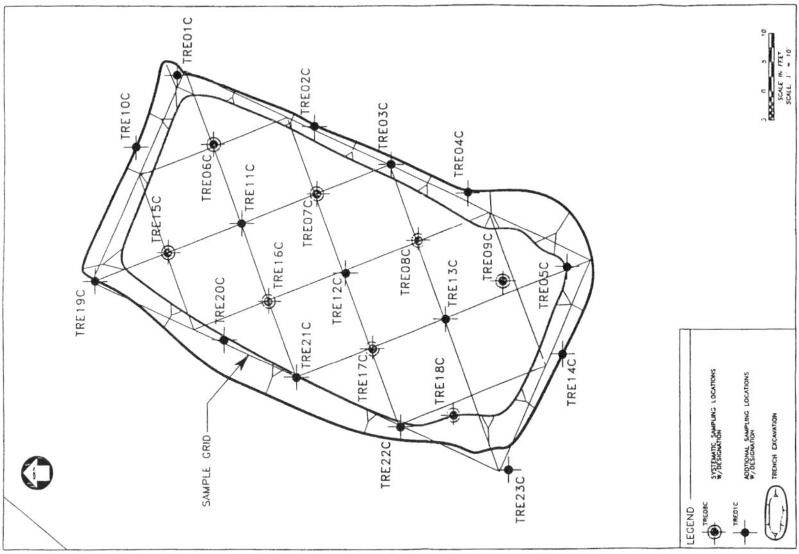

Figure 2. **Trench E - sampling grid, weapon storage area, Loring Air Force Base, Caribou, Maine.**

Sample Analysis

Samples from all 23 grid points at Trench C were analyzed for radiological components; samples from the systematic grid point numbers 6, 9, 16, and 17 were analyzed for the full suite of organic and inorganic analytes; and samples from the additional sampling grid point numbers 1, 3, 5, 10, 12, 14, 19, 21, and 23 were analyzed for inorganic analytes only. In total, 23 samples were run for radiological chemistry, 9 for inorganic analyses, and 4 for full suite at Trench C.

Split sample numbers 7, 8, 15, and 18 were collected for analysis by the University of Maine. In addition, at the direction of AFCEE, as per requests from the University of Maine representatives, an additional sample at grid point 12 was collected and submitted for analyses. Split samples from all 23 grid points (numbers 01 through 23) were submitted to the RIC for analysis.

Samples from all 23 grid points at Trench E were analyzed for radiological components; samples from the systematic grid point numbers 6, 8, 16, and 18 were analyzed for the full suite of organic and inorganic analytes; and samples from the additional sampling grid point numbers 1, 2, 3, 4, 5, 10, 11, 12, 13, 14, 19, 20, 21, 22, and 23 were analyzed for inorganic analytes only. In total, 23 samples were run for radiological chemistry, 15 for inorganic analyses, and 4 for full suite at Trench E.

Split sample numbers 6, 8, 16, and 18 were collected for analysis by the University of Maine. In addition, at the direction of AFCEE as per requests from the University of Maine representatives, 2 additional samples at grid points 10 and 12 were collected and submitted for analyses. Split samples from all 23 grid points (numbers 01 through 23) were submitted to the RIC for analysis.

Data Validation

Data validation qualifiers assigned to the analytical results were based on available QC data, and calculations from the raw data were not verified. Validation of radiological data was performed in accordance with the USEPA Guidance for Data Useability in Risk Assessment (USEPA, 1990). Validation of organic and inorganic data was performed in accordance with the most recent versions of the USEPA Contract Laboratory Program (CLP) National Functional Guidelines for Organic Data Review, and the USEPA CLP Laboratory Data Validation Functional Guidelines for Evaluating Inorganic Analyses (USEPA, 1988; USEPA, 1989).

Decontamination of Equipment

All equipment used in the exploration and the excavation of the trenches located in Areas C and E was decontaminated using EPA-approved methods. At the completion of each trench excavation, all equipment used was cleaned and wipe-sampled to check for contamination before being allowed to be removed from the EZ.

Upon completion of the excavation of Trench C and E, all equipment in the EZ was cleaned of gross contamination, scanned, and wipe-sampled to detect the presence of residual radiological contamination. If any radiological contamination was detected, the equipment was further cleaned, wiped down, and retested until the contamination was removed and the equipment verified as clean.

Contaminated residue resulting from the decontamination process (i.e., mud and/or materials removed from equipment staged on the decon pad) was collected and placed into lined rolloff boxes transported to the coal yard and placed into the last gondola for shipment to the TSDF. A submersible pump was utilized to transfer the collected decontamination rinsate from the sump of the decon pad into appropriate containers. The decontamination rinsate was then sampled, analyzed, and solidified utilizing Radsorb™ and placed into a rolloff box for disposal with the LLRW in the gondola rail cars.

Topographic Survey

An initial survey was performed on 20 June 1994 of the LLRW areas. Control points were established and surveyed. On 15 September, the outer markers for the excavations at Trenches C and E were set and surveyed. Post-excavation survey work was completed on 21 November to verify the volumes of the stockpiled soils and the quantities of waste material excavated. Based on the results of these surveys, the total volume of material excavated from Trench C was calculated to be 479.1 cubic yards, of which 294.5 cubic yards was considered to be LLRW and 184.6 cubic yards overburden. The total volume of material excavated from Trench E was calculated to be 449.6 cubic yards, of which 438.2 cubic yards was considered to be LLRW and 11.4 cubic yards overburden.

CONCLUSIONS

Ogden successfully executed the statement of work (SOW) in accordance with the AFCEE-approved QPPs, achieved the objectives for the removal action, and obtained Regulatory-issued Clean Closure of the five USTs and two LLRW trenches for the USAF. Post-excavation sampling and Clean Closure laboratory analyses performed at the two LLRW trenches confirmed that the cleanup levels for the radionuclides of concern were achieved. The analytical data were validated in accordance with USEPA guidelines, and confirmed the accuracy of the laboratory results. A total of 1,460.70 tons of LLRW was excavated from the two LLRW trenches and transported to TSDF via railroad. A total of 20 gondola rail cars were utilized for the transportation. All rail cars were accepted by the TSDF for disposal, and certification of disposal was provided to AFCEE and LAFB.

Clean Closure analytical results for two Trenches were received from 18 to 28 December, 1994. With the exception of one sample, all samples that were analyzed for the radionuclides of concern were determined to be below the background criteria. Analytical results for hazardous constituents were also within the established limits identified for the IRM.

Based on an agreement reached between the USEPA and MEDEP, Clean Closure of the two LLRW trenches was to be tied to the acceptance of the RI/FS Risk Assessment for OU 1. Ogden demonstrated that all LLRW had been removed from Trench C and E, achieving the stated cleanup criteria. Ogden received notification from the LAFB Base Closure Representative August 1995 that the ROD for OU1 had been signed by the USEPA and MEDEP, and Clean Closure approved for the two LLRW Trenches.

REFERENCES

Ogden, 1994. Final Radioactive Waste Sites Operable Unit 1, Site Characterization Report for Soil and Underground Storage Tank Low Level Radioactive Waste Removal prepared for the Air Force Center for Environmental Excellence, Brooks AFB, TX, July 1994.1.

Ogden, 1995. Draft RI/RA Report for Underground Storage Tanks and Low Level Radioactive Waste Trenches at Loring Air Force Base prepared for the Air Force Center for Environmental Excellence, Brooks AFB, Texas, February 1995.

(OSHA) Occupational Safety and Health Administration. 1993. Title 29 Part 1910 of the Code of Federal Regulations. Occupational Safety and Health Standards, January.

(USEPA) U.S. Environmental Protection Agency. 1988. Region I Laboratory Data Validation: Functional Guidelines for Evaluating Organic Analyses; prepared for the USEPA Hazardous Site Evaluation Division; Washington, DC.; February 1988.

(USEPA) U.S. Environmental Protection Agency. 1989. Region I Laboratory Data Validation: Functional Guidelines for Evaluating Inorganic Analyses; prepared for the USEPA Hazardous Site Evaluation Division; Washington, DC.; June 13, 1988 modified February 1989.

(USEPA) U.S. Environmental Protection Agency. 1990. Guidance for Data Useability in Risk Assessment, Interim Final, Office of Emergency and Remedial Response, Washington, DC. October, 1990.

(U.S. NRC) U.S. NRC NUREG/CR-5849. 1992. Manual for Conducting Radiological Surveys in Support of License Termination, prepared for the U.S. NRC by ORAU-92/C57, June.

SITE ASSESSMENT

CHAPTER 27

The SPLP as an Alternative to TCLP Analysis for Metals: A Connecticut Example Using Lead

Stephen H. Brown, **Dudley W. Alleman**, **Robert A. Stewart**, and **L. Andrew White**, HRP Associates, Inc., Plainville, Connecticut

Bruce G. Hyde, Office of Development & Planning, City of New London, Connecticut

INTRODUCTION

In June 1994, HRP Associates, Inc. (HRP) began environmental investigations and remedial actions at the former New London Mills facility at 30-80 Pequot Avenue, New London, Connecticut. The study area is shown on Figure 1. The work was conducted for the City of New London, whose intent was to redevelop the site into a marine science learning center. HRP's work was completed in close cooperation with the Connecticut Department of Environmental Protection (CT DEP) Urban Sites Remedial Action Program of the Bureau of Water Management, who provided expedient review of environmental issues at the site. Consequently, remediation proceeded in a timely manner to facilitate property redevelopment in an economically depressed region.

The site was used as a manufacturing facility from the late 1800s through the early 1980s. The property was first occupied by the Brown Cotton Gin Company, which manufactured cotton gins and printing presses, and subsequently by the Babcock Printing Press Manufacturing Company. From the 1930s until the 1960s, New London Mills occupied the site and manufactured hard surface floor coverings (linoleum). The site was vacant in the early 1970s.

From circa 1977 to the mid 1980s, the Electric Boat Division Warehouse and several other businesses used the site buildings to store large quantities of paper and wood pulp. A large fire destroyed several warehouse buildings in 1985, and what was left of the fire-damaged buildings was razed shortly thereafter. The remaining two buildings were essentially vacant from 1980 to 1995, when they were demolished as part of the current site development. From the early 1980s to circa 1994, the property was used as a storage yard for Allied Crane.

Figure 1. **Location of the Study Area**

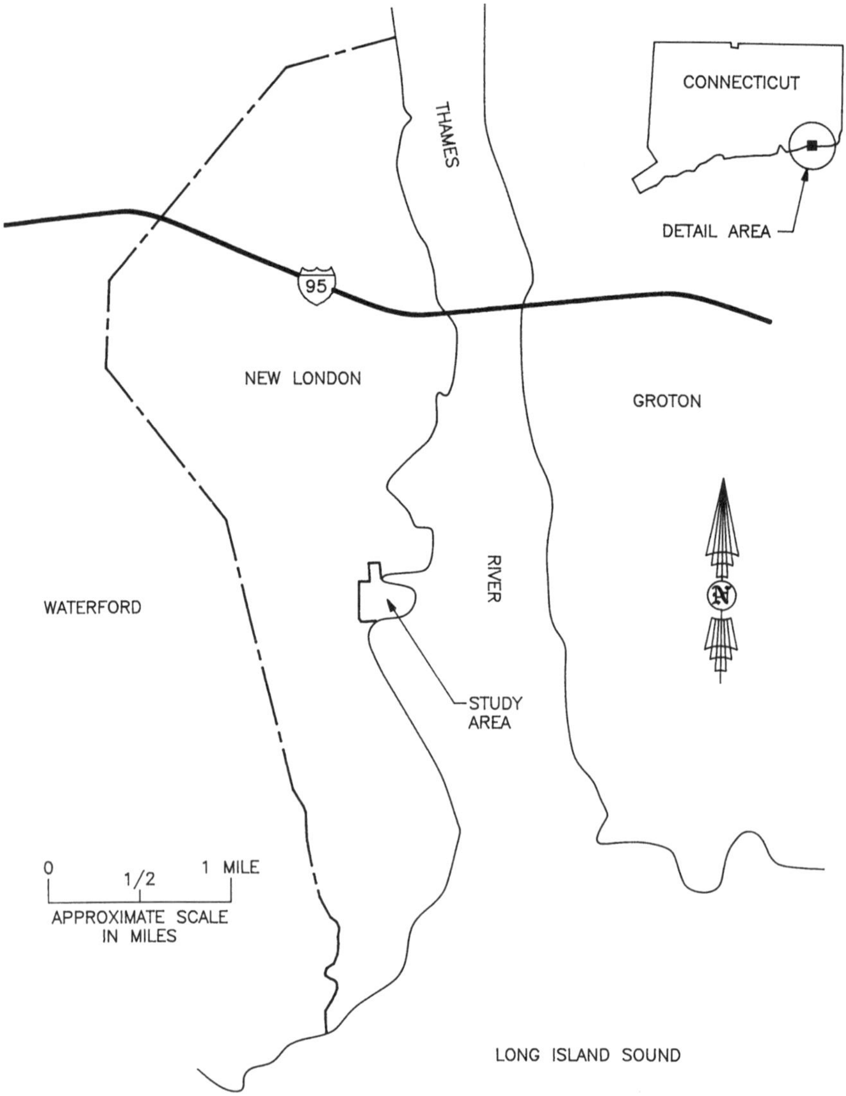

In 1987, HRP prepared a Phase I Site Assessment report for a potential site developer, followed by a Phase II subsurface screening investigation. More recently, HRP began a comprehensive site investigation in July of 1994. This investigation evolved into a CT DEP-approved plan for remedial action, which began in early 1995. The marine science learning center is scheduled to be completed in 1997.

The focus of this chapter is on the use of an alternative analytical technique for the analysis of leachable lead in vadose-zone soils at the subject site. The traditional testing methodology for metal leachability in Connecticut soils has been the Toxicity Characteristic Leaching Procedure (TCLP). Connecticut's newly adapted Remediation Standard Regulations (RSR) now allows the Synthetic Precipitation Leaching Procedure (SPLP) as an alternative. This chapter presents the results of SPLP testing on 37 soil samples for which TCLP lead analyses initially did not meet remediation standards. At the time of HRP's work (mid-1994 to late 1995), the CT DEP's remediation standards were evolving rapidly, and the New London Mills site was a trial study for the CT DEP and the City of New London. The results obtained by HRP contributed in part to the CT DEP's final adoption of the SPLP as an alternative investigative tool to the TCLP.

DESCRIPTION OF SURFICIAL DEPOSITS

As shown on Figure 2, two distinct areas are designated on the approximately 22-acre property, and each has a separate history of development. The first area is approximately 11 acres of undeveloped land which forms a small peninsula on the east side of the site. This area has been raised by as much as 10 feet with respect to the property as a whole, due to sequential filling of the Thames River in the last 50 years. The area was filled in accordance with various state and federal permits. This portion of the site is informally designated the "eastern fill area" (EFA).

The second area, on the western portion of site, is also roughly 11 acres and was developed as a complex of manufacturing buildings in the late 1800s. Many of the buildings have been razed, leaving only concrete foundations. As of early 1995, only two of the buildings and associated bituminous parking areas remained. These buildings had been essentially vacant since 1980. This part of the site is referred to as the "western manufacturing area" (WMA), and is the focus of this chapter.

Surficial deposits in the WMA consist of fill, whose origin is unknown, except where onsite sources were presumed for certain materials, such as demolition debris and slag. A light to dark brown, fine to coarse sand, with fine to coarse gravel and varying amounts of silt is present from the surface to a depth of two feet below grade (b.g.). This material contains pockets of demolition debris, consisting of bricks, scrap metal, broken concrete, and wood, typically in areas adjacent to concrete building foundations. Underlying this material from 2-5 feet b.g., the limit of exploratory test pits, is a dark brown to orange-brown fine to medium sand, with some fine to coarse gravel, and varying amounts of silt, metal, and brick. The water table was typically encountered in this interval, averaging approximately 4' b.g.

Locally in the WMA, at depths typically greater than two feet b.g., is an interval of orange-yellow to orange-brown or dark reddish brown to dark gray

Figure 2. **Base Map Showing Ground Water Contours in October, 1994**

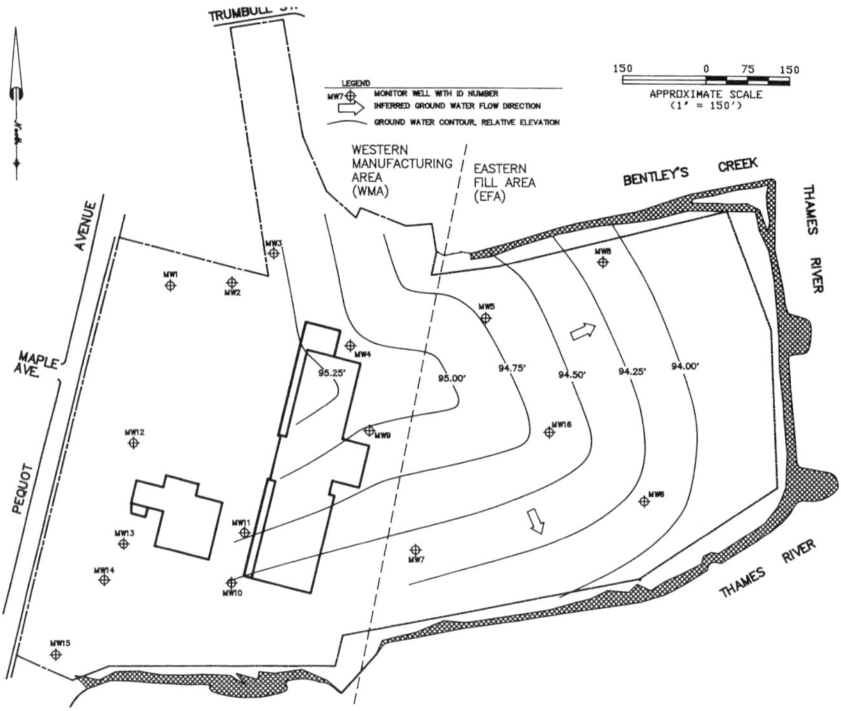

slag. Figure 3 depicts the distribution of slag in the WMA. The slag is a dense, fused mixture of silt, sand, and gravel sized particles in a fine grained ash matrix. The proportion of ash varies.

Slag is commonly formed on furnace walls or as a by-product from secondary metal smelters. High-temperature slag is mainly an ash, and at lower temperatures slag is more typically a sintered material, as described by Bryers (1986). Both types of slag deposits have been observed on the site, and are attributed to a foundry which was located on the WMA from at least 1891 to 1921, based on historical Sanborn Fire Insurance Maps.

The composition of slag depends on (1) the mineralogy and chemistry of the original coal or coke that was used as fuel, (2) the temperature of combustion, and (3) the gases that were present during combustion. In general, slag is composed of clay minerals, quartz or its high temperature polymorphs, pyrite and/or other iron sulfide minerals, elemental iron, and carbonate minerals such as calcite or dolomite. Other minerals or elements may be present in slag, depending on the collective "mineral matter" that was present in the original coal.

Figure 3. **Occurrence of Slag on the Western Manufacturing Area**

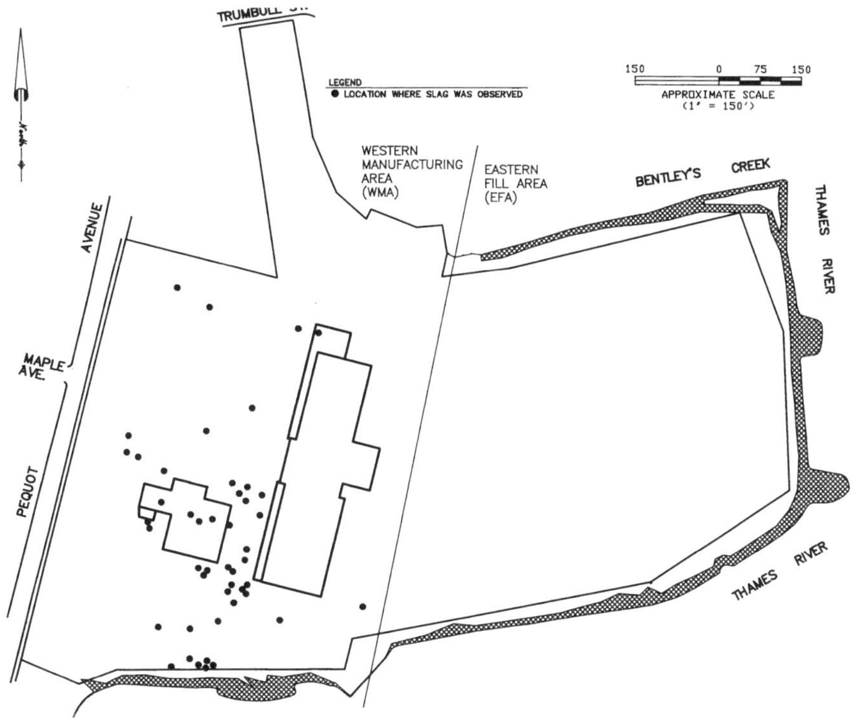

Additionally, if the slag is derived from smelting operations, the composition (including heavy metal content) would be related to the composition of the ore.

At the New London Mills site, 16 soil samples containing slag were analyzed for total lead by Mass Analysis (EPA Method 3050) to evaluate this material for risks associated with direct human exposure, as required by the DEP. The range of total lead concentrations (27.6 - 4200 ppm) for these 16 samples is interpreted to reflect smelting operations, as the range is about an order magnitude higher than that for low-and high-temperature coal ash (3.1 - 258 ppm) experimentally obtained by Ruch, et. al.(1973).

Below the orange-brown sand and slag are estuarine deposits which were encountered only on the northern portion of the WMA. These deposits correlate with estuarine sediments observed at the same relative stratigraphic position beneath the EFA, and consist of light- to medium-brown, fine sand and silt with a trace of organics. On the southern portion of the WMA, beach deposits consisting of gray, fine to coarse, well-stratified sands directly underlie the sand-and-slag interval. The apparent depth to bedrock for the WMA ranges between 10 and 25 feet b.g., based on a limited number of test boring refusals.

SCOPE OF INVESTIGATION AND INTRODUCTION TO CONNECTICUT REMEDIATION STANDARDS FOR LEAD

The CT DEP has designated groundwater quality classes and uses for the entire state. Class "GB" has been assigned to groundwater on and near the site, based on the presence of known or suspected sources of contamination in the same drainage basin, and the urban character of this part of New London. Groundwaters in a "GB" area are presumed to be unsuitable for human consumption without prior treatment.

HRP used test pits, test borings, and monitor wells to characterize soil and groundwater quality on the WMA. State-certified laboratory analyses of soil and groundwater samples included volatile and semivolatile organic compounds, and numerous inorganic constituents. Lead is the focus of this chapter. It is important to note that two rounds of groundwater analyses from a network of 16 shallow monitor wells revealed no detectable lead. However, Toxicity Characteristic Leaching Procedure (TCLP) analysis of numerous samples of fill materials collected from across the site revealed 37 locations where leachable lead concentrations exceeded the pollutant mobility criteria (PMC) of 0.15 parts per million (ppm) for Class "GB" groundwater quality areas.

The PMC for lead in "GB" area soil is 10 times the groundwater protection criteria (GWPC) for lead of .015 ppm. This standard (0.15 ppm) is then compared to a TCLP or SPLP leachability analysis following current DEP requirements. The pollutant mobility criteria is designed to protect groundwaters from the leaching of vadose-zone lead, after dilution of the leachate within the groundwater. If the PMC is exceeded, then some method of remedial action is typically necessary.

The DEP has also proposed direct exposure criteria (DEC) to evaluate potential risks associated with human exposure to accessible shallow soils. This evaluation is based on total metal content via mass analysis. Although HRP addressed this issue, the results will not be presented in this chapter, except as discussed in the context of slag characterization (above).

CT REGULATORY FRAMEWORK FOR TCLP AND SPLP TESTING

Since 1990, the CT DEP has required the TCLP method for analyzing metals for two purposes. First, the TCLP test is used to assess metal leachability in soil as a source of groundwater contamination, by comparing the TCLP results to CT DEP Pollutant Mobility Criteria (PMC) or drinking water standards (DWS). Secondly, the TCLP test is used to determine whether the levels of any RCRA metals are characteristically hazardous for waste classification purposes. As previously mentioned, in "GB" areas the CT DEP set a level of 10 times the GWPC or DWS for RCRA metals in soil as the pollutant mobility remedial threshold. Above this threshold, the contaminated soil must be remediated, which in most cases has meant removal and offsite disposal.

Since shortly after the introduction of the TCLP method for general use, consultants and regulators recognized that the TCLP more aggressively extracted certain metals than the Extraction Procedure (EP) method it replaced. Consequently, significant volumes of contaminated soil, particularly lead contaminated soils, were routinely identified on many urban sites in Connecticut, even where no point sources were obvious. At some sites, the presence of lead in the soils appeared to be related to historical fill materials. However, in many cases, the source appeared to be the general urbanization and industrial history of the site area (e.g., lead-containing stormwater runoff, vehicle emissions, and airborne industrial emissions).

Also, TCLP detections of low to moderate levels of metals in soil were rarely associated with a corresponding plume of metals in groundwater, as would be predicted by the TCLP method. This was the case at the New London Mills site, as no lead was detected in site groundwaters, even though leachable levels of lead had been identified above the PMC standard in the soils by the TCLP test. Because of this fundamental problem, and due to the large number of sites being remediated because of lead in soil, in the absence of identifiable sources, the CT DEP began to examine alternative methods for evaluating pollutant mobility, in particular the Synthetic Precipitation Leaching Procedure, or SPLP.

The SPLP is designated as EPA Method 1312, which follows EPA SW-846 protocol, as does the TCLP. A solution of acetic acid and sodium hydroxide is used as the extraction fluid for TCLP metals analysis to initially adjust the pH to 4.93±.05 for most samples. In contrast, a solution of sulfuric and nitric acids is used as the extraction fluid for SPLP metals analysis to initially adjust the pH to 4.20±.05 (east of the Mississippi River). The principal difference between the two methods when they are used for metals analysis is that the TCLP extraction fluid is buffered so that the pH remains steady during the extraction procedure, whereas the SPLP extraction fluid is not buffered. In this way, the pH of the sample has an impact for the SPLP method, as would be expected.

In summary, the TCLP proved to be unrealistically aggressive for lead, and over-predicted its mobility. Although lead was detected above PMC cleanup standards in numerous site soils at the New London Mills by TCLP, no lead was identified in groundwater from any of 16 monitor wells in the course of two sampling rounds. Therefore, in December of 1994, HRP requested and was given permission by the CT DEP to use the SPLP in place of the TCLP to determine the pollutant mobility of lead in soil and the associated cleanup requirements at the New London Mills property. Subsequently, in late December 1994, the CT DEP released the revised "Proposal for the Connecticut Clean-Up Standard Regulations," which became the eventual basis for final cleanup criteria (the RSR, mentioned above) for soil and groundwater. The RSR accepts either the TCLP or SPLP to evaluate pollutant mobility.

ANALYTICAL RESULTS

The SPLP analyses were completed on the 37 samples where previous TCLP analyses had yielded lead concentrations above the class "GB" groundwater PMC of 0.15 ppm. The SPLP samples were obtained from the same jar of soil used for the TCLP analyses, but not from the same aliquot initially analyzed by TCLP.

Of the 37 samples in which lead exceeded the PMC by TCLP, lead exceeded the PMC by SPLP in only five samples (Table 1). A semi-log plot of the TCLP and SPLP data is shown on Figure 4. Two extreme TCLP data points (samples 2 and 26) and all SPLP data below detection limits (ND) have been omitted for clarity. There is no apparent linear or logarithmic relationship between the TCLP and the SPLP data, only an average 47-fold reduction in detected lead concentrations. In only one case (sample 7) was the SPLP value higher than the TCLP value (1.02 ppm vs. 0.495 ppm), presumably due to a "hot spot" within the sample jar. As a direct result of the SPLP analyses, the volume of lead-contaminated soil requiring excavation and offsite disposal was reduced by 80%, as compared to the volume of material which would have required removal using TCLP data alone.

The TCLP/SPLP program at the former New London Mills site was a small part of the entire Remedial Investigation/Feasibility Study (RI/FS) process. Other remedial actions were proposed and accepted by the CT DEP for the Western Manufacturing Area, including removal and offsite disposal of soil contaminated with total petroleum hydrocarbons (TPHs) and polynuclear aromatic hydrocarbons (PAHs). An environmental use restriction will be used for soil in which total lead (by EPA 3050) exceeds the corresponding direct exposure criteria (DEC). Such soils will be a minimum of 2-4 feet below grade upon site devel-

Figure 4. **TCLP vs. SPLP Analytical Data From 28 Soil Samples**

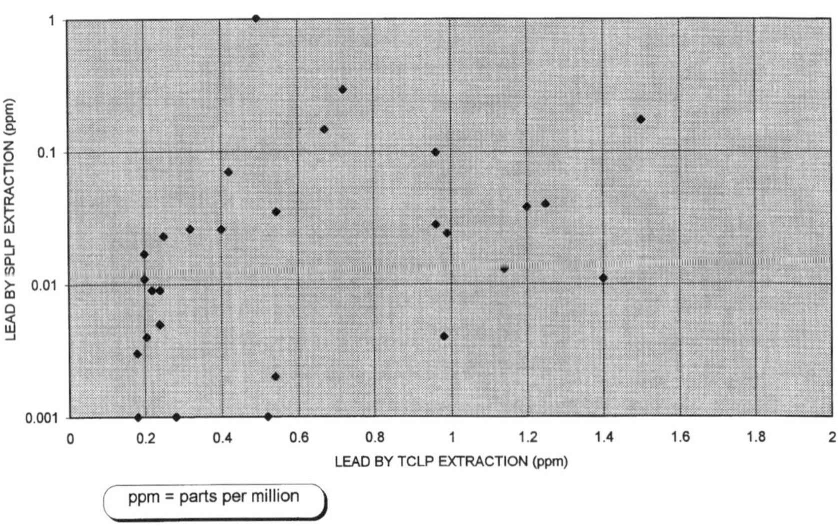

Table 1 **Summary of SPLP Results In Comparison To TCLP Values**

Sample I.D.	TCLP Lead Concentration (ppm)	SPLP Lead Concentration (ppm)
1	0.960	0.098
2	4.42	0.192
3	0.670	0.146
4	1.14	0.013
5	0.205	0.004
6	0.545	0.035
7	0.495	1.02
8	0.250	0.023
9	0.990	0.024
10	1.78	ND
11	0.255	ND
12	1.34	ND
13	0.240	ND
14	1.50	0.172
15	1.40	0.011
16	1.50	ND
17	0.54	0.002
18	1.20	0.038
19	0.72	0.291
20	0.32	0.026
21	0.24	0.005
22	0.96	0.028
23	0.56	ND
24	0.52	0.001
25	0.28	0.001
26	22.7	0.403
27	0.42	0.070
28	0.20	0.017
29	0.22	0.009
30	0.40	0.026
31	0.20	0.011
32	1.25	0.040
33	0.18	0.003
34	0.68	ND
35	0.98	0.004
36	0.24	0.009
37	0.18	0.001

opment. Ongoing groundwater monitoring for selected organic and inorganic parameters, including lead, is part of the remedial action plan for the site.

CONCLUSIONS

Based on the results of this investigation, the SPLP is an important alternative to the TCLP for metals leachability testing in Connecticut, and perhaps in other states where soil leachability testing for metals is current policy. At this time, the SPLP does not entirely replace the TCLP, particularly where a need exists for RCRA hazardous waste classification. As the SPLP is applied to additional Connecticut sites in a variety of settings, a more systematic relationship between the SPLP and the TCLP may emerge.

As a direct result of the use of the SPLP at the former New London Mills site, approximately $600,000 was saved in remedial costs, money that was better spent on performing remedial actions at the site where a threat to human health and/or the environment was at issue. Based upon the SPLP results, much of the fill material on the western manufacturing area did not require excavation and offsite disposal as part of the remedial action. This project has been a success in converting a classic example of a "brownfield" into a greenfield, whose eventual use as a marine science and learning center will become a cornerstone in revitalizing the New London area.

REFERENCES

Bryers, R. W. 1986. *Influence of Segregated Mineral Matter in Coal on Slagging in Mineral Matter and Ash in Coal*, ACS Symposium Series 301, p. 353 (or Chapter 2) (K.S. Vorres, Ed.) Washington, DC, American Chemical Society.

Ruch, R.R., Gluskoter, J.J., and Shimp, N.F., 1973. Occurrence and Distribution of Potentially Volatile Trace Elements in Coal: An Interim Report, Environmental Geology Notes, Illinois State Geological Survey, Urbana, April.

CHAPTER 28

A Novel Approach for the Determination of the Source of Hydrocarbons in Contaminated Soils

R.P. Philp, **M. Kombrink**, and **J. Allen**, School of Geology and Geophysics, University of Oklahoma, Norman, Oklahoma

INTRODUCTION

For many years there has been an ongoing interest in the correlation of environmental pollutants with the sources responsible for such spills. The types of organic compounds spilt into the environment are many and varied, and range from pesticides to chlorinated solvents, to toxic warfare reagents, to crude oils and refined petroleum products. Naturally, each of these classes of compounds reacts in a different way, once spilt in the environment. Some will be oxidized, some biodegraded, some hydrolyzed, and others rapidly adsorbed onto mineral phases, slowing down their rate of biodegradation and bioavailability. Processes such as the ones mentioned above greatly affect our ability to undertake the correlations necessary to determine the point source of any spill.

This chapter is specifically concerned with crude oils and refined hydrocarbon products spilt into the environment, and the development of a relatively novel analytical approach which can be used, in conjunction with other techniques, to undertake correlations and determination of the source for such spills. (It should be noted that, despite the fact only hydrocarbons are discussed in this chapter, it is proposed that GCIRMS will be equally applicable to other types of chemicals spilled into the environment.) From the moment an oil or refined hydrocarbon product is spilled, changes will start to occur to the hydrocarbons. Initially the changes will be dominated by evaporation and water washing, both of which tend to remove the lighter hydrocarbon components in the oil (Palmer, 1993). Within a relatively short period of time biodegradation will become dominant, and then not only will the residual light components be removed, but progressively heavier components are also removed in a fairly specific and predictable manner (Peters, 1993). Ultimately, all of the saturate hydrocarbons will be removed, leaving an unresolved complex mixture of components that are very resistant to biodegradation. With this scenario in mind, it becomes extremely difficult to undertake correlations between unaltered and altered samples. Traditionally, techniques such as gas chromatography (GC) or gas chromatography-mass spectrometry (GCMS) have been used to obtain fingerprints of various compounds' classes for correlation purposes (Philp, 1993). Two samples coming from the same source, or which are genetically related to each other, may be expected to have the same fingerprint, but once the fingerprint has been altered as result of weathering it becomes difficult to undertake correlations with any degree of confidence.

Recent advances have led to the development of a novel analytical technique; namely, combined gas chromatography-isotope ratio mass spectrometry (GCIRMS), which appears to be a potentially powerful correlation tool (Brenna, 1994; Hayes et al., 1990; Freeman et al., 1990; Murray et al., 1994). GCIRMS in effect provides the ability to determine the isotopic composition of individual components present in complex mixtures which are resolvable by gas chromatography. Prior to this development, use of carbon isotopic values was limited to determination of bulk isotopic values for unfractionated samples or individual fractions such as total saturate, aromatic, or polar fractions of crude oils (Kvenvolden et al., 1993a; 1993b). There are, of course, limitations to the GCIRMS approach, and in particular, these limitations are related to the fact that in order to get good reproducible and reliable data one needs to have good GC resolution. In other words, if two components are co-eluting, it will be difficult to obtain reliable isotopic values for either of these components. In the situation with well-resolved components, reliable isotopic data will be available for the individual components.

The purpose of this chapter is to demonstrate that GCIRMS has a potential role to play in the correlation of hydrocarbons which are contaminants in the environment with their suspected sources, the basic premise being that as long as individual components, such as *n*-alkanes, can be detected in the weathered samples, the isotopic composition of these individual components in unweathered and weathered samples should be virtually identical.

In the following discussion, examples of this technique will be discussed along with its application to the correlation of several weathered and unweathered samples. It is important to emphasize that this technique will not completely replace the more conventional techniques of GC and GCMS. Instead, GCIRMS should be used in conjunction with those techniques to provide supporting evidence, and in many cases additional evidence, for any correlations that are being made between samples.

EXPERIMENTAL

Samples

For the purposes of this preliminary study, samples of crude oils, fuel oils, gasoline samples, and tar balls were obtained from a variety of sources. Specific locations of samples will not be given since they are not necessary for the discussion. Samples were obtained in both unweathered and weathered condition. In some cases the samples were weathered naturally, and in other cases the samples were weathered artificially in the laboratory.

Gas Chromatography

Prior to analyses by GCIRMS, the samples were analyzed by gas chromatography. For the most part, the samples were typically analyzed, without any fractionation, using a DB-1 fused silica capillary column (0.25 mm i.d.; 1.0 μm film thickness), and temperature programmed from 40 to 300°C at 4°C/min.

Gas Chromatography-Mass Spectrometry

Where necessary, samples were analyzed by GCMS using a Finnigan TSQ 70 combined with a Varian 3400 GC. Temperature conditions were the same as described above, and the system was normally used in the GCMS mode, monitoring commonly used biomarker ions such as m/z 191 and m/z 217 for determination of terpane and steranes, respectively.

Gas Chromatography-Isotope Ratio Mass Spectrometry

GCIRMS was performed using a Finnigan MAT 252 Isotope Ratio Mass Spectrometer coupled with a Varian 3400 gas chromatograph. GC conditions were as described above.

RESULTS AND DISCUSSION

GCIRMS has been used in a number of applications in the past two or three years since its development in the late 1980s. In particular, there have been a number of American Chemical Symposia dedicated to the use of this technique specifically in geochemical applications. Papers from the first of these symposia have been published in *Organic Geochemistry* (Vol. 21, Nos. 6/7, 1994). The second one was held at the ACS meeting in Chicago in August 1995, and at this time only abstracts from that meeting are available (210th American Chemical Society National Meeting, Chicago, August 1995, abstracts only). However, these two meetings provide an excellent account of the state of this technique. It is not proposed to go into an extensive literature review of GCIRMS in this chapter; rather, the interested reader is referred to the proceedings and abstracts of those two symposia for more detailed information.

As part of the development and application of any new technique or approach, it is necessary to undertake a certain amount of method development. In the next few paragraphs, a brief discussion will be given to illustrate some of the work that has been undertaken to demonstrate the reliability of GCIRMS as a technique for the correlation of weathered and unweathered oil, or oil-derived samples. Initially one oil sample was analyzed several times to demonstrate

Figure 1. **Examples showing (a) the reproducibility of the GCIRMS technique, and (b) the fact that there is little or no isotopic fractionation occurring as a result of HPLC separation of the oil into saturate and aromatic fractions.**

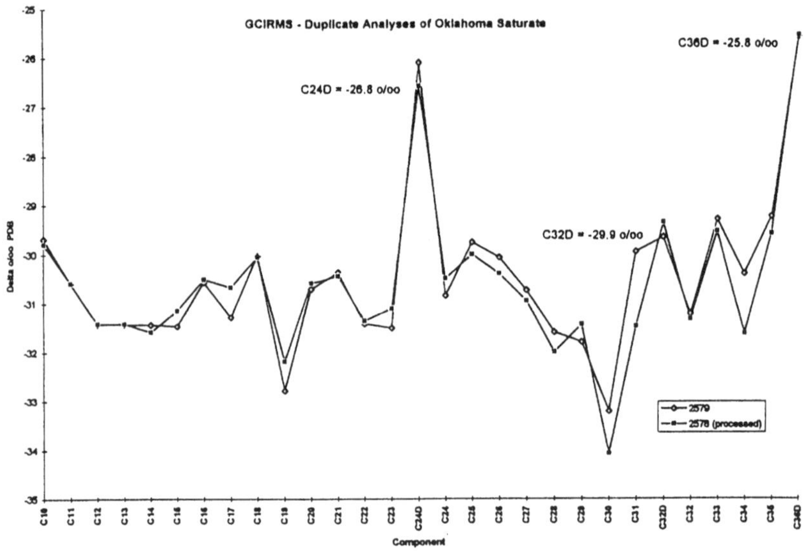

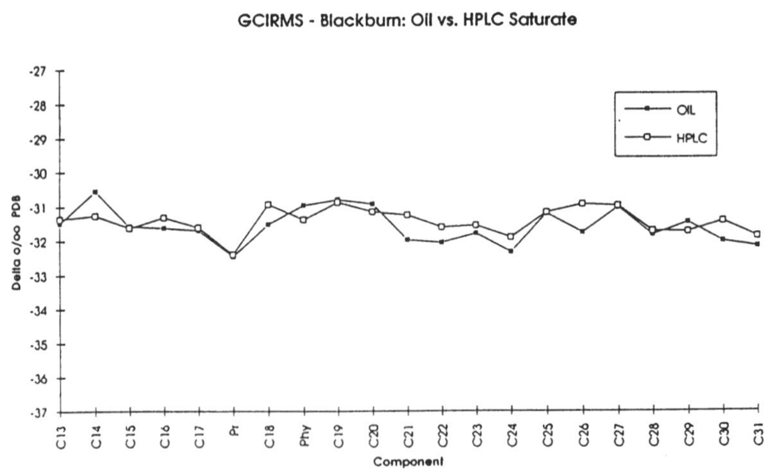

the reproducibility of the technique. Results from this part of the work are shown in Figure 1a. The data shown in this figure are obtained from the analysis of the oil by GCIRMS, following which the major peaks in the chromatogram are selected, and isotopic values for each of these components calculated and plotted onto the spread sheet, as shown in this figure. The slight discrepancies that can be observed for the isotopic compositions of the higher carbon numbered alkanes result from the very low concentration of these compounds in the original sample.

Crude oils are very complex mixtures of a wide variety of compound classes, including saturate, aromatic, and various polar compounds. Oils can be fractionated into the compound classes by a variety of chromatographic techniques, such as HPLC, column chromatography, or thin layer chromatography. In certain situations, sample size may preclude our ability to reliably fractionate samples into individual fractions, making it necessary to analyze unfractionated samples directly by GCIRMS. Hence, following the reproducibility checks it was necessary to determine whether or not there was any isotopic fractionation during the separation of a crude oil into saturate and aromatic fractions by HPLC. If not, and if the situation arose, a comparison could be made, for example, between the isotopic composition of *n*-alkanes present in whole oils vs. the corresponding values for the *n*-alkanes in the associated saturate fractions. Results of such a comparison are shown in Figure 1b, which shows the isotopic composition of individual components present in a whole oil, and the corresponding values for the same components in the saturate fraction separated by HPLC prior to analysis by GCIRMS. As can be seen from this plot, there is virtually no difference between the whole oil and the values for the saturate hydrocarbon fraction. It should also be noted in Figure 1a, and other figures, that various standard components are included, and it is essential that such standard components are included in any study of this type. These standards, which are totally deuterated *n*-alkanes with carbon numbers of 24, 32, and 36, are of known isotopic composition, and more importantly, do not coelute with any of the undeuterated components. The isotopic compositions of these standards have been determined individually using the batch inlet system of a separate isotope ratio mass spectrometer. Hence their values are known, and if these values are not achieved during the GCIRMS analysis of a sample, then the complete set of GCIRMS data must be discarded and the sample rerun. In other words, these components serve as internal quality controls and their values should be obtained reproducibly to within +/- 0.3%.

From the data shown above, and other more extensive laboratory studies, it has been clearly established that GCIRMS is reproducible, and that with the addition of internal standards of known isotopic composition, the effects of weathering on crude oils and hydrocarbon products can be investigated. For the purposes of this chapter, a number of crude oil samples and refined hydrocarbon products, weathered and unweathered, have been examined by GC, GCMS, and GCIRMS. The first example makes use of an oil from the Middle East which has been weathered by evaporation and compared with its unweathered counterpart. The gas chromatograms of the two samples are shown in Figure 2, from which it can be seen that the evaporation has removed the more volatile components. Analyses of these two samples by GCIRMS to determine the isotopic

Figure 2. **Gas chromatograms for unweathered and weathered samples of Middle East Crude oils.**

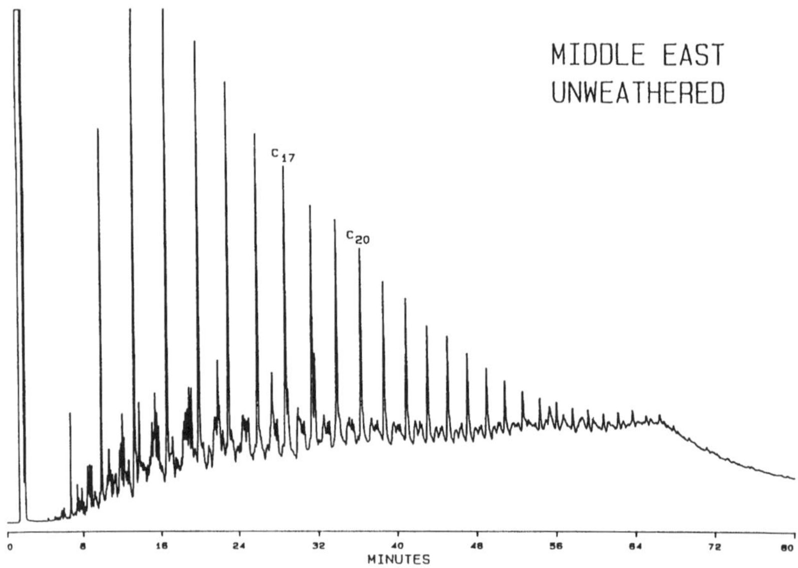

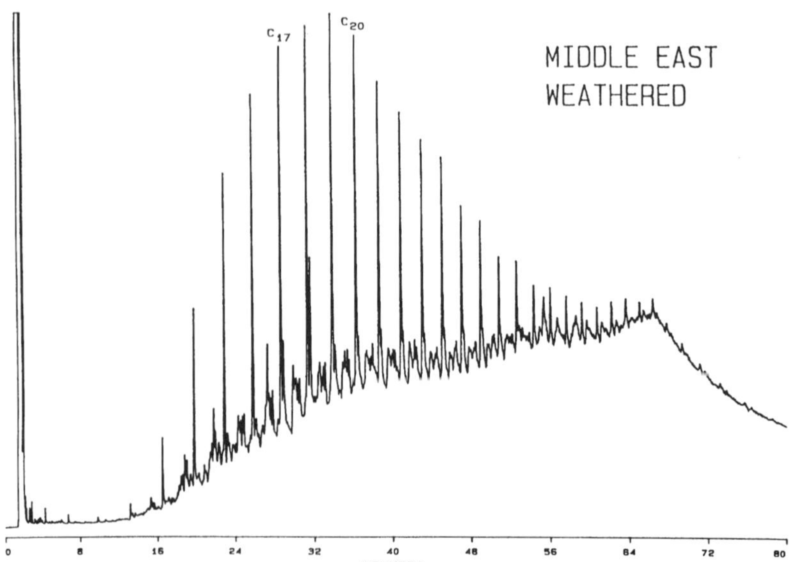

Figure 3. **Comparison between the isotopic data for the individual n-alkanes in the unweathered and weathered samples shown in Figure 2.**

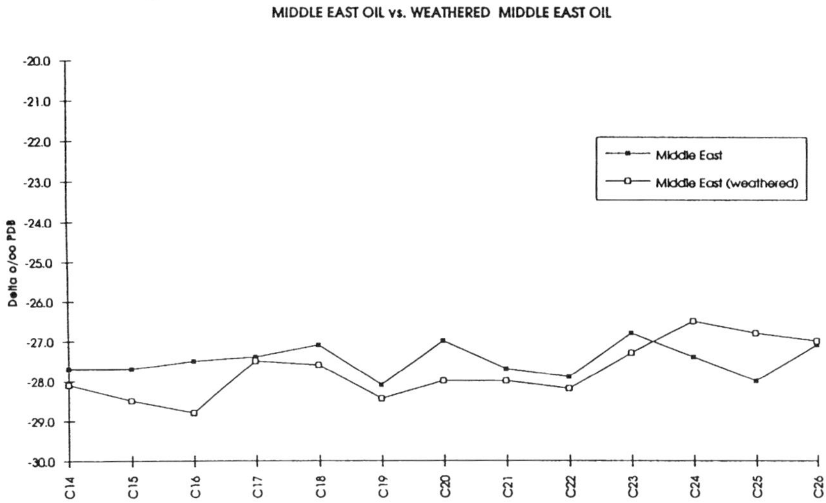

compositions of the individual *n*-alkanes resulted in the data shown in Figure 3. It is clear from these data that little change has occurred to the isotopic composition of the *n*-alkanes as a result of the evaporation process.

Similar results and conclusions can be drawn from samples of fuel oils, and other hydrocarbon products, that have been weathered naturally and compared with their unweathered counterparts. Gas chromatograms and the corresponding isotopic data for three additional sets of weathered and unweathered samples are shown in Figures 4-6. The major conclusion to be drawn from these sets of data is that, in general, the effects of weathering do not appear to have a significant effect on the isotopic composition of the major individual components. This in turn reinforces the notion that the GCIRMS has the potential to be a very useful tool for correlation purposes, particularly if used in conjunction with GC and GCIRMS to determine the source of crude oils and refined hydrocarbon products that have been spilt into the environment.

Crude oil samples which are heavily weathered and subjected to biodegradation initially lose their lower molecular weight *n*-alkanes and then, with time, increasingly higher carbon number alkanes will be removed. This can lead to a relative concentration effect whereby at a certain stage of the biodegradation process, higher carbon number components may appear to be present in relatively high concentrations. For example, in Figure 7a and b, hydrocarbons from two heavily weathered tar balls have been characterized by high temperature gas chromatography (HTGC). HTGC in itself is a relatively new technique, and permits one to readily determine the distribution of hydrocarbons in carbon

Figure 4. **Gas chromatograms and GCIRMS data for an unweathered crude oil and the same sample after several days of weathering.**

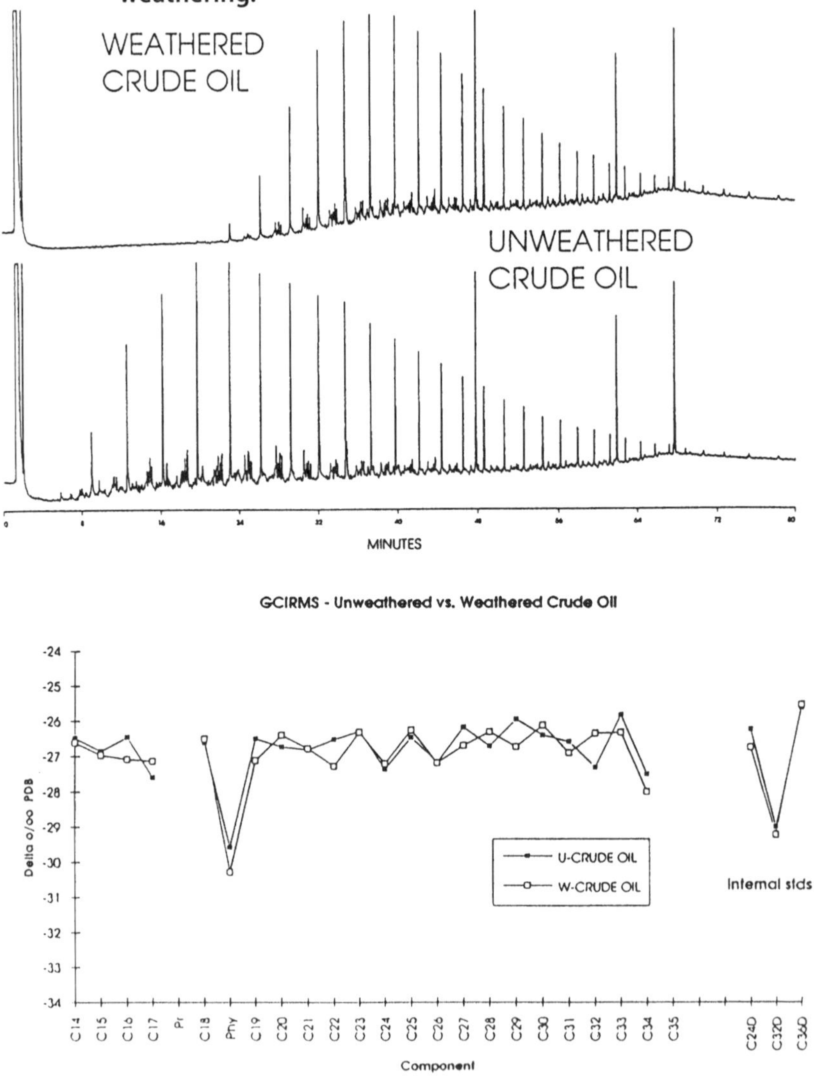

number ranges (i.e., $>C_{35}$) that are not normally detected by conventional gas chromatography (Philp, 1994). This is an important development, particularly for characterization of weathered oil samples where such concentration effects may be expected to occur. Results from the analyses of these two samples by GCIRMS produced the data shown in Figure 8a. The strong similarity between the isotopic composition of these two samples leads to the proposal that these two tar ball samples are derived from the same source, probably in this case the

Figure 5. **Gas chromatograms and GCIRMS data for an unweathered light fuel oil, and the same sample after several days of weathering.**

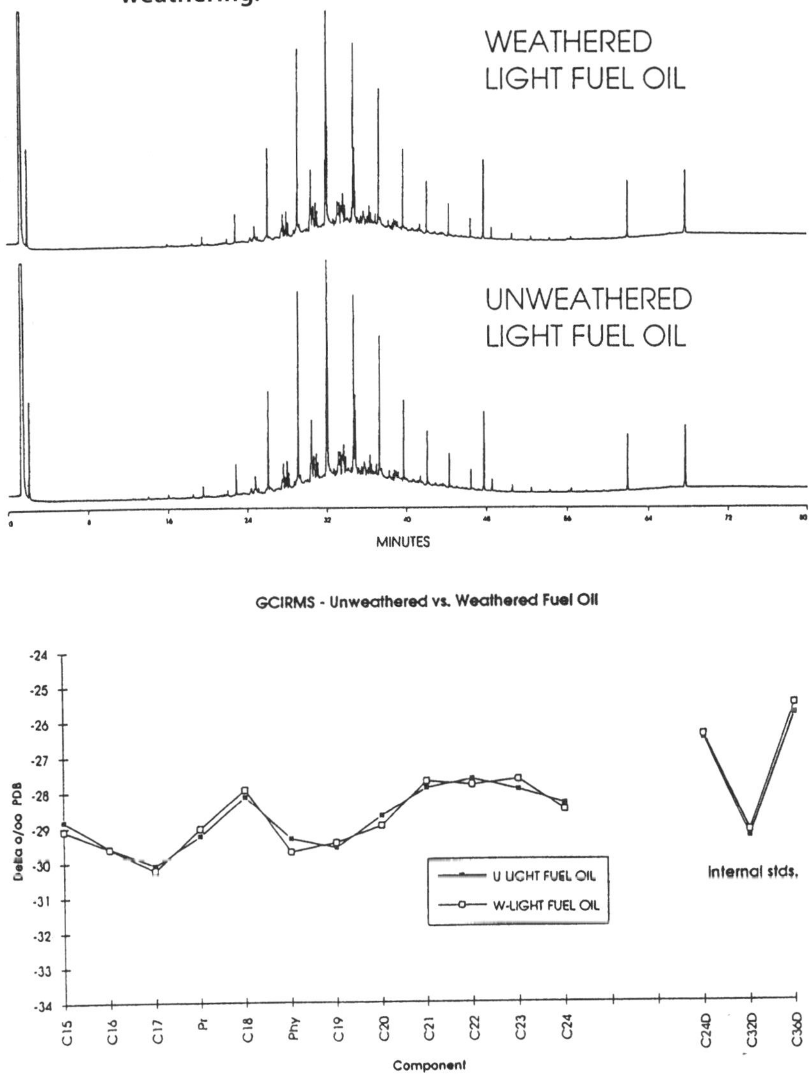

same tanker, or two tankers carrying oil from the same source. Confirmation of this conclusion comes from GCMS analysis of the same samples and determination of the distribution of the commonly used biomarkers-steranes and terpanes (Figures 9a and b and Figures 10a and b). The similarity between these biomarker distributions for the two samples supports the conclusions based on the isotope distributions.

Figure 6. **Gas chromatograms and GCIRMS data for an unweathered #1 fuel oil, and the same sample after several days of weathering.**

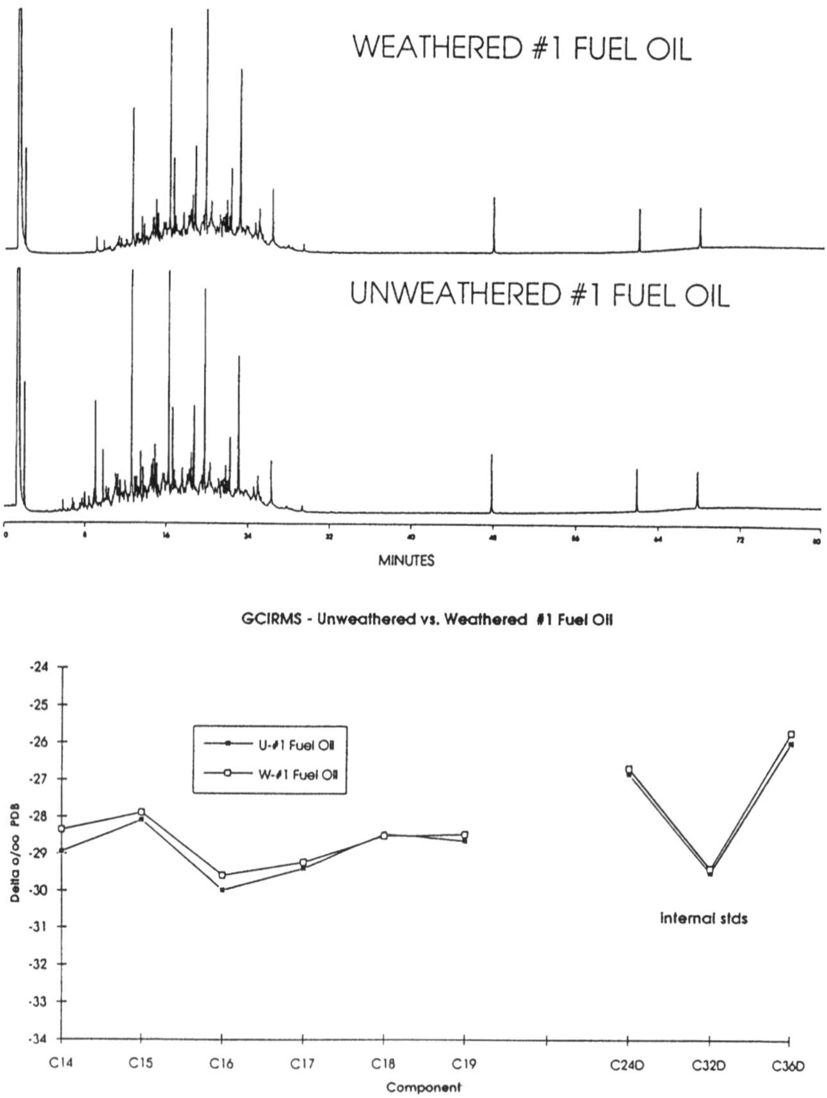

Gas chromatograms of the hydrocarbons from a second set of weathered crude oil samples are shown in Figures7c and d. These samples, while not as weathered as the samples shown in Figures 7a and b, clearly have been weathered to the extent where the low molecular weight hydrocarbons have been removed, and the initial stages of biodegradation have commenced. GCIRMS data

Figure 7 a-d. **Gas chromatograms showing the hydrocarbon distributions of tar samples that have been weathered to varying degrees.**

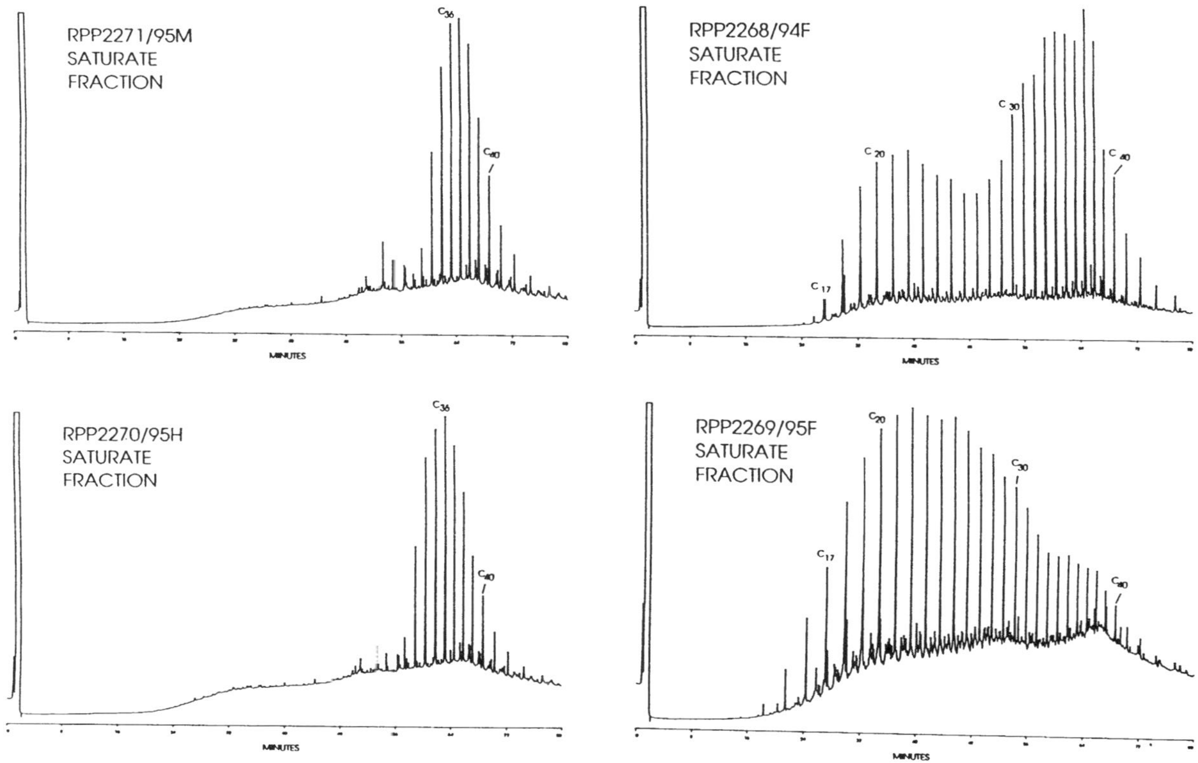

Figure 8 a & b. **The carbon isotope distributions for the individual n-alkanes in the two sets of tar ball samples.**

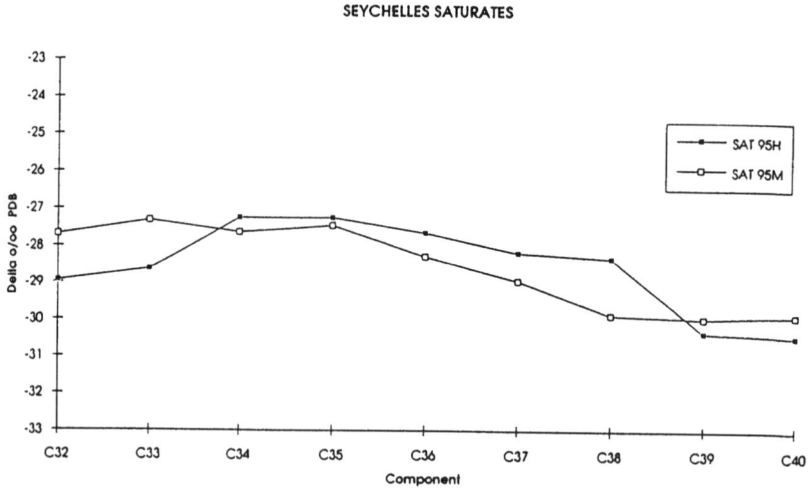

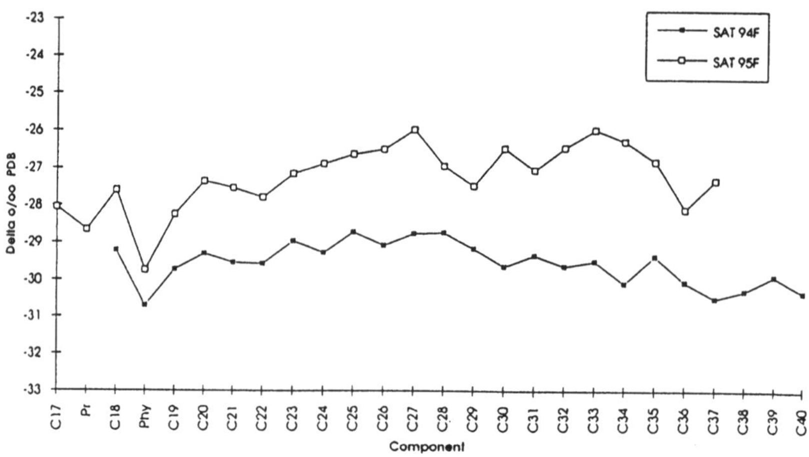

shown in Figure 8b for these two samples suggests that these samples are not directly related to each other, as seen from the significant differences in their isotopic compositions, which are far greater than the differences that may be expected from experimental errors or nonreproducibility of the technique. Sterane and terpane biomarker data for these two samples (Figures 9c and d and Figures 10c and d) shows that while there are a number of basic similarities in these samples, one sample contains abundant proportions of the biomarker called 18α(H)-oleanane, characteristic of a contribution from angiosperm-derived organic matter. Again, the isotopic composition combined with the biomarker data clearly support the idea that in this case these two samples are not derived from identical sources, and that the differences between the samples do not result from biodegradation or weathering, but differences in origin.

Figure 9 a-d. **GCMS data showing the m/z 191 chromatograms for the two sets of tar ball samples shown in Figure 7 a-d.**

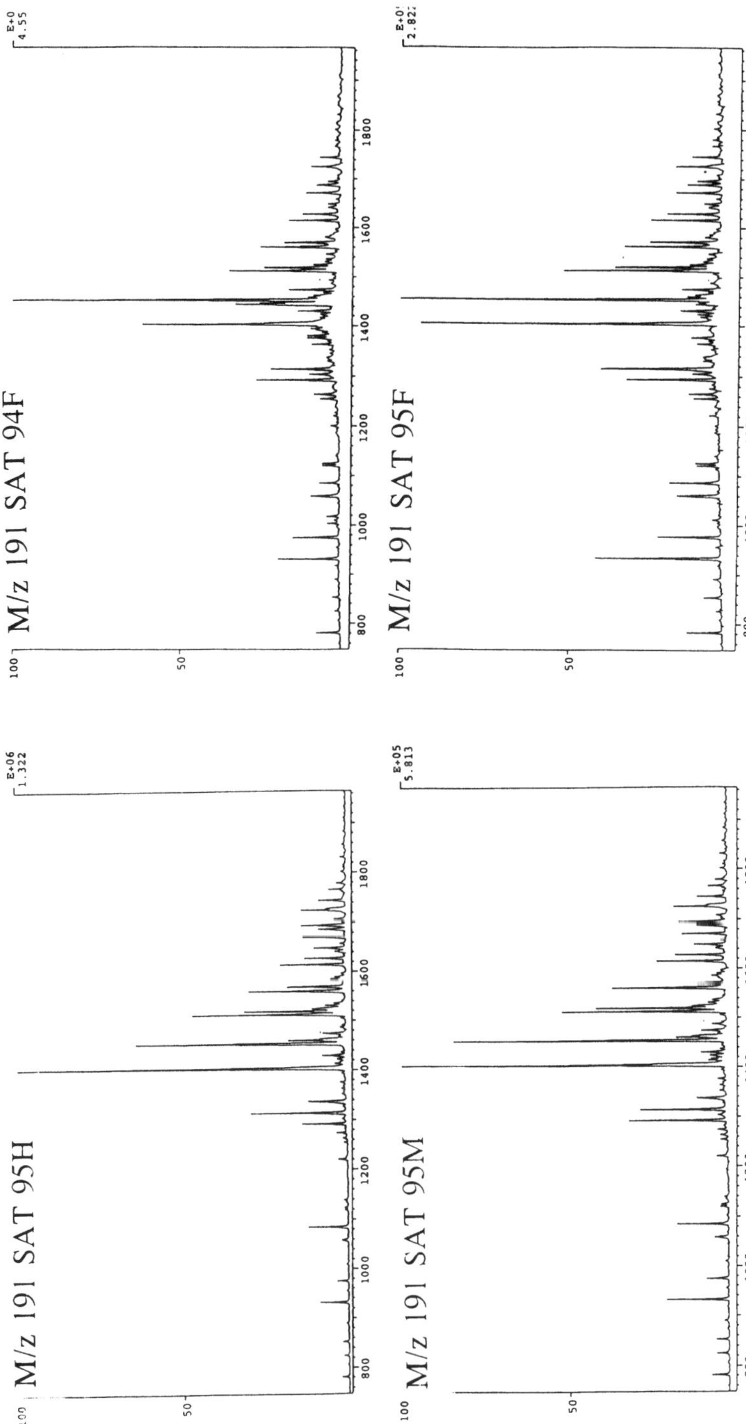

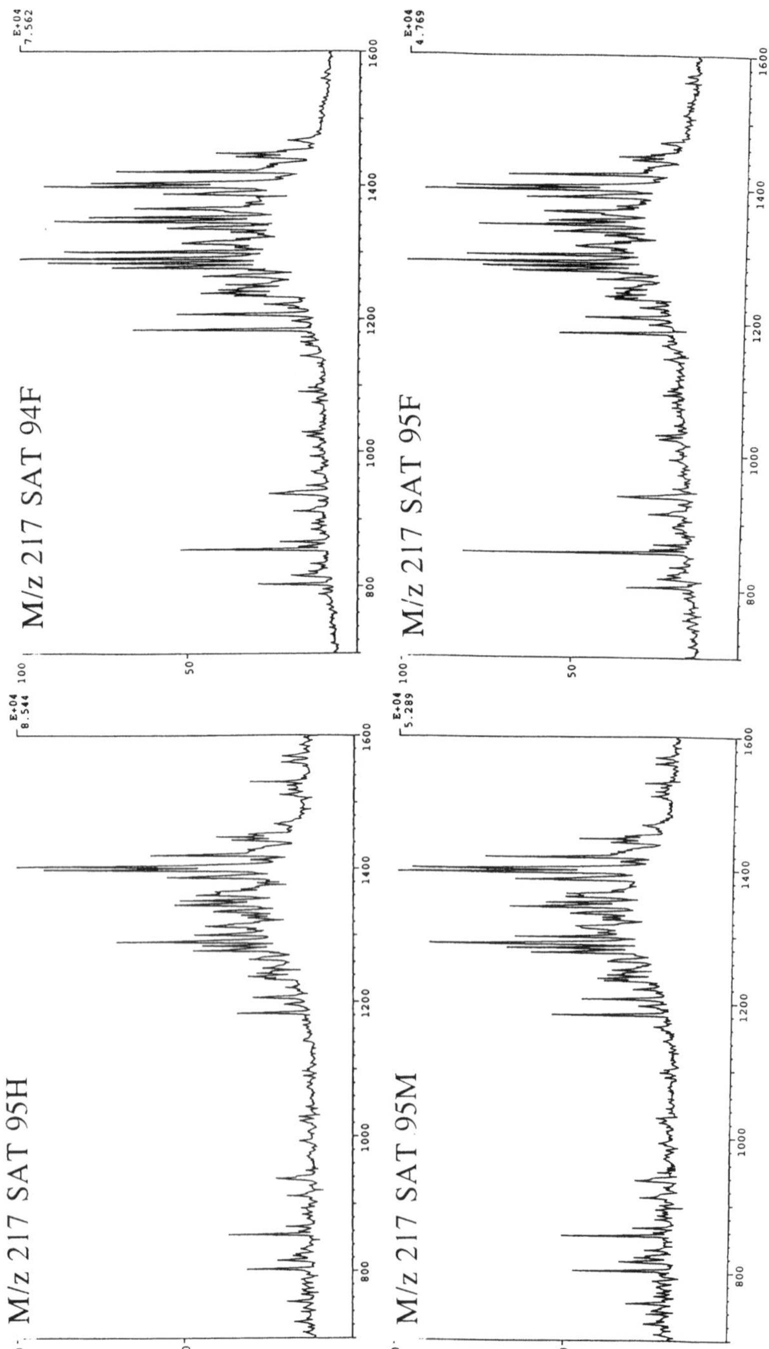

Figure 10 a-d. **GCMS data showing the m/z 217 chromatograms for the two sets of tar ball samples shown in Figure 7 a-d.**

Figure 11. **Carbon isotope data for the major components of the two gasoline samples, showing differences in the carbon isotope composition of certain components between the two samples.**

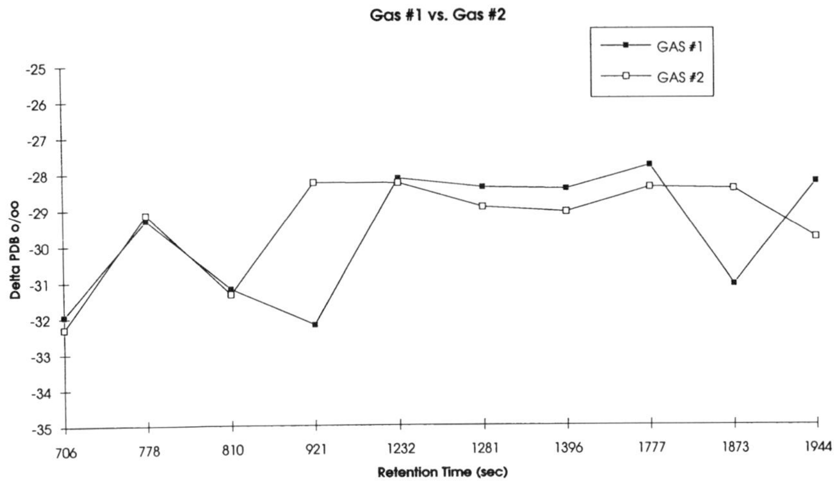

A slightly different example demonstrates the application of GCIRMS to the characterization of more volatile gasoline samples. Two gasoline samples collected from local service stations on opposite corners of an intersection were analyzed initially by gas chromatography, resulting in virtually identical chromatograms, both qualitatively and quantitatively. However, analyses of the same two samples by GCIRMS provided the data shown in Figure 11, where the isotopic composition of a selected number of components are presented in the same format as above. The components used to construct this diagram have not been identified, since for the purpose of this example it is only necessary to ensure that the same components are being monitored for each sample, and this can be done on the basis of retention time data. Despite the very close similarity of the GC data, it can be seen from Figure 11 that there are, indeed, a number of differences between the two samples, particularly the two components with retention times of 921 and 1873 sec. Based on the isotopic data alone, it is possible to distinguish these two gasoline samples which were virtually identical on the basis of gas chromatography. This example illustrates the potential of GCIRMS for determination of the source of samples responsible for problems associated with leaking underground storage tanks.

The results described in this chapter have all been concerned with the characterization of whole oils or their saturate hydrocarbon fractions. In many cases the hydrocarbon samples may be extensively weathered to the extent that virtually all the saturate hydrocarbons have been removed, making it impossible to use them for correlation purposes. In such a situation it is necessary to utilize the more resistant aromatic hydrocarbons for correlation purposes. The aromatic fractions can be isolated by HPLC and then analyzed by GCIRMS in

Figure 12. **GCIRMS chromatogram to illustrate the complexity of the aromatic fractions present in crude oils and the problems associated with the determination of the isotopic compositions for the individual components.**

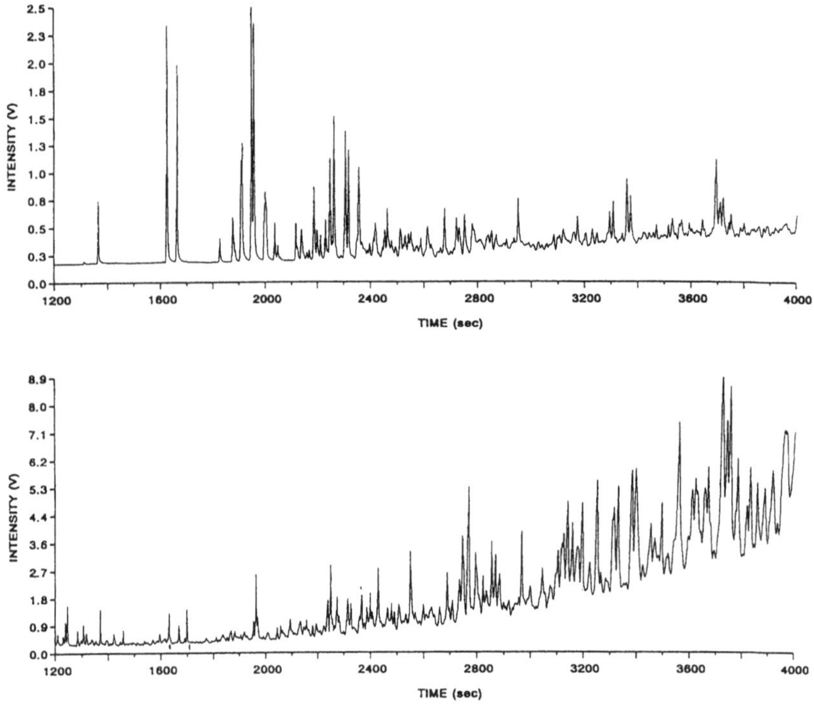

the manner as described above. It should be emphasized that a potential problem is the complexity of the aromatic fraction, as shown in Figure 12. There are many co-eluting or closely eluting components in these fractions and, ideally, complete resolution is required to get the most reliable isotopic data. However, there are still certain components such as naphthalenes and phenanthrenes that may be useful for correlation purposes. In order to partially eliminate the problems concerned with coelution, the two major pairs of phenanthrene isomers can be considered as one peak in each case, and a combined isotope value obtained for these two components. This is satisfactory, since if the two isomers coelute in one sample, they will coelute in the sample with which it is being correlated; therefore, carbon isotope values are not necessarily required for the individual components in this situation. The values obtained for these phenanthrene isomers plus various naphthalene isomers are shown in Table 1. It can be seen that there is a reasonably good correlation between samples that are known to be related. The discrepancies observed here are simply related to the complexity of the mixture, and the errors associated with removing the large isotopic contribution from the underlying unresolved complex mixture of components in the sample.

SUMMARY

The purpose of this chapter has been to introduce the relatively novel analytical concept of GCIRMS as a potential monitoring and correlation tool for use in a variety of environmental problems. The examples used in this particular study are related to environmental problems caused by contamination from crude oils or refined hydrocarbon products. It is proposed that ultimately the technique can be used for other environmental problems. It is not suggested that GCIRMS should be used as a stand-alone technique. Rather, it raises the degree of confidence to which correlations are made to a higher level than those simply based on GC and GCMS data. This is particularly important when the samples being examined are weathered or heavily biodegraded. The successful use of GCIRMS depends on the premise that with weathering and biodegradation, the isotopic composition of individual components in a mixture will not be greatly affected. Hence, as long as individual components remain in the weathered sample, then isotopic compositions can be determined and used for correlation purposes.

With very heavily weathered samples, the aromatic fractions may be utilized for correlation purposes, since these components are more resistant to biodegradation. However, a potential problem with the aromatic fractions is their great complexity and the need for good resolution of individual components in order to obtain reliable isotopic data. This problem can be rectified to a large extent by additional fractionation of the aromatic fraction, since it is already known that such fractionation will not affect the isotopic compositions of the individual components.

In brief, GCIRMS has the potential to become a very powerful correlation tool of use in environmental problems. It will be of most use when used in conjunction with other analytical techniques such as GC or GCMS. It will be of greatest additional value in the cases of samples that are heavily biodegraded to the extent that changes have started to occur to the biomarker distributions, making correlations difficult or impossible by conventional techniques.

ACKNOWLEDGMENTS

The work described in this proposal was supported in part by a grant from the U.S. Coast Guard administered through the U.S. Department of Transportation (USDOT-ID# 57-94-G-00079). The GCIRMS system was purchased with grants from the NSF and DOE-BES University Instrumentation programs, plus additional funds from the University of Oklahoma.

REFERENCES

Brenna, J.T. 1994. High Precision Gas Isotope Ratio Mass Spectrometry: Recent Advances in Instrumentation and Biomedical Applications, *Acc. Chem. Res.*, **27**, 340-346.

Freeman, K.H., Hayes, J.M., Trendel, J.M., and Albrecht, P.A. 1990. Evidence from Carbon Isotope Measurements for Diverse Origins of Sedimentary Hydrocarbons, *Nature*, **343**, 254-256.

Hayes, J.M., Freeman, K.H., Popp, B.N., and Hoham, C.H. 1990. Compound Specific Isotope Analyses: A Novel Tool for Reconstruction of Ancient Biogeochemical Processes, *Org. Geochem.* **16**, 1115-1128.

Kvenvolden, K., Hostettler, F.D., Rapp, J.B., and Carlson, P.R. 1993a. Hydrocarbons in Oil Residues of Prince William Sound, Alaska. *Mar. Pollut. Bull.* **26**, 24-29.

Kvenvolden, K., Threlkeld, C.N., Warden, A., and Carlson, P.R. 1993b. Forensic Geochemistry-Carbon Isotopic Identification of Two Sources of Oil Residues in Prince William Sound, Alaska. In: *Organic Geochemistry-Poster Session Volume*, pp. 735-738 (Oygard, K., Ed.) Oslo, Norway, F. Hurtigtrykk.

Murray, A.P., Summons, R.E., Boreham, C.J. and Dowling, L.M. 1994. Biomarker and *n*-Alkane Isotope Profiles for Tertiary Oils: Relationship to Source Rock Depositional Setting. *Org. Geochem.* **22**(3-5), 521-542.

Palmer, S. 1993. Effect of Biodegradation and Water Washing on Crude Oil Composition. In: *Organic Geochemistry-Principles and Applications* pp. 511-534 (Engel, M.H., and Macko, S.A., Eds.) New York, Plenum Press.

Peters, K.E. and Moldowan, J.M. 1993. *The Biomarker Guide.* Englewood Cliffs, NJ, Prentice Hall.

Philp, R.P. 1993. Oil-Oil and Oil-Source Rock Correlations: Techniques. In: *Organic Geochemistry*, pp.445-460 (Engel, M.H. and Macko, S.A., Eds.), New York, Plenum Press.

Philp, R.P. 1994. High Temperature Gas Chromatography for the Analysis of Fossil Fuels. In: The Uses of Chromatography for Analysis of Synthetic and Fossil Fuels, *J. High Res. Chromatog.* **17**(6), 398-406.

BIOREMEDIATION

CHAPTER 29

Vapor Extraction, Air Sparging, and Bioventing in Combination Form a Technically- and Cost-Effective Scenario to Remediate Petroleum Hydrocarbons

David A. Brown and **James N. Baker**, P.G. Parsons Engineering Science, Inc., Liverpool, New York

Michael P. Mailloux, Chevron U.S.A. Products Company, Inc., Atlanta, Georgia

INTRODUCTION

When the appropriate site conditions exist, air sparging, vapor extraction, and bioventing can be combined to form a technically- and cost-effective scenario to remediate petroleum hydrocarbon contaminated soils. A former Gulf Terminal in Upstate New York meets these conditions. The site geology consists of highly permeable sands and gravels with only trace amounts of silt. The groundwater table is approximately 15 feet below the ground surface, which provides an ideal vadose zone. The site contaminants are petroleum fuel residuals, primarily from the former storage and transfer of gasoline distillates. A series of pilot studies were conducted at the site in August and September of 1994 to determine the validity of the proposed technologies. Based on the pilot study results, it was determined that the combined technologies of soil vapor extraction, air sparging, and bioventing could be used to effectively remediate the site. Using the pilot study data as the design basis, Parsons ES designed and installed a full-scale remediation system to address both the vadose and phreatic zone contaminants. The SVE portion of the system was placed into operation in April of 1995, and to date has removed over 12,000 pounds of petroleum hydrocarbons, including over 30 pounds of benzene. The overall costs for remediating the site including pilot studies, detailed design, system installation, and one year of operation are estimated at $5.60 per cubic yard for the estimated 35,000 cubic yards of contaminated soil at the site. The pilot study, full-scale operational results, and projected remediation costs are the focus of this chapter.

PROJECT BACKGROUND

Site Location

This former Gulf Oil Terminal site is located in upstate New York. Gulf Oil operated the Terminal from 1939 to 1981. In the mid-1980s, Chevron acquired the property when Gulf Oil holdings became a part of Chevron. The site occupies an area of approximately one acre, and formerly contained an inactive petroleum storage and distribution facility.

Site History

Site investigation activities began in 1987 with the installation of six monitoring wells to determine if free phase hydrocarbons were contaminating groundwater beneath the site. No measurable free phase hydrocarbons were found in the wells; however, total benzene, toluene, ethylbenzene, and xylenes (BTEX) concentrations were in excess of 8,600 ppb in a groundwater sample from monitoring well MW-2 (Figure 1). Results of the soil sampling indicated concentrations of BTEX and polynuclear aromatic hydrocarbons (PAHs) at 981 and 160 mg/kg, respectively, in soils beneath the former loading rack (Figure 1). The results also indicated that the soils and groundwater south of the site had no detectable levels of petroleum contamination, and therefore, had not been impacted by the contamination within the property boundaries.

Based on these data and subsequent site investigations conducted in 1989, 1993, and 1994 in preparation for transfer of the property, the New York State Department of Environmental Conservation (NYSDEC) issued a Stipulation Agreement requiring Chevron to conduct additional environmental activities, including an offsite investigation and remedial efforts.

Site Geology and Hydrogeology

The unconsolidated deposits underlying the site consist of a discontinuous thin layer (0 to 8.5 feet) of fill mixed with sand and gravel. This layer overlies a silt loam which grades into a fine to coarse sand and gravel deposit. The sand and gravel deposit extends to a depth of at least 30 feet. Bedrock was not encountered at the site in any soil borings conducted to date.

Groundwater flow beneath the site occurs in the unconsolidated glacial outwash deposits. Groundwater flow direction is generally to the south, or toward the Susquehanna River, which is located approximately 1000 feet south of the site. The nearest municipal water supply well field is located approximately 0.8 miles southeast of the site, which is on the opposite side of the Susquehanna River. The depth to groundwater at the site ranges from 15 to 17 feet below the ground surface (bgs) during the summer months, and fluctuates as much as eight feet during the more wet months of spring.

Figure 1. **Vapor extraction/air sparging pilot test area (former Gulf terminal, upstate New York).**

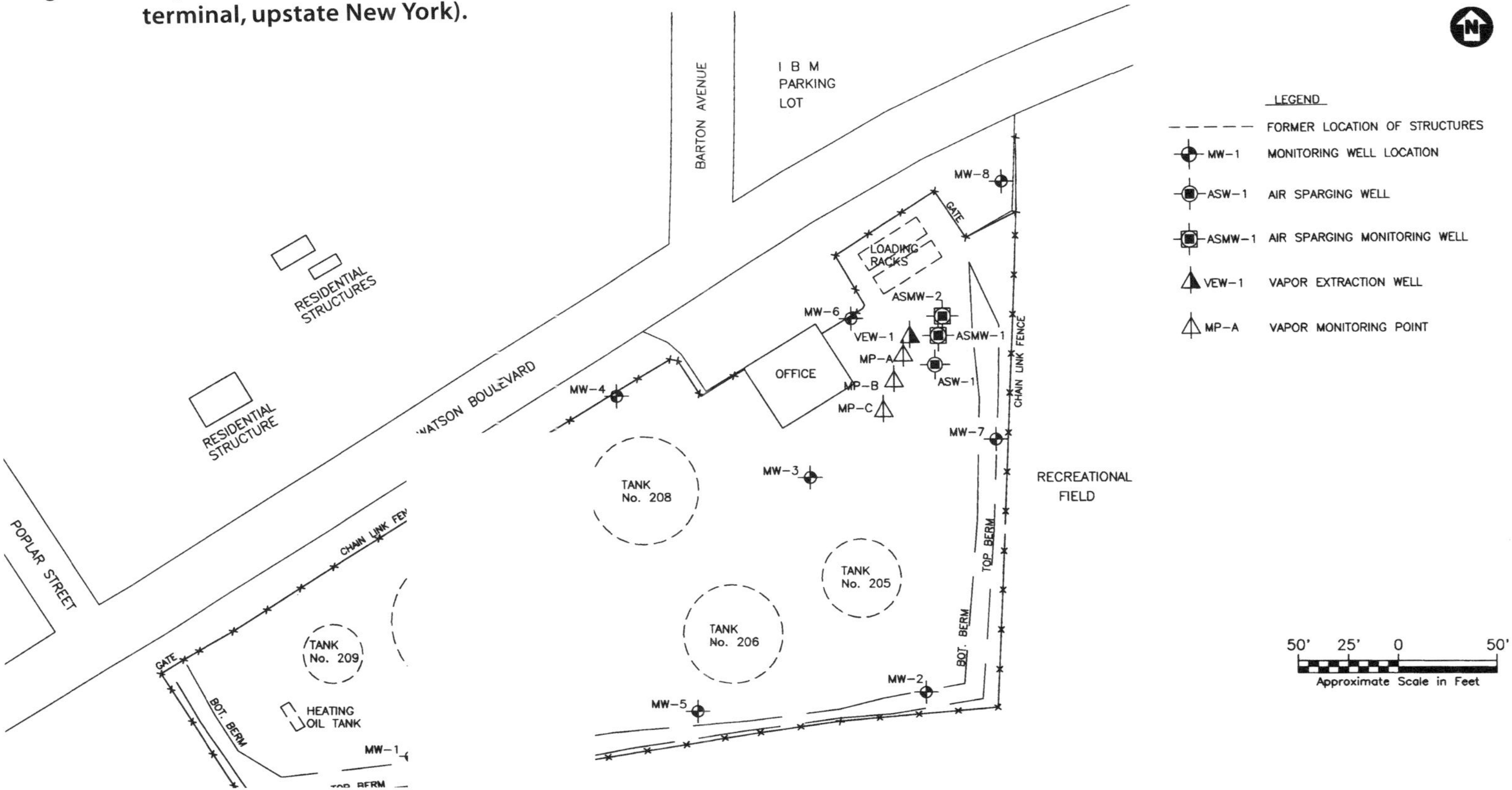

PILOT TEST

Pilot Test Objectives

The primary objectives of the pilot studies were threefold: to determine the zone of influence of a vapor extraction well installed in the vadose zone (i.e. unsaturated zone); to determine the radius of influence of aeration from an air injection point in the phreatic zone (i.e., saturated zone); and to determine whether bioventing will be effective at biologically remediating the residual hydrocarbons in the unsaturated zone soil. Residual hydrocarbons unaffected by vapor extraction may be treated by implementation of bioventing technology.

Well and Monitoring Point Installations

On July 19 and 20, 1994, Parsons ES installed one air sparging well (ASW-1), one vapor extraction well (VEW-1), two air sparging monitoring wells (ASMW-1 and ASMW-2) for saturated zone monitoring, and three soil vapor monitoring points (MP-A, MP-B, and MP-C) for unsaturated zone monitoring. The locations of the wells and monitoring points are shown on Figure 1. These wells and monitoring points were installed in the area of highest documented contamination at the site. The wells and points were located and installed to facilitate their incorporation into a full-scale SVE/AS and/or bioventing remediation system if the pilot testing proved successful.

The ASW was installed to a total depth of 29 feet bgs (11 feet below the groundwater table), and was constructed with a 2-foot long section of 2-inch diameter Schedule 40 factory-slotted PVC wellscreen set at a depth interval of 27 to 29 feet bgs. The VEW was constructed of 4-inch diameter Schedule 40 PVC with a slot size of 0.03-inch. The total depth of the VEW was 17 feet bgs, with a screened interval from 7 to 17 feet bgs. Two air sparging monitoring wells (ASMW-1 and ASMW-2) were installed at distances 15 and 25 feet laterally away from the ASW. Each ASMW was constructed with a 1.5-inch diameter by 1-foot long section of 0.020-inch slot PVC wellscreen set at 24 feet bgs. Three soil vapor monitoring points (MP-A, MP-B, and MP-C) were installed at 10, 24, and 40 feet radially away from the VEW. Each MP was constructed to provide multiple depth soil gas monitoring, with two discrete depth monitoring points set at six and ten feet bgs. Each discrete point was constructed using a 6-inch section of 1/2-inch diameter Schedule 40 PVC wellscreen with a 0.02-inch slot size. Type K thermocouples were also installed at both discrete monitoring point depths at MP-A to allow for soil temperature monitoring.

Soil, Soil Gas, and Groundwater Samples

Soil Samples

Soil samples were collected from each new well and monitoring point installed for the pilot study. The samples were analyzed for volatile organic compounds (VOCs) and PAHs, along with total iron, alkalinity, pH, total Kjeldahl nitrogen (TKN), total phosphates, percent moisture, and grain size analysis.

Groundwater Samples

Groundwater samples were collected from all eight previously installed groundwater monitoring wells, the vapor extraction well, and the air sparging well prior to the start of the pilot study. Additionally, groundwater samples from monitoring wells MW-4 and MW-5 were collected upon completion of the air sparging phase of the pilot study to determine if any lateral migration of contaminants had been caused by the air injection into the saturated zone.

The initial groundwater samples were analyzed for VOCs, PAHs, alkalinity, and total and dissolved calcium, magnesium, manganese, and iron. The samples from MW-4 and MW-5 collected upon completion of the pilot study were analyzed for VOCs only. In addition to the laboratory analyses, groundwater samples were also analyzed for four field parameters: dissolved oxygen, pH, conductivity, and temperature.

Soil Gas Samples

Prior to the pilot study, soil gas samples were collected from the VEW, the ten-foot interval from monitoring point MP-A, the six-foot intervals in MP-B and MP-C, and from monitoring wells MW-6, 7, and 8. These samples were collected to determine the initial soil gas concentrations of benzene, ethylbenzene, toluene, and xylenes (BTEX), and total volatile petroleum hydrocarbons (TPH) in the soil prior to any pilot study work. Additionally, the soil gas samples were monitored for oxygen, carbon dioxide, and total volatile hydrocarbons by use of field instruments. Soil gas samples were also collected from the vacuum blower exhaust to determine the bulk removal rate of hydrocarbons during the pilot test. Following completion of the pilot studies, a second round of soil gas samples were collected from the previously sampled monitoring points. These samples were used to determine the short-term reduction in volatile compounds from the site soils. All soil gas samples were collected in Summa canisters and analyzed by laboratory gas chromatograph/mass spectrometry.

PILOT STUDY DESCRIPTION

Three pilot studies were carried out under this program, including soil vapor extraction, air sparging/soil vapor extraction, and bioventing. The vapor extraction and air sparging pilot studies were conducted following the procedures described in Chevron's 1993-1994 Environmental Consultants Manual (1993). The bioventing pilot study followed the procedures described in the document entitled Test Plan and Technical Protocol for a Field Treatability Test for Bioventing (Hinchee et al., 1992). Each pilot study procedure is briefly described below.

Soil Vapor Extraction Pilot Study

The soil vapor extraction pilot study was conducted during the week of 1 August, 1994. A 2-hp rotary lobe vacuum blower was utilized to withdraw air from the soils at the site. The extracted air stream was treated via two granular activated carbon (GAC) adsorbers prior to being released to the atmosphere. A thermal anemometer was installed downstream of the GAC canisters to measure the extracted air flow rate. Vacuum gauges located at VEW-1 and on the blower intake were used to measure system vacuum. Additionally, Magnahelic vacuum gauges were installed at each MP to measure vacuum response in the site soils. The extracted and treated air stream was monitored during the test with a total volatile hydrocarbon (TVH) analyzer. This instrument was used to provide a gross measurement of volatile organics being extracted by the SVE system.

The pilot study was conducted by operating the vacuum blower at a steady flow rate and vacuum, and monitoring vacuum changes over time at each monitoring point. Two SVE tests were conducted at two vacuum levels (12 and 13.5 inches water vacuum). The test at 12 inches of water vacuum was discontinued after only 10 minutes, as steady state was quickly realized. The second test at 13.5 inches water vacuum also achieved steady state within 10 minutes, but was continued for a total of 16 hours. The second test was run to monitor effects of higher vacuum on system performance, and to better define the long-term removal rates of VOCs.

Air Sparging/Vacuum Extraction Pilot Study

The air sparging/vacuum extraction pilot study was conducted in September 1994. This test was conducted with a 3/4-hp rotary vane air compressor. An air rotometer was used to measure the sparge air flow rate into ASW-1. Pressure gauges were installed at ASW-1, on each air sparge monitoring well, the vapor extraction well, and on each of the three vapor monitoring points. The sparge test was run for 21 hours at an injection rate of 7.0 cfm and a steady-state pressure of 4.3 psi. After approximately one hour of air injection, the ASMWs were monitored for dissolved oxygen, and water samples were collected for analysis of pH, temperature, and conductivity. Water level readings were collected immediately after the dissolved oxygen measurements. Additionally, water level readings and dissolved oxygen concentrations were collected from the vapor extraction well and monitoring wells MW-6 and MW-7. A final round of water level readings, dissolved oxygen, pH, temperature, and conductivity were collected from the ASMWs, the VEW, monitoring wells MW-6 and MW-7, and from the air sparging well upon completion of the sparge test.

Bioventing Pilot Study

Two separate bioventing tests were conducted at the site in July and August 1994. The first test was an in situ respiration test, the objective of which was to determine the rate at which soil bacteria degrade petroleum hydrocarbons. The second test was an air permeability test necessary to determine the radius of oxygen influence and optimum air injection rate. The air permeability test was also used to determine the more efficient method (bioventing or vapor extraction) for oxygenating the vadose zone soils.

In Situ Respiration Test

The in-situ respiration tests were performed in July 1994 at discrete depth monitoring points MPA-10, MPB-6, and MPC-6. Atmospheric air was injected into each of the three MPs for 26 hours at a rate of 1.1 cubic feet per minute to oxygenate local contaminated soil. At the end of the air injection period, the air supply was cut off, and oxygen, carbon dioxide, and total volatile hydrocarbon levels were monitored for the next 42 hours. The decrease in oxygen concentrations and increase in carbon dioxide concentrations over time were used to estimate rates of bacterial degradation of fuel residuals.

Air Permeability Test

The air permeability test was conducted in August 1994 by using the VEW as the air injection well. Air was injected at a rate of 30 cubic feet per minute for two hours into the 4-inch-diameter VEW using a 2-hp rotary lobe blower unit. Magnahelic gauges were located at each MP, and pressure responses were measured over time. The pressure versus time data were input into a computer model to estimate the radius of influence. Oxygen concentrations at each MP were also monitored to verify that oxygen levels in the soil increased as the result of air injection.

PILOT STUDY RESULTS

Initial Soil, Soil Gas, and Groundwater Monitoring

Soil Sample Results

Benzene was detected at concentrations of 1,300 and 3,700 parts per billion (ppb) in the soil samples from 17 feet bgs at ASMW-1 and nine feet bgs at MP-A, respectively. Concentrations of total xylenes ranged from 47,000 to 550,000 ppb, and concentrations of ethylbenzene ranged from 12,000 to 77,000 ppb. The complete soil sample results are presented on Table 1.

Groundwater Sample Results

The results of the groundwater samples are presented in Tables 2 and 3, and show elevated concentrations (i.e., above method detection limits) of BTEX compounds and other VOCs in monitoring wells MW-2, 6, 7, and 8, the vapor extraction well, and the air sparging well. Values for many of these compounds are above the NYSDEC STARS Memo #1 guidance values. Field analysis of

Table 1. **Soil Labratory Analytical Results (Former Gulf Terminal, July 1994)**

COMPOUND	NYSDEC Guidance Values[1]	SAMPLE LOCATION (Depth in feet)						
		VEW-1 (8 to 10') & (14 to 16')[2]	ASW-1 (8 to 10') & (14 to 16')[2]	ASMW-1 (16 to 18')	ASMW-2 (14 to 16')	MP-A (8 to 10')	MP-B (8 to 10')	MP-C (8 to 10')
Soil Hydrocarbons (ppb)								
Benzene	14	<830	<8900	1300	<6200	3700	<16000	<3100
Ethylbenzene	100	12000	69000	12000	68000	77000	47000	19000
Toluene	100	<830	<8900	<1200	<6200	100000	<16000	<3100
Total Xylenes	100	47000	450000	51000	350000	550000	410000	100000
Isopropylbenzene	100	2500	TIC	1900	14000	8400	TIC	4200
n-Propylbenzene	100	5600	14000	3600	22000	17000	18000	11000
p-Isopropyltoluene	100	1900	TIC	TIC	TIC	TIC	TIC	TIC
1,2,4-Trimethylbenzene	100	54000	150000	40000	250000	230000	320000	110000
1,3,5-Trimethylbenzene	100	16000	44000	9600	63000	62000	75000	27000
n-Butylbenzene	100	9200	10000	4100	25000	19000	28000	13000
sec-Butylbenzene	100	1600	TIC	TIC	TIC	TIC	TIC	TIC
Naphthalene	200	11000	14000	6800	44000	27000	47000	19000
Methyl t-butyl ether	1000	<8300	<89000	<12000	<62000	<62000	<160000	<31000
Soil Inorganics								
Iron (mg/kg)		17600	NA	NA	NA	23000	16100	21200
Alkalinity (mg/kg as CaCO3)		<200	NA	NA	NA	<200	<200	<200
pH (units)		6.4	NA	NA	NA	6.5	6.8	6.5
TKN (mg/kg)		180	NA	NA	NA	310	150	210
Phosphates (mg/kg)		440	NA	NA	NA	360	490	380
Soil Physical Parameters								
Soil Temperature MPA-6,10 (64.2,57.3 deg.F)								
Moisture (% wt.)		13	15.4	16.6	15.2	17.5	8.1	9.3
Gravel (%)		21.6	NA	NA	NA	0.8	34.9	14.8
Sand (%)		60.7	NA	NA	NA	16.1	53.7	60.6
Silt (%)		3.0	NA	NA	NA	52.1	11.4	7.5
Clay (%)		14.7	NA	NA	NA	31.0	INC.	17.1

[1] NYDEC – Stars Memo #1 – gasoline contaminated soil
– guidance policy for groundwater quality protection (August 1992)

[2] Sample was composited from two depth intervals as insufficient sample was collected from any single split-spoon sample.

NA – not analyzed for

INC – % recorded as clay and silt

TIC – Tentatively Identified Compound.

Table 2. **Groundwater Laboratory Analytical Results (Former Gulf Terminal, July 1994)**

COMPOUND	NYSDEC GUIDANCE VALUES (ppb)[1]	VW-1	MW-1	MW-2	MW-3	MW-4	MW-4[(2)]	MW-5	MW-5[(2)]	MW-6	MW-7	MW-8	ASW-1
STARS Compounds (ppb)													
Benzene	0.7	30	ND	520	ND	ND	ND	ND	ND	560	ND	5.7	2900
Ethylbenzene	5	210	ND	1000	ND	ND	ND	ND	ND	110	1900	ND	1500
Toluene	5	ND	ND	200	ND	ND	ND	ND	ND	49	140	ND	140
Total Xylenes	5	1200	ND	5200	ND	ND	ND	ND	ND	ND	8900	ND	5200
Isopropylbenzene	5	ND	ND	ND	ND	ND	ND	ND	ND	ND	ND	ND	ND
n-Propylbenzene	5	ND	ND	ND	ND	ND	ND	ND	ND	ND	ND	ND	ND
p-Isopropyltoluene	5	ND	ND	ND	ND	ND	ND	ND	ND	ND	ND	ND	ND
1,2,4-Trimethylbenzene	5	280	ND	530	ND	ND	ND	ND	ND	ND	880	ND	650
1,3,5-Trimethylbenzene	5	70	ND	150	ND	ND	ND	ND	ND	ND	ND	6	140
n-Butylbenzene	5	ND	ND	ND	ND	ND	ND	ND	ND	ND	ND	ND	ND
sec-Butylbenzene	5	ND	ND	ND	ND	ND	ND	ND	ND	ND	ND	ND	ND
Naphthalene	10	51	ND	ND	ND	ND	ND	ND	ND	ND	290	ND	200
Methyl t-butyl ether	50	ND	ND	ND	ND	ND	ND	ND	ND	ND	ND	ND	ND
Metals (ppm)													
Dissolved Calcium	None	161	ND	70.7	125	ND	ND	ND	ND	117	46.4	42.3	94.1
Dissolved Iron	None	ND	ND	60.1	1.3	ND	ND	ND	ND	13.3	59.1	0.75	62.9
Dissolved Magnesium	None	21.3	ND	13.5	20.8	ND	ND	ND	ND	20.6	6.8	ND	15.3
Dissolved Manganese	None	8.4	ND	15.3	8.0	ND	ND	ND	ND	10.3	5.5	1.9	3.8
Total Calcium	None	163	ND	79.5	131	ND	ND	ND	ND	118	61.9	105	235
Total Iron	300[S]	4.4	ND	123	36.4	ND	ND	ND	ND	51.0	213	75.4	241
Total Magnesium	35000[G]	23.0	ND	17.7	34.0	ND	ND	ND	ND	19.8	19.1	26.1	55.2
Total Manganese	300[S]	8.5	ND	18.2	8.2	ND	ND	ND	ND	10.1	11.5	10.9	7.4

[1] NYDEC – STARS Memo #1 – gasoline contaminated soil
– guidance policy for groundwater quality protection (August 1992)
[2] Second sample, collected after air sparging test.
[S] NYS ambient water quality standards
[G] NYS ambient water quality guidance values
ND = Sample had non-detectable concentrations of this parameter.

Table 3. **Groundwater Field Results Before and After Air Sparging (Former Gulf Terminal, September 1994)**

Location	Compound	Amount Prior to Air Sparging (ppm)	Amount after 1 hr. of Air Sparging (ppm)
Chevron–ASW–1	Water Level (feet bgs)	15.51	16.70
	Dissolved Oxygen (mg/l)	0.4	6.1
	pH (units)	NA	NA
	Conductivity (umhos/cm)	NA	NA
	Temperature(deg F)	NA	NA
Chevron–VEW–1	Water Level (feet bgs)	15.33	15.05
	Dissolved Oxygen (mg/l)	0.5	4.05
	pH (units)	NA	NA
	Conductivity (umhos/cm)	NA	NA
	Temperature(deg F)	NA	NA
Chevron–ASMW–1	Water Level (feet bgs)	17.90	17.31
	Dissolved Oxygen (mg/l)	1.7	2.8
	pH (units)	6.76	7.14
	Conductivity (umhos/cm)	680	800
	Temperature(deg F)	59.6	54.7
Chevron–ASMW–2	Water Level (feet bgs)	16.52	17.27
	Dissolved Oxygen (mg/l)	1.5	2.9
	pH (units)	7.03	7.14
	Conductivity (umhos/cm)	850	800
	Temperature(deg F)	57.2	55.6
Chevron–MW–6	Water Level (feet bgs)	21.33	20.96
	Dissolved Oxygen (mg/l)	0.5	0.6
	pH (units)	NA	NA
	Conductivity (umhos/cm)	NA	NA
	Temperature(deg F)	NA	NA
Chevron–MW–7	Water Level (feet bgs)	20.90	20.90
	Dissolved Oxygen (mg/l)	0.6	0.5
	pH (units)	NA	NA
	Conductivity (umhos/cm)	NA	NA
	Temperature(deg F)	NA	NA

NA = not analyzed for.

groundwater samples indicated depleted dissolved oxygen levels in all the impacted monitoring points .

Soil Gas Sample Results

As shown in Table 4, all the soil gas monitoring locations had completely depleted oxygen levels (except background wells MW-4 and MW-5). Additionally, all points had high carbon dioxide readings (4.8 to 10.0 %), and TVH readings ranging from 2,000 ppm to greater than 20,000 ppm. These readings suggest that the indigenous microorganisms have completely depleted the naturally available oxygen supply, producing increased CO_2 concentrations, and indicating significant biological activity. In contrast, the background soil gas monitoring at two uncontaminated groundwater monitoring wells, MW-4 and MW-5, indicated near atmospheric conditions in the soil gas (i.e., greater than 20 % oxygen and less than 1.0 % carbon dioxide).

Table 4. **Initial Soil Gas Chemistry (Former Gulf Terminal, July 1994)**

Sample Location (ft)	O_2 (%)	CO_2 (%)	TVH (ppm)
Chevron–VW–1	0	7.0	16800
Chevron–MPA–6	0	6.8	20000+
Chevron–MPA–10	0	6.5	20000+
Chevron–MPB–6	0	6.5	20000+
Chevron–MPB–10	0	6.5	18400
Chevron–MPC–6	0	6.0	20000+
Chevron–MPC–10	0	6.9	16800
Chevron–MW–4	18.2	1.0	20
Chevron–MW–5	20.9	0.5	20
Chevron–MW–6	0	10.0	4000
Chevron–MW–7	0	8.5	12500
Chevron–MW–8	0	6.0	11000

Results of the laboratory analyses (Table 5) of soil gas samples from the monitoring points indicated significant concentrations of BTEX and TPH compounds. Concentrations of BTEX compounds ranged from 30 to 815 parts per million (ppm), and TPH concentrations ranged from 3,300 to 26,000 ppm. Monitoring well MW-6 had the highest concentrations of both BTEX and TPH.

Soil Vapor Extraction Pilot Study Results

Evaluations of the vapor extraction test data from the two tests were conducted according to the Chevron Protocol. Application of 12.0 inches of water vacuum and 116 cfm soil vapor flow in the first test produced an observed influence of 0.40 inches water vacuum at monitoring point MPC-10, located 40 feet from the vapor extraction well. The highest vacuum of 0.63 inches water was measured after only 5 minutes into the test at monitoring point MPA-6, located 10 feet from the VEW. During the second test, application of 13.5 inches water vacuum and 130 cfm soil vapor flow at the VEW produced an observed influence of 0.30 inches water at monitoring well MW-7, located approximately 55 feet from the VEW. The highest vacuum measured during the second test was 0.75 inches water measured after 15 minutes into the test at MPA-6.

The vacuum data from each SVE pilot test was normalized (i.e., the observed vacuum divided by the applied vacuum at the extraction well) and plotted on a semi-log scale against the normalized radial distance from the VEW (i.e., radial

Table 5. **Soil Gas Laboratory Analytical Results Before and After Vapor Extraction (Former Gulf Terminal, August 1994)**

Location	Compound	Amount Prior to Vapor Extraction (ppm)	Amount after 21 hours of Vapor Extraction(ppm)
Chevron–VW–1	TPH	14000	7600
	Benzene	52	< 2.1
	Toluene	< 2.2	< 2.1
	Ethyl Benzene	36	38
	Total Xylenes	70	99
Chevron–MPA–10	TPH	18000	7300
	Benzene	45	< 3.5
	Toluene	13	< 3.5
	Ethyl Benzene	54	24
	Total Xylenes	190	100
Chevron–MPB–6	TPH	22000	5700
	Benzene	44	< 2.8
	Toluene	130	25
	Ethyl Benzene	91	7.2
	Total Xylenes	500	130
Chevron–MPC–6	TPH	26000	14000
	Benzene	55	< 6.4
	Toluene	250	180
	Ethyl Benzene	100	69
	Total Xylenes	410	300
Chevron–MW–6	TPH	8200	1100
	Benzene	< 1.2	< 0.46
	Toluene	9.4	< 0.46
	Ethyl Benzene	12	8.9
	Total Xylenes	12	23
Chevron–MW–7	TPH	5400	3200
	Benzene	< 0.67	< 0.40
	Toluene	< 0.67	< 0.40
	Ethyl Benzene	10	8.1
	Total Xylenes	20	22
Chevron–MW–8	TPH	3300	2200
	Benzene	< 0.95	< 0.76
	Toluene	< 0.95	< 0.76
	Ethyl Benzene	14	12
	Total Xylenes	22	19
* Chevron–VEW–1	TPH	NS	10000
	Benzene	NS	< 1.8
	Toluene	NS	< 1.8
	Ethyl Benzene	NS	58
	Total Xylenes	NS	140

* sample collected from blower exhaust
TPH = total petroleum hydrocarbons
ppm = parts per million by volume
NS = no sample collected

distance divided by depth to groundwater at the VEW). Figure 2 presents the normalized plot for the first test, and Figure 3 shows the plot for the second test. A straight line fitted through the normalized points represents the spatially averaged radial distribution of vacuum induced at the extraction well. The 0.1-inch of water vacuum line shown on the plot represents the minimum required vacuum measurement considered valid for evaluating vacuum distribution. Chevron has determined that minor variations in low vacuum measurements (i.e., below 0.1-inch water vacuum) tend to scatter data significantly on a loga-

Figure 2. **Normalized vacuum plot VE Test 1 (former Gulf terminal, August 1994).**

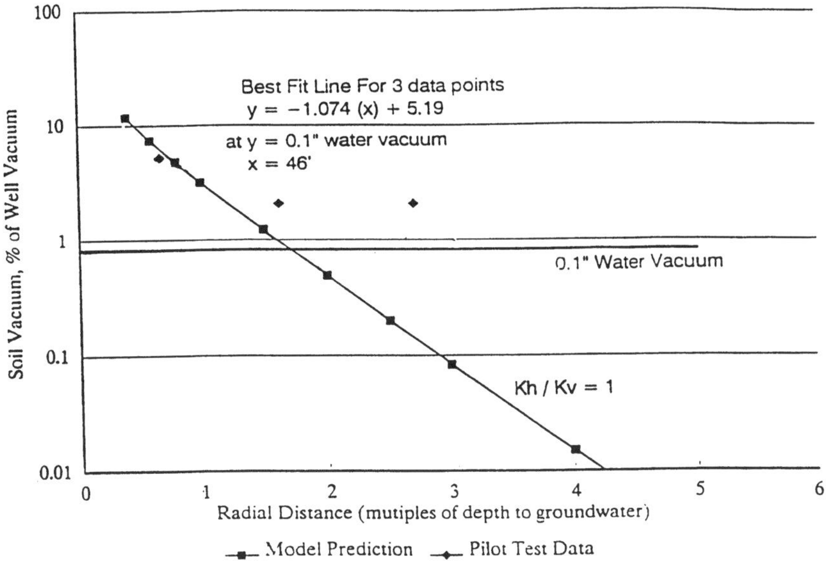

Well Head Vacuum = 12 in. H2O
Depth to Groundwater = 14.8 ft.
Extracted air flow = 116 cfm
Total Test Time = 10 min.

Point	Stabilized Pressure	% Well Vacuum	Radial Distance (ft)
MPA	0.63	5.25	10
MPB	0.25	2.08	24
MPC	0.25	2.08	40

rithmic plot, and are therefore less reliable than higher vacuum measurements made nearer the extraction well. The Kh/Kv = 1 line shown on the plots represents the predicted vacuum response line in the case where horizontal permeability equals vertical permeability. In general, most sites have horizontal permeability greater than vertical permeability (i.e., a flatter slope than the Kh/Kv = 1 line). However, Chevron's Protocol utilizes Kh/Kv = 1 as the minimum standard criteria to determine if a sufficient pressure drop with distance is observed during pilot tests. Based on these normalized plots, vacuum extraction appeared to be applicable at the site (i.e., observed data points are above the 0.1-inch water

Figure 3. **Normalized vacuum plot VE Test 2 (former Gulf terminal, August 1994).**

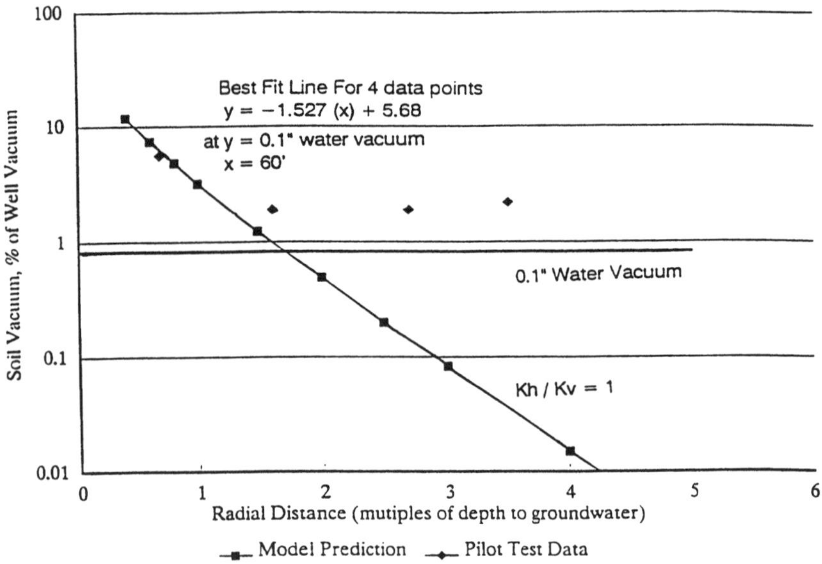

Well Head Vacuum = 13.5 in. H2O
Depth to Groundwater = 14.8 ft.
Extracted air flow = 130 cfm
Total Test Time = 60 min.

Point	Stabilized Pressure	% Well Vacuum	Radial Distance (ft)
MPA	0.75	5.6	10
MPB	0.25	1.9	24
MPC	0.25	1.9	40
MW−7	0.3	2.2	52

line, and a straight line through the data points has a flatter slope than the Kh/Kv = 1 line).

The field data was then entered into Chevron's Soil Vent Model to estimate the required well spacing for the full-scale SVE system. Additional data including depth to groundwater, soil vapor flow rate, level of hydrocarbon contamination, and extraction well construction data are entered into the model to predict well spacing and remediation time. Based on the model output, extraction wells spaced 60 feet apart would achieve an estimated minimum air flow rate of 0.5 pore volume per day throughout site soils. Additionally, approximately 50

Table 6. **Soil Vapor Monitoring Results for Benzene (Former Gulf Terminal)**

Location	Compound	Amount Prior to Vapor Extraction (ppm)	Amount after 21 hours of Vapor Extraction(ppm)
Chevron–VW–1	TPH	14000	7600
	Benzene	52	< 2.1
	Toluene	< 2.2	< 2.1
	Ethyl Benzene	36	38
	Total Xylenes	70	99
Chevron–MPA–10	TPH	18000	7300
	Benzene	45	< 3.5
	Toluene	13	< 3.5
	Ethyl Benzene	54	24
	Total Xylenes	190	100
Chevron–MPB–6	TPH	22000	5700
	Benzene	44	< 2.8
	Toluene	130	25
	Ethyl Benzene	91	7.2
	Total Xylenes	500	130
Chevron–MPC–6	TPH	26000	14000
	Benzene	55	< 6.4
	Toluene	250	180
	Ethyl Benzene	100	69
	Total Xylenes	410	300
Chevron–MW–6	TPH	8200	1100
	Benzene	< 1.2	< 0.46
	Toluene	9.4	< 0.46
	Ethyl Benzene	12	8.9
	Total Xylenes	12	23
Chevron–MW–7	TPH	5400	3200
	Benzene	< 0.67	< 0.40
	Toluene	< 0.67	< 0.40
	Ethyl Benzene	10	8.1
	Total Xylenes	20	22
Chevron–MW–8	TPH	3300	2200
	Benzene	< 0.95	< 0.76
	Toluene	< 0.95	< 0.76
	Ethyl Benzene	11	12
	Total Xylenes	22	19
* Chevron–VEW–1	TPH	NS	10000
	Benzene	NS	< 1.8
	Toluene	NS	< 1.8
	Ethyl Benzene	NS	58
	Total Xylenes	NS	140

* sample collected from blower exhaust
TPH = total petroleum hydrocarbons
ppm = parts per million by volume
NS = no sample collected

to 200 pore volumes removed were estimated to achieve cleanup of an estimated 1,000 mg/kg gasoline from the site soils.

Approximately one hour after completion of the second vapor extraction test on 2 August 1994, soil gas samples were collected from the vapor extraction well, monitoring points MPA-10, MPB-6, and MPC-6, and groundwater monitoring wells MW-6, 7, and 8 (Table 5). The O2 and CO2 had returned to their pretest conditions of 0.0% O2 and greater than 6.0% CO2. Total volatile hy-

Figure 4. **Change in D.O. from baseline air sparging test (former Gulf terminal, upstate New York).**

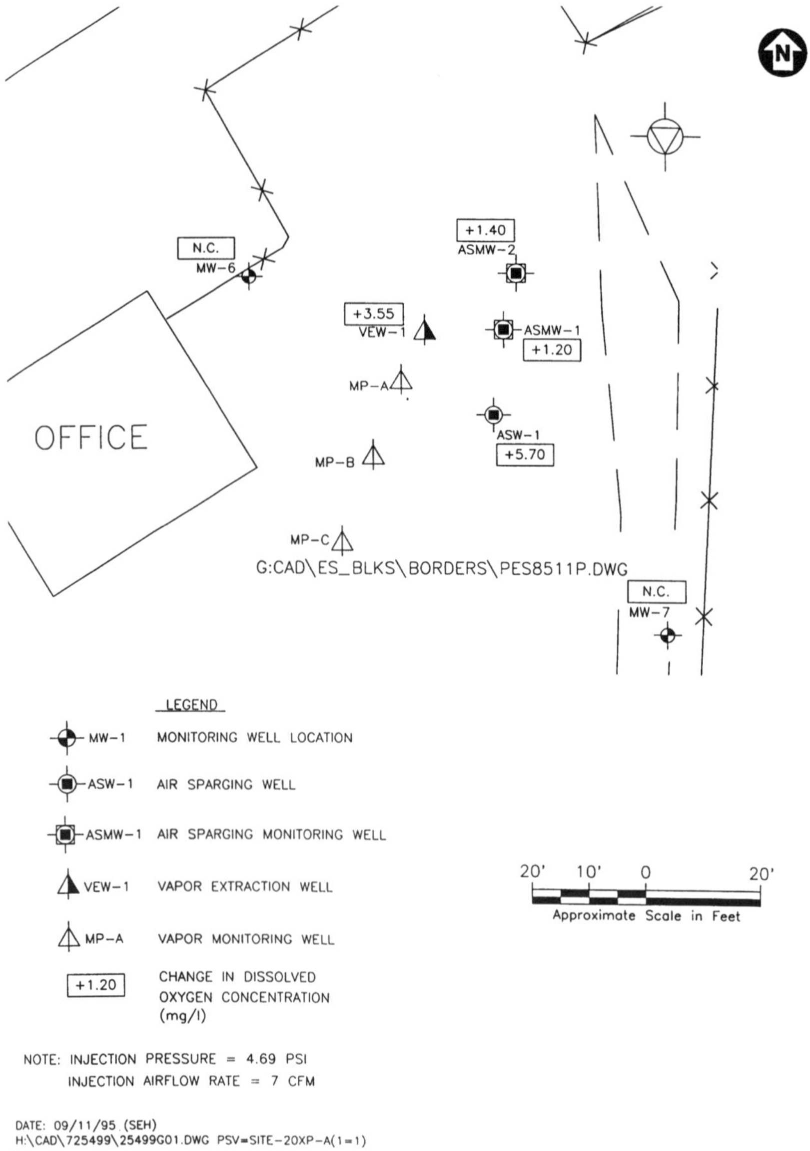

drocarbon readings also returned to near pretest conditions with concentrations ranging from 1,800 to 13,600 ppm (compared to 8,000 to greater than 20,000 ppm prior to the pilot study). Laboratory results indicated decreased concentrations of BTEX and TPH as compared to the pretest concentrations. The decreased concentrations do not, however, indicate that significant soil remediation occurred during the pilot study. The site soil and soil gas did not have adequate time to reequilibrate after the second SVE test.

Air Sparging/Vapor Extraction Pilot Study Results

The results of the pressure monitoring at the two ASMWs, located at 15 and 25 feet from the ASW, indicated a maximum pressure response of 26% and 7% of the injection well gauge pressure, respectively.

Dissolved oxygen measurements presented on Figure 4 show the dissolved oxygen concentrations in the ASMWs, and VEW-1 and ASW-1 all increased significantly during the course of the test. The largest increase was seen in VEW-1. The large increase at VEW-1 may be due to its screen interval being set much higher than the screen intervals of the two ASMWs (17 feet bgs, compared to 24 feet bgs). What likely occurred is that the higher screen interval intercepted a greater portion of the zone of effective oxygen influence. In high permeability soils, the zone has a much steeper angle from the injection well, and the deeper screens at the ASMWs (24 feet bgs) may have intercepted a smaller portion of this zone.

All of the monitored points had some degree of water level response as measured during the air sparging test. The greatest response was seen at ASMW-1 (0.88-foot rise over equilibrium). Significant responses were also seen at groundwater monitoring wells MW-6 and MW-7 (0.043-foot and 0.10-foot, respectively). Both of these groundwater wells are approximately 50 feet from the injection well. The maximum water level responses were measured at approximately one hour into the sparge test. Later water level measurements were nearly equal to the former equilibrium conditions seen before the sparge test. The water level response at the injection well (ASW-1) was measured approximately 30 minutes after completion of the sparge test. This may account for the rather small water level response at ASW-1, due to the groundwater nearly reaching its former equilibrium level within the 30 minute time frame.

Based on the data collected during all of the air sparging tests, the radius of effective oxygen influence is estimated to be approximately 25 feet. Significant increases in dissolved oxygen concentrations were recorded at both ASMWs and at VEW-1. Significant pressure increases were also seen at the two ASMWs, and significant water level responses were seen at distances as great as 50 feet.

Bioventing Pilot Study Results

Air Permeability Test

Steady-state pressure levels were achieved at all MPs within 10 minutes of test initiation. Additionally, steady-state pressures were achieved at groundwater monitoring wells MW-6 and MW-7 within 10 minutes of test initiation.

The steady-state test method was utilized to determine the soil's permeability and estimated radius of influence (Sellers and Fan, 1991). Using this test method, an air permeability value of 73.6 darcys was calculated for this site. Soils with permeabilities greater than 10 darcys are considered to be highly permeable to air flow (Johnson et al., 1990). The radius of pressure influence is estimated to be approximately 45 feet at an injection rate of 30 cfm and 3.5 inches water pressure.

After one hour of air injection at VEW-1, the oxygen concentrations in monitoring points MPA-6 and MPA-10 had increased from 0.0% to 4.3% and 1.5%, respectively. No increases were seen at monitoring points MPB and MPC. At two hours, the oxygen concentrations at MPA-6 and 10 had increased to 6.5% and 10%, respectively. No increases were seen at either MPB or MPC in two hours. These data were sufficient to determine the efficacy of air injection (i.e., bioventing) for delivering oxygen to the unsaturated zone soils.

In-Situ Respiration Test

In-situ respiration tests were performed at the discrete monitoring points MPA-10, MPB-6, and MPC-6. These points were chosen based on their low initial oxygen readings (0.0 %), high carbon dioxide readings (greater than 6%), and high TVH readings (greater than 20,000 ppm). The MPs were monitored for 42 hours after cessation of the air injection period, during which time the oxygen concentrations in all three points returned to 0.0%. The measured oxygen losses were then used to calculate biological oxygen utilization rates. The results of the in situ respiration testing for monitoring point MPA-10 are presented in Figure 5.

Oxygen loss was linear at every point tested during the respiration test. Oxygen utilization rates observed at the site were very consistent, and ranged from 0.0092% to 0.0195% per minute.

Based on the rates of oxygen utilization, an estimated 2,075 mg of petroleum hydrocarbons per kilogram of soil can be degraded each year. This value is the average of the fuel consumption rates calculated for the three points at which respiration tests were conducted. The depth interval-specific fuel consumption rates were calculated using observed oxygen utilization rates, estimated air-filled porosities, and a conservative stoichiometric ratio of 3.5 mg of oxygen consumed for every 1 mg of fuel biodegraded.

Figure 5. **Respiration test MPA-10 (former Gulf terminal, July 1994).**

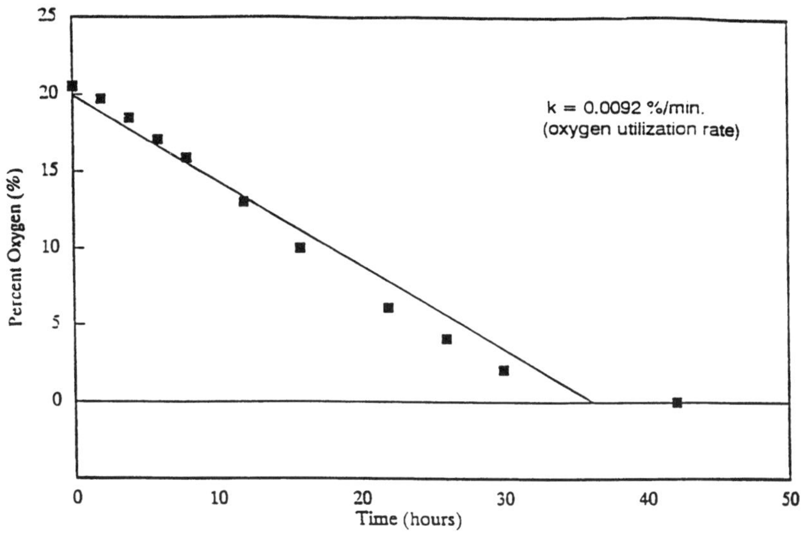

FULL-SCALE SYSTEM DESIGN AND IMPLEMENTATION

Based on the geologic factors and the type of contaminants at the site (i.e., high permeability sand and gravel soils and weathered gasoline contaminant), the technologies of vapor extraction, air sparging, and bioventing appeared to be ideally suited for remediating both the soil and groundwater at the site. The pilot studies carried out in July, August, and September of 1994 support the use of these technologies as the most cost-effective and technologically-effective remedies for the site. A combination of these technologies appeared to remediate both the soil and groundwater of volatile and semivolatile petroleum hydrocarbons.

System Design

Based on the pilot study results and size of the site (approximately 300 feet by 125 feet), the full-scale system design consisted of 10 combination vapor extraction/bioventing wells and six air sparging wells. Based on an SVE effective radius of at least 50 feet and approximately a 30% overlap, 10 SVE wells properly located would be sufficient to remediate the entire east end of the property, and a large portion of the offsite property to the east and south. The vapor extraction/bioventing wells were manifolded to a single air supply header which was configured at the blower for operation in either the vapor extraction mode or bioventing mode. The six air sparging wells were manifolded to a single air supply header which was fed by a rotary vane air compressor. Air sparging would be conducted simultaneously with the SVE system at six locations where groundwater contamination is the highest. The six air sparging wells would not impact the entire east end of the property, however. Four ASWs would be located in the area of highest groundwater contamination, and would primarily operate for source reduction in that area. A secondary benefit of these wells would be to increase dissolved oxygen levels in the groundwater and create a plume of oxygenated water moving downgradient to the south with the natural groundwater movement. The two remaining ASWs would be located in the southeast corner of the property. These wells would also oxygenate the groundwater for further biodegradation of contaminants and prevent/reduce the potential for offsite migration of site contaminants.

The SVE/AS system was connected to a programmable controller to not only operate the system, but also provide safety features, including the following:

- explosivity meter to detect elevated vapor concentrations in the compressor,
- high level shutoff switch in moisture separator tanks that, if activated, would shut the system down,
- a telemetry system to monitor air temperature and flow rate.

It is anticipated that the combined SVE/AS portion of the treatment scheme would only be in operation until such time as the VOC emissions are reduced to the point where bioventing can be initiated (typically when the soil gas hydrocarbons in the vapor extraction exhaust fall below 1,000 ppm). After this occurs, it is expected that the SVE system would be shut down, and bioventing would begin to complete remediation of the site soils.

System Installation

Installation of the full-scale system began in March 1995 with the construction of five additional air sparging wells (ASWs) and nine combination vapor extraction/bioventing wells (Figure 6). The air sparging wells were constructed of 2-inch diameter, Schedule 40 PVC with a 2-foot long section of 0.02-inch slotted screen set at depths of 25 or 30 feet bgs. The nine additional soil vapor extraction/bioventing wells (VEWs) were installed to depths of approximately 15 feet. Each VEW was constructed with a 10-foot long section of 4-inch diameter Schedule 40 PVC with a slot size of 0.04 inch. All delivery piping was installed approximately 24 to 36 inches belowground to prevent vandalism to the system.

The sparging/vapor extraction and bioventing system also included the installation of three additional groundwater and soil vapor monitoring points (MPs). The MPs were constructed of 1½-inch and ½-inch diameter Schedule 80 PVC, with 12-inch and 6-inch screen intervals set at four discrete depths. The depth intervals set at 6 and 10 feet were constructed of ½-inch diameter by 6-inch long PVC wellscreen with 0.02-inch slots, and will be used to monitor the vadose zone soils. The 17- and 22-foot depth monitoring points were constructed of 1.5-inch diameter by 12-inch long PVC wellscreen with 0.02-inch slots, and will be used to monitor the groundwater in the contaminated area.

Figure 6. **In-situ remediation system construction drawing, site location and system layout (former Gulf terminal, upstate New York).**

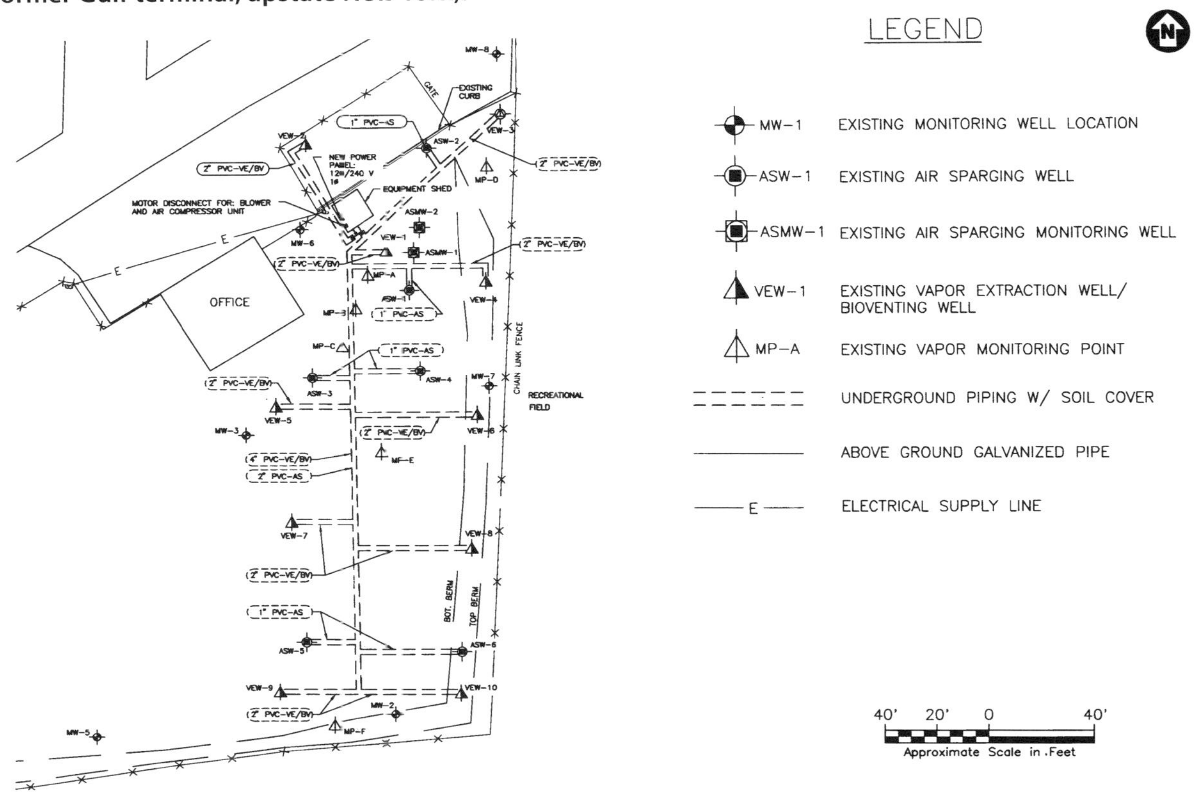

The VEW header was connected to an explosion-proof vacuum/blower rated at 300 cfm at 30 inches water vacuum. The vacuum/blower unit was installed with a 40-gallon moisture separator (including an overfill sensor), in-line air filter, pressure relief valves, 20-foot exhaust stack with a continuous explosive vapor monitor, and various valving to convert the system from the vapor extraction to bioventing mode.

The ASW header was connected to a rotary vane, explosion-proof, 100% oil-less air compressor. The compressor is able to deliver up to 50 cubic feet per minute of air at a pressure of 10 pounds per square inch. The air compressor is configured with an in-line air filter and various valves and gauges to control the delivery pressure.

SYSTEM STARTUP AND FIRST FOUR MONTHS OPERATION

Installation of the full-scale system was completed in April 1995. System startup was initiated immediately, beginning with operation of the SVE system only. Initially, only the SVE system was started in order to meet the NYSDEC air emission limit of 0.22 pound benzene per day. Based on the pilot study data, operation of both the SVE and air sparging systems would have exceeded this limit.

Soil vapor samples were collected from the SVE system during the initial startup and first three months of operation of the system from 19 April 1995 through 21 July 1995. All samples were analyzed for BTEX and total volatile petroleum hydrocarbons (TPH) via modified Method TO-3, which included GC/MS analysis. The results of the analyses for benzene are presented on Table 6.

As shown on Table 6, the initial stack gas samples indicated that the system was discharging greater than 0.22 pound benzene per day until 4 May 1995. The total air flow rate from the SVE wells was decreased on 4 May in order to meet

Table 6. **Soil Vapor Monitoring Results for Benzene (Former Gulf Terminal)**

DATE	BENZENE CONCENTRATION (ppmv)	FLOWRATE (CFM)	BENZENE EMISSION (lbs./day)
19-APR-95	0	0	0.000
19-APR-95	22	240	1.510
20-APR-95 am	60	240	4.118
20-APR-95 pm	57	240	3,912
21-APR-95	47	240	3.226
26-APR-95	16 (1)	240	1.098
04-MAY-95	2.2 (2)	240	0.151
15-MAY-95	7.7	240	0.528
22-MAY-95	1.9	240	0.130
21-JUN-95	1.4	240	0.096
21-JUL-95	2.2	240	0.151

(1) Flowrate for vapor extraction system reduced by approximately 40% on 26 April 95.
(2) Flowrate for vapor extraction system reduced by approximately 50% on 3 May 95.

the air emission limit. After that date, the SVE system was operating below the 0.22 pound benzene per day limit. A transient exceedance of the 0.22 pound per day limit was indicated by a 15 May 1995 sample. All samples collected after 15 May 1995 indicated that the SVE system is operating below the emission limit.

Overall, it is estimated that after 90 days operation of the SVE system, over 30 pounds of benzene were removed from the site soils. Additionally, approximately 42 pounds of toluene, 60 pounds of ethylbenzene, 180 pounds of xylene, and over 12,000 pounds of TPH were removed by SVE alone. Efforts are currently under way to estimate the amount of fuel removed by biodegradation by correlation to the mass of CO2 emitted from the SVE stack.

PROJECT COSTS

The site measures approximately 300 feet by 125 feet, and contamination has been found to an average depth of 25 feet. The total volume of contaminated soil based on these numbers is approximately 35,000 cubic yards. The total per yard costs of the remediation including the pilot study, system design, installation, and one year of operation is approximately $5.60. If remediation extends out to two years, the unit costs are increased to $6.20 per cubic yard.

CONCLUSIONS

The site conditions at the former Gulf Terminal are ideally suited for air based in-situ treatment. The site soils consist of highly permeable sand and gravel deposits. The site contaminants are relatively volatile and biodegradable gasoline residuals. Based on very positive pilot study results and the mass of contaminants removed in the first three months of operation, it appears that the site is well on its way to remediation.

The SVE system has been the sole technology operating so far, due to the low air emission limit set by the NYSDEC. After three months operation, the emissions from the SVE system have been reduced to approximately one-third of the 0.22 pound benzene per day limit. It is estimated that the emission rate will fall below one-quarter of the limit by mid-September. At that point, the air sparging system will be activated to begin remediation of the phreatic zone. Because the approach to phreatic zone treatment is not overly aggressive (i.e., source reduction of the most contaminated zone and a biosparging approach for the remainder of the site's phreatic zone), remediation of this zone is expected to require more than one year of operation.

The SVE and AS systems will be operating concurrently in order to assure capture of the AS gases injected into the phreatic zone. Once the emission rate for the combined SVE and AS systems falls below the 0.22 pound per day limit, the SVE system will be turned over to bioventing treatment, while the AS system air injection rate will be reduced to follow a biosparging type approach.

Based on the results to date and the overall anticipated costs including one year of operation, the combined in situ technologies have been highly effective for removing site contaminants in an extremely cost-effective manner. However, good engineering judgment must always be exercised prior to installing a similar system at other petroleum contaminated sites. This former Gulf Terminal site has the optimal conditions for this type of treatment. Site by site analysis and pilot studies are the only way to determine the applicability of these technologies.

REFERENCES

Chevron U.S.A. Products Company. 1993. 1993-1994 Environmental Consultants Manual.

Hinchee, R.E., Ong, S.K., Miller, R.N., Downey, D.C., and Frandt, R. 1992. Test Plan and Technical Protocol for a Field Treatability Test for Bioventing, Revision 2.

Sellers, K., and Fan, C.Y. 1991. Soil Vapor Extraction: Air Permeability Testing and Estimation Methods, Proceedings of the 17th RREL Hazardous Waste Research Symposium.

Johnson, P.C., Stanley, C.C., Kemblowski, M.W., Byers, D.L., and Colthart, J.D. 1990. A Practical Approach to the Design, Operation, and Monitoring of In Situ Soil-Venting Systems, *Groundwater Monitoring Rev.* Spring, p. 159.

CHAPTER 30

Innovative Biological Solid-Phase (or Land) Treatment of Petroleum Hydrocarbons (PHCS) at the United States Navy's Craney Island Fuel Terminal in Portsmouth, Virginia

Paul E. Flathman, OHM Remediation Services Corp., Findlay, Ohio

Gregory P. Hatchett, Department of the Navy, Atlantic Division, Naval Facilities Engineering Command, Norfolk, Virginia

Jeffrey C. Weatherly, OHM Remediation Services Corp., Glen Allen, Virginia

John H. Carson, Jr., OHM Remediation Services Corp., Findlay, Ohio

James M. Wrona, OHM Remediation Services Corp., Glen Allen, Virginia

Douglas E. Jerger, Paul R. Lear, Patrick M. Woodhull, OHM Remediation Services Corp., Findlay, Ohio

James L. Brown, OHM Remediation Services Corp., Trenton, New Jersey

INTRODUCTION

In 1993, the United States Navy constructed and first tested a 600 ft x 1,000 ft (180 x 300 m), 14-acre (5.6 hectare), solid-phase (or land) treatment facility at the Fleet and Industrial Supply Center (FISC), Craney Island Fuel Terminal in Portsmouth, Virginia. The facility was designed to biologically remediate large volumes of petroleum hydrocarbon (PHC)-contaminated soils obtained from Craney Island and other nearby Navy facilities. In 1994, under the U.S. Navy's Atlantic Division's (LANTDIV) Contract N62470-93-3032, Delivery Order 0016, OHM Remediation Services Corp. was contracted to remediate approximately 18,000 yd^3 (14,000 m^3) of PHC-contaminated soil transported to the facility from Site 13 at the Navy's FISC, Yorktown Fuels Division, Yorktown, Virginia.

Soil at the Yorktown facility had become contaminated with Fuel Oil 535. This was a heavy, special grade fuel oil that had been produced for the Navy by ARCO Petroleum Products Company. PHCs from tank still bottoms had contaminated loamy sand soil at the facility, and the PHCs had weathered in place for 14 years prior to excavation and treatment. Most of the weathered product was contained in hard, hydrophobic clods up to several inches (centimeters) in diameter. As determined by USEPA Method 418.1 (Total Recoverable

Petroleum Hydrocarbons, Spectrophotometric, Infrared), total PHC (TPH) concentration in the soil averaged 1,200 mg/kg. As specified in the Virginia Solid Waste Management Regulations, the cleanup level at the site was 50 mg/kg TPH. A risk-based treatment variance for industrial locations later established the cleanup level at 1,000 mg/kg TPH.

The project objectives were to reduce TPH concentration to 50 mg/kg in a 120-day treatment period, and to compare the effectiveness of three innovative treatment products to conventional treatment methods by being able to detect a 20% or greater difference in treatment with 80% confidence. Innovative treatment areas (ITAs) A (bacterial enzyme), B (selected bacterial mixture), and C (oleophilic nutrient/selected bacterial mixture) were constructed at the facility to evaluate three innovative products for increasing the rate of biological treatment of the PHCs. The efficacy of these innovations was compared to a conventional solid-phase control (Area D) after 120 days of treatment.

FIELD IMPLEMENTATION

Solid-phase (or land) treatment of hazardous waste is a managed technology that involves the controlled treatment of a waste in the upper soil zone. This treatment technology relies on the dynamic physical, chemical, and biological processes that result in the biodegradation, immobilization, or transformation of a hazardous waste to an environmentally acceptable level (Brown and Deuel, 1980; Cookson, 1995; LaGrega et al., 1994; Loehr et al., 1979). Solid-phase treatment has been widely used throughout the world to treat petroleum industry (ERT, 1983) and other industry wastes to acceptable levels (Parr et al, 1983; Sims et al., 1986a; Sims et al, 1986b).

In the 14-acre (5.6 hectare) treatment cell, the 18,000 yd^3 (14,000 m^3) of PHC-contaminated soil were spread to an average depth of 9.6 in.(24 cm). ITAs A (bacterial enzyme), B (selected bacterial mixture), and C (oleophilic nutrient/selected bacterial mixture) were constructed at the facility to evaluate three products for increasing the rate of biological treatment of the PHCs. The efficacy of these innovations was compared to a conventional solid-phase control (Area D) after 120 days of treatment. Twelve tons (11 metric tons) of diammonium phosphate (DAP), $(NH_4)_2HPO_4$, was initially added to the soil in all areas of the treatment cell as an immediately available source of nitrogen and phosphorus to support microbial growth on the PHCs. On Days 30, 60, and 90, additional 12-ton (11 metric-ton) quantities of DAP were also added to the entire site. A schematic of the treatment cell is presented in Figure 1. A Caterpillar SS-250 Soil Stabilizer and Reclaimer (Caterpillar, Inc.; Peoria, IL) and a Scat 482B Compost and Bioremediation Turner (Skat Engineering; Delhi, IA) were used to reduce the size of the clods, to mix the soil amendments, and to make the PHCs more available for aerobic biological treatment.

Figure 1. Schematic of the land treatment cell, and triangular grid sampling on Day 0.

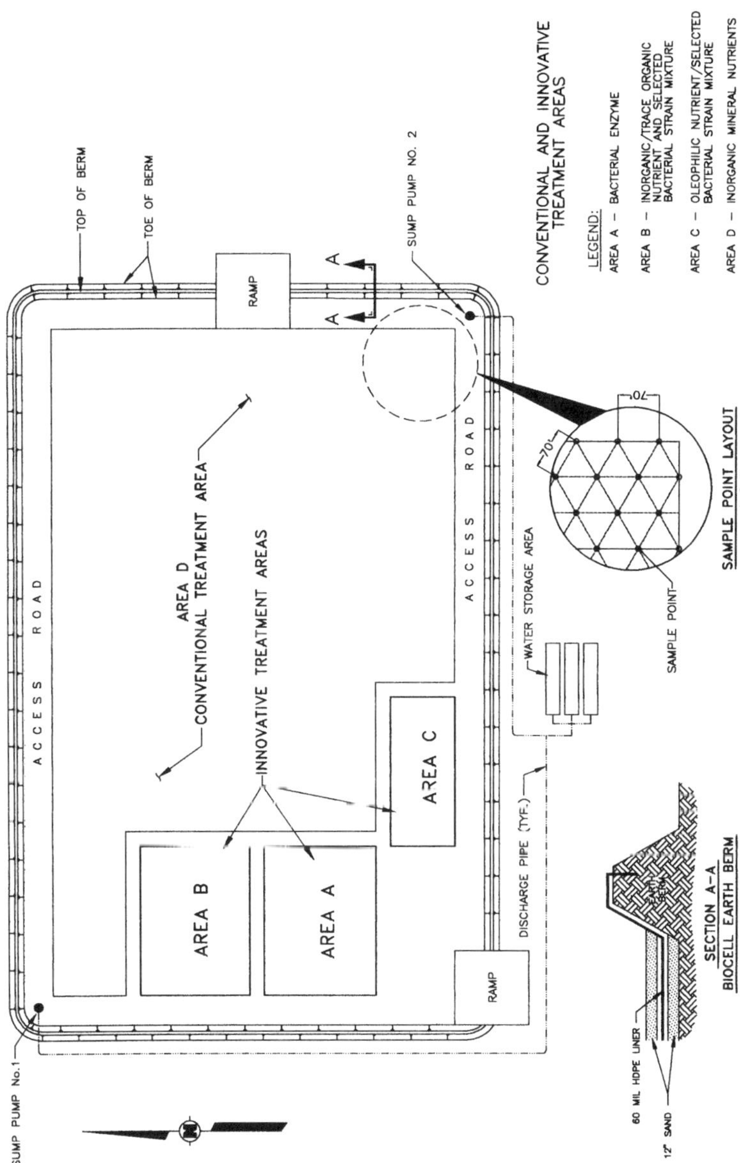

A sampling and analysis plan was developed which had the following objectives:

- to validate the attainment of the 50 mg/kg TPH cleanup level in soil at the site as required in the Virginia Solid Waste Management Regulations,
- to compare the effectiveness of three innovative treatment products to conventional treatment methods by being able to detect a 20% or greater difference in treatment with 80% confidence,
- to assure the maintenance of soil conditions conducive to enhanced microbial growth on PHCs, and
- to establish that leachate discharged from the treatment cell to the wastewater treatment plant at Craney Island would not adversely affect plant operations.

To meet those objectives, samples were analyzed according to USEPA (1983; 1986; 1992), American Society of Agronomy/Soil Science Society of America (ASA/SSSA) (Klute, 1986; Page, 1982) and American Public Health Association's *Standard Methods* procedures (Table 1) (APHA, 1985).

With the exception of ITA C, baseline sampling was performed on Day 0 prior to startup of the treatment cell. Interim samples were collected on Days 30, 60, and 90, and final samples were collected from the treatment cell on Day 120. In ITA C, baseline sampling was performed on Day 30. Innovative treatment in Area C was, therefore, evaluated over a 90-day period. Baseline samples in each treatment area were collected along an equilateral triangular grid (Figure 1) according to USEPA procedures (Barth and Mason, 1984; Mason, 1983). The maximum information from a fixed number of samples in a fixed area is obtained by maximizing the average distance between adjacent sample locations. The equilateral triangular grid affords the highest average spacing between sample locations and provides the maximum information about contaminant distribution.

In each ITA (i.e., Areas A, B, and C), a total of 13 composite samples were collected with each composite sample comprised of 10 sampling points. In the conventional treatment area (CTA), i.e., Area D, a total of 20 composite samples were obtained with each composite sample comprised of 6 sampling points. In addition, two duplicate samples were collected and analyzed for TPH. One duplicate sample was collected from the CTA, and a second duplicate sample was collected from one of the ITAs. Duplicate samples were generated by repeating the sample collection process for the preparation of one composite sample. Samples were collected using a 22 1/3 ft (6.8 m) spacing in each of the ITAs and a 70 ft (21 m) spacing within the CTA. Starting points and directional orientation were selected at random. Samples were collected using an AMS Model 317.04 soil auger 2 1/4 in.(5.7 cm) in diameter and 12 in.(30 cm) in length which included a 6.0 in.(15 cm) cylinder for containing the soil (Forestry Suppliers, Inc., Jackson, MS). Any clods found in composite samples were pulverized in the field, and each sample was thoroughly mixed. A representative subsample for analysis was obtained by successive coning and quartering until the desired sample volume was obtained. In addition to TPH, baseline samples were also analyzed for BETX, TOX, TOC, pH, available mineral nutrients, bacterial population density, and soil moisture content according to the analytical methods presented in Table 1.

When baseline sampling was completed on Day 0, the soil was placed into windrows, the sampling plan was modified, and interim sampling within the windrows occurred on Days 30, 60, and 90; final sampling occurred on Day 120 (Figure 2). Total windrow length was determined for both the ITAs and the CTA.

Table 1. **Analytical Methods for Monitoring Analyses**

Parameter	Analytical Method	
	Soil	Leachate
Total Petroleum Hydrocarbons (TPH)	USEPA 418.1	USEPA SW-846 M8015
Benzene, Ethylbenzene, Toluene, and o-, m-, and p-Xylene (BETX)	USEPA SW-846 8020	USEPA SW-846 8020
Oil and Grease	Not Analyzed	USEPA 413.1
Total Organic Halogens (TOX)	USEPA SW-846 9022	USEPA 450.1
Total Organic Carbon (TOC)	USEPA SW-846 9060	USEPA 415.1
pH	ASA/SSSA 12-2.6	USEPA 150.1
Available Mineral Nutrients		
Ammonium-Nitrogen (NH_4-N)	ASA/SSSA 33-3/ 33-4	USEPA 350.2
Nitrate-Nitrite-Nitrogen (NO_3-NO_2-N)	ASA/SSSA 33-3/ 33-4	USEPA 352.1
Phosphate-Phosphorus (PO_4-P)	ASA/SSSA 24-5.1/ 24-5.3	USEPA 365.2
Potassium (K)	ASA/SSSA 13-3.3.1.2/ 13-3.3.3	USEPA SW-846 6010
Aerobic Heterotrophic Bacterial Population Density	SM 9215 B	Not Analyzed
Soil Moisture Content	ASA/SSSA 21-2.2	Not Analyzed

From this total length, 12 and 40 ft (3.7 and 12 m) were subtracted from each ITA and from the CTA, respectively. The distance between each sampling point was calculated by dividing this length by the number of sampling points required. A total of 26 and 120 sampling points were required for each ITA and for the CTA, respectively. On Day 30, sampling respectively started at 3 and 10 ft (0.91 and 3.0 m) for each ITA and for the CTA. Sampling point locations on Days 60 and 90 were established by adding 3 and 10 ft, respectively (0.91 and 3.0 m), to each prior sampling point location for the ITAs and for the CTA. A total of 26 sample points at 19 ft (5.8 m) increments were established for each of the ITAs. For every 13 samples collected in each ITA, one composite sample was prepared, for a total of 2 composite samples in each ITA. A total of 120 sample points at 234 ft (71.3 m) intervals were established in the CTA. For every 20 samples collected in the CTA, one composite sample was prepared for a total of 6 composite samples. In addition, a duplicate sample was also collected from the CTA and analyzed for TPH. The duplicate sample was generated by repeating the sample collection process for the preparation of one composite sample. Vertical 6 in.(15 cm) samples were collected from the center of each 3 1/2 ft (1.0 m) high windrows by employing the soil auger used for baseline sample collection on Day 0.

Except for the number of composite samples prepared, final sampling on Day 120 consisted of utilizing the same number of sampling points within the ITAs and within the CTA. As on Days 0, 30, 60, and 90, a total of 26 and 120 sampling points were collected in each ITA and in the CTA, respectively. For every 2 samples collected within each ITA, one composite sample was prepared for a total of 13 composite samples. For every 6 samples collected within the CTA, one composite sample was prepared for a total of 20 composite samples. As with baseline sampling performed on Day 0, two duplicate samples were collected and analyzed for TPH. One duplicate sample was collected from the CTA, and a second duplicate sample was collected from one of the ITAs.

On a daily basis, soil moisture content in discrete samples collected from the treatment cell was determined using a Speedy® Moisture Tester (Soiltest; Lake Bluff, IL). If the soil moisture content was less than 10% on a wet-weight basis, additional water was provided to the soil through an irrigation system constructed at the site. On a weekly basis, pH and screening analyses for available mineral nutrients were performed on eight composite samples collected from the treatment cell. Two of the composite samples were from the ITAs, and the remaining six samples were from the CTA. Available mineral nutrient analyses (i.e., NH_4-N, NO_3-NO_2-N, and PO_4-P) were performed colorimetrically by nesslerization, cadmium reduction, and ascorbic acid, respectively, following aqueous extraction of nitrogen, acid-fluoride extraction of phosphorus, and flocculation of the solids (Hach Company, Loveland, CO). On Days 30, 60, 90, and 120, in addition to TPH, samples were also analyzed for BETX, TOX, TOC, pH,

Figure 2. Schematic of windrows constructed in the land treatment cell for sampling on Days 30, 60, 90, and 120.

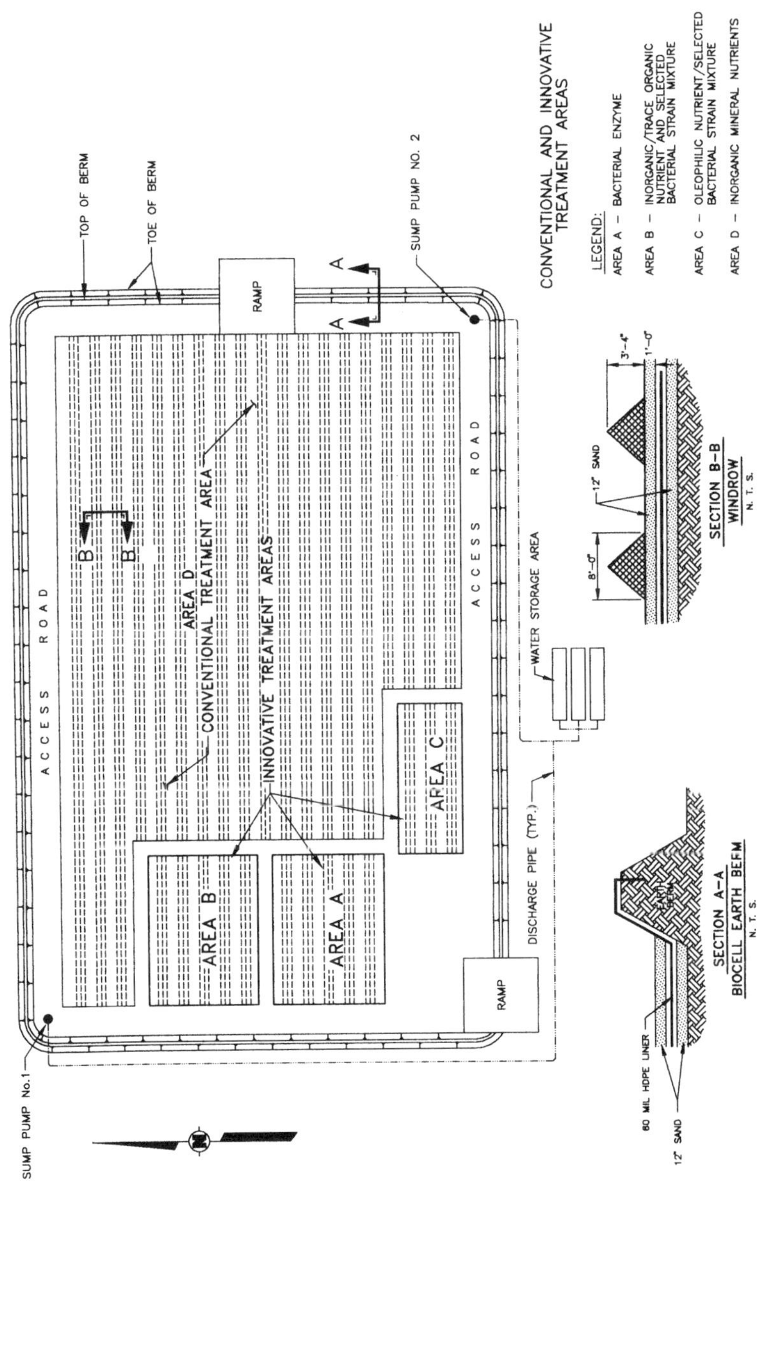

available mineral nutrients, bacterial population density, and soil moisture content according to the analytical methods presented in Table 1.

RESULTS AND DISCUSSION

BETX and TOX were either not detected or were detected at concentrations below regulatory concern in the soil and in the two leachate samples analyzed. In one leachate sample, oil and grease was quantified at a concentration of 130 mg/L. In general, other compounds analyzed in the leachate samples (Table 1) were either not detected or were detected at such low levels that wastewater treatment plant operations at Craney Island would not be affected. At an initial concentration of 1,200 mg/kg, TPH in soil generally accounted for only a small percentage (i.e.,~11%) of the TOC quantified (i.e., 11,000 ± 3,800 mg/kg, $x \pm s$, $n = 7$). Soil pH remained within an acceptable range (i.e., pH 6.9 to pH 8.1, geometric mean (GM) = 7.6, n = 50) for biological treatment. Mineral nutrients were always available to support microbial growth on available organics. In the 50 samples analyzed, GMs for NH_4-N, NO_3-NO_2-N, PO_4-P, and K concentrations (and concentration ranges) in the soil were, respectively, 11 (4 to 81), 30 (4 to 430), 22 (8 to 220), and ~47 (<20 to 130) mg/kg. Bacterial population density remained relatively high (i.e., 10^5 to 10^7 cfu/g dry-weight, GM = 1.9×10^6 cfu/g dry-weight) in the samples analyzed. These analyses were performed prior to each of the four 12-ton (11 metric-ton) applications of DAP to the treatment cell which would increase the NH_4-N concentration in the soil by ~20 mg/kg.

Performance data have indicated that biodegradation of most chemicals in soil can be modeled using a first-order reaction rate or a mixture of first-order rates. First-order kinetics generally apply where the available concentration of the chemical being degraded is low relative to the biological activity in the soil (Carson and Gupta, 1993). Biological treatment of TPH was modeled according to the following integrated form of the first-order model:

$$C(s,t) = C(s,0)e^{-k(s)t}$$

where

$C(s,t)$ = average TPH concentration (mg/kg) in treatment area s at time t (days)

$C(s,0)$ = average TPH concentration (mg/kg) in treatment area s at time 0 (days)

$k(s)$ = first-order rate constant (day^{-1}) for area s

t = time (days)

s = treatment area.

The first-order model was fit to the TPH data using the method of generalized linear interactive modeling (GLIM) for the purpose of estimating initial

TPH concentrations and first-order rate constants for each area (Statistical Sciences, 1994; Venables and Ripley, 1994; Dobson, 1983). The above first-order model is linear in the log scale and, including errors, can be expressed according to the following equation:

$$\ln[C(s,t)] = \ln[C(s,0)] + [-k(s)]\cdot t = K_0(s) + K_1(s)\cdot t + \varepsilon$$

where

$K_0(s) + K_1(s)\cdot t$ is the linear predictor

ε is a random variable with mean 0, corresponding to random error

"ln" is the link function which links the linear predictor with the predicted value in the scale of the observations.

GLIM models are a generalization of ordinary linear models that can incorporate several features often found in real modeling situations, namely:

- nonlinear (but transformable to linearity) relationships between independent and dependent variables
- nonconstant error variance which is a function of the conditional expected value of the independent variable
- possibly nonnormal error distribution, even for studentized residuals.

A GLIM model, rather than ordinary nonlinear least squares regression, was required, because the error variance (i.e., variance of random deviations of observed concentrations from the "true" values) in TPH data was not constant, but was an increasing function of average TPH concentration (Figure 3). This phenomenon is quite common in environmental and other earth sciences data, particularly when the data are collected over a spatial domain (Carson and Gupta, 1993; Cressie, 1991), and is sometimes referred to as the "proportional effect" (Isaaks and Srivastava, 1989).

Table 2 presents estimates of the contrasts in the first-order rates of treating the PHCs in the ITAs compared to the CTA control (Area D). Although the fitted first-order rate constants for ITAs A and B were slightly higher than that for the CTA, the differences were not statistically significant at the 0.01 level. The *p*-values for the rate differences for ITAs A and B were 0.13 and 0.31, respectively. The rate contrast between ITA C and the CTA was highly significant ($p = 2.7 \times 10^{-6}$). The Z-scores presented in Table 2 have a distribution closely approximated by that of a standard normal deviate. Their values are referred to tables of the standard normal distribution for the purpose of estimating their associated *p*-values (i.e., the probability that a Z-score as large or larger could be generated by chance under the null hypothesis of no difference in rates).

Based on initial results, the model was refit by assuming the same rate of TPH degradation (but different initial TPH concentrations) for ITAs A and B as for the CTA. Random variation was modeled as having a distribution symmetric about 0, and a standard deviation proportional to the TPH concentra-

Figure 3. **Standard deviation versus sample mean of TPH concentration grouped by treatment area and sampling date with the area of each circle proportional to the square root of the number of samples.**

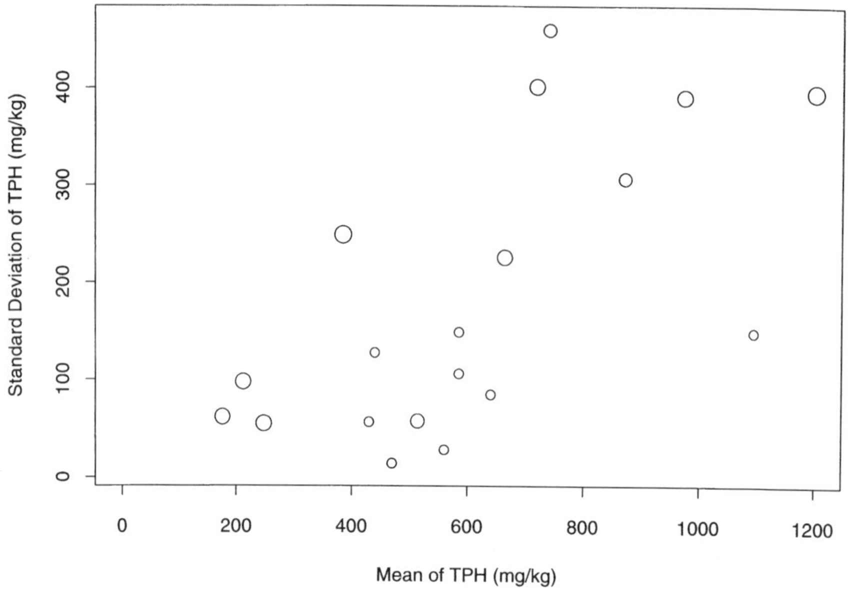

Table 2. **Comparison of the First-Order Rates of Biological Treatment in the ITAs Compared to the CTA**

Treatment Area	Treatment	Difference in Rates (Day-1)	Standard Error	Z-Score	p-Value
A	Bacterial enzyme	0.00206	0.00184	1.12	0.131
B	Inorganic / trace organic nutrient and selected bacterial strain mixture	0.00091	0.00184	0.496	0.310
C	Oleophilic nutrient / selected bacterial strain mixture	0.00989	0.00218	4.56	0.000002744

tions predicted by the model. Studentized residuals from a GLIM model are ordinary residuals divided by the standard error of prediction and corrected for transformation bias. Studentized residuals from this model were plotted against model predictions for the purpose of assessing whether the error variance was correctly modeled, and whether higher order model terms might be required (Figure 4). Inspection of Figure 4 indicated the absence of any trends which would indicate lack of fit. The studentized residuals were distributed symmetrically about the zero line, and their dispersion appeared to be independent of the predicted values.

Studentized residuals versus normal quantiles were plotted to assess the shape of the error distribution (Figure 5). The studentized error distribution appeared to be symmetric (but not necessarily normal) except for the extreme tails, which exhibited some skewness to the right. Extreme points on the plots in Figures 4 and 5 were labeled according to their corresponding area. Figures 4 and 5, therefore, indicated that there were no patterns which would suggest that the model was deficient. The mean residual deviance (analogous to the mean squared error) for this model was 0.162, which indicated good explanatory power.

The estimated first-order rate for ITA C was approximately twice that of the pooled estimate rate for ITAs A and B and for the CTA (Table 3). Test results for TPH and first-order model predictions with 95% confidence limits for ITAs A, B, and C and for CTA D as a function of time are presented in Figures 6 through 9, respectively. As a result of delays in initial sampling of ITAs A and B and in laboratory analysis of samples from those areas, as presented in Figures 6 and 7, initial sample collection was considered to be on Day 7. The delay in laboratory analysis was accounted for by applying the rule that biological reaction rates approximately double or halve for each respective 10°C increase or

Figure 4. **Diagnostic checking of the first-order model: studentized residuals versus fitted values predicted by the first-order model to assess stabilization of variance and appropriateness of the final model.**

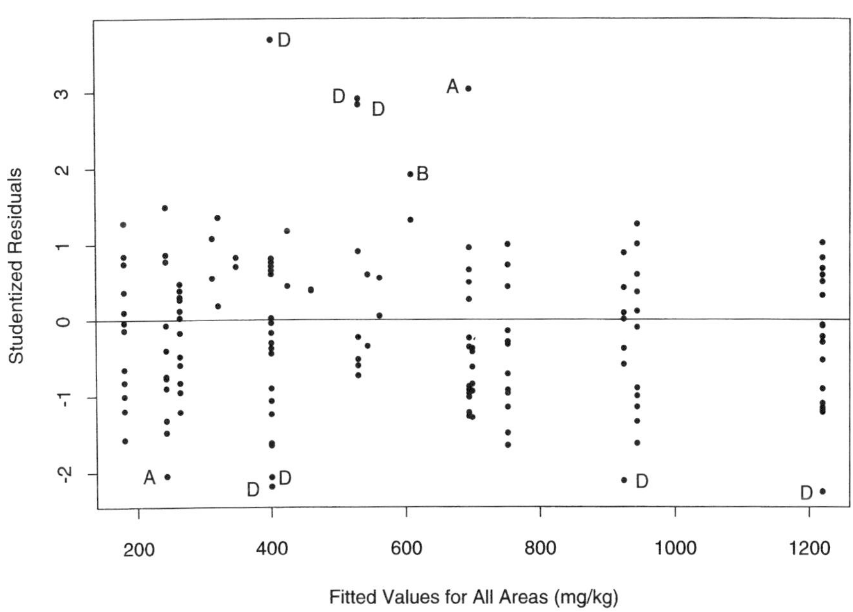

Figure 5. **Diagnostic checking of the first-order model: studentized residuals versus normal quantiles to assess shape of the error distribution.**

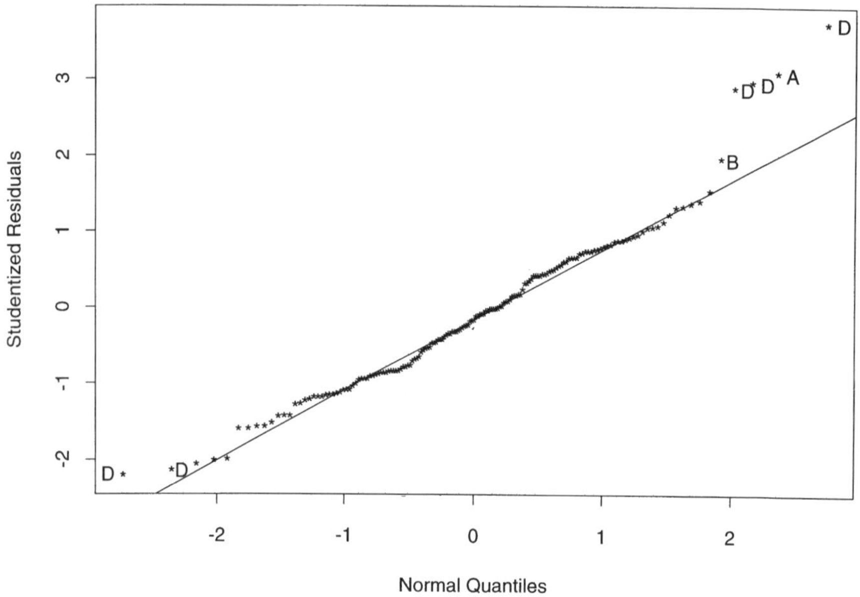

Table 3. **First-Order Rates of Biological Treatment in the Innovative and Conventional Land Treatment Areas**

Treatment Area	Treatment	K (Day-1)	Standard Error	95% Lower Confidence Limit	95% Upper Confidence Limit
A	Bacterial enzyme	0.00928	0.000803	0.0077	0.0109
B	Inorganic / trace organic nutrient and selected bacterial strain mixture				
D	Conventional land treatment				
C	Oleophilic nutrient / selected bacterial strain mixture	0.0184	0.00146	0.0155	0.0213

Figure 6. **TPH concentrations and first-order model prediction with 95% confidence limits presented as a function of time for ITA A, the bacterial enzyme.**

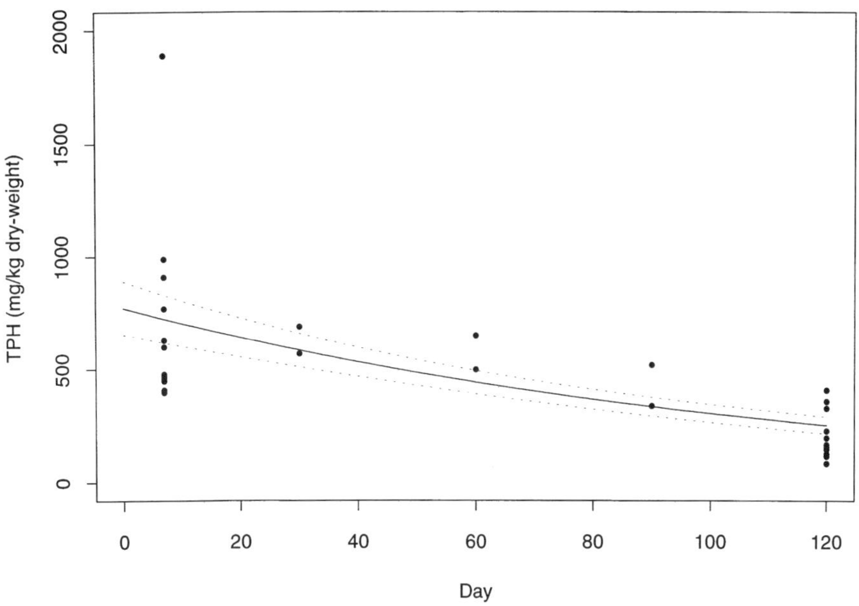

Figure 7. **TPH concentrations and first-order model prediction with 95% confidence limits presented as a function of time for ITA B, the inorganic / trace organic nutrient and selected bacterial strain mixture.**

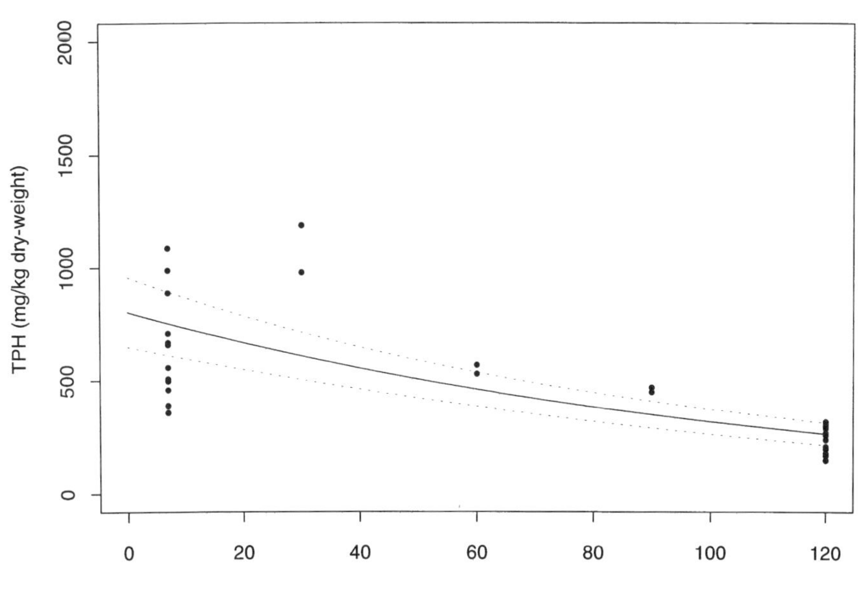

Figure 8. **TPH concentrations and first-order model prediction with 95% confidence limits presented as a function of time for ITA C, the oleophilic nutrient/selected bacterial strain mixture.**

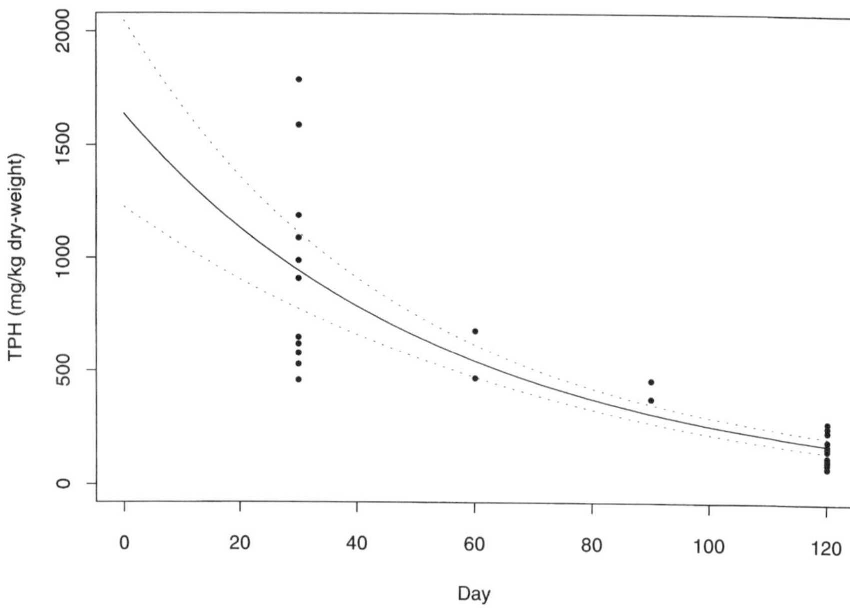

Figure 9. **TPH concentrations and first-order model prediction with 95% confidence limits presented as a function of time for the CTA.**

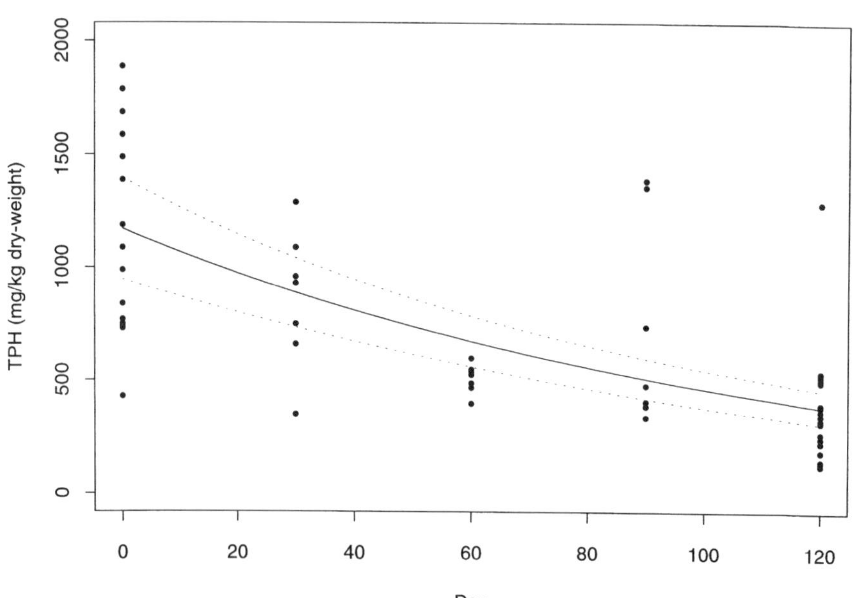

decrease in temperature. The rule was applied to the difference between soil temperature in the treatment cell and the temperature in the laboratory's sample storage refrigerator. The correction was somewhat conservative in that it might have slightly inflated the estimated first-order rates for ITAs A and B. Other than this, the comparison of first-order rates was not affected by the difference in time for initial sample collection in each area.

Table 4 presents the percent reduction for each area, based on the final model and on the results of the initial and final sampling events. The extent of PHC treatment in ITA C was substantially greater than in any of the other areas. With greater than an 80% reduction in PHC concentration in ITA C after 90 days of treatment compared to 67% reductions in the other areas after 120 days of treatment, the benefit of the oleophilic nutrient/selected bacterial mixture for remediation of PHC-contaminated soil was demonstrated.

The oleophilic nutrient/selected bacterial strain mixture evaluated in ITA C was developed by Eugene Rosenberg and others at Tel-Aviv University in Tel-Aviv, Israel, through the financial support of Makhteshim Chemical Works Ltd (Beer-Sheva, Israel) (Rosenberg et al., 1992). The product, known as System E.T. 20 (formerly MCW.B 20) is listed on the USEPA's National Contingency Plan (NCP) Product Schedule. The oleophilic nutrient, known as F-1, is a modified urea-formaldehyde polymer containing 18% N and 10% P as P_2O_5. The selected bacterial mixture in System E.T. 20, obtained using enrichment culture techniques, are those organisms that used crude oil as a carbon and energy source for growth, and used F-1 as the source of nitrogen and phosphorus to support that growth. The selected microflora and the F-1 adhered to the oil/water interface, as observed microscopically and by the fact

Table 4. **TPH Reduction in the ITAs and in the CTA**

			Reduction in TPH (Percent)	
Treatment Area	Treatment	Treatment Time (Day Number)	Estimated From Initial and Final Data	Estimated From Model
A	Bacterial enzyme	0 to 120	69.7	67.1
B	Inorganic/trace organic nutrient and selected bacterial strain mixture	0 to 120	62.8	67.1
C	Oleophilic nutrient/ selected bacterial strain mixture	30 to 120	80.8	80.9
D	Conventional land treatment	0 to 120	67.1	67.1

that crude oil degradation proceeded even when the water phase was removed and replaced seven times with unsupplemented water (i.e., in a simulated open system). In a field trial performed on an oil-contaminated sandy beach between Haifa and Acre, Israel, in the summer of 1991, 86% degradation of pentane extractables was reported after 28 days of enhanced biological treatment, compared to 15% in an untreated control.

CONCLUSIONS

Innovative treatment in Area C (oleophilic nutrient/selected bacterial strain mixture) was more effective in treating TPH than innovative treatment in Areas A and B and conventional treatment in Area D. In ITA C, ~80% reduction in TPH was obtained over a 90-day treatment period (i.e., in only 75% of the time allowed for the other treatments). In Areas A, B, and D, ~67% reduction in TPH was obtained over a 120-day treatment period. The first-order rate of biological treatment of TPH in ITA C was about twice that obtained in the other areas. The effectiveness in treating TPH in ITAs A and B was not significantly different than in treating TPH in CTA D.

ACKNOWLEDGMENTS

The authors would like to acknowledge Paul M. Cavanaugh, Field Chemist (OHM), for his onsite role in efficiently guiding project operations.

REFERENCES

APHA (American Public Health Association), American Water Works Association, and Water Pollution Control Federation. 1985. *Standard Methods for the Examination of Water and Wastewater*, 16th ed. Washington, DC, American Public Health Association.

Barth, D.S. and Mason, B.J. 1984. Soil Sampling Quality Assurance User's Guide, EPA-600/4-84-043. Environmental Monitoring Systems Laboratory, U.S. Environmental Protection Agency, Las Vegas, NV.

Brown, K.W. and Deuel, L. 1980. Hazardous Waste Land Treatment, SW-874. Municipal Environmental Research Laboratory, Cincinnati, OH, U.S. Environmental Protection Agency.

Carson, J.H., Jr., and Gupta, A.K. 1993. Multivariate Non-Normal Statistics in Site Characterization and Evaluation. In: *Multivariate Environmental Statistics*, pp. 75-98. (Patil, G.P. and Rao, C.R., Eds.), New York, Elsevier Science Publishers B.V.

Cookson, J.T., Jr. 1995. *Bioremediation Engineering: Design and Application.* New York, McGraw-Hill, Inc.

Cressie, N.A.C. 1991. *Statistics for Spatial Data.* New York, John Wiley & Sons.

Dobson, A.J. 1983. *Introduction to Statistical Modeling.* New York, Chapman and Hall.

ERT (Environmental Research and Technology). 1983. Land Treatment Practices in the Petroleum Industry. Washington, DC, American Petroleum Institute.

Isaaks, E.H. and Srivastava, R.M. 1989. *Applied Geostatistics.* New York, Oxford University Press.

Klute, A., Ed. 1986. *Methods of Soil Analysis, Part 1, Physical and Mineralogical Methods*, 2nd Edition. Madison, WI, American Society of Agronomy and Soil Science Society of America.

LaGrega, M.D., Buckingham, P.L., and Evans, J.C. 1994. *Hazardous Waste Management.* New York, McGraw-Hill, Inc.

Loehr, R.C., Jewell, W.J., Novak, J.D., Clarkson, W.W., and Friedman, G.S. 1979. *Land Application of Wastes*, Volumes 1 and 2. New York, Van Nostrand Reinhold Company.

Mason, B.J. 1983. Preparation of Soil Sampling Protocol: Techniques and Strategies, EPA-600/4-83-020. Environmental Monitoring Systems Laboratory, U.S. Environmental Protection Agency, Las Vegas, NV.

Page, A.L., Ed. 1982. *Methods of Soil Analysis, Part 2, Chemical and Microbiological Properties*, 2nd Edition. Madison, WI, American Society of Agronomy and Soil Science Society of America.

Parr, J.F., Marsh, P.B., and Kla, J.M., Eds. 1983. *Land Treatment of Hazardous Wastes.* Park Ridge, NJ, Noyes Data Corporation.

Rosenberg, E., Legmann, R., Kushmaro, A., Taube, R., Adler, E., and Ron, E. 1992. Petroleum Bioremediation - A Multiphase Problem. *Biodegradation* 3:337-350.

Sims, R.C., Sims, J.L., Sorensen, D.L., Doucette, W.J., and Hastings, L.L. 1986a. Waste/Soil Treatability Studies for Four Complex Industrial Wastes: Methodologies and Results, Vol. 1, Literature Assessment, Waste/Soil Characterization, Loading Rate Selection, EPA/600/6-86/003a. Robert S. Kerr Environmental Research Laboratory, Office of Research and Development, U.S. Environmental Protection Agency, Ada, OK.

Sims, R.C., Sims, J.L., Sorensen, D.L., Doucette, W.J., and Hastings, L.L. 1986b. Waste/Soil Treatability Studies for Four Complex Industrial Wastes: Waste Loading Impacts on Soil Degradation, Transformation, and Immobilization, EPA/600/6-86/003b. Robert S. Kerr Environmental Research Laboratory, Office of Research and Development, U.S. Environmental Protection Agency, Ada, OK.

Statistical Sciences, Inc. 1994. S-PLUS for Windows, Version 3.2 Supplement. Seattle, WA: StatSci: a Division of MathSoft, Inc.

U.S. Environmental Protection Agency. 1983. Methods for Chemical Analysis of Water and Wastes, EPA-600/4-79-020. Environmental Monitoring and Support Laboratory, Office of Research and Development, Cincinnati, OH.

U.S. Environmental Protection Agency. 1986. Test Methods for Evaluating Solid Waste, Physical/Chemical Methods, 3rd Edition, SW-846. Office of Solid Waste and Emergency Response, Washington, DC.

U.S. Environmental Protection Agency. 1992. Test Methods for Evaluating Solid Waste, Physical/Chemical Methods, 3rd Edition, SW-846, Update 1. Office of Solid Waste and Emergency Response, Washington, DC.

Venables, W.N. and Ripley, B.D. 1994. *Modern Applied Statistics with S-Plus.* New York, NY: Springer-Verlag.

CHAPTER 31

Bioventing in Soil Piles to Augment Landfarming of Hydrocarbon-Impacted Soil

Marilyn K. Guthrie, ARCO Products Company, 4 Centerpointe Drive, Suite 100, La Palma, CA 90623

Terence R. Regan, **Michael T. Paczkowski**, and **Derek Sharp**, Geraghty & Miller, Inc., One Corporate Drive, Andover, MA 01810

INTRODUCTION

ARCO Products Company operated a terminal in East Providence, Rhode Island, between the 1920s and the mid-1980s, when the terminal was used for the bulk storage of gasoline, kerosene, and fuel oils. Total petroleum hydrocarbon (TPH) concentrations as high as 10,000 mg/kg and total benzene, toluene, ethylbenzene, and xylene (BTEX) concentrations as high as 17 mg/kg in soil have been detected at the site. Since 1988, landfarming using a tractor-mounted, horizontal auger has been the primary remedial technology used to treat the hydrocarbon-impacted soil to reach the cleanup criteria of 50 mg/kg for TPH and 0.25 mg/kg for BTEX. Generally, excavation and stockpiling were the primary construction activities during the winter, and landfarming and backfilling during the summer. In an effort to increase productivity and extend the soil treatment season, plans were prepared to construct a low-cost soil pile to biovent soil stockpiles during the winter months, beginning in late 1992.

The decision to construct soil piles was based on (1) the desire to maintain progress and momentum at the site throughout the year, rather than mobilizing and demobilizing each season; (2) the hypothesis that ambient temperature is the limiting factor for landfarming, but oxygen is the limiting factor in a soil pile over the winter, since the pile's mass would regulate temperatures within the pile enough to sustain biological activity; and (3) the cost of installing and maintaining bioventing equipment is low, and the potential benefits are great, relative to the cost of either constructing an ordinary stockpile, or demobilizing during the winter. The availability of space at the site for landfarming, excavating, and/or stockpiling soil at any given time was also considered, so that a balance was maintained between untreated soil available for remediation, and treated soil available for backfilling and site grading.

OBJECTIVES

The objectives of the soil-pile construction in the ARCO Kettle Point remediation program are to (1) maintain progress toward site closure throughout the year, including the winter months when landfarming is not feasible; (2) extend the soil-treatment season from the current 6- or 7-month season; and (3) provide pretreatment to the soil, thereby reducing the time needed to landfarm the soil to reach the cleanup criteria during the subsequent summer months.

Accomplishing these objectives allows treatment of a greater volume of soil in a given time period, relative to landfarming alone. This results in a lower cost per yard for soil treatment, an overall reduction in remediation costs, and a reduction in the time required to complete the entire soil remediation program at the site.

SOIL-PILE CONSTRUCTION

Construction of soil pile ESP92-1 began in September 1992 using impacted soil excavated from three former tank areas at the site. Pile design included three adjacent modules, approximately 15 ft (4.6 m) high, and 110 ft (33 m) by 185 ft (56 m) at the base, each with its own blower and offgas treatment system (Figure 1). In Modules 1 and 2, a series of slotted pipes installed horizontally at the 5 ft (1.5 m) and 10 ft (3 m) elevations within the pile were connected via a manifold to a 5-horsepower regenerative blower. The blowers extracted air from one layer of pipes in the pile, and reinjected the extracted air into the other layer. The reinjected air was heated by friction as it passed through the blower. Module 3 was constructed with vertical slotted injection and extraction points manifolded together on top of the pile and connected to the blower. Quick-connect fittings were used to facilitate reversing the injection/extraction flow direction within the pile.

Approximately 5 to 10 % of the air extracted from the pile was treated by the offgas treatment system, consisting of two vapor-phase carbon canisters, prior to discharge into the atmosphere. Accordingly, a similar amount of make-up air was obtained from the atmosphere and injected into the pile. This exchange of ambient air was used to replenish oxygen used by the microorganisms within the pile during biodegradation. Based on their review of the soil pile design, the Rhode Island Department of Environmental Management (RIDEM) determined that no written permits were required to construct or operate the soil pile. Operational parameters, including oxygen, carbon dioxide, soil temperature, and pressure/vacuum, as well as organic vapor concentrations, were periodically measured in the field to monitor pile performance.

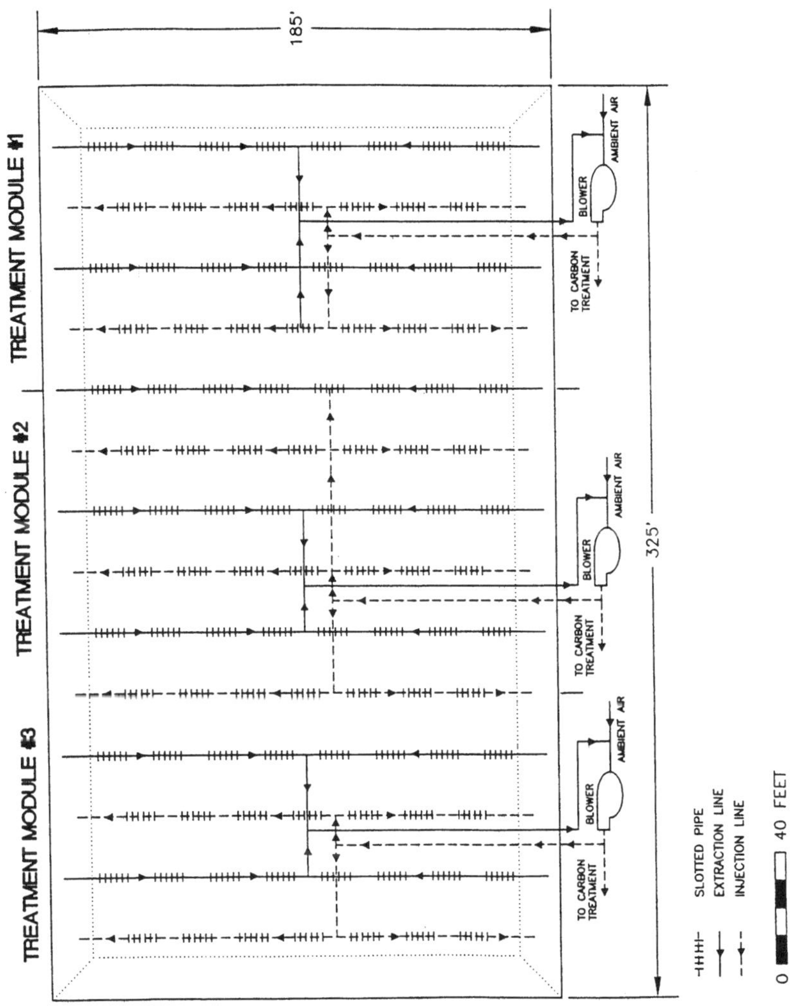

Figure 1. **Engineered soil pile ESP92-1 plan view.**

When completed, ESP92-1 contained approximately 42,750 yd^3 (32,700 m^3) of soil in the three modules. Monitoring data collected during operation of ESP92-1 was used to assess the level of biodegradation activity within the pile, and adjust the air injection and extraction accordingly. On average, all three modules showed relatively high oxygen levels and relatively low carbon dioxide levels while the blowers were operating, but when the blowers were shut down, the oxygen levels slowly decreased while a corresponding increase in the carbon dioxide levels was observed. Pressure or vacuum was also measured at the monitoring points installed between the injection and extraction lines to monitor airflow within the pile.

SOIL PILE PERFORMANCE AND RESULTS

In the spring and early summer of 1993, dismantling of ESP92-1 began concurrently with the startup of the 1993 landfarming season. To assess soil pile performance, preliminary indicator soil samples were collected from a portion of Module 1 of ESP92-1 for analysis of TPH and BTEX. These results were then compared to the concentrations measured in soil samples collected as Module 1 of ESP92-1 was being constructed. The comparison showed that TPH and BTEX concentrations in that portion of Module 1 of ESP92-1 had decreased substantially during pile operation, and that concentrations in 2,700 yd^3 (2,060 m^3) of the 2,900 yd^3 (2,220 m^3) sampled during the indicator sampling round were below the cleanup criteria. Due to these favorable results, Modules 2 and 3 were operated longer than originally planned, and in August 1993, compliance soil samples were collected directly from the remainder of Module 1 and from all of Module 2 of the soil pile. Samples were not collected from Module 3 until later in the year because Module 3 was constructed and started operation later than Modules 1 and 2.

By collecting compliance samples from the soil within the soil pile rather than sampling the soil after moving it into treatment cells for landfarming, handling costs were minimized. Compliance samples were collected from within the soil pile by subdividing the pile into cells containing 200 yd^3 (153 m^3) or less (the compliance sampling-frequency requirement), drilling a boring through the center of the cell from the top of the pile to the base, and collecting a composite soil sample from the boring for analysis of BTEX and TPH. The soil that had reached the cleanup criteria was moved into an excavation area for backfilling and compaction; soil with BTEX or TPH levels still above the cleanup criteria was moved into a treatment cell for additional treatment via landfarming. After soil pile operation for approximately 187 days, BTEX and TPH concentrations in Modules 1 and 2 had decreased by 94 and 49 %, respectively, and approximately 70 % of the soil analyzed contained BTEX and TPH concentrations below their respective cleanup criteria (Figures 2 and 3).

Overall, the soil that was "pretreated" in soil piles showed faster degradation rates for TPH than soil that was not pretreated in a soil pile. Figure 4 is a plot of the ratio of TPH concentration at a given time to the initial TPH concentration (Ct/C0) versus treatment time in days. As shown on Figure 4, soil that was moved directly from an excavation into a treatment cell for landfarming (i.e., nonpretreated soil) had slower TPH degradation rates than the soil that was pretreated in a soil pile and then landfarmed. The half-life for pretreated soil was approximately 8 days, compared to a half-life of approximately 17 days for the nonpretreated soil. Although no microbial counts were measured in any of the soil, the faster landfarming treatment times observed in the pretreated soil may be due to the establishment of microorganisms within the soil pile before the soil was loaded into treatment cells for landfarming. By comparison, soil placed into treatment cells for landfarming directly from an excavation may not have had the same opportunity for the microorganisms to establish growth prior to the start of landfarming.

BENEFITS AND COST SAVINGS

The performance of soil pile ESP92-1 exceeded the primary objective for building the soil pile, which was to pretreat the impacted soil excavated during the winter of 1992-1993 via bioventing to decrease the time needed to remediate the soil via landfarming the following summer. Soil samples collected from ESP92-1 following bioventing showed that approximately half of the soil sampled had TPH and BTEX concentrations below the cleanup criteria. The performance of the soil pile resulted in cost savings that have helped reduce the overall soil

Figure 2. **Decrease in BTEX concentrations (soil pile ESP92-1).**

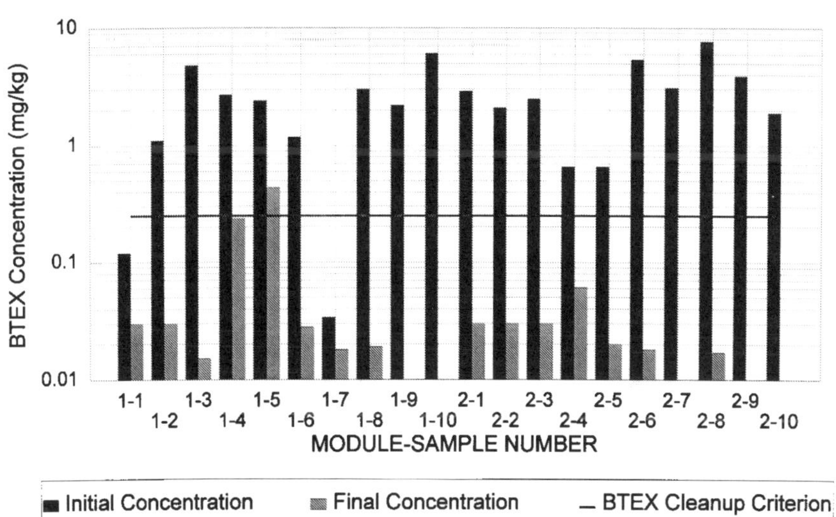

Figure 3. **Decrease in TPH concentrations (soil pile ESP92-1).**

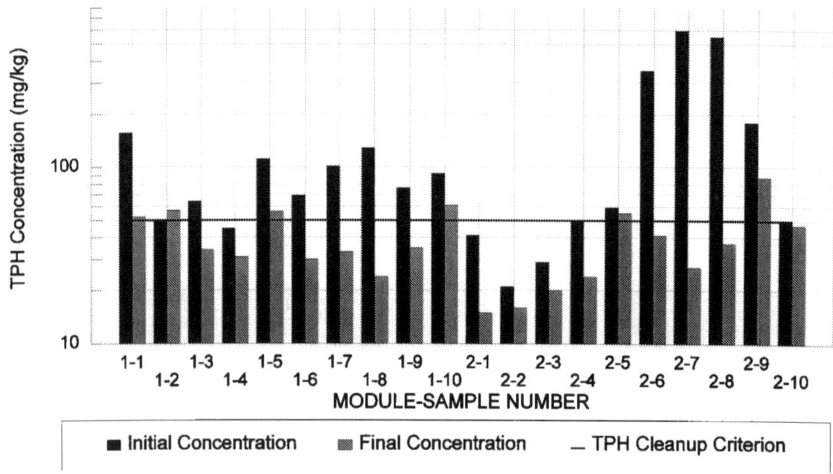

remediation cost to approximately \$42/yd^3. Some of the benefits and/or cost savings realized from soil pile construction and operation are as follows:

1. Soil that reached the cleanup criteria in the soil pile did not have to be loaded into treatment cells and landfarmed, then unloaded into an excavation for backfilling and compaction. Instead, the soil was moved directly from the soil pile into an excavation for backfilling, saving soil-handling costs and also the cost of operation and maintenance of the machinery for tilling the soil. Reduced soil handling saved between \$165,000 and \$215,000 in contractor costs.
2. Due to favorable monitoring results, bioventing of soil in ESP92-1 continued through July 1993, rather than moving the soil into treatment cells for landfarming as originally planned. Between April and October 1993, excavation continued in areas originally scheduled for excavation beginning in November 1993, putting the excavation work and the entire remediation program ahead of schedule. The decreased remediation time resulted in an estimated \$150,000 to \$225,000 in savings in reduced project management, construction management, database management, update/status meetings, and site monitoring activities.
3. The impacted soil excavated during the summer of 1993 was moved directly into treatment cells for landfarming, rather than into an untreated soil stockpile, saving soil handling costs. Reduced soil handling saved between \$100,000 and \$150,000 in contractor costs.
4. The area excavated during the summer of 1993 was used to backfill treated soil directly from landfarming operations, eliminating the need to construct a 16,300 yd^3 (12,500 m^3) stockpile of treated soil as originally planned to maintain a balance between availability of treated and untreated soil. The reduced soil handling saved approximately \$57,000 in contractor costs.

Figure 4. Comparison of landfarming TPH degradation rates for pretreated versus nonpretreated soil.

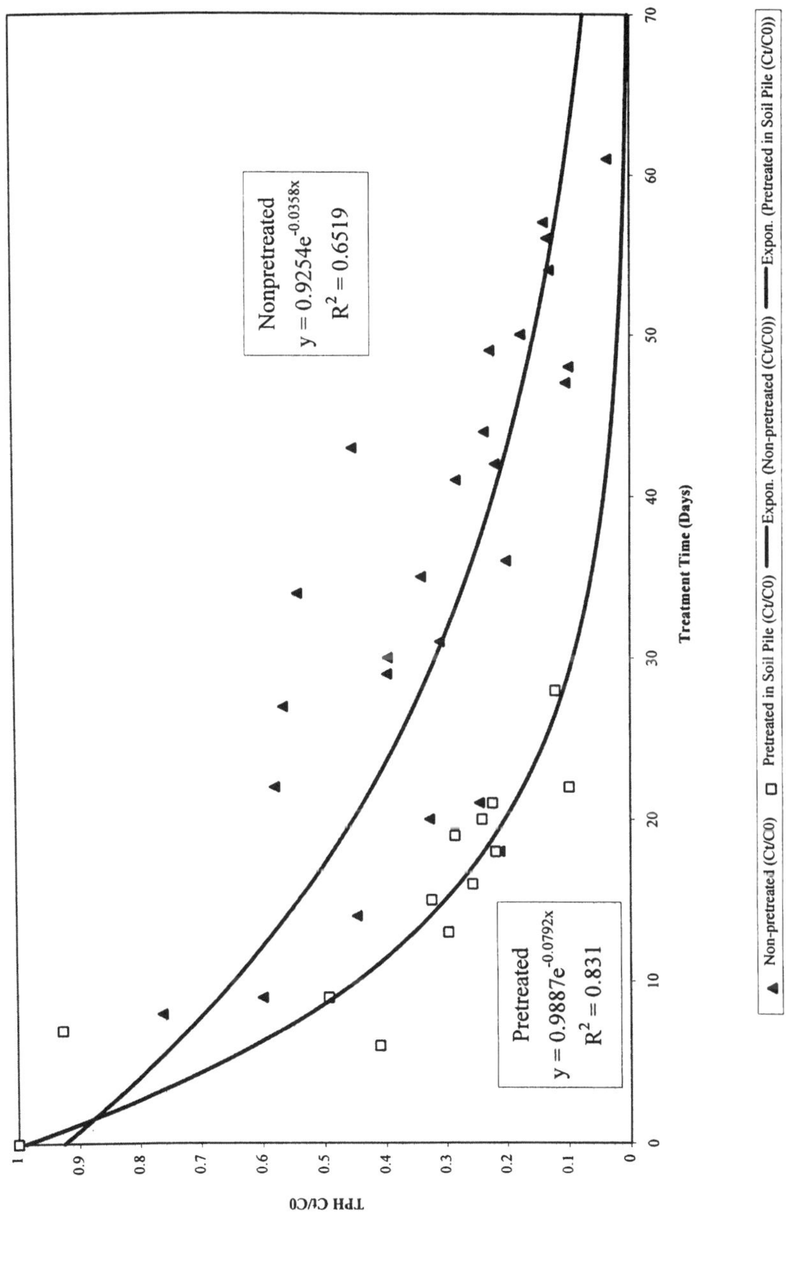

The costs for constructing and operating soil pile ESP92-1 are estimated at approximately $150,000, including capital costs for equipment and piping, contractor costs for soil pile construction, and operation and monitoring costs for one year. As described above, the total cost savings realized from bioventing in soil piles in 1992 is estimated between $472,000 and $647,000. Therefore, the net cost savings realized as a result of having constructed and operated soil pile ESP92-1 is between $325,000 and $497,000. Over the course of site remediation, it is estimated that bioventing in soil piles will save between $1M and $1.5M, and decrease the time required to remediate the site by as much as 3 years.

CONCLUSIONS

The positive results and overall cost savings achieved using bioventing in soil piles to augment landfarming at Kettle Point demonstrate the effectiveness of combining these two treatment technologies to implement a low-cost, highly efficient ex-situ soil bioremediation program. For landfarming programs conducted in cold-weather climates, bioventing in soil piles in the winter months can reduce TPH and BTEX levels significantly. TPH and BTEX levels in a large fraction of the soil placed into soil piles at ARCO Kettle Point reached the cleanup criteria. For the remainder that did not reach the cleanup criteria in the soil pile, observed treatment times were faster during landfarming compared to soil that had not been pretreated in a soil pile prior to landfarming. The decreased landfarming time allowed processing and throughput of a larger volume of soil in a given time, thereby reducing overall treatment time and cost to approximately $42/yd^3$. Enhanced treatment combining bioventing and landfarming techniques at Kettle Point is estimated to result in a total cost savings of up to $1.5M, and shorten remediation time by up to 3 years, compared to landfarming alone.

CHAPTER 32

A Basic Biocell Approach for Small-Scale Petroleum-Contaminated Soil Projects

Susan M. Lasdin and **Christopher M. O'Neill**, New York State Department of Environmental Conservation, Division of Spills Management, Albany, New York

BACKGROUND

As the number of petroleum-contaminated sites requiring cleanup in New York state increases, so does the desire for better, faster, and more cost-effective ways to investigate and remediate these sites. This situation has become a more significant concern as the option of landfilling petroleum-contaminated soils has decreased due to the closing of more and more landfills, especially to petroleum-contaminated media, and because landfilling is becoming more costly for the few which remain in operation. The New York State Department of Environmental Conservation (DEC) has been investigating alternatives to landfilling which meet environmentally-protective limits for contaminants in air, water, and soil. DEC is also seeking alternatives which lend themselves to the development of a generic design standard for small-scale projects. Development of such standards is desirable, as it will reduce the time lag and the overall costs of a spill cleanup of small-scale projects, by providing the design work for a responsible party (RP) or contractor to adapt similar projects and eliminate the need for more costly professional support. Development of generic design standards also assists many RPs and contractors who are interested in using innovative remedial technologies but lack personal knowledge and experience in those technologies.

Alternatives to soil disposal have developed from necessity, yet have also fostered the beneficial reuse or onsite treatment of soils. Reuse options for treated soils are described in DEC's Spill Technology and Remediation Series (STARS) Memo #1: Petroleum-Contaminated Soil Guidance Policy, and include reuse as a construction material, returning soil to the original excavation, placing the soil elsewhere onsite, and reuse offsite at a preapproved location. Onsite treatment options which are being used in New York state by DEC include thermal treatment, soil venting, cold-mix asphalt manufacturing, and bioremediation. Soil type, contaminant type, soil quantity, or contaminant concentration, are some of the factors which may limit small-scale uses of some of these onsite soil remediation techniques. Bioremediation, however, appears to be a viable alternative for the development of a design standard which can be adapted to many different small-scale projects.

The design standard which DEC is developing is described in a paper, "Generic Biocell and Biopile Designs for Small-Scale Petroleum-Contaminated Soil Projects" (The Paper). The Paper includes designs for a biocell and a biopile, using indigenous microorganisms or using commercially-available biological blends, and apply to small-scale (approximately 30 - 100 yd^3) petroleum-contaminated soils. This volume of soil, which is typical of the removal of an underground storage tank, was chosen recognizing that a standard approach to cleaning up this amount of soil could reduce costs for many responsible parties who do not have the resources to hire professionals with experience in bioremediation, reduce the project duration time as compared to less active approaches, and this would be a manageable quantity of soil for application of a generic design standard.

Other design conditions which are discussed in The Paper include maintaining neutral pH and moderate moisture, and operating at temperatures above 40°F. Nutrient requirements, also discussed in The Paper, include nitrogen, phosphorous, and oxygen, and were calculated based on the average composition of the contaminants in the soil and the amount of soil to be treated by solving the chemical reactions associated with hydrocarbon degradation. The Paper includes calculation and tabulation of nutrient requirements for varying levels of contamination and amounts of soil.

The same principles found in The Paper can be used for smaller or larger quantities. A less elaborate design may be appropriate for smaller quantities, and the use of an experienced professional consultant is recommended for larger quantities. It is important to emphasize that this guidance is intended exclusively for small scale ex-situ projects. It should not be extrapolated to design larger and more complex sites.

PURPOSE

A field project was conducted to test the bioremediation procedures described in The Paper. Based on the results of the field test, The Paper is being expanded and incorporated into a formal NYS DEC guidance document identifying guidelines and recommended practices to be used by RPs, contractors, DEC staff, etc., for bioremediation of petroleum-contaminated soil quantities of 100 yd^3 or less. The intent of this effort is to develop a cost-effective, simple, "do-it-yourself" remediation technique.

SCOPE

The field project consisted of 5 biocells and 2 biopiles designed and set up according to The Paper, as follows:

(1) 50 yd^3 cell of gasoline-contaminated soil
(1) 50 yd^3 cell of fuel oil-contaminated soil
(1) 50 yd^3 pile of gasoline-contaminated soil
(1) 50 yd^3 pile of fuel oil-contaminated soil
(1) 25 yd^3 control cell of gasoline-contaminated soil
(1) 25 yd^3 control cell of fuel oil-contaminated soil
(1) 50 yd^3 bioaugmentation cell of fuel oil-contaminated soil.

Extensive microbial and contaminant sampling was conducted to evaluate the effectiveness of the generic approach. Graphs of Total Petroleum Hydrocarbons (TPH) results and microbial count results are included in the Results section of this chapter.

The cells and piles were constructed and operated from the beginning of September 1993 through mid-November 1993. The project was shut down until spring 1994 because microbial activity is minimal at soil temperatures below 40°F. The final operation of the project was conducted from June 1994 to October 1994.

Total project costs, including construction and materials, soil mobilization and spiking, contracted labor, and sample analysis, are estimated at approximately $320,000. These costs do not include DEC man-hours and travel related to planning and collecting samples, evaluating results, and report preparation. The project funds consisted of contributions from the USEPA (approx. $120,000), Stewart's (approx. $180,000), and LVL (approx. $20,000). Approximately $80,000 was used for sample analysis. The remaining $240,000 was used for soil mobilization, construction materials and labor, project maintenance, and sample collection. A discussion of costs expected for an actual remediation project is included in this report.

DISCUSSION

Issues which have been evaluated during the field project include aeration of the cells; nutrient application; control and monitoring of moisture, temperature, and pH; and reuse of leachate. Success of operation and completion of this project was monitored by measuring moisture, pH, TPH, total concentrations, microbes, and soil type at regular intervals throughout the project.

Additional issues which are being addressed before a NYS DEC guidance document can be developed include development of a generic sampling procedure and scheme; discussion of the benefits of bioremediation vs. the long-term liability of landfilling; and cost analyses.

Project Description

The project was conducted at a remote site in Saratoga Springs, NY, in cooperation with an RP (Stewart's) who was willing to participate in the project to develop a generic design for bioremediation of small amounts of petroleum-contaminated soil.

In addition to testing the design feasibility and application, and determining the cost of implementing the generic design, 7 treatment areas were selected for construction to test bioremediation of gasoline versus fuel oil, use of a cell versus a pile, and bioaugmentation versus biostimulation. As illustrated in Figure 1, the cells were bermed and lined, and filled with 18 to 24 inches of soil. The biocell soils were initially fertilized, and sprayed with water and tilled regularly to moisten and oxygenate the soil. The cells were also covered with a tent to protect the soil from precipitation.

As illustrated in Figure 2, the piles were bermed and lined, and mounded with 4 to 5 feet of soil. The biopile soils were initially fertilized, and sprayed with water regularly to moisten the soil. Air was pumped into the pile through flexible-perforated piping wound through the pile. The piles were also covered with polyethylene to protect against precipitation. Old tires were use to hold the cover down and also to allow air space between the top of the pile and the cover.

A total of 300 yd^3 of contaminated soil was divided for treatment into 5 biocells and 2 biopiles, as illustrated in Figure 3. Area 1 was a fuel oil-contaminated soil cell, approximately 22 feet by 50 feet by 1.5 feet. This area was designed to test biostimulation of fuel oil-contaminated soil in a cell.

Figure 1. **Design specifications for biocell.**

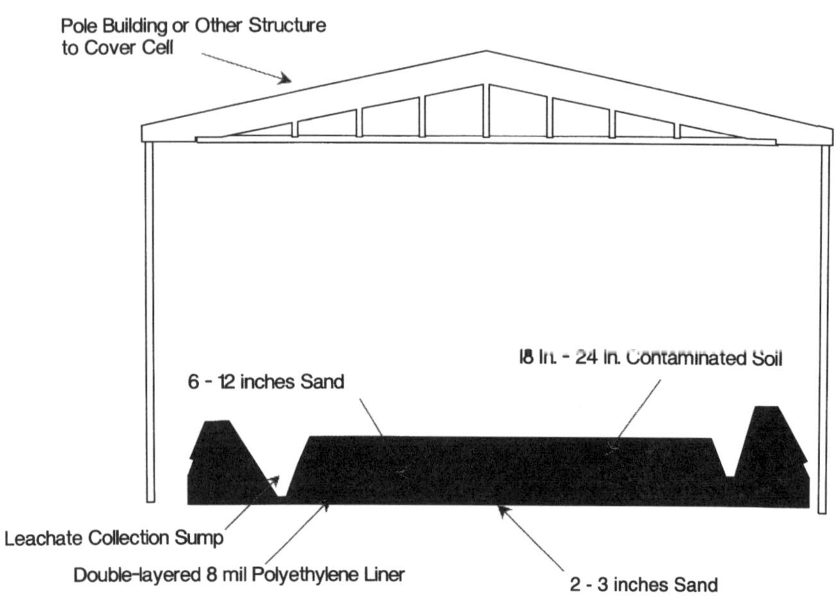

Figure 2. **Design specifications for biopile.**

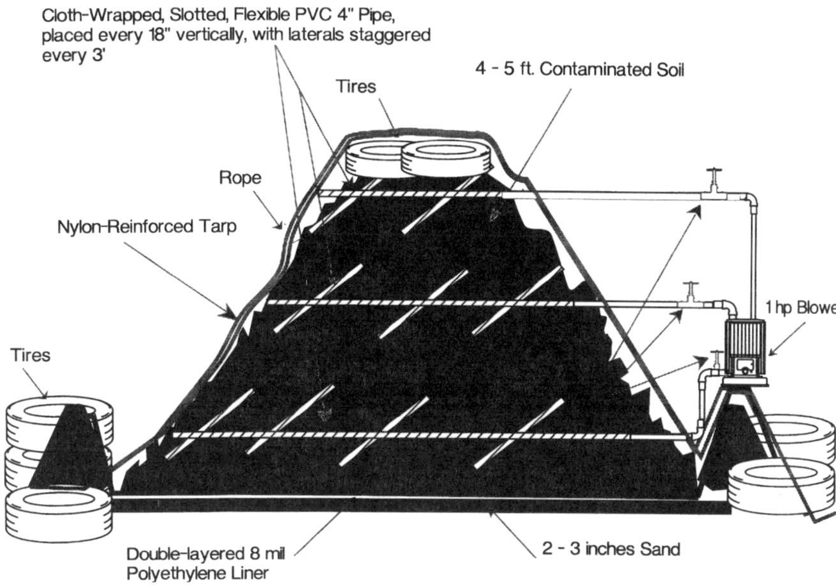

Biostimulation is a process where natural bioremediation is enhanced by adding nutrients, oxygen, and moisture to stimulate the existing bacterial population to utilize the petroleum food source.

Area 2 was also a fuel oil-contaminated soil cell of the same size; however, this soil was augmented with bacteria which was delivered to the site freeze-dried and fermented in warm water prior to application to the site several times per week. This cell was designed to test bioaugmentation of fuel oil-contaminated soil in a cell. Bioaugmentation is a process where bioremediation is assisted by adding "off-the-shelf" bacterial blends. Fuel oil was chosen for bioaugmentation because it is more difficult to degrade the heavier, more complex petroleum hydrocarbons found more predominantly in fuel oil than in gasoline.

Area 3 was a fuel oil-contaminated soil cell, approximately 25 feet by 25 feet by 1.5 feet, and Area 4 was a gasoline-contaminated soil cell, the same size as Area 3. These cells were designed as control cells to represent what many "do-it-yourselfers" have done. The soil was spread out in treatment cells, a small amount of fertilizer was added, the soil was tilled once, leaving the cell unattended and open to the environment.

Area 5 was a gasoline-contaminated soil cell, approximately 25 feet by 50 feet by 3 feet. This area was designed to test biostimulation of gasoline-contaminated soil in a cell. Area 6 was a fuel oil-contaminated soil biopile, and area 7 was a gasoline-contaminated soil biopile, both approximately 20 feet by 22 feet

Figure 3. **Biocell and biopile project site.**

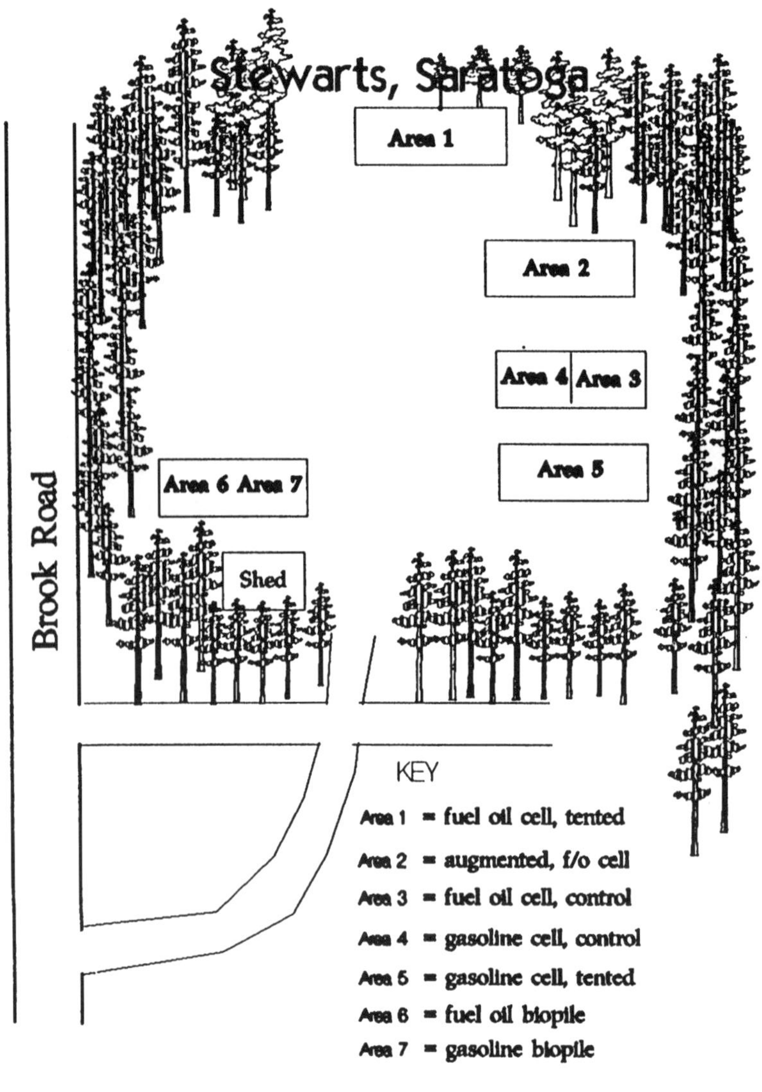

by 3 feet. Area 6 was designed to test biostimulation of fuel oil in a pile, and area 7 was designed to test biostimulation of gasoline in a pile.

Aeration

Piles

Aeration was provided to the pile by pumping ambient air through a continuous helical-structure of perforated piping covered with a geotextile material. Less contaminant removal in the piles seems to indicate that aeration may have been a problem. Perhaps the helical-structure allowed for short-circuiting of air. Short-circuiting occurs due to varying resistance to air flow such as soil settling or clogging of pipe perforations, resulting in air preferentially flowing through the path of least resistance. A manifold of piping with valves at the manifold may improve air distribution by allowing control of the direction and flow rate of air passing through the piping. Air could be directed to the most contaminated parts of the pile, and also to dry out the very wet areas, or away from areas which may be drying out from too much air. A smaller diameter pipe would also be desired to allow more surface area of soil to be in contact with the air; however, the piping used is not available in a smaller diameter. Perhaps the availability of a smaller diameter, flexible, perforated piping could be investigated for a project of this nature.

Smaller diameter piping would also facilitate easier sampling, which may contribute to solving the aeration problem. Sampling challenges were encountered when the piping was struck by the auger, and subsequent auguring locations were necessary to avoid the piping obstruction. These sampling problems also caused the geo-textile material on the piping to be torn up. The purpose of this geo-textile was to prevent soil from entering the perforations in the piping; therefore, this could have also contributed to poor air distribution in the pile.

Cells

Aeration was provided to the cells by tilling twice per week. Tilling is a very difficult, labor-intense process in saturated soil, clayey soil, and soil deeper than 10 inches. Based on the difficulty of tilling the soil with a garden tiller, it is recommended that a farm tiller should be used, or that the cells should be constructed with thinner layers of soil. If thinner layers are used (10 inches or less of soil), a suitable rototiller could be rented at a garden store or a rental center, to efficiently turn the soil layer over in the cell.

It is recommended that clayey soil be treated in a cell rather than a pile to promote the best aeration. The use of compost material to promote aeration of clayey soil by bulking the soil, should be investigated further.

An alternative to be investigated, for both the tilling problem in the cells and the aeration of clayey soils, is the use of oxygen-containing nutrient blends being marketed by various vendors in the industry. These blends, containing nutrients and oxygen in a solid form, are mixed into the soil and solubilize when moisture is added. Investigation of the cost and effectiveness of nutrient blends relative to the use of fertilizer and aeration is recommended.

Nutrient Addition

Nutrients were provided in the form of air, water, and fertilizer as discussed in The Paper. Fertilizer requirements were determined based on the volume of soil to be treated and TPH, using Tables A1: Amount of Fertilizer Required to Treat Gasoline-Contaminated Soil, and A2: Amount of Fertilizer Required to Treat Fuel-Oil Contaminated Soil, from The Paper. Tables A1 and A2, illustrated in

Figure 4. **Amount of fertilizer required to treat gasoline-contaminated soil.**

Table A1: Amount of Fertilizer Required to Treat Gasoline-Contaminated Soil (lbs.) (N:P = 6:1)

Total Petroleum Hydrocarbons (TPH) (ppm)	Volume of Soil (yd.³)							
	30	40	50	60	70	80	90	100
1000	59	79	99	118	138	158	177	197
2000	118	158	197	236	276	315	355	394
3000	177	236	296	355	414	473	532	591
4000	236	315	394	473	552	631	709	788
5000	296	394	493	591	690	788	887	985
6000	355	473	591	709	828	946	1064	1182
7000	414	552	690	828	966	1104	1242	1379
8000	473	631	788	946	1104	1261	1419	1577
9000	532	709	887	1064	1242	1419	1596	177
10000	591	788	985	1182	1379	1577	1774	1971
12500	739	985	1232	1478	1724	1971	2217	2463
15000	887	1182	1478	1774	2069	2365	2660	2956
17500	1035	1379	1724	2069	2414	2759	3104	3449
20000	1182	1577	1971	2365	2759	3153	3547	3941

Figure 5. **Amount of fertilizer required to treat fuel-oil contaminated soil.**

Table A2: Amount of Fertilizer Required to Treat Fuel Oil-Contaminated Soil (lbs.) (N:P = 6:1)

Total Petroleum Hydrocarbons (TPH) (ppm)	Volume of Soil (yd.³)							
	30	40	50	60	70	80	90	100
1000	56	75	93	112	131	149	168	187
2000	112	149	187	224	261	298	336	373
3000	168	224	280	336	392	448	504	560
4000	224	298	373	448	522	597	671	746
5000	280	373	466	560	653	746	839	933
6000	336	448	560	671	783	895	1007	1119
7000	392	522	653	783	914	1044	1175	1306
8000	448	597	746	895	1044	1194	1343	1492
9000	504	671	839	1007	1175	1343	1511	1679
10000	560	746	933	1119	1306	1492	1679	1865
12500	699	933	1166	1399	1632	1865	2098	2331
15000	839	1119	1399	1679	1958	2238	2518	2798
17500	979	1306	1632	1958	2285	2611	2938	3264
20000	1119	1492	1865	2238	2611	2984	3357	3730

this report as Figures 4 and 5, were developed based on contaminant concentration and the stoichiometry of the degradation equations. The derivation of these equations and the subsequent fertilizer amounts are illustrated and discussed in Appendix C of The Paper.

For the field project, we added all of the fertilizer at once, in dry pelletized form. There has been much discussion in the industry over the use of liquid fertilizer versus pellets, and periodic versus bulk application. Future investigation should be conducted to evaluate whether or not variation in the form and application of fertilizer influences the effectiveness of bioremediation.

Moisture

Piles and cells were sprayed with water as needed to maintain wet but "not puddly" conditions. Conversely, in order to prevent saturation from rain, the cells and piles were covered. The piles were covered by laying polyethylene sheeting over the soil and anchoring it down with used tires.

According to The Paper, the cover on the cells required a peaked design so that water would not collect on it. The cell cover also should have been designed to allow soil tilling operations to occur without disturbing the cover, or to be easily removed and replaced to allow for soil tilling as needed. Originally, a pole barn structure was recommended, but cost estimates were prohibitive (approximately \$3000 for a barn to cover 50 yd^3).

For our field project, tents were constructed of reinforced-nylon tarps and cable. This design still allowed some puddling of rain water on the covers and into the cells. The covers and cables were stressed and weakened and were torn in some places, requiring regular maintenance. This made it difficult to remove the cover for tilling, and left some parts of the cell exposed to the rain. Other alternatives which should be investigated are portable garages or surplus army tents or any other viable structure. These structures should be cheaper than a pole barn, stronger than the tarpaulin tents, and tall enough that they will not have to be removed for tilling. Perhaps an alternative may be to leave the cell uncovered until rain becomes a problem, or perhaps the project could be active during the "dry" season and shut down when conditions are too wet and cold for bioremediation to occur.

Another contribution to maintaining proper moisture was leachate control. Because there was inadequate sump space to contain leachate until the sump could be pumped, the sump overflowed onto the cell and pile soil, saturating much of the treatment soil. This may have contributed to lower removal rates than expected.

Future designs could include a downward-sloped treatment area foundation, and a moat around it which is large enough to contain the amount of leachate anticipated before the moat can be pumped.

There was also concern regarding the complete distribution of moisture throughout the piles, or alternatively, regarding the possible drying-out of the soil by the pumped-in air. However, moisture measurements performed on several rounds of the soil samples indicate similar moisture was maintained in the piles as in the cells.

Temperature

Bioremediation of contaminated soil operates best at a soil temperature between 40° and 100°F. Black covers will absorb heat to keep the pile warm, and most clear polyethylene which was used during the project degraded to some extent due to exposure to weather and sunlight, while the black polyethylene remained intact.To prolong the life of bioremediation processes in the Northeast, the use of black polyethylene covers is recommended.

pH

The maintenance of neutral pH (pH=7) was also investigated. Liming may be necessary for acidic conditions ($pH<7$). The pH of alkaline soil ($pH>7$) should be reduced by the acidic by-products generated from the biological activity in the soil. As biological activity and resulting pH reduction continue, monitoring may be necessary to identify if acidic conditions occur and subsequent liming is necessary. Based on pH monitoring during three months of the project, which indicated that neutral pH was maintained, pH monitoring does not seem necessary after measuring and adjustment at the start of the project, for a project of this size and duration.

The use of lime to reduce the pH of the soil was also investigated. It has been reported that lime may be harmful to most hydrocarbon-degrading bacteria, and magnesium sulfate may be used as a substitute. Since magnesium sulfate is apparently difficult to obtain, and since we have limited data discouraging the use of lime, we will continue recommending its use. Further investigation of the use of lime for pH control is recommended in a project where pH control is a concern.

Contaminated Leachate

Collection and disposal versus reuse of the contaminated leachate which may collect in the sump areas of the cells or piles was investigated. If the leachate is collected, it would have to be analyzed for proper disposal. Depending on the level of contamination, disposal options may include discharging to a municipal sewer system with prior approval from the sewage treatment plant owner/operator, or to a regulated disposal facility. An acceptable alternative to disposal is reuse of the leachate for moistening the cells or piles.

Contaminated water was reused on all piles and cells except for the control cells. This project did not provide sufficient data to substantiate whether or not reuse of contaminated water has an effect on the efficiency of bioremediation. However, there was no obvious adverse effect which would preclude the reuse of contaminated water during the bioremediation process. Therefore, it is recommended that contaminated leachate can be simply reused for moistening the cells or piles for a project of this nature. Any excess leachate collected would need to be disposed of properly, as described in The Paper.

SAMPLING

Extensive microbial and contaminant sampling was conducted to evaluate project effectiveness for the following parameters: microbe counts, soil sieve, % solids, pH, TPH and contaminant-specific analyses for the five cells and two piles used in the field project.

Samples were collected randomly and analyzed, in accordance with NYSDEC's STARS Memo #1: Petroleum-Contaminated Soil Guidance Policy. Samples were collected from random locations based on a 2-D grid in the cells and a 3-D grid in the piles. A computer program designed by Christopher O'Neill, P.E. of DEC, was used to determine the sample locations for this project. A copy of this program, "BASIC Program for Random Sampling Locations, is included in another DEC Report, "Field Instrument Evaluation Project," (O'Neill, 1995) and briefly described below. More information on sampling grids can be found in NYS DEC's Sampling Guidelines and Protocols manual.

In order to determine random sample locations using a grid, the dimensions of the cells (x,y) and the pile (x,y,z) are entered as the upper limits of the grid. Random numbers between 0 and 1 were then used to determine each dimension in the grid. Since a 5-digit random number table was used, each random number was divided by 100,000 to obtain an adjusted random number between 0 and 1. One adjusted random number was then multiplied by the upper limit for each dimension to determine the random sample location.

For example, treatment area #1 was 23 feet wide (x) and 28 feet long (y). A random number of 78261 would be divided by 100,000 and multiplied by 23 to obtain an x-dimension of 18' for the random sample location.

Each sampling event consisted of collecting two grab samples and two composite samples from each of the treatment areas. The analytical results obtained from the four samples (2 grabs and 2 composites) for each soil contaminant type were averaged, to determine the value for that parameter for that particular treatment area. One set of results used zero for "nondetect" results reported by the laboratory, and another set of results used the detection limit for "nondetect" results reported by the laboratory. Figures 6 and 7 show BTEX (sum of Benzene, Toluene, Ethylene, and Xylenes) values plotted using "nondetect" equals zero and nondetect = detection limit, respectively. As these figures illustrate, there was no apparent significant difference in using zero or the detection limit for "nondetects" for BTEX. Figures 8 and 9 show the values for PAH (sum

Figure 6. **BTEX data summary: Areas 1-7.** (Non-detect equals zero)

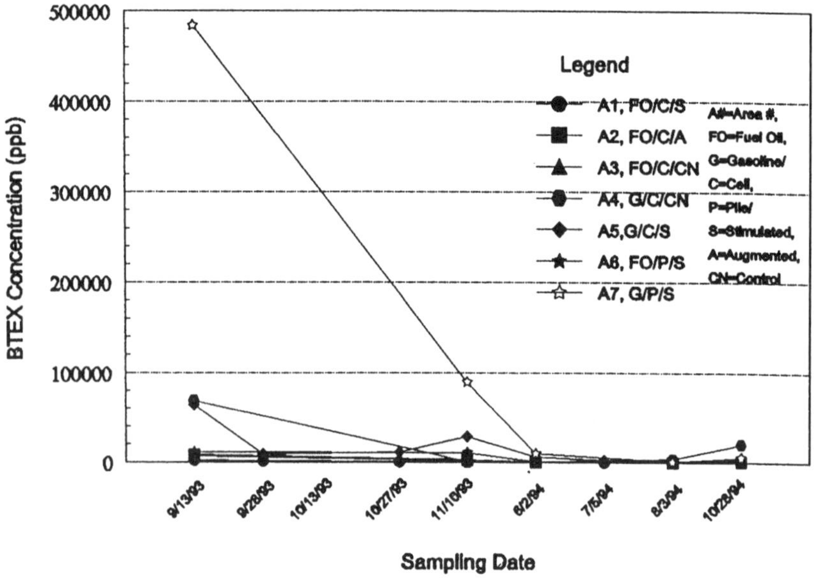

Figure 7. **BTEX data summary: Areas 1-7.** (Non-detect equals detection limit)

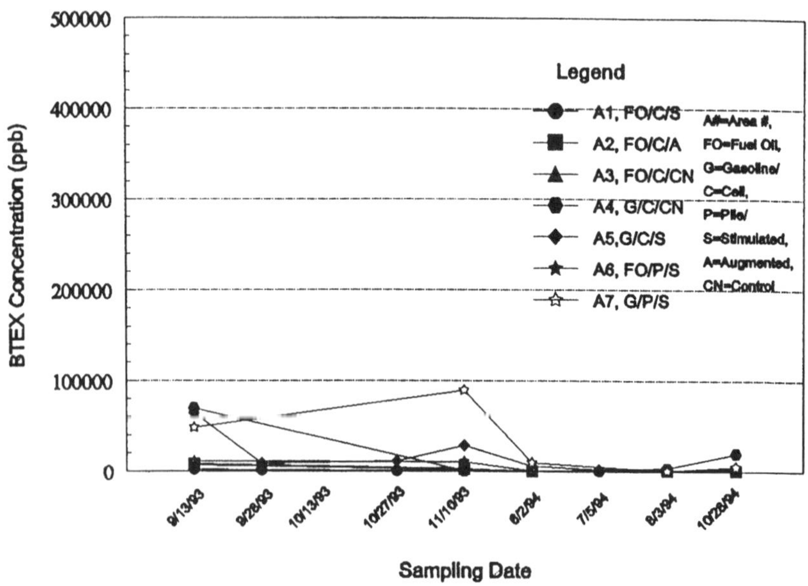

Figure 8. **PAH data summary: Areas 1, 2, 3, and 6.** (Non-detect equals zero)

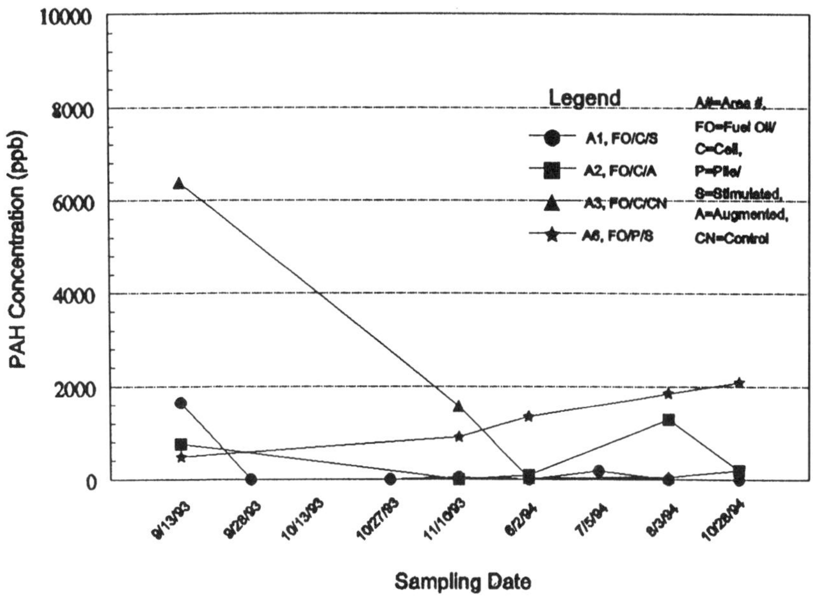

Figure 9. **PAH data summary: Areas 1, 2, 3, and 6.** (Non-detect equals detection limit)

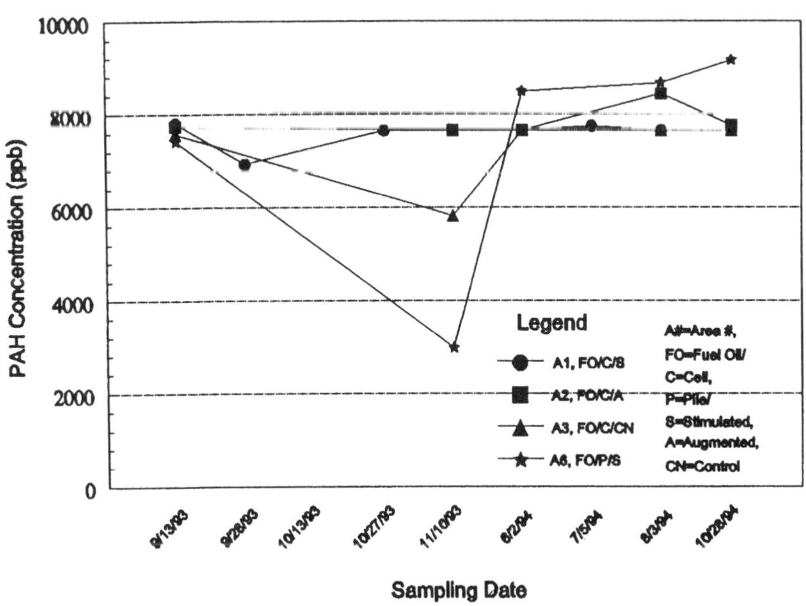

of Poly Aromatic Hydrocarbons). Although the numbers are higher when the detection limit was used, the trend for both plots was similar, indicating that using the detection limit or zero for "nondetects" does not affect the results.

The sampling schedule consisted of collecting samples for an initial TCLP (Toxicity Characteristic Leachate Procedure) and individual component analyses (EPA 8021 and 8270), monthly TPH and periodic individual component analyses, final individual component analyses, monthly microbe analyses, and periodic pH and % solids analyses. For actual remediation projects, initial TCLP analyses would probably not be performed, and periodic individual component analyses would probably not be performed until project completion was suspected. It is important to note that although sampling guidelines are available, analytical parameters and frequencies should be determined on a site-by-site basis.

RESULTS

According to literature, a microbe population, defined by the number of colony-forming units (CFU) present, of 10^6 - 10^8 is the recommended level of microbes sufficient to support contaminant degradation. As illustrated in Table 1, all areas developed microbial counts within the recommended range of (10^6 - 10^8), seeming to indicate that the system was operating satisfactorily in all the areas. Table 1 also includes colony-utilizing population (CUP), which is the percentage of the microbe population which are consuming the petroleum products. Areas 3, 4, and 6 dropped down to 10^5 CFU at the end of the project. A decrease in microbes with high contamination levels could indicate the system is not operating efficiently, since sufficient food source (contamination) is available for the microbial population to survive. A lower microbe level was expected in the control cells (areas 3 and 4) based on one-time rototilling and nutrient addition at start-up. However, these numbers were very similar to those in the

Table 1. **Microbe Counts: Colony Farming Unit (CFU) Concentration Colony Utilizing Population (CUP) %**

Sampling Area	Date of Sample							
	9/14/93	10/13/93	11/10/93	6/2/94	7/5/94	8/3/94	9/8/94	10/28/94
1 (FO/C/S)	1.5E+06	2.3E+10 70%	2.46E+08	2.2E+08 8%	5.61E+06	2.25E+06 42%	4.99E+06	1.79E+07 73%
2 (FO/C/A)	1.4E+08	1.9E+06 0%	3.19E+06	4.87E+07 4%	3.71E+07	1.52E+06 36%	4.40E+05	2.15E+06 0%
3 (FO/C/CN)	9.5E+06	3.7E+07 59%	1.35E+07	2.28E+06 12%	2.46E+06	4.25E+06 0%	4.00E+05	6.3E+05 0%
4 (G/C/CN)	1.3E+07	1.1E+09 0%	3.12E+07	9.93E+6 2%	4.86E+06	1.1E+06 0%	1.76E+07	2.4E+05 0%
5 (G/C/S)	1.4E+06	2.3E+07 28%	4.3E+06	1.96E+06 0%	4.87E+06	1.89E+06 0%	1.89E+07	1.51E+08 0%
6 (FO/P/S)	4.2E+06	3.6E+07 76%	2.45E+06	1.04E+07 0%	5.57E+06	2.59E+06 8%	1.02E+06	3.99E+05 0%
7 (G/P/S)	2.3E+07	4.2E+07 0%	1.27E+06	1.12E+06 0%	5.4E+07	4.85E+06 0%	7.41E+06	4.05E+07 76%

stimulated areas, indicating that the minimal biostimulation in the control cells was somewhat effective. For pile #6, perhaps the pile was not receiving enough oxygen or was too moist. (See discussion in Aeration and Moisture sections of this report.)

As illustrated in Table 2, soil sieve analyses indicate similar soil types in all cells. The soil types range from 44 to 60% sand, 9 to 31% gravel, and 26 to 44% silt/clay. This amount of silt/clay may have contributed to lower contaminant removal than expected, as it is more difficult to bioremediate silty/clayey soil.

Table 2. **Soil Sieve Analysis**

Sample Area	% Gravel	% Sand	%Silt/Clay
Area 1	30.6	43.5	25.9
Area 2	10.5	45.2	44.3
Area 3	18.7	45.4	35.9
Area 4	9.9	58.3	31.8
Area 5	11.9	54.9	33.2
Area 6	8.9	51.5	39.6
Area 7	10.9	59.8	29.4

Analysis results of percent solids, listed in Table 3, indicate similar moisture content in the cells and piles, ranging from 86 to 90% solids. Therefore, with the limited data collected during this project, it seems that distributing moisture in the pile may be no more difficult than in the cell.

Table 3. **% Solids**

Sample Number	% Solids
Area 6 C1	87.3
Area 7 C1	87.8
Area 1 C2	89.2
Area 2 C2	----
Area 3 C2	87.3
Area 4 C2	86.8
Area 5 C2	86.2
Area 6 C2	89.6
Area 7 C2	89.2

Table 4 illustrates that neutral pH was maintained during the three months in which it was monitored. Perhaps this means that pH monitoring is not necessary for a project of this size and duration.

At the end of the field project, none of the cells or piles met the criteria of STARS #1 for all contaminants of concern, although there was significant de-

Table 4. **pH**

Sample Data	Sample Number	pH
07/05/95	Area 1 C1	7.7
	Area 2 C1	8
	Area 3 C1	8
	Area 4 C1	7.8
	Area 5 C1	7.7
	Area 6 C1	8.6
	Area 7 C1	8
09/08/94	Area 1 C1	
	Area 2 C1	
	Area 3 C1	7.6
	Area 4 C1	7.9
	Area 5 C1	7.9
	Area 6 C1	7.9
	Area 7 C1	8.2
10/28/94		
	Area 1 C2	7.98
	Area 2 C2	----
	Area 3 C2	7.75
	Area 4 C2	7.73
	Area 5 C2	7.96
	Area 6 C2	7.73
	Area 7 C2	7.94

crease for most contaminants. This inability to meet STARS #1 criteria could be attributed to the challenges in maintaining proper aeration and moisture, as discussed in those sections of this chapter.

The TPH results show a general downward trend in the cells and no significant removal in the piles. Although TPH is not part of STARS #1 cleanup criteria, it may be appropriate to use as an inexpensive qualitative indicator of bioremediation progress.

TPH, BTEX, and PAH removal rates are illustrated in Table 5. Based on these results, bioremediation is more efficient in a cell than in a pile. Trends for benzene, BTEX, naphthalene, PAH, total volatiles, and xylene were also examined. These values varied up and down, although a generally downward trend was exhibited. This variation in results in the biocells could be attributed to varying depth of samples collected, and to the tilling problems which may have inhibited bioremediation from occurring successfully in the bottom portion of the contaminated soil layer in the cell. This variation may also be evidence for the necessity of three-dimensional random sampling, to account for varying degrees of contamination existing in different locations in the soil cells as well as the piles.

Table 5. **% Removal**

Area	TPH	BTEX	PAH
1 (FO/C/S)	88%	76%	100%
2 (FO/C/A)	53%	91%	
3 (FO/C/CN)	54%	97%	100%
4 (G/C/CN)	59%	71%	NA
5 (G/C/S)	8%	98%	NA
6 (FO/P/S)	0%	74%	
7 (G/P/S)	0%	89%	NA

The overall data presented in Figures 6 and 10 are a comparison of BTEX removal for all the areas. For Figure 10, the y-axis was limited to 40,000 ppb to magnify the lower section of the graph for easier evaluation of this section of the graph. These figures illustrate that the rate of BTEX removal was the same for areas 1, 2, and 3 for the first five sampling events. These areas were fuel oil cells which were biostimulated, bioaugmented, or a control, respectively. If the fourth data point in area 6, fuel oil in a pile which was biostimulated, was eliminated, then area 6 would also show the same removal rate for BTEX. For the gasoline cells and pile, however, there was no relation in the results except for the period of June 1994 to August 1994, when there were similar removal rates for areas 5 and 7. The gasoline areas showed fluctuations in BTEX removal, although overall there was a net removal. This is not what might be expected, because BTEX are volatiles and are the major components of gasoline. A pos-

Figure 10. **BTEX data summary: Areas 1-7 (y-axis limited to 40,000 ppb).**

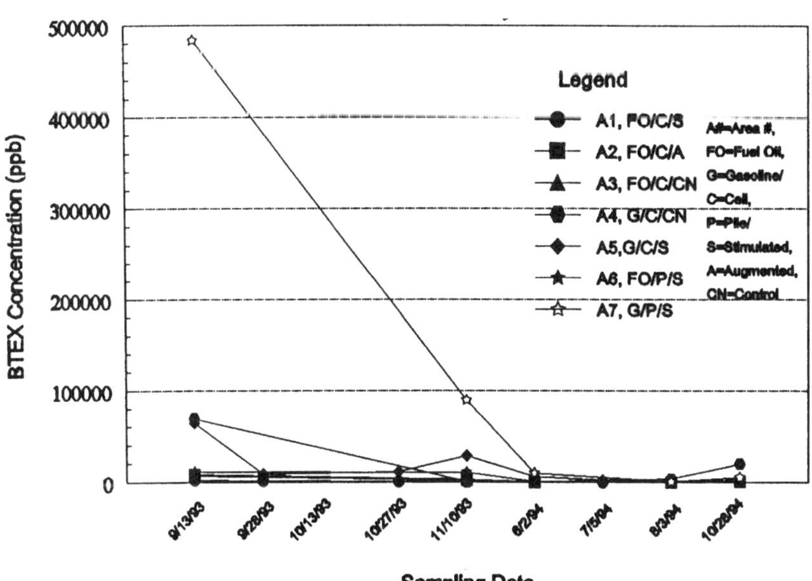

sible explanation for this is error in collecting samples. If not, it might be possible to conclude that biodegradation of fuel oil is more efficient than gasoline. Perhaps the nonhomogeneity of the soil contributed to the lack of relationship in removal rates. More frequent and larger numbers of samples may allow better data interpretation and relationships in removal rates.

The overall data presented in Figure 8 are a comparison of PAH data for areas 1, 2, 3, and 6. Area 1 shows faster removal; however, there were more data points for area 1 than any of the others. If data for the same data points are compared, then areas 1, 2, and 3 had similar removal rates. These areas were fuel oil cells which were biostimulated, bioaugmented or a control, respectively. Area 6 showed an increase in PAH. A possible explanation for this is either an error in collecting samples or the random sample encountered a "hot spot." Once again, more intense sampling might actually show a similar removal rate for area 6 as for areas 1, 2, and 3. If not, it might be possible to conclude that biodegradation of fuel oil is more efficient in a cell than in a pile.

The overall data presented in Figure 11 are a comparison of TPH Method 418.1 analysis results for all areas. Area 1 displayed a significant downward trend. The other areas fluctuated, with much less overall change in TPH. Given the fluctuations in Figure 8, this may support the current concern in the industry that TPH is not a good indicator of petroleum hydrocarbon removal in gasoline and fuel oil-contaminated soils.

Figure 11. **TPH data summary: Areas 1-7.**

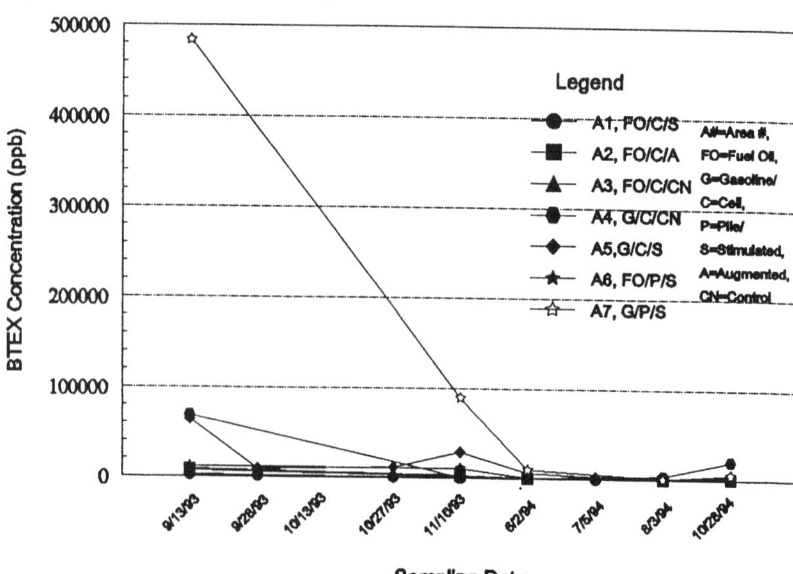

OBSERVATIONS AND RECOMMENDATIONS

The following recommendations, which were addressed in the preceding discussion, were based on observations of the apparent effects of variations in aeration, moisture, temperature, pH, and the reuse of leachate, on the results of the bioremediation project.

Aeration

Piles

To enhance aeration in the piles, a manifold of piping with valves at the manifold branches may improve air distribution by allowing control of the direction and flow rate of air passing through the piping. Perhaps the availability of a smaller diameter, flexible, perforated piping could be investigated for a project of this nature, to solve aeration problems due to surface area of piping exposed to the soil, and to disturbing of the pipes during sampling.

Cells

To enhance aeration in the cells, based on the difficulty of tilling the soil with a garden tiller, it is recommended that a farm tiller should be used, or that the cells should be constructed with thinner layers of soil. If thinner layers are used (10 inches or less of soil), a suitable rototiller could be rented at a garden store or a rental center, to efficiently turn over the soil layer in the cell.

It is recommended that clayey soil be treated in a cell rather than a pile, to promote the best aeration. The use of compost material to promote aeration of clayey soil should be investigated further.

Piles and Cells

To possibly simplify the aeration and/or nutrient addition processes, investigation of oxygen-containing nutrient blends is recommended. Also, the use of liquid versus pellet form fertilizer, and periodic versus bulk application should be investigated.

Moisture

Cells

To solve the moisture problems in the cell, it is recommended that alternatives for covering a cell be investigated. Some options to investigate are portable garages or surplus army tents. These structures should be cheaper than a pole barn, stronger than the tarpaulin tents, and tall enough that they will not have to be removed for tilling. Perhaps an alternative to covering the cell may be to leave the cell uncovered until rain becomes a problem, or perhaps the project could be active during the "dry" season and shut down for the "rainy" and "cold" seasons.

Piles and Cells

To solve the leachate collection problem, future designs should include a downward-sloped cell or pile foundation, and a moat around the cell or pile which is large enough to contain the amount of leachate anticipated before the moat can be pumped. Also, it is recommended that contaminated leachate can be simply reused for moistening the cells or piles for a project of this size and duration.

To prolong the life of bioremediation processes in the Northeast, the use of black polyethylene covers is recommended. Black covers will absorb heat to keep the pile warm. Most clear polyethylene which was used during the project has degraded to some extent, while the black polyethylene remained intact.

pH

Further investigation of the use of lime for pH control is recommended. Also, pH monitoring does not seem necessary after measuring and adjustment at the start of the project, for a project of this nature.

Others

The possible effect of plant life in the control cell on passive bioremediation deserves further investigation. Even though bioremediation was not actively pursued in these cells, there was 54 - 59% removal of TPH in these cells. Perhaps the plant life-cycle contributed nutrients or moisture to these cells.

CONCLUSIONS

Based on this project, bioremediation can be accomplished in a simple, "backyard" setup. However, it appears imperative that much attention be given to the oxygen supply and moisture control. It appears that less money invested in a simpler setup, without control of these conditions, such as using an inadequate tiller or unsturdy cell covers will require much more time and money invested in monitoring and maintaining the conditions during the project because less than ideal conditions may require a longer project duration. In this field project, much money and time were spent on the repair of the tiller, on labor for longer time required to till with an inadequate tiller, and time and materials required for numerous repairs of the tents.

Also, the treatment cell required more maintenance than the piles; however, if BTEX, PAH or TPH is used as an indicator of efficiency of bioremediation and based on this project, a cell is more efficient than a pile. Based on PAH and TPH, the bioaugmentation was less effective than the biostimulation. There was no consistently better result between fuel oil and gasoline.

During this project, approximately $250/ton was spent on sample analysis. Another $800/ton was spent on design, soil mobilization, construction materials and labor, project maintenance and monitoring, and sample collection. A generic sampling plan will be included in the NYSDEC guidance document based on the results of the field project. According to this plan, sample analysis costs could be reduced approximately 60%, from $250/ton during the research project to approximately $70/ton for an actual cleanup project. In addition, costs can be reduced significantly for actual remediation projects based on what construction materials the RP may have access to, and the cost for labor to construct, maintain and monitor the project.

The guidance document will also include a simple explanation of bioremediation, much like in this chapter, and guidance with recommendations for varying operating conditions and possible alternatives to experiment with and study in the future. Finally, a discussion of the benefits, as well as a cost analysis, of bioremediation versus the long-term liability of landfilling will be included in the guidance document because it seems that it can be so much simpler, quicker, and less expensive to landfill the contaminated soil at this time.

ACKNOWLEDGMENTS

This report, which was prepared by the New York State Department of Environmental Conservation Division of Spills Management, summarizes the results of a research project designed to test DEC's guidelines and recommended practices for bioremediation of small-scale amounts of petroleum-contaminated soil. The project progressed satisfactorily due primarily to the support and cooperation of the following companies and individuals:

NYSDEC - Donald Darmer, Richard Wagner, Anthony Karwiel

Stewart's Ice Cream Shops, Inc. - Chad Fowler

Lebanon Valley Landscaping, Inc. - Bill Black, Paul Heilmann

American Spill Abatement, Inc. - George Longworth, Harold Strom, and Bill Toran

Waste Stream Technology, Inc. - Mike Barnhart and Jim Heizy

Adirondack Environmental Services - Paul Batista

Con-Test Analytical Laboratory - T.J. Lyons, Mark Simon

A special acknowledgement is given to Stewart's and their construction coordinator, Chad Fowler. They contributed significantly by subsidizing the sampling costs and providing other key services. This project exemplified the cooperation of the DEC and industry toward achieving a common goal.

Additional appreciation is extended to the DEC Central Office and Region 5 Divisions of Solid Waste and Regulatory Affairs for their support in expediting construction and operation of this project.

REFERENCES

Lasdin, Susan M., P.E., and O'Neill, Christopher M., P.E., NYSDEC, "Generic Biocell and Biopile Designs for small-scale Petroleum-contaminated Soil Project," May 1, 1995

O'Neill, Christopher, P.E., NYSDEC, "Field Instrument Evaluation Project," May 1. 1995.

RISK ASSESSMENT

CHAPTER 33

An Overview of Risk-Based Corrective Action (RBCA) for Petroleum Release Sites

John B. Gustafson and **Curtis C. Stanley**, Shell Development Company, Houston, TX

Paul C. Johnson, Department of Civil Engineering, Arizona State University, Tempe, AZ

Dennis D. Rounds, South Dakota Petroleum Release Compensation Fund, Pierre, SD

INTRODUCTION

Currently, the methods for determining "how clean is clean" are being questioned and reevaluated across the country. Both industry and environmental regulatory agencies are making efforts to focus on risk when making decisions regarding the urgency for, and extent of, corrective action. While risk-based cleanups have been encouraged, many regulators, environmental consultants, and tank owners may be uncomfortable with the practice. This is due to a number of reasons including: lack of clear guidance, the perception that problems can be "risked away," inability to translate the CERCLA risk assessment paradigm to UST activities, and concern over residual liability issues.

In response to needs expressed by regulatory agencies and industry, the American Society for Testing and Materials (ASTM) undertook the challenge to develop a standardized approach to risk-based corrective action (RBCA) for petroleum release sites. In February 1993, a diverse and balanced task group was established by ASTM Subcommittee E50.01 on Storage Tanks, part of the E50 committee on Environmental Assessment. The task group is comprised of representatives from the U.S. Environmental Protection Agency (USEPA), state regulatory agencies, state cleanup funds, environmental consulting firms, and petroleum, insurance, and banking industries. The task group prepared draft documents that were sent out for review and balloting according to ASTM requirements. ASTM ES 38-94, "Guide for Risk-Based Corrective Action at Petroleum Release Sites," was approved as an ASTM Emergency Standard by the ASTM

Committee on Standards in mid-1994, and as an ASTM standard, E1739-95, in September of 1996.

Since its formal approval (and even during the draft phase), this document has received considerable attention. Discussions regarding its incorporation into many state and local regulatory programs are currently ongoing. In addition, ASTM, USEPA, and various industries are collaboratively developing awareness and training workshops related to this document. There is still, however, considerable confusion regarding the content and intended use of this standard, and how it relates to other agency and industry risk-assessment related activities. In this chapter we provide an overview of the E1739-95 standard.

WHAT IS RISK-BASED CORRECTIVE ACTION?

"Risk-Based Correction Action" (RBCA - pronounced like "Rebecca") is a term that is used quite liberally in the environmental industry. To many, RBCA is synonymous with "risk assessment," the scientific process for quantifying risks associated with exposure to chemicals in the environment. However, RBCA is intended as a new philosophy or framework for managing contaminant release sites. In this new approach, decisions related to urgency of response, target cleanup levels, and remedial measures are based on current and reasonable potential risks to human health and environmental resources. It is this broader definition of RBCA that is utilized in the ASTM E1739-95 "Guide for Risk-Based Corrective Action at Petroleum Release Sites." As we shall see, applying the RBCA process to a given site may or may not involve the preparation for a formal risk assessment.

HOW DOES RBCA DIFFER FROM CURRENT CORRECTIVE ACTION PROGRAMS?

Many of the components of the RBCA process are similar to those already practiced under current state and federal corrective action programs (site assessment, remedial measure selection, etc.). However, the philosophy underlying traditional and RBCA approaches is quite different.

Historically, the focus of corrective action programs has been to reduce the amount of contaminants present at a given site, with the ultimate goal being the achievement of background, or very low criteria (such as Maximum Contaminant Levels in ground water). On the other hand, RBCA decisions are dictated by current and potential risks posed by a site. Corrective action goals are then based on reducing these risks to some low, acceptable level. This may be achieved by reduction of contaminant concentrations, but it may also involve reducing the potential of exposure through the application of institutional controls, point-of-use water treatment, and the natural attenuation of contaminants.

ARE THERE OTHER APPROACHES FOR MANAGING CONTAMINANT RELEASE SITES?

So far, we have mentioned two approaches for managing contaminant release sites. In the traditional approach, the goal is to reduce contaminant concentrations below prescribed numerical standards, independent of site characteristics. In the RBCA approach, the goal is to reduce current and reasonable potential risks of adverse impacts to below some low, acceptable level. There is a third approach advocated by some, and this approach is often referred to as the "technology-based limits" approach. Here the goal is to reduce contaminant concentrations as much as is "technically feasible," given current technologies and practical constraints. Each of these three approaches has benefits and limitations. The more traditional approach is the easiest to consistently manage; however, the numerical standards are applied independent of site characteristics. The RBCA approach is better equipped to handle the diverse nature of sites, but assumes that we can indeed assess risks posed by sites. The technology-based approach is attractive as it provides an outlet for those parties who want to turn-off ineffective remedial systems. However, detractors will argue that this approach also forces the installation of remedial systems at all sites not meeting generic numerical criteria, and it inherently assumes that the most appropriate technologies will be applied correctly at every site. In addition, not all parties will easily agree on what "technically feasible" means, and that the approach will be applied inconsistently, depending more on the financial resources of the responsible party than risks posed by the site.

WHAT IS RISK ASSESSMENT, AND HOW HAS IT BEEN USED?

As stated above, the RBCA process is based on considerations of risk and exposure. While we shall see that this does not imply that formal risk assessments are prepared for each site, it is useful to review the sequence of risk/exposure assessment activities prescribed by most guidance documents (e.g., Risk Assessment Guidance for Superfund, USEPA 1989); these include:

a. identifying chemicals of concern
b. receptor identification
c. exposure analysis
d. dose-response analysis
e. risk and sensitivity/uncertainty quantification
f. risk management.

These basic activities are performed for all human health and ecological risk/exposure assessments, independent of whether the site of concern is a relatively small underground storage tank release site or a large-scale mixed-waste Superfund site. By their nature, these activities require a multidisciplinary approach involving people with a range of expertise, including site assessment, fate and transport of compounds in the environment, and human health and ecological effects. The degree of complexity required for steps (a) - (f) varies greatly, depending on the goal of the assessment and complexity of the site and surroundings.

The use of formal risk assessment has been a major component of the CERCLA program for some years, but its acceptance and use at the state and local level has been limited to date. There are many reasons for this: (a) most existing guidance has been written for complex assessments focused on large-scale sites, (b) in some cases regulators do not feel adequately trained to review formal risk assessments, (c) policy does not explicitly allow their use, or (d) it is perceived that the multitude of possible approaches will make the review process less manageable. In addition, in many cases there is a distrust resulting from the perception that a risk assessment is necessarily a complex mathematical exercise in which parameters can be adjusted to get the answer of choice. Risk assessment has also been viewed by some as an opportunity to delay implementing corrective action. In the past, risk assessments have been primarily used to justify the closure of sites where technology limitations have been encountered. These negative perceptions are obviously a hurdle to the implementation of RBCA; however, through the cooperative application of RBCA at demonstration sites, both regulators and responsible parties are beginning to realize the benefits of the RBCA approach.

CAN RISK ASSESSMENT BE INCORPORATED INTO THE CORRECTIVE ACTION PROCESS?

Considering the experience gained to date, the diverse nature of sites likely to be encountered, and the questions that must be answered (categorization immediacy of response, target levels, compliance zones, etc.), it is clear that a new risk assessment paradigm is needed. For this reason, the ASTM "Guide for Risk-Based Corrective Action at Petroleum Release Sites" is based on a "tiered" approach to risk and exposure assessment, where each tier refers to a different level of complexity. Assessment and remedial activities are appropriately tailored to site-specific conditions and risks. This flexibility allows RBCA to be more technically-defensible, protective, and hence cost-effective than traditional approaches under which all sites conform to uniform standards and procedures. While the RBCA process outlined in Figure 1 is not limited to a particular class of compounds, the E1739-95 standard emphasizes the application of RBCA to petroleum fuel releases.

The RBCA process is based on three tiers of possible activities, where the user begins at the first tier and then progresses to higher tiers, if warranted. As we shall see, by progressing through each tier, the activities of subsequent tiers become more focused and efficient. A discussion of each tier appears below.

Figure 1. **Risk-based corrective flowchart.**

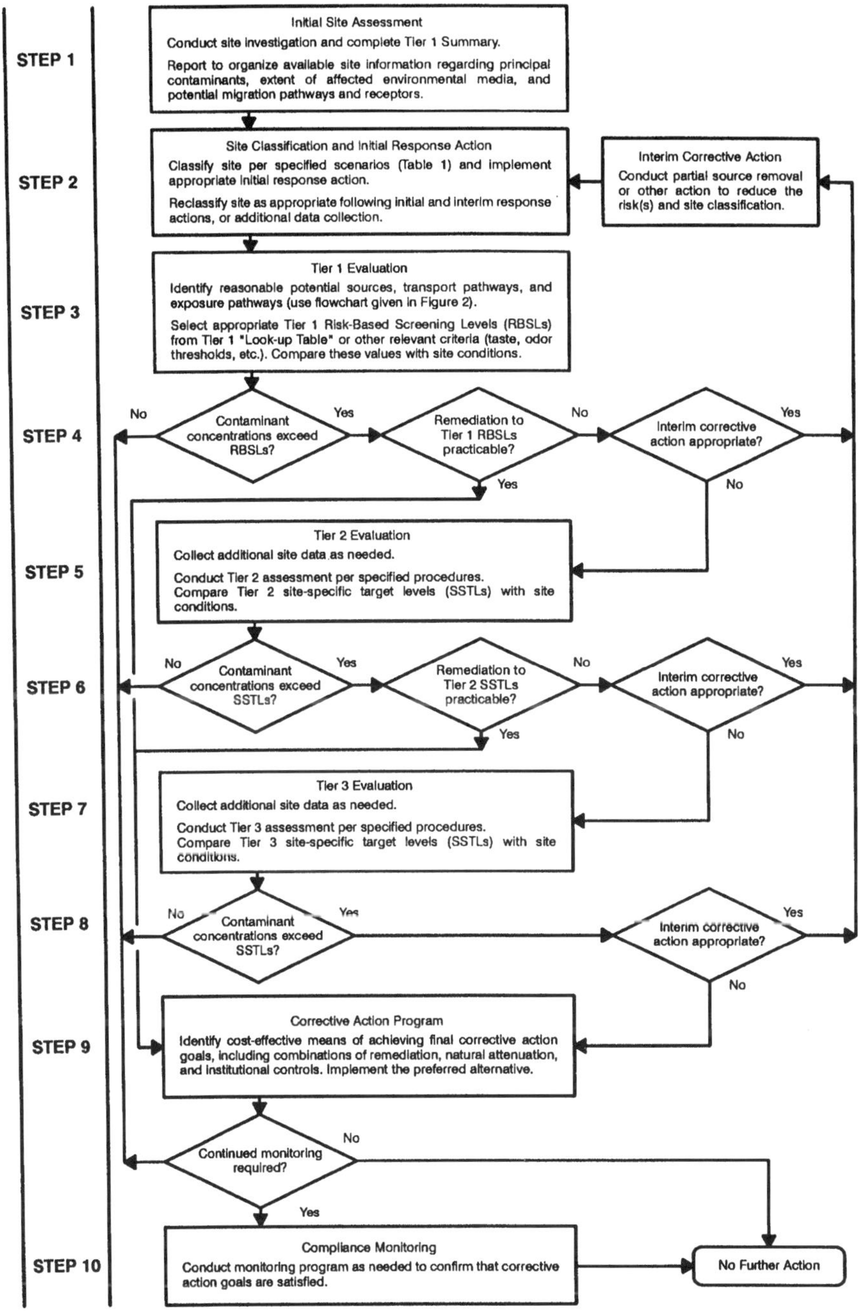

Tier 1: Conservative, But Efficient

In Tier 1, sites are assigned a Classification based on information collected from historical records, a visual inspection, and minimal site assessment data. The user is required to identify contaminant sources, obvious environmental impacts (if any), the presence of potentially impacted humans and environmental resources (e.g., workers, residents, water bodies, etc.), and potentially significant transport pathways (e.g., groundwater flow, volatilization, etc.). Sites are classified in order to ensure that resources are immediately applied toward those sites posing the greatest threats to human health and environmental resources. In some states, the classification process is also used to identify sites where agency oversight is not necessary and the corrective action is self-directed, or conducted under the oversight of "licensed site professionals."

Associated with each site classification is a prescribed immediate response action to ensure that human health and safety and environmental resources are protected. These actions range from emergency response activities necessary for sites posing short-term threats, to monitoring programs appropriate for those sites having a low potential for current or future impacts.

In addition, as part of Tier 1, a Look-Up Table containing screening level concentrations, is used to determine if site conditions satisfy the criteria for a quick regulatory closure, or if conditions exist that warrant a more site-specific evaluation of corrective action goals. Groundwater, soil, and vapor concentrations may be presented in this table for a range of potential pathways, exposure scenarios (residential, industrial, etc.), and types of contaminant sources (e.g., gasoline, crude oil, etc.). The values in this table may be a combination of health/risk-based screening levels, aesthetic (taste, odor, etc.) criteria or ecologically-based criteria. These values are applied consistently from site to site, but are"evergreen" in that they will change as new methodologies and parameters are developed. It is urged that these levels be established based on realistic exposure scenarios and the latest scientific evidence available.

Tier 2 and Tier 3: Site-Specific Corrective Action Goals - Increasing Site-Specific Data

Tiers 2 and 3 provide the user with options for determining site-specific target levels (SSTLs). In most cases, the decision to move to a higher tier is based on answers to the following questions:

- Are the assumptions used in a lower tier appropriate, relative to site-specific conditions?
- Are the goals established from a higher tier's analysis likely to be less costly to achieve?
- Is the cost for additional analyses acceptable, relative to the cost required to achieving the lower tier's goals?

It is important to note that the goal of all tiers is to achieve similar levels of protection. The only difference is that in moving to higher tiers, the user is able to develop more cost-effective corrective action plans because the conservative assumptions of earlier tiers are replaced with more realistic site-specific assumptions. However, additional site assessment data may be required. While both Tiers 2 and 3 involve developing site-specific goals, the major distinction between Tiers 2 and 3 is that Tier 2 analyses tend to be consistent with the level of site characterization data most often available (at UST sites), and Tier 3 often involves a much more significant increase in site-specific data requirements. Tier 2 analyses may involve the use of screening-level predictive models coupled with site assessment and monitoring data, and sometimes Tier 2 site-specific target levels are derived from the same equations used to calculate Tier 1 risk-based screening levels, except that site-specific parameters are used in the calculations. At other sites, the Tier 2 analysis may involve applying Tier 1 or Tier 2 risk-based screening levels at more probable points of exposure. These points are often referred to as "alternative points of compliance," and may be physically located at property boundaries or the edge of areas where access has been restricted by physical or institutional barriers. Tier 3 analyses often involve the use of complex numerical models and probabilistic analyses.

Corrective Action Selection

If exceedances of the selected target levels occur and corrective action is necessary, the user develops a corrective action plan in order to reduce the potential for adverse impacts. One option is to utilize traditional remediation processes to reduce contaminant concentrations below the target levels. Another equally viable option is to achieve exposure reduction (or elimination) through the use of institutional or physical barriers. Again, the goal here is risk reduction.

WHAT IS CONTAINED WITHIN THE ASTM GUIDE FOR RISK-BASED CORRECTIVE ACTION APPLIED AT PETROLEUM RELEASE SITES?

In general, the ASTM standard describes the framework for building a streamlined and technically-defensible "risk-based" corrective action program. While the ASTM standard focuses on petroleum release sites, the process is generally applicable to any contaminant and release scenario.

The main body of the ASTM standard describes a logical sequence of activities and decisions to be followed from the time contamination is suspected until regulatory closure is achieved. At all times, the decisions and required actions are based on ensuring that human health and beneficial uses of environmental resources are protected. The process or "framework" has been developed in such a way that resources are more effectively allocated.

Appendices are also included that contain supporting information on topics such as: (a) characteristics of refined petroleum fuels, (b) the approach used to develop the example Tier 1 Look-Up Table, (c) the role of predictive modeling in the RBCA process, (d) institutional controls, and (e) example applications of the RBCA process.

WHERE DO WE GO FROM HERE?

The ASTM standard describes the philosophy and basic framework for a risk-based corrective action process. It needs to be"customized" before application in a particular state or region. This will at a minimum involve developing, or adopting: (a) a site classification system, (b) a Tier 1 Look-Up Table, and (c) user-friendly ("cookbook") guidance written for the general public. It is envisioned that this customization of the E1739-95 RBCA process will be accomplished through collaboration between regulatory agencies,m responsible parties, the public, and other interested parties. The ASTM standard was written in such a way that, in many cases, only minor modifications to current programs are needed to conform with the ASTM approach.

CHAPTER 34

Soil Decreases the Dermal Bioavailability of Arsenic in a Chemical Mixture in Pig

Mohamed S. Abdel-Rahman, Gloria A. Skowronski, and **Abdelrazak M. Kadry**, Pharmacology and Toxicology Department, New Jersey Medical School, University of Medicine and Dentistry of New Jersey, Newark, New Jersey

Rita M. Turkall, Pharmacology and Toxicology Department, New Jersey Medical School, and Clinical Laboratory Sciences Department, School of Health Related Professions, University of Medicine and Dentistry of New Jersey, Newark, New Jersey

INTRODUCTION

Arsenic is a natural component of the earth's crust; therefore, it can be released to the environment. Low levels of arsenic from natural sources can be found in all environmental media, as well as in food; however, releases from human activities (e.g., metal smelting, chemical production and use, coal combustion, and waste disposal) can lead to substantial environmental contamination. Most releases of arsenic are to land or soil, primarily in the form of pesticides or solid wastes (ATSDR, 1993). The majority of dumping sites in the USA contain heavy metals, including arsenic. Arsenic is now the fifth most frequently found substance at National Priorities List (NPL) sites (ATSDR, 1995).

Regardless of the route of exposure to arsenic, the skin is a critical organ of arsenic toxicity, due to local absorption and binding to sulfhydryl-group-containing proteins (Hostynek et al., 1993). Chronic exposure to arsenic in industry, and its subsequent accumulation in the skin, has resulted in dermal lesions, Bowen's disease (precancerous dermatitis), and various forms of cancer (Hostynek et al., 1993; Montgomery and Waisman, 1941; Holmqvist, 1951; Birmingham et al., 1965; Magos, 1991; Cebrian et al., 1983). Skin contact with arsenicals by professional pesticide applicators has been reported to result in hyperkeratosis, necrosis, and ulceration of the skin (Hostynek et al., 1993; Leoncini, 1933; Tamponi, 1935). Available evidence also supports a causal role in the etiology of squamous cell cancers and basal cell carcinoma from exposure to arsenical medications, in addition to high levels of arsenic in drinking water (Hostynek, et al., 1993; Philipp, 1985; Dickerson, 1980).

Quantitative data are available in the literature on the dermal bioavailability of arsenic alone (Dutkiewicz, 1977; Wester et al., 1993) or adsorbed to soil (Wester et al., 1993). However, most exposures to chemicals are to mixtures, rather than to a single compound (Mumtaz et al., 1993). Since bioavailability studies are lacking following dermal exposures to arsenic in a mixture, and because quantitative studies to determine if the dermal penetration of arsenic in a mixture is influenced by a soil matrix are important in the health risk assessment process, this study was conducted to investigate the effect of two soils (sandy and clay) on the bioavailability of arsenic in a chemical mixture. In addition to arsenic, the chemical mixture consisted of chloroform, toluene, and phenanthrene. The chemicals included in the mixture were selected based on their frequency of detection at or near hazardous waste disposal sites, their known toxic potential, and their being representative of the various chemical classes (Yang, 1994).

Pig skin, which is morphologically and functionally similar to human skin, was used as the animal model in these studies because the availability of human skin is limited, and it is also difficult to control the gender, anatomical site, and condition of the donor skin (Meyer et al., 1978; Kao, 1990). Furthermore, in previous studies we demonstrated excellent correlation of the in vitro penetration of phenol or nickel through pig skin with the in vivo and in vitro penetration of these compounds reported in the literature for human skin (Skowronski et al., 1994; Abdel-Rahman et al., 1995).

METHODS

Chemicals, Animals and Soil

Arsenic in the form of H_3AsO_4 (Aldrich Chemical Co., Milwaukee, WI) was used as a carrier and labeled with ^{73}arsenic (Los Alamos National Laboratory, Los Alamos, NM).

Male York-Hampshire pigs (40-60 lb) were purchased from Cook College Farm, Rutgers University, New Brunswick, NJ.

Two soils were examined. The Atsion sandy soil containing 90% sand, 2% clay, and 4.4% organic matter was collected from an aquifer drill site of the Cohansey sand formation near Chatsworth in south central New Jersey. The Keyport clay soil containing 50% sand, 22% clay, and 1.6% organic matter was collected from the Woodbury formation near Moorestown in southwestern New Jersey. Both soils are representative of soil types widely distributed in the eastern, southern, and southwestern United States (USDA, 1977); USDA, 1972).

Penetration Studies

Whole pig skin was obtained from the costo-abdominal areas of sacrificed pigs. After the skin was washed with water and lightly shaved, it was placed in ice-cold HEPES-buffered (25 mM) Hanks' balanced salt solution (HHBSS), pH 7.4, containing 1% dextrose and gentamicin sulfate (50 mg/L), and was transported on ice to the laboratory, where it was immediately prepared for diffusion cells according to Bronaugh and Stewart (Bronaugh and Stewart, 1985). The skin was used 2-3 hours after the sacrifice time.

Excised skin was cut to a thickness of 200 μm with a dermatome (Padgett Electro-Dermatome model B, Padgett Instruments Inc., Kansas City, MO). Circular pieces of dermatomed skin were then mounted into Teflon flow-through diffusion cells modified for volatilization determinations (Crown Bio Scientific, Inc., Somerville, NJ). The exposed skin surface area was 0.64 cm^2. Diffusion cells were maintained at 35°C (yielding a skin surface temperature of 32°C) in an aluminum block holder, which was heated by a circulating water bath. The dermal side of each skin sample was bathed with receptor fluid at a flow rate of 3 mL/h by a multichannel peristaltic cassette pump (Manostat, New York, NY). The receptor fluid was HHBSS containing gentamicin sulfate (50 mg/L) and 10% fetal bovine serum (Sigma Chemical Co., St. Louis, MO). The HHBSS receptor fluid was aerated continuously with medical grade oxygen. Skin samples were allowed to equilibrate for 1 h before treatment. One tenth μCi ^{73}arsenic alone or in a mixture was administered to the epidermal surface of the skin in a total volume of 5-20 μL, or after the addition of 30 mg of sandy or clay soil. The chemical doses were 2.5 μg arsenic, 5.2 μg phenanthrene, 4.3 mg toluene, and 7.5 mg chloroform. Perfusate was collected in scintillation vials at 15 min intervals up to 1 h, at 1.5 and 2 h, then at 2 h intervals up to 16 h postdosing, using an automated fraction collector.

At the conclusion of the 16-hour study, a gentle stream of air was allowed to flow over the skin surface for 1 hour. Charcoal tubes (SKC Inc., Eighty-Four, PA), which were attached to the upper chambers of the diffusion cells, trapped chemical volatilizing from the skin surface. Radioactive compound was extracted from charcoal by shaking with 1 mL 6N hydrochloric acid. Unabsorbed chemical was washed off the skin surface once with 1 mL of a 1% aqueous soap solution (Ivory Liquid, Procter & Gamble, Cincinnati, OH), and twice with 1 mL distilled water. Cell tops were rinsed in the same manner. Skin samples were completely solubilized in Solvable (Packard Instruments Co., Inc., Meriden, CT) to determine the binding capacity of the skin. Radioactivity in receptor fluids, charcoal extracts, skin and cell washes, and skin digests was counted in Formula 989 liquid scintillation cocktail (Packard) by liquid scintillation spectrometry (LS 7500, Beckman Instruments Inc., Fullerton, CA). Sample quench was corrected by using the H-ratio method.

Data Analysis

All data were reported as the mean ± SEM. Statistical differences between treatment groups were determined by one-way analysis of variance (ANOVA) with Scheffe's test. A p value <0.05 was considered significant.

RESULTS

Figures 1 and 2 demonstrate the effect of sandy and clay soils, respectively, on the penetration rate of ^{73}arsenic alone or in a mixture through male pig skin in vitro. When arsenic was administered to skin as a mixture with chloroform, toluene, and phenanthrene, the rate of penetration was significantly increased, compared to arsenic alone from 30-45 min after treatment through the end of the study (Figure 1). For the mixture, the maximum rate of penetration (0.88 ± 0.20% initial dose/h) was observed between 1 and 1.5 h after dosing, while for

Figure 1. **The effect of sandy soil on the penetration rate of ^{73}arsenic alone or in a mixture through male pig skin in vitro.** Skin was treated with 0.1 μCi ^{73}arsenic alone or in a mixture with or without 30 mg soil. The chemical doses were 2.5 μg arsenic, 5.2 μg phenanthrene, 4.3 mg toluene, and 7.5 mg chloroform. Values (means + S.E.M.) represent the percent of the initial dose per hour collected in the receptor fluid at the indicated time interval (plotted at the midpoint of the collection period) from 13-14 replicates per treatment from 3 pigs.

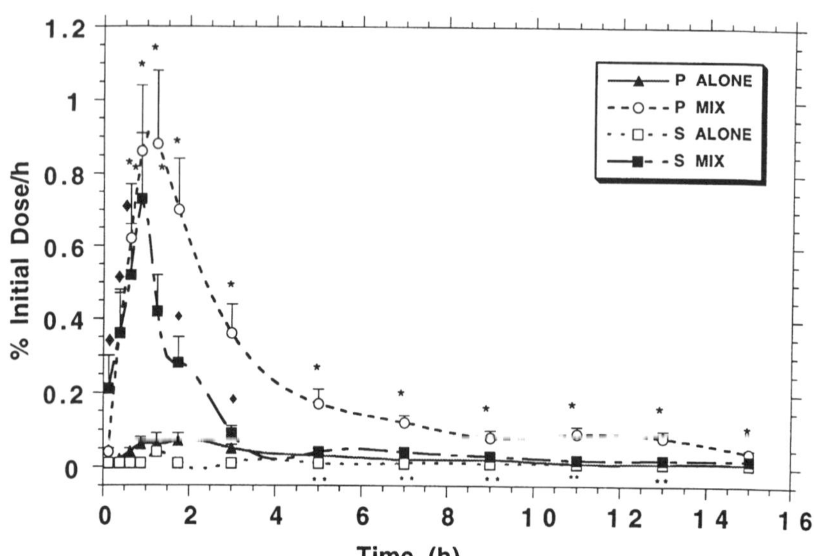

*Significantly different from arsenic alone, ANOVA (p<0.05).

**Significantly different from arsenic mixture.

◆ Significantly different from sandy soil-adsorbed arsenic alone.

the chemical alone, the maximum rate of penetration (0.07 ± 0.02% initial dose/h) was seen between 1.5 and 2 h. There were no significant differences in the rate of penetration between arsenic alone or adsorbed to sandy soil. However, the rate of dermal penetration for the arsenic mixture adsorbed to sandy soil was faster than the chemical alone adsorbed to sandy soil from 0-15 min through 45 min to 1 h after treatment. Furthermore, the penetration rate of the sandy soil-adsorbed mixture was less than the mixture without soil from the 1.5 to 2 h collection period through the 12-14 h collection period. Peak penetration for sandy soil-adsorbed arsenic mixture was 0.73 ± 0.18% initial dose/h between 45 min and 1 h. Clay soil was more effective than sandy soil in reducing the rate of penetration for the mixture (Figure 2). For each of the collection periods, the rate was no more than 0.03% initial dose/h when the mixture was adsorbed to clay soil. Moreover, the rate of penetration was the same for arsenic alone or adsorbed to clay soil.

Figure 2. **The effect of clay soil on the penetration rate of ^{73} arsenic alone or in a mixture through male pig skin in vitro.**
Skin was treated with 0.1 μCi ^{73}arsenic alone or in a mixture with or without 30 mg soil. The chemical doses were 2.5 μg arsenic, 5.2 μg phenanthrene, 4.3 mg toluene, and 7.5 mg chloroform. Values (means + S.E.M.) represent the percent of the initial dose per hour collected in the receptor fluid at the indicated time interval (plotted at the midpoint of the collection period) from 13-14 replicates per treatment from 3 pigs.

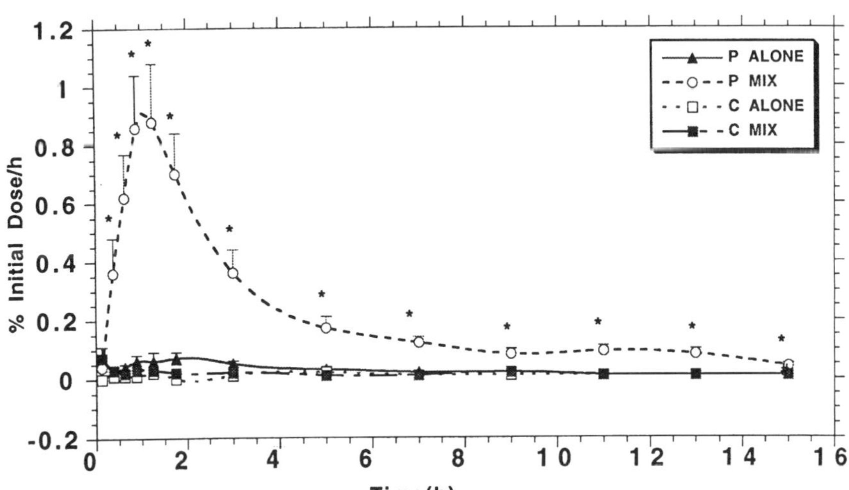

*Significantly different from arsenic alone, ANOVA ($p < 0.05$).
**Significantly different from arsenic mixture.

The cumulative penetration of arsenic describes the total amount as percent of initial arsenic which permeated skin within a designated time interval. A plot of the cumulative penetration of the arsenic mixture versus arsenic alone with or without soil is illustrated in Figures 3 and 4. There was a significant increase in the dermal bioavailability of arsenic when exposure occurred as a mixture rather than as the chemical alone. This was supported by the significantly higher cumulative percentage of arsenic penetrating skin from 45 min to 16 h after treatment. At the end of the 16-h study, the total cumulative percentage of radioactivity in the receptor fluid for the mixture was eight-fold more (3.1% initial dose) than with arsenic alone (0.4% initial dose). Both soils significantly decreased the dermal penetration of arsenic as a mixture into receptor fluid. Although the cumulative percentage of radioactivity in receptor fluid from 0-2 h was significantly higher for the sandy soil-adsorbed mixture (0.8%) than for arsenic alone adsorbed to sandy soil (0.03%), from 4-16 h significantly less ra-

Figure 3. The effect of sandy soil on the cumulative penetration of ^{73}arsenic alone or in a mixture through male pig skin in vitro. Skin was treated with 0.1 µCi ^{73}arsenic alone or in a mixture with or without 30 mg soil. The chemical doses were 2.5 µg arsenic, 5.2 µg phenanthrene, 4.3 mg toluene, and 7.5 mg chloroform. Values (means + S.E.M.) represent the percent of the initial dose collected in the receptor fluid from 0 time until the indicated time from 13-14 replicates per treatment from 3 pigs.

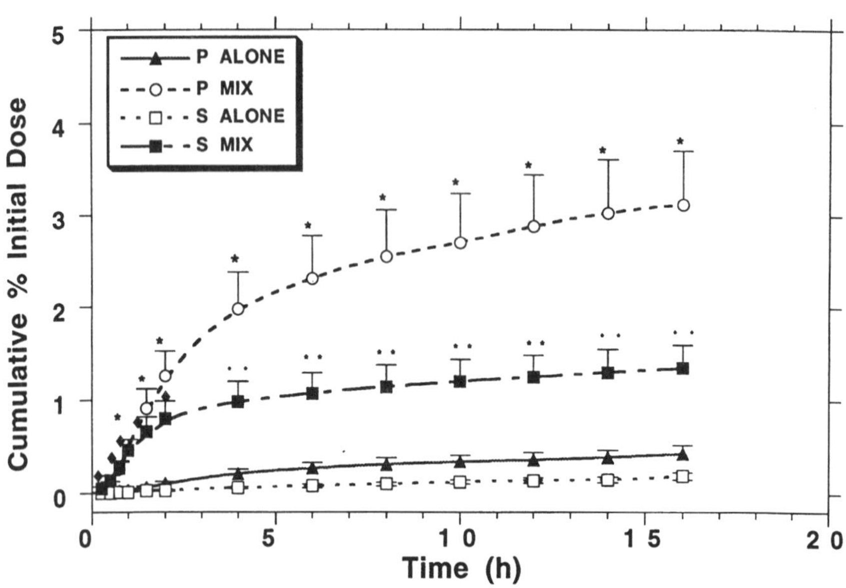

*Significantly different from arsenic alone, ANOVA ($p < 0.05$).

**Significantly different from arsenic mixture.

◆ Significantly different from sandy soil-adsorbed arsenic alone.

dioactivity from topical application of sandy soil-adsorbed arsenic mixture was collected in the receptor fluid than from the mixture in the absence of sandy soil (Figure 3). At 16 h there was more than a two-fold difference in the amount of radioactivity in the receptor fluid from arsenic mixture in the presence of sandy soil (1.4%) compared to arsenic mixture without soil (3.1%). With clay soil, the quantity of radioactivity in the receptor fluid for the mixture was significantly reduced from 30 min through 16 h versus the mixture in the absence of clay soil. Fifteen-fold more radioisotope penetrated the skin into receptor fluid within 16 h for the arsenic mixture (3.1%) than after adsorption to clay soil (0.2%) (Figure 4).

A summary of the amount of arsenic-derived radioactivity that penetrated pig skin into receptor fluid, as well as the binding capacity of arsenic to pig skin is shown in Tables 1 and 2. In the mixture, significantly more radioactivity re-

Figure 4. **The effect of clay soil on the cumulative penetration of ^{73}arsenic alone or in a mixture through male pig skin in vitro.** Skin was treated with 0.1 µCi ^{73}arsenic alone or in a mixture with or without 30 mg soil. The chemical doses were 2.5 µg arsenic, 5.2 µg phenanthrene, 4.3 mg toluene and 7.5 mg chloroform. Values (means + S.E.M.) represent the percent of the initial dose collected in the receptor fluid from 0 time until the indicated time from 13-14 replicates per treatment from 3 pigs.

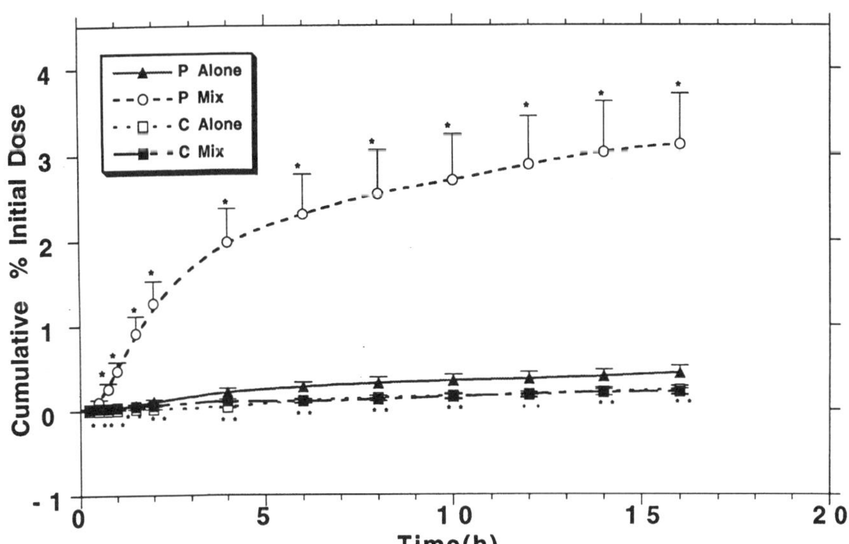

*Significantly different from arsenic alone, ANOVA ($p < 0.05$).

**Significantly different from arsenic mixture.

mained bound to skin (56.3% initial dose) compared to arsenic by itself (44.2%) (Table 1). The quantity of radioactivity bound to skin from the mixture when adsorbed to sandy soil was 50% less than that of the arsenic in a mixture. However, the bound radioactivity from the sandy soil-adsorbed mixture was threefold that of arsenic alone adsorbed to sandy soil (9.8%). Total penetration of arsenic (receptor fluid + bound to skin) was significantly increased when it was applied as a mixture to skin (59.4%) than when treatment was with chemical alone (44.6%). The total penetration of the sandy soil-adsorbed mixture was decreased (29.7% initial dose) relative to the mixture without soil, but increased versus arsenic alone adsorbed to sandy soil (10% initial dose). At the same time, significantly less radioactivity was found in the skin wash of the mixture (23.6% initial dose) than with arsenic alone (60.7%). However, more radioactivity was easily washed off the skin with soap and water after treatment with sandy soil-adsorbed mixture (46.4% of the dose). The binding capacity of arsenic in a mixture to skin was only 8.2% of the dose when adsorbed to clay soil (Table 2). Similarly, the total penetration of arsenic was significantly decreased to 8.4% when the mixture was adsorbed to clay soil. The majority of the compound, either alone or in the mixture, remains bound to the clay soil. No volatilization from the skin surface was detected from any of the treatments. Less than 5% of the administered dose was recovered in the cell washes (data not shown).

DISCUSSION

As previously reported for nickel (Abdel-Rahman et al., 1995), arsenic also binds to skin after dermal exposure. The affinity of arsenic for sulfhydryl groups leads to its accumulation in skin (Hostynek et al, 1993). In another animal study, Dutkiewicz measured the in vivo percutaneous absorption of inorganic arsenic by immersion of rat tails in solutions of radiolabeled sodium arsenate, $Na_2H^{74}AsO_4$ for 1 hour (Dutkiewicz, 1980). An initial accumulation of arsenic was observed in the skin, while uptake of arsenic into blood or tissues was undetectable for up to 24 h. This was followed by continuous, slow transport from the skin into the blood stream and to other tissues. The rate of skin absorption was 1.14 to 33.1 $\mu g/cm^2/h$ for 0.01 to 0.2 M aqueous solutions. Rat skin is much more permeable than pig skin or human skin, and can lead to a higher estimate of the absorbed dose. However, both studies (rat and pig) suggest that dermal exposure to arsenic leads initially to arsenic covalently binding to skin, but even after exposure ends, the bound arsenic may slowly penetrate into the systemic circulation and become distributed to tissues.

The in vitro penetration of arsenic-73 as H_3AsO_4 through human skin from water or soil was studied by Wester et al. (1993). Penetration was 1.9% and 0.8% of the dose from water and soil, respectively, when receptor fluid and skin concentrations were added. Most of the arsenic in their studies was recovered in the skin surface wash. In the previous study, human cadaver skin was used after

Table 1 **Summary of the Effect of Sandy Soil on the Penetration of ^{73}Arsenic Alone or in a Chemical Mixture Through Male Pig Skin in Vitro**

	Arsenic Alone	+ Sandy Soil	Arsenic Mixture	+ Sandy Soil
Receptor Fluid	0.4 ± 0.1[a]	0.2 ± 0	3.1 ± 0.6[b]	1.4 ± 0.2[c]
Bound To Skin	44.2 ± 2.8	9.8 ± 1.6[b]	56.3 ± 2.4[b]	28.3 ± 1.8[c,d]
Total Penetration	44.6 ± 2.8	10.0 ± 1.6[b]	59.4 ± 2.1[b]	29.7 ± 1.9[c,d]
Skin Wash	60.7 ± 3.4	76.1 ± 3.2[b]	23.6 ± 1.6[b]	46.4 ± 3.0[c,d]

[a] Values (means ± S.E.M.) represent the percent of the initial dose (n=13-14 replicates per treatment from 3 pigs).

[b] Significantly different from arsenic alone, ANOVA (p<0.05).

[c] Significantly different from arsenic mixture.

[d] Significantly different from sandy soil-adsorbed arsenic alone.

Table 2 **Summary of the Effect of Clay Soil on the Penetration of ^{73}Arsenic Alone or in a Chemical Mixture Through Male Pig Skin in Vitro**

	ARSENIC ALONE	+ CLAY SOIL	ARSENIC MIXTURE	+ CLAY SOIL
Receptor Fluid	0.4 ± 0.1[a]	0.2 ± 0	3.1 ± 0.6[b]	0.2 ± 0[c]
Bound To Skin	44.2 ± 2.8	5.8 ± 0.8[b]	56.3 ± 2.4[b]	8.2 ± 0.5[c]
Total Penetration	44.6 ± 2.8	6.0 ± 0.8[b]	59.4 ± 2.1[b]	8.4 ± 0.5[c]
Skin Wash	60.7 ± 3.4	37.1 ± 3.1[b]	23.6 ± 1.6[b]	32.1 ± 1.5
Bound To Soil		50.0 ± 0		50.0 ± 0

[a] Values (means ± S.E.M.) represent the percent of the initial dose (n=13-14 replicates per treatment from 3 pigs).

[b] Significantly different from arsenic alone, ANOVA (p<0.05).

[c] Significantly different from arsenic mixture.

storage under refrigeration for 5 days in Eagle's minimum essential medium. With cadaver skin, elapsed time from death to harvest of tissues and treatment, and storage of the cadaver, can contribute to variability in skin permeability (Kao, 1990). The present study was conducted on pig skin immediately after it was collected and transported in HHBSS from the farm to the laboratory. Using viable skin may explain why more arsenic was absorbed and became bound to pig skin, compared to human cadaver skin.

This investigation demonstrated that a chemical mixture of chloroform, toluene, and phenanthrene increases the dermal bioavailability of arsenic, but adsorption to either of two soils can lower the penetration of arsenic in the mixture. The combination of chloroform and toluene may be responsible for the increased permeability of arsenic in the mixture, since chloroform has been shown to completely destroy stratum corneum in human volunteers (ATSDR, 1993b), and toluene can cause skin damage due to its degreasing action (USEPA, 1983). Further studies are needed to study the mechanism of action for the observed penetration differences between arsenic alone and in a mixture.

The adsorption of arsenic present in soils depends on the type and amounts of sorbing components of soil (Yan-Chu, 1994). In this study, arsenic was more strongly adsorbed by clay soil. Clay minerals and their composition are the most fundamental factors affecting the biological toxicity of heavy metals in soils (Yan-Chu, 1994; Wu et al., 1991). Galba and Polacek found that the sorption of arsenate in soil increased with increasing clay particles, especially particles less than 0.001mm in diameter (Yan-Chu, 1994; Galba and Polacek, 1973).

In conclusion, dermal exposure to arsenic is a serious health hazard because arsenic forms a reservoir in skin from which it can eventually be released into the blood. In a mixture, the dermal bioavailability is significantly increased compared to arsenic alone. Both soils, clay more so than sandy soil, decreases the permeability of the arsenic mixture. As a result, the potential health risk from dermal exposure to arsenic is reduced if the metal is adsorbed to soil.

ACKNOWLEDGMENT

This research was supported in part through funding from the Hazardous Substance Management Research Center and the New Jersey Commission on Science and Technology.

REFERENCES

Abdel-Rahman, M.S., Skowronski, G.A., Kadry, A.M., Turkall, R.M., and Botrous, M.F. 1995. Soil Decreases the Dermal Bioavailability of Nickel in Male Pig *in Vitro*, In: *Hydrocarbon Contaminated Soils*, Volume V, (Kostecki, P.T, Calabrese, E.J. and Bonazountas, M., Eds.), Amherst, MA, Amherst Scientific Publishers.

(ATSDR) Agency for Toxic Substances and Disease Registry. 1993a. *Toxicological Profile for Arsenic*, p. 99. U.S. Department of Health and Human Services, Public Health Service, Atlanta, GA.

(ATSDR) Agency for Toxic Substances and Disease Registry. 1993b. Toxicological Profile for Chloroform, p. 44. U.S. Department of Health and Human Services, Public Health Service, Atlanta, GA.

(ATSDR) Agency for Toxic Substances and Disease Registry. 1995. *HazDat Database*, U.S. Department of Health and Human Services, Public Health Service, Atlanta, GA.

Birmingham, D.J., Key, M.M., Holaday, D.A., and Perone, V.B. 1965. An Out-Break of Arsenical Dermatoses in a Mining Community, *Arch. Dermatol.* 91, 457.

Bronaugh, R.L., and Stewart, R.F. 1985. Methods for *in Vitro* Percutaneous Absorption Studies. IV. The Flow-Through Diffusion Cell, *J. Pharm. Sci.* 74, 64.

Cebrian, M.E., Albores, A., Aguilar, M., and Blakely, E. 1983. Chronic Arsenic Poisoning in the North of Mexico, *Hum. Toxicol.* 2, 121.

Dickerson, O.B. 1980. Arsenic, In: *Metals in the Environment*, p. 1. (Waldron, H.A., Ed.), London, Academic Press.

Dutkiewicz, T. 1977. Experimental Studies on Arsenic Absorption Routes in Rats, *Environ. Health Perspec.* 19, 173.

Galba, J., and Polacek, S. 1973. Sorption of Arsenates under Kinetic Conditions in Selected Soil Types, *Acta Fytotech* 28, 187.

Holmqvist, I. 1951. Occupational Arsenical Dermatitis. A Study Among Employees at a Copper Ore Smelting Including Investigations of Skin Reactions to Contact with Arsenic Compounds, *Acta Derm. Venereol.* 31(Suppl.), 26,1.

Hostynek, J.J., Hinz, R.S., Lorence, C.R. , Price, M., and Guy, R.H. 1993. Metals and the Skin, *Crit. Rev. Toxicol.* 23, 171.

Kao, J. 1990. Validity of Skin Absorption and Metabolism Studies, In: *Methods for Skin Absorption*, p. 191, (Kemppainen, B.W. and Reifenrath, W.G., Eds.), Boca Raton, FL, CRC Press.

Leoncini, F. 1933. Contributo alla conoscenza delle dermatosi cutanee da arsenicismo professionale, *Zentralbl. Haut-Geschlechtskr.* 44, 188.

Magos, L. 1991. Epidemiological and Experimental Aspects of Metal Carcinogenesis: Physicochemical Properties, Kinetics, and the Active Species, *Environ. Health Perspect.* 95, 157.

Meyer, W., Schwarz, R., and Neurand, K. 1978. The Skin of Domestic Mammals as a Model for Human Skin with Special Reference to the Domestic Pig, *Curr. Probl. Dermatol.* 7, 39.

Montgomery, H., and Waisman, M. 1941. Epithelioma Attributable to Arsenic, *J. Invest. Dermatol.* 4, 365.

Mumtaz, M.M., Sipes, I.G., Clewell, H.J., and Yang, R.S.H. 1993. Symposium Overview. Risk Assessment of Chemical Mixtures: Biologic and Toxicologic Issues, *Fund. Appl. Toxicol.* 21, 258.

Philipp, R. 1985. Arsenic Exposure: Health Effects and the Risk of Cancer, *Rev. Environ. Health* 5, 27.

Skowronski, G.A., Kadry, A.M., Turkall, R.M., Botrous, M.F., and Abdel-Rahman, M.S. 1994. Soil Decreases the Dermal Penetration of Phenol in Male Pig *in Vitro*, *J. Toxicol. and Environ. Health*, 41, 467.

Tamponi, M. 1935. Dermatosi professionale da arsenico, *Zentralbl. Haut-Geschlechtskr.* **52**, 98.

(USDA) U.S. Department of Agriculture. 1972. National Cooperative Soil Survey: Official Series Description, Keyport Series, Soil Conservation Service, Washington, DC.

(USDA) U.S. Department of Agriculture. 1977. National Cooperative Soil Survey: Official Series Description, Atsion Series, Soil Conservation Service, Washington, DC.

(USEPA) U.S. Environmental Protection Agency. 1983. Health Assessment Document for Toluene, Office of Health and Environmental Assessment, Washington, DC, EPA-600/8-82-008F.

Wester, R.C., Maibach, H.I., Sedik, L., Melendres, J., and Wade, M. 1993. In Vivo and in Vitro Percutaneous Absorption and Skin Decontamination of Arsenic from Water and Soil, *Fund. Appl. Toxicol.* 20, 336.

Wu, L.S., Gu, Z.L., Xie, S.Q., and Zhou, D.Z. 1991. Factors Affecting Biological Toxicity of Heavy Metals in Soils and Their Regulation, *Huanjing Kexue* **12**, 12.

Yan-Chu, H. 1994. Arsenic Distribution in Soils, *Adv. Environ. Science Technol., Arsenic in the Environ.* 26, 17.

Yang, R.S.H. 1994. Toxicology of Chemical Mixtures Derived from Hazardous Waste Sites or Application of Pesticides and Fertilizers, In: *Toxicology of Chemical Mixtures*, p. 83, (Yang, R.S.H., Ed.), San Diego, CA, Academic Press.

CHAPTER 35

The Use of Terrestrial Toxicity Tests To Assess Toxicity and Efficacy of Remediation Activities on Contaminated Soils

David A. Pillard, ENSR Corporation, Fort Collins, Colorado

INTRODUCTION

A prerequisite for the cleanup of contaminated sites is the setting of remediation guidelines that, if met, will not pose unacceptable risk to organisms exposed to the site, human or otherwise. In most historical cases, remediation guidelines have been numerical concentrations established using a variety of methods, which can be thought of as absolute or relative (Jessiman et al., 1992). Absolute methods set cleanup standards for soils without considering site-specific factors. Relative methods take into account characteristics of the site as well as properties of the chemicals of potential concern (COPCs); therefore, an attempt is made to develop a more realistic assessment of the risk posed at a particular site.

Regardless of whether site-specific factors are employed, numerical cleanup standards require remediation to given chemical concentrations. Dependence on numerical levels in the cleanup of contaminated soils carries with it the same disadvantages faced in early efforts to reduce or eliminate ecological risk from effluents entering aquatic systems. One goal set by the Federal Water Pollution Control Act (often referred to as the Clean Water Act) was to prohibit the discharge of toxic pollutants in toxic amounts. Early NPDES (National Pollutant Discharge Elimination System) permits focused on meeting numerical levels; few required toxicity testing, which involves exposing test organisms to test materials and observing acute and chronic effects (Foster, 1985). However, toxicity testing has now become a critical component of most NPDES permits, since it became apparent that achieving predetermined chemical limits would not necessarily eliminate toxicity of effluents. Similarly, numerical soil cleanup guidelines may be under- or over-protective, depending upon possible unknown toxicity of additives in formulated mixtures, interactive effects of contaminant mixtures, natural variations in environmental levels, sublethal effects, and chemical and physical properties of soil.

It is beyond the scope of this chapter to provide a detailed review of all studies of compounds or contaminated soils where terrestrial toxicity tests have been employed, let alone the myriad of additional studies where soil extracts have been examined using traditional aquatic test organisms. The purpose of this chapter is to (1) discuss the advantages of toxicity tests in site assessment, (2) provide a general overview of toxicity tests that have been used to assess soil toxicity, concentrating primarily on those using terrestrial organisms, (3) describe relative sensitivities of plants and animals to various materials, and (4) discuss case studies of contaminated soil assessment.

RATIONALE FOR USING TOXICITY TESTS

An important advantage of toxicity tests is that they do not require an *a priori* knowledge of COPCs. The identification of the proper analytical methods to measure chemical concentrations in soil requires some insight into what chemicals may be present at a site in concentrations that pose a potential risk. It is often necessary to rely on historical information, or preliminary analytical scans of a minimal number of samples. Since historical information may not be complete or accurate or the sites may have been used for a variety of activities, the *a priori* identification of COPCs may be next to impossible. In addition, toxic metabolites may be formed during biodegradation that would not be detected during analysis (Hund and Transpurger, 1994). Roberts and Dorough (1984), for example, found that the degradation products of some pesticides were more toxic to earthworms than the parent compound.

Additives in commercial formulations of compounds can cause toxicity that would not be predicted from studies of the pure, active ingredient. Pillard (1995) found that formulated mixtures of ethylene and propylene glycol airplane deicers were significantly more toxic to aquatic organisms than pure ethylene and propylene glycol. Similarly, formulated glycol mixtures were also more toxic to lettuce (*lactuca sativa*) (Pillard and DuFresne, in preparation). Esteron 99, a commercially-available 2,4-D formulation was found to be more toxic than nearly pure 2,4-D to algae, *Daphnia*, and Microtox®, although root elongation was not affected (Miller et al., 1985; Thomas et al., 1986). Chemicals that are "inert" and added to mixtures to improve the efficacy of end-use products may, in fact, be active toxicants, or increase the bioavailability of the active toxicant. These inert additives would not likely be quantified during chemical analysis of water or soil collected from a hazardous waste site. Although it is sometimes common to complete a priority pollutant scan on environmental samples, doing so, more times than not, proves inconclusive (Miller et al., 1985). Samoiloff et al. (1983) found that the most toxic fraction of sediment from a lake contained none of the priority toxic chemicals; therefore, priority pollutant analysis would not have indicated the presence of any toxic materials at toxic levels.

Another benefit of toxicity tests is that, by exposing organisms directly to soils or soil extracts, it is possible to evaluate environmental factors that can modify toxicity. Many organic or inorganic materials show fluctuating levels of toxicity under varied conditions. In water, for example metal toxicity is greatly affected by hardness. Similarly, the chemical and physical characteristics of soil can affect both the toxicity and bioavailability of metals and organic compounds. Lanno et al. (1993) studied the effects of four organic compounds on *Lumbricus terrestris* using different soil types, ranging from very low organic matter (sand) to high organic matter and clay content. The soil with the highest organic matter and clay content showed the least toxicity to worms. A loam which had the same organic matter content as sand, but had a greater silt and clay content, was more toxic than the high organic matter soil, but less toxic than sand. The availability of metals may also be affected by soil texture as well as pH (Monenco Consultants Limited, 1984).

TEST ORGANISMS

Numerous organisms are available for use in soil toxicity tests. The organisms may be grouped into two categories: (1) aquatic organisms used to test eluates (elutriates), and (2) terrestrial organisms that can be used to test soils directly. Table 1 provides a summary of many of the organisms used to test soils. Some of the organisms that are used to test bulk soils directly (earthworms and plants) can also be used to test eluates by adding the eluate directly to an artificial soil or to filter paper to which the organism is exposed. Adding a liquid solution directly to a nontoxic control soil or to filter paper, or suspending plants at the surface of solution, has also been used to assess the effects of given concentrations of pure or formulated compounds (Roberts and Dorough, 1984; Pillard and DuFresne, in preparation; Wong and Bradshaw, 1982; Neuhauser et al., 1985). Many studies use an eluate to assess the effects of contaminants in soil. The eluate is considered representative of the soluble and, therefore, bioavailable portion of toxic materials. Such an assumption, however, may be incorrect. It is possible in some situations that insoluble materials are not released during extraction procedures, but could still affect organism performance (Thomas et al., 1986).

Animals

The most common animal used in soil toxicity tests is the earthworm, *Eisenia foetida*. *E. foetida* is a small to medium size worm reaching a length of approximately 100 mm (Fender, 1985). *Lumbricus terrestris*, a large worm (up to 350 mm), has also been used in testing. *E. foetida*, however, is much easier to culture than *L. terrestris*. Its life cycle is short, going from juvenile to reproducing adult in seven to eight weeks, and it survives readily in laboratory cultures with minimal care (Greene et al., 1989). In addition, since *E. foetida* is smaller than

Table 1. **Some Organisms Used in Toxicity Tests of Soils or Soil Extracts**

Common Name	Scientific Name	Common Name	Scientific Name
Aquatic Organisms Used to Test Soil Extracts (Eluates)			
Water Flea	*Daphnia magna*	Fathead Minnow	*Pimephales promelas*
Water Flea	*Daphnia pulex*	Alga	*Selenastrum capricornutum*
Water Flea	*Ceriodaphnia dubia*	Bacteria (Microtox®)	*Photobacterium phosphoreum*
Animals Used in Soil-Contact Studies			
Earthworm	*Eisenia foetida*	Earthworm	*Lumbricus rubellus*
Earthworm	*Lumbricus terrestris*		
Plants Used in Soil-Contact Studies			
Lettuce	*Lactuca sativa*	Corn	*Zea mays*
Radish	*Raphanus sativus*	Onion	*Allium cepa*
Cucumber	*Cucumis sativus*	Oat	*Avena sativa*
Cabbage	*Brassica oleracea*	Perennial Ryegrass	*Lolium perenne*
Soybean	*Glycine max*	Tomato	*Lycopersicon esculentum*
Carrot	*Daucus carota*	Mustard	*Brassica alba*
Chinese Cabbage	*Brassica campestris*	Rape	*Brassica napus*
Cress	*Lepidium sativum*	Turnip	*Brassica rapa*
Fenugreek	*Trifolium ornithopodioides*	Rice	*Oryza sativa*
Mungbean	*Phaseolus aureus*	Sorghum	*Sorghum bicolor*
Red Clover	*Trifolium pratense*	Wheat	*Triticum aestivum*
Vetch	*Vicia sativa*	Estuarine Marsh Grass	*Spartina alterniflora*
Freshwater Marsh Grass	*Echinochloa crusgalli*		

L. terrestris, less space is needed for culturing and less soil and smaller test chambers can be used during testing. Another species, *L. rubellus*, which has also been studied, is smaller than *L. terrestris* and has a shorter life cycle. However, maintaining healthy cultures of *L. rubellus* may be difficult (Roberts and Dorough, 1984). *E. foetida* is recommended for use in toxicity testing by the International Workshop on the Ecotoxicology of Earthworms (Greig-Smith, 1992).

Although questions have been raised as to the relative sensitivity of *Eisenia foetida*, studies indicate that *E. foetida* is generally as sensitive as other worms. Heimbach (1985) examined the toxicity of several pesticides to both *E. foetida* and *Lumbricus terrestris*, and found that both species showed similar sensitivities. Roberts and Dorough (1984) found that *L. rubellus* may be slightly more sensitive than *E. foetida*, but control mortality of *L. rubellus* was greater than for *E. foetida*.

Plants

Unlike soil animals, where only a few species have been used to test soil toxicity, several plant species have been studied, which has resulted in an extensive and varied database. The U.S. Environmental Protection Agency and the Organization for Economic Cooperation and Development (OECD) recommend several different species for testing, both dicots and monocots (Fletcher, 1991). Fletcher et al. (1985) reviewed the PHYTOTOX database for data related to the sensitivity of plants to herbicides. They pared down the available information by eliminating studies that did not have comparable data, and found that adequate data to make toxicological comparisons (for pesticides) were only available for a few species (e.g., oat, wheat, corn, sorghum, and soybean). Exclusion from this list does not, however, mean that a plant is not a good test species. Lettuce (*Lactuca sativa*) has become one of the more common plants species used in toxicity testing. Wang (1987) found that of cucumber (*Cucumis sativus*), lettuce, and millet (*Panicum miliaceum*), lettuce was the most sensitive to metals, although millet was more sensitive to organics. Lettuce has also been found to be more sensitive than radish (*Raphanus sativus*) or ryegrass (*Lolium perenne*) when tested using several commercially-available chemicals (Gorsuch et al., 1990).

The natural chemical and physical characteristics of soil will affect plant survival and growth, regardless of the presence of any material that might be considered toxic. Lettuce, for example, shows poor performance if soil is very wet (Pillard, 1996). Therefore, studies using soils from wetlands, for example, utilize species that are specifically adapted to very wet conditions. Walsh et al. (1991) used a marsh grass - *Echinochloa crusgalli* - to test liquid effluents and sediment on the survival and growth of seedlings. *E. crusgalli* is a freshwater marsh plant, but saltwater/estuarine marsh plants have also been studied, including *Spartina alterniflora* (Walsh et al., 1991). At some locations, it may be desirable to conduct tests using native plants found at, or near, the site. Using native plants provides a site-specific assessment. However, the test results may not be applicable to other locations, and there may be little historical data in the literature for comparison. Healthy, viable seeds of native plants can also be difficult to obtain.

TEST METHODS

The use of plants to evaluate the effects of chemicals goes back several decades and has its origins in the agricultural sciences before being adopted for toxicological investigations. Many different types of tests have, therefore, been performed over the years, and the methods are still quite varied. Like standardized aquatic toxicity tests, consistent techniques have been developed for soil toxicity tests. The U.S. Environmental Protection Agency (USEPA) recommends using *Eisenia foetida* and *Lactuca sativa* in direct assessments of soil toxicity at hazardous waste sites (Warren-Hicks et al., 1989). Water-extracts of soil (eluates) can also be evaluated using water-column organisms (e.g., *Daphnia magna*) or root elongation methods. A standard set of test protocols for conducting some soil toxicity tests has been developed by the USEPA and provides initial guidance for conducting soil tests (Greene et al., 1989). Additional test methods may be derived directly from the scientific literature. Following is a short description of some of the more basic tests.

Earthworm Survival Using *Eisenia foetida*

The earthworm survival study involves exposing the test organism to soil for 14 days and counting surviving organisms at test termination. To minimize differences between test soils or between a test soil and a control soil, all soils are hydrated to a set moisture level. This requires knowing (1) how much water is already in the soil (moisture fraction, MF), and (2) how much water the soil can hold at saturation (water holding capacity, WHC). Moisture fraction is determined by drying a known amount of soil at 100°C and measuring weight loss. Water holding capacity is determined by adding distilled water in excess to a known quantity of dried soil, letting the soil drain, and measuring the increase in weight (Greene et al., 1989). Test and control soils are typically hydrated to 75% WHC. Test chambers (e.g., quart or pint canning jars) are kept under continuous light. Test temperature will vary with the species being tested. The worms generally stay in the soil to avoid the light, although in very toxic soil they may leave the soil and climb on the side of the test chamber. The chambers are covered to prevent escape of the worms, although some provision is made for exchange of air (e.g., a hole in the jar lid). Organisms are not fed during the study.

Controls for an earthworm survival test may consist of either a field-collected soil obtained from a reference location or an artificial soil prepared in the laboratory. A reference soil collected near the hazardous waste site, but not impacted by the suspected toxicants, can provide relevant information on the site-specific ecological impacts of toxicants. In an ideal situation, the reference soil would be identical to the test soil except for the suspected toxicant. Therefore, any variations due to pH, organic content, grain size, etc., would be eliminated. However, it is often difficult to find an appropriate reference soil that mimics the characteristics of the test soil and is not contaminated. Even in a supposedly "clean" area, it is possible that unknown contaminants may affect test outcome. Because of the potential problems with field-collected reference soils, a laboratory-made artificial soil may prove to be an appropriate alternative. Although different researchers may use different recipes, a basic mixture consists of an organic source (e.g., sphagnum peat), silica sand, and kaolinite clay. Calcium carbonate is added to adjust the pH.

Like any toxicity test using environmental samples, earthworm studies may be on 100% site soil (a "screening" procedure), or a series of dilutions are set up using the control soil to dilute the site soil to a predetermined set of concentrations (e.g., 100, 50, 25, 12.5, and 6.25 percent site soil). Diluting the soil allows for the determination of a nontoxic level of soil, as indicated by the No Observed Effect Concentration (NOEC) or the LC_{50} (concentration at which 50 percent of the organisms in a test die). The use of a multiple-concentration test may be critical at hazardous waste sites to identify levels of contamination, and thus establish the most toxic areas.

While the earthworm survival study is effective in observing acute effects, chronic tests may also be performed that examine sublethal effects. The easiest sublethal effect to measure is weight change. Typically, a significantly lower weight (relative to the control) would be considered an indication of toxicity. Gibbs et al. (in press), however, found that higher weights may not necessarily indicate lower toxicity. A better alternative to growth may be reproduction. *Eisenia foetida* is hermaphroditic, with all worms having male and female reproductive organs. Because this species has a short life-cycle and is relatively prolific, two sexually-mature worms should produce cocoons and offspring unless stressed. Gibbs et al. (1996) describes an efficient method for measuring reproduction as a toxicity endpoint in *E. foetida.*

Seed Germination Using Bulk Soil

Seeds of the test species (e.g., lettuce) are placed on test (site) and control soil (16-40 mesh silica sand) which is covered with a coarse-grain sand. After the test chambers (usually plastic petri dishes) are hydrated, they are placed in a dark bag or box which is sealed to reduce water loss. After 48 hours, the test chambers are removed from the dark bag and transferred to a clear box or bag, which is sealed and placed back into an environmental chamber for an additional 72 hours. Seeds are considered germinated if the young seedling has visibly emerged through the cover sand.

Greene et al. (1989) suggests hydrating all chambers to 85% of the WHC. While seeds germinate well at 85% WHC, lettuce, cucumber, radish, and perennial ryegrass will also germinate well at moisture levels as low as 45% WHC (Pillard, 1996). Since germination can be reduced due to overwatering (lettuce is especially susceptible), a lower level of hydration may be preferable.

Although recommended solely as a germination study in Greene et al. (1989), this bulk soil study can also be used to measure root elongation, which may be a more sensitive sublethal endpoint. Because some soils can bind tightly together and thus tear roots, soils must be carefully loosened before removing plants for measurement. Soil is rinsed off the roots, and they are measured to the nearest mm. Shoot length can also be measured, although it may not be as sensitive as root elongation (Wong and Bradshaw, 1982; Gorsuch et al, 1990).

Root Elongation Using Soil Eluates

For situations where only soil extracts are to be tested, a root elongation test may be appropriate. To obtain extracts, soils are mixed with deionized laboratory water and combined in a continuous mixer for at least 48 hours. The water is then centrifuged and applied to filter paper in petri dishes. Seeds (usually 5) are placed on the filter paper, and the test chambers are kept in the dark for 5 days. Root length is measured to the nearest mm. Controls in this study consist of filter paper hydrated only with deionized laboratory water or an equivalent. Although sometimes thought to be a good measure of materials that are bioavailable, the eluate test has also been criticized as not representative of actual environmental conditions. Root elongation directly in bulk soil may be a preferable way to examine this sublethal endpoint.

Plant Survival and Early Growth

Plant survival and early growth tests examine effects on plants after they have germinated and achieved 3 to 5 days' growth. Seeds are germinated, *en masse*, in clean silica sand and transferred to test chambers (e.g., Styrofoam cups) containing site soil. Test duration is variable, depending upon the species being tested. *Echinochloa crusgalli* (a freshwater marsh grass), for example, is tested for 14 days, while *Spartina alterniflora* (an estuarine marsh grass) is tested for 21 days (Walsh et al., 1991). Test chambers are held at under a 16-hour light: 8-hour dark cycle; temperature is species-specific. The moisture level of the test chambers must be checked regularly, and water added to prevent them from drying out. Lack of nutrients can significantly affect plant growth and may produce false positive results; therefore, a nutrient solution (e.g., Hoagland's) is typically added to the test chambers at test initiation and periodically during the study. The artificial soil used in the survival and plant growth study is similar to the artificial soil used in the earthworm study. It consists of peat (usually sieved to remove large particles), fine, medium, and coarse sand, kaolinite clay, and silt (if available). Calcium carbonate is added to adjust the pH of this typically acidic mixture.

In addition to plant survival, final measurements may include seedling height and dry weight of the entire plant, or of plant parts. Separating belowground biomass from aboveground biomass is important because different chemicals may adversely affect different plant parts. Additionally, it is sometimes difficult to remove all soil particles from the roots, which can skew results and potentially mask toxicity.

SENSITIVITY OF TEST SPECIES

There is often a significant difference among test species as to their sensitivity to various toxicants. The decision of what organisms to use, as well as test methodology, may depend upon the suspected contaminants as well as the goals of the study. Thomas et al. (1986) used bioassays to characterize contamination at Rocky Mountain Arsenal in Colorado. Their study showed that, using soil extracts, algae and *Daphnia* were more sensitive to copper, cadmium, and zinc than either root elongation or earthworm survival. However, root elongation was the most sensitive endpoint when testing 2,4-D. Miller et al. (1985) also determined that algae (*Selenastrum capricornutum*) and *Daphnia* were the most sensitive to copper, cadmium, and zinc, although lettuce was the most sensitive of the higher plants tested. The greater sensitivity of higher plants to 2,4-D is likely to be related to the mode of action of the toxicant. 2,4-D is an auxin inhibitor, and algae have no auxin growth regulators. Fletcher et al. (1985), in their review of the PHYTOTOX database, determined that oat and wheat generally gave the most sensitive response to a wide range of plant herbicides; corn and sorghum were the least sensitive of the monocots, and cucumber was the most sensitive dicot. Wang (1987) found that lettuce was the most sensitive to cadmium, copper, hexavalent chromium, manganese, nickel, and zinc. Millet, however, was more sensitive to the phenol compounds (1986). Wong and Bradshaw (1982) tested nine metals and determined that copper was the most toxic to rye grass (*Lolium perenne*), with an EC_{50} of 0.02 ppm; root growth was found to be more sensitive than shoot growth. Gorsuch et al. (1990) also found that root elongation was more sensitive than plant height when seeds were exposed to 26 commercially-available chemicals; germination was not noticeably affected.

Earthworms are generally considered insensitive to a variety of toxicants (Miller et al, 1985). Thomas et al. (1986) found earthworm survival to be an insensitive endpoint when testing metals and 2,4-D, but survival was sensitive to diphenylamine. Roberts and Dorough (1984), however, reported that some metals were very toxic to *Eisenia foetida*, as were carbamate insecticides, when presented to the organism on filter paper. It is possible that when presented in a soil matrix toxicity is ameliorated by the chemical and physical characteristics of the soil. In another test exposing *E. foetida* to chemicals on filter paper, phenols were found to be the most toxic of the organic chemicals tested, while phthalates and PAHs were not very toxic (Neuhauser et al., 1985). Heimbach (1985) found carbofuran, methidathion, endosulfan, and aldicarb to be some of the more toxic materials to both *Lumbricus terrestris* and *E. foetida*.

APPLICATION OF TERRESTRIAL TOXICITY TESTS AT HAZARDOUS WASTE SITES

Despite the prevalence of chemical-based cleanup criteria at hazardous waste sites, terrestrial toxicity tests have been employed at several locations. One of the most studied waste sites in the western United States is Rocky Mountain Arsenal (RMA) near Denver, Colorado. RMA was used for weapons production and storage, and encompasses areas contaminated with metals and organic compounds. Thomas et al. (1986) collected water and soil samples from locations at RMA. Tests were run using both eluates and whole soil. In studies using eluates, algae (presumably *Selenastrum capricornutum*) was the most sensitive at some sites, while root elongation (using lettuce) was not affected. At other locations, however, algal growth was actually stimulated by the eluate extracted from site soil, while root elongation was significantly decreased, indicating different toxicants present at different locations. In addition, in some soil samples, lettuce germinated directly in soil was affected, while root elongation in water extracts prepared from the same soil was not, indicating that the phytotoxic component that impaired germination may not have been water soluble, and thus was not present in the eluate.

Seed germination and root elongation were used to tests soils at the Bunker Hill Superfund site (Linder et al., 1990). High levels of metals at this mine tailings waste site prompted the use of zeolites as amendments to sequester contaminants such as lead, cadmium, and zinc. Germination of a bush bean (*Phaseolus vulgaris*) and tall fescue (*Festuca arundinacea*), however, actually decreased with increasing zeolite concentration. Similarly, Lanno et al. (1993) found that although concentrations of parent organic compounds decreased in contaminated soils, toxicity, as measured by earthworm survival and seed germination, actually increased.

Remediation of contaminated soils and the subsequent decline in concentrations of parent compounds is not necessarily indicative of reduced toxicity, and therefore is not a good measure of ecological risk. Cleanup activities, such as the addition of chemical amendments, may reduce concentrations of the initial contaminants, but actually compound ecosystem risk by increasing the concentrations of toxic components either as additives or degradation products. If the risk is short-term and a necessary step in the elimination of toxic materials, then the risk may be acceptable. Toxicity tests stand out as an appropriate way of monitoring that risk. In addition, because test organisms react differently to different toxicants, and there is sometimes little correlation between species, it is recommended that a suite of tests with different organisms be used, rather than a single test (Hund and Transpurger, 1994; Roberts and Dorough, 1984; Miller et al., 1985; Fletcher et al., 1985).

Although laboratory tests have more commonly been used to assess soil toxicity, some researchers have performed in situ investigations by placing organisms in containers directly at the waste site. Callahan et al. (1991) studied soil toxicity at the Baird and McGuire Superfund Site in Holbrook, MA. This mixed upland/wetland forested area was used to mix pesticides, and is contaminated with chemicals such as DDT, DDE, DDD, and chlordane. Plastic test chambers were laid out in transects throughout the suspected contaminated area (and a reference area). *Lumbricus terrestris* was used in the study; five in each chamber. Survival was determined, and pesticide body burden was also measured. Nwosu et al. (1991) used in situ seed germination studies (with cucumber and wheat) to assess contamination at a study plot sprayed with 3,5-dinitro-N^4,N^4-dipropyl-sulfanilamide (Surflan™). In both studies, differences in organism response were indicative of the degree of contamination. Because soil is not removed from the waste site, there is a reduced chance of chemical or physical changes that could modify soil toxicity. However, onsite studies are subject to often significant variations in temperature, moisture, and light. Lack of experimental control may compromise results and make comparisons with other studies difficult.

SUMMARY

Long-term experience with toxicity tests, primarily aquatic but with a growing proportion of terrestrial studies, has indicated that assessment of chemical concentrations alone is not sufficient to define risk to organisms, populations, or ecosystems. Although numerous cleanup criteria exist for soils (Beyer, 1990), studies have shown that these may be over- or under-protective of the soil ecosystem. Terrestrial tests, although not as standardized as aquatic tests, have nevertheless been successfully used and appropriate test protocols developed. Toxicity tests can be an important component of waste site assessment, and complement traditional chemical criteria for initial site assessment as well as site monitoring and remediation.

REFERENCES

Beyer, W.N. 1990. Evaluating Soil Contamination, United States Department of the Interior, Biological Report 90(2).

Callahan, C.A., Menzie, C.A., Burmaster, D.E., Wilborn, D.C., and Ernst, T. 1991. On-Site Methods for Assessing Chemical Impact on the Soil Environment Using Earthworms: A Case Study at the Baird and McGuire Superfund Site, Holbrook, Massachusetts, *Environ. Tox. Chem.* 10, 817.

Fender, W.M. 1985. Earthworms of the Western United States. Part I. Lumbricidae, *Megadrilogica* 4, 93.

Fletcher, J.S. 1991. Assessment of Published Literature Concerning Pesticide Influence on Nontarget Plants, In: Plant Tier Testing: A Workshop to Evaluate Nontarget Plant Testing in Subdivision J Pesticide Guidelines, p. 28 (Fletcher, J.D. and Ratsch, H., Eds.), United States Environmental Protection Agency, EPA/600/9-91/041.

Fletcher, J.S., Muhitch, M.J., Vann, D.R., McFarlane, J.C., and Benenati, F.E. 1985. PHYTOTOX Database Evaluation of Surrogate Plant Species Recommended by the U.S. Environmental Protection Agency and the Organization for Economic Cooperation and Development, *Environ. Tox. Chem.* 4, 523.

Foster, R.B. 1985. Environmental Legislation. In: *Fundamentals of Aquatic Toxicology: Methods and Applications*, Chapter 20. (Rand, G.M. and Petrocelli, S.R., Eds.) Washington, DC, Hemisphere Publishing Corp.

Gibbs, M.H., Wicker, L.F., and Stewart, A.J. 1996. A Method for Assessing Sublethal Effects of Contaminants in Soils to the Earthworm, *Eisenia foetida*, Environ. Tox. Chem. 15, 360.

Gorsuch, J.W., Kringle, R.O., and Robillard, K.A. 1990. Chemical Effects on the Germination and Early Growth of Terrestrial Plants, In: *Plants for Toxicity Assessment, ASTM STP 1091*, p. 49 (Wang, W., Gorsuch, J.W., and Lower, W.R., Eds.), Philadelphia, PA, American Society for Testing and Materials.

Greene, J.C., Bartels, C.L., Warren-Hicks, W.J., Parkhurst, B.R., Linder, G.L., Peterson, S.A., and Miller, W.E. 1989. Protocols for Short Term Toxicity Screening of Hazardous Waste Sites, United States Environmental Protection Agency, EPA/600/3-88/029.

Greig-Smith, P.W. 1992. Recommendations of an International Workshop on Ecotoxicology of Earthworms, Sheffield, UK (April 1991), in *Ecotoxicology of Earthworms*, p. 247 (Greig-Smith, P.W., Becker, H., Edwards, P.J., and Heimbach, F., Eds.) Andover, UK, Intercept.

Heimbach, F. 1985. Comparison of Laboratory Methods, Using *Eisenia foetida* and *Lumbricus terrestris* for the Assessment of the Hazard of Chemicals to Earthworms, *J. Plant Dis. Protect.* 92, 186.

Hund, K. and Transpurger, W. 1994. Ecotox-Evaluation Strategy for Soil Bioremedation Exemplified for a PAH-Contaminated Site, *Chemosphere* 29, 371.

Jessiman, B., Richardson, G.M., Clark, C. and Halbert, B. 1992. A Quantitative Evaluation of Ten Approaches to Setting Site-Specific Cleanup Objectives, *J. Soil Contam.* 1, 39.

Lanno, R.P., Stephenson, G.L., and Wren, C.D. 1993. The toxicity of Four Chemicals to *Lumbricus terrestris* in Three Natural Soils, presented at the 14th Annual Meeting, Society or Environmental Toxicology and Chemistry, Houston, TX, November 14-18.

Lanno, R.P., Wren, C.D., and Stephenson, G.L. 1993. Earthworm and Seed Germination Toxicity Tests to Assess the Efficacy of Bioremediation of Chlorophenol and PAH-Contaminated Soils, presented at 14th Annual Meeting, Society of Environmental Toxicology and Chemistry, Houston, TX, November 14 to 18.

Linder, G., Greene, J.C., Ratsch, H., Nwosu, J., Smith, S., and Wilborn, D. 1990. Seed Germination and Root Elongation Toxicity Tests in Hazardous Waste Site Evaluation: Methods Development and Applications, In: *Plants for Toxicity Assessment, ASTM STP 1091*, p. 177 (Wang, W., Gorsuch, J.W., and Lower, W.R., Eds.), Philadelphia, PA, American Society for Testing and Materials.

Miller, W.E., Peterson, S.A., Greene, J.C., and Callahan, C.A. 1985. Comparative Toxicology of Laboratory Organisms for Assessing Hazardous Waste Sites, *J. Environ. Qual.* 14, 567.

Monenco Consultants Limited. 1984. Development of Clean-Up Criteria for Soils at the Pincher Creek Gas Plant, prepared for Gulf Canada Resources, Inc., Report 10.

Neuhauser, E.F., Loehr, R.C., Malecki, M.R., Milligan, D.L., and Durkin, P.R. 1985. The Toxicity of Selected Organic Chemicals to the Earthworm, *Eisenia foetida*, *J. Environ. Qual.* 14, 383.

Nwosu, J.U., Ratsch, H., and Kapustka, L.A. 1991. A Method for On-Site Evaluation of Phytotoxicity at Hazardous Waste Sites, In: *Plants for Toxicity Assessment: Second Volume, ASTM STP 1115*, p. 333 (Gorsuch, J.W., Lower, W.R., Wang, W., and Lewis, M.A., Eds.), Philadelphia, PA, American Society for Testing and Materials.

Pillard, D.A. 1995. Comparative Toxicity of Formulated Glycol Deicers and Pure Ethylene and Propylene Glycol to *Ceriodaphnia dubia* and *Pimephales promelas*, *Environ. Tox. Chem.* 14, 311.

Pillard, D.A., 1996. Adjusting Hydration Water Volume to Decrease Preparation Time in Seed Germination Studies, Bull. Contam. Toxicol. 56, 599.

Pillard, D.A. and DuFresne, D. Phytotoxicity of Formulated Glycol Deicers and Pure Ethylene and Propylene Glycol, in preparation.

Roberts, B.L. and Dorough, H.W. 1984. Relative Toxicities of Chemicals to the Earthworm *Eisenia foetida*, *Environ. Tox. Chem.* 3, 67.

Samoiloff, M.R., Bell, J., Birkholz, D.A., Webster, G.R.B., Arnott, E.G., Pulak, R., and Madrid, A. 1983. Combined Bioassay-Chemical Fractionation Scheme for the Determination and Ranking of Toxic Chemicals in Sediments, *Environ. Sci. Technol.* 17, 329.

Thomas, J.M., Skalski, J.R., Cline, J.F., McShane, M.C., Simpson, J.C., Miller, W.E., Peterson, S.A., Callahan, C.A., and Greene, J.C. 1986. Characterization of Chemical Waste Site Contamination and Determination of Its Extent Using Bioassays, *Environ. Tox. Chem.* 5, 487.

Walsh, G.E., Weber, D.E., Simon, T.L., and Brashers, L.K. 1991. Toxicity Tests of Effluents with Marsh Plants in Water and Sediment, *Environ. Tox. Chem.*, 10, 517.

Walsh, G.E., Weber, D.E., Simon, T.L., Brashers, L.K., and Moore, J.C. 1991. Use of Marsh Plants for Toxicity Testing of Water and Sediment, In: *Plants for Toxicity Assessment, STP 1115*, p. 341 (Gorsuch, J.W., Lower, W.R., Wang, W., and Lewis, M.A., Eds.), Philadelphia, PA, American Society for Testing and Materials.

Wang, W.C. 1986. Comparative Toxicology of Phenolic Compounds Using Root Elongation Method, *Environ. Tox. Chem.* 5, 891.

Wang, W. 1987. Root Elongation Method for Toxicity Testing of Organic and Inorganic Pollutants, *Environ. Tox. Chem.* 6, 409.

Warren-Hicks, W., Parkhurst, B.R., and Baker, Jr., S.S. 1989. Ecological Assessment of Hazardous Waste Sites: A Field and Laboratory Reference, United States Environmental Protection Agency, EPA/600/3-89/013.

Wong, M.H. and Bradshaw, A.D. 1982. A Comparison of the Toxicity of Heavy Metals, Using Root Elongation of Ryegrass, *Lolium perenne*, *New Phytol.* 91, 255.

CHAPTER 36

Selection of Ecological Toxicity Benchmarks for an Ecological Risk Assessment of Terrestrial Wildlife

Kristin G. Lawrence, Jennifer C. Kotanchik, Kenneth G. Symms, and **Denice H. Wardrop**, Environmental Standards, Inc., Valley Forge, Pennsylvania

INTRODUCTION

The ecological risk assessment (ERA) process, like the human health risk assessment process, is based on two major elements: characterization of exposure and characterization of ecological effects. As noted in the Environmental Protection Agency (EPA's) Framework for Ecological Risk Assessment (USEPA, 1992a), the EPA realized that, while the traditional risk assessment paradigm (i.e., hazard identification, exposure assessment, dose-response toxicity assessment, and risk characterization) is generally applicable, ecological risk assessment differs from human health risk assessment in three points of emphasis: (1) ecological risk assessment endpoint receptors can be characterized at a population or community level rather than an organism level; therefore, characterization of effects may be more complex than for individual species; (2) receptors can include species from such a wide taxonomic range (e.g., earthworms to mink) that consequently, no one set of measurement endpoints can be generally applied; and (3) possible effects of nonchemical stressors may be included. In order to accommodate these differences, the EPA set forth a unique paradigm (Figure 1) for conducting an ERA (USEPA, 1992). The basic components are described briefly below.

Problem Formulation

Problem formulation presents a preliminary characterization of exposure and effects and the examination of data needs, issues, and objectives in order to define the scope, objectives, and feasibility of the assessment. This step is comparable to planning issues that are addressed at the beginning of all risk assessments. Early involvement of risk assessment personnel allows significant tailoring of the sampling plan to accommodate specific ecological risk assessment hypotheses. In addition, information compiled during the problem formulation can be used to select ecologically-based endpoints.

Figure 1. **Ecological Risk Assessment Paradigm.**

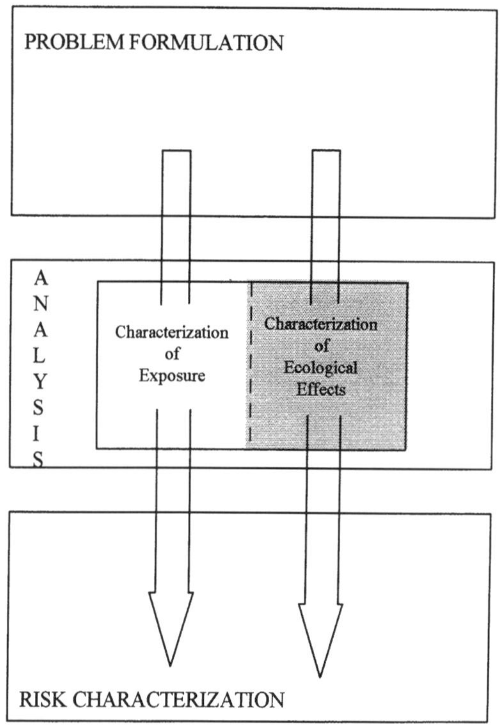

Analysis

Analysis is a characterization of exposure and ecological effects. The analysis phase explicitly addresses the two main requirements of risk: the inherent ability of a stressor to cause adverse effects, and the potential for co-occurrence of a stressor and an ecological component. This step is comparable to the hazard identification, exposure assessment, and dose-response assessment units in the traditional paradigm. Therefore, the analysis phase consists of the following subsections: selection of chemicals of concern (COCs), selection of receptors of concern (ROCs), selection of assessment and measurement endpoints, exposure assessment, and ecological effects assessment.

Risk Characterization

Risk characterization evaluates the likelihood of adverse ecological effects associated with exposure to the identified stressors, including a discussion of uncertainties and the strengths and weaknesses of the assessment. This step is comparable to risk characterization in the traditional paradigm.

This chapter will focus on the characterization of ecological effects component of the Analysis phase (Figure 1). The ecological effects assessment quantitatively links concentrations of contaminants to adverse effects in receptors (USEPA, 1991). Literature reviews, field studies, and/or laboratory tests provide the dose-response information; however, most ERAs do not require all three types of studies (1992b). Literature reviews generally provide sufficient toxicity information to derive a toxicity benchmark and complete an ERA (Maughan, 1993).

DERIVATION OF TERRESTRIAL TOXICITY BENCHMARKS

The relationship between the stressor and the assessment and measurement endpoints identified during problem formulation is analyzed in the characterization of ecological effects (USEPA, 1992a). In this process, a toxicity benchmark value, defined here as an endpoint indicating a specific toxic effect in a particular species based on laboratory investigations, is derived. These toxicity values are required for comparison to intake concentrations (estimated in the Exposure Assessment) in order to characterize ecological risk via the quotient method. It is essential that the outputs of the exposure assessment and the effects assessment are comparable. That is, it is necessary to determine if the receptors are directly exposed to the COC (i.e., direct contact with water, soil or sediment), or if they are exposed through indirect pathways such as accumulation in food chains (USEPA, 1994a). It is equally important to consider the route of exposure: will the receptor ingest, inhale, or dermally absorb the COC. These factors should be periodically considered during the ERA in order to ensure the critical exposure pathways are complete (Maughan, 1993). For every complete exposure route, a toxicity benchmark should be derived for each respective COC and ROC (USEPA, 1994a).

A large quantity of benchmarks for aquatic species have been reported in the literature, but information for terrestrial species is sparse. Therefore, the selection of benchmarks is largely constrained by the availability of published data and by the need to match the benchmarks reported in the literature to the ecological ROC at a particular site. A number of sources of toxicity information exist, but these sources vary with regard to the availability and strength of supporting evidence. The following protocol has been established for the determination of toxicity indices; it defines both a hierarchy of sources to be consulted and the methodology for the determine toxicity values. This protocol has been developed in the absence of any available EPA methodology, but is intended to represent both the best sources and methods currently available and those in current use by the EPA. Toxicity values for the COC at a given site can be obtained by utilizing methodology similar to that utilized to develop toxicity reference doses (RfDs) in a human health risk assessment:

1. Toxicity values should first be obtained from the following Oak Ridge National Laboratory draft Screening Benchmarks for Ecological Risk Assess-

ment databases (Opresko, Sample, and Suter, 1994). This source currently supersedes all other information sources.

2. Wildlife-specific toxicity values for compounds that are not available in the Oak Ridge databases are derived using a variety of sources for uncorrected toxicity values, such as:
   a. IRIS (Integrated Risk Information System) database (IRIS, n.d.). This database contains the Reference Doses (RfDs) and Cancer Slope Factors (CSFs), which have been verified by the EPA's RfD and Carcinogen Risk Assessment Verification Endeavor (CRAVE) workgroups.
   b. ATSDR (Agency for Toxic Substances and Disease Registry). Toxicological Profiles (ATSDR, 1990). ATSDR is in Athens, Georgia, and is part of the U.S. Department of Health and Human Services.
   c. Hawley's Condensed Chemical Dictionary (Sax and Lewis, 1987).
   d. RTECS (Registry of Toxic Effects of Chemical Substances) (NIOSH, n.d.). A continuously updated electronic database maintained by the National Institute for Occupational Safety and Health (NIOSH). Accessed through TOXNET, National Library of Medicine, Bethesda, Maryland.

Note: When utilizing the Quotient Method in an ERA, the most appropriate endpoint is the highest No-Observable-Adverse-Effect-Level (NOAEL) published from a well-conducted study (USEPA, 1994A; DeSesso and Price, 1990). As noted in the EPA Guidance notes (USEPA, 1994b), chronic tests have greater resolution and, therefore, are more useful than subchronic or acute toxicity tests. All toxicity values presented in the Oak Ridge Database (Opresko et al., 1994) are based on chronic NOAELs; however, these values are not available for all constituents. If the toxicity value available is not a chronic NOAEL, conversion to a NOAEL is performed using the methodology outlined in Figure 2. This methodology was derived by conjoining procedures found in Ford, Applehans, and Ober (1992) and U.S. EPA's Ecological Risk Assessment Guidance for Superfund: Process for Designing and Conducting Ecological Risk Assessments (USEPA, 1994a). If there is no NOAEL, LOAEL, LD50, or other toxicity data for a given constituent, quantitative structure-activity relationships (QSARs) can be utilized (USEPA, 1992a).

3. If there is no species-specific information for the ROC at a given site, wildlife-specific toxicity values subsequently can be derived from the NOAEL using the body-weight correction methodology presented in the Oak Ridge database (Opresko et al., 1994).
4. Some baseline studies have been conducted using taxonomic extrapolation. Hoffman et al. (1990) compared the differences in the rate of xenobiotic metabolism by vertebrates and found the relative rate of metabolism to be, in general, mammals > birds > fish, and that bioaccumulation tends to be inversely related to rate of xenobiotic metabolism. The EPA (USEPA, 1994c)

Figure 2. **Deviation of No Observable Adverse Effect Levels.**

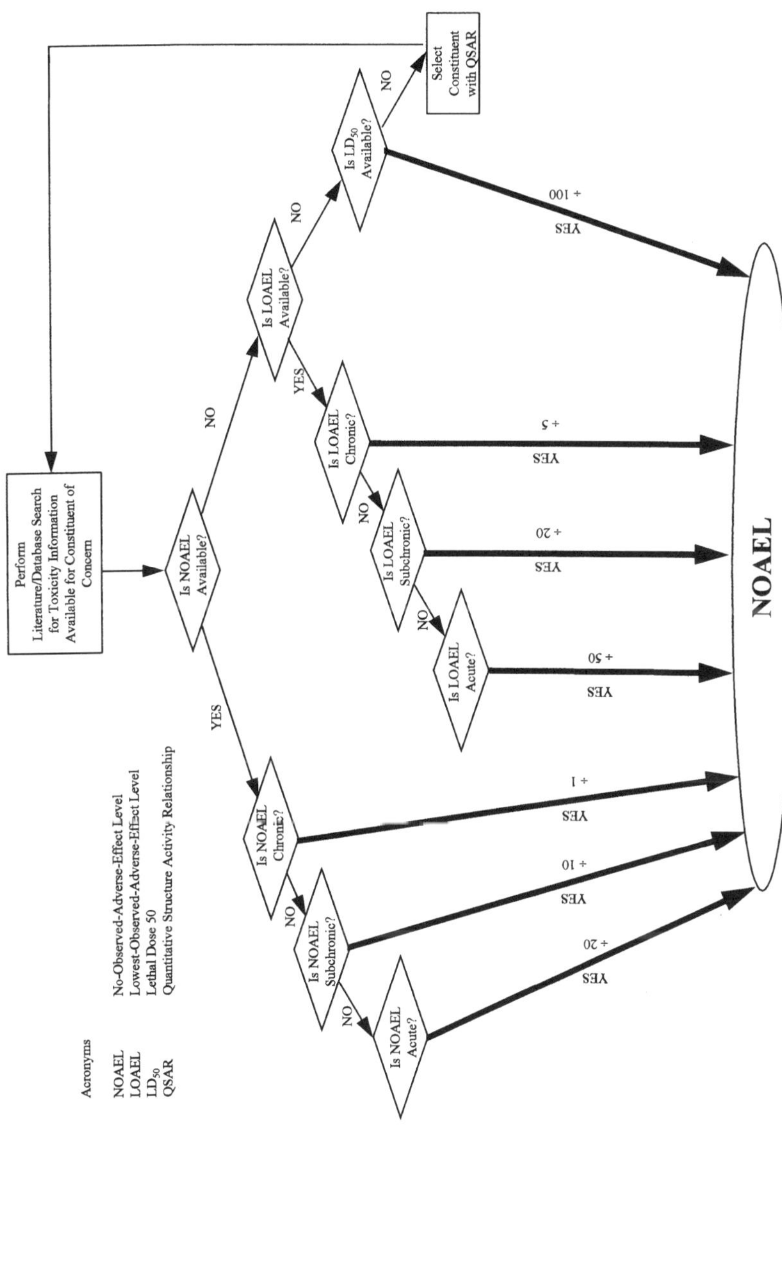

presented a study comparing rats with starlings, starlings with red-winged blackbirds, and mallard ducks with bullfrogs which indicated that variation between classes was one to two orders of magnitude greater than variation among classes. While these studies set a good foundation for a potential to extrapolate between classes, no methodology currently is endorsed by the EPA for deriving toxicity values from one class to another. In the absence of any guidance, an additional safety factor of 10 is, therefore, often applied to the available NOAEL in the study to convert to the appropriate class toxicity value.

5. If an endangered species is selected as a ROC, an additional safety factor of 10 should be added (Gaudet et al., 1995).

A generic example of toxicity benchmarks developed for constituents, in order that potential risks posed by all constituents can be quantitatively evaluated with the objective of minimizing inherent uncertainty in the process, is presented in Table 1.

Table 1. **Toxicity Benchmarks for the Raccoon**

Chemicals	Benchmark Toxicity Value	Safety Factor	Surrogate Toxicity Value	Source	Reference
Pyrene	10.60	100	2700	LD50 for rat. Converted using Opresko's equation #4	Sax & Lewis, 1987
Cobalt	1.96	NA	5	NOAEL for rat based on reproduction. Converted using Opresko's equation #4	ATSDR, 1990
Copper	6.06	NA	11.71	NOAEL for mink based on reproduction. Converted using Opresko's equation #4	Opresko et al., 1994
Magnesium	10	100	1000	Subchronic NOAEL for Japanese quail based on physiology.	Opresko et al., 1994

UNCERTAINTIES

As with human health, the effects characterization portion of the assessment has the most uncertainty. Several types of extrapolations necessary in this step compound uncertainties. For example, taxonomic extrapolation creates uncertainty as to whether an untested species exhibits the same sensitivity to a given stressor as a tested species (USEPA, 1992a). In addition to extrapolation between taxa, pronounced variation in response to stressors has been reported for laboratory and field organisms of the same species; similar reports of variation apply to laboratory organisms that are exposed to the field and field organisms that are used in the laboratory (IRIS, n.d.). Extrapolation between responses in order to derive a NOAEL from a LOAEL or an LD50 also contribute to uncertainty (USEPA, 1992a). Likewise, by utilizing a NOAEL as an endpoint, the focus is on an individual level (Ford et al., 1992), rather than a population level.

For ecological risk assessment this presents an overly conservative approach (Mayernik and Fehrenkamp, 1992) in that it places judgment in the hands of the risk assessor rather than the risk manager. However, at the present time most EPA guidance as well as other publications recommend this approach (USEPA, 1994a; DeSesso and Price, 1990; Ford et al., 1992). Uncertainty also occurs when equivalence is assumed between short-term mortality and long-term depauperation, or by assuming that reduced individual fecundity equates to reduced population size (USEPA, 1992a). These assumptions do not incorporate synergistic effects between individuals and populations (e.g., competition and trophic levels). Although these uncertainties deal more directly with risk characterization, they are still intimately linked with the characterization of ecological effects.

REFERENCES

ATSDR. 1990. Agency for Toxic Substances and Disease Registry, Toxicological Profiles, United States Department of Health and Human Services, Atlanta.

DeSesso, J.M., and Price, F.T. 1990. *General Guidance for Ecological Risk Assessments at Air Force Installations*, Brooks Air Force Base, The Mitre Corporation.

Ford, K.L., Applehans, F.M., and Ober, R. 1992. Development of Toxicity Reference Values for Terrestrial Wildlife, in *Proceedings of the HMC/Superfund Conference*, Washington, DC, 1992, p. 803.

Gaudet, C.L., Power, E.A., Milne, D.A., Nason, T.G.E., and Wong, M.P. 1995. A Framework for Ecological Risk Assessment at Contaminated Sites in Canada, Part I: Review of Existing Approaches, *Hum. Ecol. Risk Assess.*, **1**(2), 43.

Hoffman, D.J., Rattner, B.A., and Hall, R.J. 1990. ES & T Series: Wildlife Toxicology, *Environ. Sci. Technol.*, 24(3), 276.

IRIS. United States Environmental Protection Agency, Integrated Risk Information System, National Library of Medicine, Bethesda, MD.

Maughan, J.T. 1993. *Ecological Assessment of Hazardous Waste Sites,* New York, Van Nostrand Reinhold.

Mayernik, J.A., and Fehrenkamp, K. 1992. A New Model for Conducting Quantitative Ecological Risk Assessments at Hazardous Waste Sites, in *Proceedings of the HMC/Superfund Conference*, Washington, DC, 1992, p. 813.

NIOSH. National Institute for Occupational Safety and Health, Registry of Toxic Effects of Chemical Substances, National Library of Medicine, Bethesda, MD.

Opresko, D.M., Sample, B.E., and Suter, G.W. 1994. Toxicological Benchmarks for Wildlife: 1994 Revision, Oak Ridge National Laboratory, Oak Ridge, TN.

Sax, N.I., and Lewis, R.J. 1987. *Hawley's Condensed Chemical Dictionary*, 11th edition, New York, Van Nostrand Reinhold.

USEPA. 1991. Office of Solid Waste and Emergency Response, ECO Update, Volume 1, Number 2, United States Environmental Protection Agency, Washington, DC, 1991, p. 5.

USEPA. 1992a. Risk Assessment Forum, Framework for Ecological Assessment, United States Environmental Protection Agency, Washington, DC.

USEPA. 1992b. Office of Solid Waste and Emergency Response, ECO Update, Volume 1, Number 4, United States Environmental Protection Agency, Washington, DC, 1992, p. 7.

USEPA. 1994a. Environmental Response Team, Ecological Risk Assessment Guidance for Superfund: Process for Designing and Conducting Ecological Risk Assessments, United States Environmental Protection Agency, Edison, NJ.

USEPA. 1994b. Office of Solid Waste and Emergency Response, ECG Update, Volume 2, Number 1, United States Environmental Protection Agency, Washington, DC, 1994, p. 3.

USEPA. 1994c. Risk Assessment Forum, Ecological Risk Assessment Issue Papers, United States Environmental Protection Agency, Washington, DC.

CHAPTER 37

Cost-Effectiveness of Risk-Based Closures at UST Sites

Karen M. Scruton, Parsons Engineering Science, Inc., Atlanta, Georgia

James N. Baker, Parsons Engineering Science, Inc., Liverpool, New York

INTRODUCTION

Given the 1998 deadline for upgrading, replacing, or closing undergroun.. tor-age tank (UST) systems, many owners at sites throughout the country will need to comply with the regulations in the most cost-effective manner possible. Sites that do not pose a significant health threat to humans or the environment and, therefore, can be closed with "no further action" (NFA) required need to be identified and eliminated from further remedial activities as quickly as possible to allow the allocation of limited funds to sites that do pose a health risk and will require further remedial action.

In order to achieve NFA at a site, the risk to human and ecological receptors must be shown to be acceptable. Traditionally, a baseline risk assessment has been completed to determine the extent of risk at a site. Even at UST sites, which usually have only a small amount of data, the risk assessment can cost $10,000 or more, depending on the complexity and number of sources of contamination at the UST site.

Over the past several years, Parsons ES has established a streamlined risk-based method for evaluating UST sites for a major transportation client. This approach has resulted in agency acceptance of higher site-specific cleanup goals and NFA results at sites located in Alabama, California, Georgia, Massachusetts, New York, Ohio, Oregon, South Carolina, Tennessee, Virginia, Washington, and West Virginia.

The Environmental Protection Agency (EPA) is currently moving in this direction. Recent EPA guidance (USEPA, 1995) encourages the use of risk-based decision-making at UST sites, based on an approach established by the American Society for Testing and Materials (ASTM) in 1994 (ASTM, 1994).

TRADITIONAL BASELINE RISK ASSESSMENT

The traditional baseline risk assessment is primarily based on the USEPA's Risk Assessment Guidance for Superfund (RAGS) (USEPA, 1989a; 1989b) and subsequent supplements for both human health and ecological assessments.

The human health risk evaluation is comprised of the following five major steps:

1. Site Characterization
2. Data Evaluation
3. Exposure Assessment
4. Toxicity Assessment
5. Risk Characterization

The ecological evaluation is comprised of the following three main sections (Menzie et al., 1993):

1. Problem Formulation
2. Exposure Characterization
3. Risk Characterization

Figures 1 and 2 present flowcharts outlining these steps and the degree of complexity required to perform a human health and ecological risk assessment, respectively. Although many of these steps will be abbreviated, given the small amount of data usually available at the UST site, all steps of the process must be evaluated. The cost of performing a baseline risk assessment at a UST site, therefore, can be $10,000 or more for the human health risk assessment alone. This amount may be doubled if an ecological risk assessment is required. This amount is unjustified at sites where a screening evaluation of the data shows that the concentrations detected will probably lead to a "no further action" conclusion.

STREAMLINED RISK EVALUATION

A streamlined method of evaluating potential risks to human health and the environment has been used by Parsons ES at UST sites for a major transportation client. The streamlined approach focused on identifying potential human and ecological receptors and performing a screening evaluation of the chemicals detected in site media. Figure 3 provides a schematic of the steps involved in the risk-based streamlined approach. The following major steps are involved in the streamlined approach:

1. Site Characterization
2. Data Evaluation
3. Screening Evaluation
4. Recommendations

Figure 1. Human health risk assessment process.

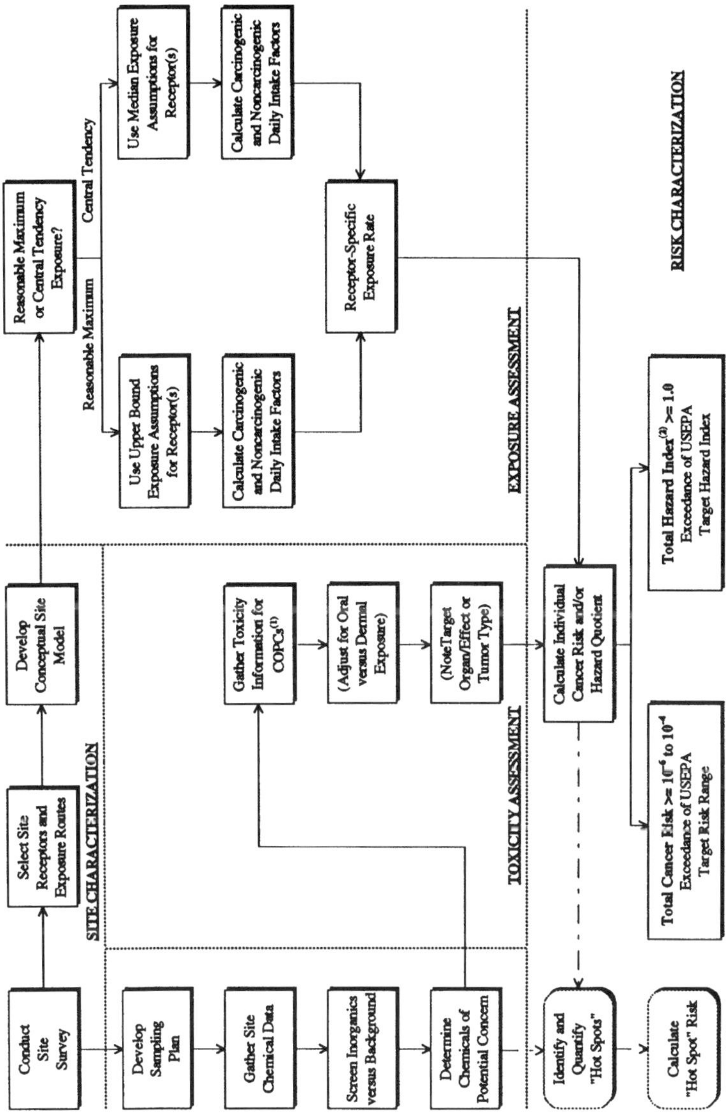

1. Chemical(s) of potential concern. 2. Sum of individual hazard quotients.

Figure 2. **Ecological risk assessment process.**

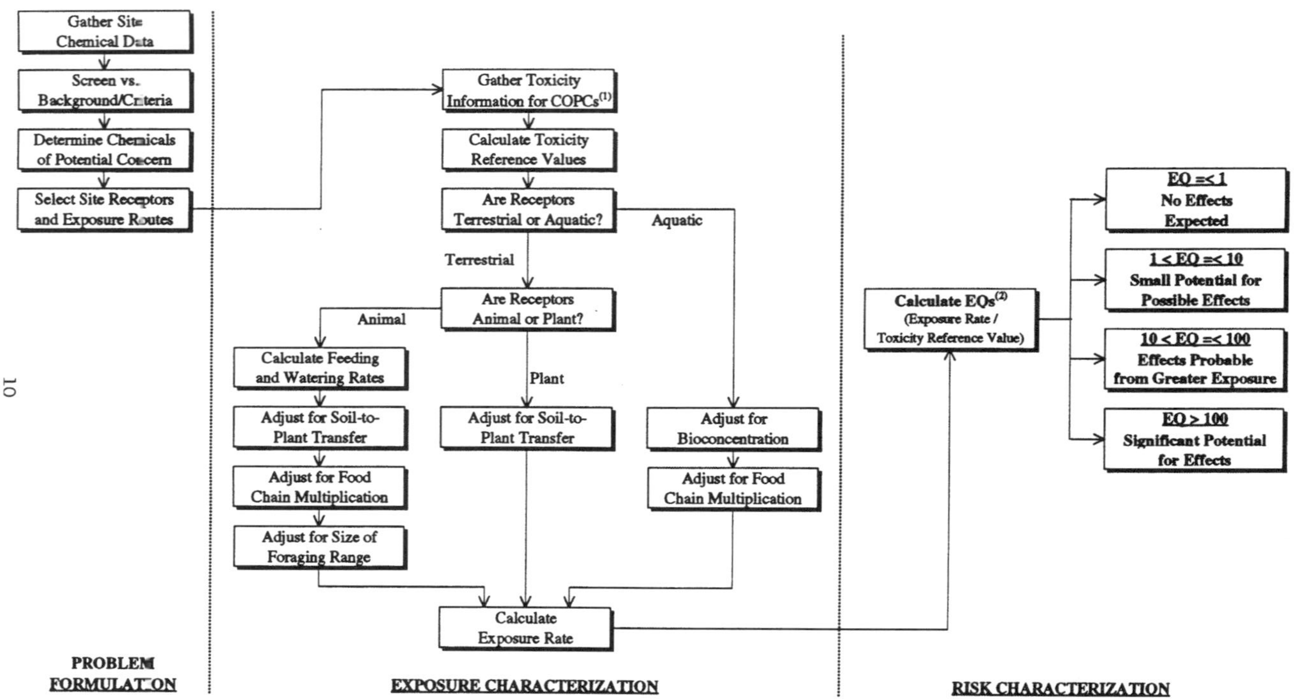

Based upon Menzie et al., 1993.
1. Chemical(s) of potential concern.
2. Ecological quotient.

10

The most significant step in this process is the screening evaluation, which includes identification of applicable or relevant and appropriate requirements (ARARs). The ARARs may be federal or state standards, such as Maximum Contaminant Levels (MCLs) in groundwater or state-specific soil cleanup levels, or may be risk-based cleanup levels. If federal or state ARARs are not available for a particular medium, risk-based cleanup levels are derived using the most recent federal USEPA (USEPA, 1991), regional USEPA, or state guidance. Consequently, the resultant ARARs used in the screening evaluation are site-specific.

If the screening evaluation determines that the maximum detected site concentration exceeds ARARs, a baseline risk assessment is recommended. If site concentrations do not exceed ARARs, a "no further action" alternative is recommended.

Figure 3. **Streamlined risk-based screening process.**

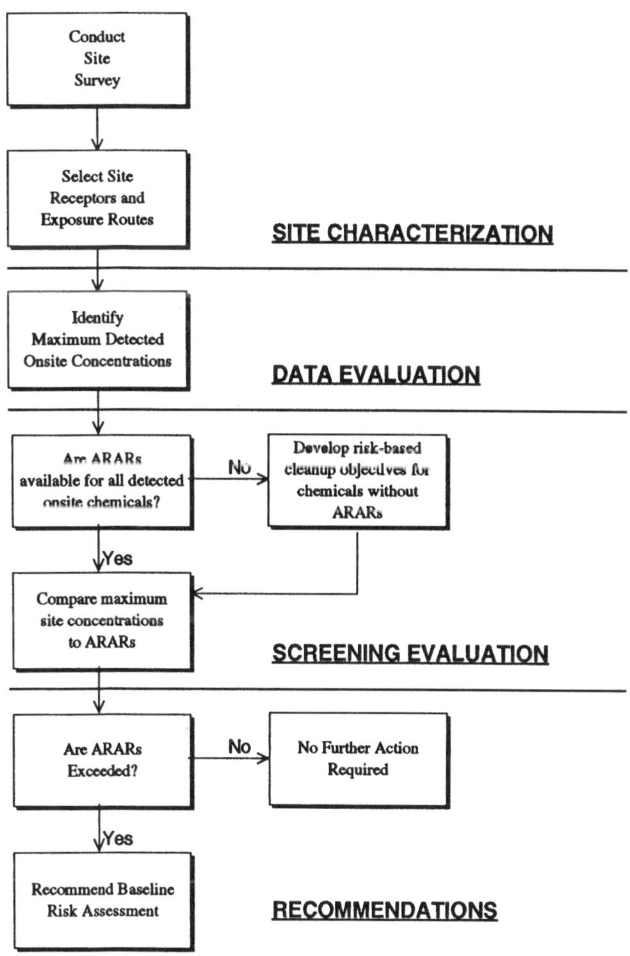

CURRENT DIRECTION OF USEPA REGARDING RISK-BASED CLOSURES

Many USEPA regions and individual states are currently accepting risk-based cleanups. Some regions, such as Regions 3 and 10, have issued region-specific risk-based cleanup levels which are based on USEPA methodology (USEPA, 1991). Numerous states also have developed their own cleanup criteria.

In 1994, the American Society for Testing and Materials (ASTM, 1994) issued an Emergency Standard Guide for Risk-Based Corrective Action (RBCA) Applied at Petroleum Release sites. This guide provides a tiered approach which involves increasingly complex levels of data collection and risk analysis. In Tier 1, conservative, generic risk-based screening levels (RBSLs), aesthetic criteria, and other appropriate standards are used to screen maximum concentrations of site contaminants. When the Tier 1 RBSLs are not appropriate, Tier 2 provides a method to derive site-specific target levels (SSTLs) based on site-specific conditions. When the Tier 2 SSTLs are considered inappropriate, Tier 3 provides methodology to derive site-specific SSTLs based on more sophisticated site-specific analysis (i.e., use of Monte Carlo simulations). If the SSTLs at an appropriate tier are exceeded, a selected corrective action plan should be implemented. This involves determining the most cost-effective option for achieving the final corrective action levels.

The USEPA has recently (USEPA, 1995) provided a directive encouraging the use of a risk-based decision-making approach in UST corrective action programs based on the ASTM emergency standard.

The USEPA is currently looking at establishing a standardized tiered approach to risk assessment for all hazardous waste sites, where the lower tiers deal with comparison of maximum site concentrations to generic or site-specific risk-based cleanup levels. If cleanup levels defined in the lower tiers are exceeded, a higher-tiered risk assessment must be completed. A higher-tiered risk assessment employs the traditional USEPA baseline risk assessment methodology described above. Parsons Engineering Science, Inc. is currently working with the EPA to identify and analyze the various tiered methods employed by USEPA regions and states in order to establish a standardized risk-based tiered method to be used throughout all regions and states.

SUMMARY AND CONCLUSIONS

Prior to release of the ASTM emergency standard for risk-based closure of UST sites, Parsons ES had performed streamlined risk-based evaluation at UST sites for a major transportation client that resulted in "no further action" outcomes.

Within three years of the submittal of the first of these streamlined risk-based closures, ASTM released their Emergency Standard Guide for Risk-Based Corrective Action (RBCA) Applied at Petroleum Release Sites (ASTM, 1994). This guide provides a tiered approach to risk assessment that utilizes screening methodology to determine whether detected chemicals are chemicals of potential concern at a site without performing a traditional baseline risk assessment.

The USEPA has recently provided a directive encouraging the use of a risk-based decision-making approach in UST corrective action programs based on the ASTM emergency standard. The USEPA is currently looking at establishing a standardized tiered approach to risk assessment. Parsons Engineering Science, Inc. is working with the EPA to identify and analyze the various tiered methods employed by EPA regions and states in order to establish a standardized risk-based tiered method to be used throughout all regions and states.

REFERENCES

(ASTM) American Society for Testing and Materials. 1994. Emergency Standard Guide for Risk-Based Corrective Action Applied at Petroleum Release Sites. Designation: ES 38-94. July, 1994.

Menzie, C., Cura, J., Freshman, J., and Svirsky, S. 1993. Evaluating Ecological Risks and Developing Remedial Objectives at Forested Wetland Systems in New England. In: *Application of Ecological Risk Assessment to Hazardous Waste Site Remediation*, pp. 89-100. Workshop Proceedings. USEPA Science Advisory Board.

(USEPA) United States Environmental Protection Agency. 1989a. Risk Assessment Guidance for Superfund: Volume I - Human Health Evaluation Manual (Part A). Interim Final. Office of Emergency and Remedial Response. EPA/540/1-89/002. December, 1989.

(USEPA) United States Environmental Protection Agency. 1989b. Risk Assessment Guidance for Superfund: Volume II - Environmental Evaluation Manual. Interim Final. Office of Solid Waste and Emergency Response. EPA-540-1-89-001. March, 1989.

(USEPA) United States Environmental Protection Agency. 1991. Risk Assessment Guidance for Superfund: Volume I - Human Health Evaluation Manual (Part B, Development of Risk-Based Preliminary Remediation Goals). Interim. Office of Research and Development. EPA/540/R-92/003. December, 1991.

(USEPA) United States Environmental Protection Agency. 1995. Use of Risk-Based Decision-Making in UST Corrective Action Programs. OSWER Directive 9610.17. Office of Solid Waste and Emergency Response. March 1, 1994.

CHAPTER 38

A Scientific Approach to Assessing Risks of Soil Contaminants at Day-Care Facilities in Detroit, Michigan

Kenneth G. Symms, Jennifer C. Kotanchik, Thomas Seery, Kristin G. Lawrence, and **Robert J. Fares**, Environmental Standards, Inc., Valley Forge, Pennsylvania

INTRODUCTION

Risk assessment is a process utilized by the U.S. Environmental Protection Agency (EPA) and most regional, state, and local regulatory agencies to evaluate potential adverse health effects associated with chemical substances in the environment and to provide a basis for determining remedial action objectives. For exposure to chemical carcinogens, which usually drives the remedy, the National Oil and Hazardous Substances Pollution Control Plan (NCP) requires that lifetime cancer risk after cleanup should fall within the range of one in ten thousand (10^{-4}) to one in a million (10^{-6}) (USEPA, 1993a). Typically, EPA and state regulators require remediation to levels that afford protection to the 10^{-6} risk benchmark, even for an implausible, hypothetical individual who could be maximally exposed.

The 100 fold degree of discretion would be a major factor in the cost of a site cleanup, if it were properly exercised by regulating agencies in accordance with the intent and provisions of the NCP. However, the conservatism of setting acceptable risk to the extreme of the range is compounded by other factors in the risk assessment process which are even more biased, such as the scenarios and conditions of exposure that are typically assumed for a hypothetical individual who is to be protected. For example, some regulatory agency project managers may insist that a risk assessment of a specific site include the assumed future residential-use or day-care facility scenarios, even if the possibility of these future site uses is nonexistent, in spite of the fact that the most current guidance calls for realism and site-specificity in deriving the exposure scenarios. In addition, regulatory agencies often require the use of standard (default) exposure parameters, such as the amount of soil children could inadvertently ingest day after day at a day-care center in Detroit, Michigan, for example.

The forced use of highly conservative exposure factors which are propagated through a risk assessment results in estimates which are implausible and unrealistic. The 1991 Regulatory Program Report (Executive Office of the President of the United States) recognized the problem of extreme conservatism in risk assessments (1991), as reflected in the following excerpt:

> Unfortunately, risk assessment practices continue to rely on conservative models and assumptions that effectively intermingle important policy judgements within the scientific assessment of risk. Policymakers must make decisions based on risk assessments in which scientific findings cannot be readily differentiated from embedded policy judgements. This policy environment makes it difficult to discern serious hazards from trivial ones, and distorts the ordering of government's regulatory priorities.... The continued reliance on conservative (worst-case) assumptions distorts risk assessment, yielding estimates that may overstate likely risks by several orders of magnitude.

In this same document, the Office of Management and Budget (OMB) highlighted the following reasons (Executive Office, 1991), among others, why use of conservative assumptions was problematic:

> Risk Assessments with such extreme conservative biases do not provide decision-makers with the information they need to formulate an efficient and cost-effective regulatory strategy, and a perverse and unfortunate outcome of using upper-bound estimates based on compounded conservative assumptions is that it may lead to regulate insignificant risks and ignore more serious risks.

To illustrate this, a risk assessment was performed of native, unpolluted, pristine land in Detroit, Michigan, under the assumption of a day-care facility scenario utilizing typically employed default exposure factors. For example, risk assessors typically assume that children ingest 200 mg of contaminated soil each day for 350 days/year (residential scenario) or 250 days/year for six years (day-care scenario), even in areas such as Detroit, Michigan, where snow and ice cover the ground for many months each year. It was further assumed that children with an average body weight of 16 kg absorb naturally occurring minerals dermally from soil adhering to an exposed skin surface area of 3,910 cm^2 with an adherence factor of 1.45 mg soil per cm^2 of exposed skin,[3] and an absorption factor of 1%. Under these assumptions and applying the most recently EPA-published cancer slope factors for arsenic, the estimated total cancer risk posed to children resulting from exposure to the average natural soil concentrations of arsenic (9.8 mg/kg) in lower Michigan is about 1×10^{-5}. This level of incremental cancer risk is generally unacceptable to regulatory reviewers.

Are children at day-care centers in Detroit or elsewhere in the country incurring unacceptable cancer risks because of naturally-occurring arsenic in soil?

Should regulatory agencies require any property which could theoretically be utilized for residential or day-care centers to rigorously remediate to residues of arsenic in soil that are "cleaner than clean"? If regulatory reviewers of risk assessments require the forced use of extremely conservative assumptions, is there any alternative risk assessment methodology that is at once realistic and consistent with EPA's own published guidelines?

The purpose of this chapter is to demonstrate that by adhering to the procedures and approaches detailed in EPA's Guidelines for Exposure Assessment (USEPA, 1992a), and other relevant guidance documents, conservative but realistic expressions of risk are generated which show that children at day-care centers are not exposed to added risks exceeding the *de minimis* benchmark of 10^{-6} from naturally occurring elements in soil. Exposure and risk assessments are intended by the EPA to be based on the best scientific information available and on sound professional judgment in the absence of specific data. The concatenation of worst-case assumptions to derive exaggerated and meaningless risk estimates is inconsistent with the EPA's own technical and policy guidelines.

THE RISK ASSESSMENT PROCESS

Several guidance documents have been published by the EPA to aid toxicologists in the performance of the baseline risk assessment. One EPA document, Risk Assessment Guidance for Superfund, Volume I, Human Health Evaluation Manual, provides specific guidance on the use of appropriate exposure scenarios in the risk assessment (USEPA, 1989b).

The basic outline of the process involves four steps:

1. Data evaluation
2. Exposure assessment
3. Toxicity assessment
4. Risk characterization

Data evaluation is the process of analyzing site-sampling results, and identifying the contaminated media (soil, sediment, groundwater, surface water, and air) and constituents of concern to human health. The sampling results are compiled, sorted by environmental medium, and analyzed statistically to generate upper 95% confidence limits of the mean concentrations for each substance in each medium. The statistical analysis may be performed on the data as a whole or on subsets of data if the site is large or activities at the site warrant subsets. The EPA recommends using the lower of the 95th percent upper confidence limit of the mean concentration, or the maximum concentration of a chemical in any given medium, as the best estimate of reasonable maximum exposure concentration (USEPA, 1989b). Reasonable maximum exposure is defined by the EPA as the highest exposure that is reasonably expected to occur at a site (1989b). Where a substance thought to be present in a medium is not actually detected in a sample, the EPA requires that risk assessors conservatively assume the chemical to be present, nonetheless, at a concentration equivalent to one-half its quantitation limit.

Exposure assessment identifies actual and potential exposure routes in each relevant medium. Examples of exposure routes include drinking groundwater and ingesting soil. The population at actual or potential risk of exposure is incorporated into each exposure route; e.g., residents drinking the groundwater and construction workers ingesting soil. An exposure equation is developed for each route and exposure parameters are chosen (e.g., ingestion rate, body weight, etc.). The results of the equations present an estimation of reasonable maximum daily intakes averaged on an annual basis for noncarcinogens and a lifetime basis for carcinogens. The EPA provides guidance on how to chose exposure parameters that characterize reasonable maximum exposure (USEPA, 1989b). The EPA acknowledges that professional judgment is required in this process, and that the specific values provided in their guidance "should be regarded as general recommendations, and could change based on site-specific information" (USEPA, 1989b).

To develop site-specific factors that estimate upper-bound exposure and dose, EPA's Guidelines for Exposure Assessment specifies that not all exposure parameters should be maximized (USEPA, 1992a), or the resulting assessment will be much too conservative. Rather, the assessor should judge which exposure parameters are on the expected distribution; i.e., parameters that represent the actual population and not just create a fictitious worst-case individual. These requirements indicate that it is the EPA's intended policy to avoid characterizing the worst possible scenario. The EPA's technical experts have attempted to convey these concepts to regional offices and risk managers, as illustrated in the following published excerpts:

> The Guidelines do not encourage the use of worst-case assessments, but rather the development of realistic assessments based on the best data available. A technical support document and a substantial section of the Guidelines currently discuss evaluation of uncertainty in order to produce objective assessments using the best (not the worst-case) estimates available either for preliminary or in-depth exposure assessments. (*Federal Register*, 51 FR 34053, Sept. 24, 1986).

> By maximizing the parameters in a scenario for exposure, the [risk] assessor is looking at the top end of the distribution of exposures in a population (if indeed the worst case actually exists in the population). A legitimate use of worst-case scenarios is to determine if the exposure or risk is low enough at this extreme so as to dismiss concern for this scenario. It is not legitimate to use a worst-case scenario to prove that there in fact, exists a concern in a real population. In constructing a worst-case scenario, the assessor has usually added assumptions or used particular data points that bring into question whether the sce-

nario actually represents the real world. If the exposure or risk value estimated by a worst-case scenario is high enough to cause concern, the assessor must reevaluate the parameters used and perform reality checks before deciding a problem really exists. (*Federal Register*, 53 FR 48846, Dec. 2, 1988).

An example of an exposure scenario that does not follow these requirements is a child day-care exposure scenario utilizing standard EPA default values. These child default values include a soil ingestion rate of 200 mg/day, an exposure frequency of 250 days/year, an exposure duration of 6 years and a body weight of 16 kg. When the naturally occurring average concentration of arsenic (7.2 mg/kg) detected in soils throughout the conterminous United States (Schacklette and Boerngen, 1984) is applied to this standard EPA child ingestion exposure scenario, a carcinogenic risk of about 1×10^{-5} is calculated. This risk exceeds the EPA's "point of departure" carcinogenic risk benchmark of 1×10^{-6} which would likely trigger regulatory action for a site under investigation.

Toxicity assessment determines chemical dose-response relationships based on existing toxicity studies and information supplied by the EPA in their online database, Integrated Risk Information System (IRIS) (1995), and their annually updated Health Effects Assessment Summary Tables (HEAST) (USEPA, 1993b). Chemical dose-response relationships are quantified for noncarcinogenic and carcinogenic chemicals, and are termed risk reference dose (RfD) values and cancer slope factors (CSF), respectively. RfD values are estimates of safe human doses of chemicals with wide margins of safety incorporated. The values are based on experimentally derived no-observed-adverse-effect-level (NOAEL) in animals and, rarely, in humans. The NOAEL is adjusted with safety factors to account for possible differences in sensitivity between animals and humans (when utilizing an animal NOAEL), human variability, and short-term versus long-term study length. If a NOAEL is not available in the scientific literature, the RfD value is derived from the lowest level that has exhibited an effect. An additional safety factor converts this dose to a NOAEL. The application of multiple safety factors may not be appropriate for every chemical investigated and, therefore, results in very conservative RfD values. In addition, a RfD value is a point estimate and not a range of probabilities like cancer slope factors. Intakes, therefore, are judged on a bright-line basis, and not on a continuum between acceptability and unacceptability (Williams, 1993). If a substance, therefore, falls below the line of acceptability, it must be cleaned up to that line (Environmental Health Letter, 1993).

Individual RfD values are derived for oral and inhalation exposure to a chemical. In addition, chronic RfD values are derived for lifetime or long-term exposures, and subchronic RfD values are derived for less-than-lifetime exposures (two weeks to seven years) (USEPA, 1989a).

The EPA calculates cancer slope factors using a linearized multistage mathematical model to extrapolate the results of tumor incidence data from experiments with laboratory animals to low level chemical exposure in humans. A few of the EPA's published cancer slope factors are based on human data. The EPA's published cancer slope factors are conservative and represent upper-bound estimates of lifetime cancer risk. The lower-bound risk, which is never utilized in the risk assessment, may be substantially lower and may even be zero (Williams, 1993).

In the determination of tumor incidence data, the EPA administers a maximum tolerated dose (MTD) of a substance to laboratory animals, and the resulting tumors are counted. Carcinogenicity of a substance is then derived by comparing the tumor count with the administered dose. One compelling argument against this methodology is that such a large, toxic dose of any substance, whether it is carcinogenic or not, may increase the rate of spontaneous tumors which also appear at significant rates in animals which were not given test chemicals. That is, it may be the dose that is carcinogenic and not the chemical, *per se*. By using this approach, the EPA apparently is focusing on the most extreme experiments instead of performing studies that may have more scientific potential for estimating human risk.

The EPA has developed a classification scheme for potential carcinogenic chemicals (USEPA, 1989b), based on the weight of evidence for producing tumors in test animals given very high doses or in some cases, on epidemiological data (e.g., workers exposed to chemical agents for many years) (Table 1).

Cancer slope factors are derived for chemicals in groups A through C. Arsenic, for example, is classified as a group A carcinogen. The slope factor is used in risk assessments to estimate an upper-bound lifetime probability of an individual developing cancer as a result of exposure to a carcinogen (USEPA, 1989b).

The EPA has not published an oral cancer slope (potency) factor for arsenic in IRIS or the most recent update to HEAST. However, a footnote appears beside the "NA" (not applicable) entry for the unit risk for arsenic which states "a unit risk of 5E-5 $(\mu g/L)^{-1}$ has been proposed by the Risk Assessment Forum and this recommendation has been scheduled for SAB review"(USEPA, 1988).

Table 1. **EPA Weight-of-Evidence Classification System for Carcinogenicity**

Group	Description
A	Human carcinogen
B1	Probable human carcinogen; limited human data are available
B2	Probable human carcinogen; sufficient evidence in animals and inadequate or no evidence in humans
C	Possible human carcinogen
D	Not classifiable as to human carcinogenicity

The unit risk of 5E-5 $(\mu g/L)^{-1}$, proposed by the EPA's Risk Assessment Forum in 1988, is commonly utilized as a default value by risk assessors quantifying risks associated with oral exposure to arsenic at Superfund sites. The CSF derives to 1.75 units of cancer risk per mg/kg/day of exposure averaged over a lifetime, assuming 2 L of water ingestion per day for a lifetime by a 70 kg individual. Any such use of this interim, unverified value for estimating cancer risk was intended to be qualified with important caveats that have generally been ignored. The EPA's Risk Assessment Council reviewed the Forum's interim proposed value and concluded (USEPA, 1988):

> . . . in making case-specific risk management decisions, program offices should be aware of qualities and uncertainties of a carcinogenic risk estimate for ingested inorganic arsenic that might mitigate their concerns compared to estimates of risks for other carcinogens. In the Council's view, these qualities and uncertainties could, in a specific risk management situation, modify one's concern downward as much as an order of magnitude. In such instances, the management document must clearly articulate this fact and state the factors that influenced such a decision.

Indeed, there are numerous factors at issue in the derivation of the interim, proposed CSF of 1.75 risk units per mg/kg/day for ingestion of inorganic arsenic including: (1) the validity of the Taiwan study; (2) the negative epidemiological findings in several U.S. communities; (3) the very high treatability and survival rate of nonmelanomatous skin cancer; (4) the possible role of arsenic in human nutrition (i.e., the "essentiality" issue); and (5) the likelihood of a threshold mechanism of action and pharmacokinetic considerations (Science Advisory Board, 1989).

Because many regulatory reviewers insist upon use of an oral CSF of 1.75 $(mg/kg/day)^{-1}$ for arsenic in risk assessments of Superfund sites, this value is applied in this assessment of naturally occurring arsenic in surface soils.

Risk characterization is a comparison of estimated average daily chemical intakes calculated in exposure assessment, with acceptable daily intake levels derived in toxicity assessment. The magnitude of risk from exposure to a noncarcinogenic chemical is expressed as a hazard quotient. Hazard quotients are calculated by dividing daily chemical intake levels by the chemical specific oral or inhalation RfD value. The sum of hazard quotients is termed the hazard index. If the hazard index equals one or below, no remedial activity is required to protect the population from potential noncarcinogenic health hazards associated with exposures to site contaminants. If the hazard index is considerably greater than one, a potential could exist for noncancer health effects. Although these intakes have a greater probability of causing an adverse effect, they should not be assumed, by definition, to be unacceptable. Cancer risks are determined by multiplying the average daily lifetime chemical intake by cancer slope factors. The EPA has established a risk range between 1×10^{-4} to 1×10^{-6} as the overall risk-based cleanup goal for Superfund sites (USEPA, 1989b), although the EPA commonly requires target risk adequate for the protection of human health as one-in-a-million (i.e., 1×10^{-6}). A cancer risk equal or below 1×10^{-6} indicates that no remedial activity is required to protect the population from potential carcinogenic health risks associated with exposures to site contaminants.

Currently, the EPA and the Superfund process are under scrutiny to determine if regulations have been followed to produce realistic baseline risk assessments and not worst-case scenarios. The Hazardous Waste Cleanup Project (HWCP) (1993), a coalition of major industry trade associations, has recently completed three papers addressing the federal Superfund waste site cleanup program. One paper entitled, " Exaggerating Risk: How EPA's Risk Assessments Distort the Facts at Superfund Sites Throughout the United States," reports that risk assessments often rely on unreasonable, unrealistic, theoretical, worst-case exposure assumptions instead of relevant and plausible site-specific data (USEPA, 1988). The result of multiplying each worst-case assumption magnifies the conservatism of the risk assessment output. The EPA's own guidelines recommend that all available data be utilized, especially site-specific information, and using ranges of values rather than single values for each assumption in the calculations (USEPA, 1988).

Critics of the risk assessment process are also quick to point out that the U.S. government has spent a significant amount of time and money studying how to improve the process. Yet, the EPA continues to rely on their outdated methods (Environmental Health Letter, 1993).

Monte Carlo Simulation

Monte Carlo simulation is one approach which can, if employed properly, scientifically address overly conservative risk assessments. This statistically and scientifically more valid approach avoids the multiplicative exaggerations of combining and propagating upper-bound exposure assumptions. The methodology has recently been encouraged by the EPA's own Science Advisory Board. Monte Carlo simulation considers not only the worst-case assumptions, but also the most probable, as well as the entire range of the variables involved in the estimations of exposure. Rather than inputting a single upper-bound value (such as a soil ingestion rate of 200 mg/day for a child), the computerized analysis is choreographed to run thousands of calculations (iterations), each time selecting values at random within the specified ranges of a "probability density function" (PDF) for each major variable. Thus, the full ranges of model parameter assumptions can be combined to calculate the entire probability distribution of the exposure, rather than just the upper-bound, single-point estimation or default values. In this framework, a complete distribution of risk is derived. This probability distribution of risk will afford any upper-bound estimate desired, such as the 95th percentile. The 95th percentile is what the EPA hopes to achieve when it recommends the reasonable maximum exposure (RME) approach. However, when a series of conservative point estimates are multiplied, the resulting estimate is often closer to the 99th percentile or greater. The 95th percentile obtained by Monte Carlo simulation can be orders of magnitude below the RME obtained via multiplication of default point estimates.

Sound Science in the Risk Assessment Process

The use of site-specific data and information which can improve estimates of exposure is the policy articulated by the EPA. The National Contingency Plan states that where sound scientific information is available, it is preferred over the use of conservative default assumptions (Science Advisory Board, 1989). The following section discusses a more valid approach for evaluating exposure parameters relevant to exposure to soil constituents by children at day-care centers in the Detroit area.

METHODOLOGY

As a means to gather site-specific information, a survey of day-care centers operating in the city of Detroit and its suburbs was conducted. Surveys were sent to 100 day-care centers requesting information concerning children's activities and playground conditions at their facilities.

The results of the survey were analyzed statistically. If a respondent failed to answer any question, the input value was left blank. Statistical analysis was performed on the data, and averages and standard errors were calculated for each parameter. Probability density functions were generated from the statistical analysis for input into Monte Carlo simulations.

The following comments excerpted from the EPA's Exposure Factors Handbook serve to summarize the methodology for assigning probability density functions for input variables (USEPA, 1989a):

> In general, the selection of a probability distribution to represent an input factor in the exposure models should be based upon any gathered information about that factor, theoretical arguments, and/or expert opinions. A probability distribution can be ascertained from such information as the following: general shape of the distribution, minimum, maximum, mode, mean, median, midrange and other percentiles. Available data on the probability distributions for each of the exposure factors discussed in this handbook have been presented in previous sections. When distribution data are not available, distributions can be assigned using professional judgement.

PDFs for each parameter in ingestion and dermal exposure scenarios are presented in Tables 2 and 3.

Table 2. **Model Assumptions for Soil Ingestion by Children (ages infant to 6 years) at a Day-Care Facility**

Oral Intake (mg/kg/day) = $\frac{Cs*IR*EF*ED*WF*FT*MF*ES*ME*FI*CF}{BW*AT*365}$

Model Parameter	Symbol	Units	Point Estimate	Distribution Type/Function	Model Assumptions For Probabilistic Estimate: Mean Arith.	Geom	se	min	max	Source
Arsenic concentration in soil	C_s	mg/kg	12.0	Normal	9.8		0.72	0.05	39	Range from Huron-Erie Lobe, MI
Ingestion rate	IR	mg/day	86	Lognormal		4.1	0.8			Thompson and Burmaster (1991)
Exposure frequency days spent at day-care (based on 50 weeks/year*days/week)	EF	days/year	192	50*Normal	3.85		0.36	0	5	Survey
Fraction of week spent outdoors (season-adjusted)	WF	unitless	0.75							Survey
Summer/Fall				Normal/(5days/week)	4.16		0.15			Survey
Winter/Spring				Normal/(5days/week)	3.34		0.17			Survey
Total time available at day-care (exclusive of field trips away from site)	FT	unitless	0.96	1-(Normal/(173 total hrs available per	7.01		1.60			Survey
Fraction of months in the year spent outdoors (season-adjusted) -	MF	unitless	0.68							Survey
Summer/Fall				Normal/6 months	4.59		0.17			Survey
Winter/Spring				Normal/6 months	3.54		0.21			Survey
Exposure duration	ED	years	2.90	Normal	2.9		0.19	0	6	Survey
Matrix effect for inorganics	ME	unitless	1.00	Discrete	1					Conservative Assumption
Exposed soil fraction	ES	unitless	0.06	Normal	0.06		0.02	0	1	Survey
Conversion factor	CF_1	kg/mg	1.00E-06	Discrete	1.00E-06					
Fraction ingested from contaminated source (season-adjusted) -	FI	unitless	0.58							
Summer/Fall				Normal/(2.2 hrs/day)	1.66		0.1			Survey; Juster and Stafford (1985)
Winter/Spring				Normal/(2.2 hrs/day)	0.88		0.05			Survey; Juster and Stafford (1985)
Body weight	BW	kg	13.95	Normal	13.95		3.62			EPA (1989)
Averaging time for noncarcinogenic effects	AT_{nc}	years	2.90	Normal	2.9		0.19	0	6	Survey
Averaging time for carcinogenic effects	AT_c	years	70	Discrete	70					EPA (1989)
Conversion factor	CF_2	days/year	365	Discrete	365					
Subchronic oral RfD for arsenic	RfD	mg/kg-day	3.00E-04	Discrete	3.00E-04					IRIS
Oral slope factor for arsenic	SF	mg/kg-day^{-1}	1.75	Discrete	1.75					IRIS

Table 3. **Model Assumptions for Dermal Contact With Soil by Children (ages infant to 6 years) at a Day-Care Facility**

Dermal Intake (mg/kg/day) = $\frac{Cs*SA*AdF*ABS*EF*ED*WF*ES*MF*FT*FI*CF}{BW*AT*365}$

Model Parameter	Symbo	Units	Point Estimate	Distribution Type/Function	Model Assumptions For Probabilistic Estimate: Mean Arith.	Geom	se	min	max	Source
Arsenic concentration in soil	Cs	mg/kg	12.0	Normal	9.8		0.7	0.1	39	Range from Huron-Erie Lobe, MI
Surface area available for exposure (adjusted for seasonal clothing) -	SA	cm^2	1752							
Summer/Fall				0.3863*Normal	6800		70			EPA (1991, 1986)
Winter/Spring				0.1291*Normal	6800		70			EPA (1991, 1986)
Soil adherence factor	AdF	mg/cm^2	1.45	Triangular	0.51			0	1.5	Lepow (1974), EPA (1989a,b)
Absorption factor for inorganics (0.0001/hr)*(10hr)	ABS	unitless	0.001	Discrete	0.001					Roy Smith, PhD, Region III EPA
Exposure frequency days spent at day-care (based on 50 weeks/year*days/week)	EF	days/year	192	50*Normal	3.85		0.4	0	5	Survey
Fraction of week spent outdoors (season-adjusted) -	WF	unitless	0.75							Survey
Summer/Fall				Normal/(5days/week)	4.16		0.2			Survey
Winter/Spring				Normal/(5days/week)	3.34		0.2			Survey
Total time available at day-care (exclusive of field trips away from site)	FT	unitless	0.96	1-(Normal/(173 total hrs available per month))	7.01		1.60			Survey
Fraction of months in the year spent outdoors (season-adjusted) -	MF	unitless	0.68							Survey
Summer/Fall				Normal/6 months	4.59		0.2			Survey
Winter/Spring				Normal/6 months	3.54		0.2			Survey
Exposure duration	ED	years	2.90	Normal	2.9		0.2	0	6	Survey
Matrix Effect for Inorganics	ME	unitless	1.00	Discrete	1					Conservative Assumption
Exposed soil fraction	ES	unitless	0.06	Normal	0.06		0	0	1	Survey
Conversion factor	CF1	kg/mg	1.00E-06	Discrete	1.00E-06					
Fraction contacted dermally from contaminated source (season-adjusted) -	FI	unitless	0.58							Survey
Summer/Fall				Normal/(2.2 hrs/day)	1.66		0.1			Survey; Juster and Stafford (1985)
Winter/Spring				Normal/(2.2 hrs/day)	0.88		0.1			Survey; Juster and Stafford (1985)
Body weight	BW	kg	13.95	Normal	13.95		3.6			EPA (1989b)
Averaging time for noncarcinogenic effects	ATnc	years	2.90	Normal	2.9		0.2	0	6	Survey
Averaging time for carcinogenic effects	ATc	years	70	Discrete	70					EPA (1989b)
Conversion factor	CF2	days/year	365	Discrete	365					
Subchronic oral RfD for arsenic	RfD	mg/kg-day	3.00E-04	Discrete	3.00E-04					IRIS
Oral slope factor for arsenic	SF	$mg/kg\text{-}day^{-1}$	1.75	Discrete	1.75					IRIS

RISK ASSESSMENT INPUT

The exposure parameters for the child day-care risk assessment were obtained from widely accepted studies and developed from results of the surveys. Thirty-one of the surveys were returned and completed, 4 surveys were returned unanswered, and 65 surveys were not returned. The following parameters were derived from the survey results, and together with the study parameters are detailed in the subsections below:

- Exposure frequency
- Exposure duration
- Fraction of week spent outdoors
- Fraction of months spent outdoors
- Total time at day-care
- Fraction of time children play in exposed soil
- Fraction ingested from contaminated source
- Averaging time for noncarcinogenic effects

Concentration in Soil

The risk assessment is undertaken with the purpose of demonstrating that average residues of arsenic in native surficial soils in the greater Detroit area (and the United States in general) do not pose an unacceptable risk if day-care facilities were to be constructed and operated on them. Accordingly, conservative assumptions regarding naturally occurring concentrations of arsenic in surficial soils are employed. Kesler-Arnold and O'Hearn compiled data collected during numerous environmental assessments in lower Michigan in which background soil samples were collected (1990). Average concentrations of arsenic were found to vary significantly based on soil type (e.g., clay, sand, or silt), and significant differences were noted within the three different glacial lobes. Statistical analysis of soils within each geologic lobe revealed that the Huron-Erie lobe manifested the greatest average concentration of arsenic in the background soil samples. The mean concentration of arsenic was found to be 9.8 mg/kg on the Huron-Erie locations, with a standard deviation of 8.5 mg/kg (n - 139 samples). A normal distribution was assumed for purposes of generating a probability density function for the concentration term for arsenic in surface soil in Michigan.

Ingestion Rate

The EPA's exposure default value for soil ingestion is 200 mg/day as recommended in RAGS Part A (USEPA 1989b). More accurate estimates have been determined by Binder et al. (1986) and Calabrese et al. (1989) in studies that analyzed children's stool for trace metals. The most recent of these studies was conducted by Thompson and Burmaster, who modified the Binder study data by using the actual stool weights (1991), instead of the 15-gm value used originally, and published parameterized distributions for use in Monte Carlo simulations. The result was an estimated lognormal mean soil ingestion rate from all sources of soil for young children of 4.13 mg/day with a standard deviation of 0.80. This value translates to 62 mg/day and is utilized in Monte Carlo to represent an accurate and realistic 50th percentile assumption for how much soil a child actually ingests.

Exposure Duration

The survey results revealed that children do not generally attend the same day-care facility for 6 years. The results indicate children may receive home care for some of the first 6 years of life, or children may relocate outside of the day-care region. Exposure duration for children attending day-care in the Detroit, MI area was derived from the survey results and determined to be normally distributed with an average of 2.9 years and a standard error of 0.19 years.

Exposure Frequency

According to the survey, children attend day-care on varying schedules. Some children attend day-care for 8 hours or more per day, every day of a work week. Other children attend day-care only 2 or 3 days per week or half days (4 hours or less) for 2 or 3 days per week. Therefore, it is not reasonable to assume an aggregate child attends day-care for over 8 hours per day for every day of the week, since the child has the opportunity to be at day-care for a range of time per week. The survey reveals the average child attends day-care 3.85 days per week with a maximum of 5 days per week. Fifty weeks per year was assumed to be a reasonable maximum duration an average child will attend day-care.

Fraction of the Week Spent Outdoors

Of the 4.59 months available for outdoor play during the summer and fall, the average number of days per week available for children to play outdoors is 4.16 with a standard error of 0.15. Of the 3.54 months available for outdoor play during the winter and spring months, the average number of days per week available for children to play outdoors is 3.34 with a standard error of 0.17 days per week.

Number of Months Spent Outdoors

During every year in Detroit, weather influences when children play outside at a day-care facility. The summer and fall months, May through October, are known as low-flow regime months, indicating less precipitation than the winter and spring months. During the summer and fall months there are usually fewer cold and high precipitation days than during winter and spring months, thus children are allowed outside to play for longer duration and for larger portions of each week. The survey results indicate children are allowed outdoor play on average 4.59 months, with a standard error of 0.17 months for the 6-month period of the summer and fall, and 3.54 months with a standard error of 0.21 months for the 6-month period of the winter and spring.

Total Time at Day-Care

Day-care facilities often provide children with field trip opportunities including lunch, walks in the park or neighborhood, or museum trips away from the facility. The survey results showed that children attend field trips for 7.01 hours per month. There are, therefore, approximately 173 hours available during each month that a child may attend day-care. The fraction of time spent during field trips is time spent away, divided by total time available, which equals 4 percent of each month is spent away from an average day-care facility.

Fraction of Time Children Play in Exposed Soil

Most day-care facilities that have outdoor playgrounds, have ground coverings including grass, wood chips, rubber pads, concrete, and asphalt. Twenty of 31 facilities responded that their outdoor play areas have appropriate landscaping and ground-covering and no exposed soil in the outdoor play area. The 11 positive responses ranged from 5% to 50% exposed soil cover. The percentage of exposed soil at all day-care facilities has a mean of 6.0% with a standard error of 2%. Accordingly, the average percent of time children play in exposed soil is 6%, assuming there is an equal probability that children will play on landscaped play areas and bare soil.

Matrix Effect

When chemicals are administered in solid vehicles such as food and soil, only a fraction of the ingested dose is extracted from the vehicle and subsequently absorbed in the gastrointestinal tract (USEPA, 1992b). Gastrointestinal absorption of contaminants mixed into such a medium is inhibited by physical-chemical bonding to the matrix (Hawley, 1985). Several studies have been performed to estimate the oral absorption fractions of chemicals from soil, relative to absorption from vehicles such as water.

The U.S. EPA (1992) summarized the matrix effect of soil contaminated with TCDD (dioxin) and estimated the relative bioavailability from soil to be 25% to 50% (USEPA, 1992c). Compared with corn oil as a vehicle, gut absorption of 20% to 40% TCDD was observed when administered in a soil matrix. This estimate is in good agreement with the 20% to 26% absorption from TCDD- treated soil from the work of Poiger and Schlatter 1980).

The bioavailability of inorganic compounds in soil after ingestion is not well established. Gastrointestinal absorption of cadmium is only about 5% to 8%, 1% for chromium, 5% to 15% for lead (of which only 5% is retained in the body), and sparse amounts of nickel (Casaratt and Doull, 1986). The EPA recognizes the importance of bioavailability as exemplified in its publication of two oral risk reference dose values (RfDs) for cadmium, one for cadmium in food, and one for cadmium in water. The EPA has also published two RfDs for manganese to address the bioavailability of its different forms. The separate RfDs for food and water indicate a potentially higher bioavailability of manganese from drinking water (IRIS, 1995).

In addition, highly insoluble forms of arsenic, such as arsenic trisulfide or arsenic triselenide are not absorbed well from the gastrointestinal tract (ATSDR, 1991). While solubility and bioavailability of arsenic in soil may profoundly influence uptake into the bloodstream following dermal contact or incidental ingestion of soil containing naturally occurring arsenic, there is inadequate information to define a probability density function for a matrix effect on gastrointestinal absorption. Therefore, no matrix effect or considerations of the bioavailability of natural forms of arsenic in native soil are included in this assessment.

Fraction Ingested from Contaminated Source

Children can be exposed to soil at various locations (i.e., at home, school, a friend's house, a playground) during their daily outdoor activities. The fraction ingested from contaminated source is a parameter which determines how much of a child's time spent outdoors is spent in the day-care's outdoor facilities. The general formula to determine this fraction is:

$$\frac{\text{hours spent outside at a day-care facility each day}}{\text{total hours spent outdoors each day}}$$

The survey results were used to establish how many hours a child spends outside at a day-care facility each day. The normal mean for summer and fall months was 1.66 hours outside, with a standard error of 0.1. During winter and spring months children are expected to spend a daily mean of 0.88 hours outdoors at the day-care (with a standard error of 0.05). The total number of hours a child is expected to be outside on any given day of the year is 2.2 hours (Juster and Stafford, 1985).

Body Weight

The mean body weight for children is 13.95 kg, with a standard error of 3.62 kg (USEPA, 1989a). The distribution around the mean is assumed to be normal.

Averaging Time

The averaging time for noncarcinogenic effects, according to EPA's guidance (USEPA, 1989b), is pathway-specific, and is equivalent to the exposure duration parameter derived from the survey multiplied by 365 days/year. The averaging time for carcinogenic effects is 70 years multiplied by 365 days/year (USEPA 1989a; U.S. Bureau of Census, 1990).

Surface Area Available for Exposure

Based on information summarized by the EPA, total skin surface area was calculated to have a mean of 0.68 m^2 and standard error of 0.07 (USEPA, 1991). It is assumed that children at day-care facilities will be clothed during outdoor play time. During the summer/fall term, it is assumed children will be wearing shoes, short leg coverings, and short-sleeved shirts, thus exposing the face, lower arm, hands, and lower legs, which provides a mean of 38.6% and standard error of 0.74% of the skin available for exposure (USEPA, 1989a). It was assumed children would be wearing shoes, long sleeves, and long leg coverings during outdoor play time during the winter/spring term, thus leaving the face and hands available for exposure. The percentage of skin available during the winter/spring term has a mean of 12.9% and a standard error of 0.46% (USEPA, 1989a).

Adherence Factor

Dermal absorption of contaminants associated with soil is approximated by the methodology described in the EPA's guidance (USEPA, 1989b). The minimum amount of soil that can potentially adhere to exposed skin is assumed to be negligible. The most likely quantity of soil adhering to exposed skin is 0.51 mg/cm^2, as measured on notably soiled hands of children by Lepow et al. (1974). The maximum value for soil to skin adherence factor is 1.45 mg/cm^2.[5]

Absorption Factor

The fraction of contaminant on soil that is capable of passing through the epidermis and reaching the dermis is termed the absorption factor. In general, the stratum corneum of the human skin provides an effective barrier to environmental toxins (including naturally-occurring minerals, plant and animal toxins). For most inorganic salts, percutaneous absorption is considered insignificant relative to incidental ingestion (see, for example, U.S. EPA, Ambient Water Quality Criteria Documents for various metals, 1980) (USEPA, 1980). An absorption factor of 1% over a duration of 10-hours skin exposure time is conservatively assumed in the dermal exposure scenario for inorganics (Dr. Roy Smith, Toxicologist, U.S. EPA Region III, December 31, 1991, personal communication).

RISK ASSESSMENT OUTPUT AND CONCLUSION

The ingestion and dermal exposure scenarios, utilizing the input parameters described above, are presented in Tables 2 and 3. Table 4 provides a summary of the Monte Carlo distribution of risks posed to children through ingestion and dermal exposure of soil at day-care facilities. The most likely (mean) added lifetime cancer risk posed to children by exposure to soil arsenic under typical conditions at a day-care center is estimated to be about 4×10^{-8}, or 4 cases of cancer in 100 million children so exposed. For effects other than cancer

Table 4 **Distribution of Total Risk From Exposure to Arsenic-Contaminated Soil**

Simulation Results for Arsenic Hazard

Iterations= 5000
Simulations= 1
 # Input Variables= 23
 # Output Variables= 2
Sampling Type= Monte Carlo

Minimum =	7.79E-07
Maximum =	4.18E-02
Mean =	1.95E-03
Std Deviation =	2.43E-03
Variance =	5.90E-06
Skewness =	4.60
Kurtosis =	41.15
Errors Calculated =	0
Mode =	1.04E-03
5% Perc =	2.26E-04
10% Perc =	3.17E-04
15% Perc =	4.18E-04
20% Perc =	5.15E-04
25% Perc =	6.20E-04
30% Perc =	7.23E-04
35% Perc =	8.28E-04
40% Perc =	9.53E-04
45% Perc =	1.08E-03
50% Perc =	1.22E-03
55% Perc =	1.37E-03
60% Perc =	1.56E-03
65% Perc =	1.76E-03
70% Perc =	1.99E-03
75% Perc =	2.29E-03
80% Perc =	2.73E-03
85% Perc =	3.33E-03
90% Perc =	4.34E-03
95% Perc =	6.10E-03

risk, the maximum hazard index is about 0.04, which is well below unity or a ratio of possible concern. The EPA has indicated that exposures and risks should be estimated such that 90 to 98% of the exposed population would actually incur less exposure and attendant risk (USEPA, 1992a). Utilizing the Monte Carlo simulation analysis it is estimated that the 95th percentile of added cancer risk is about 1×10^{-7}, or 1 case of cancer in 10 million children exposed to native soil at a typical day-care center.

These findings suggest that even where no unusual engineering controls have been imposed on the design and construction of a typical day-care facility, exposures to naturally occurring elements in soil are not likely to pose any significant risks to children.

REFERENCES

(ATSDR) U.S. Department of Health & Human Services. 1991. Toxicological Profile for Arsenic, Agency for Toxic Substances and Disease Registry, Atlanta.

Binder, S., Sokal, D., and Maughan, D. 1986. Estimating Soil Ingestion: The Use of Tracer Elements in Estimating the Amount of Soil Ingested by Young Children, *Arch. Environ. Health*, **41**, 341.

Calabrese, E.J., Barnes, R., Stanek, E.J., Pastides, H., Gilbert, C.E., Veneman, P., Wang, X., Laszitity, A., and Kostecki, P.T. 1989. How Much Soil Do Young Children Ingest: An Epidemiologic Study, *Regul. Toxicol. Pharmacol.* **10**, 123.

Casaratt and Doull's Toxicology, The Basic Science of Poisons, Third Edition. 1986. (Klaasen, C., Amdur, M., and Doull, J., Eds.), New York, Macmillan Publishing Company.

Environmental Health Letter. 1993. Risk Assessment Debated; Accuracy Is Questioned, *Environmental Health Letter*, **32**, 162.

Executive Office of the President of the United States. 1991. Regulatory Program Report. Washington, DC.

Federal Register. Vol. 51, No. 185, 34053, September 24, 1986.

Federal Register. Vol. 53, 48846, December 2, 1988.

Hawley, J.K. 1985. Assessment of Health Risk from Exposure to Contaminated Soil. *Risk Anal.*, **5**, (4).

Hazardous Waste Cleanup Project. 1993. Exaggerating Risk: How EPA's Risk Assessments Distort the Facts at Superfund Sites Throughout the United States. Morgan, Lewis and Bockius, Washington, DC.

IRIS (Integrated Risk Information System). 1995. A Continuously Updated Electronic Database Maintained by the U.S. Environmental Protection Agency, TOXNET, National Library of Medicine, Bethesda.

Juster, F.T. and Stafford, F.P. 1985. *Time, Goods and Well-Being*, Ann Arbor, MI, Survey Research Center, Institute for Social Research, University of Michigan.

Kesler-Arnold, K.A. and O'Hearn, M. 1990. Background Concentrations of Metals and Cyanide in Lower Michigan Soils, *44th Purdue Industrial Waste Conference Proceedings*, Chelsea, MI, Lewis Publishers.

Lepow, M.L., L. Bruckman, R.A. Rubino, S. Markowitz, M. Gillette, and J. Kapish. 1974. Role of Airborne Lead in Increased Body Burden of Lead in Hartford Children. *Environ. Health Perspect.* **7**, 99.

Poiger, H. and Schlatter, C. 1980. Influence of Solvents and Absorbents on Dermal and Intestinal Absorption of TCDD, *Food Cosmet. Toxicol.* **18**, 477.

Science Advisory Board. 1989. Review of the Arsenic Issues Relative to the Phase II Proposed Regulations, Memorandum to the U.S. EPA Administrator, Washington, DC.

Shacklette, H.T. and Boerngen, J.G. 1984. Elemental Concentrations in Soil and Other Surficial Materials of the Conterminous United States, U.S. Geological Survey Professional Paper 1270, U.S. Government Printing Office, Washington, DC.

Thompson, K.M. and Burmaster, D.E. 1991. Parametric Distributions for Soil Ingestion by Children. *Risk Anal.* **11**,(2).

U.S. Bureau of Census. 1990. Data User's Service Division.

(USEPA) United States Environmental Protection Agency. 1980. Ambient Water Quality Criteria Documents, Office of Water, Criteria and Standards Division, Washington, DC.

(USEPA) United States Environmental Protection Agency. 1988. Special Report on Ingested Inorganic Arsenic, Risk Assessment Forum, Washington, DC.

(USEPA) United States Environmental Protection Agency. 1989a. Exposure Factors Handbook, Office of Health and Environmental Assessment, Washington DC.

(USEPA) United States Environmental Protection Agency. 1989b. Risk Assessment Guidance for Superfund, Volume I, Human Health Evaluation Manual (Part A), Interim Final, Office of Emergency and Remedial Response, Washington DC.

(USEPA) United States Environmental Protection Agency. 1991. Interim Guidance for Dermal Exposure Assessment, Office of Research and Development, Washington DC.

(USEPA) United States Environmental Protection Agency. 1992a. Guidelines for Exposure Assessment, Bureau of National Affairs, Washington, DC.

(USEPA) United States Environmental Protection Agency. 1992b. New Interim Region IV Guidance (PAH Toxicity Equivalent Factors; Absorption Factors and Soil to Skin Adherence Factors), Region IV, Atlanta, February 11, 1992.

(USEPA) United States Environmental Protection Agency. 1992c. Estimating Exposure to Dioxin-Like Compounds, Office of Research and Development, Washington, DC.

(USEPA) United States Environmental Protection Agency. 1993a. EPA National Oil and Hazardous Substances Pollution Contingency plan (NCP) Under the Comprehensive Environmental Response, Compensation and Liability Act of 1980, *Federal Register*, April 16, 1993.

(USEPA) United States Environmental Protection Agency. 1993b. Health Effects Assessment Summary Tables, Annual Update, Office of Solid Waste and Emergency Response.

Williams, C. 1993. Assessing Risk: Putting Toxicity in Perspective, *Environmental Protection*, 4, 12.

Yang et al., *In Vitro* and *in Vivo* Percutaneous Absorption of Benzo(a)pyrene from Petroleum Crude-Fortified Soil in the Rat, *Bull. Environ. Contam. Toxicol.* **43**, 207.

CHAPTER 39

Risk-Based Investigation of Inorganic Constituents in Fly Ash

Daniel P. Claycomb, Kristin G. Lawrence, and **Jennifer C. Kotanchik,**
Environmental Standards, Inc., Valley Forge, Pennsylvania

INTRODUCTION

The subject site is located in southeastern Pennsylvania and consists of approximately 27 acres of undeveloped land bordering a residential development to the north, a creek to the south and west, and a local street to the east. There are two distinct topographic areas at the site. The first area is considerably higher in elevation than the rest of the site, reportedly due to the emplacement of fly ash material from a local energy-generating facility during the late 1960s and early 1970s. This first area is relatively flat and covered with low-growth vegetation and occasional small trees. Steep slopes lead down to the second area of the site, which, for the most part, borders the creek. This second area slopes gently toward the creek and is covered with large trees and other endemic vegetation.

Results of previous investigations conducted at the site in 1989 identified an area of potential environmental concern at the topographically high area of the site where the emplacement of fly ash fill material from the energy plant reportedly occurred. Subsurface soil sampling and laboratory analyses were performed at this time. By assessing the analytical results of soil samples collected during the 1989 investigation, a potential toxicological concern with respect to total arsenic levels in the fly ash fill material was identified. As area developers gained acute interest in developing the site for residential use, Environmental Standards, Inc. (hereinafter "Environmental Standards") was contracted to conduct a risk-based investigation and evaluation of the site to determine the risk posed to human health and the environment by implementing the proposed development activities.

RISK-BASED INVESTIGATIVE ACTIVITIES (PHASE I)

In May of 1994, Environmental Standards initiated the first phase of investigative activities by conducting a subsurface soil investigation. The objective of this investigation was to better characterize the site with respect to the arsenic concentrations present at various depths in onsite soils, and to confirm the results of previous investigations with respect to other constituents of potential concern. Data collected during this phase was utilized in conducting a site-specific risk assessment which was anticipated to eliminate the constituents of potential concern. The investigation consisted of the advancement of three soil borings, located at portions of the site previously identified as fill areas containing high levels of arsenic, to depths ranging between 22 and 27 feet below ground surface, utilizing the hollow stem auger method of drilling. Surface soil samples (0-6 inches) were collected at each boring location. Split-spoon soil samples were then collected in five-foot intervals from a depth of five feet to 20 feet below grade in one boring, and from a depth of five feet to 25 feet below grade in the other two borings. The three surface soil samples were analyzed for target analyte list (TAL) metals and the 14 split-spoon soil samples were analyzed for arsenic. The results of this investigation indicated that five inorganic compounds (arsenic, beryllium, cobalt, magnesium, and manganese) could pose a potential concern to human health. Also, it was observed that the arsenic levels typically increased from their lowest levels at the surface (0-6") to the highest level at the 15 to 17 feet sampling interval, and then decreased with increased depth in the remaining sampling intervals.

Utilizing the EPA Region III's recently published screening criteria for cobalt (Smith, 1994), the concentrations of cobalt detected at the site were determined to be below the EPA's residential screening values. A similar process, using the EPA's risk-based equations and Environmental Standards' internally-generated toxicity criteria for magnesium, yielded the same result; magnesium concentrations at the site were considered to be below a residential health-based risk assessment screening value (Smith, 1994). Thus, cobalt and magnesium were no longer constituents of concern, and were eliminated from further consideration.

RISK-BASED INVESTIGATIVE ACTIVITIES (PHASE II)

With the determination of specific constituents of potential concern (arsenic, beryllium, and manganese) identified during the Phase I investigation, the original risk assessment strategy, which was anticipated to eliminate all five constituents of potential concern, had to be modified. Environmental Standards evaluated alternative strategies and decided to conduct a site-specific risk assessment using a construction worker scenario. This scenario assumed that construction workers would be exposed to materials at depths up to 15 feet below ground surface (bgs) by skin contact, inhalation, and ingestion. Also, an assessment of the quality of groundwater, sediment, and surface water was recommended. A second phase of investigation to provide further characterization of the site was needed to proceed with the scenario.

Soil Sampling Program

In September 1994, Environmental Standards conducted field activities associated with the Phase II investigation. Activities included the advancement of nine soil borings to depths ranging from 15 to 35 feet bgs. A soil sample for laboratory analysis was collected at each of three previously determined sampling intervals within each soil boring. These depths were pre-selected on the basis of previous study results and anticipated risk assessment exposure scenarios. A surface soil sample was collected at a depth of 0 to 6 inches bgs with decontaminated stainless steel troughs. This sample depth was selected because this is the soil with which construction workers will be in day-to-day contact. The remaining two soil samples, collected at 6 to 8 feet bgs and at 13 to 15 feet bgs, were obtained with decontaminated split-spoon samplers. The 6 to 8 feet sample was selected because most excavating activities related to foundation and trenching work for residential structures would terminate at this depth. The 13 to 15 feet bgs sample was selected because this was the greatest anticipated depth that earth work would ever occur during construction activities.

Groundwater Investigation

As part of the planned groundwater investigation, four boring locations were selected for borings to be advanced to the soil/groundwater interface, where a representative sample of groundwater was to be obtained. SB-9 was selected as a downgradient location, SB-10 as an upgradient location, and SB-11 and SB-12 as lateral locations (see Figure 1). Because SB-9 was the first boring where a groundwater sample was to be obtained, split-spoon samples were collected in frequent sampling intervals so that possible confining stratigraphic units and any resulting perched water table identified during the previous phase of investigation could be further characterized. This technique was also used to ensure that perched water was properly sealed to prevent vertical migration down the borehole. An actual groundwater table was not believed to have been encountered before the boring was terminated at a depth of 35 feet bgs, where weathered bedrock was encountered. Advancement of the remaining soil borings revealed that groundwater was not encountered prior to the termination of the borings at the soil/weathered bedrock interface.

Surface Water and Sediment

On September 9, 1994, Environmental Standards field personnel, in accordance with EPA stream sampling protocol, collected surface water and sediment samples from three sampling locations (upstream, midstream, and downstream) along the creek, which flows adjacent to the southern and western portions of the site. The surface water samples were collected with laboratory-supplied bottles that had been submerged into the creek. The sediment samples were collected with decontaminated stainless steel troughs.

Figure 1. **Soil profile map, subject site, southeastern Pennsylvania.**

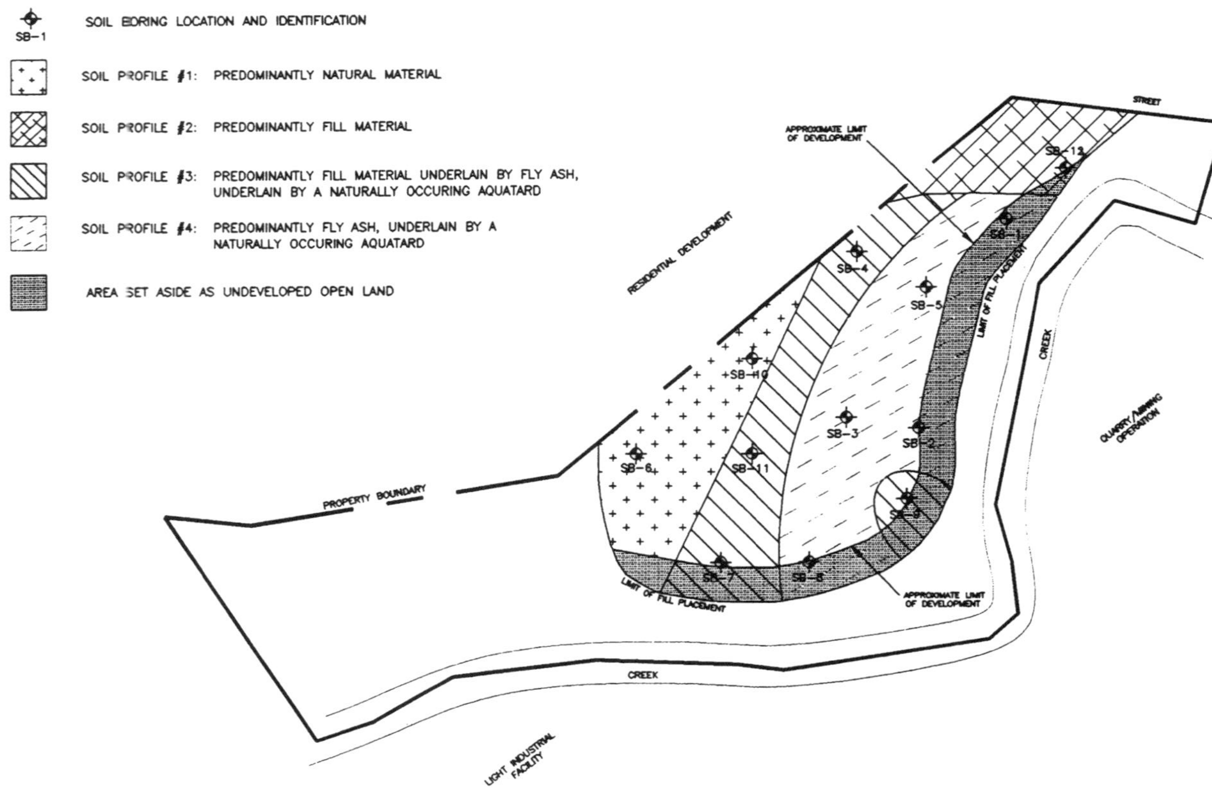

Laboratory Analysis

The soil, sediment, and surface water samples were placed into laboratory-provided containers, packed into an iced cooler for preservation, and delivered to a USEPA-approved environmental chemistry laboratory for analysis. The samples were analyzed for arsenic (SW-846 Method 7060), beryllium (SW-846 Method 6010), and manganese (SW-846 Method 6010).

Because groundwater was not encountered, no groundwater samples were collected for laboratory analysis.

SITE GEOLOGY/HYDROGEOLOGY

Environmental Standards' interpretation of subsurface conditions at the site is based on the results of the soil boring programs conducted on May 25 and September 8-9, 1994. The site soil characteristics are dependent on the amount and type of filling that occurred at different portions of the site. Distinctly different soil profiles were identified at four portions of the site during the September 1994 phase of investigation. They are described in the following profiles and in Figure 1.

Soil Profile #1

Found at the northwestern and western portions of the site, this soil profile encompassed approximately 20% of the site under investigation. Typically, coarse to fine sand with varying amounts of silt and gravel was encountered from the ground surface to depths ranging from approximately 5 to 10 feet bgs. Beneath the sand, clayey silt with weathered rock fragments was encountered to depths ranging from approximately 14 to 15 feet bgs. Saprolite and weathered bedrock were encountered in SB-10. SB-10 was terminated at approximately 23.5 feet bgs due to split-spoon refusal on bedrock. Much of the material in this profile appeared to be naturally occurring. However, some construction debris and other fill material were observed at some of the shallow sampling intervals.

Soil Profile #2

Found at the eastern portion of the site, this soil profile encompassed approximately 20% of the portion of the site under investigation. This portion of the site was comprised of predominantly fill material/construction debris.

Soil Profile #3

Found at several locations in the central and southwestern portions of the site, this soil profile encompassed approximately 20% of the portion of the site under investigation. Typically, this profile was comprised of fill material (generally, coarse to fine sand with varying amounts of silt, gravel, construction debris, and cinders) from near the ground surface to depths ranging from 6 to 12.5 feet bgs. Beneath the fill, fly ash was encountered to depths ranging from 14 to 26 feet. A naturally occurring silt and clay unit (an aquiclude) was encountered beneath the fly ash in the two soil borings that were advanced through the fly ash unit.

Soil Profile #4

Found in the central and southern portions of the site, this soil profile encompassed approximately 40% of the portion of the site under investigation. Typically, this profile consisted of fly ash from just below the ground surface to a maximum depth of 26 feet. A naturally occurring silt and clay unit (an aquiclude) was encountered beneath the fly ash in the two soil borings that were advanced through the fly ash unit.

ANALYTICAL RESULTS

Upon receipt of the analytical results, Environmental Standards tabulated and assessed the data with respect to the highest and lowest concentration of each constituent at each sampling interval, trends in constituent concentrations, and comparison to published cleanup standards and constituent concentrations in native soils. The Phase II analytical results are presented in Table 1, and a description of Environmental Standards' assessment is summarized in the following paragraphs. These data are further evaluated with regard to the site-specific risk assessment in the Conclusion section of this chapter.

Soil Samples

Arsenic was detected in 26 of the 27 soil samples collected during the second phase of investigation (four values were reported by the laboratory at concentrations below the method detection limit [BMDL]). Detected arsenic concentrations ranged from BMDL to 44 milligrams per kilogram (mg/kg) in the soil samples collected at the 0 to 6-inch interval, from BMDL to 130 mg/kg in the soil samples collected from the 6 to 8-foot interval, and from BMDL to 75 mg/kg in the soil samples collected from the 13 to 15-foot interval. The highest level of arsenic within each boring was found at either the 6 to 8-foot or 13 to 15-foot sampling interval in each of the nine borings. The Pennsylvania Department of Environmental Resources' (PA DER's) generic cleanup standard for arsenic in soil is 20 mg/kg (PA DER, 1993). Nine of the 27 soil samples analyzed for arsenic had levels in excess of this cleanup standard. However, with only one exception, these modestly elevated levels of arsenic were found in the deeper subsurface soils. Also, these modest levels of arsenic marginally exceeded published naturally occurring arsenic concentrations in soil (1 to 40 mg/kg) (Dragun, 1988).

Beryllium was detected in each of the 27 soil samples collected during the second phase of investigation. The levels of beryllium detected ranged from 0.66 to 2.8 mg/kg in the soil samples collected from the 0 to 6-inch interval, from 0.89 to 5.8 mg/kg in the soil samples collected from the 6 to 8-foot interval, and from 0.92 to 3.5 mg/kg in the soil samples collected from the 13 to 15-foot in-

Table 1. **Subject Site, Southeastern Pennsylvania**
Phase II Soil Sample Results - Inorganic Analysis (mg/kg)

Sample Location:	SB-4A	SB-4B	SB-4C	SB-5A	SB-5E	SB-5C	SB-6A	SB-6B	SB-6C	SB-7A	SB-7B	SB-7C	SB-8A	SB-8B	SB-8C
Sample Depth:	0'-0.5'	6'-8'	14'-15'	0"-6"	6'-8'	13'-15'	0"-6"	6'-8'	13'-15'	0"-6"	6'-8'	13'-15'	0"-6"	6'-8'	13'-15'
Arsenic	4.5	16	12	44	130	66	3.4	3.4	BMDL(1.9)	5.2	100	75	16	56	65
Beryllium	1.2	0.89	3.4	2.8	5.3	3.5	1.1	1.2	1.1	1.1	3.2	2.2	1.5	2.3	2.7
Manganese	1260	41	434	200	66	77.4	1620	545	391	901	75.7	111	402	70.4	74.6

Sample Location:	SB-9A	SB-9B	SB-9C	SB-10A	SB-103	SB-10C	SB-11A	SB-11B	SB-11C	SB-12A	SB-12B	SB-12C
Sample Depth:	0'-0.5'	8'-10'	13'-15'	0"-6"	6'-3'	13'-15'	0"-6"	8'-9'	13'-15'	0"-6"	6'-8'	13'
Arsenic	3.2	15	61	ND	4.7	BMDL(1.8)	1.8	12	44	BMDL(0.96)	BMDL(0.98)	2.1
Beryllium	0.66	1.1	1.9	0.75	0.54	0.92	0.86	1.1	2.1	2	1.1	1
Manganese	375	614	111	449	1050	2140	925	393	90.8	1440	499	989

mg/kg = milligrams per kilogram

BMDL (1.9) = Below Method Detection Limit (Method Detection Limit)

ND = Not Detected

terval. The highest level of beryllium at each sampling level was found in boring SB-5. PA DER has not published health-based guidelines for beryllium, but a comparison of the highest level of beryllium detected at the site (5.8 mg/kg) to naturally occurring background concentrations of beryllium reported in scientific literature (0.1 to 40 mg/kg) indicates that the levels are well within the naturally occurring ranges of beryllium (Dragun, 1988).

As with beryllium, manganese was also detected in each of the 27 soil samples collected during the second phase of investigation. The levels of manganese ranged from 200 to 1,620 mg/kg in the soil samples collected from the 0 to 6-inch interval, from 41 to 1,050 mg/kg in the soil samples collected from the 6 to 8-foot interval, and from 74.6 to 2,140 mg/kg in the soil samples collected from the 13 to 15-foot interval. The highest levels of manganese were detected in the soil samples collected from the surface sampling interval (0 to 6-inches) at seven of the nine soil boring locations, possibly due to naturally occurring concentrations in the top soil. PA DER has not published health-based guidelines for manganese, but a comparison of the highest level of manganese detected at the site (2,140 mg/kg) to published values of the naturally occurring background concentrations of manganese (100 to 4,000 mg/kg) indicates that the levels detected at the site are well within the naturally occurring ranges of manganese (Dragun, 1988).

Surface Water and Sediment Samples

Arsenic and beryllium were not detected in any of the surface water samples. Manganese was detected in each of the surface water samples at low levels, ranging from 0.0065 to 0.016 milligrams per liter (mg/L). The concentrations of manganese increased slightly from the upstream to midstream, and from the midstream to downstream surface water samples (Table 2).

Arsenic was not detected above the method detection limit in any of the sediment samples. Beryllium was detected in all sediment samples at low levels, ranging from BMDL to 0.22 mg/kg. Manganese was also detected in all sediment samples at low levels, ranging from 80.8 to 291 mg/kg. The highest levels of both beryllium and manganese in the sediment samples were detected in materials collected from the midstream sampling location (Table 3).

A site-specific risk assessment was performed utilizing the arsenic, beryllium, and manganese data collected in the Phase I and II investigations. The components of a risk assessment utilized to determine if there was an unacceptable risk were comprised of data evaluation, an exposure assessment, a toxicity assessment, and a risk characterization. Data evaluation determines what constituents are of potential concern. The exposure assessment determines which populations are exposed to the constituents of concern, and quantitatively estimates each population's exposure. In the toxicity assessment, carcinogenic and noncarcinogenic toxicity benchmarks are established for each constituent. The risk characterization combines the results from the exposure and toxicity as-

Table 2 **Subject Site, Southeastern Pennsylvania, Phase II Surface Water Results - Inorganic Analysis (mg/L)**

Sample ID	Stream-up	Stream-mid	Stream-down
Arsenic	ND	ND	ND
Beryllium	ND	ND	ND
Manganese	0.0065	0.011	0.016

Table 3 **Subject Site, Southeastern Pennsylvania, Phase II Sediment Sample Results - Inorganic Analysis (mg/kg)**

Sample ID	Sed-up	Sed-mid	Sed-down
Arsenic	BMDL (0.85)	BMDL (1.1)	BMDL (1.0)
Beryllium	BMDL (0.11)	0.22	0.16
Manganese	80.8	291	147

mg/L = milligrams per liter.

mg/kg = milligrams per kilogram.

ND= Not Detected.

BMDL (0.85) = Below Method Detection Limit (Method Detection Limit).

sessments to determine if the constituents pose an unacceptable risk to the exposed populations. There may be an unacceptable risk if the carcinogenic or noncarcinogenic levels exceed a cancer risk of 1×10^{-6} (one incidence of cancer over a lifetime in a population of one million people so exposed) and a hazard index of 1, respectively. The combined cancer risks and hazard index posed to future construction workers via exposure through all pathways and all chemicals was estimated to be 9×10^{-7} and 0.5, respectively. Children playing in the creek were estimated to incur a combined cancer risk of 3×10^{-9}, and a total combined hazard index for this population was estimated to be 0.0002.

CONCLUSIONS

This risk-based project was unique in that the data collected during the two phases of investigations not only provided an assessment of subsurface geological/ hydrogeological conditions, but also augmented the performance of a risk assessment. Geological/ hydrogeological assessment of site conditions indicated that four distinct soil profiles were present at the site, and that neither perched nor groundwater were present above the bedrock/unconsolidated material interface. Assessment of the analytical data indicated that, in some instances, arsenic concentrations may exceed PA DER's soil cleanup standards and naturally occurring levels of arsenic in soil. However, the levels of beryllium and manganese in soil were well within the range of the naturally occurring levels reported for these constituents. To further determine the potential impact of these constituents of concern to human health, a site-specific risk assessment was conducted to estimate potential risks associated with development of the site. Construction workers and children ages 6 to 15 years old were considered the most sensitive populations likely to be exposed to constituents in the site soil and adjacent creek sediments and surface water. Exposure to residents was not considered because restrictive measures would reduce the probability of residents coming into contact with the fly ash to near zero. Following the proposed construction activities, the fly ash material will be covered by at least one foot of native or imported "clean" soils. Additionally, direct contact by the resident population would be limited further by a proposed 50 to 100-foot undeveloped buffer zone along the edge of the fill material, landscaped vegetative cover, paved roads and driveways, and construction of houses.

The overall assessment of the concentrations of arsenic, beryllium, and manganese in the soil, surface water, and sediment at the site indicated that there was no unacceptable risk posed to construction workers or children by the constituent levels present at the site, given future use of the property as currently planned by developers and described to Environmental Standards. Because ground water was not

Because groundwater was not encountered during investigative activities, no analytical data of groundwater from the site are available. In the absence of analytical data, an assessment of the impact of groundwater quality at the site was made, based on the following site conditions: (1) the constituents of concern are inorganic (thus, they are easily sorbed onto soil); (2) a substantial aquiclude (the naturally occurring silt and clay layer) and competent bedrock were encountered beneath the fly ash material and above groundwater, indicating a low potential for the migration of constituents to the groundwater; and (3) proposed development plans indicate that potable water will be supplied to the development by the local water authority. Given these conditions, it did not appear likely that the constituents of concern had significantly impacted the quality of groundwater beneath the site, and that minimal to no risk was posed to future residents by the groundwater quality.

REFERENCES

Dragun, J., 1988. *The Soil Chemistry of Hazardous Materials*, Chapter 3. Silver Spring, MD, The Hazardous Materials Control Research Institute.

(PA DER) Pennsylvania Department of Environmental Resources. 1993. Cleanup Standards for Contaminated Soils (Interim), Harrisburg, PA.

Smith, R. L. 1994. Risk-Based Concentration Table, Fourth Quarter 1994, U.S. EPA Region III, Philadelphia, November 8, 1994.

CHAPTER 40

Risk Based Evaluation of Potential Remedial Actions on a Flood Controlled Wetlands Area in Southern New York

Jaime N. Goncalves and **Jennifer C. Kotanchik**, Environmental Standards, Inc., Valley Forge, Pennsylvania

INTRODUCTION

Site Description

The site is located in Chemung County, New York, which is in the Southwestern Plateau Region. This region is characterized by a series of flat-topped hills and ridges with steep slopes (Soil Survey of Chemung County, 1973). Measuring approximately 1,400 feet long and 350 feet wide, the site runs roughly east to west, and encompasses an area of approximately 11.5 acres (Figure 1). The site is bounded by a flood control levee to the north and by a stream to the south. The site lies entirely within the 100-year flood plain of the stream. Natural drainage for the site is toward the stream. The surface soils consist mostly of relatively thin alluvial deposits of silts, sands, and gravels which are loose and very permeable. The alluvial deposits overlie glacial outwash sands, gravels, lacustrine silts, and glacial till which overlie bedrock. Vegetation is mostly deciduous woodland and grasses, which are typical of forested wetlands.

The site is deed restricted, allowing hiking, biking, and recreational use only. The restriction is consistent with the U.S. Army Corps of Engineers regulations regarding land use for sites that lie within a 100-year flood plain, possess wetland characteristics, and are within a flood-control structural system. The restrictions eliminate the possibility of development, or the installation of groundwater supply wells for potable purposes.

Site History

Dumping of chlorinated compounds, fats, greases, oils, and fly-ash, appears to have started in late 1940s and continued until the late 1960s. In 1986, an adjacent property was purchased for construction of a state prison facility. In July, 1987, buried drums were found on the adjacent property and an investigation

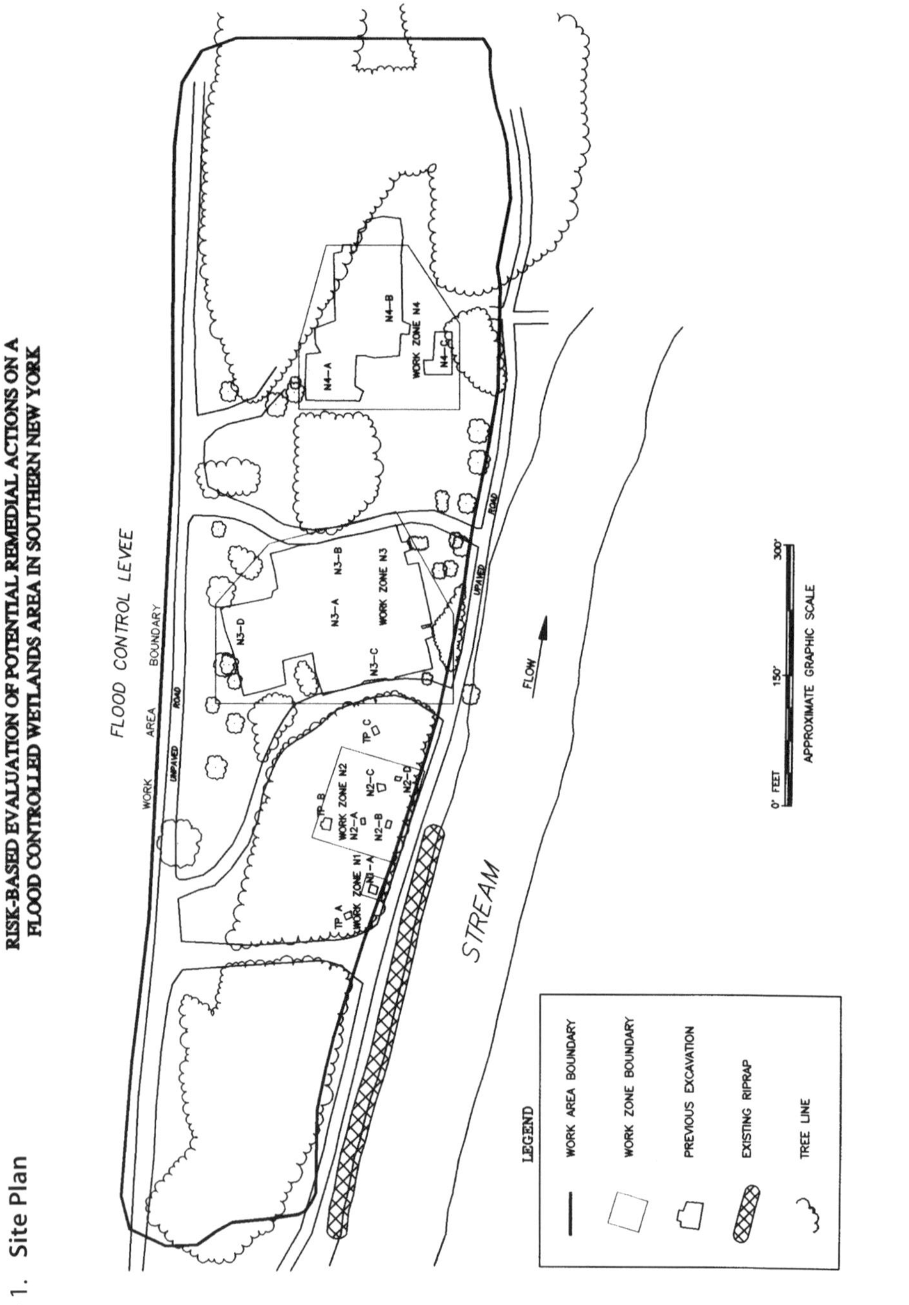

Figure 1. Site Plan

was conducted by New York State Department of Environmental Conservation (NYSDEC). Although the investigation was initially focused on the adjacent property, it quickly extended to include the site in question.

Numerous investigations and remedial actions have been undertaken at the site, culminating with the signing of the March, 1992 - Record of Decision (ROD). After the ROD was signed, the 1993 Excavation and Sampling Program and the 1994 Boring and Sampling Program were conducted, and represent the most up to date information concerning the site.

Record of Decision

After completion of the Remedial Investigation (RI), the state executed a Record of Decision (ROD) for the site in March, 1992. The ROD identified three main Remedial Action Objectives (RAOs) for the site:

1. Reduce the potential for direct exposure to waste/soils and surface soils presenting localized high potential risks.
2. Reduce the toxicity, mobility, or volume of waste constituents and soil contaminants.
3. Preserve or enhance the unique natural, environmental, ecological, and recreational resource characteristics of the site.

These objectives were identified for use in the selection of the most appropriate remedial action for the site, as well as the overall community. In addition to identifying the RAOs for the site, the ROD outlined the potential remedial actions identified in the Feasibility Study, and selected the most appropriate remedial action based on the information existing at that time. The selected remedial action, excavation of contaminated soil/waste (estimated to be approximately 5,360 cubic yards of material) with offsite disposal and groundwater monitoring, was elected on the basis of fulfilling the RAOs as well as being protective of human health and the environment. The selection was based on the data collected up to, and including, the information gathered in the RI.

Summary of Post ROD Activities

Upon execution of the ROD, plans were developed to conduct the selected remedial action at the site. Four separate work zones were established, and materials excavated. Post-excavation samples collected from the excavation walls and floors indicated that cleanup criteria would not be met in three of the four work zones. A total of 10,255 cubic yards of material was ultimately excavated and disposed of at an offsite treatment, storage, or disposal (TSD) facility when work was suspended in November, 1993. As a result of the additional information collected during the remedial activities, it was evident that additional information regarding waste distribution was required. To address the data gaps, a soil boring program was initiated in February, 1994. A total of 72 borings was installed to a depth of either 12 feet below grade or 12 feet into natural soils. The borings were installed in a 50 by 100-foot grid pattern throughout the site, except in the excavated work areas. The soil borings provided information from 285 samples. The information obtained from the 1993 and 1994 activities was used to produce a Revised Feasibility Study (RFS) that was submitted to NYSDEC.

DATA REVIEW

During the 1993 Remedial Program and 1994 Soil Boring Program, 423 composite soil samples were collected from varying depth intervals and analyzed for PCBs, cadmium, lead, and nickel. To organize the analytical data into a more manageable format, a matrix was developed using chemical-specific values and sample intervals. The matrix was divided into two sections, representing surface soils and subsurface soils. Surface soils were categorized as those soil samples collected from 0 to 3 feet below grade (bg). Subsurface soil samples were categorized as all soil samples collected greater than 3 feet below grade. The matrix was divided into four depth intervals, 0 to 3 feet bg (surface), 3 to 8 feet bg, 8 to 12 feet bg, and greater than 12 feet bg. Chemical-specific concentration categories were also developed.

The concentration categories for PCBs included 1 mg/kg; 10 mg/kg; 25 mg/kg; 50 mg/kg; and 500 mg/kg. The concentration limits for cadmium were 10 mg/kg and 510 mg/kg. The concentration categories for lead included 350 mg/kg; 400 mg/kg; and 2,000 mg/kg. The concentration categories for nickel were 500 mg/kg and 20,000 mg/kg at the surface (0 to 3 ft), and 1,000 mg/kg and 20,000 mg/kg in subsurface soils (below 3 ft).

Surface Soil Conditions

At the end of the 1993 excavation program, side wall and exit samples were collected from the North Work Areas N1, N2, N3, and N4, to determine the level of metals and PCBs remaining in the soil. Surface soil samples were collected as composite samples of the 0 - 3 foot interval. One hundred thirty-eight samples were collected during the 1993 excavation program, 45 of which were collected from the surface interval, 0-3 feet below grade.

Analytical results from the surface soils samples indicated that 64.28% contained no PCBs; 90.82% contained less than 10 mg/kg cadmium; 87.76% less than 350 mg/kg lead; and 80.61% less than 500 mg/kg nickel. Table 1 summarizes the distribution of analytical data of those samples classified as surface samples from both the 1993 and 1994 sampling events.

Subsurface Soil Conditions

In addition to surface soil samples, side wall and exit samples were collected from the work areas to evaluate the level of metals and PCBs remaining in the subsurface soil. Subsurface composite samples were collected for this purpose. The intervals ranged roughly from 3.5 to 8 feet bg, 8 to 12 feet bg, and 12 to 15 feet

TABLE 1. **Distribution of 1993 and 1994 Soil Sample Results, Southern New York Site**

PCBs							
Surface	ND	< =1 mg/kg	>1 - 10 mg/kg	>10 - 25 mg/kg	>25 - 50 mg/kg	>50 - 500 mg/kg	> 500 mg/kg
0 to 3 ft.	41	22	14	11	7	3	0
Subsurface							
>3 to 8 ft.	102	30	26	13	8	11	3
>8 to 12 ft.	70	13	5	6	7	7	2
> 12 ft.	10	0	0	4	4	4	0

Cadmium				
Surface	ND	< 10 mg/kg	>10 - 510 mg/kg	> 510 mg/kg
0 to 3 ft.	50	39	9	0
Subsurface				
>3 to 8 ft.	132	44	17	0
>8 to 12 ft.	97	9	4	0
> 12 ft.	20	2	0	0

Lead					
Surface	ND	< 350 mg/kg	>350 - 400 mg/kg	>400 - 2,000 mg/kg	> 2,000 mg/kg
0 to 3 ft.	31	55	0	10	2
Subsurface					
>3 to 8 ft.	88	73	4	20	8
>8 to 12 ft.	72	31	0	5	2
> 12 ft.	17	2	0	3	0

Nickel				
Surface	ND	< 500 mg/kg	500 - 20,000 mg/kg	> 20,000 mg/kg
0 to 3 ft.	32	47	19	0
Subsurface	ND	<1,000 mg/kg	1,000 - 20,000 mg/kg	> 20,000 mg/kg
>3 to 8 ft.	82	93	18	0
>8 to 12 ft.	74	35	1	0
> 12 ft.	17	4	1	0

ND = Compound was not detected by the analytical method employed for sample analysis.

bg. During the 1994 Soil Boring Program, 72 borings were installed across the site using a 50 foot by 100-foot grid system, similar to previous investigations. Samples were collected in varying intervals that roughly matched those of the 1993 excavation program. In all, a total of 325 samples were collected.

Analytical results from the subsurface soil samples indicated that 78.77% contained less than 10 mg/kg PCBs; 93.54% less than 10 mg/kg cadmium; 87.08% less than 350 mg/kg lead; and 93.85% less than 1,000 mg/kg nickel. Table 1 provides a more detailed summary of the distribution of analytical data of those samples classified as subsurface samples from both the 1993 and 1994 sampling events.

Groundwater

As part of the RI and subsequent investigations, seven monitoring wells were installed on the property to characterize the overburden (shallow and unconfined) aquifer beneath the site. The wells ranged from 25 to 30 feet deep. The water table was encountered between 5 to 15 feet below grade. Because of seasonal fluctuations in the water table, a 20-foot section of screen was installed in each well. The top of the screen was set from between 2 to 5 feet above the water level at the time of installation.

During the RI, two rounds of groundwater sampling were conducted. The samples were analyzed for volatile organic compounds, semivolatile organic compounds, pesticides/PCBs, and metals. The results indicated that no volatile organics, semivolatile organics, or pesticide/PCB compounds were present in the groundwater. Dissolved metals were detected at levels which did not exceed NYSDEC's Class GA standards (NYSDEC, 1991a). Total metal analytes were detected at levels which exceeded the Class GA standards (NYSDEC, 1991a); however, this was attributed to fine particulates suspended in the sample during the purging and sampling process. This was confirmed during the subsequent sampling events (Rounds 4, 5, and 6), where a lower purge rate was used, which reduced the turbulence in the monitoring well prior to sampling. In the latter rounds of sampling, Class GA standards (NYSDEC, 1991a) were exceeded only for total iron and manganese in both upgradient and downgradient wells. The results indicated that there was no contamination of the groundwater from the wastes and soils on the site.

Subsequently, unfiltered groundwater samples were collected during August 1994, at which time PCBs were detected in monitoring well MW 14 at concentrations above the Class GA standards (NYSDEC, 1991a). PCBs were also detected in monitoring wells MW 15 and MW 16 (0.071 ppb and 0.076 ppb, respectively) at concentrations well below the Class GA standards (NYSDEC, 1991a). To confirm the presence of PCBs, another sampling event was planned for MW 14; however, a natural high water event (100-year flood) occurred, which destroyed monitoring well MW 14 before a confirmatory sample could be collected. A replacement well for MW 14, MW 14R was installed adjacent to MW 14. Filtered and unfiltered samples were collected from MW 14R and submitted for analysis for PCBs and metals (nickel, lead and, cadmium). The analytical results from MW-14R indicated that PCBs (0.14 ppb) were detected in the unfiltered sample. However, the analytical results from the filtered sample were well below the laboratory detection limits, suggesting that the PCBs detected in the unfiltered sample were adsorbed to fine particulates entrained in the sample during the collection procedure and were not present in the groundwater.

Surface Water

One required element of the original RI was to determine if soils and wastes onsite were having an effect on surface water quality. To evaluate water quality, surface water samples were collected upstream and downstream of the site, and downstream of the storm water outfall on the northeast corner of the prison.

No volatile or semivolatile organic chemicals, pesticides, or PCBs were detected during the two rounds of sampling. Inorganic chemical concentrations were not significantly different at downstream locations when compared to the upstream sampling location.

Sediment

Two rounds of sediment samples were collected at the same sample locations as the surface water samples described in the above section. No volatile organic compounds, pesticides, or PCBs were detected in any of the samples. Semivolatile target compounds and Tentatively Identified Compounds were detected in the two rounds of sampling in both the upgradient and downgradient sampling points at comparable levels. Metals analyses of the downgradient samples showed no appreciable difference from the upgradient sample.

IDENTIFICATION OF ARARs/SCGs

Remedial actions at the site must comply with federal Applicable, Relevant and Appropriate Requirements (ARARs) and New York State Standards, Criteria, and Guidelines (SCGs) or other environmental statutes, unless a waiver can be justified. These statutes include those established by the USEPA and other federal agencies, as well as those established by New York state agencies, if they are promulgated, more stringent than the federal standards, and are identified in a timely manner. Potentially applicable federal ARARs fall within three general categories: Chemical-Specific, Action-Specific, and Location-Specific. Each category is briefly described below.

Chemical-Specific ARARs

Chemical-specific requirements are usually technology or risk-based numerical limitations or methodologies that, when applied to site-specific conditions, result in the establishment of acceptable concentrations of a chemical that may be found in, or discharged to, the ambient environment. Information regarding the use of chemical-specific ARARs in risk assessments can be found in Risk Assessment Guidance for Superfund Volume I— Human Health Evaluation Manual Part A, Interim Final (USEPA, 1989a) and Risk Assessment Guidance for Superfund, Volume II— Environmental Evaluation Manual, Interim Final (USEPA, 1989b).

Action-Specific ARARs

Action-specific ARARs are usually technology or activity-based requirements or limitations on actions taken with respect to hazardous substances. These requirements typically define acceptable treatment, storage, and disposal procedures for hazardous substances during the implementation of the response action. The requirements generally set performance or design standards for specific activities related to managing hazardous wastes at sites similar to this site.

Location-Specific ARARs

Location-specific ARARs are the restrictions placed on the concentration of hazardous substances or the performance of activities solely because they occur at a special location. These requirements relate to the geographical or physical position of a site rather than the nature of the materials or the proposed remedial action. These requirements limit the type of remedial action that can be implemented, and may impose additional constraints on a cleanup action. For instance, the restrictions caused by flood plains and wetlands are among the most common location-specific potential ARARs for municipal landfill sites (USEPA, 1991).

NYSDEC has elected to refer to its soil ARARs as SCGs. The list of NYSDEC SCGs examined during development of this document was dated August 1991, and appeared as Appendix E in NYSDEC's DRAFT document titled Cleanup Policy and Guidelines (NYSDEC, 1991).

REMEDIAL ACTION OBJECTIVES

The three RAOs identified in the feasibility study (FS) were still considered valid and current for the RFS. The RAOs are identified and discussed in additional detail below. Existing data indicated that surface water and sediment quality were not affected by the site; therefore, RAOs do not specifically concern themselves with these media.

Data collected from the monitoring well network suggested that filtered groundwater samples did not contain compounds in any health-based concentration of significance. In addition, it is not possible to consume site groundwater because the site is prohibited, via deed restriction, from ever having a potable supply well installed on the site. The RAOs ,therefore, do not specifically address groundwater.

Reduce Direct Contact Exposure

The first RAO was to reduce the potential for direct exposure to waste/soils and surface soils presenting localized high potential risk. Several remedial actions could potentially achieve this objective, and a key function of the RFS was to review these technologies in light of the new information obtained during 1993 and 1994 site activities.

Reduce Material Toxicity, Mobility and/or Volume

This RAO is stipulated in the 1988 National Oil and Hazardous Substances Pollution Contingency Plan (NCP) (USEPA, 1988b) and is a general goal of waste management regulatory programs nationwide. Action had already been taken to achieve this RAO to a significant degree through the excavation program completed during 1993. The 1993 Excavation Program, in and of itself, therefore, satisfied this RAO.

Preserve or Enhance Unique Resource

The site possesses certain attributes that made consideration of potential remedial actions somewhat complex when compared to typical waste sites. Wetlands, the existence of Army Corps of Engineer flood plain management structures built in 1949, and the fact that the site lies within the 100-year floodplain were all location-specific features which had to be accounted for in the remedial alternative evaluation process. Thus, the final RAO of the remedial program considered for the site had to account for its unusual setting.

RISK ASSESSMENT

The risk assessment was performed to help evaluate the different remedial alternatives in light of the RAOs. To accomplish these objectives, consideration of alternative cleanup goals was essential. Without such consideration, the remedial actions would have been, in many cases, contradictory. For example, if the extent of excavation required to meet the cleanup goals established in the original ROD resulted in major destruction of wetlands, forested habitat, and natural erosion control, it would be appropriate to consider alternative cleanup technologies, provided they could also achieve adequate reduction in direct-contact risks and other factors evaluated in the remedy evaluation process as detailed in NYSDEC's May 15, 1990 Technical and Administrative Guidance Memorandum (TAGM) Hazardous Waste Remediation (HWR) 90-4030 (NYSDEC, 1990).

Indeed, protection of wetlands and floodplains is a requirement under state and federal statutes (*viz.*, ARARs). According to an EPA guidance document entitled Conducting Remedial Investigation/Feasibility Studies for CERCLA Municipal Landfill Sites (USEPA, 1991):

> Remediation of .... landfill sites located next to [or on] wetland areas will have to be implemented in a manner which minimizes the destruction, loss or degradation of wetlands [40 CFR 6.302(a)]. Additionally, the Clean Water Act Section 404 prohibits discharge [intentional, accidental or inevitable] of dredged or fill material into a wetland area ...
>
> Remediation of landfill sites located within floodplains (for example, lowlands, and relatively flat areas adjoining inland and coastal waters) will have to be carried out to the extent possible, to avoid adverse effects, and to preserve natural and beneficial value of the floodplain (40 CFR 6.302(b)). For example, remedial actions should not be designed and constructed in a manner that destroys the usefulness of a floodplain, thereby potentially causing adjacent areas to become flooded.

In order to achieve the recommended cleanup goals established in the March, 1992 ROD, massive excavation, deforestation, devegetation, and long-term disruption of the existing habitat would be required because the PCB contamination alone extended to a significant depth in the soil column. In such circumstances, the EPA recommends waivers and alternative options, even if the cleanup goals are legitimate ARARs. For example, the EPA's CERCLA Compliance with Other Laws Manual (USEPA, 1988a) notes:

> This waiver could be used to "salvage" a remedial action option that would cause greater environmental damage or health risks solely because that option had to meet all ARARs, especially where one ARAR causes the problem. For example, attaining the ambient concentration level for PCBs spread throughout river sediment might require widespread dredging of the sediments, causing an unacceptable release of the pollutant to the water body and damaging or disrupting the ecosystem. Waiving the ARAR for ambient PCB concentrations in the sediment would eliminate the need to conduct such harmful dredging ... Remedies protective of human health and the environment but not meeting all ARARs should be compared to the remedy meeting ARARs that causes the minimum adverse impacts.

Development of Alternative Cleanup Goals

Recommended soil cleanup objectives must afford adequate protection to human health and the environment. However, recommended soil cleanup objectives must also consider site-specific conditions and probable routes of exposure for the constituents of concern. The NCP (USEPA, 1988b) states:

> Alternatives shall be developed that protect human health and the environment by eliminating, reducing, and/or controlling risks posed through each pathway by a site. The number and type of alternative to be analyzed shall be determined at each site, taking into account the scope, characteristics, and complexity of the site problem that is being addressed.

The site should perpetually remain a forested floodplain, and the only conceivable human use of the site was recreational. In fact, the site was deed-restricted from all uses other than recreational hiking and biking. The only route of exposure of potential concern at the site was direct contact with waste/soil. While EPA-recommended methodologies for performing risk assessments had considerably evolved since the RI report was submitted, the performance of another risk assessment to calculate recommended cleanup objectives for this exposure route was unnecessary because the EPA had already developed health-based soil cleanup goals for PCBs and metals of interest, based on direct-contact with soils for various exposure populations. The EPA's Guidance on Remedial Actions for Superfund Sites with PCB Contamination (USEPA, 1990) recommends a cleanup goal of 10 to 25 mg/kg of PCBs for remote areas or areas where land use is industrial, not residential.

> In remote areas or areas where land use is industrial, a more appropriate concentration at which to start analysis may be 10 to 25 mg/kg, since direct exposure is less frequent than for residential land use and higher concentrations will be protective.

In addition, the EPA's PCB Spill Policy describes the level of cleanup required for PCB spills occurring after May 4, 1987 (the effective date of the Toxic Substance Control Act, TSCA). This policy is not a regulation and only applies to recent spills (reported with 24 hours of occurrence). However, the EPA states "as a codified policy representing substantial scientific and technical evaluation it [EPA's PCB Spill Policy] has been considered in developing the guidance cleanup levels" (USEPA, 1990). The policy recommends a cleanup level of 25 to 50 mg/kg for sites in industrial or other reduced-access areas.

The EPA Region III quarterly distributes a risk-based concentration (RBC) table with recommended screening levels for numerous constituents.

> The table contains reference doses and carcinogenic potency slopes (obtained from IRIS through July 1, 1994, HEAST through November March 1994, the Superfund Health Risk Technical Support Center, and other EPA sources) for nearly 600 chemicals. These toxicity constants have been combined with "standard" exposures scenarios to calculate RBCs - chemical concentrations corresponding to fixed levels of risk (i.e., a hazard quotient of 1, or lifetime cancer risk of 10^{-6}, whichever occurs at a lower concentration) in water, air, fish tissue, and soil" (USEPA, 1994a).

Dr. Roy Smith, Senior Toxicologist in Region III explains, "simply put, RBCs are risk assessments run in reverse. For a single contaminant in a single medium, under default exposure assumptions, the RBC corresponds to the target risk or hazard quotient" (USEPA, 1994a). Region II has no RBC equivalent, but it is clear that constituent toxicities and exposure methodologies can transect administrative EPA Regional boundaries. The health-based soil action levels for cadmium and nickel were derived assuming an industrial-use scenario, which was far more likely to result in repeated exposures to soil constituents than the recreational use mandated for the property. Region III's RBC in industrial soil for cadmium is 510 mg/kg (USEPA, 1994a). In the EPA's Guidance on Residential Lead-Based Paint, Lead-Contaminated Dust, and Lead-Contaminated Soil (USEPA, 1994b), the Agency recommends a level of no-action for lead in areas where contact by children is less likely or infrequent, of 2,000 mg/kg in bare soil. EPA Region III derived a health-based soil action level for nickel based on industrial exposure of 20,000 mg/kg (USEPA, 1994a).

Current Exposure Concentration for Direct Contact with Soil

The current exposure concentration for direct contact with soil was calculated following the EPA's guidelines in Risk Assessment Guidance for Superfund, Volume I, Human Health Evaluation Manual (Part A) (USEPA, 1989a). The 95% upper confidence limits (UCLs) of the mean concentrations for the constituents of concern at the site were as follows:

- Cadmium 4.78 mg/kg
- Lead 287 mg/kg
- Nickel 362 mg/kg
- PCBs 18.7mg/kg

In accordance with current methodologies for conducting risk assessments (USEPA, 1989a), the exposure point concentration was determined by calculating the 95% UCL of the mean concentration. This was significant because the EPA's development of a 25 mg/kg of PCBs in soil concentration as protective of human health under an industrial land-use scenario was derived by assuming all soil contains 25 mg/kg. Thus, if the 95% UCL of the average concentration of PCBs is significantly less than 25 mg/kg, then no unacceptable risks would be posed to workers if the property were to be used as an industrial setting. Likewise, the 95% UCLs of the mean concentrations of cadmium, lead, and nickel were well below the cleanup goals described in the above section. Therefore, it was evident that no unacceptable health hazards were posed by the current levels of constituents present in soils at the site.

DEVELOPMENT OF REMEDIAL ALTERNATIVES

Overview

To develop potential remedial alternatives for the site, individual technologies and groups of technologies/processes were evaluated in general terms of effectiveness and implementability. This evaluation determined the applicability of specific technologies and process options in terms of their ability to attain the site-specific RAOs. From the set of remaining technologies and processes, remedial alternatives were developed. The final alternatives incorporated different combinations of technologies. The alternatives developed represent different levels of potential for achieving the RAOs established for the site.

Based on this evaluation of technologies and process options, seven remedial alternatives were most applicable to this site. The seven alternatives fall within four major groups of alternatives, which were generally:

1. actions completed to date (excavation) with improved erosion control and continued groundwater monitoring
2. capping
3. removal
4. removal and capping

Attainment of Remedial Action Objectives

As previously discussed, the site possesses certain attributes that made consideration of potential remedial actions somewhat complex when compared to typical waste sites. Under normal conditions, scoring of remedial alternatives using the methodology contained in NYSDEC's TAGM HWR-90-4030 (NYSDEC, 1990) would have indicated how well the remedial actions satisfy the RAOs. However, ecological functional values were not reflected in the scoring methodology presented in the TAGM (NYSDEC, 1990). This scoring method was at odds with attaining the RAO of preserving or enhancing the unique natural resources at the site.

The TAGM (NYSDEC, 1990) scoring methodology did not consider the potential impact of the *remedial actions themselves* on the environment (or human health, for that matter). In the case of a forested floodplain environment whose unique functions require that the vegetation remain substantially intact, scoring remedial actions based on residual risks alone was technically deficient if the remedial actions resulted in removal of the native cover. According to the scoring methodology, a remedial action which completely removed this vegetative structure and caused a loss in such functions as flood storage and habitat could receive a "perfect" score in this category and thus, have appeared to be the best alternative.

It was, therefore, desirable to counter-balance this incomparability and evaluate remedial alternatives by considering the site-specific RAOs in conjunction with the performance standard. To this end, the remedial alternatives were evaluated based on their ability to achieve the site-specific RAOs.

It was apparent that maximizing the RAO specifying reduction in volume would have minimized achieving the RAO specifying a goal of preserving or enhancing unique resources. The relative importance of each RAO necessarily determined the optimum balance in selecting the most appropriate remedy. For example, one capping alternative, which involved placement of a modified RCRA cap over all areas with PCB or lead concentrations in surface soil exceeding 1 mg/kg and 350 mg/kg, respectively, would satisfy RAO No. 1 (reduce direct contact exposure). However, it would not reduce exposure potential any more than another capping alternative which called for a gravel/soil cap. Although the gravel/soil cap alternative would not reduce water infiltration to the same extent as the RCRA cap alternative, the soil constituents of concern were essentially immobile, and reduction in subsurface mobility, accordingly, would not afford added benefit at this site. While the performance standard of an RCRA-type cap is generally regarded as superior to other cap types, from an RAO perspective, it offered no additional reduction in mobility.

Evaluation of the RAO to Preserve or Enhance the Unique Natural, Environmental, Ecological, and Recreational Resource Characteristics of the Site

This objective was consistent with provisions of federal laws regarding wetlands and floodplains. However, it was not possible to meet this objective and also meet an objective of achieving SCGs as cleanup goals or achieving maximum waste volume reduction. All remedial alternatives that involved capping the entire site would dismiss this RAO. Each remedial alternative involving further excavation would compromise, to some extent, the site-specific objective of preserving the resource characteristics of the site: the lower the cleanup objectives, the greater the extent of deforestation and habitat destruction required.

The importance of this RAO had to be weighed against other RAOs to determine which objectives merited the greatest consideration. If a remedy resulted in maximum reduction in the direct contact, exposure potential which could be effectively managed, while at the same time accomplished reduction in volume or mobility, and also preserved the natural resource characteristics of the site, that remedy would achieve the maximum balance of stated objectives and could assure adequate protection to human health and the environment. The most balanced alternative, cap areas not heavily vegetated with a 3-foot gravel/soil layer, represented the best choice for achieving this balance. If the RAO for preserving the natural quality of the local environment/recreational value of the property was not an important consideration, then the gravel/soil cap alternative represented the best choice because it would accomplish the maximum degree of future protection.

Evaluation of Cost vs. Benefit of Remedial Alternatives

Benefits realized by various remedial alternatives can be measured in several ways. For example, the greatest benefit to be accomplished in any remedy is to afford adequate protection to human health and the environment. Indeed, this benefit is in every sense the "prime directive" of environmental regulation. Other desirable benefits include: restoration of a listed site to predisposal conditions; reduction of volume, mobility, or toxicity to the maximum extent possible, as discussed above; and several other factors detailed in TAGM-HWR-90-4030 (NYSDEC, 1990) as key elements of the remedial alternative evaluation process. While cost is not an overriding consideration per se, it is, nevertheless, particularly important when alternatives do not differ substantially in achieving other objectives. For example, one alternative calling for excavation of all material containing PCBs or lead in excess of 10 mg/kg and 350 mg/kg, respectively, would adequately reduce direct contact exposure opportunities. Another alternative (cap areas not heavily vegetated with a 3-foot gravel/soil layer), however, would likely be equally, if not more, effective in reducing direct contact exposure potential (RAO Number 1). When one considers the massive excavation which already took place, both alternatives accomplish reduction in volume (RAO Number 2); and, although the most expensive alternative would result in greater reduction in volume, the most balanced alternative would result in greater reduction in erosion/mobility. Thus, both alternatives accomplished the major objective of protecting human health, but the difference in cost was more than $21,000,000. Furthermore, the most balanced alternative satisfied the third RAO for preserving the natural recreational quality and ecology of the site, while continued excavation necessitated complete deforestation. In this case, the additional cost involved in implementing continued excavation over other alternatives was not justified by the benefit of risk reduction (essentially none) and the abandonment of the third RAO (preserving the natural environmental resource and erosion-controlling nature of the mature trees on the site).

Recommended Alternative

The information collected and analyzed during RFS development resulted in the identification of a suitable remedial action for the site. The highest scoring alternative, when viewed under the strict terms of the NYSDEC TAGM (1990) quantitative protocol and the RAOs as stipulated in the ROD, called for development restrictions, groundwater monitoring, limited excavation of PCB-affected soils, cleaned areas covered with gravel/soil cap, erosion control measures maintained, and flora and fauna maintained. This alternative was also recommended because it most effectively achieved the optimum balance among the NYSDEC RAOs stated in the March 1992 ROD. In addition, the remedy could be implemented so that it adequately achieved each of the seven criteria established in NYSDEC's TAGM (1990) and was consistent with the Draft October 1991 Cleanup policy (NYSDEC, 1991).

CONCLUSION

Risk assessment has been used in the past to provide justification that a selected "remedial goal" was protective of humans and the environment. Today, the process is run in "reverse" by providing information as to the minimum amount of materials needed to be removed to provide adequate protection to the environment and humans. By coupling the risk assessment process with the process of cost-benefit analysis, responsible parties can select the remedial action that will not only provide adequate safety to the public, but also be the most cost-efficient. Recent events in Congress suggest that the federal government has begun to move in this direction, while some states have been reluctant to validate this merging of philosophies. It will be interesting to see how the state governments react as more and more responsible parties elect to conduct this type of analysis.

REFERENCES

(NYSDEC) New York State Department of Environmental Conservation. 1990. Technical and Administrative Guidance Memorandum: Selection of Remedial Actions at Inactive Hazardous Waste Sites. Division of Hazardous Waste Remediation, Albany, NY. TAGM-4030.

(NYSDEC) New York State Department of Environmental Conservation. 1991. Cleanup Policy and Guidelines. Volume I and Volume II - Appendix. Draft. Cleanup Standards Task Force, Albany, NY.

(NYSDEC) New York State Department of Environmental Conservation. 1991a. Water Quality Regulations for Surface Waters and Ground Waters. Division of Water, Albany, NY. 6NYCRR.

Soil Survey of Chemung County, New York. U.S. Department of Agriculture. Soil Conservation Service, issued September, 1973.

(USEPA) United States Environmental Protection Agency. 1988a. CERCLA Compliance with Other Laws Manual. Office of Emergency and Remedial Response, Washington DC. OSWER Directive 9234.1-01.

(USEPA) United States Environmental Protection Agency. 1988b. EPA National Oil and Hazardous Substances Pollution Contingency Plan (NCP) Under the Comprehensive Environmental Response, Compensation and Liability Act of 1980.

(USEPA) United States Environmental Protection Agency. 1989a. Risk Assessment Guidance for Superfund, Volume I, Human Health Evaluation Manual (Part A). Interim Final. Office of Emergency and Remedial Response, Washington DC. EPA/540/1-89/002.

(USEPA) United States Environmental Protection Agency. 1989b. Risk Assessment Guidance for Superfund, Volume II, Environmental Evaluation Manual. Interim Final. EPA/540/1-89/001.

(USEPA) United States Environmental Protection Agency. 1990. Guidance on Remedial Actions for Superfund Sites with PCB Contamination. Office of Emergency and Remedial Response, Washington, DC. EPA/540/G-90/007.

(USEPA) Environmental Protection Agency. 1991. Conducting Remedial Investigations/Feasibility Studies for CERCLA Municipal Landfill Sites. Office of Emergency and Remedial Response, Washington, DC. EPA/540/P-91/001.

(USEPA) United States Environmental Protection Agency. 1994a. Risk-Based Concentration Table. Region III, Philadelphia, PA.

(USEPA) United States Environmental Protection Agency. 1994b. Guidance on Residential Lead-Based Paint, Lead-Contaminated Dust, and Lead-Contaminated Soil. Office of Prevention, Pesticides and Toxic Substances, Washington, DC.

CHAPTER 41

Background Levels of Priority Pollutant Metals in Rhode Island Soils

Timothy M. O'Connor, Division of Site Remediation, Rhode Island Department of Environmental Management, Providence, Rhode Island

INTRODUCTION

The goal of this study was to investigate the occurrence, geochemical abundance, and scatter of the metals regulated by hazardous materials programs in Rhode Island soil.

Due to the natural occurrence of metals, it is necessary for environmental site investigations to attempt to determine the levels of metals in soil beyond the influence of the site being investigated. These levels are referred to as background. These results are then used as control samples, and compared to other site samples and available literature to determine the degree and extent of environmental contamination at the site.

The problem with this situation is that available literature is very limited, and the cost of analyzing samples discourages many project managers from taking what they perceive to be unnecessary samples. Also, one or two attempts at taking representative background samples can be insufficient, and can lead to over- or understating the extent of contamination. Therefore, it is often difficult for regulators, consultants, or responsible parties to address the detection of priority pollutant metals at a site. This project's goal is to attempt to solve this problem by utilizing existing site data from both the federal and state hazardous waste and hazardous materials regulatory programs.

This study relies entirely upon data which were generated under one of two regulatory authorities. The first program is under the Comprehensive Environmental Response Compensation and Liability Act, commonly referred to as Superfund. Superfund is a federal program which is run by the United States Environmental Protection Agency (USEPA). The purpose of this program is to investigate and clean up the nation's worst hazardous waste disposal sites.

The second regulatory program which generated samples used in this study was the Rhode Island Department of Environmental Management's Division of Site Remediation, which regulates the investigation and cleanup of sites similar to those in the federal Superfund program. This state program follows the general framework of the federal program, but is much smaller in size. Although these samples are generated by another organization, they are included in this study because the goals of both the state and federal programs are very similar, and the same USEPA laboratory procedures were required for both programs.

This study compiles a comprehensive database of background levels of priority pollutant metals throughout the State of Rhode Island, utilizing 106 sample points. Figure 1 shows an outline of Rhode Island, and the approximate locations of the sampling points. In addition to the gathering of analytical results, the average land usage in the vicinity of the site was also recorded. The data were then statistically evaluated with the goal of establishing statewide background levels as well as investigating to what degree man's activities influence these levels.

PARAMETERS AND METHODOLOGIES

The analytical results presented in this study included only total recoverable levels of the metals. Samples were taken from approximately the top two feet of surficial soil in the vicinity of the site investigated.

It is standard operating procedure for environmental investigators to remove approximately the top two inches of soil prior to collecting a background sample. This procedure is an attempt to limit the effects of potential pollutant sources such as automobile emissions, road runoff, or other very common sources of soil contamination.

The data used throughout this study were obtained from analytical results through USEPA laboratory methods. These methods are described in great detail in the EPA document entitled: EPA Methods for the Chemical Analysis of Water and Wastes, and Test Methods for Evaluating Solid Waste (SW-846). The EPA laboratory method numbers are: antimony (6010), arsenic (7060), beryllium (6010), cadmium (6010), chromium (6010), copper (6010), lead (6010 and 7421), mercury (7470 and 7471), nickel (6010), selenium (7740), silver (6010), thallium (7840, 7841 and 6010), and zinc (6010).

Although all of these data were generated using consistent EPA methodologies, the individual quality assurance and quality control procedures followed by the laboratories did vary. This variance is due to the overlapping jurisdiction of the state and federal hazardous materials programs. The federally investigated sites utilized the USEPA's Contract Laboratory Program (CLP). This program was initiated at the same time as the Superfund program. Its purpose is to provide consistent, high quality, legally defensible analytical results to support the Superfund program on a nationwide basis. These program laboratories utilize the same USEPA laboratory analytical methodologies as standard laboratories, but the CLP also mandates strict procedures for documentation of quality assurance and quality control.

Figure 1. **Sample locations.**

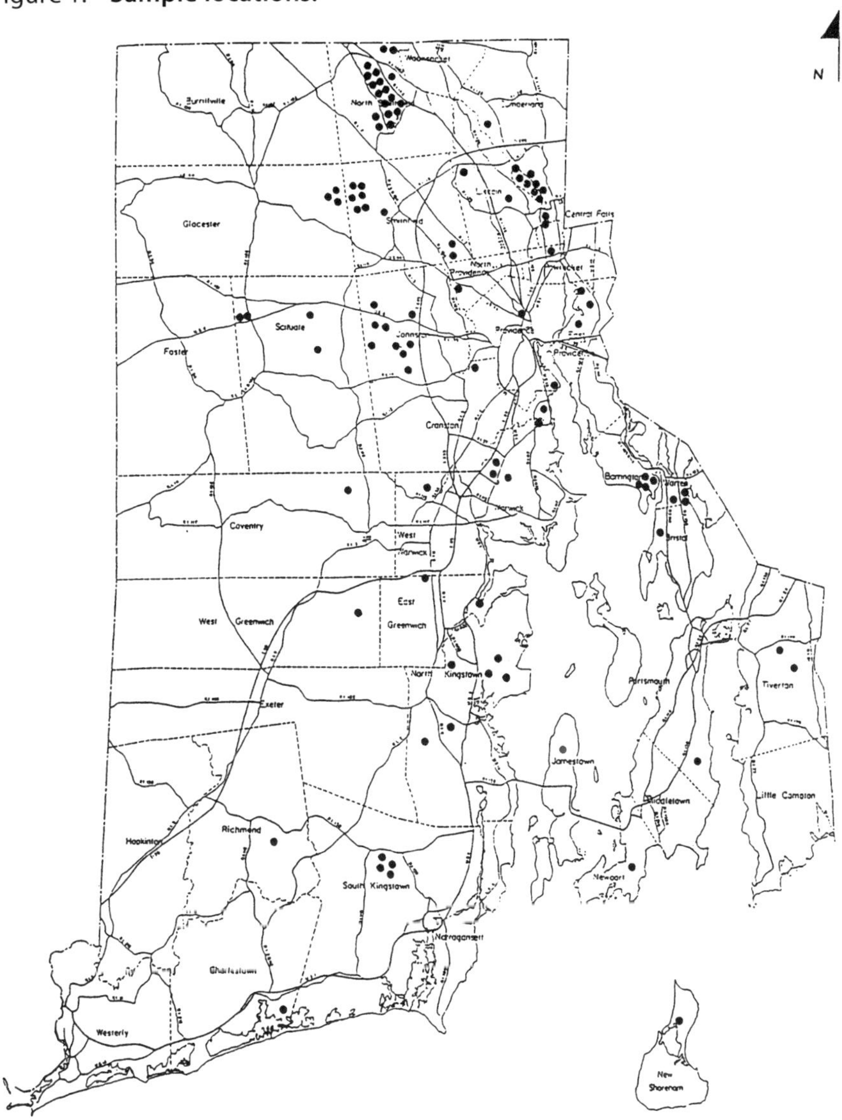

CLP data are accepted nationwide as the highest quality data available (USEPA, 1988). This program is not available, however, for use by state personnel or private contractors. Thus, data generated by a second quality assurance and quality control program is incorporated into this study. This second program is one which meets the requirements of the state, but which does not include as heavy an emphasis on validation procedures and legal defensibility as the CLP.

The land usage type was primarily determined by reviewing the report which contained the analytical results, and was supplemented by interviews with state personnel familiar with the site. In order to make this study more useful, land usage was broken down into two general categories. These were:

- High Density: used to indicate industrial parks, urban areas, areas of heavy, historic development (such as the former Naval Base, Quonset Point) and densely populated residential areas. Fifty-nine sampling points (56%) were considered to be in a high density land usage area.
- Low Density: used to indicate areas with low density residential development and very limited commercial or industrial usage. Forty-seven sampling points (44%) were considered to be in a low density land usage area.

FREQUENCY OF OCCURRENCE

A total of 106 samples were included in this study. However, due to the variety of site characteristics and the organization taking the samples, not all samples were analyzed for all parameters. Each metal was analyzed in at least 84 of the 106 samples.

On a statewide and land usage basis, Table 1 indicates a metal's frequency of occurrence in samples for which it was analyzed. Very few trends are apparent in this analysis, but several points are noteworthy. Arsenic, cadmium, copper, and mercury occur at significantly higher rates in high density areas than low density, suggesting that their occurrence is likely influenced by land usage. Barium, beryllium, and nickel, however, occur at significantly lower rates in high density areas than in low density areas, suggesting that their occurrence is not impacted by man's usage, and is likely a soil property.

On a statewide basis, six metals occurred in less than 20% of the samples analyzed for that parameter. These were antimony, cadmium, mercury, selenium, silver, and thallium. Review of the occurrence data suggested that the presence of these metals could be more the result of man's impact than natural occurrence.

STATISTICAL ANALYSIS

The goal of this study was to investigate the occurrence, geochemical abundance, and scatter of the metals regulated by hazardous materials programs. Prior to proceeding, two decisions were necessary. The first regarded how to handle a nondetect sampling result, and the second regarded the validity of statistically analyzing sampling results for metals which occurred infrequently.

For the purposes of calculating averages and deviations, this study assumed a level of half of the method detection limit (as reported in the certificate of analysis for sample results) was present for results reported as nondetectable. Based on this assumption, all metals which did not occur in at least 20% of the samples were not statistically analyzed, as the sample detection limits would have disproportionately weighted the results. These metals (antimony, cadmium, mercury, selenium, silver, and thallium) are reported in Table 2, along with their respective ratio of occurrence and range of detected levels. Using the 20% cutoff criterion left beryllium, which occurred at 47.7%, as the lowest occurring metal carried forward in the study.

Table 1. **Frequency of Occurrence Statewide and Land Usage Comparison**

	Statewide	High Density	Low Density
Antimony	5.9% (5)	7.0% (3)	4.8% (2)
Arsenic	76.4% (81)	81.4% (48)	70.2% (33)
Barium	82.4% (75)	74.5% (38)	92.5% (37)
Beryllium	47.7% (42)	35.6% (16)	60.5% (26)
Cadmium	14.4% (14)	20.0% (10)	8.5% (4)
Chromium	90.6% (96)	93.2% (55)	87.2% (41)
Copper	75.3% (73)	80.0% (44)	69.0% (29)
Lead	94.2% (98)	94.9% (56)	93.3% (42)
Mercury	17.4% (17)	19.6% (10)	14.9% (7)
Nickel	59.2% (58)	41.8% (23)	81.4% (35)
Selenium	14.1% (13)	15.2% (7)	13.0% (6)
Silver	4.2% (4)	5.9% (3)	2.3% (1)
Thallium	4.8% (4)	2.3% (1)	7.3% (3)
Zinc	92.6% (88)	96.3% (52)	87.8% (36)

Note: number in parentheses indicates the actual number of positive detections. References 4-80.

Table 2. **Range of Concentrations for Low Occurrence Metals**

Metal	Ratio Of Occurrence	Range Detected (Mg/Kg)
Antimony	5: 85	2.00 - 5.90
Cadmium	14: 97	0.22 - 3.50
Mercury	17: 98	0.02 - 0.50
Selenium	13: 92	0.05 - 1.10
Silver	4: 95	0.20 - 1.30
Thallium	4: 84	0.43 - 0.94

Source: References 4-80.

The analytical results for the remaining eight, more commonly-occurring, metals in the study (arsenic, barium, beryllium, chromium, copper, lead, nickel, and zinc) were studied in greater detail. These data were compiled into a series of frequency bar graphs which compare ranges of analytical results to their frequency. These graphs, as shown on Figures 2 through 9, indicate that the analytical results for all of the metals are positively skewed and lognormal.

Due to the lognormal distribution, the geometric data were selected as the most representative of geochemical abundance (Gilbert, 1987). The bar graphs also indicate that several of the results may fall outside the expected range of variation, and are thus suspect. Utilizing the geometric data, a 95% confidence interval was determined, and results falling outside of the interval were eliminated. As the data were positively skewed, only the upper bound of the confidence intervals was necessary. This value was calculated as the product of the geometric average and the geometric deviation to the second power.

For comparative purposes, arithmetic average and deviation, geometric average and deviation and the geometric data with 95% confidence were calculated on a statewide and land usage basis for each of the commonly occurring metals. These results are depicted in Tables 3 and 4.

The results of the categorical analysis of these data by land usage were somewhat surprising. Comparing the results of the land usage data with the statewide data, one expects a comparison of potential pollutant levels (in this case, metals) would be the largest for high density areas and lowest for low density areas. This was not consistently the case.

Geometric average levels detected in high density areas were greater than the low density average for arsenic, copper, and lead, and significantly less than the low density average for barium, chromium, and nickel.

The results for beryllium and zinc did not fit into either of the above categories. The results for beryllium showed no variance when comparing land usages to the statewide average using the geometric data and a slight variance in the arithmetic data. The zinc arithmetic average indicated a level for high density usage above the low density average, while the geometric data indicated a level lower than the statewide average for the same category.

This inconsistency of the results with regard to land usage is likely the result of most samples containing very little, if any, actual surficial soil. It is standard operating procedure for environmental investigators when taking background samples to remove the top two inches of soil in an effort to avoid ubiq uitous contamination such as automobile emissions or road runoff. This procedure appears to have impacted the results of this investigation with regard to investigating the relationship between land usage and the geochemical abundance of a metal.

Figure 2. **Arsenic statewide frequency distribution.**

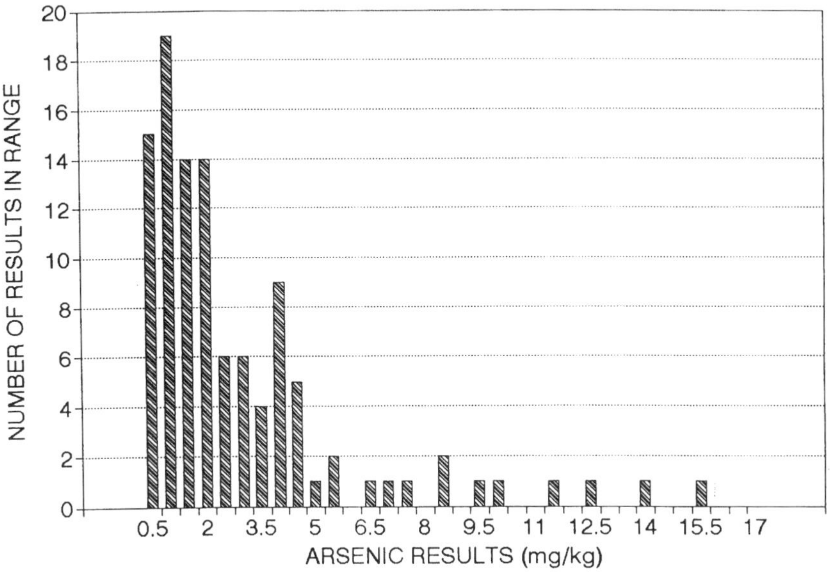

Figure 3. **Barium statewide frequency distribution.**

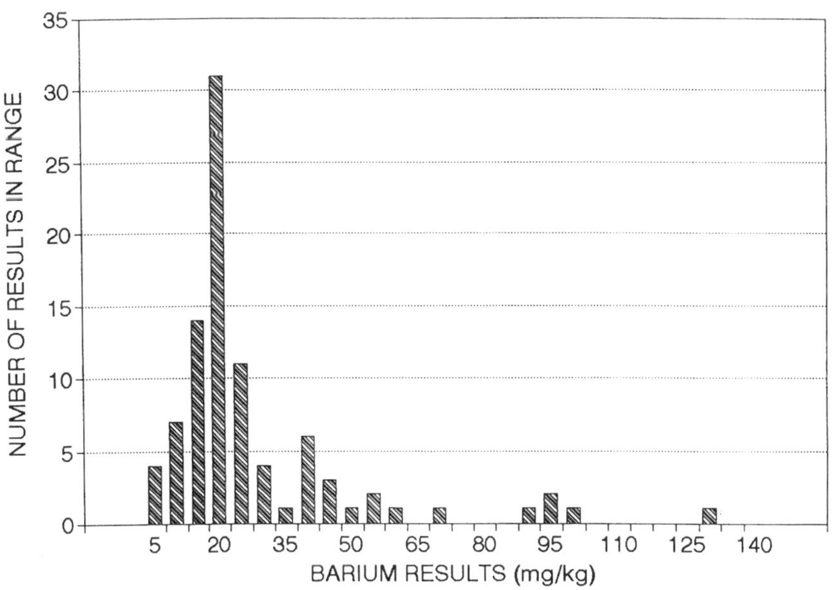

Figure 4. **Beryllium statewide frequency distribution.**

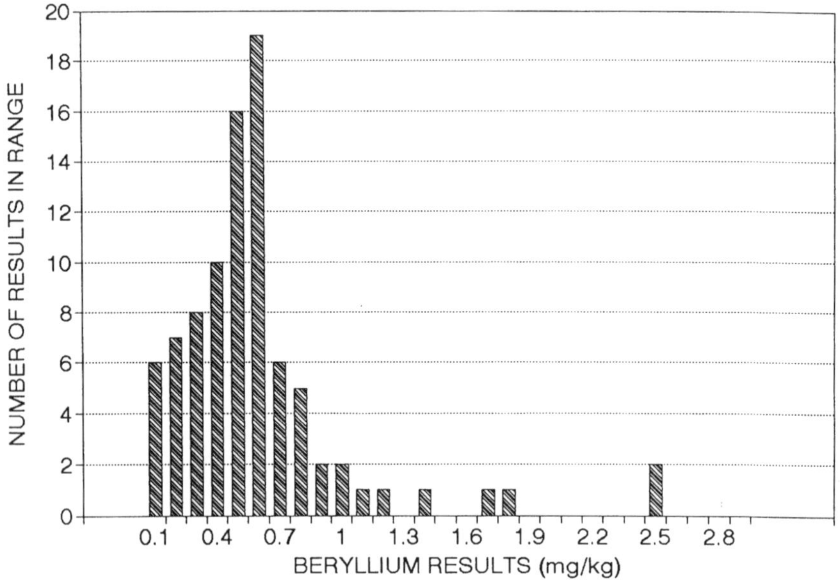

Figure 5. **Chromium statewide frequency distribution.**

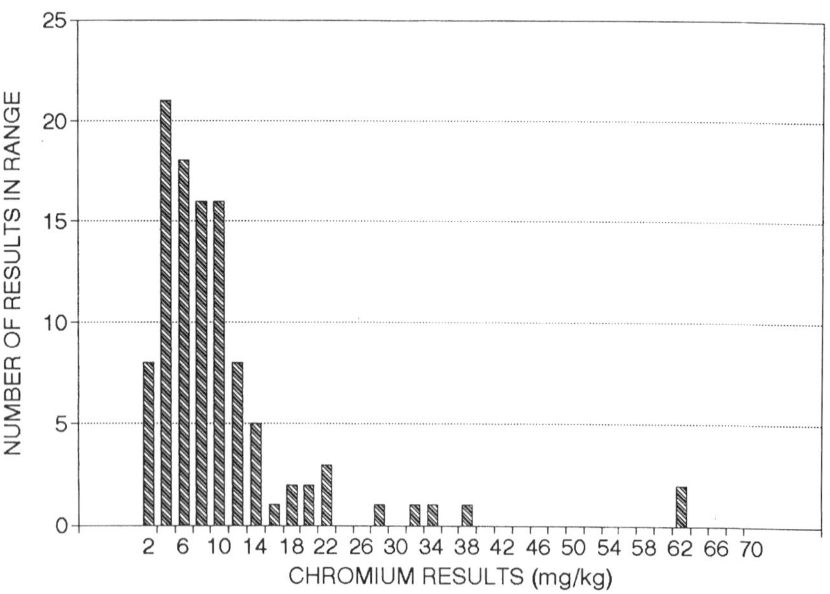

Figure 6. **Copper statewide frequency distribution.**

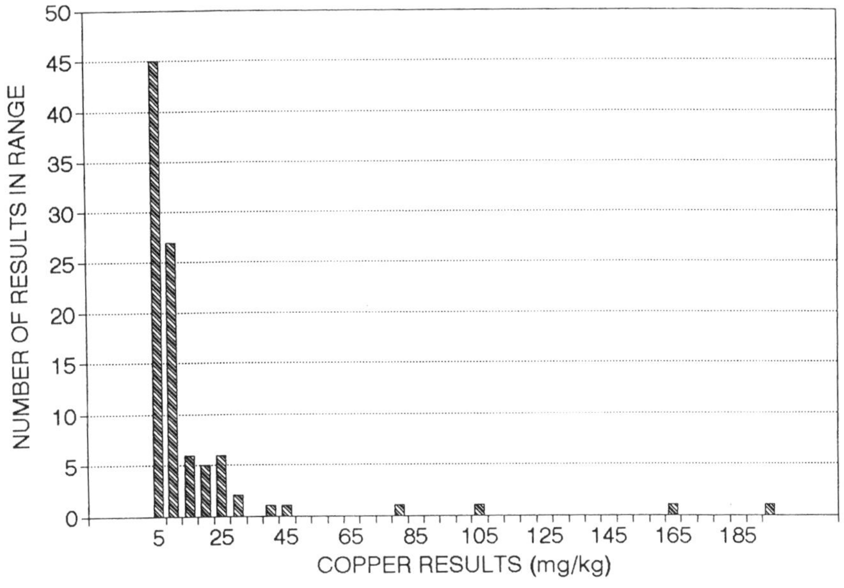

Figure 7. **Lead statewide frequency distribution.**

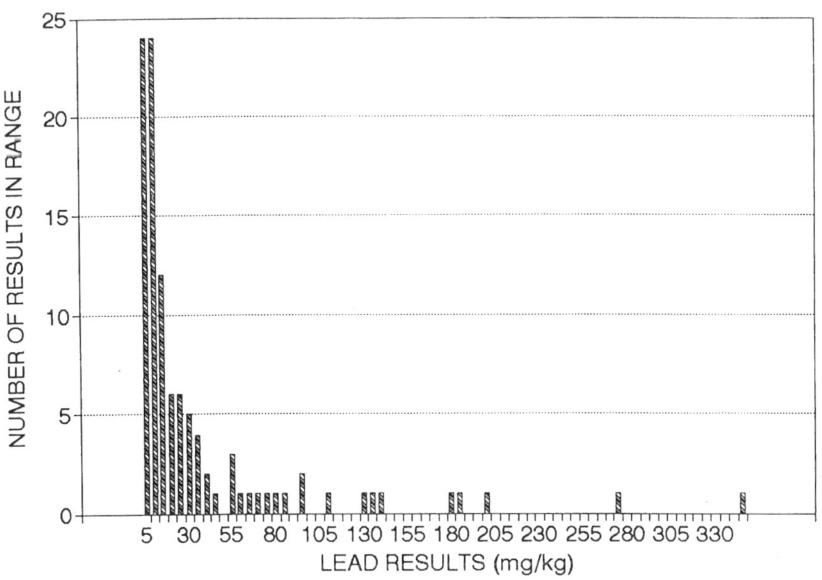

Figure 8. **Nickel statewide frequency distribution.**

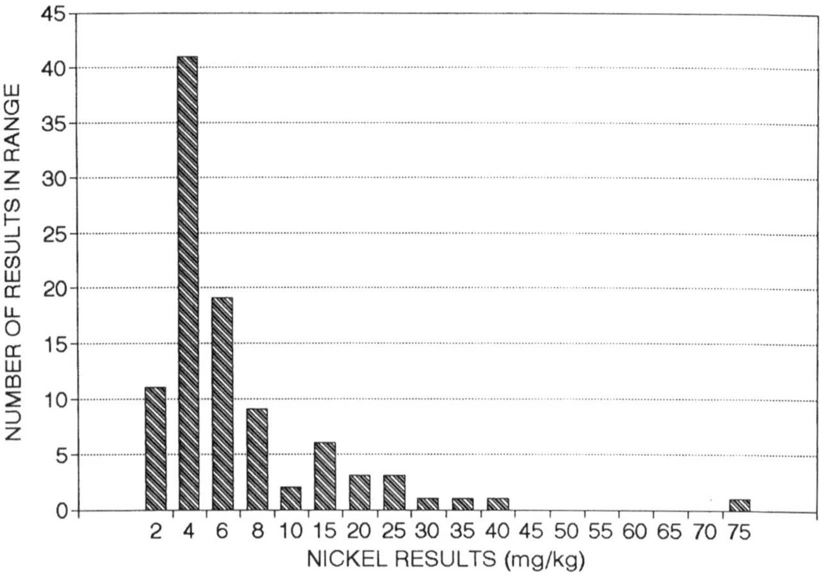

Figure 9. **Zinc statewide frequency distribution.**

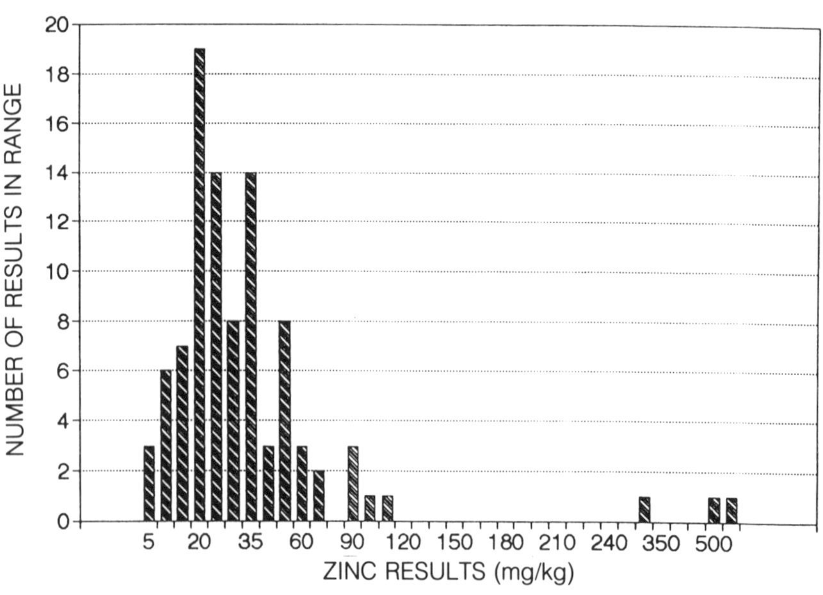

Table 3. **Statistical Summary Table**
Arithmetic Data (all results in mg/kg)

Land Usage	Statewide		High Density		Low Density	
Metal	AVE.	DEV.	AVE.	DEV.	AVE.	DEV.
Arsenic	2.71	2.93	2.74	2.86	2.66	3.01
Barium	25.80	21.61	24.15	24.34	27.91	17.31
Beryllium	0.55	0.43	0.54	0.42	0.57	0.44
Chromium	9.20	9.73	8.08	6.99	10.61	12.19
Copper	13.56	28.22	15.28	35.24	11.30	14.28
Lead	33.50	55.54	39.55	66.00	25.58	36.14
Nickel	6.81	9.33	6.03	10.63	7.80	7.21
Zinc	41.94	73.44	48.44	94.84	33.37	22.85

Source: References 4-80.

Table 4. **Statistical Summary Table**
Geometric Data (all results in mg/kg)

Land Usage	Statewide		High Density		Low Density	
Metal	Ave.	Dev.	Ave.	Dev.	Ave.	Dev.
Arsenic	1.67	1.53	1.80	1.41	1.52	1.68
95 % Ci	1.19	1.29	1.25	1.15	1.12	1.44
Barium	19.57	16.61	17.01	16.76	23.40	16.43
95 % Ci	NA	NA	NA	NA	NA	NA
Beryllium	0.43	1.47	0.43	1.44	0.43	1.51
95 % Ci	0.38	1.50	0.38	1.46	0.38	1.54
Chromlum	6.53	4.72	6.08	4.78	7.14	4.65
95 % Ci	NA	NA	NA	NA	NA	NA
Copper	6.41	4.69	6.59	4.66	6.19	4.72
95 % Ci	5.98	4.31	5.81	4.31	NA	NA
Lead	13.91	11.36	15.66	11.25	11.92	11.50
95 % Ci	NA	NA	NA	NA	NA	NA
Nickel	4.37	3.02	3.73	3.15	9.33	2.84
95 % Ci	4.24	2.92	3.53	3.06	NA	NA
Zinc	25.27	22.06	24.38	22.10	26.49	22.01
95 % Ci	NA	NA	NA	NA	NA	NA

95 % Ci = statistical recalculated data after application of 95 % confidence interval.

NA = all values in data set fell within 95 % confidence interval.

Source: References 4-80.

OTHER INVESTIGATIONS

Only one investigation of similar scope could be located. This was by the United States Geological Survey and entitled: Elemental Concentrations in Soils and Other Surficial Materials of the Conterminous United States (USGS Professional Paper 1270) (Schlette and Boerngen, 1984). This investigation began in 1961 and continued in several phases until 1981. Samples were collected by USGS scientists when traveling to project sites and during field work for other studies. A total of 1,318 samples were taken as part of this investigation and analyzed for either 35 or 46 elements, depending upon the time period the sample was taken. All 14 metals which were part of this study were included in the USGS study. The USGS study included 28 samples in New England, one of which was near the Rhode Island/Connecticut border.

The results of the USGS study are presented in Table 5. The data from this study also indicated that the metals concentrations were positively skewed and lognormal. For this reason it was concluded that the geometric average and deviation were more representative of the geochemical abundance than the arithmetic data. Arithmetic averages were also calculated for comparative purposes.

The USGS found that three metals (cadmium, silver, and thallium) were not detected at sufficient frequency to allow averages and deviations to be calculated. This finding is consistent with this study, as all three of these metals were detected at frequencies too low to evaluate the levels on a statewide basis (cadmium 14.4%, silver 4.2%, thallium 4.8%).

Table 5. **Summary of USGS Results for the Eastern United States** (mg/kg)

Metal	Freq.	Average		Geometric Deviation	Range
		Arith.	Geom.		
Antimony	23.7%	0.76	0.52	2.38	<1 - 8.8
Arsenic	98.9%	7.40	4.80	2.56	<0.1 - 73
Barium	100%	420	290	2.35	10 - 1,500
Beryllium	32.2%	0.85	0.55	2.53	<1 - 7
Chromium	100%	52	33	2.60	1 - 1,000
Copper	98.1%	22	13	2.80	<1 - 700
Lead	78.0%	17	14	1.95	<10 - 300
Mercury	100%	0.12	0.81	2.52	0.01 - 3.4
Nickel	82.0%	18	11	2.64	<5 - 700
Selenium	84.1%	0.45	0.30	2.44	<0.1 - 3.9
Zinc	98.1%	52	40	2.11	<5 - 2,900

Source: References 4-80.

Comparison of the remaining three low occurrence metals in the Rhode Island study with the USGS findings was noteworthy. Antimony was detected at 5.9% in Rhode Island, which is relatively consistent with the USGS finding of 23.7%. However, the results of this comparison for mercury [17.3% (RI) vs. 100% (USGS)] and selenium [14.1% (RI) vs. 84.1% (USGS)] show that these two metals occur at significantly higher rates in the eastern United States as a whole than they do in Rhode Island.

In general, geometric averages for the eight common Rhode Island metals were all substantially lower than the USGS results. The exception was lead, with 13.91 mg/kg (RI) versus 14 mg/kg (USGS). Barium had the greatest differential, 19.56 mg/kg (RI) to 420 mg/kg (USGS). A review of the entire data set indicates that almost all of the Rhode Island results did fall within the range of the USGS results.

SUMMARY AND CONCLUSION

The findings of this investigation indicate that of the 14 metals most commonly regulated by hazardous materials programs in Rhode Island (antimony, arsenic, barium, beryllium, cadmium, chromium, copper, lead, mercury, nickel, selenium, silver, thallium, zinc), six occur only infrequently in Rhode Island soil (antimony, cadmium, mercury, selenium, silver, thallium). Their occurrence is likely the result of anthropogenic effects. Occurrence of the more common metals in Rhode Island soils (arsenic, barium, beryllium, chromium, copper, lead, nickel, zinc) is more likely the result of nature.

For the common metals, distribution of the geochemical abundance levels is lognormal and positively skewed. For all metals in this investigation, results were consistently lower than the results found for the eastern United States in a comparable study by the United States Geological Survey.

The results of the categorical analysis of these data by land usage indicate that arsenic, copper, and lead occur at greater levels in high density areas than low density. This result suggests that their concentration is slightly influenced by land usage. Barium, chromium, and nickel, however, occur at significantly lesser levels in high density areas than in low density areas, suggesting that their concentration is not impacted by man, and is likely entirely a soil property. The results for beryllium and zinc showed little variance by land usage.

During this investigation there was a large amount of thought given to the idea of applying a confidence interval to the data and eliminating the data points which did not fall within this interval. As an original premise of this study was that the samples in it were representative of background conditions at a given site, it would seem arbitrary to then eliminate values which did not fall within an expected range. In reviewing the data set, however, it was obvious that certain values were very suspicious. Therefore, a 95% confidence interval was established using the geometric data. Applying the confidence interval to the statistical results for the common metals allowed data to be eliminated for four of the metals. The results of this application were most notable for arsenic, which had its statewide geometric average drop almost 30%.

In conclusion, eight metals regulated by hazardous material programs are common in Rhode Island soil. These are arsenic, barium, beryllium, chromium, copper, lead, nickel, and zinc. Six metals regulated by the hazardous material programs are not common in Rhode Island soil. These are antimony, cadmium, mercury, selenium, silver, and thallium. The geometric average with the application of the 95% confidence interval is the best estimate of the geochemical abundance of priority pollutant metals in Rhode Island soil.

REFERENCES

ABB Environmental Services Inc., Remedial Investigation Report Primary Source Area (OU1), June 1993.

AET, Results of Sampling at Amtrol Inc., September 14 1987.

CDM/FPC, Draft Remedial Investigation Report of GSR Landfill, 1993.

CDM/FPC, Site Inspection Report of Truk Away Landfill, 1993.

CRA, Site Assessment Report, Former Dart Industries Facility, 1992.

EA, Screening Site Inspection Report of Jamestown Landfill, 1991.

ESS, Certificate of Analysis, Foster Nike Site, August 4, 1993.

ESS, Certificate of Analysis, Ninigret NWR, December 10, 1992.

ESS, Certificate of Analysis, North Scituate Town Dump, June 15, 1993.

ESS, Certificate of Analysis, Photek, December 8, 1992.

GHR, Remedial Investigation Report of Stamina Mills Site, January 1990.

Gilbert, R.O., 1987. *Statistical Methods for Environmental Pollution Monitoring*, New York, Van Nostrand Reinhold.

GZA, Engineering Evaluation of Contamination Blue Beach Disposal Site, December 1993.

GZA, Final Chemical Data Acquisition Plan for Engineering Evaluation of Contamination Kiefer Park Tank Farm November 1992.

GZA, Remedial Investigation Report of Central Landfill, 1992.

Metcalf and Eddy, Engineering and Evaluation of Contamination Old Quarry Disposal Site, August 1992.

Metcalf and Eddy, Remedial Investigation Report of Rose Hill Regional Landfill, 1993.

NUS Corp., Screening Site Inspection Report of M. Earl Adams, 1989

NUS Corp., Screening Site Inspection Report of Bible Speaks, 1989.

NUS Corp., Screening Site Inspection Report of Barrington Landfill No. 1, 1990.

NUS Corp., Screening Site Inspection Report of Barrington Landfill No. 2, 1990.

NUS Corp., Screening Site Inspection Report of Barrington Landfill No. 3, 1990.

NUS Corp., Screening Site Inspection Report of Barrington Landfill No. 4, 1990.

NUS Corp., Screening Site Inspection Report of BB & S Lumber, 1990.

NUS Corp., Screening Site Inspection Report of Centredale Manor, 1990.
NUS Corp., Screening Site Inspection Report of Cece Macera Landfill, 1990.
NUS Corp., Screening Site Inspection Report of Comortgage, 1990.
NUS Corp., Screening Site Inspection Report of C.J. Wheldon Trucking Company, 1990.
NUS Corp., Screening Site Inspection Report of Cumberland Landfill, 1988.
NUS Corp., Screening Site Inspection Report of Derecktor Farm, 1989.
NUS Corp., Screening Site Inspection Report of Dupraw Dump, 1990.
NUS Corp., Screening Site Inspection Report of East Greenwich Landfill, 1990.
NUS Corp., Screening Site Inspection Report of Elm Tree Dump, 1990.
NUS Corp., Screening Site Inspection Report of Former Firestone Landfill, 1989.
NUS Corp., Screening Site Inspection Report of Greenwood Avenue Disposal Area, 1990.
NUS Corp., Screening Site Inspection Report of G Tech, 1990.
NUS Corp., Screening Site Inspection Report of Hartford Avenue Gravel Pits, 1989.
NUS Corp., Screening Site Inspection Report of Industrial Machine Corporation Property No. 1, 1990.
NUS Corp., Screening Site Inspection Report of Industrial Machine Corporation Property No. 2, 1990.
NUS Corp., Screening Site Inspection Report of H & H Rack, 1988.
NUS Corp., Screening Site Inspection Report of Liquid Carbonic, 1990.
NUS Corp., Screening Site Inspection Report of W.J. Lynch Paint and Varnish Company, 1989.
NUS Corp., Screening Site Inspection Report of Mays Marshall and Meier Company, 1989.
NUS Corp., Screening Site Inspection Report of North Kingstown Landfill No.2, 1990.
NUS Corp., Screening Site Inspection Report of Patriarca Property, 1990.
NUS Corp., Screening Site Inspection Report of Polytop, 1990.
NUS Corp., Screening Site Inspection Report of Precision Chrome Plating, 1990.
NUS Corp., Screening Site Inspection Report of Providence Gas #1, 1989.
NUS Corp., Screening Site Inspection Report of R & R Jewelry Inc., 1988.
NUS Corp., Screening Site Inspection Report of Red Oak Drive Disposal Area, 1989.
NUS Corp., Screening Site Inspection Report of Scituate Town Landfill, 1990.
NUS Corp., Screening Site Inspection Report of Sel Rex, 1990.
NUS Corp., Screening Site Inspection Report of Smithfield Town Landfill, 1990.
NUS Corp., Screening Site Inspection Report of Tiverton Town Dump, 1990.
NUS Corp., Screening Site Inspection Report of J. Vinagro Landfill, 1989.
NUS Corp., Screening Site Inspection Report of Warren Town Landfill, 1989.

Resource Control Associates, Phase I Soils Evaluation Report and Proposed Phase II Sampling Plan for the Carol Cable Property, August 1989.

RIDEM, Expanded Site Inspection Report of X-Ray Products, 1992.

RIDEM, Screening Site Inspection Report of Central Falls Dump, 1992.

RIDEM, Screening Site Inspection Report of Electrolizing Inc., 1990.

RIDEM, Screening Site Inspection Report of Fay Field, 1991.

RIDEM, Screening Site Inspection Report of Fields Point City Dump, 1992.

RIDEM, Screening Site Inspection Report of James Fraza Property, 1990.

RIDEM, Screening Site Inspection Report of Hilo Dump, 1992.

RIDEM, Screening Site Inspection Report of Lonsdale Narrows, 1992.

RIDEM, Screening Site Inspection Report of Anthony Macera Property, 1993.

RIDEM, Screening Site Inspection Report of New Shoreham Landfill, 1990.

RIDEM, Screening Site Inspection Report of Pine Hill Road Dump, 1991.

RIDEM, Screening Site Inspection Report of Portsmouth Town Dump, 1992.

RIDEM, Screening Site Inspection Report of Providence Metalizing, 1992.

RIDEM, Screening Site Inspection Report of Smithfield Chemical Industrial Dump, 1991.

RIDEM, Screening Site Inspection of West Warwick Landfill, 1990.

Shacklette, H.T. and Boerngen, J.G. 1984. Elemental Concentrations in Soils and Other Surficial Materials of the Conterminous United States, USGS Professional Paper 1270.

TRC Corp., Site Inspection Report of Brickyard Pond, 1993.

TRC Corp., Site Inspection Report of Carrol Cable, 1993.

(USEPA) U.S. Environmental Protection Agency. 1988. Users Guide to the Contract Laboratory Program.

Weston, Site Inspection Report of New England Wood Preserving Company, 1993.

Weston, Site Inspection Report of Nyman Manufacturing Company, 1993.

Weston, Site Inspection Report of West Greenwich Town Landfill, 1992.

Woodward Clyde, Final Pre-Design Engineering Report II Davis Liquid Waste Site, October 1993.

CHAPTER 42

Cleanup of Contaminated Soil—Unreal Risk Assumptions: Contaminant Leaching

Arnold Schiffman, Schiffman Associates, LLC, Ewing, NJ

INTRODUCTION

Background

"Preliminary remediation goals" or risk-based concentrations (in essence, a generic or site-specific requirement or standard) for the cleanup of contaminated soils and other media were set forth by the United States Environmental Protection Agency (USEPA) in Risk Assessment Guidance for Superfund (RAGS): Part B, Development of Risk-based Preliminary Remediation Goals (1991). A risk-based concentration (Co) is back-calculated by:

$$Co = \frac{RT * BW * AT}{EF * ED * TF * CR} \qquad (1)$$

Where:

- Co = Concentration of a chemical (e.g., mg/kg of soil)
- CR = Contact/intake rate (e.g., mg/d)
- EF = Exposure frequency (d/yr)
- ED = Exposure duration (yr)
- BW = Body weight (kg)
- AT = Averaging time or average exposure period (d)
- RT = A risk target of excess individual lifetime cancer risk or a target hazard index (unitless)
- TF = A toxicity factor such as a cancer slope factor, shown as SF, $[(mg/kg\text{-}d)^{-1}]$ or a chronic reference dose, shown as 1/RfD, (mg/kg-d)

A key assumption in the USEPA risk assessment paradigm is that there is a constant concentration of the contaminant in media (USEPA 1991; 1989), or contaminant mass is infinite and conservative in space and time.

Contaminant Concentration and Mass in Contaminated Soil

The exposure pathways often used to set concentration limits for chemical contaminants in soil are: ingestion of the soil, dermal contact with the soil, inhalation of soil particles, volatilization of contaminants from the soil to the atmosphere, leaching of contaminants from the soil to groundwaters, runoff of contaminated soil to streams, and uptake of contaminants by plants. Except for groundwater, these pathways are largely limited to the first meter of the soil column. A contaminant mass balance is not considered because the regulatory paradigm is based on concentration. To emphasize the importance of mass balance, the term "infinite and conservative contaminant mass" is used to describe the constant concentration assumption.

Infinite contaminant mass in soil means that all of this mass is available for each pathway. Conservative contaminant mass means that the contaminant is not degraded during transport to each pathway or during the exposure duration. In reality, contaminant mass transported from the soil to other media is lost to the soil, and many organic contaminants degrade to carbon dioxide and water. Thus, the assumption of infinite and conservative contaminant mass is a physical impossibility when used to determine cleanup requirements for contaminated soil. Physical impossibilities are not conservative assumptions, and the premise of this chapter is that a risk-based requirement is unacceptable when based on a physically impossible assumption.

This chapter presents methods to correct risk-based soil concentrations for leachable contaminant mass after these concentrations are first determined assuming constant concentration. Soil pathways for ingestion, direct contact, volatilization, soil erosion, and plant uptake are corrected for contaminant mass lost from soil by leaching. Leaching is used as an example of a mass loss mechanism to demonstrate the inappropriateness of the constant concentration assumption. Corrected concentrations determined by the methods set forth in this chapter are up to several orders of magnitude greater than risk-based concentrations calculated assuming constant concentration.

CORRECTING FOR THE INFINITE CONTAMINANT MASS ASSUMPTION

Finite Mass Derived from a Risk-based Concentration

Assuming infinite and conservative contaminant mass, a risk-based standard for a single exposure pathway defines a finite quantity of contaminant mass. For example, a risk-based concentration for ingestion of a carcinogenic chemical in residential soil is calculated as 1 milligram of chemical per kilogram of soil. For ingestion of a unit mass of soil per day, 1 mg/kg represents a "safe level" referenced to a target risk. A finite contaminant mass in soil is now defined as 1 mg/kg of contaminant in a volume of soil. Thus, a cubic meter of soil weighing 1,500 kg will contain 1,500 mg of contaminant. The same rationale applies to noncarcinogens. The major difference is the assumption of no dose threshold for carcinogens versus a dose threshold (reference dose) for noncarcinogens. Contaminant mass below the reference dose does not contribute to any "adverse health effect." For a carcinogen, contaminant mass below the "safe level" does contribute to risk of cancer.

Contaminant-Loss Mechanisms for Organic Chemicals

There are two main mechanisms for loss of an organic chemical contaminant from soil at a contaminated site: (1) physical removal of the contaminated soil itself by wind and water erosion or ingestion by humans and animals; (2) removal of the contaminant from the soil through leaching by infiltrating rainwater, volatilization to the atmosphere through the soil surface, and degradation by biological and chemical transformations (Jury and Valentine, 1989). Loss of contaminant mass by leaching is a major loss mechanism and is used to correct the constant concentration assumption. However, volatilization (Saunders, 1995) and degradation (Schiffman, in press) of contaminants can be more significant than leaching, depending upon chemical-specific characteristics.

A Leaching Correction Factor for Surface Soil

Risk-based concentrations for carcinogens are derived from a cumulative dose of chemical over an exposure period. Although the dose is usually expressed as a fixed amount or concentration per day, this concentration can vary above or below the fixed value. The effect of changing exposure period or dose is linear. Half the exposure period or half the dose means half the risk. Thus, a risk-based concentration can be corrected for known losses of contaminant mass by increasing this concentration to account for the mass lost. A similar rationale applies to noncarcinogens, except there is a threshold dose or concentration below which there are no adverse health impacts. A corrected risk-based concentration is calculated by determining a new concentration that compensates for mass lost by leaching during the exposure duration. This correction factor is applicable to contaminated soil pathways for ingestion, direct contact, erosion, plant uptake, and volatilization, and assumes no other contaminant-mass losses.

Assumptions for Derivation of a Correction Factor for Leaching

A representative elementary volume (REV) (Bear and Verruijt, 1987) of soil of one cubic meter (surface expression of one square meter by one meter deep) is used to describe contaminant concentrations. Within the REV, soil is isotropic and homogeneous, and local concentrations are volume-averaged. The concentration leaving the REV is the ratio of solute mass to the volume of water passing a cross-section of one square meter. This volume of water is recharge, or the quantity of rainwater and snowmelt that infiltrates through the unsaturated zone and reaches the water table. The REV contains a single chemical contaminant at a maximum level not exceeding residual saturation for that chemical. Residual saturation is the volume of liquid chemical held in soil by capillary forces against the force of gravity. Residual saturation is usually described as the presence of "separate phase" chemical (including solids). The minimum contaminant level is the chemical mass sorbed to the soil particles at a pore water concentration of the chemical greater than zero. Mass removal is by contaminant dissolution and desorbtion into recharge water. Recharge water enters the REV at zero concentration of the contaminant, and concentration of the chemical in the recharge water is limited by solubility.

During leaching of a separate phase organic chemical from soil, the dissolved concentration of the chemical in the recharge water quickly rises to near-saturation levels (solubility limit). Anderson et al. (1992) demonstrated near-saturation levels in laboratory experiments where water flowed around and through a section of sand containing a separate phase organic liquid. Once the chemical dissolves in water, there is equilibrium partitioning to soil particles and sorption and desorbtion are instantaneous. Furthermore, the ratio of the mass of chemical sorbed to the soil to the mass dissolved in water (the distribution coefficient) is approximated by a linear isotherm. Thus, under local equilibrium conditions, dissolution of any liquid chemical to pore water and partitioning between dissolved contaminant and soil particles is instantaneous and linear. In addition, there are no losses of chemical by degradation or volatilization.

Chemical leaching can be estimated by zero-order kinetics for separate phase chemical and by either zero-order or first-order kinetics for sorbed chemical. In reality, leaching of chemicals from soil is a complex process that combines characteristics of zero-order, first-order, and second-order kinetics. However, any shortcomings in correcting risk-based soil cleanup criteria for contaminant mass lost from soil pale in comparison to the constant concentration assumption.

Zero-Order Kinetics (Rate Constant)

Under zero-order kinetics, leaching will reduce contaminant mass in the REV to zero over a finite time (Tz). Furthermore, the half-life of a contaminant ($Tz/2$) in the REV is dependEnt on its initial concentration (Co) where higher values for Co give longer half-lives. A concentration (C) at time (t) is given by $C = Co - kt$ where k is a soil loss-rate constant equal to Co/Tz. The time (Tz) for a contaminant mass, as defined by a risk-based back-calculation, to reach zero concentration in the REV is given by:

$$Tz = \frac{BD * REV * Co}{R * SOL} * Rtd \tag{2}$$

Where:

Tz = time for contaminant mass to reach zero (days)
BD = bulk density of the soil (kg/m^3)
REV = one cubic meter (m^3)

Co = risk-based concentration, criterion or standard (mg/kg)
R = average recharge (l/day)
SOL = chemical-specific solubility (mg/L)
Rtd = a retardation factor (unitless)

Retardation is derived from equilibrium sorption phenomena and describes the apparent slower movement of a solute through soil compared to the velocity of water. The time (*Tz*) for contaminant mass removal is overestimated when separate phase chemical is present at levels that provide sufficient soluble mass to continuously exceed the capacity of all sorption sites in the soil (no retardation). *Tz* is approximated when small quantities of separate phase chemical are present. The retardation factor is defined by (Bouwer, 1991):

$$Rtd = 1 + Kd * \frac{\rho}{\theta s} = 1 + Foc * Koc * \frac{\rho}{\theta s}$$

Where:
Kd = a distribution or partition coefficient (mL/g)
ρ = dry bulk density of the soil (g/cm^3)
θs = total porosity or saturated volumetric water content (decimal percent)
Foc = fraction of organic carbon in the soil
Koc = a distribution coefficient relative to organic carbon

Natural recharge is a measure of average yearly water transport through the unsaturated zone and a measure of total water flux through the REV. It integrates variability in soil properties, topography, climate, and water uptake by plants into a single empirical value. Groundwater recharge is commonly estimated from climate, soil, and land use factors, or from stream hydrographs (Rutledge and Daniel, 1994). Other empirical methods are also used to estimate recharge (Lerner et al.,1990). Where irrigation water is significant, all of this water, or any in excess to the amount used by irrigated plants, can be added to natural recharge.

First-Order Kinetics (Rate Constant)

The USEPA used first-order kinetics to account for contaminants lost from soil, by mechanisms such as leaching, in Methodology for Assessing Health Risks Associated with Indirect Exposure to Combustor Devices that also contains a discussion of the kinetics of contaminant loss from soil (USEPA, 1990a). Under first-order kinetics, leaching does not reduce contaminant mass to zero over a finite time, and the half-life of a contaminant in the REV (residence time) is independent of its initial concentration (Co). The concentration (C) at time (t) is given by $C = Co * e^{-kt}$ where *k* is a soil loss-rate constant. A value for *k* can be determined by assuming no separate phase chemical, plug (saturated) flow, and each pore volume of recharge water passing through a REV contains the contaminant at a concentration determined by its partition coefficient (Kd):

$$k = \frac{-\ln\left(\frac{Csa}{Cs}\right)}{t}$$

$$Ct = Cs + Cw;\ Kd = \frac{Cs}{Cw};\ Cw = \frac{Ct}{1+Kd};\ Ct = Cs + \frac{Cs}{Kd}$$

After leaching by one pore volume of water:

$$Cta = Ct - \frac{Cs}{Kd};\ Cta = Cs;\ Csa = Cta - Cw;\ Csa = \left(1 - \frac{1}{1+Kd}\right)$$

Where:

Ct = initial total contaminant concentration
Cw = initial concentration in water
Cs = initial concentration in soil ("Co")
Cta = total concentration after leaching
Csa = concentration in soil after leaching ("C")

$$k = -\ln\left(1 - \frac{1}{1 + (Kd * \frac{\rho}{\theta s})}\right)\frac{R}{\theta s Z} \qquad (3)$$

Where:

k = a loss rate constant due to leaching (yr^{-1})
Kd = a distribution or partition coefficient (mL/g)
θ s= total porosity (saturated volumetric water content)
ρ = dry bulk density of the soil (g/cm^3)
R = average recharge (cm/yr)
Z = thickness of REV (cm)
(R/sZ) = pore volumes per unit time

The above analyses assume most leaching of surface soil occurs under saturated conditions from recharge during heavy rainfall events, or recharge consisting largely of irrigation water. Under unsaturated flow conditions, leaching can be envisioned as incremental losses of contaminant by soil moisture at residual saturation (unsaturated volumetric water content). For leaching under unsaturated conditions, the rate constant, k_u is given by (USEPA, 1990a):

$$k_u = \left(\frac{1}{1 + (Kd * \frac{\rho}{\theta})}\right) * \frac{R}{\theta Z} \qquad (4)$$

Where θ equals soil volumetric water content, and the other terms are the same as Equation 3. The unsaturated volumetric water content is estimated by (USEPA, 1988; Clapp and Hornberger, 1978):

$$\theta = \theta s\left(\frac{R}{Ks}\right)^{\left(\frac{1}{2b+3}\right)}$$

Where:

θ = volumetric water content
θ s = total porosity
R = average recharge (cm/yr)
Ks = saturated hydraulic conductivity of the soil (cm/yr)
b = a soil-specific moisture retention exponent (unitless)

In this chapter, k is used instead of k_u as values for θs are more readily available than θ. Figure 1 shows the ratio of k to k_u compared to the distribution coefficient for a wide range of saturated and unsaturated porosity. k varies from k_u by a factor of about two only for the most mobile chemicals. k is within about 10% of k_u for values of $(Kd * \rho)$ greater than 2, and k approaches k_u for values of $(Kd * \rho)$ greater than 10.

Figure 1. **Ratio of first-order leaching rate constants (*k/ku*) using assumptions of saturated flow (*k*) and unsaturated flow (*ku*) compared to partition coefficients (*Kd*). Ratio is shown for range of saturated (θ_S) and unsaturated (θ) conditions.**

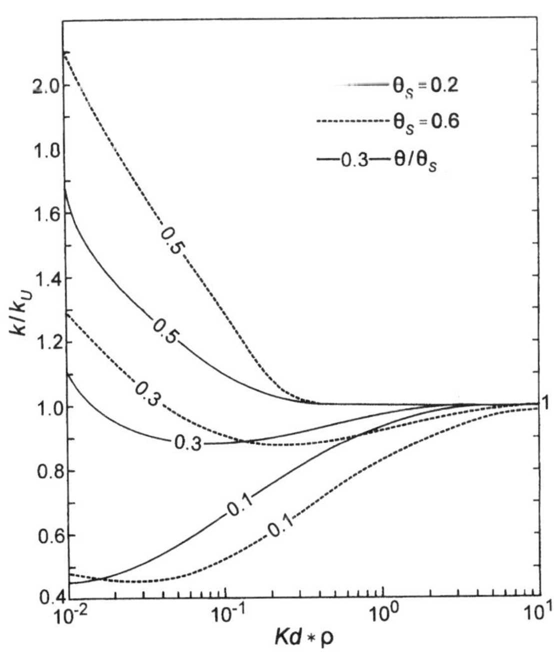

Both zero-order and first-order kinetics make for simple models of chemical mass removal by leaching with recharge water. The simplicity of this approach is consistent with the level of sophistication used in the RAGS-B risk paradigm, and appropriate for showing the need to correct for the constant concentration assumption. More complex leaching models are available and extensively described in the literature (Donigian and Rao, 1989; USEPA, 1994a). Zero-order kinetics is described first as it is applicable in the presence of separate phase chemical. In addition, nonlinear isotherms are the norm for many organic chemicals (Brusseau, 1994), and leaching half-lives in soil can be dependent upon initial soil concentration. Most soil cleanup criteria are set at low levels (1 to 100 mg/kg) that are perhaps more appropriate for analyses under first-order kinetics. However, these low levels can be artifacts of the constant concentration assumption.

ZERO-ORDER LEACHING CORRECTION FACTORS

A Zero-Order Leaching Correction Factor for Carcinogens (No Threshold)

Under an assumption of infinite contaminant mass, cumulative mass over an exposure period is equal to the risk-based concentration (*Co*) times an exposure period (*Te*), where *Te* equals *ED* x 365. Under conditions of zero-order leaching, average annual contaminant mass lost from soil to recharge-water can be disaggregated to an average daily value to keep units consistent with Equation 1. Mass loss over time is a change in contaminant concentration per REV of soil described by the zero-order equation:

$$C = Co - kt.$$

Where:

t = time (days) from zero to Te
C = a concentration at time = t (mg/kg)
Co = the risk-based concentration at t = 0
k = Co/Tz (a loss-rate constant where Tz/2 = half-life)
Tz = time to zero concentration, from Equation 2

As shown in Figure 2, this cumulative mass (area under line *Co-Tz*) is less than the cumulative mass defined using the infinite mass assumption (*Co* x *Te*). Compensating for mass lost by leaching requires a concentration higher than *Co*. This higher concentration (*Cn*) defines a cumulative mass (area under line *Cn-Ct*), over a duration of *Te* and with a loss rate of *k*, which must equal *Co* x *Te*. From Figure 2, a unitless zero-order leaching correction factor for carcinogens ($ZLCF_C$) is defined by:

Figure 2. **Diagram showing contaminant (carcinogen) mass at a concentration of *Co* over an exposure period (*Te*), assuming infinite mass (constant concentration), compared to the same concentration with a constant rate of mass loss to zero mass at time = *Tz*. Correcting for mass lost for a chemical carcinogen (no threshold) gives a concentration (*Cn*) that provides sufficient mass to compensate for mass lost.**

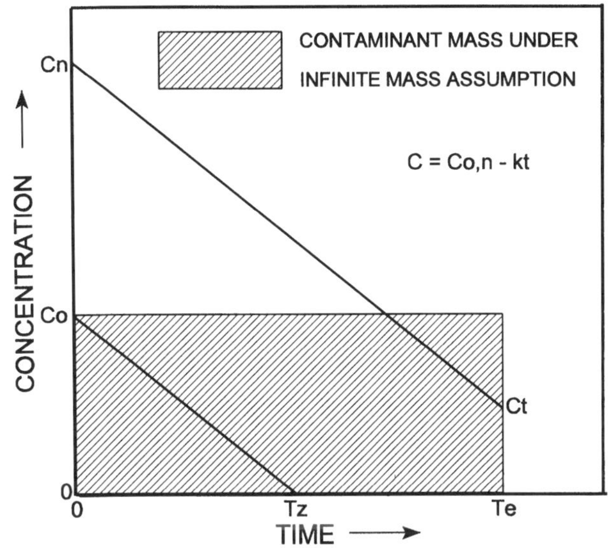

At boundary conditions:

t = 0; C = Cn: t = Te; C = Ct

Cn > Ct >= 0 (no threshold)

For: Ct = Cn - kTe; k = Co/Tz

[area under plane figure Cn,Ct,Tc,0] = [area of infinite mass]

[(Cn - Ct) x Te/2 + (Ct x Te)] = [Co x Te]

Te/2 x (Cn -(Cn - Co x Te/Tz) + (Cn - Co x Te/Tz) x Te = Co x Te

(Co x Te/Tz)/2 + Cn - (Co x Te/Tz) = Co

Cn = Co + [(Co x Te/Tz)/2]:

$$ZLCF_C = \frac{Cn}{Co} = 1 + \frac{Te}{2Tz} \qquad (5)$$

The risk-based concentration (*Co*), for carcinogens, from Equation 1 is corrected for finite mass by $Co \times ZLCF_C$. This ZLCF is applicable to all soil exposure pathways except groundwater. As concentration decreases with time, the daily dose is not an equal increment of the cumulative dose. However, this cumulative dose is equal to the cumulative dose under the infinite mass assumption.

A Zero-Order Leaching Correction Factor for Noncarcinogens (Threshold)

A ZLCF for noncarcinogens is developed similarly to that for carcinogens. The difference is the constraint of a threshold, as defined by a reference dose. When the level of a contaminant drops below the risk-based concentration determined from a reference dose, there is no longer an adverse health effect.

This analysis is based on a literal and rigid interpretation of a threshold. If short-term dose rates below the RfD can be averaged with rates above the RfD over an averaging time, then the ZLCF for noncarcinogens is the same as the ZLCF for carcinogens.

As shown in Figure 3, a loss of contaminant mass is caused by leaching at a rate defined by line *Co-Tz.* The minimum contaminant mass required for an adverse health effect, under the infinite mass assumption, is shown by *Co x Te.* Compensating for mass lost by leaching requires a concentration higher than Co. This higher concentration (*Cn*) defines a cumulative mass (area under line *Cn-Ct*), over a duration of *Te* and with a loss rate of *k*, which must meet the constraint of a threshold. The minimum corrected concentration is found when *Ct* equals *Co.* From Figure 3, a unitless zero-order correction factor for noncarcinogens ($ZLCF_{NC}$) is defined by:

Figure 3. **Diagram showing contaminant (noncarcinogen) mass at a concentration of *Co* over an exposure period (*Te*), assuming infinite mass (constant concentration), compared to the same concentration with a constant rate of mass loss to zero mass at time = *Tz*. Correcting for mass lost for a noncarcinogen gives a concentration (*Cn*), at a constant rate of mass loss to a threshold concentration (*Ct*) at time = *Te*.**

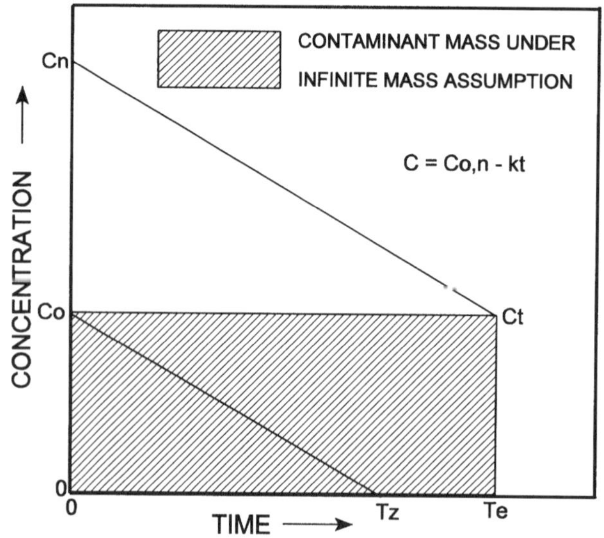

At boundary conditions:

t = 0; C = Cn
t = Te; C = Ct;
Ct = Co (threshold)
Ct = Cn - kTe; k = Co/Tz; Ct = Co
Co = Cn - Co x Te/Tz
Cn = Co + Co x Te/Tz

$$ZLCF_{NC} = \frac{Cn}{Co} = 1 + \frac{Te}{Tz} \qquad (6)$$

The risk-based concentration (Co), for noncarcinogens, from Equation 1 is corrected for finite mass by $Co \times ZLCF_{NC}$. Correcting for a significant loss of contaminant mass requires a concentration above the threshold. This is inconsistent with the paradigm for establishing soil cleanup "standards" for noncarcinogens. The magnitude of this concentration is dependent on the length of the exposure duration. The physical impossibility of constant concentration can be balanced against the reality of concentrations exceeding the threshold by using short exposure durations. This problem becomes insignificant for values of the $ZLCF_{NC}$ near one.

APPLICATION OF ZERO-ORDER LEACHING CORRECTION FACTORS (ZLCF)

Procedures

A ZLCF was calculated for 99 organic chemicals commonly used by the USEPA and state agencies to evaluate remediation of contaminated soils. Chemical-specific properties were obtained from textbook compilations (Montgomery and Welkom, 1990; Howard et al., 1990) and USEPA manuals and draft documents (USEPA 1990b; 1994b). Typical values for surface-soil properties were used (results change as properties change): porosity of 43%, bulk density of 1.51 g/cm^3, and 2% organic carbon content. Chemical-specific values for the $ZLCF_C$ and $ZLCF_{NC}$, based on leaching from soil are given in Table 1. Each chemical was evaluated as a "carcinogen" and "noncarcinogen" as the leaching algorithms are applicable to most endpoints. From Equations 2 and 5, a ZLCF was calculated for each chemical for risk-based concentrations ("*Co* or soil standards or criteria") of 1, 10, 100, 1,000, and 10,000 milligrams of chemical per kilogram of soil. The risk scenario is residential soil with an exposure duration (ED) of 70 years. Calculations were made for typical high-end and low-end values of recharge. The high-end value of 0.5 m/yr represents many humid areas in the United States, and the low-end value of 0.05 m/yr typifies many semiarid regions (Heath, 1984).

Table 1. **Factors to Correct Soil Cleanup for Chemical Mass Lost to Leaching**

Chemical Name	CAS No.	Solubility mg/l	Koc ml/g	SOL/Koc	ZLCFc[1] 70 years	ZLCFnc[1] 1 year	Half-Life days	FLCFc[2] 70 years	FLCFnc[2] 1 year	Soil Sat[3] mg/Kg
Acenapthtene	83329	4.13e+00	4.90e+03	8.43e-04	1.1e+00	1.0e+00	7.5e+04	1.1e+00	1.0e+00	4.1e+02
Acetone	67641	6.04e+05	4.60e-01	1.31e+06	6.8e+06	1.9e+05	6.3e+01	2.4e+02	5.6e+01	6.6e+04
Acrylonitrile	107131	7.50e+04	2.22e+00	3.38e+04	7.5e+05	2.1e+04	1.1e+02	1.5e+02	1.0e+01	1.1e+04
Aldrin	309002	7.84e-02	4.84e+04	4.84e+04	1.0e+00	1.0e+00	7.4e+05	1.0e+00	1.0e+00	7.6e+01
Anthracene	1201217	5.37e-02	2.12e+04	2.53e-06	1.0e+00	1.0e+00	3.2e+05	1.0e+00	1.0e+00	2.3e+01
Benzene	71432	1.78e+03	5.70e+01	3.12e+01	4.1e+03	1.2e+02	9.8e+02	1.8e+01	1.3e+00	2.3e+03
Benzo(a)anthracene	56553	1.28e-02	3.57e+05	3.59e-08	1.0e+00	1.0e+00	5.5e+06	1.0e+00	1.0e+00	9.1e+01
Benzo(a)pyrene	50328	1.94e-03	9.16e+05	2.12e-09	1.0e+00	1.0e+00	1.4e+07	1.0e+00	1.0e+00	3.6e+01
Benzo(b)flouranthene	205992	4.33e-03	8.83e+05	4.90e-09	1.0e+00	1.0e+00	1.3e+07	1.0e+00	1.0e+00	7.6e+01
Benzo(ghi)perylene	191242	7.00e-04	1.24e+06	5.65e-10	1.0e+00	1.0e+00	1.9e+07	1.0e+00	1.0e+00	1.7e+01
Benzo(k)fluoranthene	207089	4.30e-03	4.71e+05	9.13e-09	1.0e+00	1.0e+00	7.2e+06	1.0e+00	1.0e+00	4.1e+01
Benzoic Acid	65850	3.13e+03	4.07e+01	7.69e+01	9.4e+03	2.7e+02	7.3e+02	2.4e+01	1.4e+00	2.9e+03
Benzyl Alcohol	100516	4.00e+03	1.06e+01	3.77e+02	2.7e+04	7.6e+02	2.6e+02	6.7e+01	2.7e+00	1.2e+03
Bis(2-chloroethyl)ether	111444	1.18e+04	7.60e+01	1.55e+02	2.2e+04	6.2e+02	1.3e+03	1.4e+01	1.2e+00	1.9e+04
Bis(2-chloroisopropyl)ether	108601	1.70e+03	9.37e+01	1.81e+01	2.6e+03	7.5e+01	1.5e+03	1.1e+01	1.2e+00	3.4e+03
Bis(2-ethylhexl)pthalate	117817	3.96e-01	8.74e+04	4.53e-06	1.0e+00	1.0e+00	1.3e+06	1.0e+00	1.0e+00	6.9e+02
Bromodichloromethane	75274	3.97e+03	5.40e+01	7.35e+01	9.6e+03	2.8e+02	9.3e+02	1.9e+01	1.3e+00	4.8e+03
Bromoform	75252	3.21e+03	1.26e+02	2.55e+01	3.8e+03	1.1e+02	2.0e+03	8.7e+00	1.1e+00	8.4e+03
Bromomethane	74839	1.45e+04	9.00e+00	1.61e+03	1.0e+05	2.9e+03	2.3e+02	7.4e+01	3.0e+00	4.9e+03
Butanol	71363	7.47e+04	5.00e+00	1.49e+04	6.4e+05	1.8e+04	1.6e+02	1.0e+02	4.8e+00	1.5e+04
Butanone,2 (MEK)	78933	2.23e+05	1.73e+00	1.29e+05	2.3e+06	6.6e+04	9.8e+01	1.6e+02	1.3e+01	3.0e+04
Butylbenzyl pthalate	85687	2.58e+00	3.41e+04	7.57e-05	1.0e+00	1.0e+00	5.2e+05	1.0e+00	1.0e+00	1.8e+03
Carbazol	86748	7.21e-01	2.44e+03	2.95e-04	1.0e+00	1.0e+00	3.7e+04	1.3e+00	1.0e+00	3.5e+01
Carbon Tetrachloride	56235	7.92e+02	1.64e+02	4.83e+00	7.3e+02	2.2e+01	2.6e+03	6.8e+00	1.1e+00	2.9e+03
Carbon disulfide	75150	2.67e+03	5.20e+01	5.13e+01	6.6e+03	1.9e+02	9.0e+02	1.9e+01	1.3e+00	1.4e+04
Chlordane	57749	2.19e-01	5.13e+04	4.27e-06	1.0e+00	1.0e+00	7.8e+05	1.0e+00	1.0e+00	2.2e+02
Chloro-m-cresol,4	59507	3.85e+03	6.26e+02	6.15e+00	9.9e+02	2.9e+01	9.7e+03	2.2e+00	1.0e+00	4.9e+04
Chlorobenzene	108907	4.09e+02	2.04e+02	2.00e+00	3.1e+02	9.8e+00	3.2e+03	5.5e+00	1.1e+00	1.7e+03
Chloroform	67663	7.96e+03	5.60e+01	1.42e+02	1.9e+04	5.4e+02	9.6e+02	1.8e+01	1.3e+00	1.0e+04
Chloromethane	74873	6.34e+03	6.00e+00	1.06e+03	5.2e+04	1.5e+03	1.8e+02	9.3e+01	4.1e+00	2.1e+03
Chlorophenol,2	95578	2.15e+04	3.91e+02	5.50e+01	8.8e+03	2.5e+02	6.1e+03	3.1e+00	1.0e+00	1.7e+05
Chrysene	218019	1.94e-03	3.12e+05	6.22e-09	1.0e+00	1.0e+00	4.8e+06	1.0e+00	1.0e+00	1.2e+01
Cresol, p-	106445	2.26e+04	7.65e+01	2.95e+02	4.1e+04	1.2e+03	1.3e+03	1.4e+01	1.2e+00	3.7e+04
DDD	72548	7.33e-02	8.94e+04	8.20e-07	1.0e+00	1.0e+00	1.4e+06	1.0e+00	1.0e+00	1.3e+02
DDE	72559	1.92e-02	8.64e+04	2.22e-07	1.0e+00	1.0e+00	1.3e+06	1.0e+00	1.0e+00	3.3e+01
DDT	50293	3.41e-03	2.37e+05	1.44e-08	1.0e+00	1.0e+00	3.6e+06	1.0e+00	1.0e+00	1.6e+01

Chemical Name	CAS No.	Solubility mg/l	Koc ml/g	SOL/Koc	ZLCFc[1] 70 years	ZLCFnc[1] 1 year	Half-Life days	FLCFc[2] 70 years	FLCFnc[2] 1 year	Soil Sat[3] mg/Kg
Di-n-butyl pthalate	84742	1.03e+01	1.69e+04	6.39e-04	1.1e+00	1.0e+00	2.6e+05	1.0e+00	1.0e+00	3.7e+03
Di-n-octyl pthalate	117840	3.00e+00	9.80e+08	3.06e-09	1.0e+00	1.0e+00	1.5e+10	1.0e+00	1.0e+00	5.9e+07
Dibenz(a,h) anthracene	53703	6.70e-04	1.80e+06	3.72e-10	1.0e+00	1.0e+00	2.8e+07	1.0e+00	1.0e+00	2.4e+01
Dibromochloromethane	124481	3.44e+03	7.20e+01	4.78e+01	6.6e+03	1.9e+02	1.2e+03	1.5e+01	1.2e+00	5.3e+03
Dichlorobenzene,1,2	95501	1.25e+02	3.76e+02	3.32e-01	5.4e+01	2.5e+00	5.9e+03	3.2e+00	1.0e+00	9.5e+02
Dichlorobenzene,1,3	541731	1.23e+02	2.33e+03	5.28e-02	9.7e+00	1.2e+00	3.6e+04	1.3e+00	1.0e+00	5.7e+03
Dichlorobenzene,1,4	106467	7.30e+01	5.16e+02	1.41e-01	2.4e+01	1.6e+00	8.0e+03	2.5e+00	1.0e+00	7.6e+02
Dichlorobenzidine,3,3	91941	3.52e+00	2.44e+03	1.44e-03	1.2e+00	1.0e+00	3.7e+04	1.3e+00	1.0e+00	1.7e+02
Dichloroethane,1,1	75343	5.16e+03	5.20e+01	9.92e+01	1.3e+04	3.7e+02	9.0e+02	1.9e+01	1.3e+00	6.1e+03
Dichloroethane,1,2	107062	8.31e+03	3.80e+01	2.19e+02	2.6e+04	7.5e+02	6.8e+02	2.6e+01	1.4e+00	7.2e+03
Dichloroethene,1,2 (cis)	156592	4.94e+03	2.90e+01	1.70e+02	1.9e+04	5.4e+02	5.4e+02	3.2e+01	1.6e+00	4.6e+03
Dichloroethene,1,2 (trans)	156605	8.03e+03	5.00e+01	1.61e+02	2.1e+04	5.9e+02	8.7e+02	2.0e+01	1.3e+00	1.3e+04
Dichloroethylene,1,1	75354	3.00e+03	6.50e+01	4.62e+01	6.2e+03	1.8e+02	1.1e+03	1.6e+01	1.3e+00	4.6e+03
Dichlorophenol,2,4	120832	4.93e+03	1.46e+02	3.38e+01	5.1e+03	1.5e+02	2.3e+03	7.5e+00	1.1e+00	1.5e+04
Dichloropropane, 1,2-	78875	2.68e+03	5.83e+01	4.60e+01	6.1e+03	1.8e+02	1.0e+03	1.8e+01	1.3e+00	3.4e+03
Dichloropropene,1,3	542756	1.55e+03	2.60e+02	5.96e+00	9.3e+02	2.8e+01	4.1e+03	4.4e+00	1.1e+00	8.2e+03
Dieldrin	60571	1.87e-01	1.09e+04	1.72e-05	1.0e+00	1.0e+00	1.7e+05	1.1e+00	1.0e+00	4.1e+01
Diemethylphenol,2,4	105679	6.25e+03	1.26e+02	4.96e+01	7.4e+03	2.1e+02	2.0e+03	8.7e+00	1.1e+00	1.6e+04
Diethyl pthalate	84662	8.83e+02	8.20e+01	1.08e+01	1.5e+03	4.4e+01	1.4e+03	1.3e+01	1.2e+00	1.5e+03
Dimethyl pthalate	131113	4.19e+03	4.60e+01	9.11e+01	1.1e+04	3.3e+02	8.1e+02	2.2e+01	1.4e+00	4.3e+03
Dinitrophenol,2,4	51285	5.80e+03	1.00e+02	5.80e+01	8.4e+03	2.4e+02	1.6e+03	1.1e+01	1.2e+00	1.2e+04
Dinitrotoluene,2,4	121142	2.35e+02	5.10e+01	5.59e+00	7.2e+02	2.2e+01	8.8e+02	2.0e+01	1.3e+00	3.2e+02
Dinitrotoluene,2,6	606202	1.05e+03	4.20e+01	2.50e+01	3.1e+03	8.9e+01	7.5e+02	2.3e+01	1.4e+00	9.9e+02
Endosulfan	115297	2.31e-01	7.38e+02	3.13e-04	1.1e+00	1.0e+00	1.1e+04	2.0e+00	1.0e+00	3.4e+00
Endrin	72208	2.46e-01	1.08e+04	2.28e-05	1.0e+00	1.0e+00	1.7e+05	1.1e+00	1.0e+00	5.3e+01
Ethylbenzene	100414	1.73e+02	2.21e+02	7.83e-01	1.2e+02	4.5e+00	3.5e+03	5.1e+00	1.1e+00	7.9e+02
Fluoranthene	206440	2.32e-01	4.91e+04	4.73e-06	1.0e+00	1.0e+00	7.5e+05	1.0e+00	1.0e+00	2.3e+02
Fluorene	86737	1.86e+00	7.96e+03	2.34e-04	1.0e+00	1.0e+00	1.2e+05	1.1e+00	1.0e+00	3.0e+02
Heptachlor	76448	2.73e-01	6.81e+03	4.01e-05	1.0e+00	1.0e+00	1.0e+05	1.1e+00	1.0e+00	3.7e+01
Heptachlor epoxide	1024573	2.68e-01	7.24e+03	3.70e-05	1.0e+00	1.0e+00	1.1e+05	1.1e+00	1.0e+00	3.9e+01
Hexachlorobenzene	118741	8.62e-03	3.75e+04	2.30e-07	1.0e+00	1.0e+00	5.7e+05	1.0e+00	1.0e+00	6.5e+00
Hexachlorobutadiene	87683	2.54e+00	6.99e+03	3.63e-04	1.1e+00	1.0e+00	1.1e+05	1.1e+00	1.0e+00	3.7e+02
Hexachlorocyclopentadiene	77474	1.53e+00	9.59e+03	1.60e-04	1.0e+00	1.0e+00	1.5e+05	1.1e+00	1.0e+00	3.0e+02
Hexachloroethane	67721	4.08e+01	1.83e+03	2.23e-02	4.6e+00	1.1e+00	2.8e+04	1.3e+00	1.0e+00	1.5e+03
Indo(1,2,3-cd)pyrene	193395	1.70e-02	4.36e+06	3.90e-09	1.0e+00	1.0e+00	6.7e+07	1.0e+00	1.0e+00	1.5e+03
Isophorone	78591	1.20e+04	3.00e+01	4.00e+02	4.5e+04	1.3e+03	5.6e+02	3.1e+01	1.6e+00	8.4e+03

Chemical Name	CAS No.	Solubility mg/l	Koc ml/g	SOL/Koc	ZLCFc[1] 70 years	ZLCFnc[1] 1 year	Half-Life days	FLCFc[2] 70 years	FLCFnc[2] 1 year	Soil Sat[3] mg/Kg
Lindane	58899	4.20e+00	1.38e+03	3.04e-03	1.5e+00	1.0e+00	2.1e+04	1.5e+00	1.0e+00	1.2e+02
Methoxychlor	72435	8.84e-02	7.79e+04	1.13e-06	1.0e+00	1.0e+00	1.2e+06	1.0e+00	1.0e+00	1.4e+02
Methyl isobutyl ketone	108101	1.55e+01	2.12e+04	7.31e-04	1.1e+00	1.0e+00	3.2e+05	1.0e+00	1.0e+00	6.6e+03
Methylene chloride	75092	1.74e+04	1.60e+01	1.09e+03	9.5e+04	2.7e+03	3.4e+02	5.0e+01	2.1e+00	7.7e+03
Methylphenol,2	95487	2.77e+04	5.40e+01	5.13e+02	6.7e+04	1.9e+03	9.3e+02	1.9e+01	1.3e+00	3.3e+04
N-Nitrosodi-n-propylamine	621647	1.46e+04	1.70e+01	8.57e+02	7.7e+04	2.2e+03	3.6e+02	4.8e+01	2.0e+00	6.4e+03
N-Nitrosodiphenylamine	86306	3.74e+00	3.27e+02	1.14e-02	2.8e+00	1.1e+00	5.1e+03	3.6e+00	1.1e+00	2.5e+01
Napthalene	91203	3.11e+01	9.64e+02	3.23e-02	6.2e+00	1.1e+00	1.5e+04	1.7e+00	1.0e+00	6.0e+02
Nitrobenzene	98953	1.92e+03	1.31e+02	1.47e+01	2.2e+03	6.3e+01	2.1e+03	8.4e+00	1.1e+00	5.2e+03
PCBs	1336363	3.10e-03	4.52e+05	6.86e-09	1.0e+00	1.0e+00	6.9e+06	1.0e+00	1.0e+00	2.8e+01
Pentachlorophenol	87865	1.34e+01	5.67e+02	2.36e-02	4.8e+00	1.1e+00	8.8e+03	2.3e+00	1.0e+00	1.5e+02
Phenol	108952	9.08e+04	2.20e+01	4.13e+03	4.1e+05	1.2e+04	4.4e+02	4.0e+01	1.8e+00	4.9e+04
Pyrene	129000	1.37e-01	6.82e+04	2.01e-06	1.0e+00	1.0e+00	1.0e+06	1.0e+00	1.0e+00	1.9e+02
Styrene	100425	2.57e+02	9.12e+02	2.82e-01	4.7e+01	2.3e+00	1.4e+04	1.8e+00	1.0e+00	4.7e+03
Tetrachloroethane, 1,1,1,2-	630206	1.10e+03	3.99e+02	2.76e+00	4.4e+02	1.4e+01	6.2e+03	3.0e+00	1.0e+00	8.9e+03
Tetrachloroethane,1,1,2,2	79345	3.07e+03	7.90e+01	3.89e+01	5.4e+03	1.6e+02	1.3e+03	1.3e+01	1.2e+00	5.2e+03
Tetrachlororthylene	127184	2.32e+02	3.00e+02	7.73e-01	1.2e+02	4.5e+00	4.7e+03	3.9e+00	1.1e+00	1.5e+03
Toluene	108883	5.58e+02	1.31e+02	4.26e+00	6.3e+02	1.9e+01	2.1e+03	8.4e+00	1.1e+00	1.5e+03
Toxaphene	8001352	6.79e-01	5.01e+02	1.36e-03	1.2e+00	1.0e+00	7.8e+03	2.5e+00	1.0e+00	6.9e+00
Trichlorobenzene,1,2,4	120821	3.07e+01	1.54e+03	1.99e-02	4.3e+00	1.1e+00	2.4e+04	1.4e+00	1.0e+00	9.5e+02
Trichloroethane,1,1,1	71556	1.17e+03	9.90e+01	1.18e+01	1.7e+03	5.0e+01	1.6e+03	1.1e+01	1.2e+00	2.6e+03
Trichloroethane,1,1,2	79005	4.40e+03	7.60e+01	5.79e+01	8.0e+03	2.3e+02	1.3e+03	1.4e+01	1.2e+00	7.2e+03
Trichloroethylene (TCE)	79016	1.18e+03	9.40e+01	1.26e+01	1.8e+03	5.2e+01	1.5e+03	1.1e+01	1.2e+00	2.4e+03
Trichlorophenol,2,4,5	95954	9.65e+02	1.40e+03	6.89e-01	1.1e+02	4.2e+00	2.2e+04	1.5e+00	1.0e+00	2.7e+04
Trichlorophenol,2,4,6	88062	7.53e+02	2.83e+02	2.66e+00	4.2e+02	1.3e+01	4.4e+03	4.1e+00	1.1e+00	4.3e+03
Vinyl acetate	108054	2.24e+04	5.00e+00	4.48e+03	1.9e+05	5.5e+03	1.6e+02	1.0e+02	4.8e+00	8.5e+03
Vinyl chloride	75014	2.73e+03	1.10e+01	2.48e+02	1.8e+04	5.1e+02	2.6e+02	6.5e+01	2.6e+00	2.7e+03
Xylenes	1330207	1.86e+02	2.60e+02	7.15e-01	1.1e+02	4.2e+00	4.1e+03	4.4e+00	1.1e+00	1.0e+03
p-Chloroaniline	106478	3.36e+03	4.10e+01	8.20e+01	1.0e+04	2.9e+02	7.3e+02	2.4e+01	1.4e+00	3.1e+03

SOIL PARAMETERS: porosity = 0.43; bulk density = 1.51g/cm^3; fraction organic carbon = 0.02

1. Zero-order Leaching Correction Factor for *Co* or soil criterion of 1 mg/Kg
2. First-order Leaching Correction Factor based on half-life or residence time in soil calculated from soil loss constant *k*
3. Soil saturation limit or maximum concentration of a chemical in soil without any separate phase chemical present

Results and Discussion

The ZLCF is related to the term of chemical solubility (*SOL*) divided by the organic carbon partition coefficient (*Koc*). Values of $ZLCF_C$ versus *SOL/Koc* (g^2/L^2) are plotted in Figure 4. Data point scatter is related to variability in chemical-specific parameters. Figure 4 is a nomogram (conversion chart for point values) used to find a ZLCF for a carcinogenic chemical, given a soil standard or criterion (*Co*) determined from Equation 1. Soil concentrations are shown

Figure 4. **Nomogram for determining a contaminant (carcinogen) zero-order mass-loss correction factor, due to leaching ($ZLCF_C$), for risk-based soil concentrations determined using the infinite mass assumption. Exposure time (*Te*) is 25,550 days (70 years). Soil properties given as: bulk density = 1.51 g/cm^3, porosity = 0.43, fraction organic carbon = 0.02.**

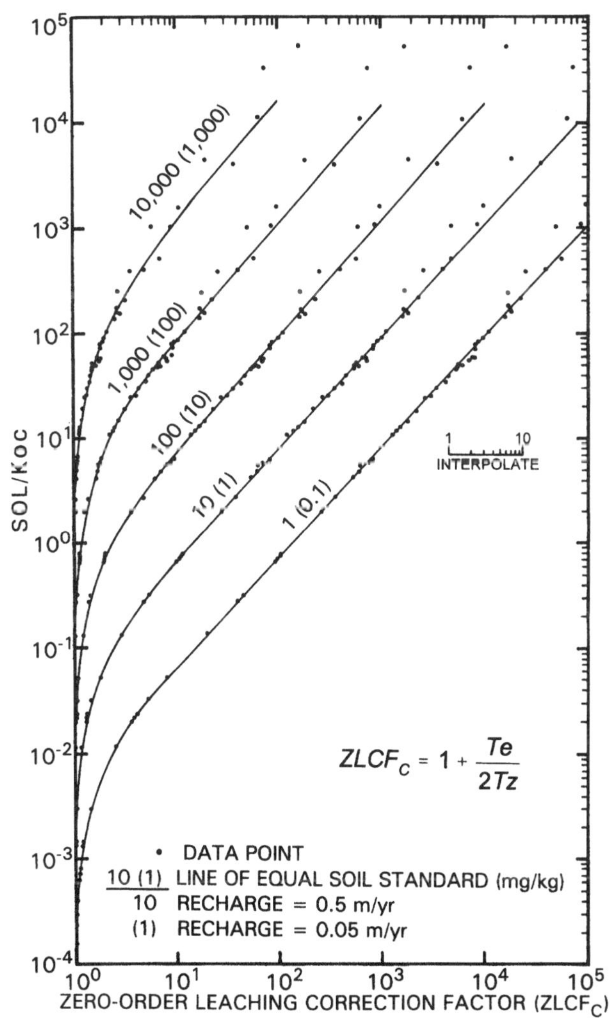

in powers of ten, and other values of *Co* can be interpolated. *Co* is multiplied by the $ZLCF_C$ to obtain a standard corrected for mass lost by leaching. Equations 2 and 5 are used for soil parameters different from the values used for Figure 4.

Figure 4 shows that chemicals with very low water solubility and related high partition coefficients do not leach from the soil at a rate sufficient to require a significant correction to the risk-based concentration ($ZLCF_C \approx 1$). Thus, these chemicals are not affected by mass lost to leaching when setting contaminated soil standards or criteria. At the other extreme, chemicals with very high solubility and related low partition coefficients leach so rapidly that the corrected concentrations (*Co* x ZLCF) exceed levels that are physically possible. Therefore, these chemicals should play no practical role in standards for soil contamination involving pathways other than groundwater. This is intuitive and self-evident.

Between these two extremes are many chemicals for which large changes to the intake and risk parameters of Equation 1 result in little change to the corrected risk-based concentration (soil standard or criterion). For example, a soil standard of 1 mg/kg is calculated from Equation 1, using an excess cancer risk (target risk) of 10^{-6}. The value of SOL/Koc for this chemical is given as 10. From Figure 4, the ZLCF is about 1,200. Thus, the soil concentration required for this chemical to provide a cumulative contaminant mass equivalent to the 1 mg/kg concentration calculated under the infinite mass assumption is 1,200 mg/kg (1 mg/kg x 1,200). For target risks of 10^{-5} and 10^{-4}, calculated risk-based concentrations are equal to 10 mg/kg and 100 mg/kg, respectively. From Figure 4, the corresponding $ZLCF_C$ values are about 120 and 12, respectively. The corrected concentrations (1 mg/kg x 1,200, 10 mg/kg x 120, and 100 mg/kg x 12) for all target risk levels are about the same—1,200 mg/kg. This result is counterintuitive but valid if leaching is governed by zero-order kinetics.

The assumptions used for zero-order kinetics state that the rate of chemical mass lost to infiltrating rainwater does not change with increasing concentration of the chemical in surface soil. However, the time (*Tz*) for all of the chemical to be lost from a REV of soil changes in proportion to the initial concentration. For chemicals within the "linear" range of Figure 4, a doubling of initial concentration doubles the time it takes to lose all contaminant mass (Equation 2). Thus, the ZLCF is decreased by half as the risk-based concentration (*Co*) is doubled—and the corrected concentration (*Co* x ZLCF) remains about the same. As shown in Table 1, an order of magnitude change to a standard for many chemicals (e.g., from a change in risk level or intake rate) produces a much smaller change in the corrected concentration. As a specific example, New Jersey uses a residential direct contact soil cleanup value of 23 mg/kg for TCE (Cohen and Mercer, 1993). This value is health-based, using an incidental soil ingestion exposure pathway and a target risk level of 10^{-6}. For a range of target risk levels of 10^{-4} to 10^{-7}, the corresponding soil cleanup criteria are 2,300 mg/kg, 230 mg/kg, 23 mg/kg, and 2.3 mg/kg, respectively. From Equations 2 and 5, an exposure duration of 70 years, and the soil parameters from Table 1, the corresponding corrected soil cleanup criteria are about 4,090 mg/kg, 2,030 mg/kg, 1,820 mg/kg, and 1,800 mg/kg, respectively.

The ZLCF provides a method to categorize organic carcinogenic chemicals into three groups for setting soil cleanup standards or criteria: chemicals unaffected by the infinite mass assumption, chemicals so strongly affected as to be inappropriate, and chemicals so significantly affected as to require correction. When the time for a chemical to completely leach from the soil (*Tz*) is equal to the exposure duration (*Te*), the ZLCF is equal to 1.5. As *Tz* increases beyond *Te*, the ZLCF approaches one. Thus, risk-based soil standards for chemicals where *Tz* is greater than *Te* are relatively independent of the infinite contaminant mass assumption. For zero-order leaching, recharge can be measured in a minimum period of one day, and the ZLCF at *Tz* equals one is about 1.3×10^4 and 5.5×10^3 for exposure periods of 70 and 30 years, respectively. As *Tz* approaches zero, the ZLCF approaches infinity. Thus, chemicals that leach this rapidly ($Tz<1$) from surface soils represent a negligible risk for soil pathways other than groundwater. Risk-based concentrations for chemicals with values of *Tz* less than *Te* and greater than one must be corrected to account for finite contaminant mass.

Figure 5 is a nomogram for noncarcinogenic chemicals based on Equation 6. The results are analogous to Figure 4, but the most important use of this nomogram is to identify chemicals with a $ZLCF_{NC}$ equal to about one for the exposure period of concern. These chemicals are not significantly affected by the infinite mass assumption. Zero-order leaching correction factors for an exposure period of one year can be used as minimum correction factors, and are shown in Table 1.

Equations 5 and 6 are applicable to other contaminant mass-loss processes that can be described by zero-order kinetics. A correction factor is calculated by determining the time for contaminant mass to reach zero concentration and using this value as Tz in Equations 5 or 6. The combined effect of several mass-loss mechanisms is determined from a Tz for each loss mechanism (e.g., Tz_1, Tz_2, Tz_3), and calculating a composite *Tz* by:

$$Tz = \frac{1}{\frac{1}{Tz_1} + \frac{1}{Tz_2} + \frac{1}{Tz_n}} \tag{7}$$

FIRST-ORDER LEACHING CORRECTION FACTORS (FLCF)

A concentration of a chemical in soil, after leaching, can be described by a first-order equation:

$$C = Co * e^{-kt} \tag{8}$$

Figure 5. **Nomogram for determining a contaminant (noncarcinogen) zero-order mass-loss correction factor, due to leaching ($ZLCF_{NC}$), for risk-based soil concentrations determined using the infinite mass assumption. Exposure times (*Te*) are 2,190 days (6 years) and 365 days. Soil properties given as: bulk density = 1.51 g/cm^3, porosity = 0.43, fraction organic carbon = 0.02.**

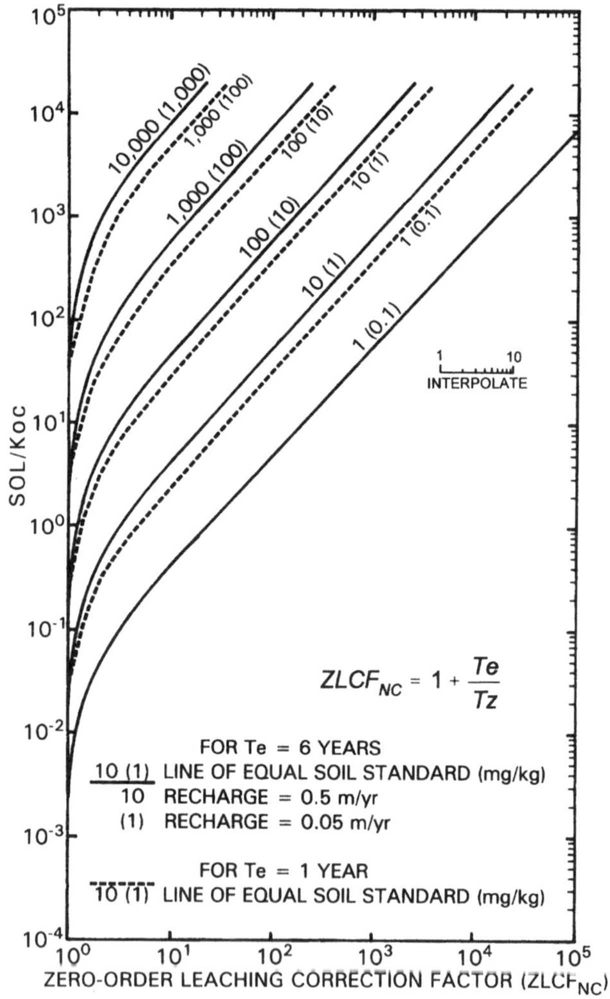

Where:

t = time (days)

C = a concentration at time = t (mg/L)

Co = a risk-based concentration at t = an initial time

k = a rate constant (days^{-1}) calculated as the natural log of 2 divided by half-life for a specific chemical.

Cumulative contaminant mass is given as:

$$CMd = Co * \int_{t=0}^{t=Te} e^{-kt}\,dt \qquad (9)$$

Where:

Te = ED x 365 (exposure period for mass loss)

CMd = cumulative contaminant mass after mass lost by leaching

For a given organic-chemical in soil, the solution to equation 9 for cumulative contaminant mass after leaching (CMd) and available for human uptake is given by:

$$CMd = Co * \frac{1 - e^{-kTe}}{1 - e^{-k}} \qquad (10)$$

First-Order Leaching Correction Factor for Carcinogens (No Threshold)

As shown in Figure 6, this cumulative mass (CMd or area under curve at *Co*) is less than the cumulative mass defined using the infinite mass assumption (*Co x Te*). A concentration higher than *Co* is required to compensate for mass lost by leaching. This higher concentration (*Cn*) defines a cumulative mass (area under curve *Cn-Ct*), with a decay rate of *k* over a duration of *Te*, which must equal *Co x Te*. From Figure 6, a first-order leaching correction factor for carcinogens ($FLCF_C$) is defined as:

$$Cn * \frac{1 - e^{-kTe}}{1 - e^{-k}} = Co * Te; \quad FLCF_C - \frac{Cn}{Co} - Te\left(\frac{1 - e^{-k}}{1 - e^{-kTe}}\right) \qquad (11)$$

A risk-based concentration (*Co*) from Equation 1, for carcinogens, is corrected for leachable mass by *Co* x $FLCF_C$.

First-Order Leaching Correction Factor for Noncarcinogens (Threshold)

An FLCF for noncarcinogens is developed similarly to that for carcinogens. The difference is the constraint of a threshold as defined by a reference dose. Problems inherent with this interpretation of a threshold are the same as for zero-order leaching. As shown in Figure 7, cumulative mass with leaching (area under curve at *Co*) is less than the cumulative mass defined using the infinite mass assumption (*Co x Te*). To compensate for mass lost by leaching requires a concentration higher than *Co*. Under the constraint of a threshold, *Ct* must equal

Figure 6. **Diagram, for carcinogens, showing contaminant mass at a concentration of *Co* over an exposure period (*Te*), assuming infinite mass (constant concentration), compared to the same concentration with an exponential rate of mass loss by leaching. Correcting for mass lost for a chemical carcinogen (no threshold) gives a corrected concentration (*Cn*) that provides sufficient mass to compensate for mass lost.**

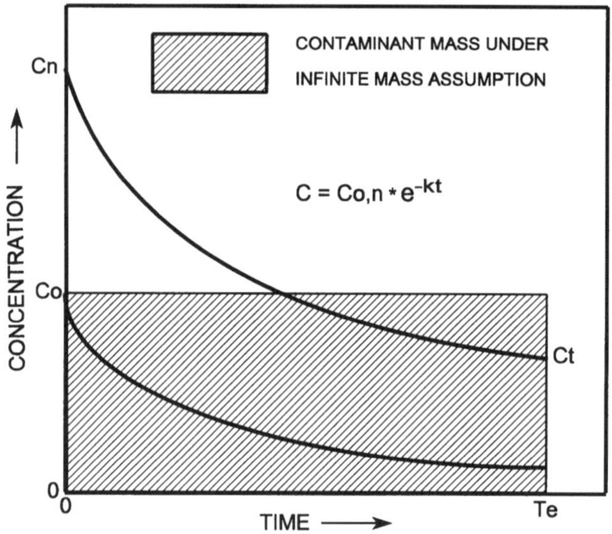

Co at time equal to Te. From Figure 7, a first-order leaching correction factor for noncarcinogens ($FLCF_{NC}$) is defined as:

$$Ct = Cn * e^{-kTe}; \quad Ct = Co$$
$$FLCF_{NC} = \frac{Cn}{Co} = e^{kTe} \tag{12}$$

A risk-based concentration (*Co*) from Equation 1), for noncarcinogens, is corrected for degradable mass by *Co* x $FLCF_{NC}$. As in the case for zero-order leaching, a $FLCF_{NC}$ significantly greater than one is inconsistent with the risk paradigm used to set cleanup standards or criteria for noncarcinogens.

APPLICATION OF FIRST-ORDER LEACHING CORRECTION FACTORS (FLCF)

Procedures

An FLCF was calculated from a loss rate constant (*k*) derived from Equation 3. Data for soil properties are from Table 1. The loss rate constant is directly proportional to the recharge rate and a rate of 0.5 m/yr was used.

Figure 7. **Diagram, for noncarcinogens, showing contaminant mass at a concentration of *Co* over an exposure period (*Te*), assuming infinite mass (constant concentration), compared to the same concentration with an exponential rate of mass loss by leaching. Correcting for mass lost for a noncarcinogen, to a threshold value (*Ct*), gives a corrected concentration (*Cn*) that provides sufficient mass to compensate for mass lost.**

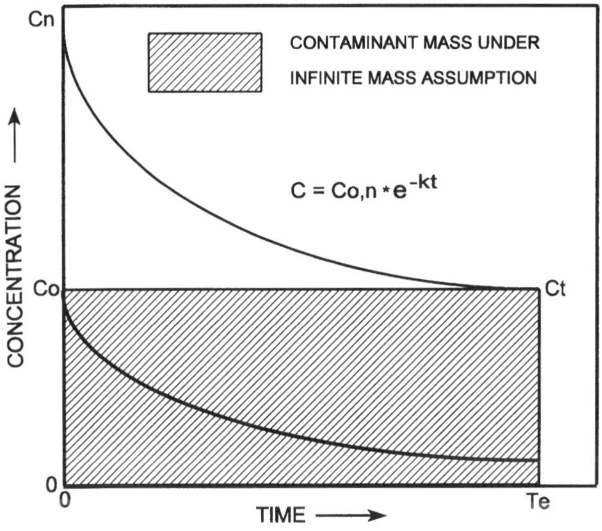

Results and Discussion

Figure 8 is a nomogram for determining $FLCF_C$ values (carcinogens) for exposure durations of 30 and 70 years. Chemical-specific values for the $FLCF_C$ based on first-order leaching are given in Table 1. The first-order loss rate constant (k) is given by Equation 3. Half-lives (measure of residence time in soil) are calculated from the rate constant k by: *half-life* = *0.693/k*. Time steps or units used for k (e.g., 1/year, 1/month, 1 /day) affect the value of the $FLCF_C$ as Equation 9 assumes a continuity of mass loss. Recharge is taken as 0.5 m/year disaggregated to even monthly increments (0.042 m/month), and half-lives are in days for comparison of time steps.

Mass loss by monthly recharge is a reasonable representation of reality (compared to daily or yearly recharge) for this simple model. For large rate constants (short half-lives), contaminant mass (Figure 6) can approach zero during a time step, and the value for the $FLCF_C$ approaches the exposure duration (Te). The $FLCF_C$ is not applicable under these conditions. Selection of the time step is an important consideration for a discontinuous loss process such as leaching by recharge water. Figure 8 shows the effect of different time steps. For example, use of a yearly recharge rate would exclude use of the $FLCF_C$ for several chemicals listed

Figure 8. **Nomogram for determining a contaminant (carcinogen) first-order mass-loss correction factor, due to leaching ($FLCF_C$), for risk-based soil concentrations determined using the infinite mass assumption. Values shown for *Te* = 30 years and 70 years.**

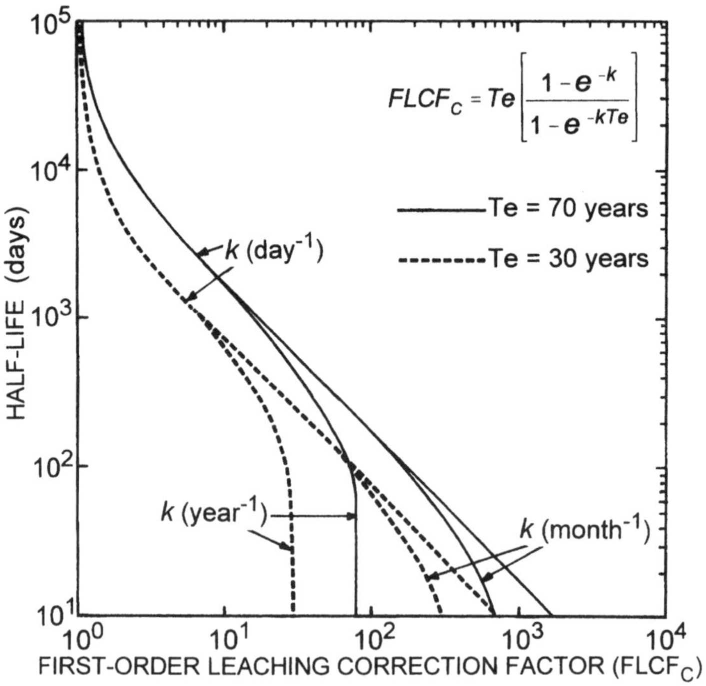

(half-lives of less than about 200 days), although the value for the time step does not significantly affect the $FLCF_C$ for chemicals with longer half-lives.

Half-lives of ten thousand days give an FLCF of about two after 70 years, and a factor of two is significant for remedial actions. As a specific example, New Jersey uses a residential direct-contact (ingestion) soil cleanup value of 6 mg/kg for pentachlorophenol (NJDEP, 1994). Accounting for mass lost by leaching gives a corrected value of 14 mg/kg ($FLCF_C = 2.3$).

Figure 9 shows $FLCF_{NC}$ values (noncarcinogens) for exposure durations of one year and six years. Six years was selected as representative of soil ingestion by children, and one year is taken as a practicable short exposure duration. The $FLCF_{NC}$ is based on a threshold concentration (Equation 12), and is not affected by the time steps used for the loss rate constant. Chemicals with values of the $FLCF_{NC}$ of about one are not significantly affected by the assumption of constant concentration. For an exposure duration of six years, only noncarcinogenic chemicals with half-lives greater than 4,000 days would be unaffected by the assumption of constant concentration. For other chemicals, the value for the $FLCF_{NC}$ is so large that the idea of a correction factor is inconsistent with the paradigm used to establish soil cleanup standards or criteria for noncarcinogens.

Figure 9. **Nomogram for determining a contaminant (noncarcinogen) mass-loss correction factor, due to leaching ($FLCF_{NC}$), for risk-based soil concentrations determined using the infinite mass assumption. Values shown for *Te* = 6 years and 1 year.**

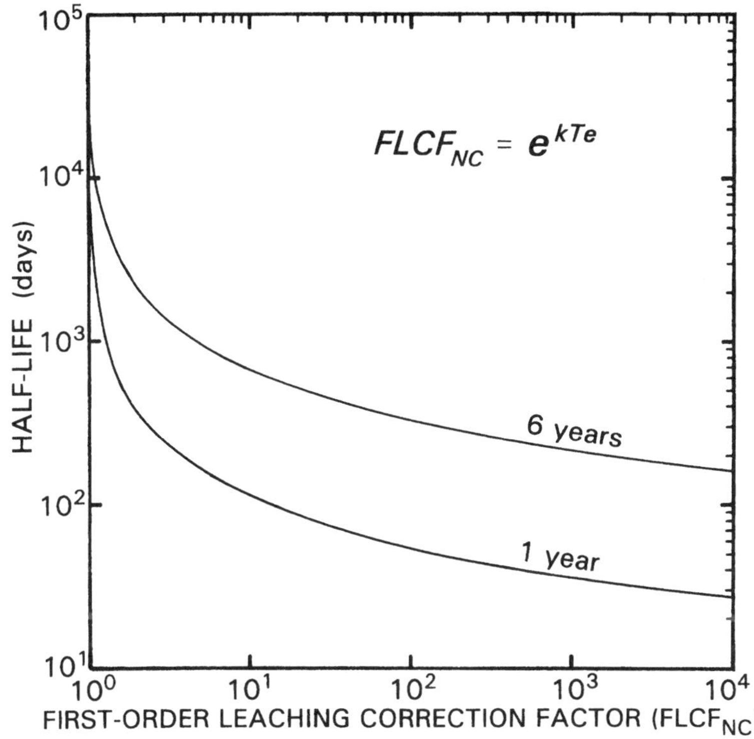

However, the $FLCF_{NC}$ for an exposure duration of one year can be considered a minimum correction factor (Table 1).

Equations 5 and 6 and Figures 8 and 9 are applicable to other contaminant mass loss processes which can be described by first-order kinetics. Where values for soil loss constants can be determined for more than one process such as leaching (k_{sl}), degradation (k_{sg}), and volatilization (*ksv*), a combined soil loss constant (k_s) assuming like time steps for loss rates is given as: $k_s = k_{sl} + k_{sg} + k_{sv}$ (USEPA, 1990a). The value for k_s can be substituted for *k* in Equations 5 and 6 or used in Figures 8 and 9.

COMPARISON BETWEEN ZERO-ORDER AND FIRST-ORDER CORRECTION FACTORS

Zero-order and first-order correction factors are comparable for chemicals that do not rapidly leach from soil. For highly mobile chemicals, the ZLCF is greater than the FLCF. The magnitude of this difference is dependent on the risk-based concentration (*Co*) as the ZLCF decreases as *Co* increases (Figures 4 and 5), while the FLCF is independent of *Co*. Table 2 compares first-order and zero-order correction factors for contaminant mass loss by leaching, for a cross-section of chemicals, based on soil parameters in Table 1 and *Co* = 10 mg/kg. These factors are comparable for low chemical solubilities, but differ as solubility increases. This uncertainty in correcting for the constant concentration assumption suggests that site-specific leaching data are needed to establish risk-based soil standards.

Table 2. **Correction Factors for Chemical Mass Lost by Leaching Under Zero-Order and First-Order Kinetics**

Chemical Name	Solubility mg/L	FACTOR Order Zero	First
PCBs	0.004	1	1
Heptachlor	0.3	1	1.1
Toxaphene	0.7	1	2.5
Pentachlorophenol	13	1.4	2.3
Hexachloroethane	41	1.4	1.3
1,4, Dichlorobenzene	73	3.3	2.5
Tetrachlororthylene	232	13	3.9
Carbon Tetrachloride	792	74	6.8
Trichloroethylene	1180	180	11
Benzene	1780	410	18
Vinyl chloride	2730	1800	65
1,2, Dichloroethane	8310	2600	26
Methylene chloride	17400	9400	50

LIMITATIONS TO CORRECTED RISK-BASED CONCENTRATIONS

Physical Constraints

Correcting a risk-based concentration for finite and conservative contaminant mass sometimes leads to high concentrations that are physically impossible. This situation implies that the calculated risk-based concentration is not possible, as this concentration could not persist for the exposure duration. Examples are corrected concentrations of a contaminant in soil exceeding available pore space or approaching total soil mass. A corrected concentration may also approach a physically possible limit and still be inappropriate. For example, a concentration that exceeds residual saturation and thus is within the range of drainable porosity. The limit for residual saturation of common organic contaminants is about 10% to 20% of porosity for soils (Cohen and Mercer, 1993). However, some of these chemicals are solids at normal soil temperatures, and could occupy most available pore space.

Corrected concentrations can exceed the lower limit of residual saturation (called soil saturation) and some separate phase chemical (liquid or solid) will be present. Several pathways for risk-based concentrations (e.g., volatilization from soil to air) are based upon partitioning theory and are invalid in the presence of any soil saturation (USEPA, 1989). Using the method proposed by Feenstra et al. (1991), upper-bound limits for soil saturation were calculated and are shown in Table 1. This limit is derived from partitioning theory and calculated from the sum of the maximum concentration of a chemical that can be present in soil pore water, the maximum concentration sorbed to soil, and the maximum concentration in soil gas. Although separate phase chemical can exist at levels lower than those shown in Table 1, concentrations above these values mean that some separate phase chemical must be present in the soil. Values for soil saturation are useful in evaluating corrected risk-based concentrations, as the presence of separate phase chemical influences correction factors calculated under zero-order and first-order kinetics. Corrected concentrations under zero-order kinetics may be underestimated when significant separate phase chemical is present, and will be substantially underestimated under first-order kinetics. For example, a risk-based standard (*Co*) of 500 mg/kg for 1,1 dichloroethane with an $FLCF_C$ of 19 is corrected to 9,500 mg/kg. This corrected concentration is greater than the soil saturation limit of 6,100 mg/kg (Table 1).

Rate-Limited Sorption and Desorption

The local equilibrium assumption is not valid for many chemicals under field conditions (Brusseau, 1994). Desorption rates are often slower than adsorption rates. However, desorption rate constants for many chemicals are inversely proportional to equilibrium partition coefficients (Brusseau and Rao, 1991). Thus, chemicals with high partition coefficients (i.e., high values for *Koc*) desorb at a slower rate than chemicals with low partition coefficients. This factor minimally affects the ZLCF and FLCF, as chemicals with high partition coefficients do not significantly leach from soil.

Effective Porosity

All pore spaces in soil are not available for movement of fluids. Soil has dead-end pores and stagnant water. Porosity available to flow of water is termed effective porosity. Effective porosity is less than total porosity. Thus, a quantity of chemical can be trapped as separate or dissolved phase, and be resistant to leaching. The field consequence is detection of contaminants in soil that should have been leached out. In addition, for some analyses of fate and transport of chemicals in soil, effective porosity may be a more appropriate value to use than total porosity.

Heterogeneities

The flow of recharge water through a REV of soil assumes a continuous, homogeneous porous media. Under field conditions, surface soils contain fractures, worm burrows, plant roots, and other heterogeneities that provide channels for water flow. These channels can have flow and transport properties significantly different from average soil.

Organic Carbon

Percent organic carbon in the soil is assumed to govern sorbtion properties of the soil. Other substances in soil, such as precipitates of iron oxyhydroxides, can affect sorption properties of the soil.

Critical Pathways

The usual methodology applied to determine a soil contaminant standard or criterion evaluates risk from exposure by multiple pathways. A standard is selected from the pathway that yields the lowest contaminant concentration (critical pathway). Direct contact with soil is often not the critical pathway compared to groundwater or air. However, this situation is based on the assumption of constant concentration. Correcting for contaminant mass lost can alter this relationship.

Other Limitations

There may be acute human health effects at higher concentrations resulting from application of a leaching correction factor. Concentrations may be at an explosive or corrosive level or a level of acute toxicity.

CONCLUSIONS AND OBSERVATIONS

Risk-based concentrations for organic chemical contaminants in soil, calculated assuming constant concentration, can be corrected for contaminant mass lost by leaching using the methods presented here. This correction is conservative, as it represents a single mass-loss mechanism. The selection of zero-order or first-order kinetics to correct for leaching is less significant than recognizing the importance of including mass-balance in algorithms used to determine risk to human health from contaminated soil.

Risk-based concentrations for direct contact with chemicals of very low water solubility and very high soil/water partition coefficients are unaffected by leaching, as they persist in soil over the long exposure durations used in the USEPA's risk paradigm. Chemicals with very high water solubility and low partition coefficients leach so rapidly that they are not available for most contaminated-soil exposure routes other than groundwater. For other organic chemicals, long exposure durations of 25 to 70 years used in calculating risk-based concentrations for carcinogens dominate their fate and transport. For these chemicals, zero-order correction factors yield corrected values essentially independent of risk levels (e.g., one-in-one million) and intake parameters such as soil ingestion rates. First-order correction factors give cleanup goals or standards several times to a few hundred times greater than those calculated assuming constant concentration. In both cases, the number of chemicals for which mass-loss by leaching is significant decreases as exposure durations become smaller.

For noncarcinogens, the risk paradigm of a level of daily exposure to a chemical not expected to cause deleterious effects over a lifetime is incompatible with any significant loss of chemical mass. For this case, contaminant mass loss needs to be incorporated into the risk algorithms. Mass loss can be directly accounted for, or the risk algorithm can be adjusted in a manner similar to "uncertainty factors" and "modifying factors" used to establish a Reference Dose.

The problem of constant concentration versus contaminant mass loss is analogous to debate over reasonable values for soil ingestion rates. A daily soil-intake rate equal to zero or equal to body weight is not real. Likewise, an assumption of infinite mass is not real for most organic chemicals. A physically impossible assumption is not a conservative assumption.

Given long exposure durations, contaminants will eventually leach from soil—the question is, at what rates? As unreal assumptions give unreal results, the problem is one of selecting appropriate parameters for leaching using either a deterministic or probabilistic approach. Given appropriate parameters, many chemical contaminants in soil will likely not be significant to human health, using the USEPA risk paradigm in RAGS-B, as levels required to account for finite contaminant mass do not practically occur. For these chemicals in soil, toxicological endpoints with shorter exposure durations, such as reproductive and developmental effects, may be more appropriate for determining remedial levels for contaminated soil.

REFERENCES

Anderson, M.R., Johnson, R.L., and Pankow, J.F. 1992. Dissolution of Dense Chlorinated Solvents into Ground Water: 1. Dissolution from a Well-Defined Residual Source, *Ground Water* **30**, 250-256.

Bear, J. and Verruijt, A. 1987. *Modeling Groundwater Flow and Pollution*, pp. 17-21. Norwell, ME, Kluwer Academic Publishers.

Bouwer, H. 1991. Simple Derivation of the Retardation Equation and Application to Preferential Flow and Macrodispersion, *Ground Water* **29**, 41-46.

Brusseau, M.L. and Rao, P.S. 1991. Influence of Sorbate Structure on Nonequilibrium Sorption of Organic Compounds, *Environ. Sci. Technol.* **25**, 1501-1506.

Brusseau, M.L. 1994. Transport of Reactive Contaminants in Heterogeneous Porous Media, *Rev. Geophys.* **32**, 285-313.

Clapp, R.B. and Hornberger, G.M. 1978. Empirical Equations for Some Soil Hydraulic Properties, *Water Resour. Res.* **14**, 601.

Cohen, R.M. and Mercer, J.M. 1993. *DNAPL Site Evaluation*, pp. 4-16 to 4-22. Boca Raton, FL, C.K. Smoley, CRC Press, Inc.

Donigian, A.S., Jr., and Rao, P.S.C. 1989. Overview of Terrestrial Processes and Modeling. In: *Vadose Zone Modeling of Organic Pollutants*, pp. 3-36 (Hern, S.C. and Melancon, S.M., Eds.) Chelsea, MI, Lewis Publishers, Inc.

Feenstra, S., Mackay, D.M., and Cherry, J.A. 1991. A Method for Assessing Residual NAPL Based on Organic Chemical Concentrations in Soil Samples, *Ground Water Monitoring Rev.* **11**, 128-136.

Heath, R.C. 1984. Ground Water Regions of the United States, U.S. Geological Survey Water Supply Paper 2242, p. 19.

Howard, P.H., Boethling, R.S., Jarvis, W.F., Meylan, W.M., and Michalenko, E.M. 1990. *Fate and Exposure Data for Organic Chemicals, Volume I, Large Production and Priority Pollutants, Volume II, Solvents*, Chelsea, MI, Lewis Publishers, Inc.

Jury, W.A. and Valentine, R.L. 1989. Transport Mechanisms and Loss Pathways in Soil. In: *Vadose Zone Modeling of Organic Pollutants*, pp. 37-60 (Hern, S.C. and Melancon, S.M., Eds.), Chelsea, MI, Lewis Publishers, Inc.

Lerner, D.N., Issar, A.S., and Simmers, I. 1990. *Groundwater Recharge: a Guide to Understanding and Estimating Natural Recharge*, International Contributions to Hydrology: Vol. 8, Hannover, West Germany, International Association of Hydrogeologists, Verlag Heinz Heise.

Montgomery, J.H. and Welkom, L.M. 1990. *Ground Water Chemicals Desk Reference*, Chelsea, MI, Lewis Publishers, Inc.

(NJDEP) New Jersey Department of Environmental Protection. 1994. Guidance Document for the Remediation of Contaminated Soils, Site Remediation Program, June.

Rutledge, A.T. and Daniel, C.C., III. 1994. Testing an Automated Method to Estimate Ground-Water Recharge from Streamflow Records, *Ground Water* **32**, 180-189.

Saunders, P.F. 1995. Calculation of Soil Cleanup Criteria for Volatile Organic Compounds as Controlled by the Soil-to-Groundwater Pathway: Comparison of Four Unsaturated Soil Zone Leaching Models, *J. Soil Contam.* **4**, 1-24.

Schiffman, A. Contaminated-Soil Cleanup Standards: Unreal Risk Assumptions-Contaminant Lost to Degradation, Proceedings of the Hazardous Materials Control Resources Institute, SUPERFUND XVI Conference, Washington, DC, in press.

(USEPA) U.S. Environmental Protection Agency. 1988. Superfund Exposure Assessment Manual, Office of Solid Waste, EPA/540/1-88/001, April.

(USEPA) U.S. Environmental Protection Agency. 1991. Risk Assessment Guidance for Superfund, Volume 1, Human Health Evaluation Manual, (Part B, Development of Risk-Based Preliminary Remediation Goals) Interim, Office of Emergency and Remedial Response, EPA Publication 9285.7-01B, December.

(USEPA) U.S. Environmental Protection Agency. 1989. Risk Assessment Guidance for Superfund, Volume 1, Human Health Evaluation Manual, (Part A) Interim Final (Office of Emergency and Remedial Response, EPA/540/1-89/002, Chap. 6, December).

(USEPA) U.S. Environmental Protection Agency. 1990a. Methodology for Assessing Health Risks Associated with Indirect Exposure to Combustor Emissions, Interim Final, Office of Health and Environmental Assessment, EPA/600/6-90/003, January.

(USEPA) U.S. Environmental Protection Agency. 1990b. Basics of Pump-and-Treat Ground-Water Remediation Technology, Appendix A-Chemical Data, EPA/600/8-90/003, March.

(USEPA) U.S. Environmental Protection Agency. 1994a. Evaluation of Unsaturated/Vadose Zone Models for Superfund Sites, Office of Research and Development, EPA/600/R-93/184.

(USEPA) U.S. Environmental Protection Agency. 1994b. Technical Background Document for Draft Soil Screening Level Framework, Part 5: Chemical-Specific Parameters, Office of Solid Waste and Emergency Response, Review Draft, July.

RADIONUCLIDES

CHAPTER 43

Evaluating Human Health Risk from Radioactive Contamination in Soils

Roy Eckart, University of Cincinnati, Cincinnati, Ohio

Roy Tan, Tetra Tech, Inc., Falls Church, VA

INTRODUCTION

This chapter is one of a series of papers investigating the differences in methodology between the Environmental Protection Agency (EPA) and the Department of Energy (DOE) related to the human health risk assessment due to radionuclides in soil. Previous work (Eckart and Tan, 1992; Tan and Eckart, 1993) compared the methodology for the gaseous pathways, such as plumes and inhalation. For the food ingestion pathway, the specific goals are:

- characterize the radiological human health risk assessment methodologies and default parameters that are developed by DOE or EPA or their contractors, and are being used in the DOE sites,
- compare the DOE and EPA representative methodologies and their default parameter values so as to distinguish the fundamental differences between them,
- quantify the differences in risk assessment results for both the DOE and EPA representative methodologies and the default parameter values that are applied to a typical DOE site with common radionuclides, and
- analyze the importance of the differences in methodologies and parameter values between the DOE and EPA.

The purpose of this research is to help to develop an understanding of the fundamental differences between the DOE and EPA risk assessment methodology. The major issue is how these differences impact on human health risk assessment results.

BACKGROUND

There are many former weapons plant sites and other DOE sites that will be remediated and restored. The cleanup level will be established by a regulatory agency, most likely the EPA. In most cases, the cleanup level will be specified as either a health risk level or an annual dose limit. One of the most important steps is to utilize a mathematical model to predict the potential health risk to people exposed to the residual radioactive contamination in the soil.

This exposure is characterized by environmental pathways such as gaseous transport of radionuclides (which leads to inhalation), ingestion pathways (food and water intake), and direct radiation exposure. All of the pathways have been mathematically modeled and developed into computer programs. More than 100 computer models (or programs) plus extensive documentation have been developed for the human health risk assessment process for hazardous and radioactive waste sites (Moskowitz et al, 1992). These models and documents are used in all analysis work, ranging from the source characterization, fate and transport, to human health effects. The DOE and the EPA are playing a major role in developing these methodologies. These agencies have established radiological health effect risk assessment methodologies for determining risk from soil containing radioactive material. These methodologies are in the form of computer programs or extensive documentation.

Although both the EPA and the DOE utilize the pathway analysis technique to evaluate the potential human health risks from radioactive materials in the environment, there are significant differences in the implementation of the methodologies and parameter values. The preliminary work on the comparison between the DOE and EPA risk assessment methodology and parameter values has shown that significant differences exist between the two methodologies and the default parameter values (Eckart and Tan, 1992; Tan and Eckart, 1993). The differences in risk or dose results can be more than an order of magnitude.

To avoid the problems introduced by the different risk assessment results between the DOE and the EPA, it is important to understand the details of the differences in the methodologies and parameter values between the DOE and EPA. The understanding of the fundamental differences and their impacts on the assessment results could help determine the appropriate uses of the DOE and EPA methodologies and parameter values. This chapter may be an important step toward establishing a realistic and consistent risk assessment methodology and the supporting parameter values.

COMPARISON OF METHODOLOGIES AND PARAMETERS

There are more than 100 computer models that are being used to support site remediation, restoration, and cleanup. These models are also used to manage site restoration activities, assist in low-level waste management, and plan for high-level waste management. More than 40 of these models were developed with the sponsorship of the U.S. DOE or the U.S. EPA. These models can also be used to characterize sites, model the movement of radioactive and hazardous waste, and model the transport of material through single or multiple pathways. Many of the models also predict the effect of hazardous substances on human health or the ecology.

SELECTION OF REPRESENTATIVE METHODOLOGIES

In order to ensure the best possible comparison, the following criteria for selecting the DOE and EPA representative risk assessment methodologies were specified:

1. The methodology must be recognized by the DOE or EPA.
2. The methodology must be widely used for DOE sites.
3. The methodology must be able to model the physical conditions and exposure scenarios of the typical radioactive waste site.
4. The methodology should be applicable for multimedia environmental transport analysis (vs. single transport pathway analysis).
5. The methodology must be applicable for complete radiological human health risk assessment. The methodology must model the transport of radionuclides existing at the contaminated site through the environment, and determine human dose or health risk.

Based on these criteria, the following four methodologies were selected for the DOE/EPA risk analysis comparison.

DOE Methodology

RESRAD (Yu et al, 1993)
GENII (Napier et al, 1988)

EPA Methodology

Risk Assessment Guidance for Superfund (RAGS) (USEPA, 1989)
PRESTO (USEPA, 1987)
This paper focuses on the ingestion pathways:

- plants grown in the soil
- meat from cattle grazing on the land
- milk from cows grazing or fed from the land
- drinking water from the site
- soil ingestion
- aquatic (fish, shellfish, etc.)

This paper compares the DOE and EPA risk assessment methodology for these ingestion pathways.

DOE METHODOLOGY

The average annual effective dose equivalent is determined by Equation 1:

$$H_{E,ip}(t) = DSR_{ip} C\ S_i(0) \tag{1}$$

where $H_{E,ip}(t)$ is average annual effective dose equivalent received at time t, $DSR_{ip}(t)$ is the dose/source ratio for the ith principal radionuclide and p^{th} environmental pathway, and $S_i(0)$ is initial concentration of the ith principal radionuclide in a uniformly contaminated zone.

The dose-to-source ratio (DSR) is used to determine the dose to humans from a unit source of radioactivity in the soil. It is calculated by Equation 2:

$$DSR_{ip}(t) = \sum_{R=0}^{k} DCF_{R}p\ C\ ETF_{R}p(t) \cdot SF_{R}(t) \tag{2}$$

where DCF_{lp} is the dose conversion factor for the ith principal radionuclide and p^{th} exposure pathway

$ETF_{lp}(t)$ is the source factor

k is the total number of the principal decay products of the ith principal radionuclide

F is a sequence number for the jth principal decay products.

The Environmental Transport Factor (ETF) accounts for the transport of the radioactive material through the environmental pathways to humans. For example, it would be the quantity of radioactive material absorbed by plants through the roots that ends up in the edible portion of the plant and the quantity of the particular plant that is consumed by humans. Simply stated, the ETF represents the actual quantity of the radioactive material ingested or inhaled by humans.

The Environmental Transport Factors, $ETF_{ipq}(t)$, for plant, meat and milk pathways, are computed as:

$$ETF_{ipq} = FA_p \bullet FCD_{ipq}(t) \bullet \Sigma(DF_{pk} \bullet FSR_{ipqk}(t)) \tag{3}$$

where $ETF_{ipq}(t)$ is environmental transport factor for the ith principal radionuclide and pq^{th} environmental pathway at time t

FA_p is area factor

$FCD_{ipq}(t)$ is cover and depth factor

DF_{pk} is dietary factor, i.e., annual consumption of kth food class for the pth food pathway

$FSR_{ipqk}(t)$ is food/soil concentration ratio.

Each of the parameters in Equations 1-3 can be found in the RESRAD manual (Yu et al., 1993) and are discussed in detail.

The drinking water ingestion subpathway is defined by the amount of drinking water actually consumed from a contaminated well onsite. The environmental transport factor (ETF) for the drinking water sub-pathway is given by Equation 4:

$$ETF_{i7}(t) = DF_7 \bullet FDW \bullet [WSR_{i1}(t) \bullet FD1 + WSR_{i2}(t)\ x\ (1 - FD1)] \quad (4)$$

where DF_7 is annual intake of drinking water
FDW is fraction of drinking water from site
WSR_{i1} is water/soil concentration ratios for well water
WSR_{i2} is water/soil concentration ratios for surface water
$FD1$ is fraction of well water used for drinking.

EPA METHODOLOGY

The detailed equations for the EPA methodology can be found in the previously cited references (USEPA 1987; 1989). Only a few typical equations are cited in this chapter.

In the EPA methodology, the ingestion pathway includes contributions from the ingestion of vegetables, ingestion of milk and meat from cattle or goats, and ingestion of contaminated ground or surface water. The cattle and goats are assumed to consume contaminated water and vegetation. The vegetables can be assumed to have been grown either on- or offsite. Storage time is considered in the intake calculations for the ingestion of all vegetation, milk, and meat. The average storage and transport times are input parameters, and radionuclide(s) are decayed for this time period.

The general equation for the ingestion of radionuclides in food and water is given by:

$$Q_{ing} = C_{lv}\ C\ U_{lv} + C_p \bullet U_p + C_{cm} \bullet U_{cm} + C_{gm} \bullet U_{gm} + C_f \bullet U_f + C_w\ Uw \quad (5)$$

where Q_{ing} is annual ingestion of the radionuclide
C_{lv} is the radionuclide concentration in leafy vegetable intake
C_p is the radionuclide concentration in cows' milk
U_{cm} is the annual intake of cows' milk
C_{gm} is the radionuclide concentration in goats' milk
U_{gm} is the annual intake of goats' milk
C_f is the radionuclide concentration in animal flesh
U_f is the annual intake of animal flesh
C_w is the radionuclide concentration in water
U_w is the annual intake of water.

BASIC DIFFERENCES

The EPA does not consider:

- soil ingestion pathway
- soil cover depth
- size of the contaminated area

These are important parameters in the analysis. For example, children ingest soil, either inadvertently or due to pica. Cattle ingest large quantities of soil during grazing. Adults ingest soil inadvertently due to poorly washed fruits and vegetables. Soil ingestion can be an important pathway.

The soil cover depth affects the quantity of radioactive material that is available for soil ingestion, resuspension, or plant root uptake. The size of the contaminated area is fundamentally important to the quantity of plant produce, meat, and milk consumed from the site.

The DOE does not consider:

- radioactive decay between harvest, or slaughter, and consumption

This time delay would be very important for short half-life radionuclides.

RISK ASSESSMENT ANALYSIS

Seven radionuclides were evaluated in this analysis:

americium-241
radium-226
uranium-238
cesuim-137
strontium-90
plutonium-239
thorium-241

These radionuclides are typical of radionuclides encountered at midwest sites. We assumed uniform distribution of these radionuclides in the contaminated layer of soil.

EXPOSURE SCENARIO

A common, realistic exposure scenario was developed for this analysis. It is a typical midwest site; 150,000 square meters were used to store radioactive waste and by-products from a production operation. There is an aquifer located under the site and a nearby river. The vadose zone has several layers of topsoil, sand and silt, sand, and brown sand. PRESTO cannot model multilayered vadose zone profile. It can only accept one type of vadose zone. In order to fully explore the different methodologies, three activity scenarios were compared: residential, agricultural, and commercial. Significant differences appeared in the occupancy and shielding factors and exposure duration for the three scenarios.

PARAMETER VALUES

One of the difficulties in any analysis is determining parameter values. It is particularly difficult in the ingestion pathway, since some of the parameters relate to crop yield and quantities of food consumed by individuals. Thus, many analysts will rely on default values built into the computer program.

Table 1 compares the important parameter values for the ingestion pathway.

Table 1. **Ingestion Pathway Input Parameter Values**

Parameter Description	Unit	DOE	EPA
Ingestion rate of fruit, vegetable & grain	kg/yr	160	119
Ingestion rate of meat	kg/yr	63	35
Ingestion rate of milk	kg/yr	92	140
Ingestion rate of drinking water	L/yr	510	700
Ingestion of soil	g/yr	36.5	36.5
Daily intake of fodder by beef cattle	kg/d	68	50
Daily intake of water by milk cow	L/d	160	60
Retention fraction for air	-	0.25	0.20
Wet-weight crop yield for leaf vegetables	kg/m2	1.5	0.65
Wet-weight crop yield for forage	kg/m2	1.1	0.67
Time of exposure for vegetables	day	60	30
Weathering removal constant for vegetation	yr-1	20	18.4

DOSE CONVERSION FACTOR (DCF)

The dose conversion factor (DCF) and the slope factor are based on a detailed analysis of the dose to human tissue and organs from inhalation or ingestion of specific radionuclides. These values are widely reported and utilized in health risk studies. The DCF and the Slope Factor determine the dose or risk due to the ingestion or inhalation of a unit of a radionuclide (usually a picocurie). Table 2 lists the dose conversion factors and slope factors that were used in this study.

Table 2. **Dose Conversion and Slope Factors for Ingestion Pathway**

Radionuclide	f1[a]	Dose Conversion Factor (mrem/pCi) DOE	EPA	Slope Factor (1/pCi)
Am-241	1.0E-03	4.50E-03	3.45E-03	2.40E-10
Cs-137	1.0E+00	5.00E-05	4.45E-05	2.80E-11
	1.0E-03	4.30E-03	3.43E-04	2.30E-11
Pu-239	1.0E-05	5.80E-05	n/a	n/a
Ra-226	2.0E-01	1.10E-03	7.40E-04	1.20E-10
	3.0E-01	1.40E-04	5.62E-06	3.60E-11
Sr-90	1.0E-02	2.20E-05	n/a	n/a
Th-232	2.0E-04	2.80E-03	2.16E-04	1.20E-11
	5.0E-02	2.5E-04	7.90E-06	2.80E-11
U-238	2.0E-03	3.80E-05	n/a	n/a

[a] Fraction of an element entering the GI tract that reaches body fluids

RISK ASSESSMENT RESULTS FOR THE INGESTION PATHWAY

Table 3 presents the results of this study for the ingestion pathway in terms of the dose-to-source ratio or the risk-to-source ratio. Both results are therefore based on a unit (pCi) of radioactivity in the soil.

Table 3 **Dose Risk for Plant Ingestion at t = 0 Per Unit Source (1 pCi/g)**

		Dose (mrem/yr)		Risk (1/yr)	
Radionuclide	f1[a]	DOE	EPA	DOE	EPA
Am-241	1.0E-03	1.72E-01	1.64E-01	9.17E-09	1.14E-08
Cs-137	1.0E+00	7.54E-02	8.31E-02	4.22E-08	5.23E-08
	1.0E-03	1.65E-01	1.01E-02	8.83E-09	6.77E-09
Pu-239	1.0E-05	2.20E-03	n/a	n/a	n/a
Ra-226	2.0E-01	1.66E+00	9.79E-02	1.81E-07	1.59E-08
	3.0E-01	1.58E+00	9.64E-02	4.06E-07	6.18E-07
Sr-90	1.0E-02	2.48E-01	n/a	n/a	n/a
Th-232	2.0E-04	1.07E-01	6.96E-03	4.59E-10	3.87E-10
	5.0E-02	2.37E-02	2.14E-03	2.65E-09	7.58E-09
U-238	2.0E-03	3.59E-03	n/a	n/a	n/a

It is important to note that the dose or risk actually varies with time. Figure 1 presents an example of this change for uranium-238 during the 1000 year horizon. The magnitude of the difference in results will change as a function of time. We have only reported this difference during some initial time period (t=0).

Figure 1. **Dose to source ratio for plant ingestion - Sr-90.**

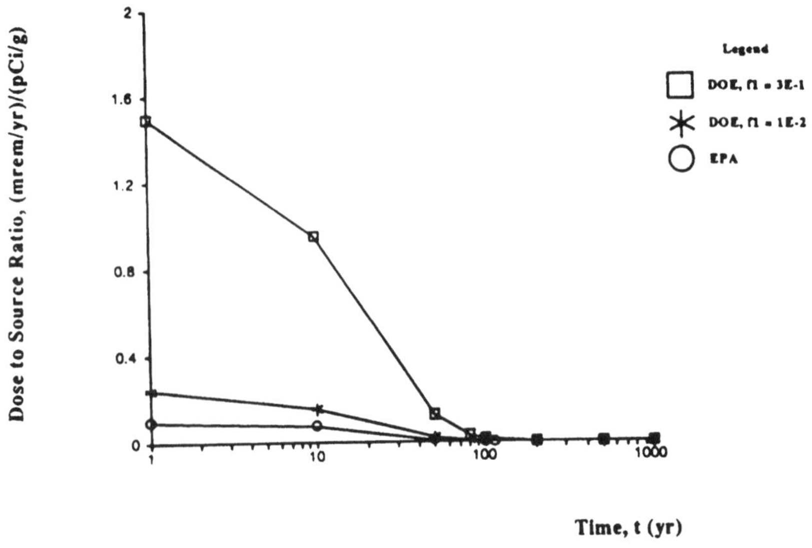

CONCLUSIONS

The results of this research project indicate that there are significant differences between the DOE and the EPA in dose and risk calculations. We believed that the EPA methodology was always more conservative than the DOE methodology. That was not the case. In some pathways, the EPA is more conservative, while in other pathways, the DOE results are the most conservative. What is clear in all cases is that there is no consistency between the DOE and the EPA in dose and risk assessment. The differences in the plant ingestion pathways are mainly due to site area, dose conversion factor, dust deposition, dust retention, and plant productivity. RESRAD offers the option to change the DCF value depending on the f1 value (fraction of an element entering the GI tract that reaches body fluids) for a radionuclide. Similar differences appear in the meat and milk pathways. Only DOE methodology considers the effect of the size of the contaminated area through the area factor FA (<1.0) and the partial radionuclide absorption through the foliage to food radionuclide transfer coefficient Tiv (<1.0) in foliar deposition. On the other hand, only EPA methodology considers the radioactive decay during plant-growing season and during the time period between the harvest and consumption.

Both RESRAD (DOE methodology) and PRESTO (EPA methodology) are generic computer programs that can be used to perform risk assessment for sites contaminated with radionuclides. RESRAD has more flexibility in modeling the site, allowing for multiple unsaturated zones and accounting for variable size and shape in sites modeled. PRESTO has more flexibility in modeling exposure pathways. It allows for modeling of exposure on- or offsite.

The risk assessment results indicate that the dose and risk results from the EPA methodology are different from the corresponding values from the DOE methodology. However, the differences in dose and risk results vary with pathways and radionuclides. The DOE results may be higher than the results from the EPA for some pathways and radionuclides, and EPA results may be higher for other pathways and radionuclides.

These differences in dose and risk are caused either by different assumptions and different computing equations in the methodologies or different parameter values, or both. Default parameter values play very important roles in the risk assessment. Many of them are not consistent between DOE and EPA. To reduce the departures in risk assessment results between the DOE and EPA, employing consistent parameter values is a key step. In addition, as users of these methodologies, site-specific parameters must be used for each program in order to have meaningful results.

REFERENCES

Eckart, R. and Tan, Z. 1992. Comparison and Analysis of Parameter Values Between the DOE and EPA Radiological Risk Assessment Methodologies, Radiological Environmental Assessment Group, Department of Mechanical, Industrial and Nuclear Engineering, University of Cincinnati, Cincinnati, OH.

Moskowitz, P.D. et al. 1992. Computer Models Used to Support Cleanup Decision-Making at Hazardous and Radioactive Waste Sites, *Risk Anal.*, **12**, 4.

Napier, B. et al. 1988. GENII - The Hanford Environmental Radiation Dosimetry Software System, Volume 1: Conceptual Presentation, U.S. Department of Energy, PNL-6584.

Tan, Z. and Eckart, R. 1993. Comparison of the DOE and EPA Risk Assessment Methodologies and Default Parameter Values for the Air Exposure Pathway, *Transactions of American Nuclear Society*, **68**, Part A.

U.S. Environmental Protection Agency (EPA), Office of Radiation Programs. 1987. Low-Level and NARM Radioactive Wastes, Model Documentation PRESTO-EPA-CPG, Methodology and Users Manual, Washington, DC, EPA 520/1-87-026.

U.S. Environmental Protection Agency (EPA), Office of Emergency and Remedial Response. 1989. Risk Assessment Guidance for Superfund: Human Health Evaluation Manual, Volume I, (Part A, Interim Final), Washington, DC, EPA/540/1-89/002.

Yu, C. et al. 1993. Manual for Implementing Residual Radioactive Material Guidelines Using RESRAD, Version 5, Argonne National Laboratory, Argonne, IL.

REGULATORY

CHAPTER 44

A Simple Method for Determining Treatment Needs for Vapor Phase Remedial Emissions

Mark D. Casey and **John J. Fitzgerald**, Massachusetts Department of Environmental Protection, Woburn, Massachusetts

INTRODUCTION

Remedial systems used in the cleanup of contaminated soil and groundwater can result in point source discharges of organic vapors to the ambient air. Air strippers and soil vapor extraction system are the most common technologies of this nature. The discharge of contaminants in this manner can result in significant risks to public health and/or may create nuisance odor conditions at nearby receptors. This chapter presents a simplified screening approach to evaluate potential adverse impacts that may be associated with such discharges, in order to help determine the need for and/or extent of requisite offgas control systems.

BACKGROUND

Decisions regarding the need for, and extent of, offgas control systems for remedial air emissions are usually predicated on the regulatory requirements in effect at any given location. Depending upon regional air quality conditions, population densities, and/or general regulatory climate, treatment requirements for remedial emissions generally fall into one of three categories:

- Offgas controls categorically required at all sites - In such cases, a treatment performance standard is often specified, such as 95% removal, application of Best Available Control Technologies (BACT), and/or prevention of significant health risks or nuisance odors.
- Offgas controls required if untreated emissions would present a significant risk to public health or welfare - In such cases, remedial proponents are required to evaluate the potential impacts of untreated emissions, and institute control systems as needed prior to activation of remedial systems.

- Offgas controls required after-the-fact only if odor complaints are received -In such cases, the presumption is that significant impacts are not likely to occur from such systems, given their relatively low emission rates, and controls only need to be retrofitted to remedial systems if and when a problem develops.

In all three cases, a knowledge of potential health risks and odor impacts resulting from such emissions, treated or untreated, would be necessary or desirable. While definitive information in this regard can only be obtained through a detailed and site-specific modeling and risk assessment process, a simple screening evaluation would be useful to characterize potential "worst-case" impacts.

EVALUATION OBJECTIVE AND APPROACH

The objective of this evaluation was to develop a series of equations and emission-distance graphs relating remedial air emission rates of common site contaminants to downwind increased ambient air concentrations of potential concern. This exercise involved (1) establishing a typical remedial air emission scenario utilizing "reasonable" worst-case conditions, (2) using a conservative air dispersion model to calculate downwind increased ambient air concentrations, and (3) establishing "acceptable" increased ambient air concentrations of contaminants of interest.

Twenty common environmental contaminants were evaluated. An EPA air dispersion model was used to quantitatively predict increased ambient air concentrations of emitted contaminants, at human receptors located at varying distances from an emission point. Subsequently, "acceptable" contaminant levels at the downwind receptors were set and used to obtain acceptable emission rates at the discharge stack.

The Air Model

The EPA Screen Model was selected as the air dispersion model of choice. This model is a conservative variation of the Industrial Source Complex (ISC) Model and as the name implies, is used to "screen out" emission sources that are not expected to contribute to a condition of air pollution, even under "worst case" site conditions.

Modeling Input Assumptions

Given the fact that the downwind receptor concentration is a function of many site-specific factors, modeling every possible site scenario would be virtually impossible. For this reason, a limited number of hypothetical site scenarios were created in order to arrive at a "reasonable worst case" representation of a typical gas station remediation site. A sensitivity analysis was performed on the input variables to the SCREEN Model to determine which variables had the greatest impact on the model results. During the sensitivity analysis, gas exit temperature, ambient air temperature and receptor height were held constant. Changes in stack height were found to cause only minor changes in receptor concentrations. With this in mind, the reasonable worst-case value for stack height and diameter were selected based on typical remedial system designs (i.e., a typical air stripper will generally be at least 15 feet tall and may be as large as 2 feet in diameter).

Various stack emission velocities were evaluated, corresponding to air flow rates ranging from 100 scfm to 500 scfm. Surprisingly, changes in emission velocity did not result in significant changes in downwind receptor concentrations (ground-level concentrations within 50 meters from the stack were somewhat lower at higher emission velocities, but were fairly constant at greater distances). The value of 0.5 m/s (300 scfm) was, therefore, selected as the reasonable worst case. The selected site scenario and model input parameters used in this evaluation are presented in Table 1.

Table 1. **Reasonable Worst Case Emission Scenario/SCREEN Model Input Variables**

Stack Height (m)	Stack Inside Diameter (m)	Stack Exit Velocity (m/s)	Stack Gas Exit Temp (°K)	Ambient Air Temp (°K)	Receptor Height (m)
4.5	0.6	0.5	293	293	Ground Level

Building Downwash and Meteorological Assumptions

Building Downwash and Meteorological Assumptions

Stack and/or building downwash effects can result in lower concentrations of contaminants within the "wake" region of a downwind plume, with a concomitant rise in the concentrations of contaminants in the "cavity region" (Figure 1) located close to the source. The size of the cavity region is dependent on the size and shape of nearby buildings and, therefore, modeling every conceivable scenario was deemed impossible. Instead, a range of building shapes and sizes expected at and around typical remediation sites were evaluated. In each case, the "cavity" length was less than 20 meters and, therefore, as a conservative assumption and simplifying condition, no stack or building downwash effects were incorporated in the modeling exercise. The application of this screening analysis is being limited to receptor locations greater than 20 meters from a remedial air emission.

Finally, the model was set to calculate downwind receptor concentrations based on worst-case meteorological conditions. In this setting, the model evaluates an array of 21 different combinations of wind speeds and stability classes,

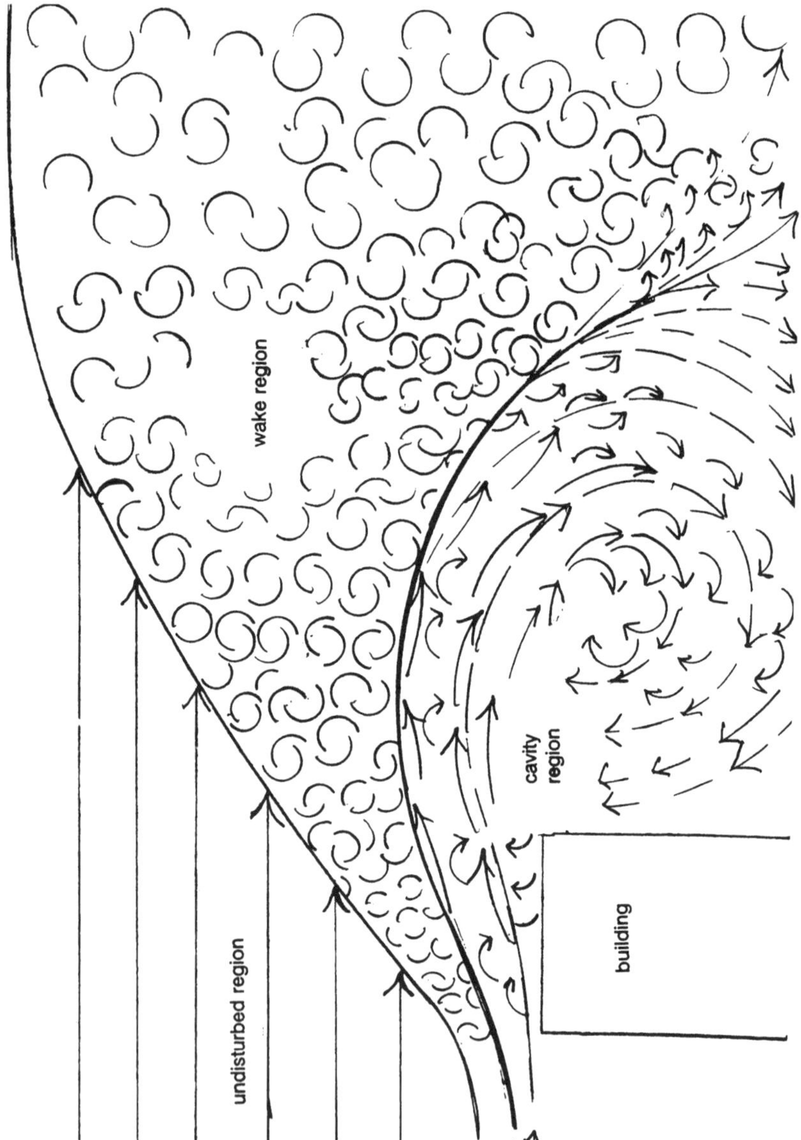

Figure 1. **Diagram of building downwash effects.**

and provides an output of the worst-case condition. This was deemed to be an acceptable degree of conservatism for a first level screening exercise.

Establishing "Acceptable" Increased Receptor Concentrations

Selection of Air Contaminants of Concern

In total, 20 volatile and semivolatile organic compounds were selected for evaluation, based upon their prevalence at contaminated sites, volatility, toxicity, and/or odor potential. These 20 compounds are listed in Table 2.

Table 2. **Health/Odor Characteristics of Targeted Contaminants**

Compound	TEL (μg/m3)	AAL (μg/m3)	Adjusted AAL (μg/m3)	Odor Threshold (μg/m3)
Acetone	160.5	160.5	2,408	550
Benzene	1.74	0.12	1.8	516
Bis (2-ethylhexyl)Phthalate	1.36	0.77	11.55	NA
Carbon-tetrachloride	85.5	0.07	1.05	10,000
Chlorobenzene	93.88	6.26	93.9	140
1,1-Dichloroethane	11.01	0.04	0.6	493,000
1,2-Dichloroethane	11.01	0.04	0.6	23,000
1,2-Dichloroethene	215.6	107.8	1,617	1,100,000
Dichloromethane	9.45	0.24	3.6	3,000
Ethylbenzene	118.04	118.04	1,773	400
Methyl ethyl ketone	32.07	32.07	481	5,580
Methyl tertiary butyl ether	714[a]	Not Available	Not available	2,040
Naphthalene	14.25	14.25	213.8	50
Phenol	52.33	52.33	785	15
Tetrachloroethylene	922.2	0.02	0.3	32,000
1,1,1-Trichloroethylene	1,038	1,038	15,570	60,000
Trichloroethylene	36.52	0.61	9.15	2,000
Toluene	10.24	10.24	153.6	6,000
Vinyl Chloride	3.47	0.38	5.7	25,000
Xylene	11.8	11.8	177	13.2

[a] Recommended TLV of 100 ppm (300 mg/m^3) divided by 420.

Selection of "Acceptable" Increased Ambient Air Concentrations

Health-based ambient air guidelines and odor threshold data were assembled for each of the targeted compounds, as displayed in Table 2. Ambient air guidance values promulgated by the state of Massachusetts were selected for the health-based component of this evaluation process; the odor threshold concentration was designated as a 50% odor recognition level contained in published literature (Verschueren, 1983).

Massachusetts has promulgated two ambient air guidance values for common air contaminants: Threshold Effect Levels (TELs) and Allowable Ambient Limits (AALs) (MA DEP, 1990). The TEL values are protective of short-term exposures, and are based upon a 24-hour averaging period. The AALs are protective of a lifetime (75 year) exposure, and are based upon a yearly averaging period. Other states have promulgated similar standards and/or guidelines.

Because it was assumed that most remedial emissions would be limited to 5 years or less, for the purposes of this evaluation, it was necessary to adjust the AAL values to account for a 5-year exposure period by multiplying each value by a factor of 75/5.

Grouping Contaminants

For developmental, implementation, and logistical reasons, the contaminants were grouped into four categories as shown in Table 3 (as opposed to having 20 separate graphs). This categorization was based upon the commonality of a "design value" established for each compound. As stated above, the health-based guidelines provided in Table 2 represent acceptable standards averaged over various time periods; the TEL over a 24-hour period and the AAL over an annual period. The SCREEN Model, however, provides maximum hourly concentrations, and therefore, a direct comparison of the TELs and AALs to the model outputs was not appropriate. In order to make the values in Table 2 comparable to hourly concentrations, the TEL was divided by 0.4, and the AAL by 0.1. No adjustments were necessary for the odor thresholds. The design value for each compound was selected as the lowest of these three values, and the contaminant groupings were established accordingly. The lowest design value for an individual compound within a group was used as the acceptable increased ambient air concentration value for all compounds within the grouping.

Modeling Process and Data Analysis

In the SCREEN Model, one inputs emission flux and site conditions, and the model calculates resultant increased air concentration at varying downwind distances. The objective of this exercise, however, was to approach this problem from the opposite end: designate "acceptable" air levels at varying distances from the emission, and then back-calculate the maximum emission rate at the source that would not exceed the designated value. Thus, it was necessary to perform

Table 3. Grouping of Targeted Contaminants

Compound	TEL/0.4 (µg/m3)	Adjusted AAL/0.1 (µg/m3)	Odor Threshold (µg/m3)	Design Value (µg/m3)	Acceptable Receptor Concentration (µg/m3)
Group1					
Benzene	4.35	18	516	4.35	
Bis(2-ethylhexyl)Phthalate	3.4	115.5	Not available	3.4	
Carbon-tetrachloride	213.8	10.5	10,000	10.5	
1,1-Dichloroethane	27.53	6.0	493,000	6.0	
1,2-Dichloroethane	27.53	6.0	23,000	6.0	
Tetrachloroethylene	2,306	3.0	32,000	3.0	0.3
VinylChloride	8.7	57	25,000	8.7	
Group2					
Dichloromethane	23.63	36	3,000	23.6	
Naphthalene	35.63	2,138	50	50	
Phenol	130.8	7,850	15	15	
Toluene	25.6	1,536	600	25.6	
Xylene	29.5	1,770	13.2	13.2	13.2
Group 3					
Chlorobenzene	234.7	939	140	140	
Methyl-ethyl-ketone	80.18	4,810	5,880	80.18	32.07
Trichloroethylene	91.3	91.5	2,000	91.3	
Group 4					
Acetone	410.4	2,480	500	401.4	
1,2-Dichloroethylene	539.1	16,170	1,100,000	539	
Ethylbenzene	295.1	17,730	400	295.1	118.04
1,1,1-Trichloroethylene	2,595	155,700	60,000	2,595	
Methyl Tertiary Butyl Ether	1,785[a]	Not available	2,040	1,785	

[a] A TLV of 100 ppm (300 mg/m^3) divided by 420 was used to obtain this value (See, Table 1)

successive iterations of the SCREEN Model in order to determine the appropriate input emission fluxes.

The above trial-and-error process was carried out for each group of compounds, and the resultant emission-distance data was plotted on semi-log graph paper. A linear regression analysis of the data was also performed to generate a line of best fit, determine correlation coefficients, and construct the Emission-Distance Graphs in Figure 2 through Figure 6.

Emission-Distance Graphs

The correlation coefficient for each of the emission-distance graphs was 0.97. Note that the line established in each graph represents the "acceptable" air concentration value within that group of compounds. Thus, emission-distance co-

Figure 2. **Emission rate vs. distance: Group 1.**

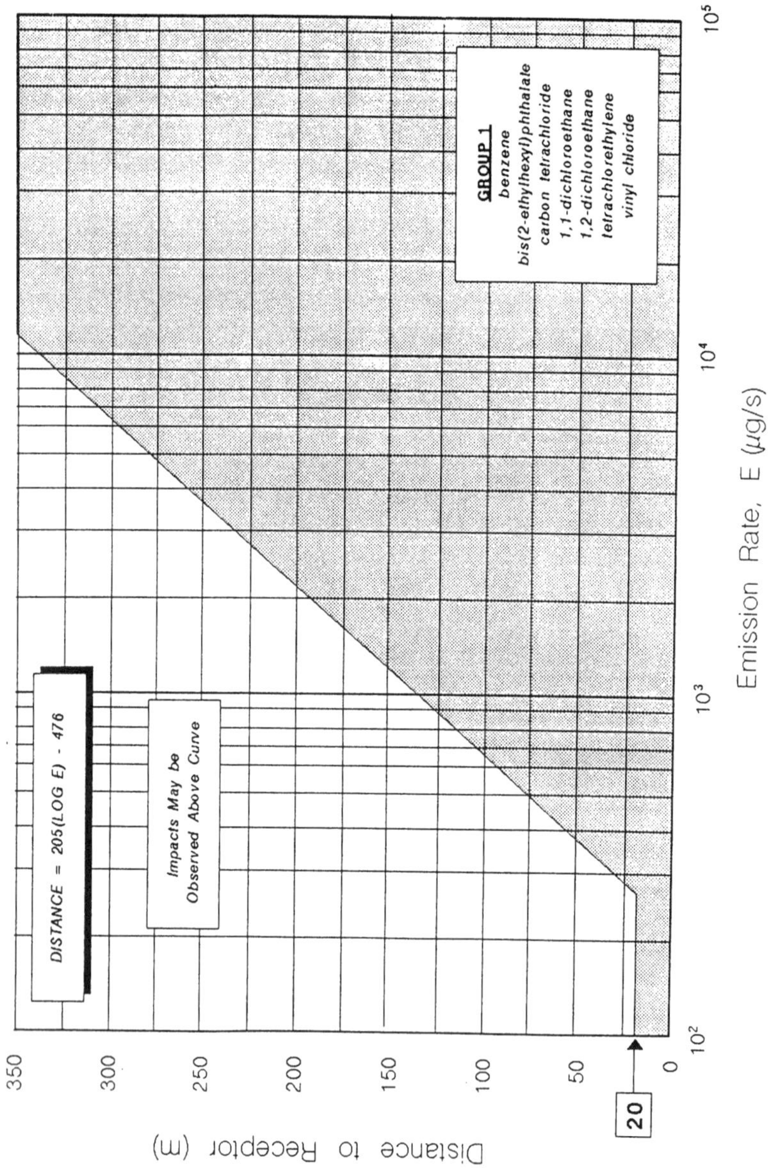

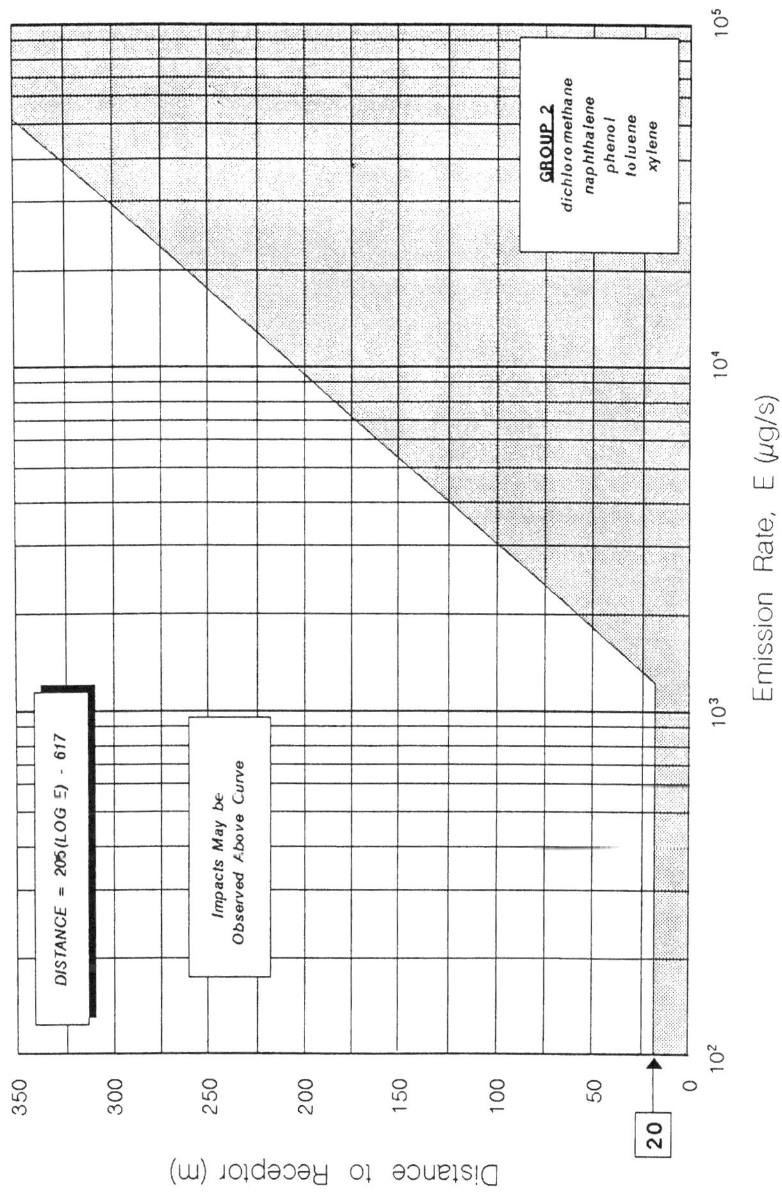

Figure 3. **Emission rate vs. distance: Group 2.**

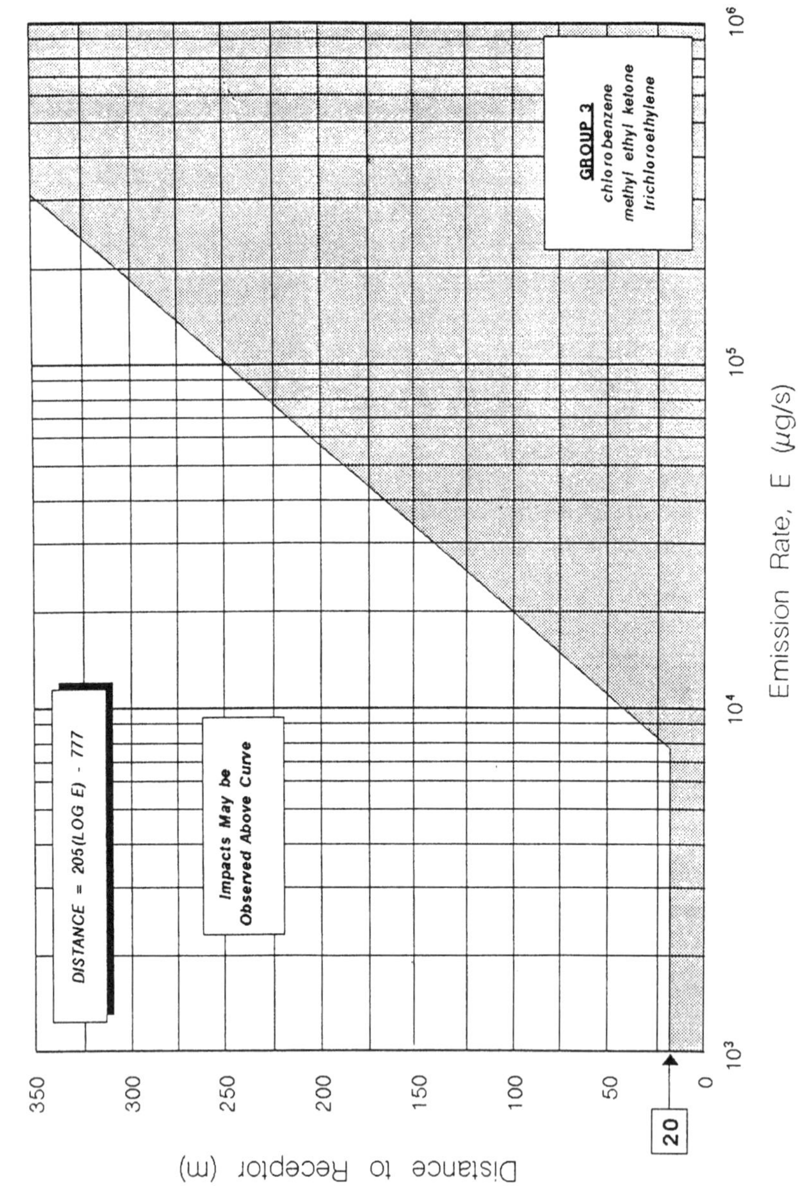

Figure 4. **Emission rate vs. distance: Group 3.**

Figure 5. **Emission rate vs. distance: Group 4.**

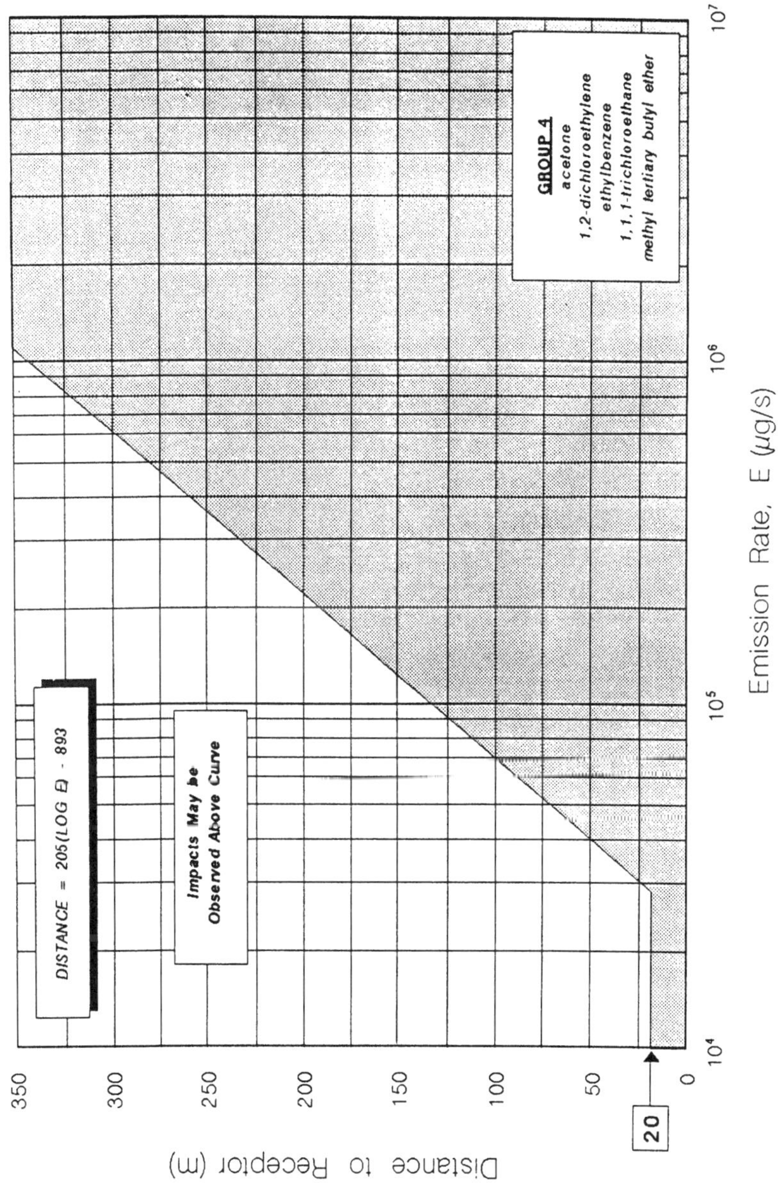

Figure 6. **Emission rate vs. distance: Group 5.**

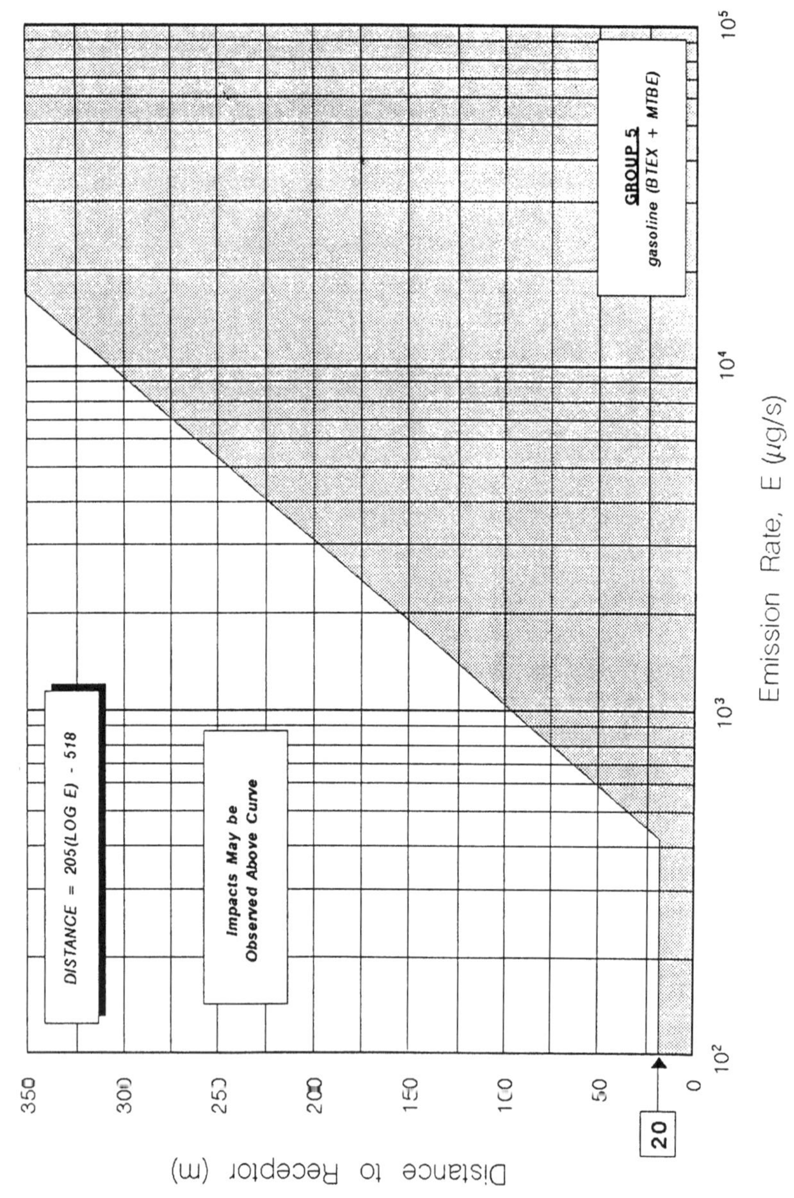

ordinates below the plotted line could exceed the "acceptable" increased air concentration at that distance, and therefore, additional site-specific analysis or vapor controls would be required. Conversely, emission-data coordinates above the plotted line would likely not result in an exceedance of the "acceptable" air concentration value. For each graph, an equation for a line of best fit is also provided for cases where graphs do not provide the necessary range or level of accuracy.

The plot in Figure 6 illustrates the "acceptable" increased air concentration for the discharge of gasoline vapors. Vapor discharges from gasoline releases were thought to present a unique concern due to the numerous compounds which make up gasoline, many of which are not adequately quantified with current analytical techniques. Gasoline components such as alkenes can cause nuisance odor conditions at downwind receptors, even though laboratory analysis of the discharge by conventional ("BTEX") analysis may indicate contaminant concentrations below acceptable levels. In addition, it has been suggested that synergistic odor conditions may result from mixtures of emitted organic vapors. A separate gasoline graph was deemed necessary to account for these conditions. The graph in Figure 6 considers the cumulative BTEX and MTBE emission rate as the independent variable for evaluating the potential impacts from the discharge of gasoline vapors.

Correlating the Emission Rate to Groundwater and Soil Gas Concentrations

For convenience, Figure 7 can be used to obtain a worst-case emission flux from an air stripper at various aqueous flow rates, based on a known groundwater influent concentration and assuming 100% mass transfer of the dissolved contaminants to the vapor phase. This correlation can be simply obtained from the following equation:

$$E = [C_w \times Q_w] / 15.84 \qquad (1)$$

where:

E = the air emission rate (μg/s)
C_w = the groundwater influent concentration (μg/L)
Q_w = the groundwater flow (pumping) rate (gal/min)

Likewise, Figures 8 and 9 can be used to obtain the emission flux from a soil vapor extraction (SVE) system at various air flow rates, based on the influent vapor concentration. This relationship is provided below:

$$E = [\, C_a \times Q_a \,] / 2118 \qquad (2)$$

where:

E = the air emission rate (μg/s),
C = the influent vapor concentration (μg/m^3),
Q_a = the air flow rate (cfm).

Figure 7. **Air stripper emission rate vs. influent groundwater concentration.**

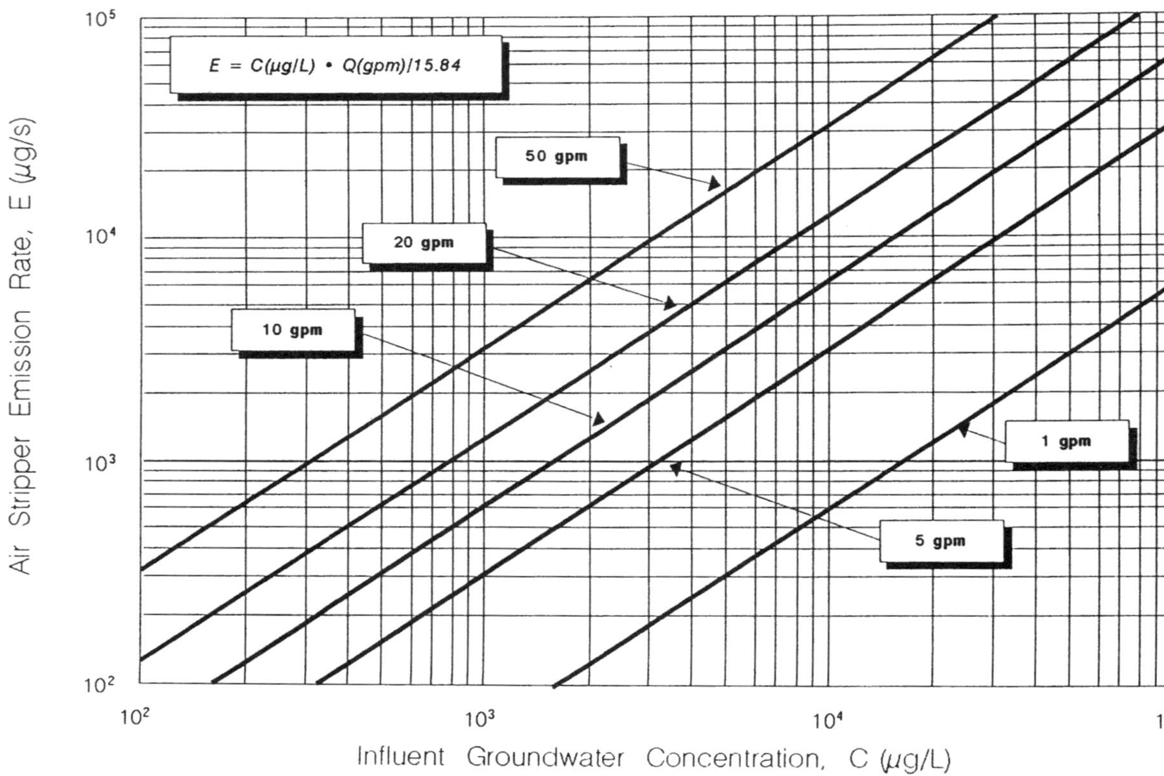

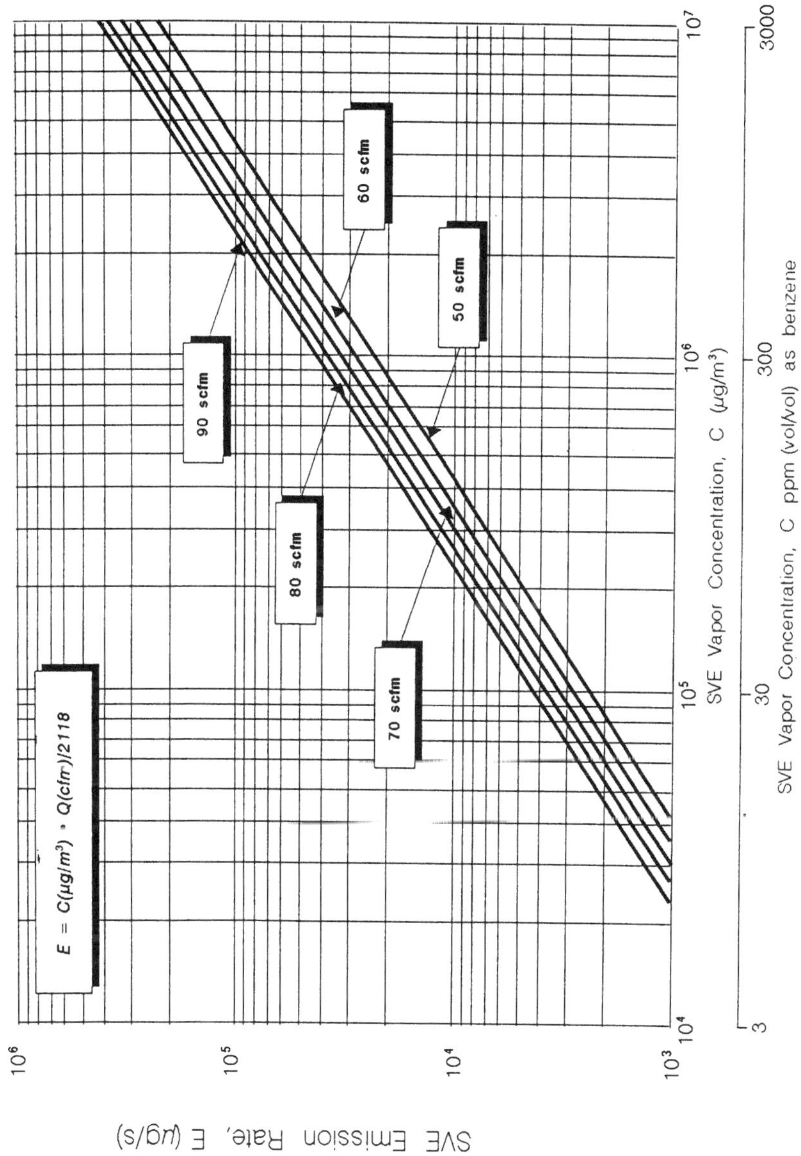

Figure 8. **SVE emission rate vs. vapor concentration.**

Figure 9. **SVE emission rate vs. vapor concentration.**

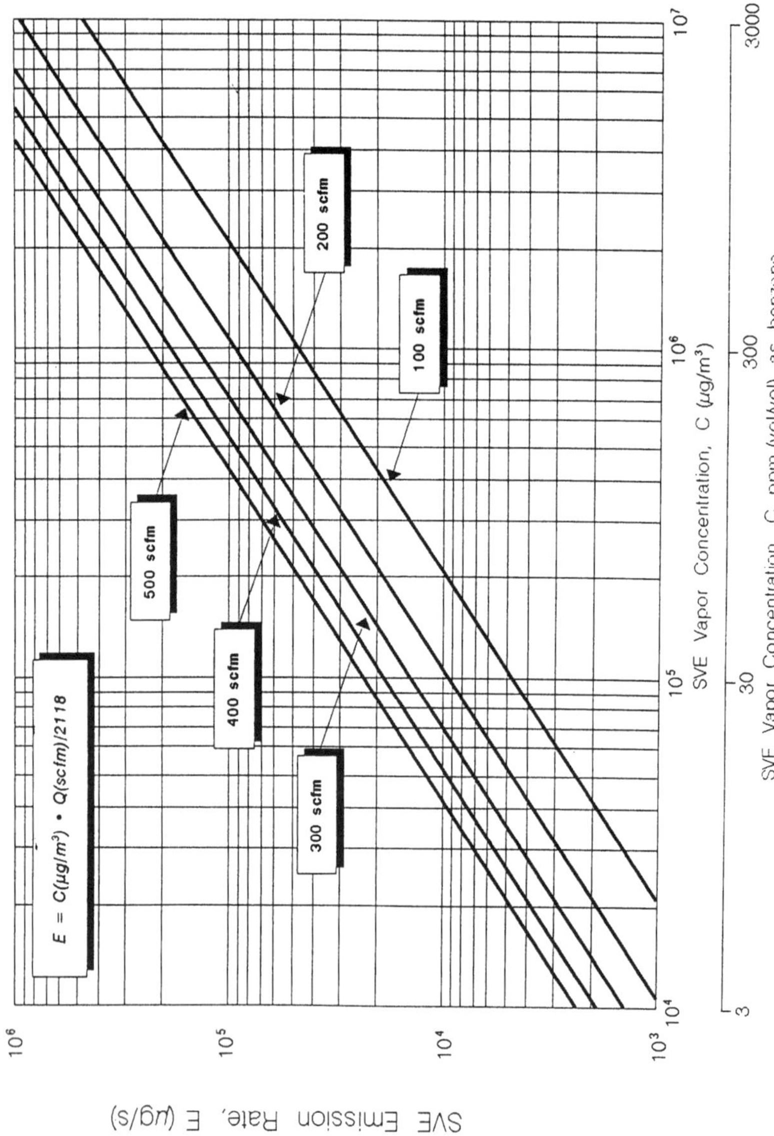

SUMMARY AND CONCLUSIONS

The goal of this modeling exercise was to develop a first-level screening tool to evaluate vapor treatment requirements on remedial discharges to the ambient air. The resultant emission-distance graphs provide a simple method for deciding when remedial air emissions are not likely to produce adverse health and/or odor impacts at nearby receptors, and when further evaluation of the issue may be necessary. As previously stated, the need for treatment of remedial air emission is typically a function of regulatory policies in effect in any given location. Where treatment of remedial emission is required, the approach presented here can be used to help determine the type and extent of necessary treatment.

Because it was developed as a screening tool, much of the rationale and assumptions used in its development are conservative, attempting to represent a reasonable worst-case site scenario. These worst-case assumptions, of course, may not be appropriate in every situation, and a more detailed evaluation may be desirable.

Moreover, because of the need to define certain simplifying and limiting factors, the use of this model is not recommended for all sites. Limiting site scenarios for the use of this model include: nonpoint source emissions; stack height less than 4.5 meters above ground level; potential receptors elevated above the ground surface (i.e., flag pole receptors); and the potential for building downwash effects. In these instances, a more complex modeling approach by qualified individuals would be appropriate to evaluate the potential risks posed by the vapor discharge.

REFERENCES

(MA DEP) *The Chemical Health Effects Assessment Methodology (CHEM)* and *The Method to Derive Allowable Ambient Limits (AAL)*, Volumes I and II. 1990. Massachusetts Department of Environmental Protection [Office of Research and Standards Publication No. 90-1].

Verschueren, K. 1983. Handbook of Environmental Data on Organic Chemicals, Second Edition, New York, Van Nostrand Reinhold Company.

CHAPTER 45

Field Instrument Evaluation Project

Christopher O'Neill, New York State Department of Environmental Conservation, Division of Spills Management, Albany, New York

INTRODUCTION

The New York State Department of Environmental Conservation conducted a research project at a petroleum spill cleanup site to compare the results obtained from several field analytical methods with each other and results from a stationary laboratory.

There was no conspicuous direct correlation discovered between results from a photoionization detector (PID), an immunoassay test kit, a gas chromatograph, and a stationary laboratory. Data comparisons involving laboratory data were limited because much less laboratory data were collected than field instrument data. In general, the PID provided higher contaminant concentrations than any other method for both fuel oil-contaminated soil and gasoline-contaminated soil. Through the use of curve-fitting and logarithmic data, several visual comparisons are noteworthy.

For gasoline-contaminated soil, curve fitting for the PID results and the immunoassay BTEX results produced parallel lines, illustrating identical trends for the data. Fitting curves to the common logarithm (LOG) of the gas chromatograph (GC) BTEX data, in parts per billion, and the LOG of the PID data, in parts per million or in parts per billion, produced parallel curves, and, therefore, identical data trends.

The fuel oil-contaminated soil raw data exhibited equivalent data trends through curve fitting of the GC BTEX data and the PID data. The stationary laboratory BTEX data, when approximated with curve fitting, resembled the immunoassay PAH fit curves, illustrating similar data trends. Additional laboratory data, especially toward the end of the project period, may have supported more comprehensive data comparisons.

Applying curve fitting approaches to the LOG of the data from fuel oil-contaminated soil, produced several visual data trend similarities, including: (1) the LOG of the immunoassay kit PAH data and the LOG of the stationary laboratory BTEX data; (2) the LOG of the GC BTEX data and the LOG of the stationary laboratory PAH data; (3) the LOG of the immunoassay kit data and the LOG of the stationary laboratory PAH data; and (4) the LOG of the PID data and the LOG of the GC BTEX data. The limited number of stationary laboratory data points adversely affected laboratory data comparisons, while the number of field instrument data points was conducive to developing and comparing data trends more confidently.

The total cost of this field and laboratory project was approximately $31,500. Additional funding and research work may provide further insight on the comparability of field and laboratory data for petroleum spill sites.

PROJECT SCOPE

The Field Instrument Evaluation Project (Project) was undertaken to use several common field instrument methods side-by-side, and splitting soil samples with a stationary laboratory, to assess the value or usefulness of the data collected, and the comparability of the data generated from each method.

The field methods and equipment used in the Project included: an HNU, Model HW101, 10.2 eV photoionization detector (HNU Systems, Inc., 1988); a Quantix Systems Portable Workstation immunoassay kit (Quantix) (Agri-Diagnostics Associates, 1992); and a portable or laboratory-grade gas chromatograph (GC).

NYSDEC personnel collected the soil samples; split the samples for each analytical method to be used; analyzed the samples with the HNU and Quantix kit; and shipped the samples to a contract laboratory for specified GC analyses and standard USEPA method analyses.

SAMPLING PROCEDURES

The soil samples were collected from petroleum-contaminated soil which was undergoing bioremediation in two lined treatment areas. One treatment area contained 50 cubic yards of soil, spread out in an 18-24 inch layer, contaminated with No. 2 fuel oil, and the other treatment area contained 50 cubic yards of soil, spread out in an 18-24 inch layer, contaminated with regular unleaded gasoline.

The specific locations for sampling each treatment area were determined using a random number table, a simple BASIC computer program, and an X-Y grid based on the dimensions of each treatment area.

Soil was dug at each sampling location using a trowel or shovel, and placed in a resealable plastic bag. (Composite samples were combined either upon digging, or from the separate plastic bags.) The samples were mixed slightly by kneading the soil in the bags. The samples were transferred to glass jars and labeled for appropriate analyses, and then placed in a sample cooler to await analysis or shipment to the laboratory.

ANALYTICAL PROCEDURES

The HNU measures the molecules present in the analyzed gas (air) which can be ionized by 10.2 electron volts of energy. The instrument has a high sensitivity to alkenes and aromatics, and negligible sensitivity to alkanes in the air being analyzed (HNU Systems, Inc., 1988). Table 1 identifies several petroleum product constituents and their respective ionization potentials (*CRC Handbook of Chemistry and Physics*, 199). Factors which may affect the results obtained from photoionization detectors include: high relative humidity; presence of alkanes and methane; and the absolute and relative abundances of detectable compounds (Robbins, 1990).

The HNU was used each week to measure the contamination present in the headspace of the soil sample jars. Specifically, the full soil sample jars were emptied to one-half full; aluminum foil was fitted across the mouth of the jar; the jar was capped with its cover; the jar was set aside, in ambient air temperature, for 10-30 minutes; the jar cap was removed; the HNU sampling probe was pushed through the aluminum foil; and the instrument reading was recorded as "Total Volatiles" in parts per million (ppm).

The immunoassay kits have a specific group of target compounds which they react with and quantify. The kits may cross-react if other compounds are present in the analyzed sample. In particular, the Quantix BTEX kit is based on the USEPA 6-component BTEX standard, and is designed to detect benzene, toluene, ethylbenzene, o-xylene, m-xylene, and p-xylene. The BTEX kit also

Table 1. **Ionization Potentials for some petroleum compounds.**

Ionization Potential For Selected Petroleum Constituents	
Petroleum Constituent	Ionization Potential (eV)
Benzene	9.24
Ethyl benzene	7.76±0.08
Toluene	8.82±0.01
o-Xylene	8.56
m-Xylene	8.58
p-Xylene	8.44
Naphthalene	8.12
sec-Butyl benzene	8.68±0.01
tert-Butyl benzene	8.68±0.01
Fluorene	8.63
Anthracene	7.55
Phenanthrene	8.1

reacts with naphthalene, 1,2,4-trimethyl-benzene, 2-methylnaphthalene, and n-propylbenzene. The Quantix PAH kit is based on the USEPA 16-component PAH standard, and is designed to detect phenanthrene, acenaphthylene, 2-methylnaphthalene, naphthalene, acenaphthene, dimethylnaphthalene, fluorene,1-methylnaphthalene, and pyrene (PSI, 1994).

The Quantix kit was used each week to measure Total PAH (polyaromatic hydrocarbons) in the soil contaminated with fuel oil, and Total BTEX (benzene, toluene, ethylbenzene, and xylenes) in the soil contaminated with gasoline. The manufacturer's procedures of soil sample extraction and colorimetric biochemistry were followed, and the instrument reading was recorded as Total BTEX, or Total PAH in parts per million (ppm).

The portable or laboratory-grade GC was used to measure benzene, toluene, ethylbenzene, and xylenes in both the fuel oil-contaminated and the gasoline-contaminated soils. Weekly samples were shipped overnight and analyzed within 24-48 hours in a climate-controlled laboratory setting. (The original plan for the Project called for a "field-grade" or portable GC; however, the lack of availability, cost and logistics of having a portable GC at the site once per week led to the decision to change to a laboratory-grade GC with 24-48 hour turnaround on analyses, and 2-3 weeks turnaround on reporting the results.) The results were reported within normal turnaround time (2-3 weeks), and were evaluated in terms of individual compound concentrations and Total BTEX (arithmetic sum of benzene, toluene, ethylbenzene, and xylenes concentrations).

Standard laboratory methods established by the USEPA (SW 846 methods) were used to measure individual volatile and semivolatile compound concentrations in the fuel oil-contaminated soil (via EPA Methods 8021 and 8270, respectively), and individual volatile compound concentrations in the gasoline-contaminated soil (via EPA Method 8021). Monthly samples were delivered to the laboratory within 24 hours for normal turnaround on analyses (approximately 7 days) and reporting (2-3 weeks). The laboratory results were tabulated and evaluated in terms of individual compound concentrations, Total BTEX (arithmetic sum of benzene, toluene, ethylbenzene, and xylenes concentrations), Total PAH (arithmetic sum of EPA Method 8270 analyte concentrations), and Total Volatiles (arithmetic sum of EPA Method 8021 analyte concentrations).

ANALYTICAL RESULTS

Each sampling event consisted of collecting two grab samples and two composite samples from both the fuel oil-contaminated soil and the gasoline-contaminated soil. The analytical results obtained from the four samples (2 grabs and 2 composites) for each soil contaminant type were simply averaged, assigning a value of zero for recorded "Nondetectable" values.

The fuel oil-contaminated soil consisted of approximately 43% sand, 31% gravel, and 26% silt/clay materials, and the gasoline-contaminated soil consisted of approximately 55% sand, 12% gravel, and 33% silt/clay materials. During the Project, the fuel oil-contaminated soil ranged in moisture content (by observation) from dry to very wet, and the gasoline-contaminated soil ranged in moisture content (by observation) from moist to very wet. The effects of soil type and moisture content on the analytical results obtained from the selected methods have not been evaluated, except for the physical collection and handling challenges associated with dry or very wet soils.

The HNU and Quantix results differed by 0.1% to 8000% over the 29 sampling events, with several data sets differing by less than 25%. The data values from the HNU and Quantix kits varied greatly from the values from the GC and stationary laboratory. The GC BTEX and the stationary laboratory BTEX data varied by 21% to 583% over the 9 laboratory sampling events, with two data sets for the gasoline-contaminated soil differing by less than 25%.

Tables 2 and 3 identify the overall ranges of data obtained from each analytical method for each contaminant type. The highest and lowest data point for each method did not necessarily correspond to the same sampling event. For example, the highest laboratory, Quantix and HNU data points were obtained from samples collected on 13 September 1993 for the fuel oil-contaminated soil, but the highest GC data point was obtained from 22 June 1994 samples.

Table 2. **Fuel oil-contaminated soil data.**

Fuel Oil-Contaminated Soil Analytical Data Ranges			
Analytical Method		Highest Data Point (Sample Date)	Lowest Data Point (Sample Date)
Lab - 8021	(Total Volatiles)	11,887 ppb (9/13/94)	30 ppb (6/2/94)
- 8270	(Total PAH)	1,650 ppb (9/13/94)	Not Detectable (Several)[a]
- 8021	(Total BTEX)	1,890 ppb (9/13/93)	7 ppb (6/2/94)
GC	(Total BTEX)	978 ppb (6/22/94)	Not Detectable (7/5/94)
Quantix	(Total PAH)[b]	558,000 ppb (9/13/93)	200 ppb (6/9/94, 6/20/94)
HNU	(Total Volatiles)[b]	61,000 ppb (9/13/93)	6000 ppb (7/5/94)

[a] 9/28/93, 10/27/93, 6/2/94, 8/3/94, 10/28/94.

[b] Instrument actually reports data in ppm units.

Table 3. **Gasoline-contaminated soil data.**

Gasoline-Contaminated Soil Analytical Ranges			
Analytical Method		Highest Data Point (Sample Date)	Lowest Data Point (Sample Date)
Lab - 8021	(Total Volatiles)	139,045 ppb *(9/13/93)*	5,090 ppb *(9/8/94)*
- 8021	(Total BTEX)	63,838 ppb *(9/13/93)*	1,001 ppb *(9/8/94)*
GC	(Total BTEX)	19,838 ppb *(10/14/94)*	275 ppb *(6/9/94)*
Quantix	(Total BTEX)[a]	308,000 ppb *(10/28/94)*	Not Detectable *(10/5/93, 10/19/93)*
HNU	(Total Volatiles)[a]	318,000 ppb *(10/28/94)*	22,000 ppb *(7/12/94)*

[a] Instrument actually reports data in ppm units.

The fuel oil-contaminated soil laboratory data for Total PAH ranged between 1650 ppb and Nondetectable (approximately < 330 ppb). Total Volatiles laboratory data ranged between 11,887 ppb and 30 ppb, and Total BTEX laboratory data ranged between 1890 ppb and 7 ppb. Figures 1 and 2 illustrate the raw data from the HNU analyses, Quantix Total PAH analyses, GC BTEX analyses, and laboratory analyses.

The overall data presented in Figure 1 suggest that at relatively higher PAH concentrations (sampling events 1-8), the Quantix data was higher than the HNU data. At relatively lower PAH concentrations, the HNU data was higher than the Quantix data. Since the HNU is capable of measuring volatile and most semivolatile (PAH) petroleum compounds, the Quantix kit may have provided higher data points due to the physical extraction of soil samples for analyses being more effective than the headspace analyses of the HNU. The Quantix and HNU data were relatively much higher than the laboratory and GC data for the entire soil concentration range. The time needed for the soil samples to be transported and analyzed by the laboratory may have allowed contaminant degradation to occur in the samples; however, the data in Figure 2 illustrate that the laboratory BTEX results were often higher than the GC BTEX results obtained from "next day" analyses. Evaluating sample handling, transportation and holding times potential effects were beyond the scope of the Project. In addition, it is worth noting that the HNU provided data points above background for the entire soil concentration range, without any "Nondetectable" results, suggesting that the lower detection limit for the Project for fuel oil contamination may be less than 30 ppb Total Volatiles in the soils.

Figure 1. **Fuel oil-contaminated soil field instrument raw data.**

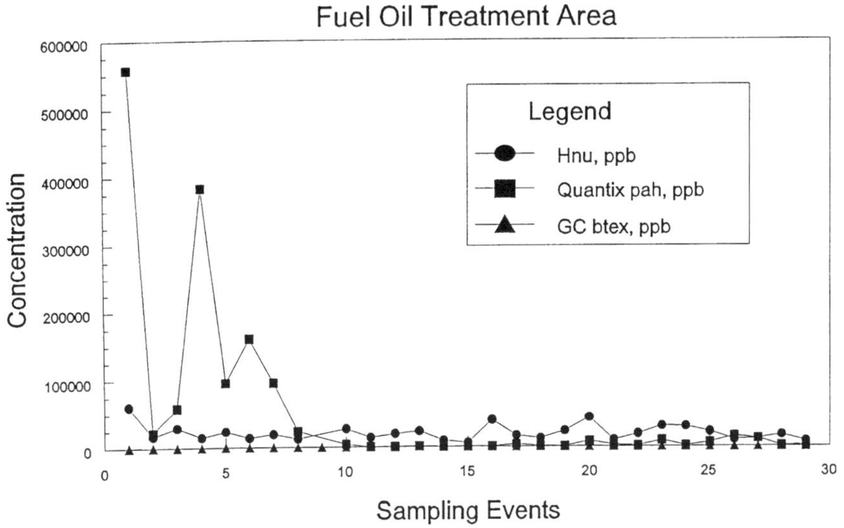

Figure 2. **Fuel oil-contaminated soil GC and stationary lab raw data.**

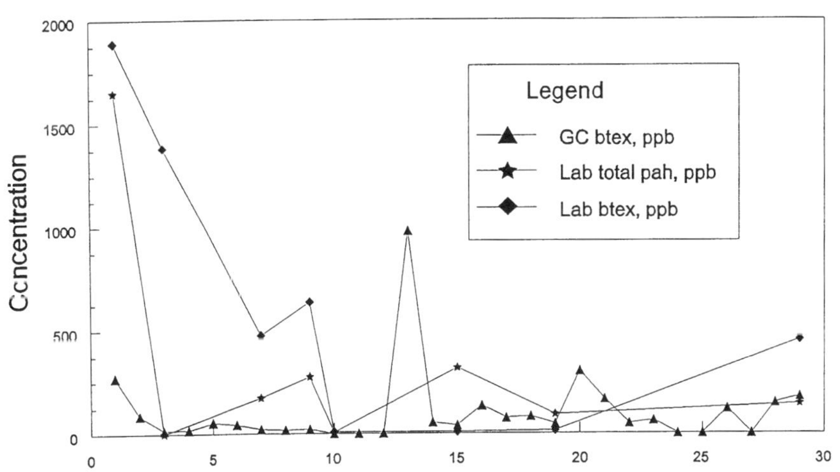

The gasoline-contaminated soil laboratory data for Total Volatiles ranged between 139,045 ppb and 5,090 ppb, and for Total BTEX ranged between 63,838 ppb and 1001 ppb. Figure 3 illustrates the raw data from the HNU analyses, Quantix Total BTEX analyses, GC BTEX analyses, laboratory Total BTEX analyses.

The overall data presented in Figure 3 suggest, in general, that the HNU results were greater than or equal to the Quantix results. For sampling events 3-7, the Quantix data, the GC data and the laboratory data were approximately equal. For sampling events 17-29, with relatively low BTEX concentrations, the

Figure 3. **Gasoline-contaminated soil raw data summary.**

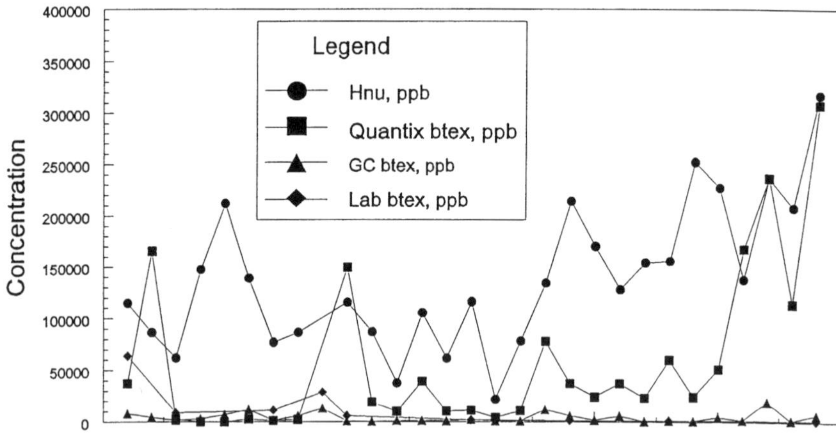

HNU and Quantix data were higher than the GC and laboratory data, with the HNU and Quantix data approximately equal on occasion. As stated previously, the transportation and holding times may have enabled contaminant degradation to occur in the samples prior to being analyzed by the GC and laboratory methods; however, potential sample degradation was not evaluated as part of the Project. In addition, the HNU provided data points above background for the entire soil concentration range, without any "Nondetectable" results, suggesting that the lower detection limit for the Project for gasoline contamination may be less than 5,000 ppb Total Volatiles in soil.

DATA COMPARISONS

The Project focused on the use of field instruments to obtain contaminant concentration data quickly and accurately for virtually immediate interpretation of the analytical results. The Project more closely resembled true field operations and situations than otherwise would be encountered in a "research-oriented" controlled atmosphere; however, using soil as the medium of interest may have impacted the ability to compare data obtained from different field and laboratory instruments.

The soil used for the Project was taken to the Project site as it was excavated from a petroleum spill site. The natural soil was nonhomogeneous; therefore, the distribution of contaminants within the soil matrix is assumed to have been nonhomogeneous as well. Even though samples were split in the field upon collection, the analytical results may have reflected different soil types with different affinities for petroleum, and different contaminant concentrations being measured.

Since soil moisture content was not controlled as part of the Project, the presence of water in the collected samples may have affected the amount of vapors generated by the wet soil, and thereby affected the HNU headspace readings. The field extractions, associated with the Quantix kits, and the laboratory extractions may have been affected by the amount of moisture in the analytical samples also. During some sampling events, the collection and handling of soil samples was complicated by the presence of excessive moisture, resulting in "mud" samples, or by the presence of little moisture, resulting in "hardpan" samples.

Time and sample degradation may have affected the comparability of analytical results the most. The HNU and Quantix analyses were performed shortly after the samples were collected, within minutes of one another. The GC analyses were generally performed within 24-48 hours later, and the laboratory analyses were performed within 7-10 days after sample collection. There may have been volatile contaminants lost to the environment, or microbial degradation within the sample containers to contribute to differences in results from the different analytical techniques.

Given the above factors which may have affected the ability to compare results from different analytical methods, there are still some comparisons which are evident from evaluating the Project data.

Other than the fact that the HNU generally provided higher contaminant concentrations than any other method for both fuel oil-contaminated and gasoline-contaminated soils, there was no conspicuous correlation between analytical results from the HNU, the Quantix kits, the gas chromatograph, and the stationary laboratory.

The data were evaluated using curve fitting computer software to "smoothen" and approximate the graphs of the raw data from each analytical method. The selected computer software was capable of approximating raw data with the following mathematical curve functions (PSI, 1994):

1. a straight line fit, in the form $y = a\,x + b$
2. a power function fit, in the form $y = a\,x^b$
3. a mono-exponential function fit, in the form $y = a\,e^{b\,x}$
4. a polynomial function fit, in the form $y = a_n\,x^n + a_{n-1}\,x^{n-1} + ... + a_1\,x + a_0$, where n is the order of the polynomial function.

Gasoline-Contaminated Soil

The raw data graphs were approximated using curve fitting to obtain the most appropriate types of curves to compare between data sets from the different analytical methods. Figure 4 illustrates the raw data from the HNU for gasoline-contaminated soil, and the approximations of the raw data curve with a straight line fit, a power function fit, and a mono-exponential fit.

Figure 5 illustrates the curve fitting for the gasoline-contaminated soil HNU raw data, the 4th-order polynomial fit, the 6th-order polynomial fit, and the

10th-order polynomial fit. Based on visual comparison of the HNU raw data and the six possible approximation curves, the straight line and 6th-order polynomial curves were selected to approximate and compare the gasoline-contaminated soil data from the field instruments and the stationary laboratory.

The raw data from the Quantix kit for the gasoline-contaminated soil is presented in Figure 6, along with the straight line fit and the 6th-order polynomial curve fit of the data.

The raw data from the GC for the gasoline-contaminated soil is shown in Figure 7. The GC Total BTEX data curve was selected for comparison with the

Figure 4. **Gasoline-contaminated soil HNU raw data and curve fitting.**

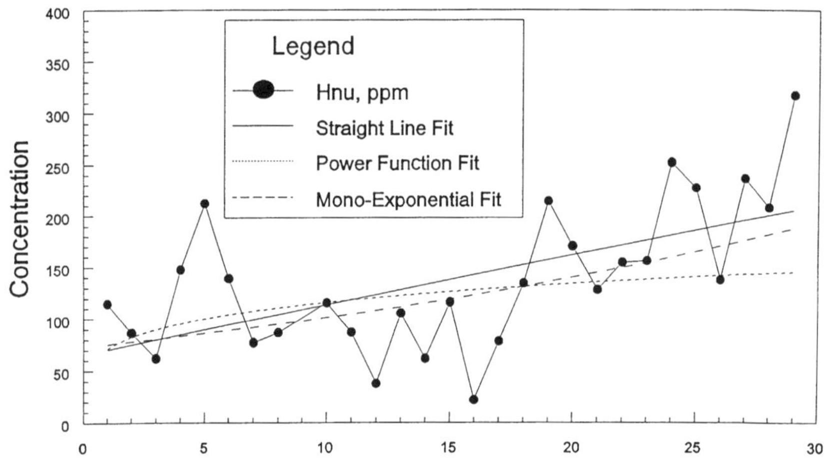

Figure 5. **Gasoline-contaminated soil HNU raw data and polynomial curve fitting.**

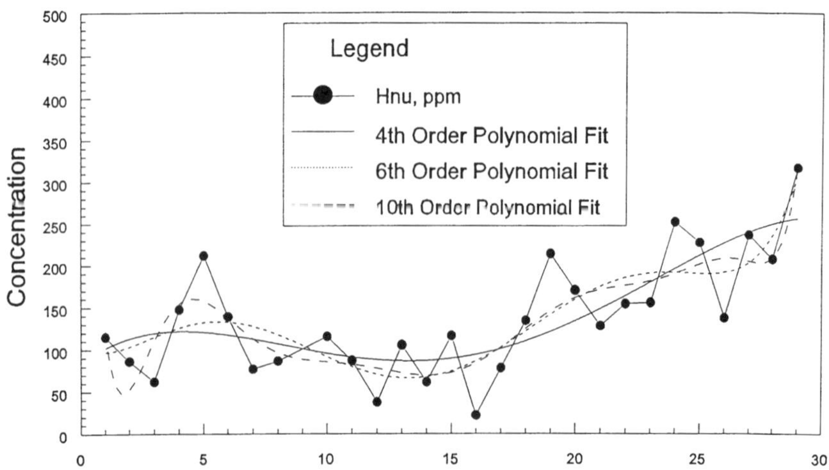

results from the other analytical methods of the Project, since the Quantix kit is designed to measure Total BTEX, and since the GC Total BTEX curve is very similar to the GC xylene and GC Total Volatiles curves.

The GC BTEX raw data, the straight line fit and the 6th-order polynomial fit are illustrated in Figure 8.

The 6th-order polynomial curves for the HNU, Quantix and GC BTEX results have been combined in Figure 9. The appropriate units are noted since the HNU and Quantix results are represented in parts per million (ppmv and

Figure 6. **Gasoline-contaminated soil Quantix BTEX raw data and curve fitting.**

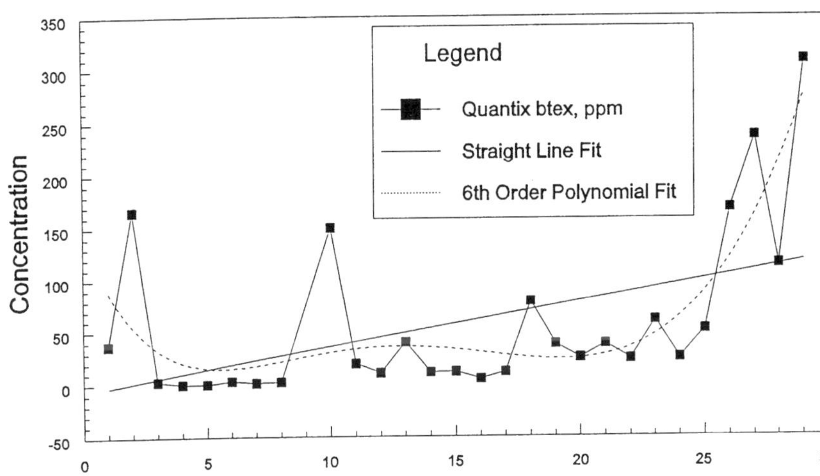

Figure 7. **Gasoline-contaminated soil GC raw data.**

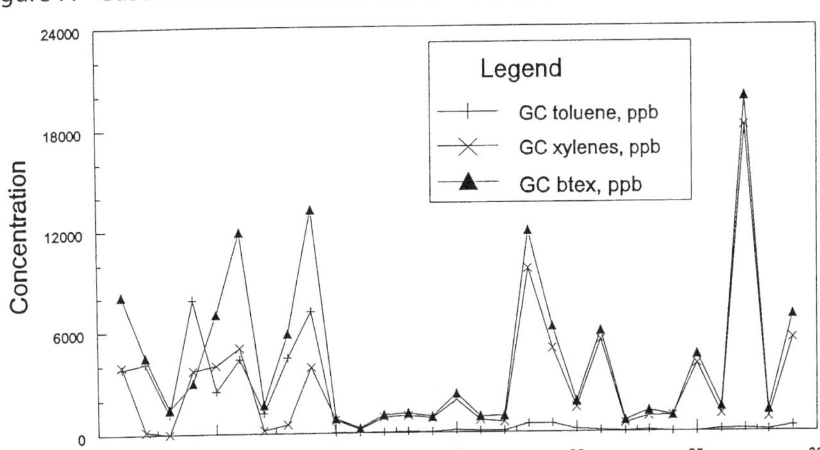

mg/kg, respectively), while the GC BTEX results are represented in parts per billion (μg/kg).

The scale used in Figure 9 and the mixture of ppm and ppb units illustrate an equivalent trend between the HNU and the Quantix 6th-order polynomial fit curves, as they are aligned essentially one on top of the other. A change in units and/or a change in scale might provide more information.

Figure 10 illustrates the 6th-order polynomial fit for the HNU, Quantix, and GC BTEX results using ppb units for all three field analyses. The overall data trends between the HNU and the Quantix kit are similar, but the HNU curve

Figure 8. **Gasoline-contaminated soil GC BTEX raw data and curve fitting.**

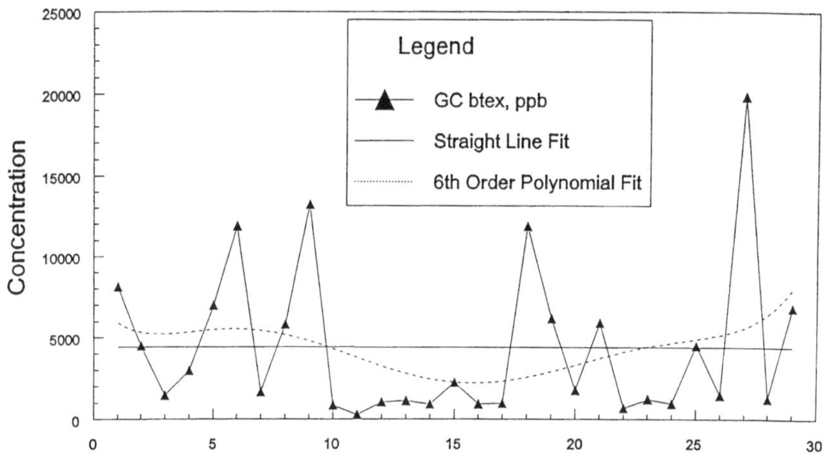

Figure 9. **Gasoline-contaminated soil field instrument data 6th-order polynomial fit curves.**

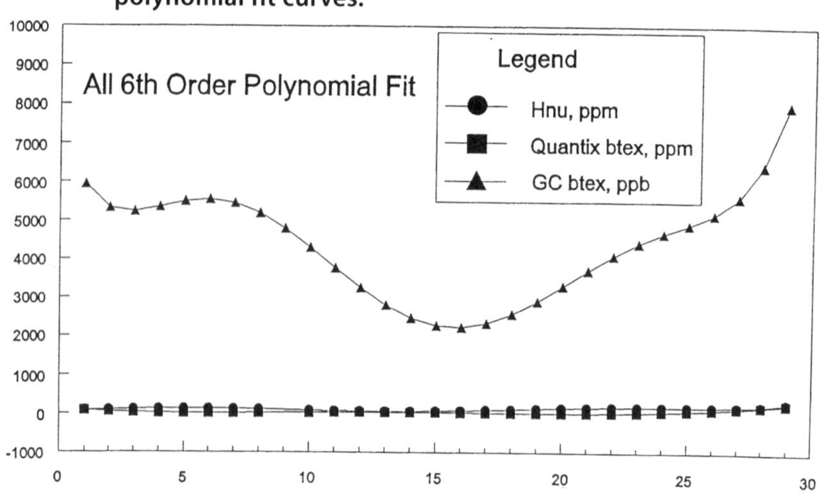

suggests a more oscillating nature. The HNU data fluctuations may be a factor of the instrument's sensitivity to ambient conditions.

The stationary laboratory raw data for Total BTEX, Total Volatiles, and Xylene for gasoline-contaminated soil is presented in Figure 11. The laboratory BTEX graph was selected for comparison with the data collected from the HNU, Quantix, and GC methods, because the BTEX data graph reflects the same trends as the Total Volatiles and Xylenes, as illustrated by similar shapes, and because the Quantix and GC methods measured BTEX in the soil samples.

Figure 10 .**Gasoline-contaminated soil field instrument data 6th-order polynomial fit curves.**

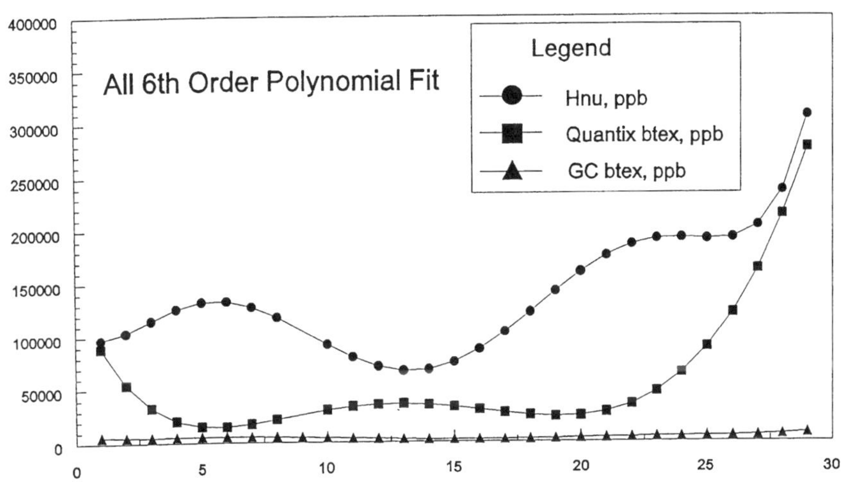

Figure 11. **Gasoline contaminated soil stationary laboratory raw data.**

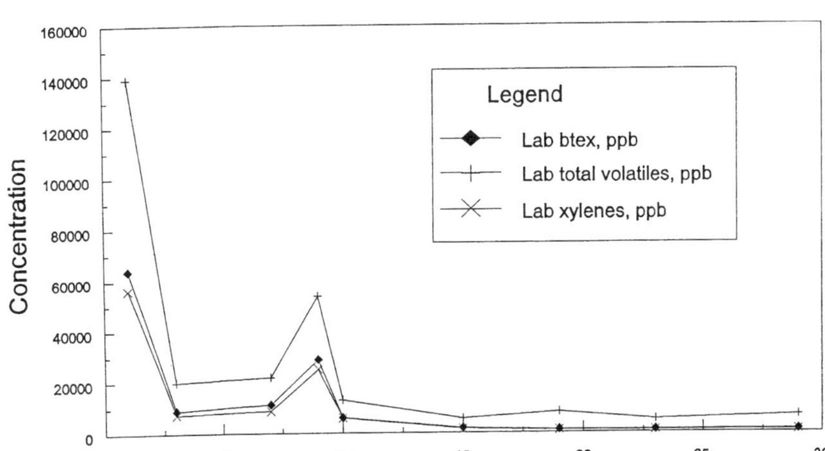

The laboratory BTEX graph, the straight line fit, and the 6th-order polynomial fit of the data are illustrated in Figure 12. The laboratory BTEX data is not amenable to a 6th-order polynomial curve, as illustrated by the valley in the 6th-order fit curve between sampling events 23 and 29. Additional laboratory data points would probably provide a database more amenable to polynomial curve fitting.

Figure 13 illustrates the straight line fit curves for the HNU, Quantix, GC, and laboratory data on gasoline-contaminated soil. The HNU and Quantix straight line fit curves are parallel lines, indicating equal data trends, while the GC and laboratory lines cross and diverge.

Figure 12. **Gasoline-contaminated soil laboratory BTEX raw data and curve fitting.**

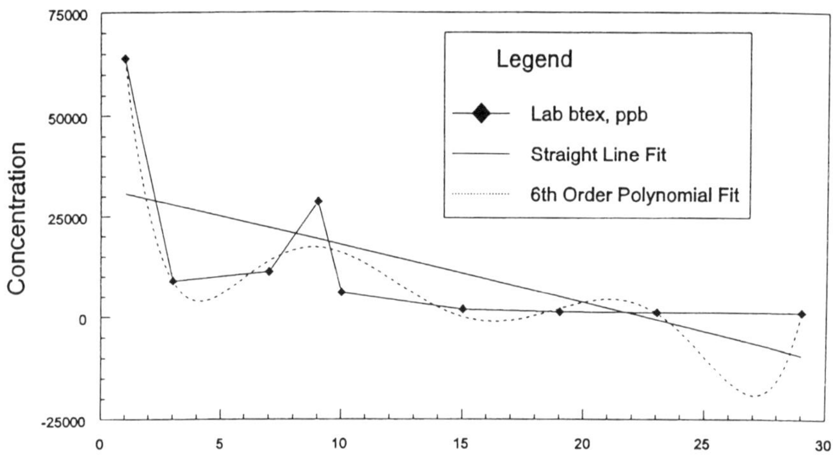

Figure 13. **Gasoline-contaminated soil straight line fit data.**

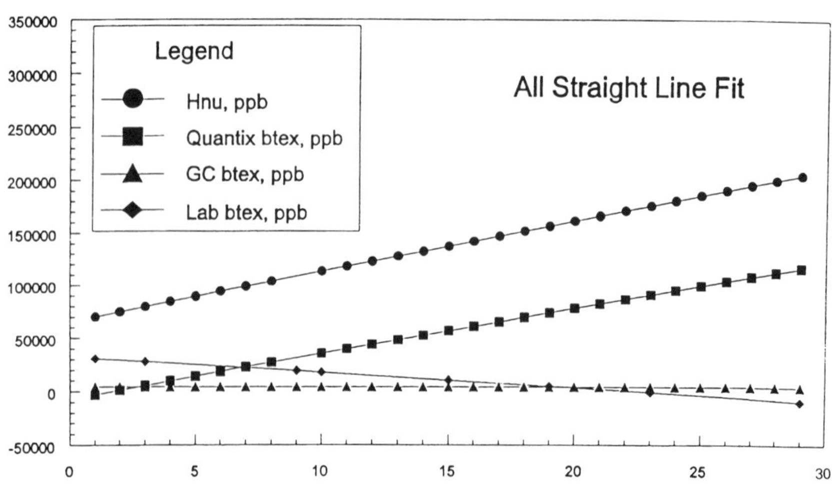

Figure 14 contains 6th-order polynomial curves fit to the HNU, Quantix, GC BTEX, and laboratory BTEX data for gasoline-contaminated soil. The mixing of units results in the overlap of the HNU and Quantix curves at the zero line. The GC curve is nearly horizontal, and is similar to the HNU and Quantix curves, but the scale and units selected for the figure limit the possible data comparisons.

All of the analytical results, as approximated by 6th-order polynomials, are illustrated in ppb units in Figure 15. The Quantix and laboratory curves are similar for sampling events 1-3, while the GC and laboratory curves converge to the zero line by sampling event 15. The HNU and Quantix fit curves plot out with

Figure 14. **Gasoline-contaminated soil 6th-order polynomial fit data.**

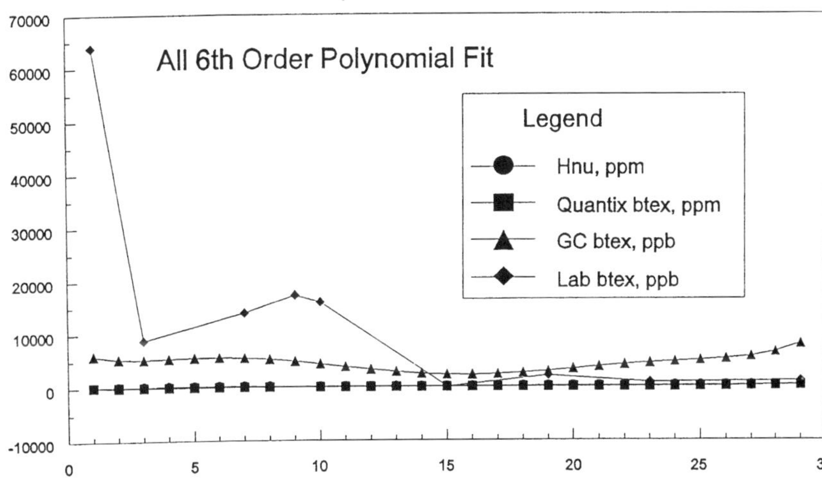

Figure 15. **Gasoline-contaminated soil 6th-order polynomial fit data.**

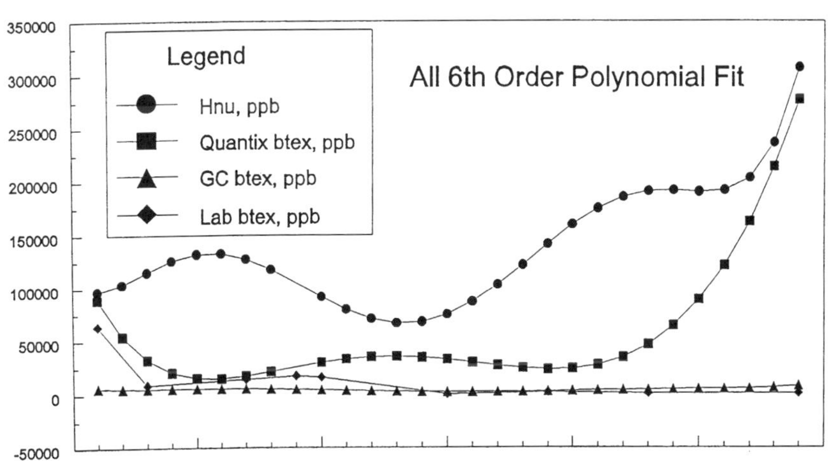

both curves starting nearly at the same point for sampling event 1 and ending at nearly the same point for sampling events 28 and 29. For sampling events 2-26, the HNU and Quantix curves are nearly mirror images of each other.

The common logarithm (LOG) of the analytical data collected from the gasoline-contaminated soil sampling events are presented in Figure 16. The LOG HNU curve and the LOG GC curve are saw-tooth curves, while the LOG Quantix curve has a larger range of data including two "Nondetectable" results at sampling events 4 and 5. The LOG laboratory curve is gently sloping to horizontal. Additional laboratory data points may have produced more of a sawtooth pattern.

Figure 16. **Common logarithm of the gasoline-contaminated soil raw data.**

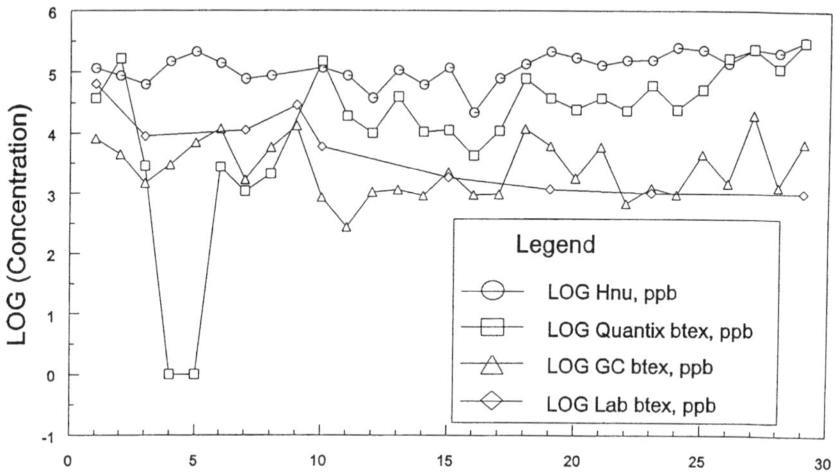

Figure 17. **Common logarithm and curve fitting of the gasoline-contaminated soil HNU raw data.**

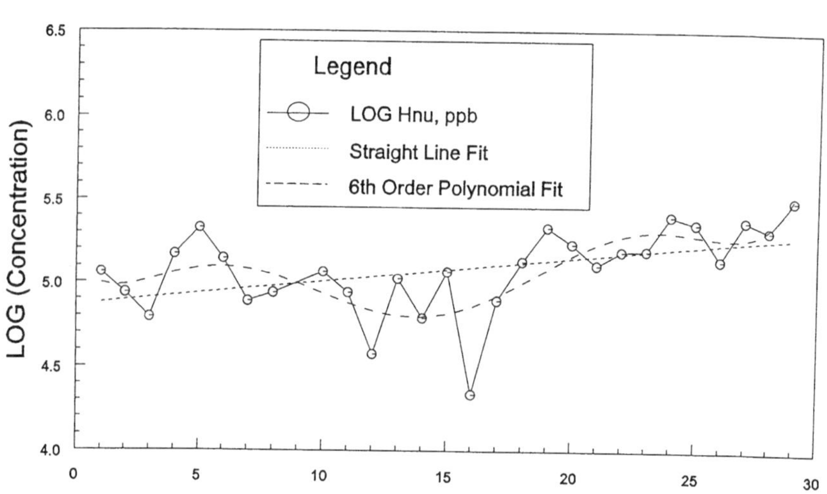

Figures 17, 18, 19, and 20 illustrate straight line and 6th-order polynomial curve fitting of the LOG data for the HNU, Quantix, GC, and laboratory data, respectively.

The straight line fit curves for the gasoline-contaminated soil LOG data are graphed in Figure 21, using ppb units on the raw data for all of the analytical methods. The LOG of the HNU and GC BTEX results, fit as a straight line, are parallel lines, indicating equal data trends, while the LOG of the Quantix and laboratory lines cross and diverge.

The 6th-order polynomial fit curves for the gasoline-contaminated soil LOG data are graphed in Figure 22, using ppb units on the raw data for all of the

Figure 18. **Common logarithm and curve fitting of the gasoline-contaminated soil Quantix raw data.**

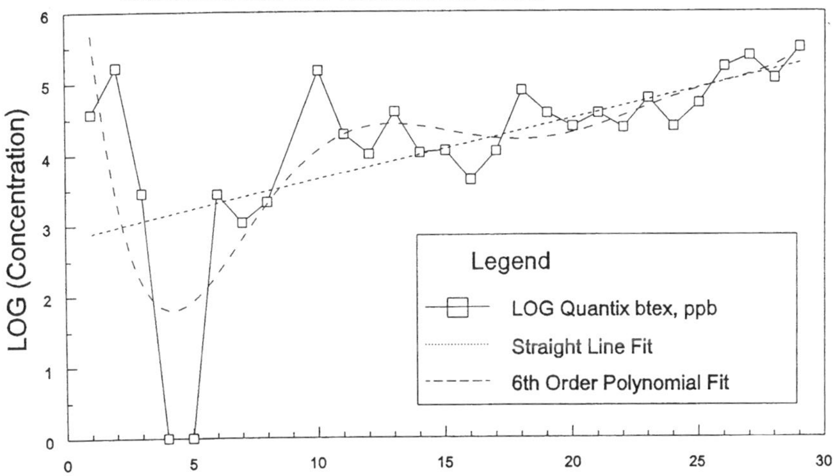

Figure 19. **Common logarithm and curve fitting of the gasoline-contaminated soil GC BTEX raw data.**

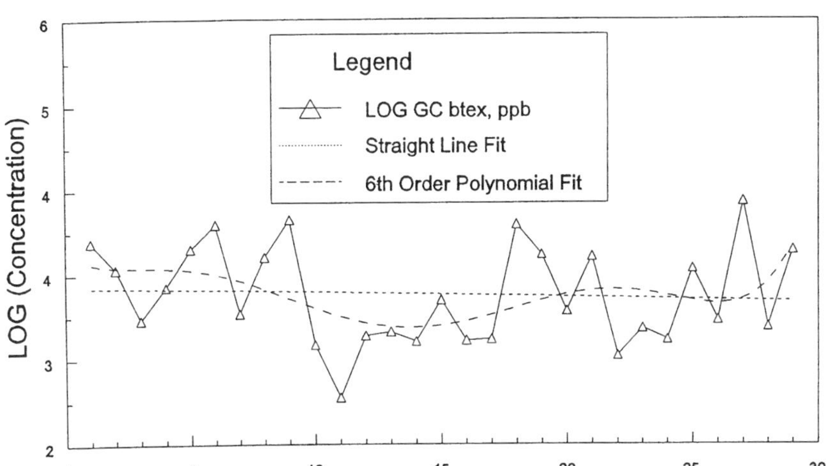

analytical methods. The LOG of the HNU and GC BTEX results, fit as a 6th-order polynomial function, are parallel curves, between sampling events 1 and 27, indicating equivalent data trends. The shape of the LOG Quantix data curve is distorted by the two "Nondetectable" data points for sampling events 4 and 5. The laboratory curve does not compare favorably to any of the other curves; however, if additional data points had been collected for laboratory analyses, more reliable comparisons would be possible.

Figure 20. **Common logarithm and curve fitting of the gasoline-contaminated soil laboratory raw data.**

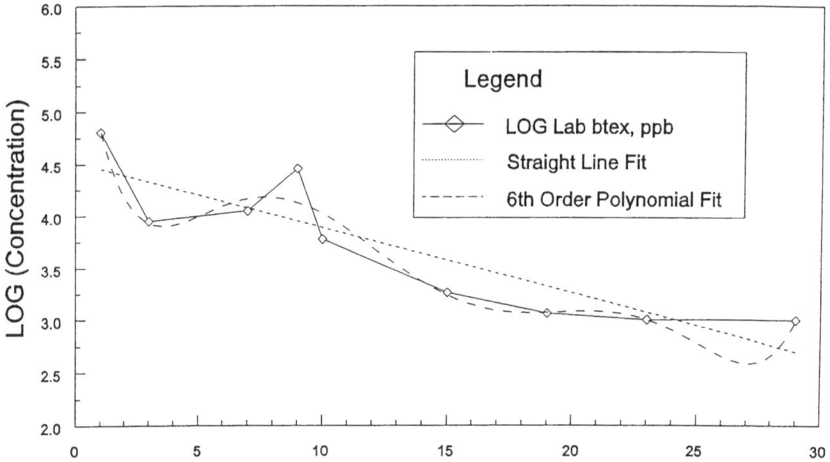

Figure 21. **Common logarithm straight line fitting of the gasoline-contaminated soil data.**

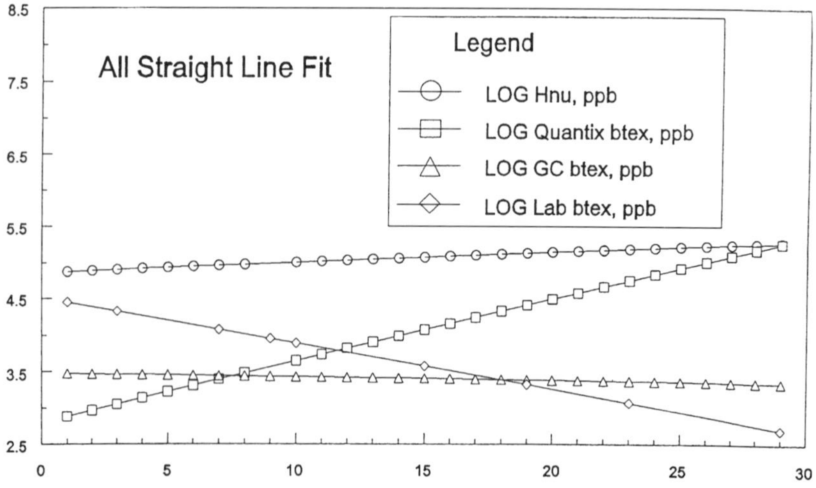

Figure 22. **Common logarithm 6th-order polynomial fitting of the gasoline-contaminated soil data.**

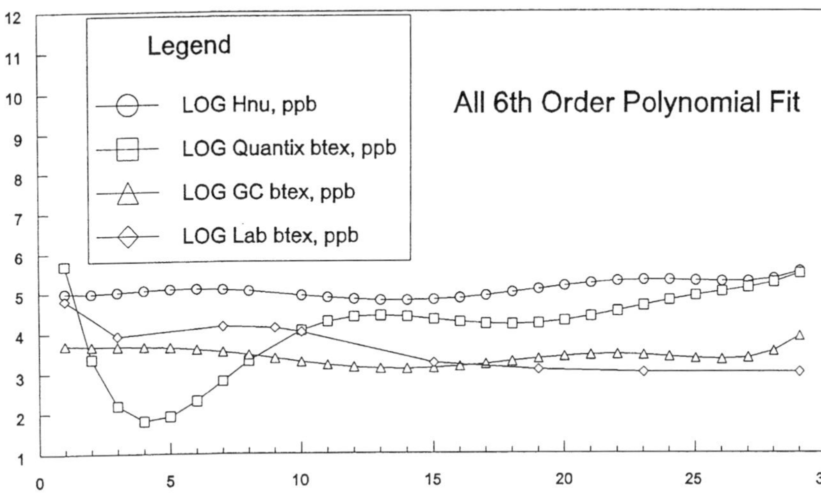

Fuel Oil-Contaminated Soil

The raw data curves for the fuel oil-contaminated soil from the HNU, the Quantix, and the GC methods are provided in Figures 23 and 24. Figure 25 identifies the laboratory data available for comparisons. Both laboratory BTEX and laboratory PAH have been included in the data comparisons, since the GC BTEX data and the Quantix PAH data are included.

The HNU and Quantix raw data are generally higher than the GC BTEX data. The HNU measured detectable levels of contamination on every sampling event during the Project, which had not been expected for fuel oil contamination.

Based on visual comparison, the straight line, power function, and 4th-order polynomial fit curves were selected to approximate and compare the fuel oil-contaminated soil data from the field instruments and the stationary laboratory.

Figures 26, 27, 28, 29, and 30 illustrate the raw data curves, and fit curves for the HNU, Quantix, GC BTEX, laboratory BTEX, and laboratory PAH, respectively.

The straight line curves for the Project analytical raw data are shown in Figures 31 and 32. The HNU and GC linear fit curves demonstrate similar data trends as parallel horizontal lines. Note, however, that the HNU and Quantix curves are based on raw data in ppm units, while the GC and laboratory data points are in ppb units. The laboratory BTEX straight line fit in Figure 31 is identical to the laboratory PAH straight line fit in Figure 32, indicating that the laboratory BTEX and laboratory PAH results have the same trend.

The 4th-order polynomial curves for the fuel oil-contaminated soil are shown in Figures 33, 34, and 35. In Figure 33, the HNU and Quantix fit curves

Figure 23. **Fuel oil-contaminated soil field instrument raw data.**

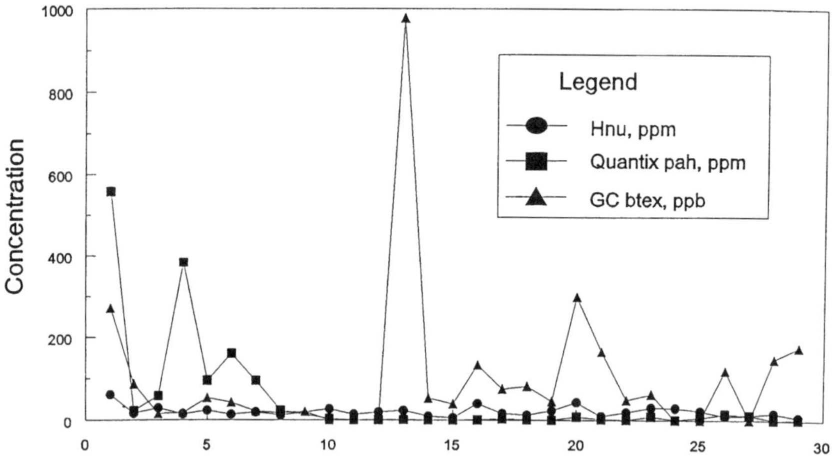

Figure 24. **Fuel oil-contaminated soil field instrument raw data.**

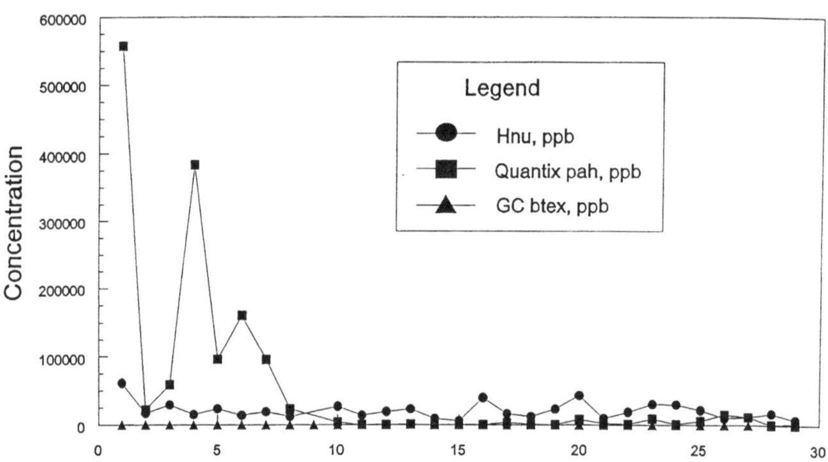

Figure 25. **Fuel oil-contaminated soil stationary laboratory raw data.**

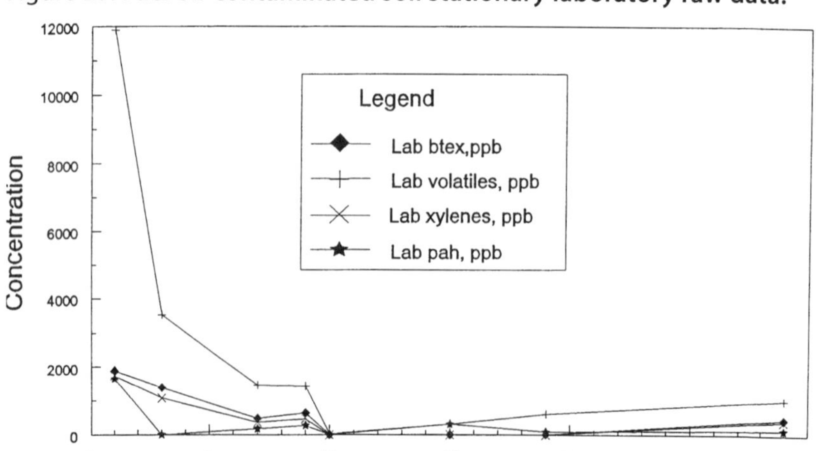

Figure 26. **Fuel oil-contaminated soil HNU raw data and curve fitting.**

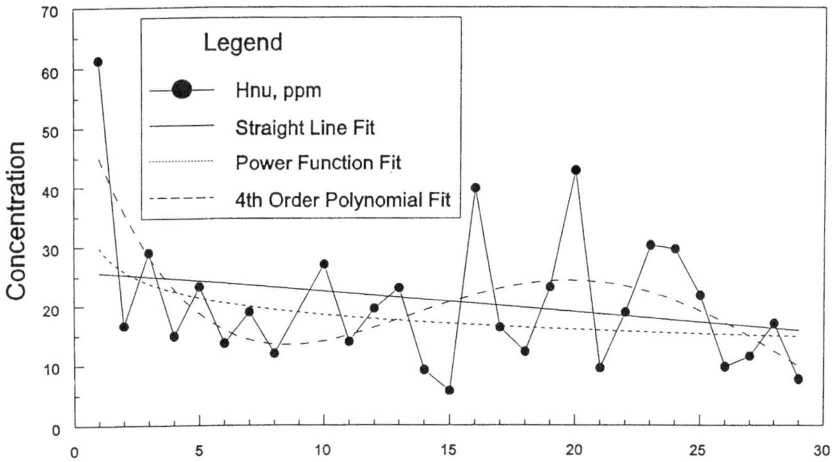

Figure 27. **Fuel oil-contaminated soil Quantix PAH raw data and curve fitting.**

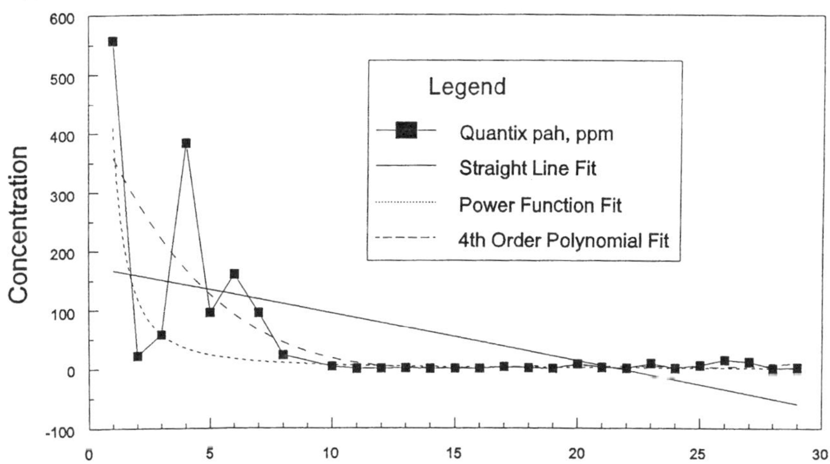

Figure 28. **Fuel oil-contaminated soil GC BTEX raw data and curve fitting.**

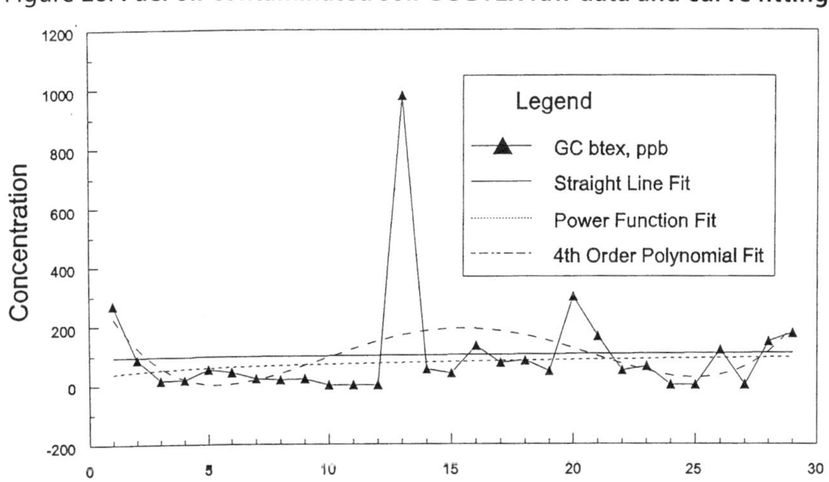

exhibit equal trends between sampling events 7 and 29, while the Quantix and GC fit curves exhibit equal trends between sampling events 1 and 5. The scale needed to accommodate the laboratory BTEX data crunches the other three curves on top of each other along the zero line. Although additional laboratory data would probably facilitate better data comparisons, the laboratory BTEX fit curve resembles the Quantix and GC curves between sampling events 1 and 7. Figure 34 shows the same data as Figure 33, without the laboratory curve, to provide better resolution of the HNU, Quantix, and GC fit curves. The GC BTEX and Quantix data, as fit with a 4th-order polynomial, illustrate similar data trends for sampling events 1-4. The HNU and Quantix fit curves are parallel between sampling events 13 and 25, and coincide on sampling events 10, 11, 28, and 29.

Figure 29. **Fuel oil-contaminated soil laboratory BTEX raw data and curve fitting.**

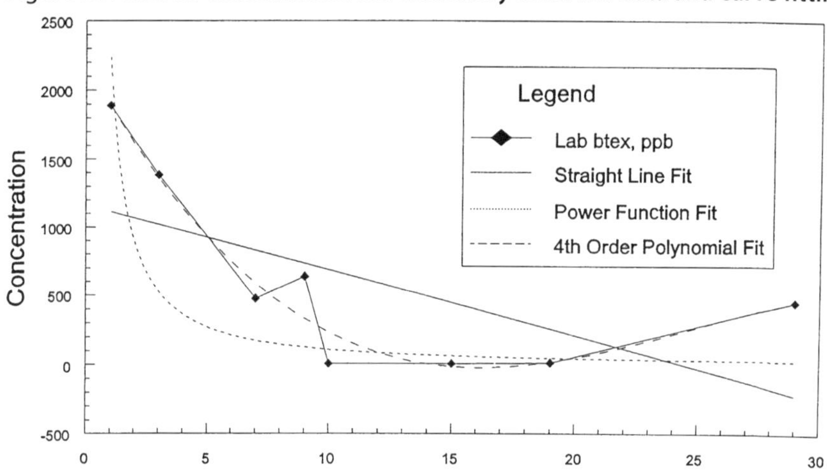

Figure 30. **Fuel oil-contaminated soil laboratory PAH raw data and curve fitting.**

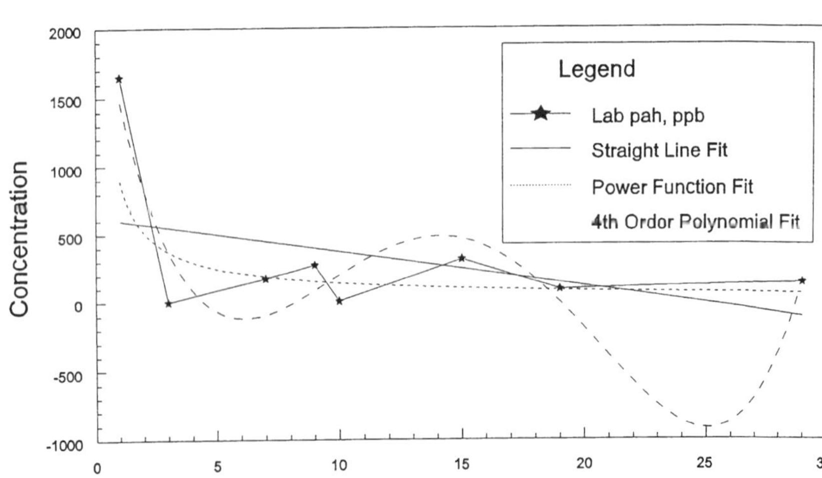

The graphical scales and units complicate a more direct comparison of the data at this time; however, the laboratory BTEX curve in Figure 33 resembles the overall trends illustrated in Figure 34 for the GC BTEX and Quantix 4th-order polynomial fit curves.

Figure 33 includes the laboratory PAH data approximated by a 4th-order polynomial. Again, the scale on the y-axis compresses the HNU, Quantix, and GC curves. Additional laboratory data points might provide better data comparisons; however, the shape of the laboratory PAH 4th-order approximation exhibits some resemblance to the GC BTEX data curve.

Figure 31. **Fuel oil-contaminated soil raw data straight line fit data.**

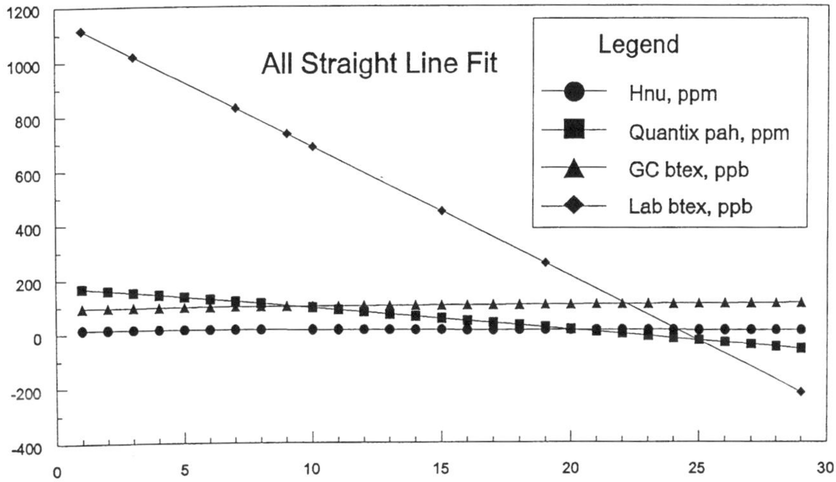

Figure 32. **Fuel oil-contaminated soil raw data straight line fit data.**

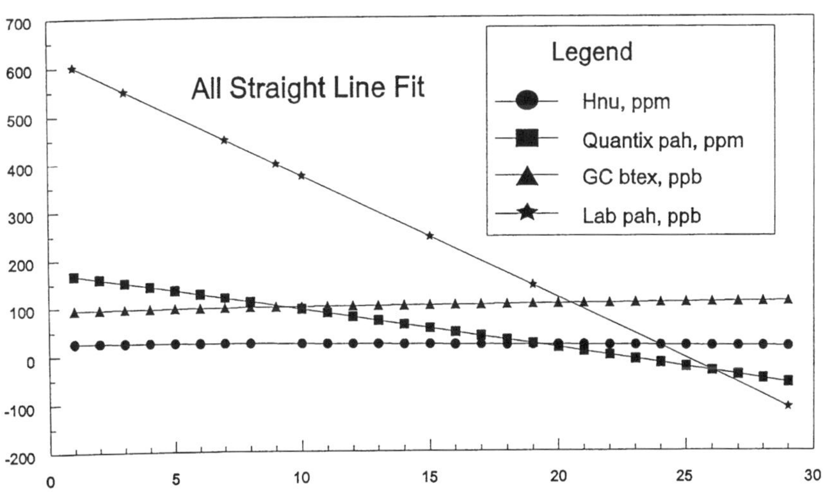

Fitting the fuel oil-contaminated soil raw data to a power function, in the form of $y = ax^b$, provides additional data comparisons. In Figure 36, all of the curves become parallel at higher sampling event numbers, corresponding to lower contaminant concentrations, and the characteristics of a power function curve. Figure 37 demonstrates a clearer view of common curve shapes and similar data trends for Quantix and laboratory BTEX data.

Figure 33. **Fuel oil-contaminated soil raw data 4th-order polynomial fit curves.**

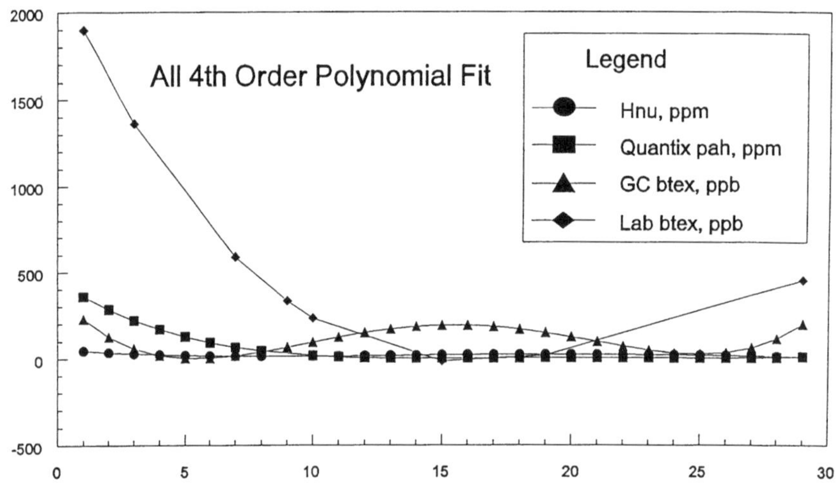

Figure 34. **Fuel oil-contaminated soil raw data 4th-order polynomial fit curves.**

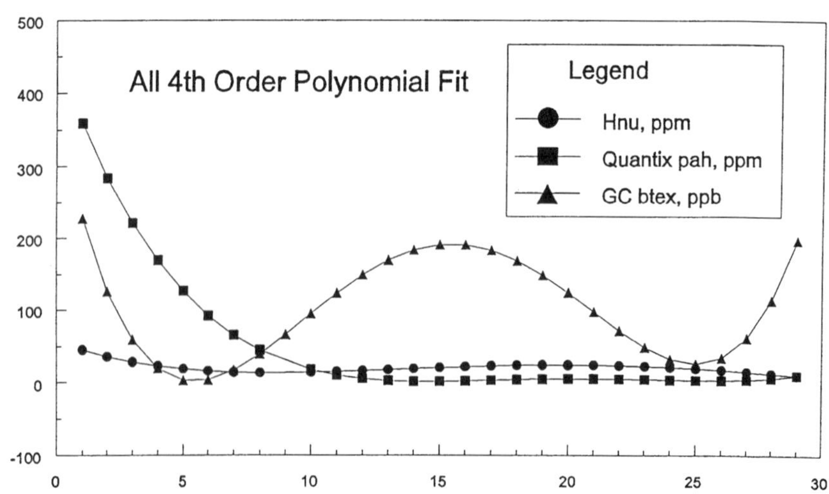

The laboratory PAH data as a power function fit curve is included with the HNU, Quantix, and GC data in Figure 38. The laboratory PAH approximation is similar to the Quantix curve, but additional data would enable better comparisons.

The fuel oil-contaminated soil data was evaluated using the common logarithm (LOG), as was the gasoline-contaminated soil data. Figures 39 and 40 contain the LOG of the HNU, Quantix, GC, and laboratory data. The LOG labo-

Figure 35. **Fuel oil-contaminated soil raw data 4th-order polynomial fit curves.**

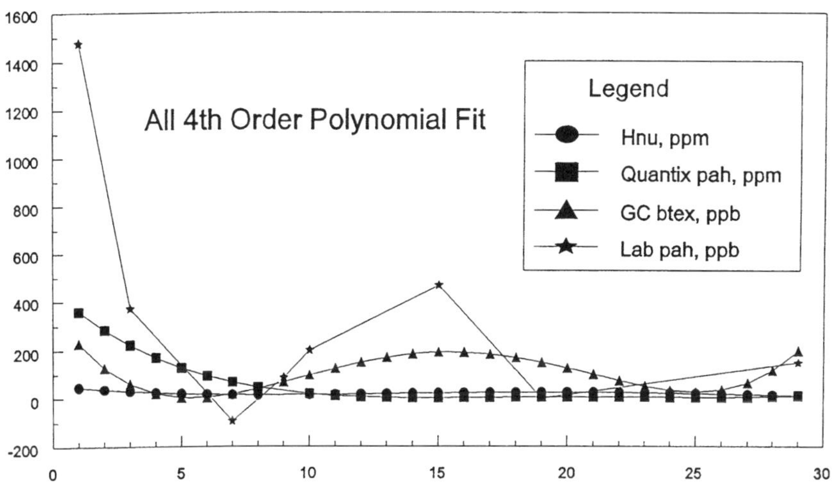

Figure 36. **Fuel oil-contaminated soil raw data power function fit curves.**

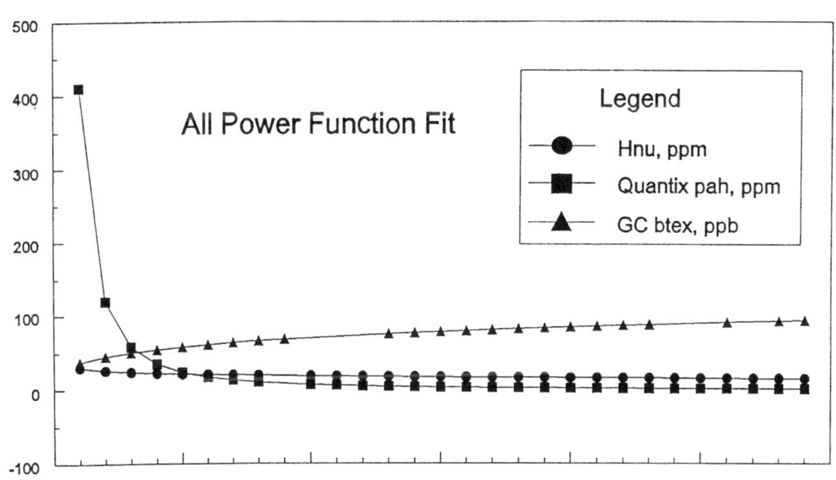

ratory BTEX curve is somewhat similar in shape to the LOG GC BTEX curve, even though the LOG GC BTEX curve is distorted due to "Nondetectable" results for sampling events 10, 11, 12, 24, 25, and 27. More laboratory BTEX data points would be helpful in comparing data curves, especially in sampling event areas 10-15 and 19-29. At several sampling points, the LOG HNU curve intersects the LOG Quantix curve. The LOG laboratory PAH curve is similar to the GC BTEX curve in shape, and coincides on some sampling events. Additional laboratory PAH data would be helpful, especially after sampling event 15.

Figure 37. **Fuel oil-contaminated soil raw data power function fit curves.**

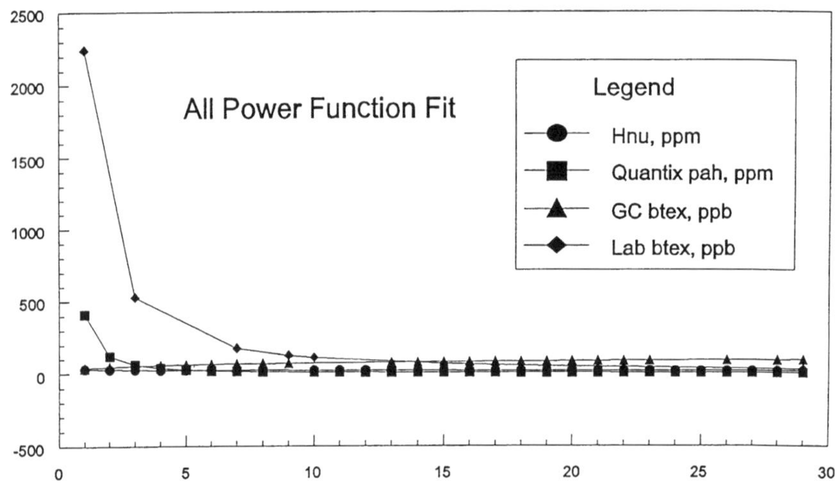

Figure 38. **Fuel oil-contaminated soil raw data power function fit curves.**

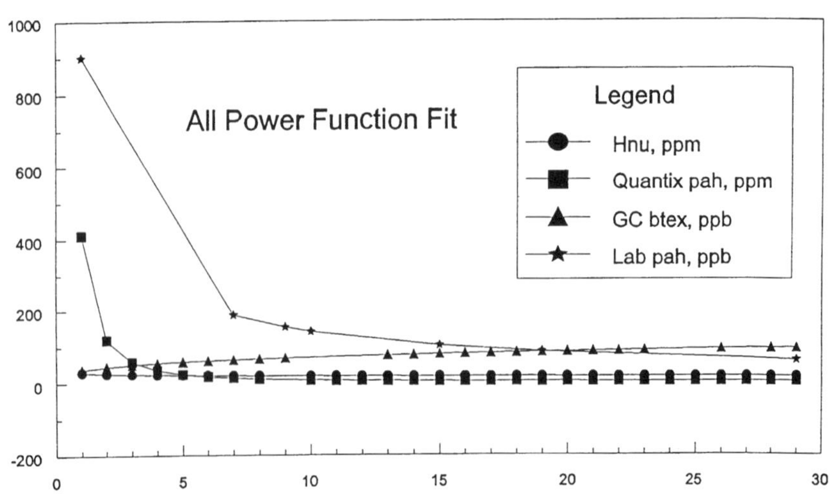

Figures 41, 42, 43, 44, and 45 illustrate the LOG data and curve fitting for the HNU, Quantix, GC BTEX, laboratory BTEX, and laboratory PAH, respectively. The LOG laboratory PAH data are visually not amenable to a 4th-order polynomial curve, as evident from the valley of the curve between sampling events 19 and 29.

Figure 39. **Fuel oil-contaminated soil common logarithm data curves.**

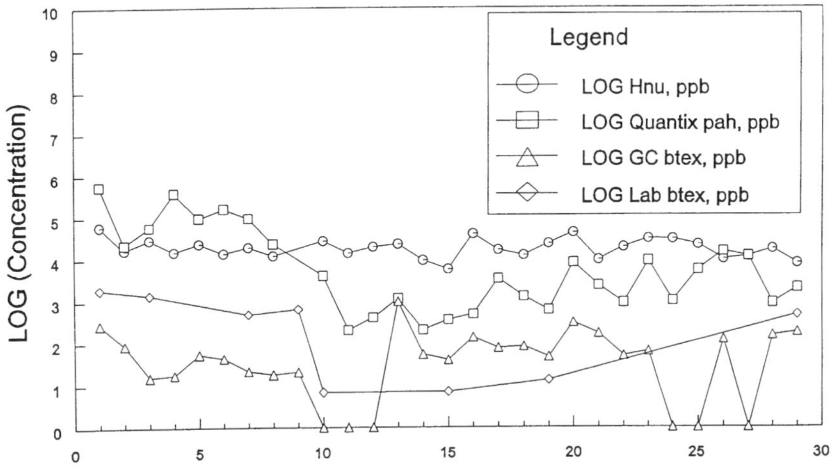

Figure 40. **Fuel oil-contaminated soil common logarithm data curves.**

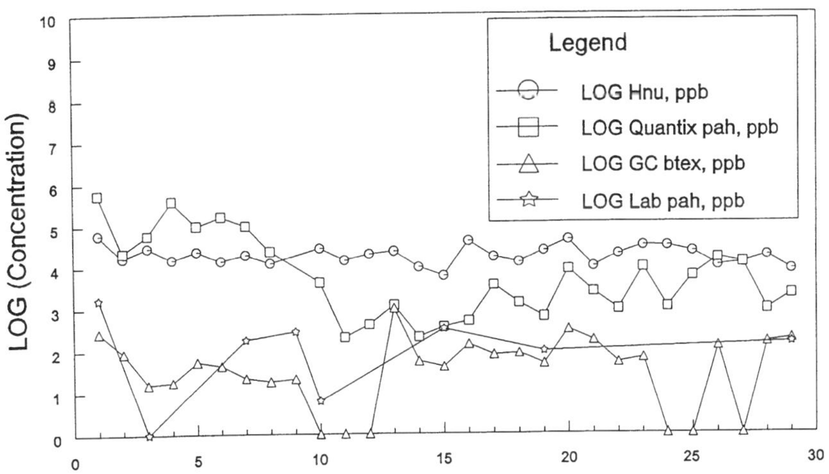

The straight line curves for the LOG data are shown in Figures 46 and 47. The LOG GC BTEX data curve and the LOG HNU data curve are parallel, indicating equivalent data trends, and the LOG Quantix data trend is equal to the LOG laboratory BTEX trend. In Figure 47, the LOG Quantix line and the LOG laboratory PAH line are converging as the sampling event number increases.

Figures 48 and 49 contain the 4th-order polynomial fit of the LOG data for fuel oil-contaminated soil. The LOG laboratory BTEX data curve in Figure 48

Figure 41. **Fuel oil-contaminated soil common logarithm HNU data and curve fitting.**

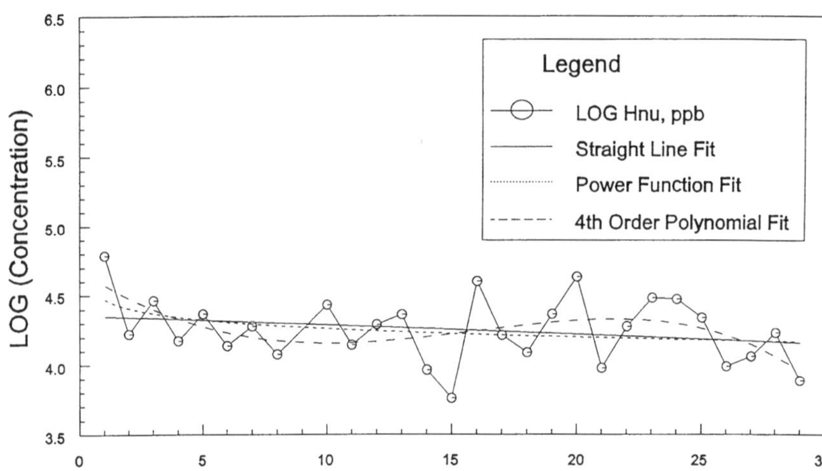

Figure 42. **Fuel oil-contaminated soil common logarithm Quantix PAH data and curve fitting.**

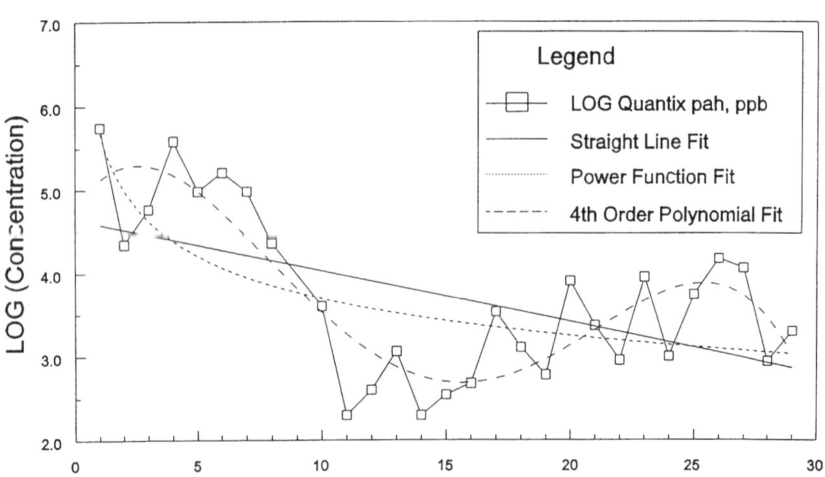

Figure 43. **Fuel oil-contaminated soil common logarithm GC data and curve fitting.**

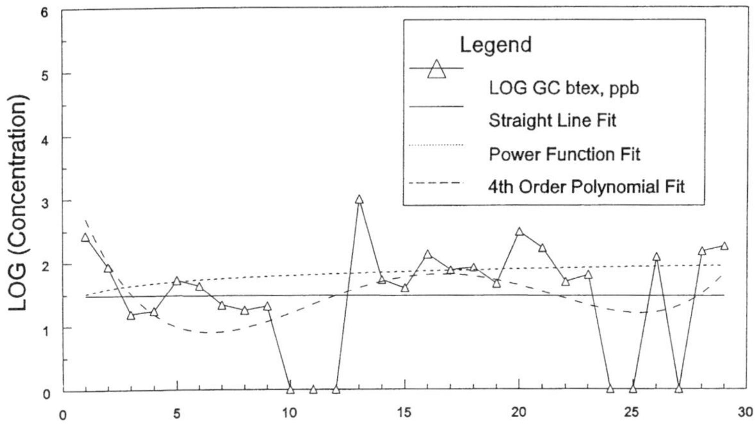

Figure 44. **Fuel oil-contaminated soil common logarithm laboratory BTEX data and curve fitting.**

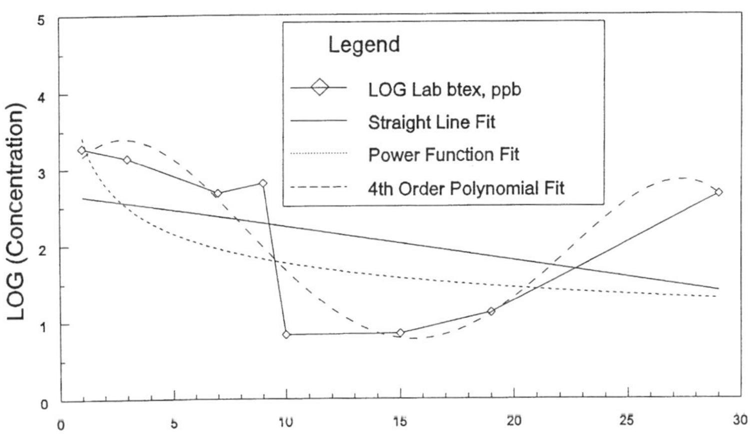

Figure 45. **Fuel oil-contaminated soil common logarithm laboratory PAH data and curve fitting.**

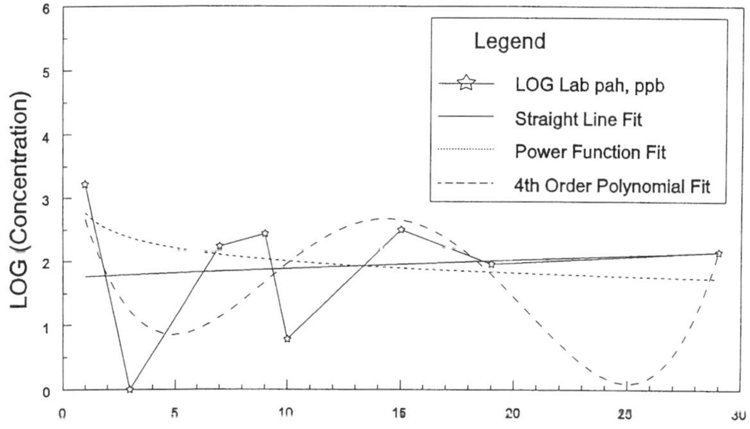

resembles the LOG GC BTEX data curve, but additional laboratory data points, especially between sampling events 19 and 29, would provide potentially a more favorable comparison between these two data trends. The LOG laboratory PAH data curve in Figure 49 resembles the LOG GC BTEX curve, but additional laboratory data points, especially between sampling events 19 and 29, would help in the curve comparison. The LOG Quantix curve and the LOG GC BTEX curve are nearly mirror images of each other.

The power function fit curves for the LOG data from fuel oil-contaminated soil are contained in Figures 50 and 51. The curves in Figure 50 suggest equiva-

Figure 46. **Fuel oil-contaminated soil common logarithm straight line fit data.**

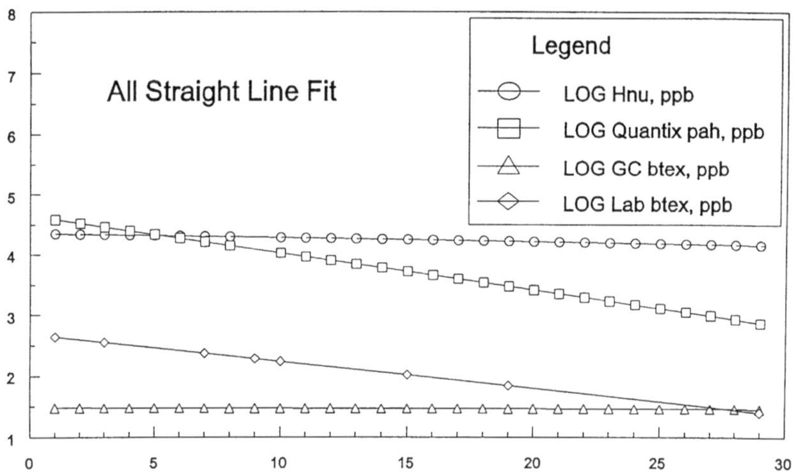

Figure 47. **Fuel oil-contaminated soil common logarithm straight line fit data.**

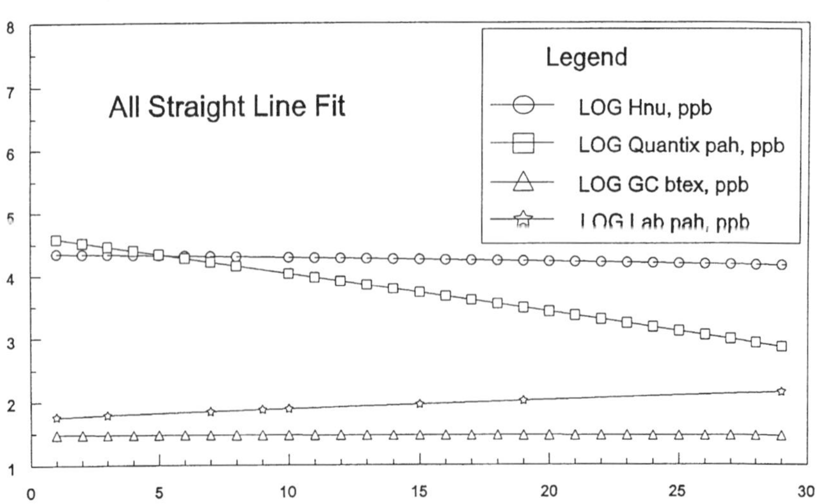

lent data trends for the LOG Quantix data and the LOG laboratory BTEX data. All four curves become parallel at higher sampling event numbers, corresponding to lower contaminant concentrations, and due to the characteristics of power function curves. Figure 51 illustrates equivalent data trends for the LOG Quantix data and the LOG laboratory PAH data. Again, all four curves become parallel at higher sampling event numbers, corresponding to lower contaminant concentrations, and due to the characteristics of power function curves.

Figure 48. **Fuel oil-contaminated soil common logarithm 4th-order polynomial fit curves.**

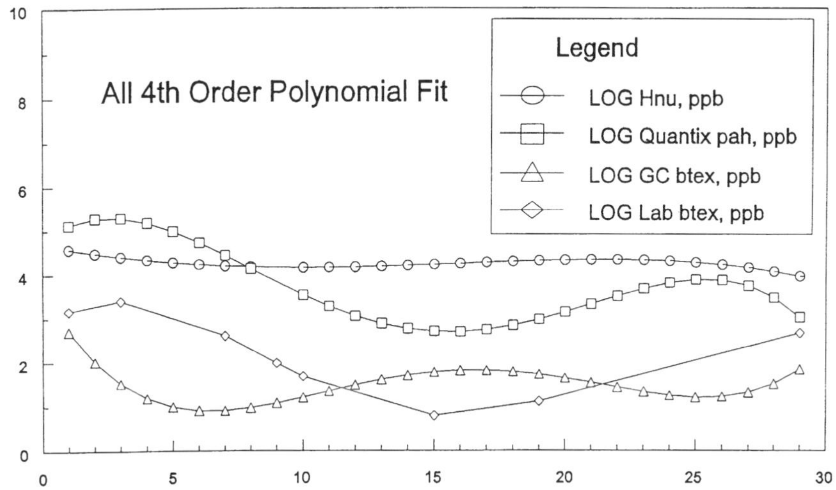

Figure 49. **Fuel oil-contaminated soil common logarithm 4th-order polynomial fit curves.**

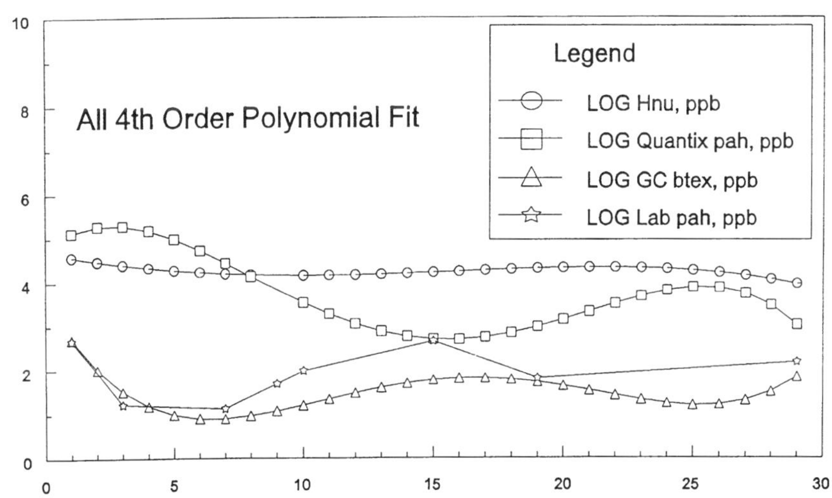

Figure 50. **Fuel oil-contaminated soil common logarithm power function fit curves.**

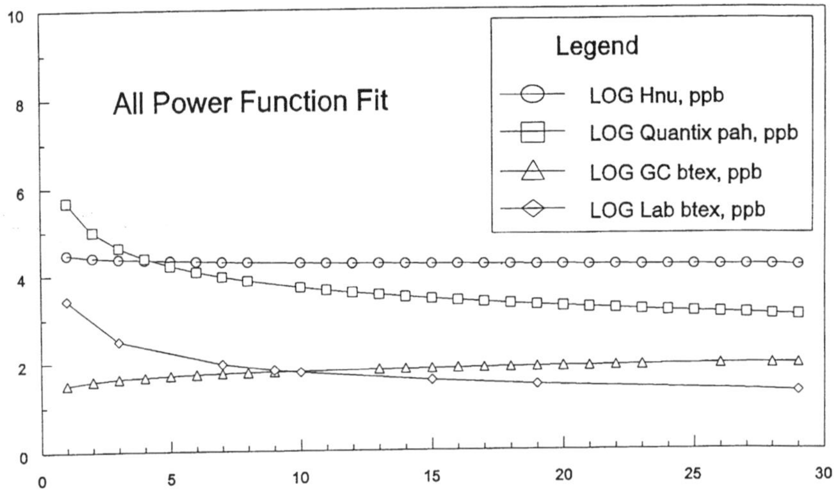

Figure 51. **Fuel oil-contaminated soil common logarithm power function fit curves.**

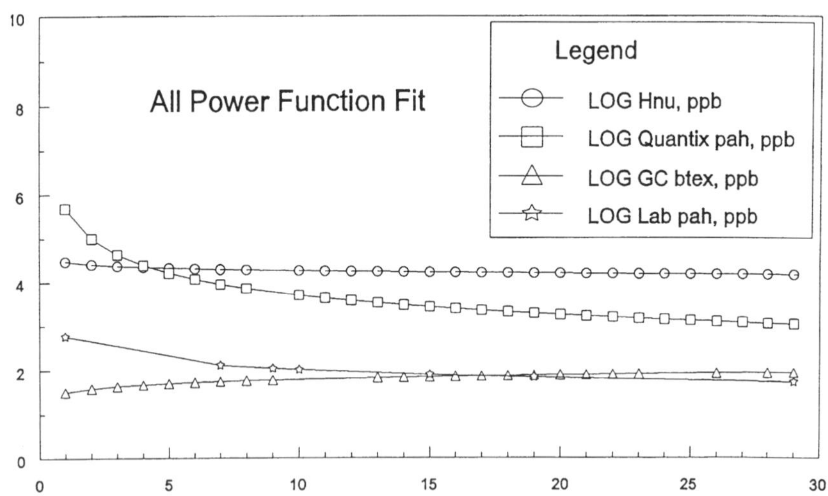

COST

The costs associated with the Project are detailed in Table 4. The analytical work from September 1993 to October 1993 totaled $9,997, while the 1994 work totaled $21,494. The Project laboratory costs would have been higher if the Project had not been combined with a bioremediation field project. The bioremediation field project accounted for laboratory results from three of the eight sampling events of this field instrument evaluation project. Furthermore, additional costs would have been incurred without the cooperation of the contractors and equipment companies involved in the Project.

Table 4. **Project cost summary.**

Field Instrument Evaluation Project Cost Summary			
Analyses	Number of Sampling Events		Sub-Total Cost
	1993	1994	
HNU	8	20	$0.00
Quantix	8	20	$9,714.16
Portable GC	9	0	$2,700.00
Lab-Grade GC	0	20	$10,452.00
Lab 8021 and 8270	2	3	$8625.00

CONCLUSIONS

Upon evaluating the results of measuring petroleum contamination using an immunoassay kit, a PID, a GC, and stationary laboratory, the immunoassay kit and PID provided higher contaminant concentration readings than the GC or the stationary laboratory, for both fuel oil-contaminated and gasoline-contaminated soils. During the Project, the PID indicated contaminant levels present in the sample headspace for all sampling events, for both fuel oil-contaminated and gasoline-contaminated soils. No direct correlations of the field and laboratory analytical results were discovered; however, the data trends for the analytical results did show some evidence of comparability.

Gasoline-Contaminated Soil

Approximations of the raw data using curve-fitting computer software illustrated equal data trends for gasoline-contaminated soil analyzed with the PID and the immunoassay kit.

Curve-fitting for the LOG of the data demonstrated equal data trends between the PID and GC BTEX results.

Fuel Oil-Contaminated Soil

An approximation of the raw data from the fuel oil-contaminated soil suggested identical trends between the PID and the GC BTEX data. The laboratory BTEX data and the laboratory PAH data also exhibited equal data trends. The laboratory PAH data, with a smaller number of data points, showed similarities to the immunoassay PAH data presented as well.

The LOG graphs of the data, as approximated through curve fitting, illustrated identical trends between the PID and the GC BTEX results. The LOG immunoassay kit curve was identical in shape to the LOG laboratory BTEX curve, demonstrating equal data trends. The LOG laboratory PAH curves were similar to the LOG GC BTEX curves, despite the limited amount of laboratory data available. Further comparisons identified equal data trends for the LOG immunoassay kit data and the LOG laboratory PAH data.

Based on this Project and the limited data set established within this Project, field instruments can be useful in obtaining contaminant information onsite, within minutes of collecting soil samples. However, the reliability and comparability of the field analyses may be affected by the soil sampling procedures, the ambient weather conditions, the types of soil being sampled, and the contaminants present in the soil being sampled. Since each field instrument has its limitations, the usefulness of field-generated data depends on regulatory approval, instrument reliability, user's confidence in the instrument, and the data needs or objectives of the user.

RECOMMENDATIONS

The Project has helped to identify several areas worthy of further evaluation.

1. Standardize sample collection and handling procedures to ensure volatile contaminant losses and cross-contamination are minimized.
2. Evaluate additional field instruments, such as flame ionization detectors, other photoionization detectors, other immunoassay techniques, and a field-grade gas chromatograph, and compare their results with each other and quick-turnaround analyses by SW846 methods in a stationary laboratory.
3. Utilize more homogeneous soils to improve the comparability of results from "split" samples.
4. Measure and record ambient weather conditions (temperature, humidity, etc.) and soil moisture content for their potential effects on results obtained from field analytical methods.
5. Utilize spiked matrix samples to compare individual instrument results to a known value.
6. Utilize duplicate samples and blank samples to improve quality assurance and quality control of the gathered data.
7. Utilize mathematical data manipulations and approximations in conjunction with graphical data evaluation to establish data trend comparisons.

ACKNOWLEDGMENTS

The Project progressed satisfactorily, due primarily to the support and cooperation of the following companies and individuals:

NYSDEC - Donald Darmer, Richard Wagner, Anthony Karwiel, Susan Lasdin
Stewarts Ice Cream - Chad Fowler
American Spill Abatement - George Longworth, Harold Strom, William Toran
Lebanon Valley Landscaping - William Black, Paul Heilmann
Quantix Systems - James Donovan, Wendy Bogle, Craig Ruggeri
Adirondack Environmental Services - Paul Batista
Con-Test Analytical Laboratory - T.J. Lyons, Mark Simon

REFERENCES

CRC Handbook of Chemistry and Physics, 71st Edition. 1990. Boca Raton, FL, CRC Press, Inc., pp. 10-212 to 10-213.

Immunoassay: The Future of Analysis for Agricultural, Food and Environmental Applications. 1992. Agri-Diagnostics Associates, 1 Executive Drive, Moorestown, NJ, 08057.

Instruction Manual, HNU Model HW-101 Portable Hazardous Waste Analyzer. 1988. HNU Systems, Inc., 160 Charlemont Street, Newton Highlands, MA, 02161-9987.

PSI-Plot Version 3 User's Manual. 1994. Poly Software International, P.O. Box 526368, Salt Lake City, UT, 84152, p. 150.

Robbins, G., and Midwest Research Institute, Soil Vapor Survey Boot Camp Manual. 1900. U.S. Environmental Protection Agency Office of Underground Storage Tanks, 401 M Street SW, Washington, DC, 20460, p. 3-9.

CHAPTER 46

A Simplified Transport-Based Methodology for Setting Site-Specific Soil Cleanup Goals for Organic Chemicals

James Ridenour, New York State Department of Health, Bureau of Toxic Substance Assessment, Albany, New York

INTRODUCTION

Experience gained by scientists, engineers, modelers, and regulators during efforts to remediate hazardous waste sites and other environmental contamination problems has led to a heightened awareness of the dynamic interrelationships among environmental media. With regard to remediation of contaminated soil, one of the most significant of these interrelationships is that between the soil and groundwater; specifically, the propensity for contaminated soil to function as a reservoir of contaminant mass which can degrade groundwater quality over the long term.

Because soil can continuously release contaminants which migrate with percolating water to underlying groundwater, questions about whether or not contaminated soil should be remediated are often decided in terms of protecting the groundwater resource. This is done partly in recognition of the potential for contaminated soil to degrade groundwater quality, but primarily because while generic soil cleanup standards have not been developed, generic groundwater standards generally have been.

Although the tie between soil contamination and groundwater quality has been recognized, determining whether soil contaminant levels will or will not cause groundwater impacts above levels of concern remains problematic. Lacking the fiscal resources and requisite experience to effectively utilize more complex mathematical models as predictive tools, regulators often resort to simpler techniques when evaluating the adequacy of proposed remedial strategies for addressing soil contamination. The most common of these approaches are built upon equilibrium partitioning.

EQUILIBRIUM PARTITIONING

The model most commonly used for calculating potential groundwater concentrations is predicated on the assumption that the contaminated soil and groundwater exist in equilibrium. The equilibrium groundwater concentration due to contact with the contaminated soil is calculated using Equation 1:

$$C_L=C_S/K_d \qquad (1)$$

where:

C_L = water concentration
C_S = soil concentration
K_d = Koc*foc
Koc = compound-specific organic carbon partitioning coefficient
foc = soil organic carbon content

The result generated using Equation 1 can be compared to the relevant standard or guidance value. If the standard is contravened, then cleanup is generally considered to be necessary. By algebraic manipulation of the partitioning equation, a soil cleanup level that will not contravene the groundwater standard can be derived, as in Equation 2:

$$C_S=C_L{}^*K_d \qquad (2)$$

In this case, C_L represents the groundwater standard or guideline.

Equilibrium partitioning methods, though they are commonly used, simple to apply, and can yield conservative estimates of potential groundwater impacts, are technically indefensible because they are insensitive to the site-specific conditions which govern contaminant transport through the unsaturated zone to the groundwater table (Figure 1).

Figure 1. **Soil:water system described by equilibrium partitioning vs. field conditions.**

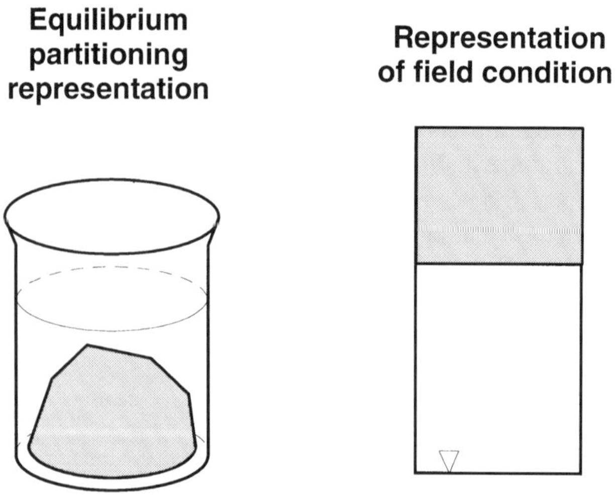

Example 1

Figure 2 depicts a generic vertical profile through a zone of contaminated soil located at some distance above the water table. The vertical extent of the contaminated zone is represented by Z1, the distance between the bottom of the source and the water table by Z2, and V1 and V2 represent the average velocity of water moving through these zones.

Consider a site as illustrated in Figure 2, where the soil is uniformly contaminated by a single organic chemical to a concentration of 20 parts per million (ppm). The soil organic carbon content (foc) is one percent and the organic carbon partitioning coefficient (Koc) for the chemical is 1000. The equilibrium groundwater concentration (C_L) is calculated by substituting these values into Equation 1, yielding a result of 2 ppm. Groundwater comparison values are generally lower than this, suggesting that soil remediation would be appropriate for this site.

Example 2

Next, consider three separate sites, all contaminated by the same contaminant and to the same level as in example 1. All three sites share the same soil conditions as the site in example 1, and the vertical dimension of Z1 at all three sites is 1 meter. Z2 at the three sites are 1, 5 and 10 meters, respectively. Using Equation 1, the estimated impact of each of these sites is also the same as that for the site in example 1.

Figure 2. **Schematic vertical profile through generic site.**

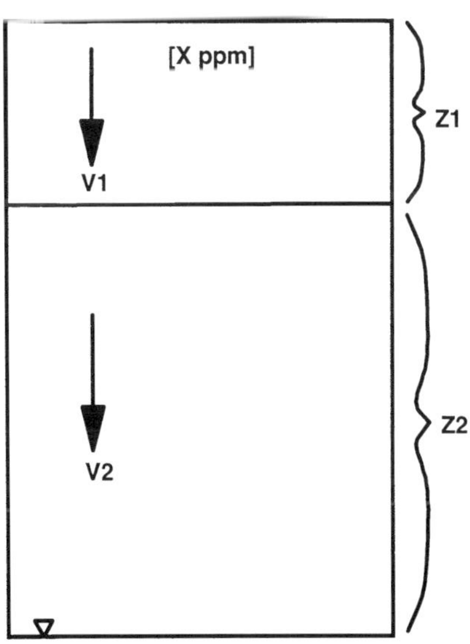

The fact that the estimated impacts are the same for all four of the cases described above, despite their different characteristics, is quite significant. In fact, regardless of the size of the contaminated zones, the depth to groundwater, the amount of rainfall, the characteristics of the soils or any other transport-related factor, as long as the initial concentration and foc are the same, the estimated impacts will be the same. This analysis also suggests that a cleanup value designed to protect groundwater quality at any one of these sites would be appropriate for them all. The influence of concentration on estimated impacts is discussed in example 3.

Example 3

Finally, consider two sites with the same soil characteristics as described above. The first consists of .0254 meters (1 inch) of soil contaminated to a concentration of 50 ppm, with 10 meters of clean soil between the water table and the source. The second consists of 10 meters of contaminated soil at a concentration of 20 ppm, located .0254 meters above the water table. Intuitively, all other conditions being the same, the second source is more likely to impact groundwater. Although the soil in source 1 is at a higher concentration, source 2 contains 160 times the contaminant mass present in source 1, and this mass is 400 times closer to groundwater. Despite this, Equation 1 predicts impacts of 5 ppm from source 1, and 2 ppm from source 2.

The preceding examples illustrate some of the problems associated with relying upon partitioning estimates for evaluating potential impacts to groundwater. Because partitioning-estimated groundwater impacts are generally conservative by several orders of magnitude, they can often lead remedial project managers to select remedies which are overly far-reaching in a mistaken belief that such actions are necessary to protect groundwater. Such remedies are not only seldom necessary, but are often extremely expensive.

Equilibrium partitioning cannot provide valid estimates of likely impacts to groundwater because it ignores the separation of the source from the water table. As Z2 increases relative to Z1, this error is magnified. Thus, the range of conditions over which partitioning estimates of groundwater impacts are appropriate is extremely narrow.

Some regulators have attempted to address this shortcoming of partitioning analyses by applying arbitrary correction factors to the partitioning estimates, yielding lower estimated concentrations. For example, the New York State Department of Environmental Conservation has adopted a methodology which calls for a generic correction factor of 100 to be applied to partitioning-estimated groundwater impacts (NYSDEC, 1990; NYSDEC, 1994). Such approaches may provide appropriate impact estimates for some sites, and conservative estimates for some others. However, if the source is large or the distance to the water table small (relative to each other), then these corrected estimates may be decidedly unconservative. In fact, the original partitioning estimates may better reflect the effect of the source on groundwater quality. In short, groundwater impacts arising from soil contamination are sensitive to transport conditions, including the velocity of percolating water, interactions between the soil and the contaminant of interest, the dispersivity of the soil, and the separation between the contaminated soil and the water table (Figure 3). It is not possible to account for the effects of these parameters on concentration in a meaningful way without using the transport equation. A simplified transport-based approach is needed that has more technical merit than either equilibrium partitioning, or partitioning as modified by an arbitrary correction factor.

APPROACH

One reason why equilibrium partitioning methods are so often employed is that alternatives that make use of more sophisticated analyses are often difficult for regulators to use. Mathematical models that incorporate various solutions to the transport equation generally fall into this category. However, using appropriate boundary conditions, the transport equation can be solved in a way

Figure 3. **Concentration vs. time as a function of depth.**

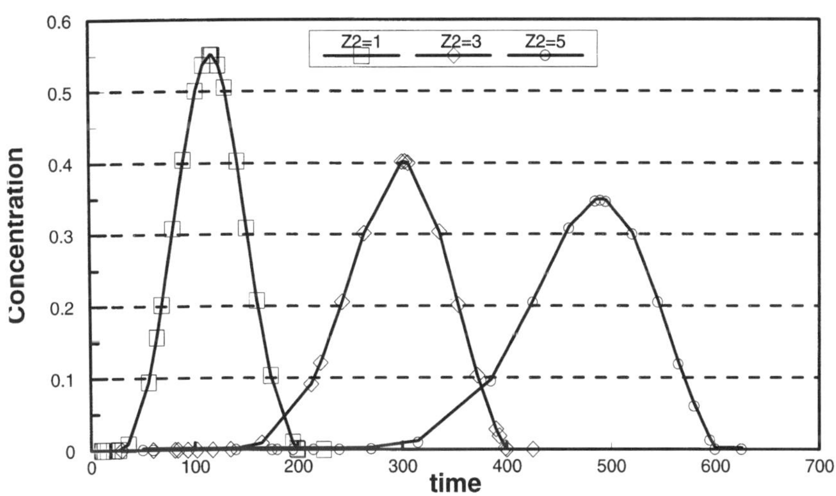

that provides a means to very easily estimate the average rate of mass release from a contaminated source. This, in turn, allows the development of a simplified method for evaluating worst-case impacts of sources at any selected depth, including that of the water table. The derivation of this approach is discussed in detail elsewhere (Ridenour, 1993; Ridenour and Chinery; Chinery and Ridenour a,b). In general terms, it involves analytical solution of the transport equation in one dimension to address the time-dependent decline of contaminant concentration within the source, and solution of the transport equation in one dimension to address dispersion of leachate as it moves through the unsaturated zone beneath the source, toward the water table.

Transport Equation

The transport equation mathematically describes the most significant processes involved in mass transport. It can be written as a first-order differential equation to describe changes in concentration with time or distance. The general form of this solution is dC/dt = -kC (USEPA, 1988). Solutions of this equation plot as linear functions on a semilog graph. The slopes of these lines, representing mass removal from contaminated sources by percolating water, are the values of k for these sources. The boundary conditions applicable to the zone below the source are, at t = 0, the pore water concentration at the water table interface = 0, and at t = 0; the time-variant concentration of leachate exiting the source can be expressed by $C_L = C_{Lo}e^{-kt}$, where C_L represents the concentration of pore water exiting the source at any time of interest, and C_{Lo} is the initial solute concentration in the source (Figure 4). Cleary et al. (1981) used these boundary conditions to generate an analytical solution to the transport equation describing the time-dependent concentration of leachate reaching the water table interface. The parameter, k, is an important term in his solution; however, his method did not provide a means of estimating k.

Estimating K1

As mentioned above, mass removal from the contaminated zone is a function of prevailing transport conditions. Under dispersive conditions, mass removal begins early, and the source is quickly depleted. Under advective conditions, the source concentration remains relatively unchanged until the contaminant mass exits the source, essentially all at once. Because of these different mass

Figure 4. **Effluent concentration as a function of time for generic source.**

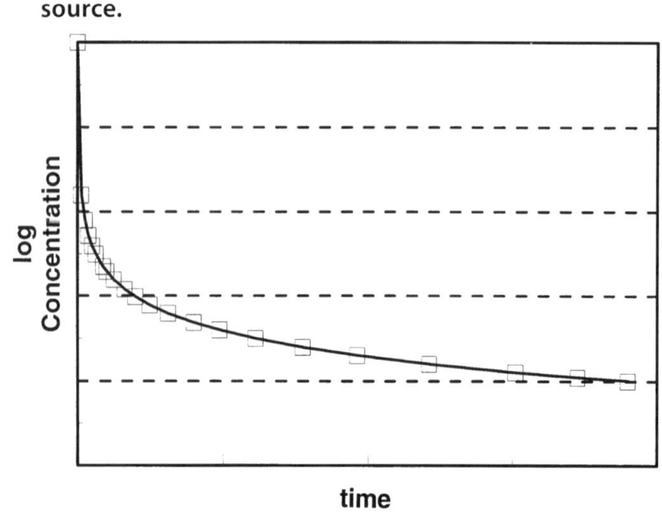

removal characteristics, "average" mass removal is best described by the early part of the concentration vs. time curve for dispersive transport, and by the later part of the curve for advective transport. These average mass removal rates for specific sources, which correspond to the k term in Cleary's solution, can be mathematically generated using the method presented by Ridenour (1993) to select the appropriate length of time over which to evaluate k for a given source. The source-specific decay rate is represented by the variable, K1. The plot of K1 vs. V/(R*Z1) (where R represents the retardation factor, a measure of the interaction of a compound with the soil), is a series of straight lines, with slopes related to the Peclet number (a measure of whether advective or dispersive forces are dominant in a transport system). Because of this linear relationship, K1 can be estimated by linear regression, using Equation 3 (Ridenour, 1993):

$$K1=1.234*Pe^{-.639}*V/(R*Z1) \tag{3}$$

where:

D = hydrodynamic dispersion coefficient (often estimated as the measured or estimated dispersivity of the medium multiplied by the fluid velocity)

Pe = column Peclet number, V*Z1/D

and other terms are as previously defined. The correlation coefficient relating the regression-generated estimates of K1 to the original source-specific values was .994, indicating exceptional agreement.

Using K1 to Estimate C_{max}

It is apparent that pore-water and groundwater concentrations below a contaminated soil source are tied to the rate of contaminant release from the source. When contaminant mass exits the source a little at a time, impacts are small. When mass exits the source as a slug, impacts are maximized. The generic concentration vs. time profile, describing contaminant concentration at any depth of interest, appears in Figure 5. Because the type curve is smooth, a bisector algorithm can be used to identify the highest contaminant concentration arriving at any depth of interest, including that of the water table. The procedure involves calculating the slopes of tangents to the effluent curve using a bisector algorithm to select the next point to evaluate, until the peak concentration is identified. Applying this approach to a large number of sources separated from modeled water tables at various distances below the sources yields results that plot as a series of arcs distinguished by the ratio of Z1/Z2. The Y-axis is the ratio of the peak concentration to the equilibrium concentration, and the X-axis variable is comprised of components describing K1, the interaction of the contaminant of interest with the soil, the dispersivity of the medium, and the average velocity of water percolating through Z2 (Ridenour, 1993).

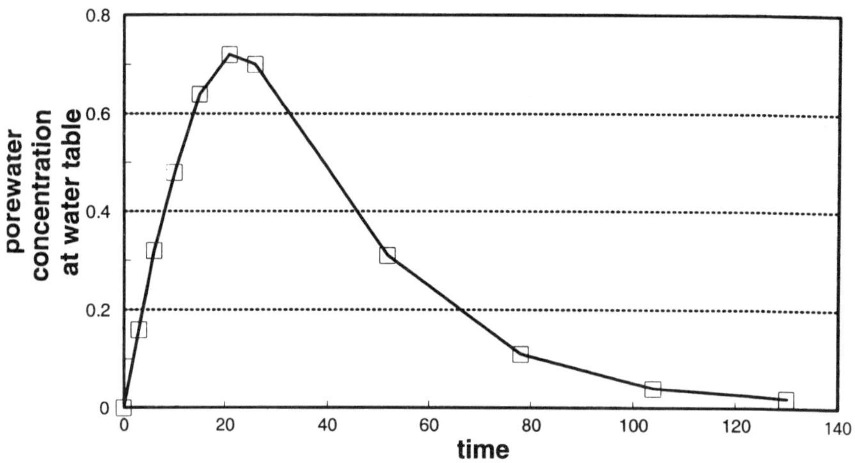

MODEL VERIFICATION

Gaber et al. (1992) conducted a column leaching study using the herbicide, Picloram. From information provided by the authors, the initial equilibrium pore water concentration in Z1 was estimated to be 0.8066 mg/L. Using Equation 3 to estimate K1 for this source, then using the graphical technique described above, the ratio of the estimated equilibrium concentration to the predicted maximum concentration is found to be approximately 0.88. By applying this ratio to the estimated equilibrium concentration, the maximum effluent concentration exiting the column was estimated to be 0.71 mg/L. The authors reported that the concentration of the applied solution was 0.99023 mg/L, and that the ratio of the maximum effluent concentration to the applied concentration was 0.73. This corresponds to a maximum effluent concentration of 0.72 mg/L. The difference between the measured and predicted values is slightly more than one percent. In contrast, applying a 100-fold correction factor to the equilibrium concentration, as recommended by some regulatory programs (NYSDEC 1990; 1994), yields a result of 0.008 mg/L, a value which represents an error of more than 8900% in comparison to the measured result. Perhaps more significantly, the 8 ppb estimate could lead project managers to conclude that a source with these conditions is of no practical significance, when they might conclude otherwise if they believed the same source might generate concentrations closer to 710 ppb.

CONCLUSION

This method recognizes that groundwater impacts are closely tied to the size of the contaminated source and the rate of contaminant release. Because it is based upon the classical transport equation, it accounts for important physical phenomena which govern the mass transport of contaminants in the unsaturated zone. Yet it remains simple to use, easy to understand, and the results are easy to interpret. Application of this method allows a quick assessment of the significance of worst-case groundwater impacts that avoids an arduous mathematical treatment without sacrificing the improved accuracy of the dispersion model. The approach builds upon simple equilibrium partitioning, and so should be easily understood and applied by the regulated community as well as by regulatory entities. It also provides an opportunity for users to conduct "what if?" analyses in short order, without the services of a consultant or modeler. Most importantly, it offers the user an analysis tool which can be used to ensure that the scope of remedial actions are adequate to protect the groundwater resource, as well as to guard against the commitment of scarce financial resources to unnecessarily conservative remedial responses. Finally, maximum groundwater impacts estimated by this method compare favorably with results generated in soil column leaching studies.

REFERENCES

Chinery, R. and Ridenour, J. A Rapid Assessment Technique for Evaluating the Significance of Contaminated Soil Sources: 2. Derivation of Average Source Decay Rate (K1) (in preparation).

Chinery, R. and Ridenour, J. A Rapid Assessment Technique for Evaluating the Significance of Contaminated Soil Sources: 3. Application of K1 to Estimate Maximum Impacts at the Water Table (in preparation).

Cleary, R., Miller, D., and Pinder, G. 1981. Groundwater Pollution and Hydrology, Princeton, New Jersey. Course materials dated July 13 to 17, 1981.

Gaber, H., Comfort, S., Inskeep, W., and El-Attar, H. 1992. A Test of the Local Equilibrium Assumption for Adsorption and Transport of Picloram, *Soil Sci. Soc. Am. J.* **56**, 1392.

NYSDEC (New York State Department of Environmental Conservation). Handling and Disposition of Excavated Petroleum Contaminated Soil, Memorandum of Understanding, Division of Water and Division of Solid Waste, March 6, 1990.

NYSDEC (New York State Department of Environmental Conservation). Division Technical and Guidance Memorandum: Determination of Soil Cleanup Objectives and Cleanup Levels, Division of Hazardous Waste Remediation, January 24, 1994.

Ridenour, J. 1993. A Rapid Assessment Technique for Evaluating Worst-Case Groundwater Impacts Arising from Finite, Temporal Contaminated Soil Sources, Graduate Thesis, Rensselaer Polytechnic Institute, Troy, NY.

Ridenour, J. and Chinery, R. A Rapid Assessment Technique for Evaluating the Significance of Contaminated Soil Sources: 1. Implications of Equilibrium Partitioning Approaches (in preparation).

USEPA (United States Environmental Protection Agency). 1988. Selection Criteria for Mathematical Models Used in Exposure Assessments: Ground-Water Models. Washington, DC.

ENVIRONMENTAL FATE

CHAPTER 47

Multispecies Screening Analysis of Soil Chromium Fate and Transport

Joshua E. Lieberman, ENSR Consulting and Engineering, Acton, Massachusetts

INTRODUCTION

The primary fate and transport processes of sorption, biodegradation, and aqueous solubility apply to both organic and inorganic contaminants, but not in the same way. Organic contaminants sorb primarily to soil organic carbon, and secondarily to soil minerals such as clays. Biodegradation or hydrolysis will transform most of these contaminants irreversibly into simpler and generally less toxic compounds within a time frame which is short, relative to that of soil-forming processes.

Inorganic contaminants, particularly metals, behave quite differently. They do tend to sorb to a variety of soil minerals, particularly clays and hydroxides, but precipitation may also be an important stabilizing mechanism under some conditions. Metals themselves are relatively little affected by degradation processes, a fact which accounts for both the high background values of many soil metals as well as the occurrence of economic mineral deposits worldwide. Metals mobility can be enhanced by organometallic complexation (e.g., methylation), forming compounds which some microbes can generate and others can destroy; however, the fate and transport of metal contaminants is largely controlled by reversible transformations between a variety of solid minerals and dissolved species according to the composition and physical state of the host soil or sediment, including parameters such as pH (acid-base index), Eh (oxidation-reduction potential, D.O. (dissolved oxygen), and temperature.

The importance of such transformations is particularly acute in the case of transition metals, whose different valence states may lead to vastly different fate and transport properties. The present chapter focuses on chromium, whose two environmentally common valence states exhibit very different characteristics. While Cr^{III} is very little soluble in water and relatively nontoxic (it may in fact be an essential nutrient), Cr^{VI} is both highly mobile and distinctly toxic to humans, animals, and vegetation (Alloway, 1990). A number of studies have shown that both oxidation and reduction of chromium are rapid, relative to other soil and groundwater processes (Palmer and Puls, 1995). This means that chromium contamination of soil or groundwater cannot be assumed to remain in one form or another. Any fate and transport analysis of chromium should therefore account for dynamic partitioning between the trivalent and hexavalent forms in response both to variations in factors such as pH, and to mobilizing processes such as pore water migration.

Two examples of chromium geochemical behavior are shown in Figure 1. Chromium species stability as a function of Eh and pH is represented in Figure 1a. This diagram shows that even in the relatively simple Cr - H_2O system, small changes in pH or Eh can alter the speciation and therefore the mobility of chromium by several orders of magnitude. Figure 1b shows the strong dependence of Cr^{III} solubility on pH when solid Fe-Cr hydroxides are present. For example, Cr^{III} in an acidic waste stream would tend to precipitate almost completely upon mixing with normal Fe-hydroxide-bearing soil.

Figure 1. **Geochemical speciation and solubility of chromium: (a) speciation as a function of Eh and pH in the system Cr - H_2O; (b) solubility of Cr^{III} as a function of pH in the presence of Cr or Fe-Cr hydroxide. The thin dashed line represents an MCL concentration of 0.1 mg/L. Adapted from Palmer and Puls (1995) and Dragun (1988).**

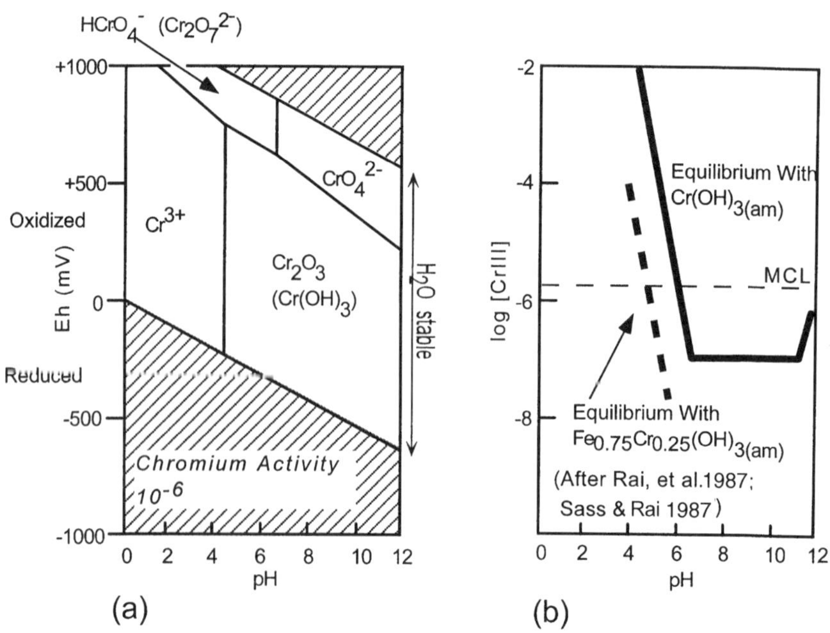

The screening fate and transport analysis presented in this chapter takes a novel approach to accounting for the transport behavior of both chromium species in contaminated soil at a former plating shop site. A factor for partitioning of total chromium between soil and pore water (Kd) is derived which accounts for dynamic partitioning between the two chromium species during pore water migration. This factor is then used as input for two simple transport models which predict the rate and concentration of chromium movement, first through unsaturated zone soils into groundwater, and then within the aquifer to hypothetical water supply wells downgradient.

The resulting predictions (and their sensitivity to input parameters) for future concentrations of groundwater chromium at potential receptors are used to evaluate remedial alternatives, including excavation, in-situ fixation, capping, and no-action. Finally, questions are raised by this analysis as to what constitutes sensible public policy for any site with metals-contaminated soils.

SITE DATA

The site considered in this chapter is a former plating shop located on the grounds of a semiconductor manufacturing facility. Acidic wastes containing a mixture of dissolved chromium (III) and (VI) were spilled or discharged to drains, and found their way into the soil underneath and around the shop. The site soils consist of mixed fine sands and clays; depth from ground surface to the water table is approximately 72 feet. Figure 2 presents the results of analyses done on soil boring samples from in and around the shop. The distribution of total chromium with soil depth is shown in Figure 2a: the highest concentrations are generally found within 10-12 feet of the ground surface, but in some borings the decrease in concentration with increasing depth seems to be leveling out. The depth trends in Cr^{VI} shown in Figure 2b are similar. The depth trends in Cr^{VI} / Cr^{III} ratio shown in Figure 2c are more equivocal, and in several borings the ratio appears either to be leveling out or increasing with depth.

FATE AND TRANSPORT METHODOLOGY

A critical component of contaminant fate and transport analysis is the exposure pathway, since without exposure to contaminants by humans, plants, or animals, a hazard cannot be determined. A conceptual water supply well exposure pathway for the site is shown in Figure 3. Under this conceptual model, chromium could be mobilized from the contaminated soil under the plating shop and transported downward into the water table by infiltrating rainwater. Once in the saturated aquifer, chromium could be transported down the hydraulic gradient by flowing groundwater to where it could be drawn into a (hypothetical) water supply well. The fate and transport analysis seeks to answer the question: if, when, and over what potential well screen locations might concentration limits for dissolved chromium be exceeded. The analysis described

Figure 2. **Sampled depth distribution of soil chromium from seven site borings: (a) total chromium; (b) Cr^{VI}; (c) ratio of Cr^{VI} to Cr^{III}. While some decrease in chromium concentration with increasing depth is evident, a similar change in the Cr^{VI} to Cr^{III} ratio is not as evident.**

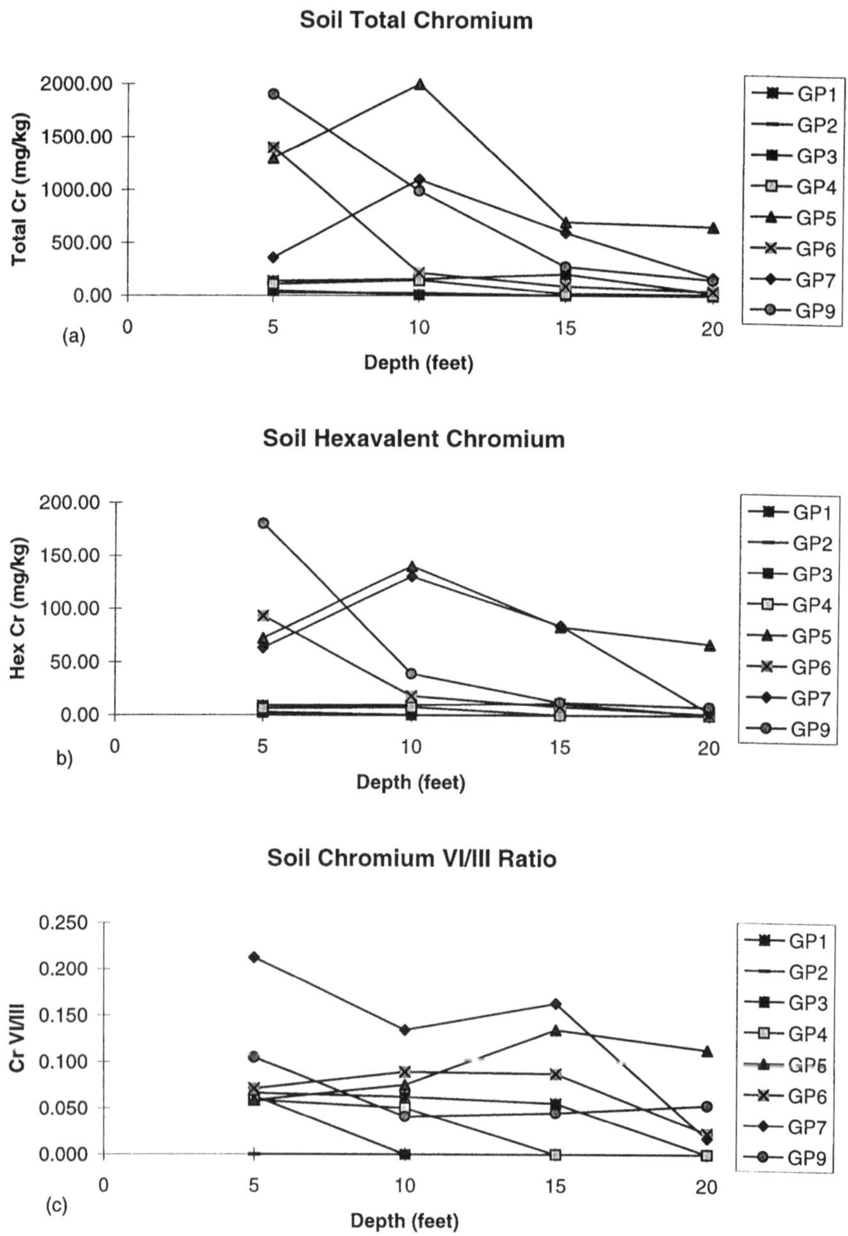

Figure 3. **Conceptual model of study site. A potential dissolved chromium exposure pathway consists of downward vadose zone movement, followed by horizontal movement in the groundwater to a hypothetical water supply wellscreen.**

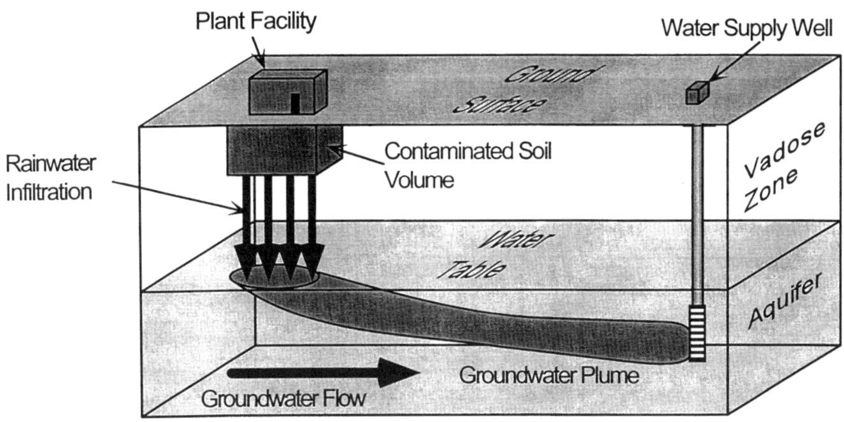

in this chapter, following the conceptual model described above, encompasses two sequential computer simulations and certain site-specific calculations which are described below.

TRANSPORT MODELING

Two computerized models were employed in this study to simulate the movement of chromium contaminant through soil and sediment pore spaces:

1. EMSOFT2 is a 1-D screening model of vadose zone (unsaturated zone and capillary fringe) contaminant transport and evaporation, which maintains a mass balance between soil, pore water, and pore gas. It solves numerical expressions developed by Jury et al. (1983, 1990). In this case, the effects of evaporation are not considered, since Cr is not volatile to any significant degree. The model also does not account for the effects of flow dispersion during migration This model provides an estimate of the timing and flux of Cr transport to the water table.

2. HPS is a pseudo-3D model of saturated zone transport (Galya, 1987) which uses an analytical solution for advective-dispersive transport from a horizontal plane source to an arbitrarily located receptor point in the aquifer. For a given flux of contaminant downward across a given area of the water table surface into a uniform saturated aquifer, this model estimates the resulting groundwater contaminant concentration at a given well screen location downgradient from the source.

Both models have been validated and employed in a large number of soil contamination studies approved by both state and federal regulators.

EMSOFT2 SETUP

For the purposes of this model, the study site has been approximated as a 1-D column extending from the surface to the water table. Table 1 presents values and sources for the model parameters which were used in the study.

Table 1. **Model Parameters for EMSOFT2 Simulation**

Parameter / Datum	Value	Comment
Chemical Species:	CrVI	
Time Period of Run:	2000+ years	or 999999 days
Depth to Ground water	72 feet	
Water-Soil Partitioning:		
Kda (CrVI) soil water	3.6 liter/kilogram	Average value reported in literature
Kdb (CrVI / CrIII) in soil	0.15 liter/kilogram	Observed average site value
Kdtotal soil Cr / water CrVI	27.6 liter/kilogram	Calculated from two values above
Water-Air Partitioning	1x10-9	Essentially no volatility
Air Diffusion Coefficient	1x10-6	Essentially no volatility
Aqueous Diffusion Coefficient.	0.432	Average literature value
Half-Life	99999 days	Does not significantly degrade with time
Porosity	0.4	Middle-upper range for sandy soil
Water Content	0.25	Reasonable estimate
Density	2	Reasonable estimate
Pore water Flux	0.0026 ft/day	Equivalent of 11.5 inches/year with no runoff
Air Boundary Layer	3.24 feet	Irrelevant since volatility is precluded
Cover Layer	0	No "clean" soil layer at surface
Contaminated Layer Thickness	19.7 feet	Assumed uniform concentration throughout layer
Contaminated Layer Cr (total) Concentration	1000 mg/kg	Average total Cr value
Equilibrium Ground water CrVI Concentration	36 mg/l	= 1000 mg/kg / 27.6 l/kg

Figure 2a gives an indication of the variation in soil chromium concentrations for several boring locations and depths in and around the former plating shop. For the purposes of the model, the initial soil contamination has been approximated as a constant 1000 mg/kg from the surface down to a depth of 20 feet. Over the distance of 52 feet to water table, dispersion, or "smearing out," of contaminant concentrations with transport, would be expected to smooth out the presently observed concentrations on the scale of 4-5 feet. EMSOFT2 does not account for dispersion explicitly, but this anticipated smoothing effect helps to make the assumption of constant Cr concentration with depth a more realistic one.

An important aspect of chromium geochemical behavior is the partitioning of total chromium between the species Cr^{VI} and Cr^{III}. A number of geochemical processes in the soil can reduce Cr^{VI} to Cr^{III} or oxidize Cr^{III} to Cr^{VI}. The ratio Cr^{VI} / Cr^{III} is plotted in Figure 2c. This ratio is normally expected to decline with depth as the soil environment becomes more oxygen-depleted and reducing. The available site data do not clearly show this trend over the limited depth interval which has been sampled. As a conservative approximation, an average value of the observed ratio (0.15) has been assumed to remain constant with depth. This approximation is equivalent to holding constant both Eh and pH, as well as the presence and composition of constituents within the soil column.

Cr^{III} is much less soluble than Cr^{VI} (Dragun, 1988), so much so that all of the dissolved Cr can be assumed to exist in the (VI) state. Therefore, Cr has been assumed to partition between the two species (VI and III) in the soil, and only exist as hexavalent Cr^{VI} in the groundwater. Partitioning is assumed to be reversible and rapid relative to the speed of transport, i.e., chromium can be oxidized as well as reduced within the soil, maintaining a constant ratio between the two species. This is also conservative, in that scarcity of oxidizing agents such as dissolved oxygen or manganese might well tend to limit the oxidation of Cr^{III} back to Cr^{VI} (and its resulting partial dissolution in the pore water).

Following this approach, the partitioning of Cr between soil and water has two coupled parts. First, Cr in the soil partitions between the Cr^{VI} and Cr^{III} species (site-specific $Kd^a \sim 0.15$). Then the soil chromium which is in the hexavalent state partitions between the soil and pore water. This partitioning ratio or Kd is estimated from Baes and Sharp (1983) ($Kd^b \sim 3.6$). The coefficient for partitioning between total soil Cr and dissolved Cr involves coupling these two partitioning reactions:

$$Kd^{total} = Kd^b + Kd^b / Kd^a \sim 27.6 \tag{1}$$

The soil is also assumed to be uniform with depth in regard to porosity (0.4), water content (0.25), dry density (2), and water flux (0.08 cm/day). The latter represents complete soaking-in of the average rainfall of 11.5 inches/year with no runoff. The depth from the surface to the water table is assumed to be a constant 72 feet.

HPS Setup

The HPS model is a pseudo-3-D model. Contaminant transport is calculated in a straight line from a segment of the water table to an arbitrarily located wellscreen. It assumes a homogeneous aquifer with constant linear groundwater flux, but does account for dispersion of contaminant in all three dimensions.

Table 2. **Model Parameters for HPS Simulation**

Parameter / Datum	Value	Comment
Chemical Species:	CrVI	
Dimensions of WT discharge	33 feet x 33 feet	Area of high chrome values near plating shop
Flux across plane into water table	0.95 ft/yr	From rainfall percolation
Influx concentration of CrVI	36 mg/l	From Kd's and calculation above
Porosity	0.4	From above
Pore velocity of horizontal flow	41.9 ft/yr	From water well measurements
Longitudinal dispersivity	10% of distance	Literature value
Transverse dispersivity	0.2% of distance	Literature value
Vertical dispersivity	0.02% of distance	Literature value
Length of well screen	10 feet	Reasonable value
Retardation	121	= 1 + Kd (soil density / porosity)

Parameters used for this model are presented in Table 2. The groundwater flow is assumed to be uniform and constant, and the aquifer is assumed to have uniform properties. The estimated hydraulic gradient at the site, from water level contours, is 0.0031. The lowest estimated hydraulic conductivity is 15 ft/day; therefore, a conservative groundwater flux is approximately 16.8 feet/year = 5.11 m/year. Accounting for the porosity of 0.4, this gives a pore velocity of 12.8 m/year. The horizontal area of contaminated soil is estimated to be a square 33 feet (10 meters) on a side. Dispersion of contaminant is calculated to be 10% of distance traveled in the direction of flow, 0.2% transverse to flow, and 0.02% in the vertical direction. The same partitioning of Cr between soil and water (Kd) has been employed as in the previous calculation.

MODEL RESULTS AND DISCUSSION

EMSOFT2 Results

Using the parameters and assumptions outlined above, Cr is estimated to reach the water table in a more or less intact 20-foot-long "slug" after 2100 years have elapsed. The concentration of Cr in the soil at the water table will remain nearly 1000 mg/kg, and the concentration of Cr (VI) in the groundwater entering the water table will remain at 36 mg/L. The velocity of Cr migration is 52/2100 = 0.025 feet/year; therefore, Cr would be released into the water table over a further period of about 800 years.

The model has essentially linear behavior in many parameters. For example, if the average contamination level were 500 mg/kg, the groundwater concentration would be about 18 mg/L. If half of the average rainfall ran off or evaporated without soaking into the site soil column, the time-to-water-table would double to about 4200 years The flux of chromium into the water table would decrease by half, but last for twice as long.

HPS Results

Predicted well water concentrations depend strongly on the distance of the wellhead from the center of the area of contamination, as shown in Figure 4 and Table 3. Given the assumptions and estimations described above, the drinking water MCL of 0.1 mg/L Cr^{VI} would be met in a typical water well more than 500 feet directly downgradient from the contaminated soil area. The concentrations of chromium in wells located more than 15 feet off the flow axis from the contaminated area would be much smaller, because of the small value of the transverse dispersion. For example, at a wellhead located 160 feet downgradient and 30 feet to the side of the contamination center, the calculated concentration of Cr^{VI} would be 0.39 mg/L.

To a good approximation, the well water concentrations of Cr expressed by the coupled EMSOFT2 and HPS models are linearly dependent on the source concentrations. For example, if the soil concentration of Cr were reduced to 500 mg/kg from 1000 mg/kg, the calculated well water concentration at 300 feet would be 0.65 mg/L instead of 1.3 mg/L.

Figure 4. **Simulated downgradient groundwater chromium concentrations. Concentrations increase to the edge of the contaminated soil zone, then decrease with distance downgradient due to the effects of dispersion and dilution. Predicted concentrations away from the axis of groundwater flow are much lower, as shown by the arrow for a location 30 feet to the side of the modeled flow axis.**

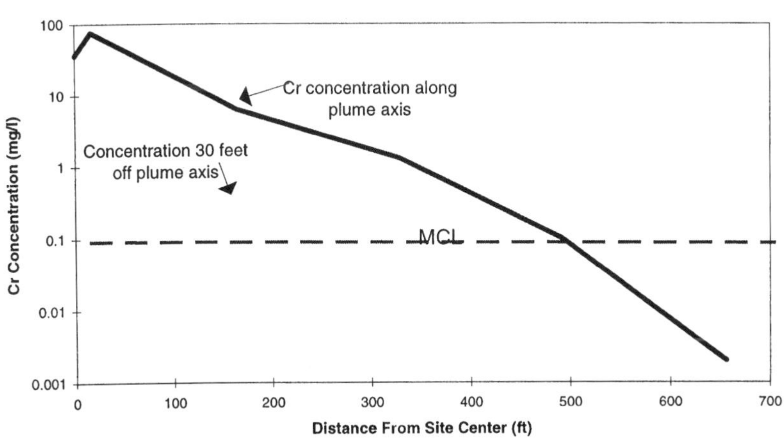

Table 3. **Results of HPS Simulation**

Distance Down-gradient (feet)	Offset From Plume Axis (feet)	Maximum Concentration (mg/l)
0	0	36
17 (edge of contamination)	0	76
164	0	6.4
164	33	0.39
330	0	1.3
495	0	0.1
660	0	0.002

CONCLUSIONS

When a simple expression for the exchange Cr^{III} <-> Cr^{VI} is included in a fate and transport model, even small proportions of Cr^{VI} in contaminated soils are shown in this chapter to be capable of resulting in Cr-contaminated groundwater after finite periods of time. For the site described in this chapter, that time period is a minimum of 2000 years. The calculated speed of migration is much slower than one would calculate considering only fractionation of Cr^{VI} between soil and pore water. An expression for distributing chromium among more than the two species described here would probably decrease this speed somewhat further, since fewer chromium species are soluble than insoluble. One could still not, however, conservatively characterize soil chromium as immobile.

This raises difficult questions for development of environmental policy and remedial guidance to cover metals-contaminated soil. The principle is widely accepted that a hazard is not mitigated simply because it is predicted to occur some ways in the future. On the other hand, the effects of most remedial alternatives cannot be predicted thousands of years into the future. For example, in-situ fixation of the hexavalent chromium, delineation of a mixing zone, or capping to limit rain infiltration would all be effective in the short term, but after several thousand years of rainfall, it is unlikely that either fixation or capping would survive; it is equally unlikely that property restrictions for a mixing zone could be preserved so long. A more reasonable likelihood is that the present subsurface geochemical parameters, representing a balance between rainwater infiltration rate and soil composition, will continue to characterize the site into the future.

There are other potential remedial alternatives. Excavation might remove the chromium from the site, but not remove the problem of storing and administering the excavated soil so as to prevent its posing a hazard. Treatment of the excavated soil to remove chromium would be possible but very expensive. Ultimately that expense ought to be weighed against the likelihood that today's environmental standards or groundwater usage will at all resemble those several thousand years from now. In other words, the calculated risk should in some way include how far in the future the hazard is predicted to occur.

Another conclusion can be drawn from this dilemma. General procedures (e.g., TCLP) and standard soil concentration limits are not very useful for intelligently determining the risks posed by metals-contaminated soil, because the fate and transport characteristics of metals are so site-specific. Only detailed information on the site-specific factors controlling the mobility of metals will yield fate and transport predictions which are both conservative and realistic. Site-specific cleanup levels can only help industry and regulators alike to avoid over-remediation of metals-contaminated soils in favor of more sensible protection for human health and the environment.

REFERENCES

Alloway, B.J. 1990. *Heavy Metals in Soils*, New York, John Wiley & Sons.

Baes, C.F. III, and Sharp, R.D. 1983. A Proposal for Estimation of Soil Leaching and Leaching Constants for Use in Assessment Models, *J. Environ. Qual.* **12**, 17.

Dragun, J. 1988. *The Soil Chemistry of Hazardous Materials*, Hazardous Materials Research Institute, Silver Spring, MD.

Galya, D. 1987. A Horizontal Plane Source Solution for Groundwater Transport. *Groundwater* **6**, 733.

Jury, W.A., Russo, D., Streile, G., and El Abd, H. 1990. Evaluation of Volatilization by Organic Chemicals Residing Below the Soil Surface. *Water Resources Res.* **26**, 13.

Jury, W.A., Spencer, W.F., and Farmer, W.J. 1983. Behavior Assessment Model for Trace Organics in Soil: I. Model Description. *J. Environ. Qual.* **12**, 558.

Palmer, C.D. and Puls, R.W. 1995. Natural Attenuation of Hexavalent Chromium in Ground Water and Soils, EPA Technical Report EPA/540/S-94/505.

Rai, D., Sass, B.M., and Moore, D.A. 1987. Chromium(III) Hydrolysis Constants and Solubility of Chromium(III) Hydroxide, *Inorganic Chemistry* **26**, 345.

Sass, B.M., and Rai, D. 1987. Solubility of Amorphous Chromium(III)-Iron(III) Hydroxide Solid Solutions, *Inorganic Chemistry* **26**, 2228.

CHAPTER 48

Estimation of an NMOC Generation Rate from a Hazardous Waste Landfill

Zhijiang Zhang and **Steven Farrick**, TRC Environmental Corporation, Windsor, Connecticut

INTRODUCTION

Landfill Gas Generation

Gases from landfills are generated as a by-product of a biological process in which microorganisms decompose organic wastes to produce carbon dioxide, methane, and other gases. Gases of concern are produced mostly by the anaerobic process, in which microorganisms generate gas in an environment deficient in oxygen. The ability of a landfill to produce gas depends on many factors, including waste composition, moisture content, pH, and nutrient availability. If the landfill is very dry, little gas will be generated (USEPA, 1994).

Initially, when waste is placed in a landfill, conditions are aerobic. Atmospheric oxygen is present, and the organic materials in the waste begin to break down, producing mainly carbon dioxide and water (Emcon Associates, 1982). These aerobic processes are exothermic (heat-producing), and the landfill temperature begins to rise. As the landfill ages and more waste is placed in it, the oxygen is depleted, and anaerobic (oxygen-deficient) conditions eventually predominate. The organic materials in the waste break down into organic acids, which then further break down into methane and carbon dioxide.

Generation rates for gases vary among landfills and within each individual landfill. Theoretically, between seven to nine cubic feet of gas can be generated per pound of dry organic matter (USEPA, 1994). Generally, methane generation rates on the order of 0.05 to 0.2 cubic feet of gas per pound of refuse per year can be anticipated for a typical municipal solid waste landfill (USEPA, 1994). However, gas generation rates can be quite variable, depending on waste composition, percent moisture, age of the material, and a number of other factors (USEPA, 1991a). As decomposition gases are produced, pressure in the landfill begins to build up. It is this pressure that drives the gas from the landfill into the soil and the atmosphere.

Potential hazards posed by landfill gas buildup and migration include explosions, asphyxiation (stress to vegetation), offsite gas migration, and disruptions of cover systems. Landfill gas migrates in response to pressure, concentration, and possibly temperature gradients. Depending on the type of nearby receptors and the nature of the landfill cover system, either passive or active landfill gas control systems might be needed to minimize potential hazards.

SITE HISTORY

The McAllister Point area of the Naval Education and Training Center (NETC) in Middletown, Rhode Island was used as the site of a sanitary and hazardous waste landfill from the mid 1950s until the mid-1970s. During this time the landfill was used for the disposal of wastes from various Naval operational and housing areas, and from the ships homeported at Newport prior to 1973. The materials reportedly disposed of at the site included domestic refuse, spent acids,

Figure 1. **Site locus plan.**

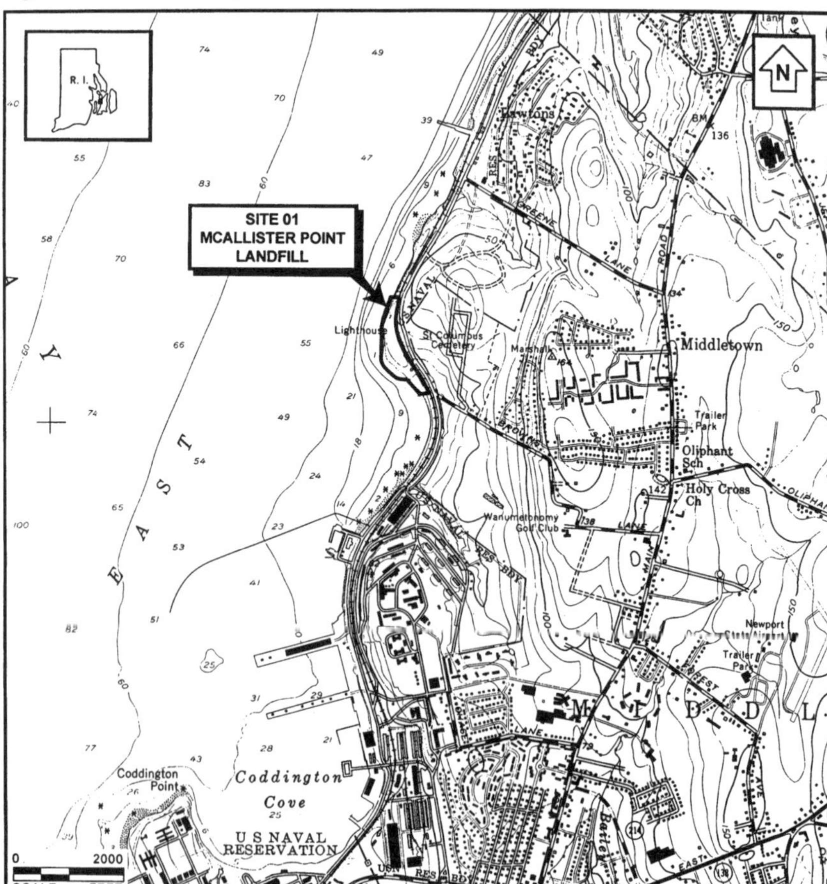

paints, solvents, waste oils, PCB-contaminated transformer oil, and construction debris. A site locus plan is provided as Figure 1.

In November 1989, the entire NETC was listed on the USEPA's National Priorities List (NPL) of abandoned or uncontrolled hazardous waste sites subject to requirements of the Comprehensive Environmental Response, Compensation, and Liabilities Act of 1980 (CERCLA) and the Superfund Amendments and Reauthorization Act of 1986 (SARA).

After listing as an NPL site, a comprehensive Remedial Investigation (RI) was conducted at the landfill (TRC Environmental Corporation, 1994a). The RI consisted of geophysical surveys, conduct of a soil gas survey, completion of soil borings and installation of monitoring wells, and collection of soil, groundwater, leachate, sediment, and landfill gas samples. Results of the investigation indicated elevated levels of both organic and inorganic contamination in soil and fill at the site. Groundwater sample results indicated the presence of aromatic volatile organic compounds (VOCs) and polycyclic aromatic hydrocarbons (PAHs) in the hundreds of parts per billion (ppb) range. Groundwater results also indicated elevated concentrations of metal analytes.

Following completion of the first phase of the RI, a Record of Decision (ROD) was signed by the Navy, EPA, and the State of Rhode Island in September 1993. Implementation of the September 1993 ROD involved the design of a cover system for the landfill. The cover system is considered a source control measure which provides containment and isolation of the landfill contents, the control of leachate generation as a result of surface infiltration, and protection against surface erosion and landfill gas migration. The cover system addresses the principal threats to human health and the environment posed by the site, and is intended to be a permanent source control remedy for the site.

Major components of the landfill cover system include:

- Capping of approximately 10 acres of the landfill with a RCRA Subtitle C multilayer cap, and capping approximately 0.8 acres with a RCRA Subtitle D soil cap
- Shore protection to prevent cap damage from wave action and storm surge events
- Landfill gas controls to manage landfill gas migration
- Surface controls to minimize erosion and manage runoff
- Use of fencing and institutional controls to control site access and future site use.

The cap provides a physical barrier to potential exposures or erosion of surficial contaminants, and provides restriction of precipitation infiltration. The ROD design implementation includes regrading of the site, improvement of drainage features, a landfill gas management system, and a reduction in grade and provision of slope protection along Narragansett Bay. Fencing and deed restrictions would be included to limit site access and future site use and development. Groundwater and storm water discharge would be monitored in accordance with federal and state regulations. The Navy would also review the remedial action, to the extent required by law, to assure that it continued to protect human health and the environment.

The McAllister Point landfill cover system has been designed to meet or exceed Resource Conservation and Recovery Act (RCRA) Subtitle C guidance as described in USEPA documents as modified according to EPA Region I review, and in accordance with accepted engineering design practices (USEPA, 1989). The cap has been designed to comply with the performance standard set forth in Section 14.12 of the Rhode Island Solid Waste Management Regulations, which requires that the cap have a maximum coefficient of permeability of 1×10^{-7} centimeters per second. The cap will consist of a top protective layer, drainage layer, low hydraulic conductivity layer consisting of a geomembrane and geosynthetic clay liner (on the top of the landfill), and a gas vent layer.

LANDFILL GAS ASSESSMENT

Both theoretical and site-specific analysis were completed to determine the need for an active or passive gas vent system at the McAllister Point Landfill. Each of these analyses are described in below.

Theoretical Analysis

Total Landfill Gas Generation Rate Assessment

To determine the type of gas control system (e.g., active or passive) to incorporate within the RCRA Subtitle C cover system at the McAllister Point Landfill, an assessment of the possible pressure impacts of landfill gas on the cap system was undertaken. This analysis involved estimating the rate of landfill gas generation, along with estimating the pressure this gas would exert on the underside of the RCRA Subtitle C cover system to be placed over the landfill.

The gross total landfill gas generation rate was assessed by the following equation:

$$Q_{LG} = R_{LG} * M \tag{1}$$

where:

Q_{LG} – gross total landfill gas generation rate (l/yr)
R_L = refuse gas generation potential (L/kg/yr)
M = mass of refuse in the landfill (kg) [density*volume]

The R_{LG} term was determined from EPA-published rates of landfill gas generation within solid waste landfills (USEPA, 1991a). Two values of R_{LG} (50 mL/kg/hr, and 2.19 mL/kg/hr) were utilized to estimate the rate of landfill gas generation at the McAllister Point landfill. In this analysis, the use of the EPA-published solid waste landfill gas generation rates is a conservative assumption, since the literature gas generation rates for active solid waste landfills would be expected to be higher than that for the construction debris, industrial waste, and solid waste observed at the inactive McAllister Point Landfill. Additionally, the literature-based solid waste landfill gas generation rate is based on production of landfill gases under ideal conditions such as a high moisture content, stable temperature, and uniform waste deposits, etc. The landfill mass term M was calculated as a product of landfill density and waste volume. For this conservative theoretical analysis, a waste density of 1.9 g/cm^3 (119 lb/ft^3) was utilized, even though lower densities have been reported (USEPA, 1991b). The total landfill volume was calculated assuming a range of landfill thicknesses from 5 to 25 feet over the entire 10.8 acre landfill. Based on these assumptions, the maximum theoretical steady state landfill gas generation rate is calculated at approximately 7.7×10^6 L/yr.

Following calculation of the total landfill gas generation rate, a conservatively estimated 50% methane, 40% carbon dioxide, 10% other gases (including 126 mg/kg total VOC concentration as determined from field testing) was used to estimate the total Non-methane Organic Carbon (NMOC) generation rate. A summary of the assumptions used and estimated theoretical NMOC generation rate based on Equation 1 is provided below:

- Landfill size: 11 acres
- Landfill thickness: variable between 5 to 25 feet
- Soil bulk density: 1.9 $gram/cm^3$ (119 lb/ft^3)
- Landfill gas composition: 50% methane, 40% carbon dioxide, 10% NMOC including all other gases
- NMOC field measurement: 126 mg/kg
- Landfill gas density: 1.28×10^{-3} $gram/cm^3$

A. • Landfill gas generation rate - 6 L/kg/yr (2.19 mL/kg/hr)(USEPA, 1991a)
Calculated NMOC emission rate is 0.62 Mg/yr (0.69 ton/yr).

B. Using the same assumptions as in (A) with the following modification:
• Landfill gas generation rate 50 mL/kg/hr (USEPA, 1991a)
Calculated NMOC emission rate is 1.90 Mg/yr (2.09 ton/yr).

EPA Tiered Estimation Methods

A second theoretical method of calculating the total landfill gas generation rate and NMOC generation rate can be conducted using the method introduced in the EPA's 1991 Tier 2 (site-specific data) Scholl-Canyon relation outlined in EPA's Air Emission Estimation Model (USEPA, 1991b). The Scholl-Canyon model assumes single stage first-order kinetic gas generation, where after a brief time lag (during which anaerobic conditions are established), gas production reaches its maximum rate. After the lag time, the gas production rate is then assumed to decrease exponentially as the organic fraction of the landfill refuse decreases. The Scholl Canyon model equation is as follows:

$$dG/dt = kL = kL_o e^{-kt} \quad (2)$$

where:

dG/dt = methane generation rate ($ft^3/lb/yr$)
k = rate constant ($year^{-1}$)
L = total refuse gas generation potential (ft^3/lb)
t = time (years)

The solution to Equation 2 may be simplified to the following form:

$$M_{nmoc}\ (Mg/yr) = RC_{nmoc}FL_0(e^{-kc} - e^{-kt}) \quad (3)$$

where:

M_{nmoc} = Mass of NMOC generated per year (Mg/yr)
R = waste deposition rate (yd^3/day)
C_{nmoc} = Concentration of nonmethane organic carbon (mg/kg)
F = unit conversion factor (7.19×10^{-9})
L_0 = total refuse gas generation potential (ft^3/lb)
k = methane generation rate constant ($year^{-1}$)
c = time since landfill closure (years)
t = time since landfill operations start (years)

In order to conduct a sensitivity analysis using Equation 3, a range of site-specific investigation results and literature values were utilized to obtain theoretical estimates of the NMOC generation rate at the McAllister Point Landfill. A summary of the ranges of input values used for the sensitivity analysis, along with the resulting NMOC generation rates, are summarized below.

C.
- Waste deposition rate of 40 cubic yards per day between mid 1950s and mid 1970s
- Waste bulk waste density of 70 lb/ft^3
- Methane generation potential (Lo) value (USEPA, 1993) of 230 m^3/Mg
- Methane generation rate constant (k) of 0.0264/yr
- NMOC field measurement of 126 mg/kg

Calculated NMOC emission rate is 0.63 Megagrams per year (Mg/yr) or 0.69 tons per year.

C1. Using the same assumptions as in (C) with the following modification:
- EPA recommended refuse density value (USEPA, 1991b) of 650 kg/m^3

Calculated NMOC emission rate is 0.37 Mg/yr (0.41 tons/yr).

D. Using the same assumptions as in (C) with the following modification:

- Methane generation rate constant (USEPA, 1991b) (k) of 0.0307/yr

Calculated NMOC emission rate is 0.65 Mg/yr (0.72 tons/yr).

D1. Using the same assumptions as in (D) with the following modification:

- EPA recommended refuse density value (USEPA, 1991b) of 650 kg/m^3 (p. G-13, EPA-450/3-90-011a)

Calculated NMOC emission rate is 0.38 Mg/yr (0.42 tons/yr).

E. Using the same assumptions as in (C) with the following modifications:

- Bulk waste density of 100 lb/ft^3
- Methane generation rate constant (k) (USEPA,1989) of 0.0307/yr

Calculated NMOC emission rate is 1.41 Mg/yr (1.55 tons/yr).

E1. Using the same assumptions as in (E) with the following modification:

- EPA recommended refuse density value (USEPA,1991b) of 650 kg/m^3

Calculated NMOC emission rate is 0.57 Mg/yr (0.63 tons/yr).

As indicated above, a range of NMOC values has been calculated to assess the relative sensitivity of the NMOC generation rate to variations in input parameters. The NMOC generation rate varied from 0.37 Mg/yr to 1.41 Mg/yr (0.41 to 1.41 ton/yr), based on the assumptions stated above.

According to the EPA's regulatory alternatives, if a landfill produces less than 25 Mg NMOC/yr, an active gas collection and control system is not required (USEPA, 1991b). The state of Rhode Island Department of Environmental Management (RIDEM) has also established a general air pollution discharge limit of five tons per year for emission sources (Air Pollution Regulation No. 9) (RIDOH/RIDEM, 1993). As noted, the maximum calculated theoretical NMOC generation rate of 1.41 Mg/yr is well below both the EPA and RIDEM threshold limits; therefore, based on two different theoretical estimates of landfill gas generation and NMOC generation potential, design and installation of an active gas control system is not warranted.

Site-Specific Analysis

Another approach to estimating the gas generation rate from the landfill was conducted using site-specific data gathered from the McAllister Point Landfill. This information was used to verify and/or modify theoretical estimates of the NMOC generation rate.

Data Collection

In order to validate results of the theoretical analysis described above under Theoretical Analysis, the composition and generation rate of site-specific landfill gases and NMOC was determined through a program of field testing.

Specific tasks which were conducted in support of the site-specific landfill gas assessment included:

- Installation of a series of vapor extraction wells and observation probes at the landfill
- Completion of a vapor extraction pilot test to assess the in-situ air permeability of materials at the landfill
- Sampling and analysis of the landfill gas.

A brief summary of each of these tasks is provided below.

Installation of Vapor Extraction Wells and Vapor Probes

A series of three vapor extraction wells and 16 vapor probes were installed on the landfill in three different configurations. Vapor extraction wells were installed to depths ranging from 13 to 16 feet below ground surface. These depths were equivalent to an elevation approximately two to three feet above the existing groundwater elevation. An 8 to 10-foot screened interval was established for each extraction well approximately 2 to 4 feet above the observed water table. Vapor probes were installed to depths ranging from 10 to 16 feet below ground surface. A uniform one-foot screen section was established at each observation probe at variable depths.

Pilot Testing

Each of the three landfill gas monitoring networks were pilot tested to assess the potential impact subsurface landfill gases may have on the air quality at the site, and on the physical composition and integrity of the proposed cap structure. Pilot tests were terminated after steady state conditions were achieved. Specifically, pilot scale testing was terminated when a steady state pressure condition, defined as no significant pressure changes observed for a period of five minutes or longer, was achieved in the vapor probes. Four pilot scale tests were conducted at the McAllister Point Landfill for durations varying between 4.25 minutes and 8 hours.

To estimate the air permeability of the landfill material, pilot scale testing results were used with the following equation modified from EPA's Hyperventilate model (USEPA, 1991c):

$$k = (Q/H)(m/p/P_w)[\ln(R_w/R_{obs})/(1-(P_{obs}/P_w)^2)]f_c \qquad (4)$$

Where:

k = intrinsic permeability (cm^2)
Q = vapor flow rate (ft^3/min)
H = vapor intake length (screened interval) (ft)
μ = air viscosity (g/cm sec)
π = 3.14159
P_w = vapor extraction well head pressure (inches of water)

R_w = vapor extraction well radius (ft)
R_{obs} = distance between extraction well and observation probe (ft)
P_{obs} = vapor observation probe pressure (inches of water)
f_c = conversion factor to obtain k in cm^2

Use of Equation 4 with a combination of field-derived data and literature values resulted in a site-specific landfill waste permeability which ranged between 2.6×10^{-8} cm^2 and 1.1×10^{-6} cm^2. The resulting derived conductivities may be then calculated from the following equation:

$$K = k\rho g/\mu \quad (5)$$

Where:
K = Air Conductivity (ft/day)
k = intrinsic permeability (ft^2)
ρ = air density (lb/ft^2)
g = gravitational constant (lb/ft^2)
μ = air viscosity (lb/ft day)

Derived conductivities varied from 7.1 ft/day to 302 ft/day for the McAllister Point Landfill.

Landfill Gas Sampling and Laboratory Analysis

To assess the landfill gas quality, a series of landfill gas samples were collected for field and quantitative analyses. Review of available data from previous site-specific soil gas surveys indicated that offgas controls would not be required for the short-term pilot scale testing. Pre-pilot scale testing calculations indicated that the expected discharge rate of both total and individual regulated VOCs (most notably benzene) would be far below RIDEM air pollution discharge standards.

During pilot scale testing, landfill gas samples were collected from a sampling port on the blower discharge pipe from each of the three extraction wells, and from each of the vapor observation wells at one of the three network configurations. Baseline landfill gas samples were collected in 3-liter Tedlar bags in accordance with EPA Method 18. Using this method, the Tedlar bag is directly connected to the 1/4" Teflon gas well sample line, which eliminates the potential for ambient air infiltration through plumbing leaks. Prior to sampling, the well was purged of two well volumes of air, using a low-flow pump configuration. Concentrations of methane, carbon dioxide, and oxygen were recorded for each sample using a hand-held infrared gas analyzer.

Landfill gas samples were collected for quantitative analysis from each of the three vapor extraction wells at the following three times during the pilot scale testing:

- At the initiation of the pilot scale test (e.g., at time t = 0 sec)
- Near the mid point of the test (variable times at t = 0.25 hr to 4.0 hr)

- Near the end point of the test (variable times at t = 1.0 to 8.0 hr).

In addition, one system blank sample was collected from the extraction system setup. Each of these samples was collected in a 3-liter Tedlar bag from a sampling port on the extraction system setup. Following collection, the samples were labeled and shipped to a qualified laboratory for analysis of volatile organic compounds (VOCs) and methane. VOCs were analyzed by a modified EPA Method T01 and T02 methodology using a GC/MS thermal desorption method, and for methane using a GC flame ionization detector methodology. Samples were field-analyzed for methane, carbon dioxide, and oxygen using the handheld infrared gas analyzer. Results of the landfill gas analysis are presented below, in the Results section.

Data Analysis

Site-specific methane concentration versus time data were plotted linearly. These data were then semilog plotted (methane concentration versus the log of the time), and then a linear regression curve fitting analysis was conducted. In this analysis, the curves follow the form of an exponential decay function where the concentration of methane approaches an asymptotic value after a long period of time. These results generally can be expressed as:

$$y_i = y_0 e^{-\alpha ti} + y_r \tag{6}$$

Where:

y_i = off-gas concentration at $t = t_i$
y_0 = static gas concentration at $t = 0$
e = natural log constant (2.718)
α = empirical constant
t_i = time at $t = t_i$
y_r = steady state gas generation rate

Based on the number of site-specific data points generated during the pilot scale testing, sufficient data exist to use best curve fitting techniques to extrapolate asymptotic values of the steady state gas generation rate y_r. A graphical depiction using site-specific data is provided as Figure 2. This graph, as well as calculations using the above relation (at times of t up to 300 hours), indicate that the percent methane in the offgas from the landfill appears to approach a value on the order of 5%, 6%, and 20%, depending on which network configuration was evaluated. In general, these values are much less than the approximate 50% methane concentration reported in the literature analysis, and confirm the conservative nature of the landfill gas assessment at this site using published values.

Figure 2. **Calculated methane levels versus time.**

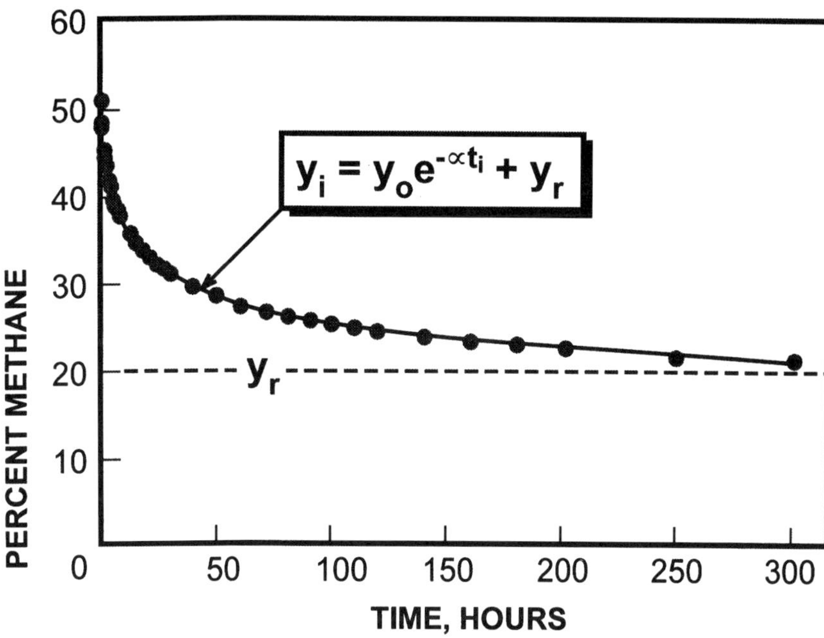

RESULTS

Landfill Gas Composition

Landfill gas samples collected during the pilot test generally indicated that concentrations of VOCs and methane from the initial gas sample (e.g., at t = 0) were highest and tended to drop with time over the course of the pilot extraction test. Minimum, maximum, and average concentrations of methane, carbon dioxide, and oxygen from the landfill gas samples are summarized below:

Parameter	Min. Conc.	Max. Conc.	Av. Conc. (n/8)
Methane (CH_4)	13.5%	56.0%	34.4%
Carbon Dioxide(CO_2)	7.2%	36.0%	17.9%
Oxygen (O_2)	not detected	15.1%	5.5%

The ratio of individual and total VOC concentrations to methane were calculated assuming a landfill gas composition of 50% methane, 40% carbon dioxide, and 10% other gases (USEPA, 1991a). This ratio was calculated as a means to determine the mass flux of individual VOCs within the landfill gas, since the rate of methane generation may be computed as described above. The assumption of 50% methane, 40% carbon dioxide, and 10% other is conservative, given the field data described above and the laboratory results. Additionally, the VOCs which comprise a portion of the 10% other gases (e.g., VOCs, O_2, etc.) are the compounds which are regulated by RIDEM and EPA, and determine the gas vent spacing analysis as discussed below. Results of these analyses were used to determine the landfill gas vent spacing by calculating the landfill area which would generate the RIDEM established discharge limit for specific regulated VOCs, as discussed below.

Landfill Gas Vent Spacing

The reported literature gas generation rate was used to estimate the possible pressure impacts on a landfill cap. This analysis was conducted by applying the ideal gas law to an assumed closed system (Richards, et al., 1962). Once again, such an analysis is anticipated to result in a conservative estimate of the pressure impacts to the cap. Results of this analysis indicate that the landfill gas will exert a net positive pressure of 1.2 psi (pounds per square inch) (15.8 psi absolute) on the underside of the cap if not allowed to escape through vents. Approximately 1.4 feet of soil cover would be required to resist the net 1.2 psi pressure. The current landfill cap design calls for a three-foot thick soil cover over the impermeable barrier layer. Therefore, even in the absence of a gas venting system, the pressure generated would not adversely impact the proposed RCRA Subtitle C cover system. Therefore, a passive gas venting design is appropriate to allow the controlled release of decomposition products of the waste material.

To ensure that landfill gas discharge is in compliance with RIDEM's air discharge regulations, the gas vent spacing was determined by calculating the landfill area (e.g., volume of landfill waste material) which would generate the RIDEM-established air discharge limits (RIDEM Air Regulation No. 9) for individual VOCs (USEPA, 1989). This was accomplished by utilizing the highest site-specific concentrations of individual VOCs detected by quantitative gas analysis using the following equation, modified from the EPA HyperVentilate model (V1.01 and V2.0) software (USEPA, 1991c).

$$A = VF/H_i = [2 \text{ x } DL_j/R_{LG}/(R_{JCH4}/\rho_S/\rho_{LG})]/H_i \tag{7}$$

Where:

A = Landfill Area where gas is generated (ft^2)
VF = Volume Factor (e.g., $[2{*}DL_j/R_{LG}/R_{JCH4}/Den_S/Den_{LG}]$) ($ft^3$)
H_i = Landfill Thickness of area i (ft)
DL_j = Discharge Limit of VOC j (lb/hr)
R_{LG} = Landfill Gas Generation Rate (mL/kg/day)
R_{JCH4} = Ratio of Individual VOC to methane (%/% or unitless)
ρ_S = Bulk Density of waste (g/cm^3)
ρ_{LG} = Landfill Gas Density (g/cm^3)

The following typical terms were used in the evaluation:

DL_j = Discharge Limit of VOCs
0.005 lb/hr benzene
1.14 lb/hr toluene
2.28 lb/hr ethyl benzene
1.26 lb/hr others

R_{LG} = Landfill Gas Generation Rate
50 mL/kg/day (EPA, 1991)
2.19 mL/kg/day (from EPA 1991 - 6 L/kg/yr)

ρ_S = Assumed Bulk Density of Waste
1.9 g/cm^3

ρ_{LG} = Landfill Gas Density
1.28×10^{-3} g/cm^3

The remaining terms within the relation (H_i and R_{JCH4}) were varied to calculate vent spacing (or area per vent), assuming differing discharge conditions. As would be expected, the vent spacing is dependent on the landfill depth, landfill gas generation rate, and RIDEM discharge limits.

As anticipated, the VOC with the lowest discharge limit, benzene, resulted in the smallest area per vent, and thus the closest required vent spacing. An area per vent of 24,498 ft^2, or a vent spacing of 157 feet was calculated with the conservative published gas generation rate (50 mL/kg/day) at the maximum landfill thickness of 25 feet. Calculations for VOCs, other than benzene, resulted in vent spacings over 2,000 feet. An approximately 150 foot landfill gas vent spacing was utilized for landfill design purposes (TRC Environmental Corporation,

Figure 3. **Gas vent system.**

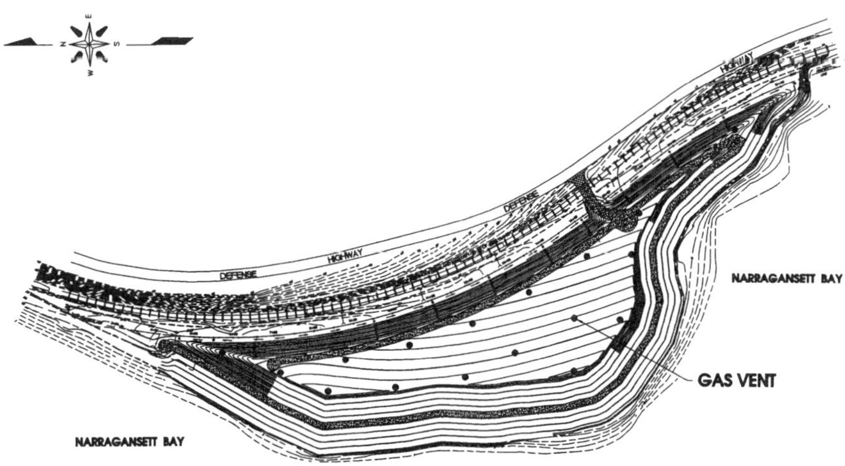

Figure 4. **Gas vent section.**

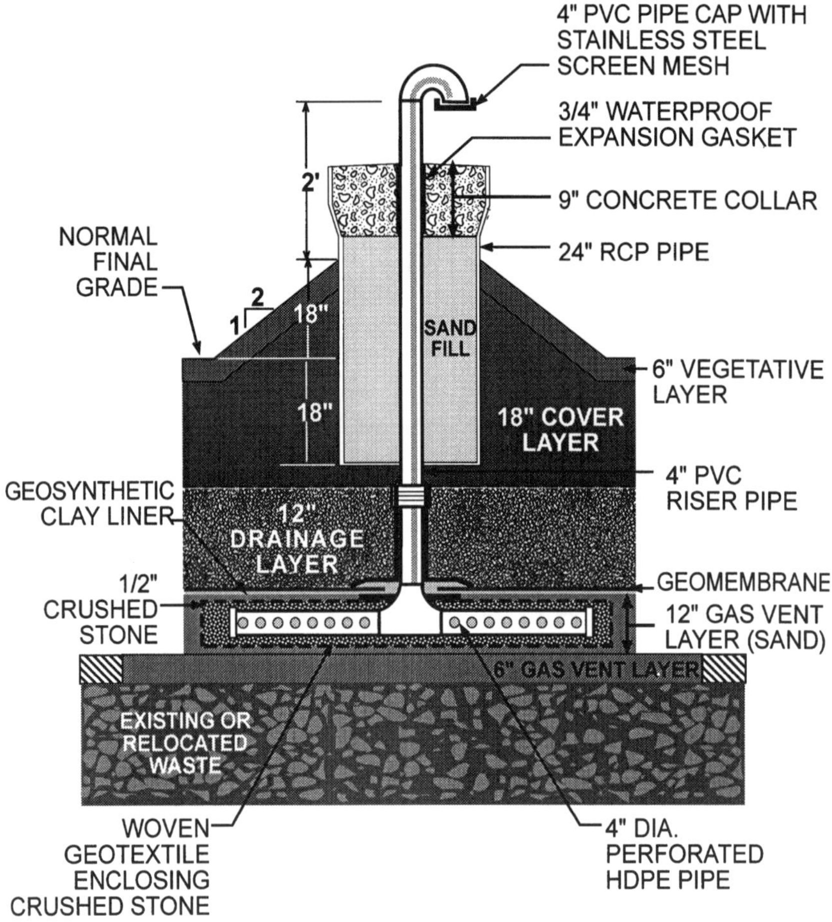

1994b). However, actual landfill vent spacings vary from approximately 100 to 260 feet. Figure 3 illustrates the vent spacing utilized at the McAllister Point landfill. In order to monitor potential gas migration away from the landfill, a series of landfill gas monitoring wells is located around the entire landward side of the landfill. A representative landfill gas vent cross section is provided as Figure 4.

CONCLUSIONS

Results of the landfill gas assessment indicate low average concentrations of regulated VOCs and methane. Through evaluation of the data and conceptual design, it was determined that a passive gas venting system is appropriate for the site. Therefore, a passive venting system has been designed at the landfill with a vent spacing established to protect the cap from pressure buildup. Furthermore, it was projected that concentrations of VOCs emitted after construction of the cap would be well below current EPA and RIDEM regulations, and therefore are not anticipated to require collection and treatment. However, given that landfill gas generation rates may deviate from that estimated, a landfill gas monitoring program will be conducted on both landfill gas vents and off-landfill gas monitoring wells to assess the need for gas control measures following construction of the cap.

REFERENCES

Emcon Associates 1982. *Methane Generation and Recovery from Landfills*, Ann Arbor, MI, Ann Arbor Science Publishers.

Richards, et al. 1962. *Modern College Physics*, Reading, MA, Addison-Wesley.

RIDOH/RIDEM. 1993. RI Clean Air Act, Air Pollution Control Regulation No. 9 & 14.

TRC Environmental Corporation. 1994a. Design Analysis for Bidding Document Submission, Cap McAllister Point Landfill, Naval Education and Training Center, Newport, Rhode Island, May.

TRC Environmental Corporation. 1994b. Phase II Remedial Investigation Report, McAllister Point Landfill, Naval Education and Training Center, Newport, Rhode Island.

(USEPA) U.S. Environmental Protection Agency. 1989. Technical Guidance Document: Final Covers on Hazardous Waste Landfills and Surface Impoundments.

(USEPA) U.S. Environmental Protection Agency. 1991a. Seminar Publication, Design and Construction of RCRA/CERCLA Final Covers, EPA/625/4-91/025, May.

(USEPA) U.S. Environmental Protection Agency. 1991b. Air Emissions from Municipal Solid Waste Landfills - Background Information for Proposed Standards and Guidelines, EPA-450/3-90-011a, March.

(USEPA) U.S. Environmental Protection Agency. 1991c. Hyperventilate Model.

(USEPA) U.S. Environmental Protection Agency. 1993. Standards of Performance for New Stationary Sources and Guidelines for Control of Existing Sources: Municipal Solid Waste Landfills, 40 CFR Parts 51, 52 and 60, June 21, 1993.[1993 EPA *Federal Register* 58, No. 117, pp. 33790-33792, June 21, 1993.]

(USEPA) U.S. Environmental Protection Agency. 1994. Seminar Publication, Design, Operation, and Closure of Municipal Solid Waste Landfills, EPA/625/R-94/008, September.

CHAPTER 49

A Numerical Method for Determining the Distribution of Nonaqueous-Phase Liquid Contamination in a Confined/Unconfined Aquifer Setting

Mark E. Kelley and **George G. Yu**, Battelle Memorial Institute, Columbus, Ohio

Mark D. White, Battelle Pacific Northwest Laboratories, Richland, Washington

Stephen H. Rosansky and **Christopher J. Perry**, Battelle Memorial Institute, Columbus, Ohio

INTRODUCTION

Measurements of free product thickness from monitoring wells commonly are used to estimate the volume of oil in a formation; however, product thickness in a monitoring well is not representative of the actual product thickness in the porous medium. Furthermore, the thickness of the NAPL (nonaqueous-phase liquid) contaminated zone in the porous medium does not determine the volume of NAPL in the formation, because the pore spaces within the NAPL contaminated zone are partly occupied by water and air. In an unconfined aquifer, the NAPL resides in the mutual presence of water below the water table and with both water and air above the water table (i.e., in the vadose or unsaturated zone). In a confined aquifer, the pores are totally saturated and the NAPL resides together with water in the pore spaces. To determine the actual volume of NAPL present in the subsurface, the spatial distribution of NAPL saturation within the contaminated zone must first be determined.

This chapter presents a numerical modeling method for calculating NAPL volume and distribution in heterogeneous porous media under confined and unconfined conditions. This method allows the site geology to be numerically simulated at any point (e.g., a monitor well) as a vertical array of nodes, each with unique properties to reflect heterogeneity within the subsurface materials. Numerical models were constructed to calculate vertical NAPL saturation profiles under equilibrium conditions (i.e., nonfluctuating water table) at monitor wells where measurements of NAPL thicknesses are available. The total NAPL volume in the stratified aquifers was determined by spatially integrating the vertical saturation profiles at each well over the area of contamination within each aquifer. The proposed method is applied to a site in the midwest United States where a fuel oil LNAPL (light nonaqueous-phase liquid) is present in a layered unconfined/confined-aquifer setting.

BACKGROUND

Lenhard and Parker (1990) developed an approach to calculate specific hydrocarbon volume in an aquifer from fluid level measurements in wells containing water and hydrocarbon. Fluid pressure distributions are inferred from well fluid levels; these fluid pressure distributions in turn are used to predict vertical fluid saturation distributions that are integrated to determine the specific hydrocarbon volume. Expressions for three-phase saturation/capillary-pressure relationships for the aquifer are required to predict fluid saturation profiles.

These expressions can be developed from two-phase retention functions according to a scaling method developed by Parker et al. (1987). The van Genuchten (1980) model describes the water saturation/capillary-pressure relationship for a two-phase air-water system as following:

$$S_e = \left[1 + (\alpha h)^n\right]^{-m} \qquad \text{for } h > 0 \qquad (1)$$

$$S_e = 1 \qquad \text{for } h \le 0 \qquad (2)$$

where S_e is the effective water saturation, h is the capillary pressure head (i.e., suction head), and α and n are empirical curve-fitting parameters for the van Genuchten model (α has the dimension L^{-1} and n is dimensionless), and $m = 1\text{-}1/n$. Lenhard and Parker (1988) applied the scaling procedure developed by Parker et al. (1987) to the van Genuchten water retention function to obtain saturation/capillary-pressure relations for multiphase systems. Their general form of the scaled van Genuchten function is

$$S^*(h^*) = \left[1 + (\alpha h^*)^n\right]^{-m} \qquad \text{for } h^* > 0 \qquad (3)$$

$$S^*(h^*) = 1 \qquad \text{for } h^* \le 0 \qquad (4)$$

where

$$S^* = \frac{S_j - S_m}{1 - S_m} \qquad (5)$$

$$h^* = \beta_{ij} h_{ij} \tag{6}$$

S_j is the saturation of the wetting fluid j for fluid system ij, S_m is the irreducible wetting-phase saturation, β_{ij} is the scaling coefficient for the two-phase system with fluids i and j, and h_{ij} is the capillary head for fluid system ij. Thus, the scaled van Genuchten function allows the calculation of the saturation of any fluid in a multiphase system when the capillary head between that fluid and its adjacent fluid is known. For example, in a three-phase air-NAPL-water system in which water is the wetting fluid, the water and total liquid saturations are independent functions of NAPL-water and air-NAPL capillary heads, and can be calculated as

$$\overline{S_w} = (\beta_{nw} h_{nw}) = S^*\left(h^*\right) \tag{7}$$

$$\overline{S_t} = (\beta_{an} h_{an}) = S^*\left(h^*\right) \tag{8}$$

where the effective water and total liquid saturations are defined by

$$\overline{S_w} = \frac{S_w - S_m}{1 - S_m} \tag{9}$$

$$\overline{S_t} = \frac{S_w + S_n - S_m}{1 - S_m} \tag{10)}$$

in which S_w is the total water saturation, S_m is the irreducible water saturation, and S_n is the NAPL saturation. Parameters β_{an} and β_{nw} are defined by Lenhard and Parker (1987) as fluid pair-dependent scaling factors:

$$\beta_{an} = \frac{\alpha_{aw}}{\alpha_{an}} \tag{11}$$

$$\beta_{nw} = \frac{\alpha_{aw}}{\alpha_{nw}} \tag{12}$$

in which α_{aw} is the surface tension of uncontaminated water, α_{an} is the surface tension of the NAPL, and α_{nw} is the interfacial tension between the NAPL and the water. Thus, it is possible to calculate the effective water saturation and total liquid saturation by substituting the terms ($\beta_{nw} h_{nw}$) and ($\beta_{an} h_{an}$) for h^* into the scaled van Genuchten function (Equations 3-4). The NAPL saturation, then, is the difference between the total liquid saturation and the effective water saturation.

$$S_n = S_t - \overline{S_w} \tag{13}$$

The capillary heads in a three-phase system can be defined as the pressure difference between phases:

$$h_{an} = h_a - h_n \tag{14}$$

$$h_{nw} = h_n - h_w \tag{15}$$

in which h_{an} is the air-NAPL capillary head, h_{nw} is the NAPL-water capillary head, h_a is the gas-phase pressure head assumed to be zero (i.e., atmospheric pressure), h_n is the NAPL-phase pressure head, and h_w is the water-phase pressure head. The capillary heads can be defined based on fluid measurements in a well:

$$h_{an} = \rho_{rn}(z - z_{an}) \tag{16}$$

$$h_{nw} = (1 - \rho_{rn})(z - z_{nw}) \tag{17}$$

where ρ_{rn} is the specific gravity of the NAPL phase, z_{an} is the elevation in the well of the air-NAPL interface, z_{nw} is the elevation in the well of the NAPL-water interface, and z is a reference elevation. Thus, well fluid levels can be related to vertical pressure distributions by the capillary heads. The fluid levels z_{an} and z_{nw} for a well containing LNAPL are illustrated in Figure 1. The fluid level z_{aw} shown on Figure 1 is the position of the air-water interface in the absence of an LNAPL. The position of z_{aw} can be calculated from the other fluid levels:

$$z_{aw} = z_{nw} + \rho_{rn}(z_{an} - z_{nw}) \tag{18}$$

Thus, well fluid-level measurements are used to calculate fluid pressure distributions in the formation adjacent to the well, which in turn are used to predict vertical fluid saturation distributions using the scaled van Genuchten equation. By vertically integrating the NAPL saturation between the top and bottom of the NAPL-contaminated zone, this method will yield the NAPL specific volume at the location of the well from which the fluid levels are measured.

PROPOSED METHOD

The proposed method uses the STOMP (Subsurface Transport Over Multiple Phases) numerical code developed at the Pacific Northwest Laboratory by White et al.(1994) and Lenhard et al. (1994) to calculate the volume of NAPL in an aquifer under equilibrium conditions. STOMP is an integrated finite-difference-based computer code for numerically modeling three-phase (i.e., air-NAPL-water) flow and transport in variably saturated porous media. The code calculates fluid saturations using the scaled form of the van Genuchten (1980) or Brooks and Corey (1964) retention functions for multiphase systems given prop-

Figure 1. **Relationship between fluid levels in a well and in the formation.**

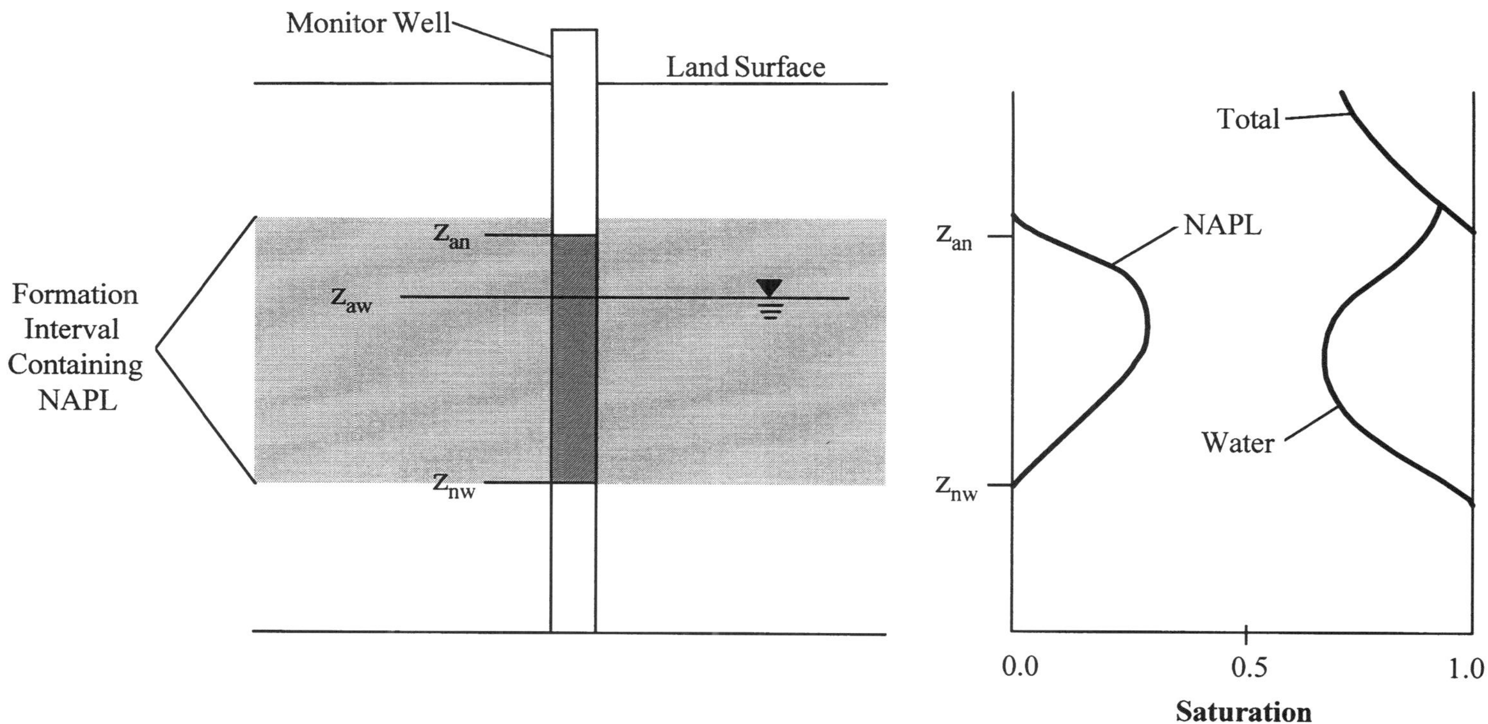

erties of the porous media and the fluids. This study shows how using a multiphase numerical modeling code such as STOMP to calculate NAPL saturation can extend the utility of the Lenhard and Parker (1990) analytical method for calculating NAPL volume by accounting for heterogeneities in the geologic media.

Model Setup

Under equilibrium conditions, a monitor well can be represented as a one-dimensional array of vertical nodes. The objective of the model simulations is to calculate the NAPL saturation profile through the formation interval containing NAPL (i.e., the interval where $S_n > 0$); therefore, the model domain must extend from the top of the NAPL-contaminated zone to the bottom of the NAPL-contaminated zone. Figure 1 illustrates the general relationship between NAPL thickness in a well and in the formation adjacent to the well. For an LNAPL, the bottom of the contaminated zone in the formation corresponds to the water-NAPL interface in the well, z_{nw}, according to the van Genuchten model. The top of the NAPL-contaminated zone in the formation occurs above the air-NAPL interface in the well, z_{an}, because of the presence of an air-NAPL capillary fringe in which $S_t < 1$ but $S_n > 0$. Therefore, the upper boundary of the NAPL-contaminated zone in the formation cannot be observed directly from well fluid levels, and must be calculated with the model. Thus, the model domain for the one-dimensional simulations was specified as the region between z_{nw} and an arbitrary elevation above z_{an}.

Solution of the scaled van Genuchten equation for NAPL saturation requires specifying a boundary condition to the problem. At the NAPL-water interface, the water pressure is equal to the NAPL pressure, which can be calculated from fluid levels in the well.

$$P_n^{nw} = P_w^{nw} \tag{19}$$

Figure 2. **Initial pressure conditions for a well with 30 cm of NAPL.**

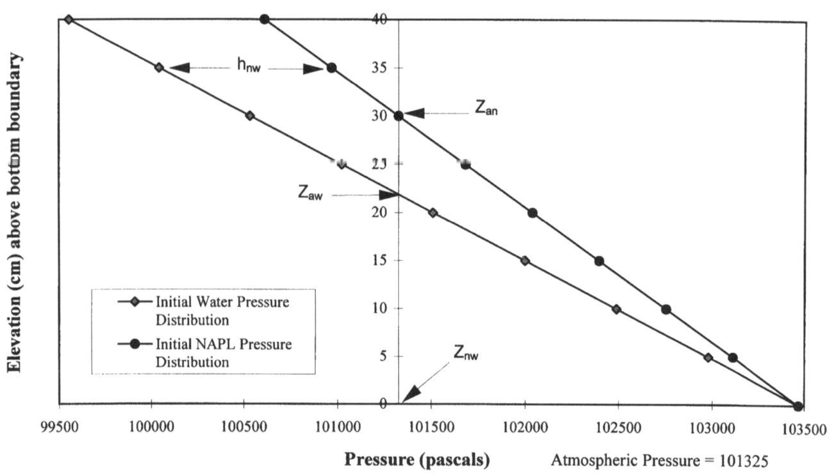

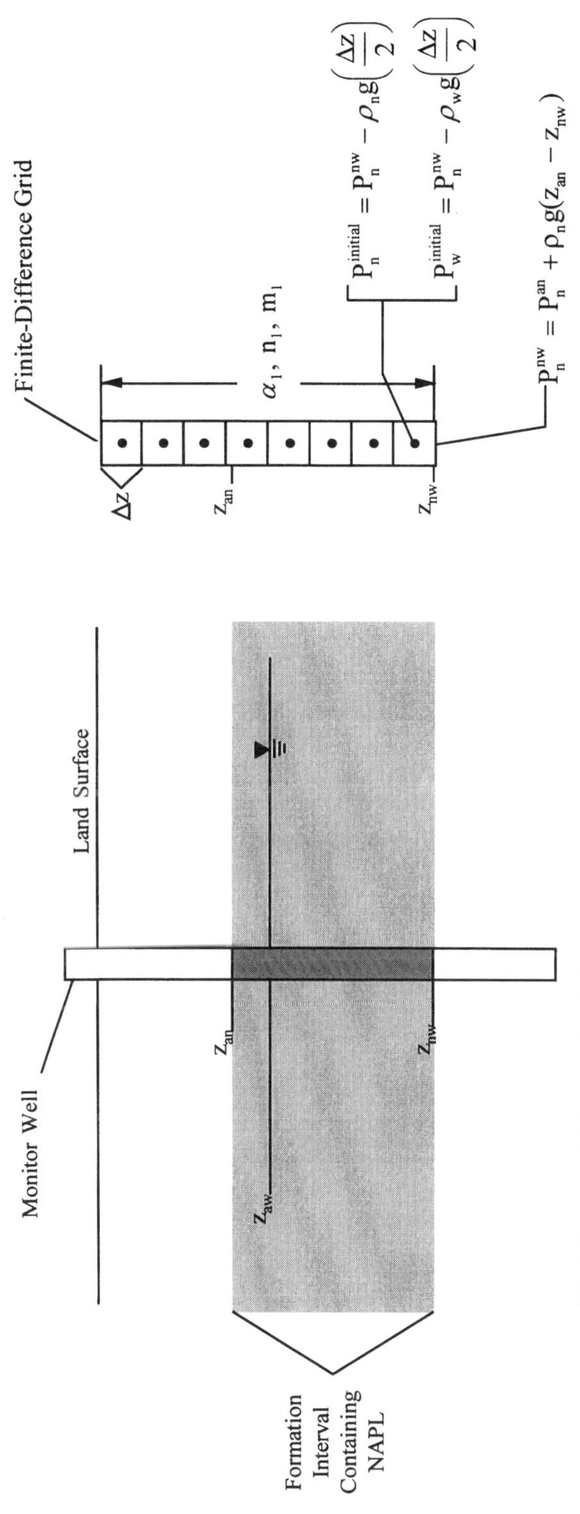

Figure 3. **Model representation of a well in an unconfined aquifer.**

Figure 4. **Model representation of a well in a confined aquifer.**

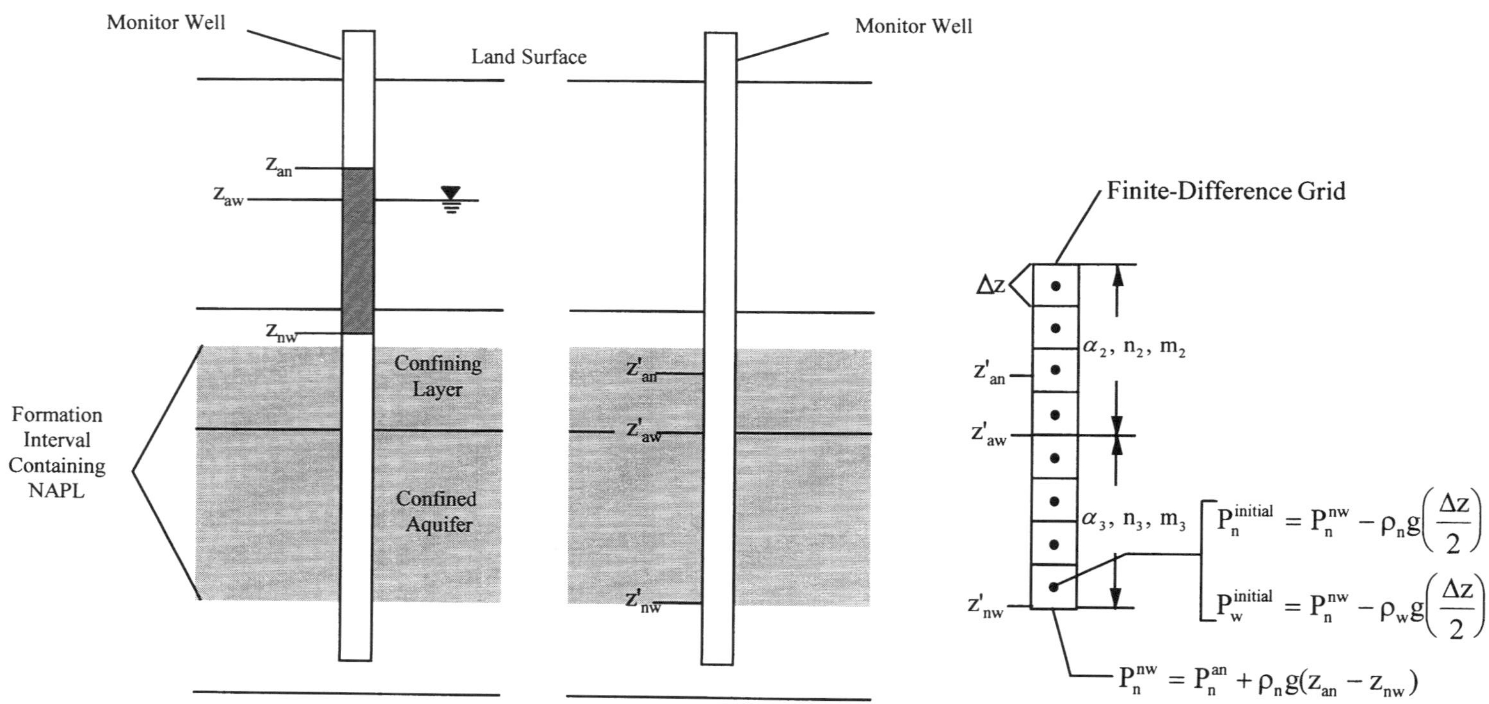

$$P_n^{nw} = P_n^{an} + \rho_n g(z_{an} - z_{nw}) \tag{20}$$

where P_n^{nw} and P_w^{nw} are the pressure of the NAPL phase and the water phase, respectively, at the NAPL-water interface in units of kg/m•sec^2 (i.e., pascals), ρ_n is the density of the NAPL phase (in units of kg/m^3), g is the acceleration due to gravity (in units of m/sec^2), and the fluid-level terms are as previously defined (in units of meters). The pressure of the NAPL at the air-NAPL interface is equal to atmospheric pressure which can be substituted into the above equation for the P_n^{an} term. Thus, the NAPL pressure at the NAPL-water interface is equal to the sum of the atmospheric pressure and the pressure due to the weight of the NAPL column. Above the NAPL-water interface, the water and NAPL pressures are dissimilar.

Initial pressure conditions must also be specified for water and NAPL for each node in the vertical one-dimensional model. Knowing the water and NAPL pressures at the water-NAPL interface, initial fluid pressures are calculated for every node in the model as:

$$P_n = P_n^{nw} - \rho_n g(z) \tag{21}$$

Figure 5. **Site Map showing Upper and Lower Sand wells that contain oil.**

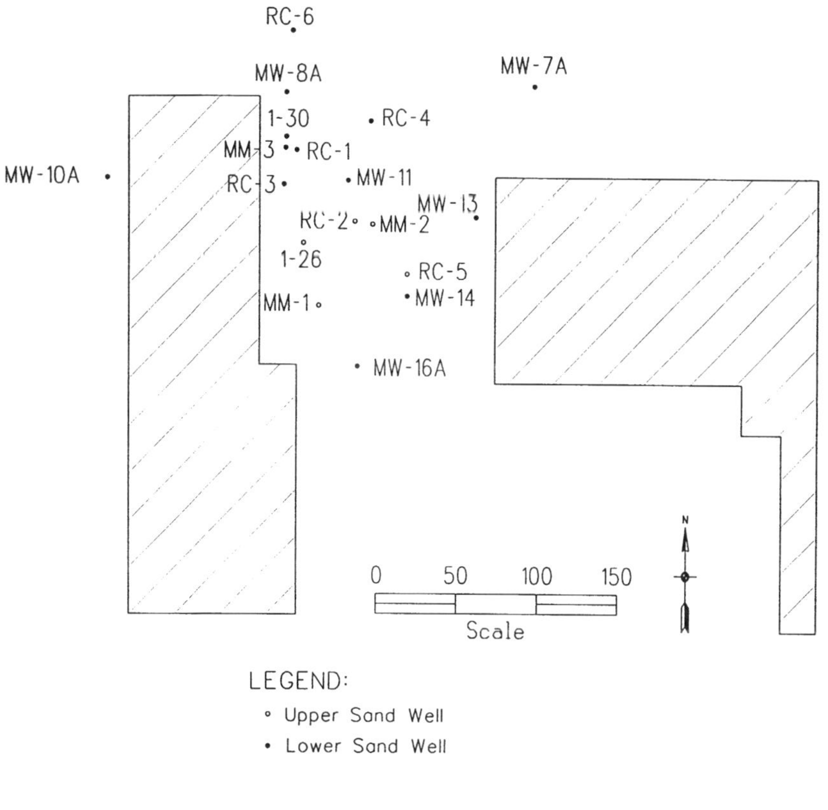

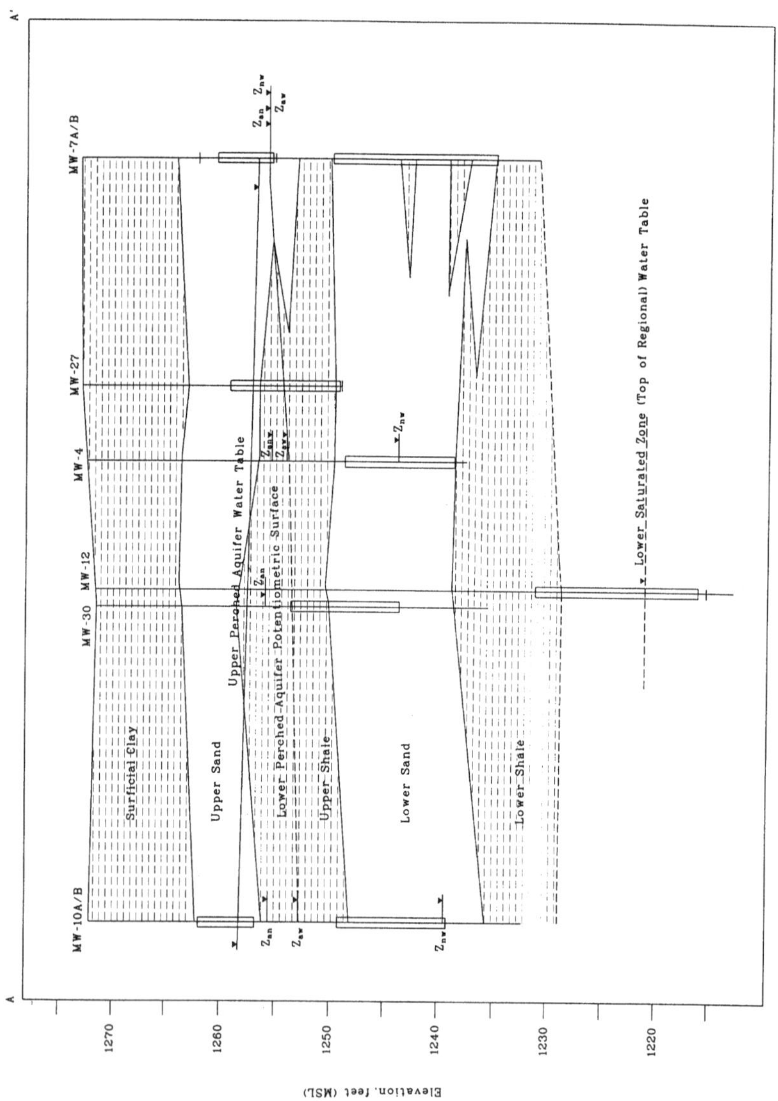

Figure 6. **Hydrogeologic cross section.**

$$P_w = P_w^{nw} - \rho_w g(z) \quad (22)$$

where z is the vertical distance from the NAPL-water interface to the node where the pressure is specified. Figure 2 illustrates how the initial water and NAPL pressure conditions would be specified for a well containing 30 cm of LNAPL having a density of 730 kg/m^3. Under these conditions, the NAPL pressure is equal to the atmospheric pressure at the air-NAPL interface, and the water pressure is equal to the atmospheric pressure at z_{aw}.

Simulation of Unconfined and Confined Aquifer Wells

In this study, one-dimensional vertical models were constructed for two types of wells containing LNAPL, those completed in an upper unconfined aquifer, and those completed in a lower confined aquifer. The model setup for the wells in the unconfined aquifer was identical to that discussed above, where air-NAPL and NAPL-water fluid levels are used to calculate a bottom boundary pressure and an initial NAPL and water pressure distribution at every node in the model. Figure 3 illustrates the one-dimensional numerical model used to simulate unconfined aquifer wells. One set of the van Genuchten α and n parameters is required to describe the characteristic moisture retention curve for the upper sand. In addition, one set of fluid-dependent scaling factors, β_{an} and β_{nw}, is required to relate the properties of the fluids to the scaled form of the van Genuchten model.

Wells completed in the confined aquifer must be treated differently from wells in the unconfined aquifer (Figure 4). Fluid levels in confined aquifer wells do not correspond to the position of the fluid levels in the formation because confining pressure acting on the aquifer causes the fluid levels in the well to rise above the aquifer where the fluids reside. In a confined aquifer, an LNAPL becomes trapped by the overlying low-permeability confining layer analogous to a natural oil deposit. To calculate the NAPL and water saturation profiles, the formation interval containing the NAPL must be determined. The position of the NAPL-containing interval is estimated by calculating a new z_{aw} called z'_{aw} that is positioned at the boundary between the confined aquifer and the overlying confining layer. Then, a new z_{nw}, called z'_{nw} is calculated and a new z_{an}, called z'_{an}, so that

$$z_{aw} - z_{nw} = z'_{aw} - z'_{nw} \quad (23)$$

$$z_{an} - z_{aw} = z'_{an} - z'_{aw} \quad (24)$$

This situation represents a confined aquifer in which the hydraulic head occurs at the boundary with the overlying confining layer. Consequently, the resulting NAPL saturation profile will straddle the aquifer-confining layer contact, thereby indicating that NAPL resides in both geologic units. In this case, two sets of van

Genuchten parameters are required to describe the characteristic moisture retention curve for each layer. Fluid-dependent scaling factors β_{an} and β_{nw} are also required to describe the behavior of the fluids.

CASE STUDY

Our method for calculating NAPL distribution was applied to a site in the midwest United States where fuel oil had been released to the subsurface over a period of approximately 50 years from a buried 250,000-gallon concrete tank. Fuel oil apparently leaked from near the base of the tank, located about 20 feet below ground surface. When the tank was installed, excavation of the tank pit breached a shale layer that separates a shallow near-surface layer of loosely consolidated sand from an underlying layer of sandstone. Head measurements show that the shale layer, where present, acts to confine groundwater in the underlying sandstone layer and causes groundwater to perch above it in the overlying sand layer. The base of the tank was emplaced below the shale layer, thereby allowing fuel oil to enter directly into the lower sandstone layer, where it was trapped by the overlying shale. As much as 16 feet of oil has been measured in monitor wells completed in the sandstone layer over a large area. The apparent thickness and areal extent of the oil in the upper sand layer is much smaller, suggesting that some fuel might have been released directly into this layer, or that a limited amount of upward seepage from the lower sand occurred, perhaps in the area where the tank breached the shale. Elsewhere, the shale layer appears to be an effective barrier preventing the upward migration of oil into the upper sand layer. Figure 5 is a site map showing the locations of wells where oil was measured in the upper and lower sand layers.

Site Hydrogeology

The near-surface geology of the site consists of five stratigraphic units that are recognizable across most of the site. Figure 6 is a cross section through the site, illustrating these stratigraphic units. Together, these units comprise a package of unconsolidated and consolidated sediments approximately 40 feet thick. Starting at ground surface, the units include a low-permeability layer of clay and sandy and clayey fill; a layer of unconsolidated silty and clayey fine to coarse sand (the Upper Sand); a low-permeability layer comprised of weathered shale and siltstone containing traces of very fine to fine sand (the Upper Shale); a layer of weakly cemented, cross-bedded sandstone comprised of fine to medium sand (the Lower Sand); and a lower shale layer that everywhere underlies the Lower Sand (the Lower Shale).

Groundwater occurs primarily in the Upper Sand and Lower Sand. Hydraulic head measurements from monitor wells completed in these layers indicate that groundwater occurs in the Upper Sand under unconfined conditions and in the Lower Sand under confined conditions. The Upper Shale acts to both perch groundwater in the Upper Sand and to confine groundwater in the Lower Sand. Heads from monitor wells completed in a regional water-bearing zone beneath the Lower Shale reveal the presence of a partially saturated zone immediately beneath the Lower Shale, indicating that groundwater in the Lower Sand is perched by the Lower Shale unit. Thus, the Lower Sand is a perched, confined aquifer. Oil has been detected in the Upper and Lower Sands, but not in the deeper regional aquifer.

Model Parameters

A model grid was developed for each well based on fluid levels measured in the well and the layer in which the well is screened. Table 1 lists the elevations of fluid levels and the geologic layer contacts for each well. Lower Sand wells are screened in the Lower Sand or across the Lower Sand/Upper Shale contact. Upper Sand wells are screened within the Upper Sand. Adjusted fluid levels (i.e., z'_{aw}, z'_{aw}, and z'_{nw}) are provided for the wells in the Lower Sand to show how the correction was made for confining conditions. Model grids for wells in the Upper Sand were constructed directly from measured fluid levels without making corrections.

The van Genuchten model parameters (i.e., α and n) were estimated for each layer using the computer code SoilProp© developed by Mishra and Parker (1993). SoilProp© converts particle-size distribution data to a soil-water retention function which is then fitted to the van Genuchten model by nonlinear regression in which α and n are the curve-fitting parameters. SoilProp© requires inputs of soil bulk density and/or saturated water content in addition to the particle size data. If one of these parameters is unknown, SoilProp© will estimate its values given a value of porosity and assuming a particle density of 2.65 g/cm^3.

Porosity values were estimated from literature values and provided as input, along with measured grain-size data and bulk densities, to determine values for the van Genuchten parameters for the three hydrogeologic layers represented in the one-dimensional models: the Upper Sand, the Upper Shale, and the Lower Sand. Given the uncertainty in porosity, van Genuchten parameters were calculated for a range of porosity for the Upper and Lower Sand. Table 2 gives the calculated van Genuchten parameters for the minimum and maximum porosity assumed for each layer. Only one set of van Genuchten parameters was calculated for the Upper Shale because we are interested primarily in determining the NAPL distribution in the sand layers.

The grain-size distribution (i.e., lithology) was assumed to be uniform throughout each layer; however, variable lithologies could have been accounted for within each layer by assigning separate van Genuchten parameters for each lithology.

Table 1. **Model Grid Parameters**

Well ID	Land Surface Elevation (ft msl)	Top of Upper Sand (ft msl)	Top of Upper Shale (ft msl)	Top of Lower Sand (ft msl)	Top of Lower Shale (ft msl)	Bottom of Lower Shale (ft msl)	Z_{an} (ft msl)	Z_{aw} (ft msl)	Z_{nw} (ft msl)	Z'_{an} (ft msl)	Z'_{aw} (ft msl)	Z'_{nw} (ft msl)
Upper Sand Wells												
MW-26	1272.42	1262.92	1254.92				1258.08		1257.94			
MM-1	1272.85						1257.08		1250.75			
MM-2	1273.40						1258.51		1255.83			
RC-2	1273.19						1257.20		1256.94			
RC-5	1274.50						1257.76		1256.17			
Lower Sand Wells												
MW-7A	1273.79	1264.79	1253.79	1250.79	1235.79	1231.79	1256.58	1256.56	1256.44	1250.81	1250.79	1250.67
MW-8A	1272.32	1262.82	1257.32	1249.32			1254.23	1253.21	1248.24	1250.34	1249.32	1244.35
MW-10A	1272.29	1262.29	1256.29	1248.29	1235.79		1255.64	1252.91	1239.59	1251.02	1248.29	1234.97
MW-11	1272.82	1263.32	1256.32	1247.82	1238.82		1255.52	1253.93	1246.14	1249.41	1247.82	1240.03
MW-13	1274.33	1264.33	1257.33	1254.33	1239.33		1256.08	1254.46	1246.54	1255.95	1254.33	1246.41
MW-14	1274.33	1264.33	1254.33	1246.33	1239.33		1256.57	1254.57	1244.80	1248.33	1246.33	1236.56
MW-16A	1272.57	1262.57	1255.57	1249.57	1238.07	1234.57	1256.18	1254.01	1243.39	1251.74	1249.57	1238.95
MW-30	1271.92	1263.92	1258.92	1250.42	1238.92		1256.32	1254.33	1244.61	1252.41	1250.42	1240.70
MM-3	1272.18			1250.00			1256.02	1254.41	1246.57	1251.61	1250.00	1242.16
RC-1	1272.15			1250.00			1254.79	1254.22	1251.45	1250.57	1250.00	1247.23
RC-3	1272.28	1262.78	1255.78	1245.78			1255.31	1253.51	1244.72	1247.58	1245.78	1236.99
RC-4	1273.05	1264.05	1257.05	1250.05	1239.05		1255.98	1253.95	1244.03	1252.08	1250.05	1240.13
RC-6	1271.98	1261.98	1255.48	1246.00			1253.41	1253.12	1251.68	1246.29	1246.00	1244.56

Table 2. **Soil Properties**

	Porosity	α	n	m
Upper Sand	0.21	9.49E-03	1.95	0.49
	0.53	4.26E-02	1.39	0.28
Upper Shale	0.31	2.22E-03	1.34	0.25
Lower Sand	0.14	1.52E-02	1.32	0.24
	0.49	4.08E-02	1.32	0.24

Note: α and n are van Genuchten parameters; m=1-1/n

Fluid-pair dependent scaling factors were calculated from measurements of surface tension and interfacial tension performed on samples of NAPL and groundwater from the site. A value of 2.3 was calculated for the air-NAPL scaling factor, β_{an}, and a value of 1.8 was calculated for the NAPL-water scaling factor, β_{nw}, based on measured values of α_{aw} (68.5 dynes/cm), α_{an} (29.7 dynes/cm), and α_{nw} (38.8 dynes/cm).

Results

Numerical models were constructed for 5 wells in the Upper Sand and 13 wells in the Lower Sand. Two simulations were performed for each well in the Upper Sand, corresponding to a high and a low porosity and a different set of corresponding van Genuchten parameters. A pair of simulations was performed for each well in the Lower Sand corresponding to the high and low porosity assumed for that layer. The porosity of the Upper Shale was held constant for simulations involving wells in the Lower Sand.

Model output includes calculated NAPL and water saturation at each node in the one-dimensional grid. A saturation profile can be plotted for each well showing the vertical distribution of NAPL in the formation adjacent to the well. Figure 7 is a calculated saturation profile for well MM-2 in the Upper Sand. The saturation profile shows an upward increase in NAPL saturation from the NAPL-water interface toward the air-NAPL interface. However, the NAPL saturation decreases above the air-NAPL interface because of the presence of air in addition to water and NAPL above this point. The upper boundary of the NAPL-contaminated zone can be observed on the saturation profile at the elevation where $S_n = 0$.

Figure 7. **Calculated saturation profile for well MM-2 in the Upper Sand (low porosity).**

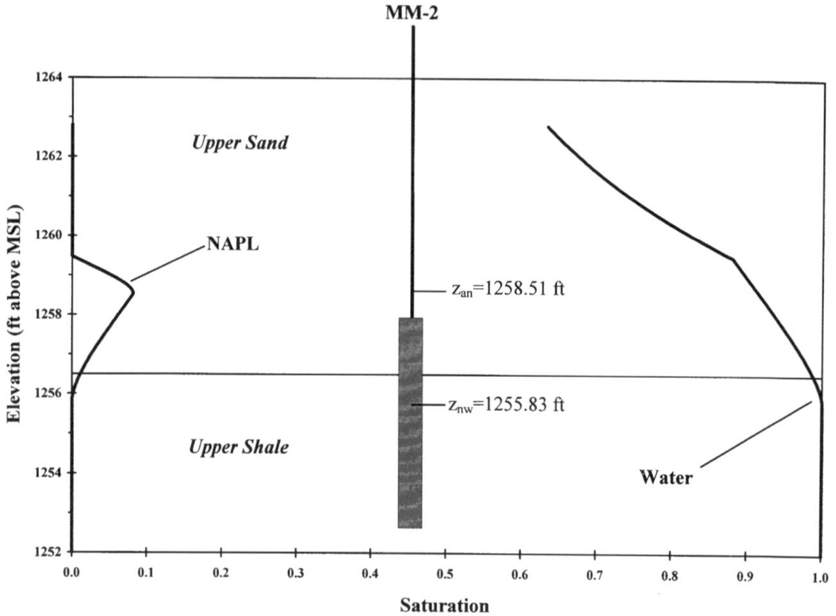

Figure 8. **Calculated saturation profile for well MW-8A in the Lower Sand (low porosity).**

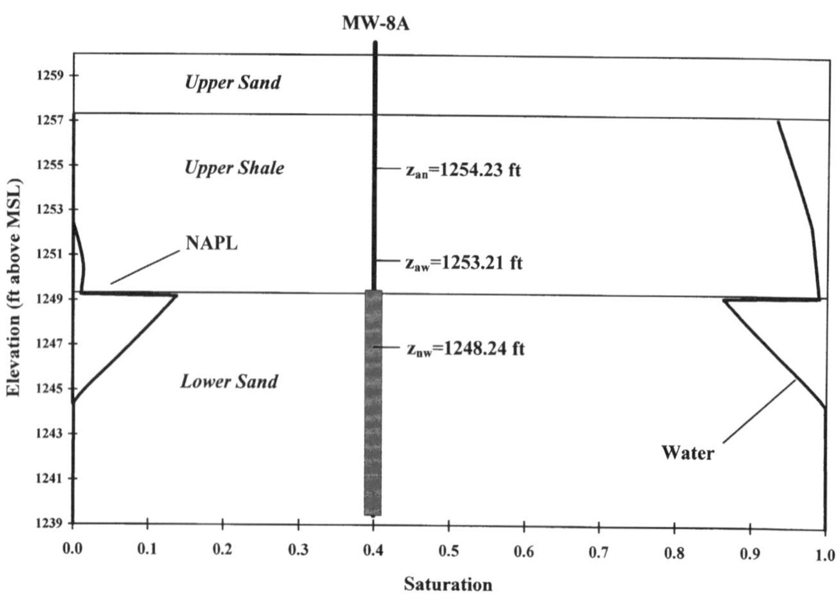

Figure 8 is the calculated saturation profile for well MW-8A in the Lower Sand. This saturation profile shows an abrupt decrease in NAPL saturation at the contact between the Lower Sand and the Upper Shale, resulting from the lithology change at this point. Above the contact, NAPL saturation increases toward the air-NAPL interface, where it reaches a local maximum in the Upper Shale. Above this point, the shale is partially saturated ($S_t < 1$), and NAPL saturation decreases upward until it reaches zero at the upper boundary of the NAPL-contaminated zone.

The specific volume of NAPL in the aquifer at each well can be calculated by summing the product of NAPL saturation and porosity for each node in the model grid as shown in Eq. 25.

$$V_s = \sum \left(\frac{S_n \cdot n \cdot \Delta V}{A_{cell}} \right) \tag{25}$$

where V_s is specific volume, S_n is NAPL saturation, n is porosity, ΔV is cell volume, and A_{cell} is cell area. Specific volume is the total volume of NAPL present per unit area of aquifer (expressed in units of cm). Table 3 gives the calculated specific volumes for each well. To derive an estimate of the total volume of NAPL in the aquifers, specific volumes must be determined for the areas between the wells. Then, once a value of specific volume is assigned to each unit area of contaminated aquifer, the values are summed to obtain an estimate of total NAPL volume in the aquifer.

Contour maps of the calculated NAPL specific volume are shown in Figure 9 for the Upper and Lower Sand in both low- and high-porosity cases. The values of specific volume shown on the Lower Sand contour map represent the sum of NAPL volume in the Lower Sand and the Upper Shale. The areal extent of NAPL in the Upper Shale is the same as in the Lower Sand because NAPL resides in both layers along the contact between the two layers. However, the volume of NAPL is greater in the Lower Sand than it is in the Upper Shale. The NAPL distribution in the Upper Sand is much more localized than in the Lower Sand and the Upper Shale. The contour plots illustrate the direct relationship that exists between porosity and specific NAPL volume.

Three-dimensional plume maps were generated to illustrate the vertical as well as horizontal distribution of NAPL in each layer. The 3D distribution of NAPL was determined from NAPL unit volumes calculated at each model node for each well. NAPL unit volume, as defined here, is obtained by multiplying the NAPL saturation at a node by the porosity at the node. Thus, NAPL unit volume defines the volume of NAPL per unit volume of aquifer. Using a gridding technique, NAPL unit volumes were then calculated at regularly spaced nodes throughout the NAPL-contaminated zone and contoured to produce the 3D plume visualizations shown in Figures 10, 11, and 12.

Table 3. **Summary of Calculated Specific Volumes**

Well ID	Oil Thickness (ft)	Oil Thickness (cm)	Upper Sand Porosity	Upper Shale Porosity	Lower Sand Porosity	Specific Volume (cm^3/cm^2)
Upper Sand Wells						
MW-26	0.14	4.27	0.21			0.000373
			0.53			0.00327
MM-1	6.33	192.94	0.21			4.11
			0.53			22.9
MM-2	2.68	81.69	0.21			0.886
			0.53			5.84
RC-2	0.26	7.92	0.21			0.00292
			0.53			0.0253
RC-5	1.59	48.46	0.21			0.166
			0.53			1.3
Lower Sand Wells						
MW-7A	0.14	4.27		0.31	0.14	0.000325
				0.31	0.49	0.00368
MW-8A	5.99	182.58		0.31	0.14	1.73
				0.31	0.49	12.5
MW-10A	16.10	490.73		0.31	0.14	17.5
				0.31	0.49	57.7
MW-11	9.38	285.90		0.31	0.14	8.41
				0.31	0.49	27.2
MW-13	9.54	290.78		0.31	0.14	4.87
				0.31	0.49	28.4
MW-14	11.80	359.66		0.31	0.14	4.75
				0.31	0.49	23.8
MW-16A	12.80	390.14		0.31	0.14	7.94
				0.31	0.49	44.9
MW-30	11.70	356.62		0.31	0.14	12.3
				0.31	0.49	39
RC-1	3.34	101.80		0.31	0.14	0.503
				0.31	0.49	4.31
RC-3	10.60	323.09		0.31	0.14	5.54
				0.31	0.49	33.2
RC-4	12.00	365.76		0.31	0.14	7
				0.31	0.49	40.3
RC-6	1.73	52.73		0.31	0.14	0.116
				0.31	0.49	1.14
MM-3	9.45	288.04		0.31	0.14	4.42
				0.31	0.49	27.6

Figure 10 is a 3D view of NAPL contamination in the Lower Sand. The upper surface of the plume in Figure 10 coincides with the boundary between the Lower Sand and the Upper Shale. The increase of NAPL saturation with elevation within the Lower Sand indicates that the maximum accumulation of NAPL occurs near the contact with the Upper Shale. Within the Lower Sand, there is also a greater accumulation of oil in the western part of the site than in the east.

Figure 11 shows the entire vertical and horizontal extent of NAPL contamination in the Upper Shale and Lower Sand. The area extent of the plume is essentially the same as that shown in Figure 10 because oil in the Upper Shale is

Figure 9. **Specific Volume Contours (cm).**

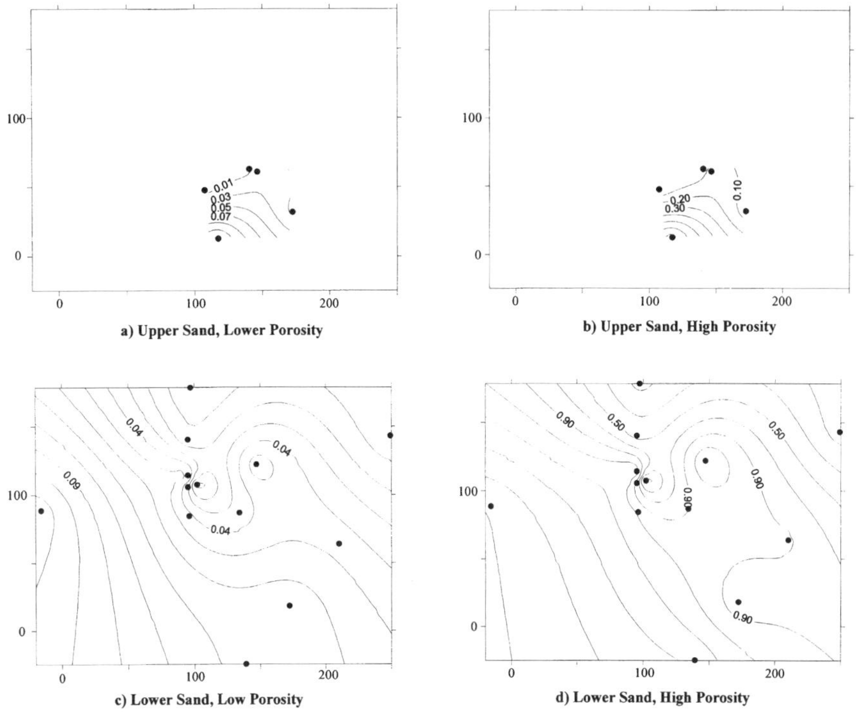

part of the same plume as oil in the Lower Sand. However, the contours in Figure 11 illustrate an upward decrease of NAPL volume within the Upper Shale in contrast to the increase observed in the Lower Sand.

Figure 12 illustrates the NAPL distribution in the Upper Sand aquifer. The areal extent of NAPL within the Upper Sand is much more localized than it is in the Lower Sand, probably because much less oil was released into this layer. Within the NAPL contaminated zone in the Upper Sand, there is an increase of contamination to the south.

CONCLUSIONS

This study presents a numerical modeling method for estimating NAPL specific volume and distribution in heterogeneous porous media under confined and unconfined conditions based on product thickness measured in monitor wells. The NAPL saturation profiles generated by this method are used to determine the NAPL distribution, volume, and plume location in the stratified aquifers. Such information is critical for the successful remediation of NAPL-contaminated sites.

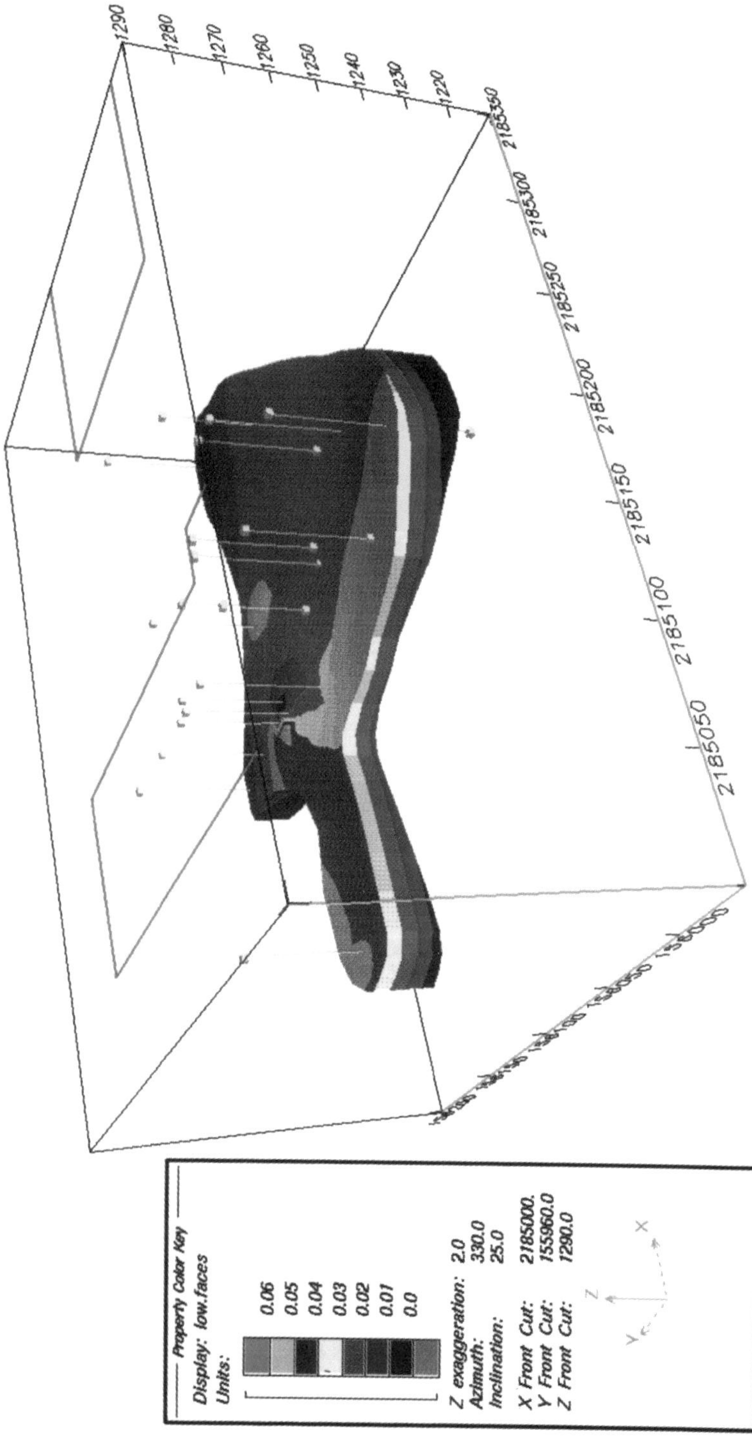

Figure 10. 3D view of NAPL distribution in the Lower Sand.

Figure 11. **3D view of NAPL distribution in the Upper Shale and Lower Sand.**

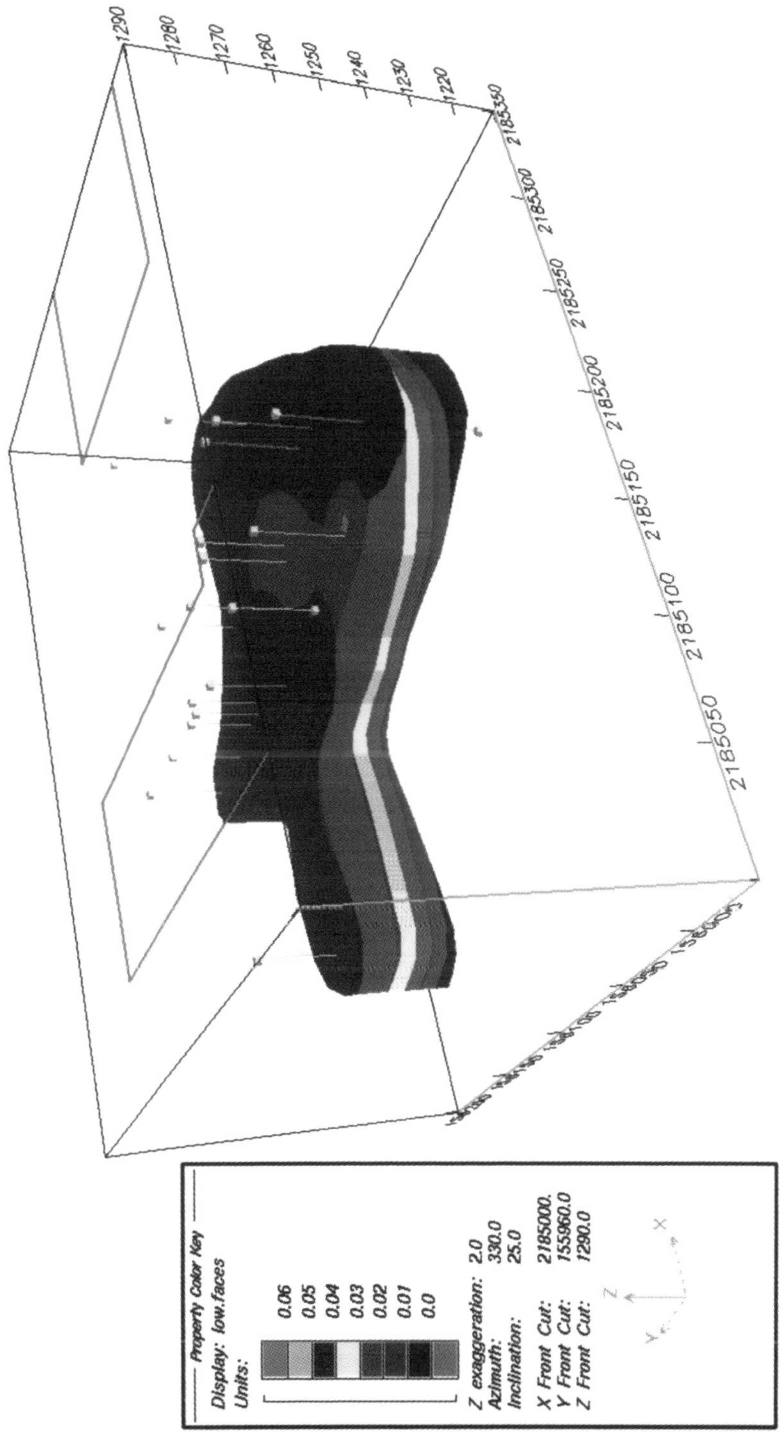

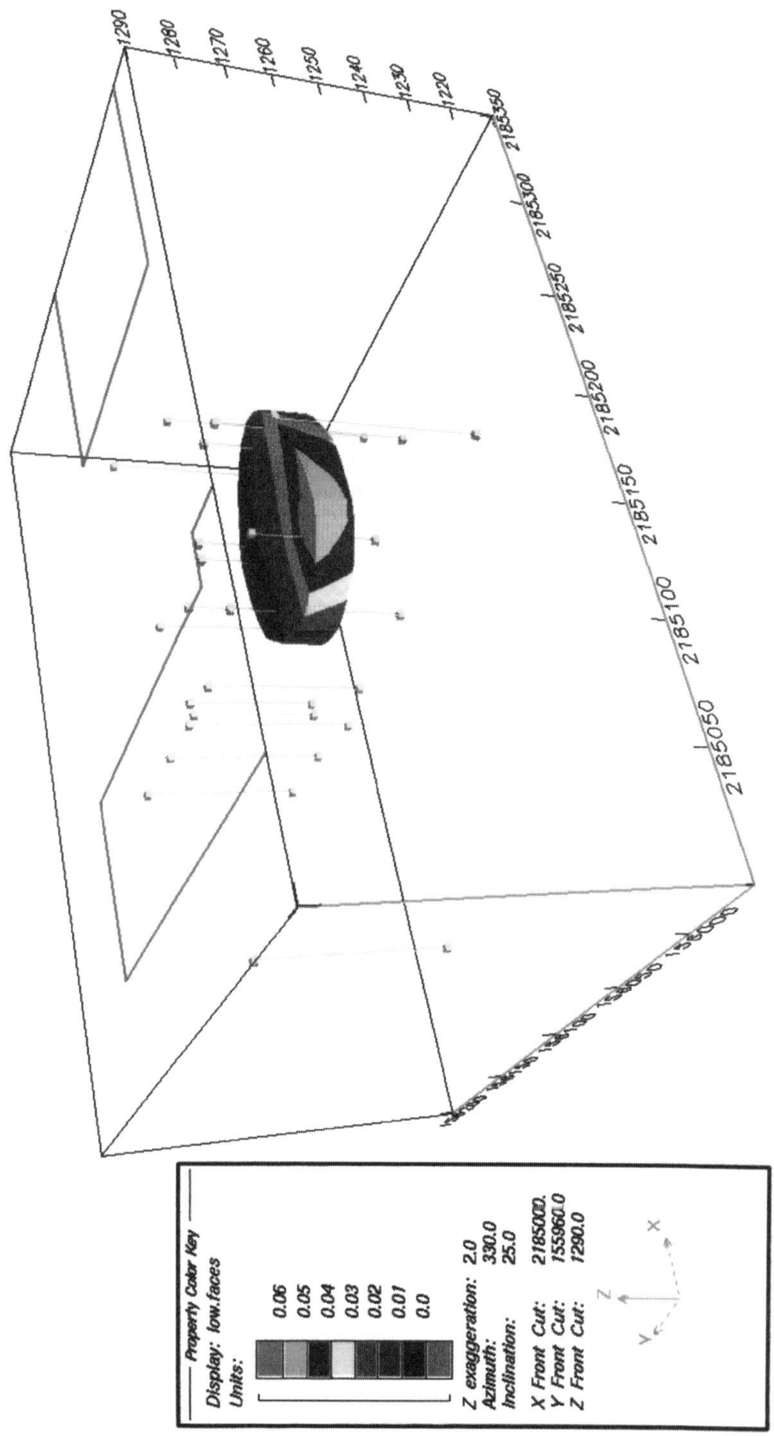

Figure 12. 3D view of NAPL distribution in the Upper Sand.

ACKNOWLEDGMENT

This study was funded by Battelle Memorial Institute through an IR&D grant in 1994. The support is gratefully acknowledged by the authors.

REFERENCES

Brooks, R.H., and Corey, A.T. 1964. Hydraulic Properties of Porous Media, Hydrology Papers No. 3, Colorado State University, Fort Collins, CO.

Lenhard, R.J., Oostrom, M., and White, M.D. 1994b. Modeling Fluid Flow and Transport in Variably Saturated Porous Media with the STOMP Simulator 2. Verification and Validation Exercises Submitted to *Water Resour. Res.*

Lenhard, R.J. and Parker, J.C. 1987. Measurement and Prediction of Saturation-Pressure Relationships in Three-Phase Porous Media Systems, *J. Contam. Hydrol.* **1**, 407.

Lenhard, R.J. and Parker, J.C. 1988. Experimental Validation of the Theory of Extending Two-Phase Saturation-Pressure Relations to Three-Fluid Phase Systems for Monotonic Drainage Paths, *Water Resour. Res.* **24**, 373.

Lenhard, R.J. and Parker, J.C. 1990. Estimation of Free Hydrocarbon Volume from Fluid Levels in Monitoring Wells, *Groundwater* **28**, 57.

Mishra, S. and Parker, J.C. 1993. User's Guide to SoilProp: A Program for Estimating Soil Hydraulic Properties and Their Uncertainty from Particle Size Data, Environmental Systems & Technology, Inc., Blacksburg, VA.

Parker, J.C., Lenhard, R.J., and Kuppusamy, T. 1987. A Parametric Model for Constitutive Properties Governing Multiphase Flow in Porous Media, *Water Resour. Res.* **23**, 618.

van Genuchten, M.Th. 1980. A closed-form equation for predicting the hydraulic conductivity of unsaturated soils, *Soil Sci. Soc. Am.* **44**, 892.

White, M.D., Oostrom, M., and Lenhard, R.J. 1994a. Modeling Fluid Flow and Transport in Variably Saturated Porous Media with the STOMP Simulator 1. Nonvolatile Three-Phase Model Description, Submitted to *Water Resour. Res.*

CLEANUP STANDARDS

CHAPTER 50

Establishing Site-Specific Cleanup Levels for Metals in Soils: A Statistical Approach

Robert A. Ellgas, Consultant, Oakland, California

INTRODUCTION

In 1993, it was reported that there were 1,200 Superfund sites, and it was estimated that there were at least 35,000 additional suspected contaminated sites in the United States with an estimated total cleanup cost of $30 billion (Cairney, 1993). By 1994, the total cleanup cost for Superfund and Resource Conservation and Recovery Act (RCRA) sites alone was estimated at $750 billion, and based on existing remedial technologies and regulatory standards, it was likely that this estimate would increase in the future (Wilson and Clark, 1994). The magnitude of these costs points to the urgent need not only for the development of innovative and cost-effective approaches to contaminated site remediation, but also for the reevaluation of the feasibility and appropriateness of achieving currently-established cleanup levels, or derived cleanup levels that are based on currently-accepted approaches to their derivation. Recent efforts in this direction are promising (National Research Council, 1994), as the U.S. Environmental Protection Agency (USEPA) and the California Regional Water Quality Control Boards (RWQCBs), for example, begin to realize the impracticality at many sites of achieving maximum contaminant levels (MCLs) in groundwater.

For soils, various methods are currently used to establish concentrations of elements and compounds that are protective of human health and the environment. These "risk-based" methods recognize that chemicals in soil can follow a number of complex pathways to reach humans, and that risk-based cleanup levels must therefore necessarily be established on a site-specific basis. The EPA, in their guidance on screening soils to establish whether existing concentrations of chemicals are acceptable (USEPA, 1994), proposes four screening levels that depend on whether the human exposure pathway is through ingestion, inhalation, or migration to groundwater (in which case EPA allows for a 1:1 dilution or a 10:1 dilution of the chemical in groundwater, depending on geochemical and hydrogeological factors that influence the fate of the chemical as it migrates from its source in the soil). The various EPA regions have also established guidance for calculating soil cleanup levels (USEPA, 1995). The Department of Toxic Substances Control at the California EPA does not publish specific soil cleanup levels, but provides guidance for how they can be calculated (CALEPA, 1992; 1994).

STATEMENT OF PROBLEM

Many metals with toxic properties occur naturally in United States soils,* and depending on the soil type and the region of the country, the naturally-occurring concentrations of these metals can vary widely (Shacklette, 1984; Dragun and Chiasson, 1991). If the methods previously mentioned for developing risk-based soil cleanup levels are applied to these metals at specific sites, it is possible that the calculated risk-based levels may actually be lower than the naturally-occurring soil concentrations of these metals. In the author's experience, examples of metals where this could be the case are cadmium, mercury, lead, and especially arsenic, for which the calculated risk-based cleanup level is often below the levels found naturally in soils.

Also, it may not be possible to calculate a risk-based cleanup level for certain metals or at certain sites due to the lack of essential data for performing the calculations. For example, the limited data on the toxicological properties of gallium makes calculation of a risk-based cleanup level problematic (CALEPA, 1994). Additionally, at sites where it is known or suspected that soils have been contaminated by metals, it may be that no site-specific hydrogeological information or soil physical and chemical information is available that would be needed to develop risk-based cleanup levels based on migration to groundwater.

Finally, it may be decided by the responsible parties to expedite soil cleanup at certain sites. The time it would take to develop risk-based soil cleanup levels for metals, and to obtain regulatory agency approval of these levels, may not meet the time limitations set for remediation.

In all these cases, the use of site-specific naturally-occurring (background) concentrations as cleanup levels could potentially be proposed as a reasonable substitute for risk-based soil cleanup levels. Site-specific background concentrations would be conservative soil remediation goals in cases where the risk-based levels would be higher than the background concentrations, but where these risk-based levels are not developed, as in either of the latter two cases mentioned above. Site-specific background levels offer advantages over background levels which could be calculated using regional data (Shacklette, 1984;

*Many organic compounds with toxic properties, in particular the polynuclear aromatic hydrocarbons (PAHs), as well as other inorganics with toxic properties (for example, fluoride), also occur naturally in certain U.S. soils. The method developed in this chapter is equally valid in application to these compounds, as well as in application to pH, but the cases described in this chapter consider only heavy metals or metalloids (i.e., arsenic).

Dragun and Chiasson, 1991) and upper prediction bounds (Hahn and Meeker, 1991), because regional data may not accurately represent the specific site.

DESCRIPTION OF METHOD

In a heterogeneous soil matrix, the naturally-occurring concentrations of metals can be expected to vary within a range that is often characteristic of the soil type and the geographic region. If a sufficient number of analytical results for metals from "background" or nonoperations areas of a site are available for the same soil type, these results can be divided into convenient concentration groups, and the group frequency can be plotted versus concentrations to obtain histograms for each metal.*

Because most soil investigation sampling programs are focused on obtaining data to delineate known or suspected areas of contamination, it is not likely that a large amount of background metals data will have been obtained from nonoperations areas. However, in many cases, the specific locations of contaminated soil are not known initially, and the first phases of sampling will likely acquire multiple measurements that are actually from uncontaminated soil. These measurements can be aggregated with any sample results from known background areas to produce the histograms described above.

A number of the analytical results used to produce histograms might actually be from areas where a chemical release has occurred. In these cases, as long as the background concentrations are sufficiently different from the concentrations representing chemical releases, and as long as there are a sufficient number of background results to perform the analysis, the method is designed to differentiate the "noise" attributed to natural background variation from the "signal" of a chemical release.

The portion of each histogram that represents the distribution of naturally-occurring concentrations of a metal can likely be modeled by a lognormal density function, which is used to model many types of environmental data, including soil contaminant data and naturally-occurring soil concentrations (Gilbert, 1987; Hwang, 1987; Mackay et al., 1983; Miller et al., 1978; Ott, 1990; Pinder and Smith, 1981). The lognormal distribution will likely be applicable if the histogram is positively skewed (meaning the mode or "peak" occurs in the lower half of the range, and a relatively long "tail" extends to the right of the mode). For a histogram that can be modeled by the lognormal distribution, the application of a log transform to the metal concentrations will allow a histogram to

*Even if the analytical results come from different soil types, the method can probably still be applied if the data are aggregated, plotted on histograms, and can be shown to follow the model described in the method. The method will probably not work for background data aggregated from soil types with very different concentration means and/or standard deviations, in which case the method should be applied to metals concentration data from each soil type where contamination is expected to be found.

be produced that can be modeled by the normal distribution. Two sample statistics of the normal distribution, the sample mean and sample standard deviation, can then be easily calculated for use in the method.

To obtain the portion of the log-transformed histogram representing background data, an "outlier screening" or "background cutoff" process is used to remove values that do not likely represent background. If sample analytical results from contaminated areas are also included on the histogram, the histogram will appear to be bimodal,* or scattered results will be found at concentrations that, by inspection, appear to be above the background distribution. Those samples representing the distribution around the higher mode, or the scattered results above the apparent background distribution, are screened out. In addition, some values at the high end of the apparent background distribution conceivably might not represent background concentrations, and therefore might conservatively be screened out on the first iteration for modeling the background distribution. This process establishes the "test background cutoff value," which is tested iteratively as described below.

At this point, another test should be performed on the log-transformed and screened data set to decide whether it has been drawn from an underlying normal distribution. Several tests are available for this purpose, depending on the size of the sample data set: the Shapiro-Wilk W test can be used for data sets where $n \leq 50$, and the D'Agostino D test can be used for data sets where n is between 50 and 1,000 (Gilbert, 1987). These tests are more powerful than the chi-square test, which is a nonparametric test commonly used for testing goodness-of-fit. It is suggested that the significance level of the W or D test not be set a priori, but that it be allowed to vary until the data fit a normal distribution. As long as the level is not too high (probably no higher than 0.3 or 0.4), it should be assumed that the data probably have come from a normally-distributed population.

If the log-transformed and screened data probably have been drawn from a normal population, the sample mean and sample standard deviation of the "test" background data are then calculated. Normal deviates (means plus varying numbers of standard deviations) can then be calculated, and then reconverted by exponentiation into nontransformed statistics and compared to the test background cutoff value. This test is performed iteratively by changing the test background cutoff value until the deviate representing a chosen percentage of the background sample data matches as closely as possible to the test background cutoff value. At this point, the test stops, and the cutoff value for background concentrations is established.

---

* The samples taken from contaminated soils could have very high concentrations, in which case the "slot size" for the histogram concentration groupings representing background concentrations might not be easily plotted on the histogram for higher concentrations. It is then convenient to group all results that will not fit on the histogram as ">." If most results fit into this group, this is where the second mode will appear.

The percentage of background data chosen to match the test background cutoff value is worth some discussion. If this percentage is set, for example, at 80%, then about 20% of the time a sample result will be concluded to represent contaminated soil, when in fact it comes from soils with background concentrations of the metal (in an hypothesis testing framework, this is called a "type I error"). This would result in what might be an unacceptably high amount of soil remediation. On the other hand, if the percentage is set, for example, at 99%, then a type I error would occur just 1% of the time. However, 99% of the background data would push the background cutoff value to about 2.6 standard deviations beyond the mean, or at a value that would probably encompass many of the data actually representing contaminated soil. Thus, a relatively high percentage of the time a sample result will be concluded to represent background, when in fact it comes from contaminated soils (a "type II error"). This type II error might be unacceptable to the oversight regulatory agency. If the upper tail of the background distribution is assumed to be an approximate mirror image of the lower tail of the contaminated soil distribution,* then the type I and type II errors will be equal where the distribution curves cross. In this case, the errors will "cancel" each other out (in other words, it will be concluded an equal percentage of the time that the sample represents either background or nonbackground, when in fact the opposite case is true). This cross point could be used to choose the percentage of background data to match the test background cutoff value, and could reasonably be argued as the point of minimum decision error. However, if the contaminated soil concentrations appear to be much higher than the background mean, it can be argued that a higher chosen percentage of the test background data truly represent background concentrations.

APPLICATION OF METHOD: TWO CASE STUDIES

The method described above has been used to establish cleanup levels for metals at two contaminated sites in California. A summary of the application of this method to these two sites is provided below. In both cases, the oversight regulatory agency (the local RWQCB and the county hazardous materials control agency, respectively) approved the cleanup levels for delineating the extent of soil remediation.

Case I: Former Semiconductor Manufacturing Facility, California

Operations at this facility included the manufacture of gallium arsenide semiconductor chips. Other metals, including chromium, copper, and lead, were also

*Unless the contaminated soil distribution is well established based on many data from known contaminated areas, this assumption cannot be tested. Also, the contaminated soil concentrations must be high enough above the background concentrations to distinguish the lower tail of the contamination distribution.

used in various manufacturing processes, and hydrofluoric acid was used for pitting and etching. Waste transport piping had the potential for leakage, so investigations focused on soils around and under waste pipe locations.

The vadose zone stratigraphy at this site consisted of nonindurated silt and clay-rich soils, and clayey sands and gravels. Silt and clay-rich soils extended from ground surface to an approximate depth of 15 feet below former grade. Interbedded silty clay and sand extended to about 29 feet below former grade, which at the time of the 1991 investigation was the approximate depth of groundwater.

During the investigation, approximately 110 analyses were performed for the 17 metals specified in the California Code of Regulations Title 22, Article 11; approximately 130 analyses were performed for fluoride; and at least 200 separate analyses were performed for arsenic and gallium. The samples for which these analyses were performed were widely distributed through all vadose zone soil types. Because of the large database in a wide variety of soil types, it was possible to apply the method described above to establish background cutoff values that could be used for soil remediation at the site.

Figures 1 and 2 are examples of the original and log-transformed histograms for arsenic in soils at the site. After applying the method iteratively to encompass 95% of the background data, the background cutoff value was calculated to be 7.9 milligrams/kilogram (mg/kg), which is shown on Figure 1. The *D* test significance level for the fitted distribution was about 0.3, meaning that the fit was not good, but that the lognormal distribution could not be ruled out.

Ninety-five percent of the background data (two normal deviates) was chosen to match the background cutoff value for several reasons. It is seen on Figures 1 and 2 that the second mode occurs at arsenic concentrations greater than 10 mg/kg. Many of these concentrations were much greater than 10 mg/kg. If a distribution were fit to the nonbackground samples, the mean would have been much higher than the background mean. This would have allowed differentiation of background "noise" from contamination "signals," and a higher percentage of the background data might have been argued to be lower than the cutoff value. However, in the interest of expediting approval of the cleanup level, 95% of the data was conservatively chosen to match the background cutoff value. It was emphasized that about 5% of the time, cleanup would be performed on soils that were not contaminated.

As a comparison to the histograms for arsenic concentration data from soils at this former semiconductor manufacturing facility, Figures 3 and 4 present original and log-transformed histograms for arsenic concentrations in soils at a natural gas compressor facility in the southeastern United States. Arsenic does not result from process operations at this facility, and therefore is not expected to be present in soils at levels above background. The "cutoff value" for background was therefore set at the highest concentration in the database. Because the background distribution including all data was fit at the 0.05 significance

Figure 1. **Histogram of arsenic concentrations in soil samples, former semiconductor manufacturing facility, California.**

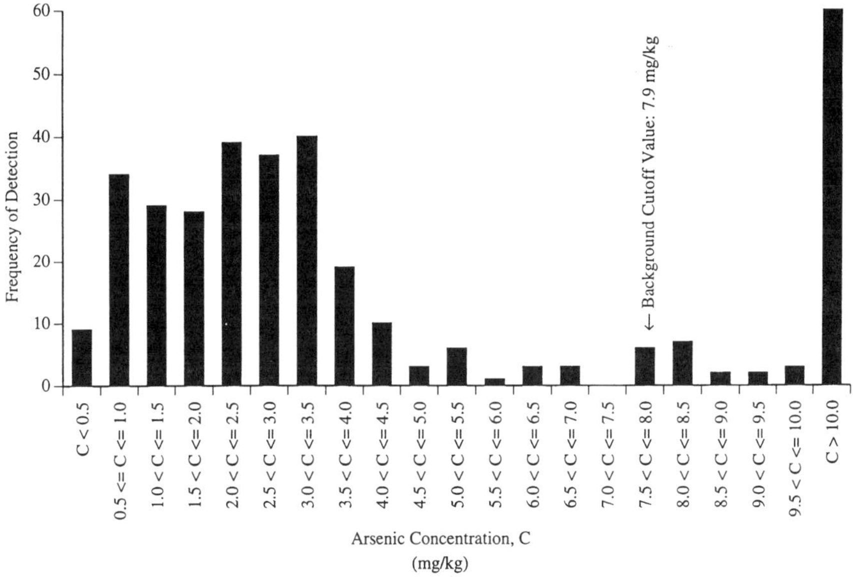

Figure 2. **Histogram of natural log of arsenic concentrations in soil samples, former semiconductor manufacturing facility, California.**

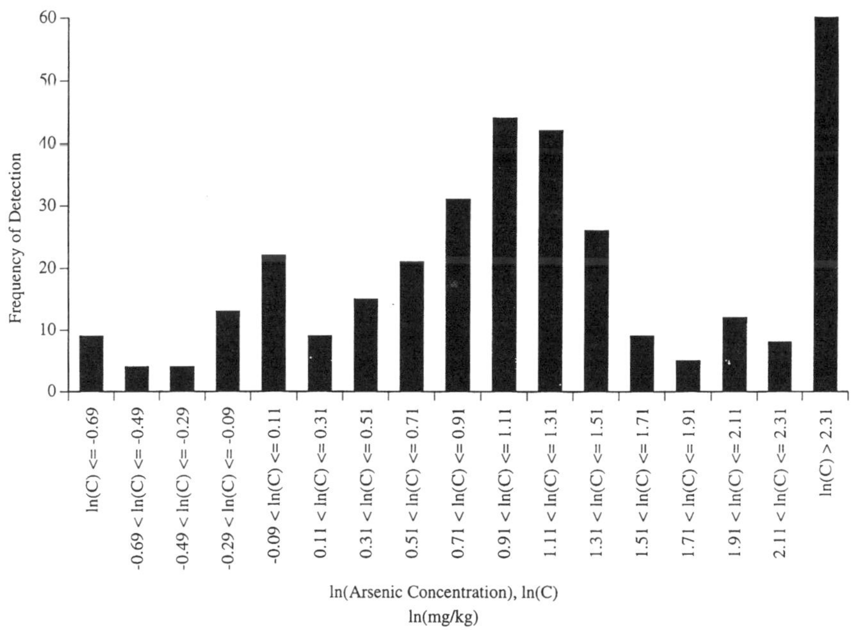

Figure 3. **Histogram of arsenic concentrations in soil samples, natural gas compressor station, southeastern United States.**

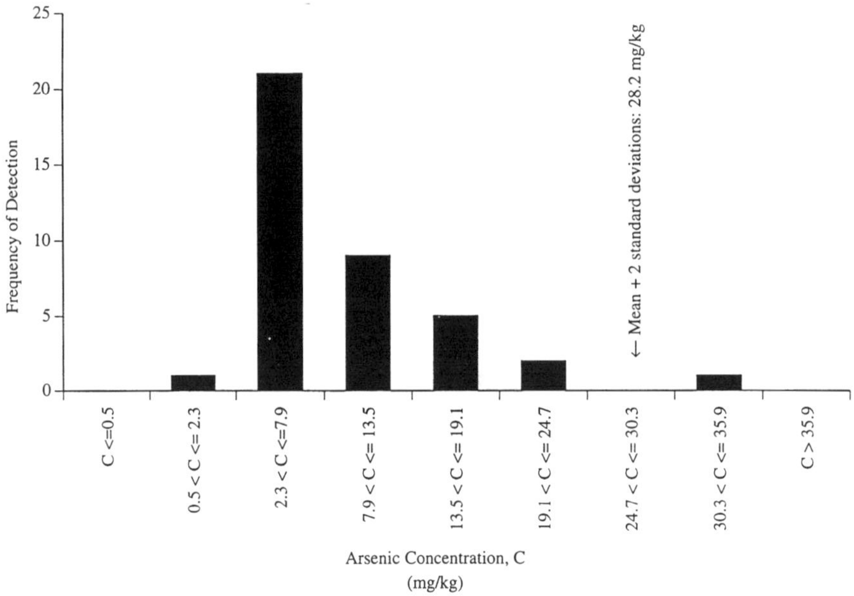

Figure 4. **Histogram of natural log of arsenic concentrations in soil samples, natural gas compressor station, southeastern United States.**

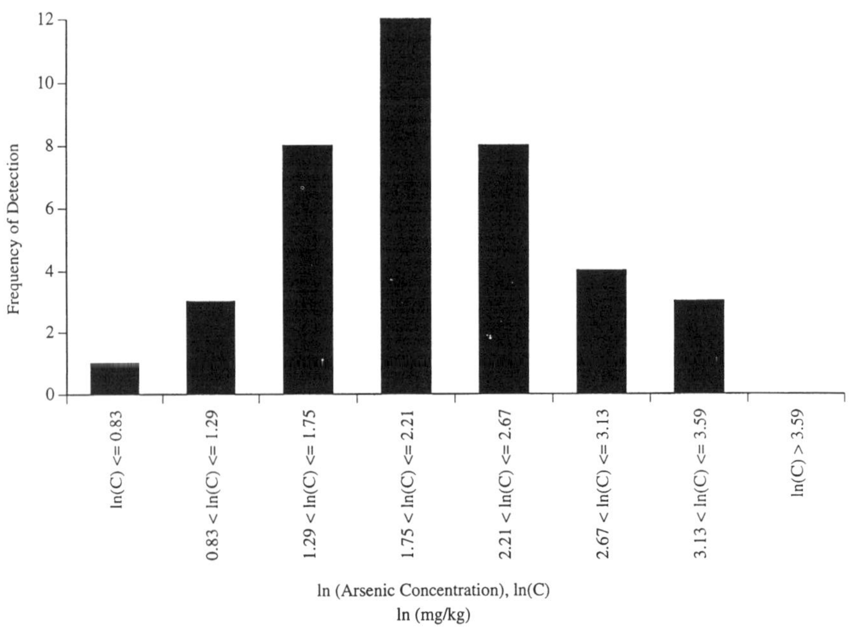

level, there is no indication of arsenic contamination at the site. As shown in Figure 3, the concentration encompassing 95% of the data (two normal deviates) is 28.2 mg/kg, which is considerably higher than the 7.9 mg/kg concentration for the California semiconductor facility. However, this matches well with the range of 1.3 mg/kg to 32 mg/kg given by Dragun and Chiasson (1991) for the specific region where the data originated.

The method was also applied to other metals at the semiconductor manufacturing site where analytical data indicated concentrations that did not fit background distributions. Table 1 summarizes the results for these metals, as well as for fluoride and pH (note that log transforms were not performed on pH data, because of pH's inherent log relationship to hydrogen ion concentration). For comparison, Table 1 also includes statistics on arsenic concentrations in "background" borings from nonoperations areas, which were required by the RWQCB. It is seen that the statistics match well with the statistics for background data from operations areas of the site.

Case II: Former Steel Specialty Products Manufacturer, California

Operations at this facility consisted primarily of light steel working and rustproof coating of premanufactured steel products. The paints used for coating dripped on the floors. When the floors were cleaned, the paint was scraped off,

TABLE 1. **Summary of Cutoff Values for Various Inorganic Chemicals and pH, Former Semiconductor Manufacturing Facility, California**

Chemical	Mean (mg/kg)	+2 Normal Deviate[a] (mg/kg)
Arsenic	2.1	7.9
Arsenic in "background" borings	2.2	6.9
Gallium	2.3	5.7
Copper	26	58
Lead	4.1	22
Chromium	47	83
Fluoride	1.4	6.0
pH[b]	7.8	+2: 9.1 -2: 6.6

Notes:

[a] +2 Normal Deviate = mean + 2 standard deviations, chosen to match the background cutoff value.

[b] Because pH values could be considered "non-background" at low values as well as high values, both positive (+2) and negative (-2) normal deviates are shown.

and scrapings were swept out the facility's back door. Soils contained chromium, lead, and zinc above background concentrations.

Soil remediation was performed within areas where a correlation was found among elevated levels of chromium, lead, and zinc, and remediation verification samples were collected and analyzed at the boundaries of these areas. Subsequent to remediation, the oversight regulatory agency wanted assurance that the remediation completed had brought soil concentrations of these metals to within background limits.

At least 50 measurements had been made of chromium, lead, and zinc concentrations in the soil during a Phase II site assessment program. Thus, the method described in this chapter was applied to obtain statistics on the background distributions of these metals.

Figures 5 and 6 present, respectively, the original and log-transformed histograms for chromium in soils at this facility. Table 2 presents the ordered sample data for chromium. After applying the method iteratively to encompass 95% of the background data (two normal deviates), the background cutoff value for chromium was calculated to be 110 mg/kg, which is shown in Table 2 and in Figure 5. The *D* test significance level for the fitted distribution was about 0.2, meaning that the fit was probably acceptable for a lognormal distribution. It is noted that the best-match cutoff value was lower than two normal deviates, so it was conservatively assumed that the cutoff value actually represented the limit on background concentrations for chromium. Statistically, more than 5% of the concentrations beyond this cutoff value would still be background concentrations.

The remediation verification sample results for soils left in place after several phases of excavation at the site were compared to background cutoff values.* Only one concentration (chromium at 200 mg/kg in one sample) potentially could have been considered above the conservatively-calculated background cutoff value (110 mg/kg). However, based on the same correlation technique mentioned above that had been used for deciding on excavation limits, it was shown that because this 200 mg/kg chromium sample had low background concentrations of both lead and zinc, this result was very likely to have been one of the more than 5% of soil concentrations of chromium that derived from the background concentration distribution. Thus, the oversight regulatory agency closed on metals remediation at this facility.

CONCLUSIONS

The statistical method described in this chapter for establishing site-specific background concentrations of metals in soils can be used for establishing soil cleanup levels for these metals under the following circumstances:

*The corresponding background cutoff values for lead and zinc were, respectively, 140 mg/kg and 450 mg/kg.

Figure 5. **Histogram of chromium concentrations in soil samples, former steel specialty products manufacturing facility, California.**

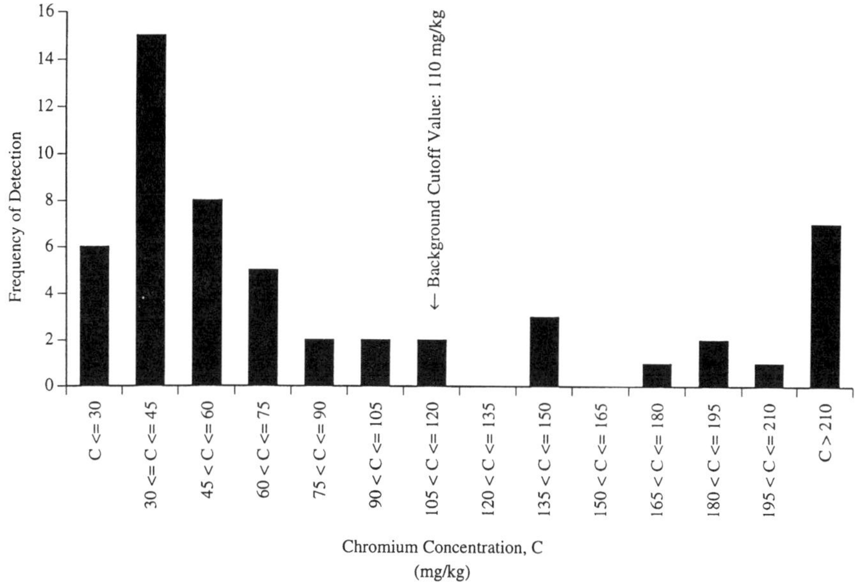

Figure 6. **Histogram of natural log of chromium concentrations in soil samples, former steel specialty products manufacturing facility, California.**

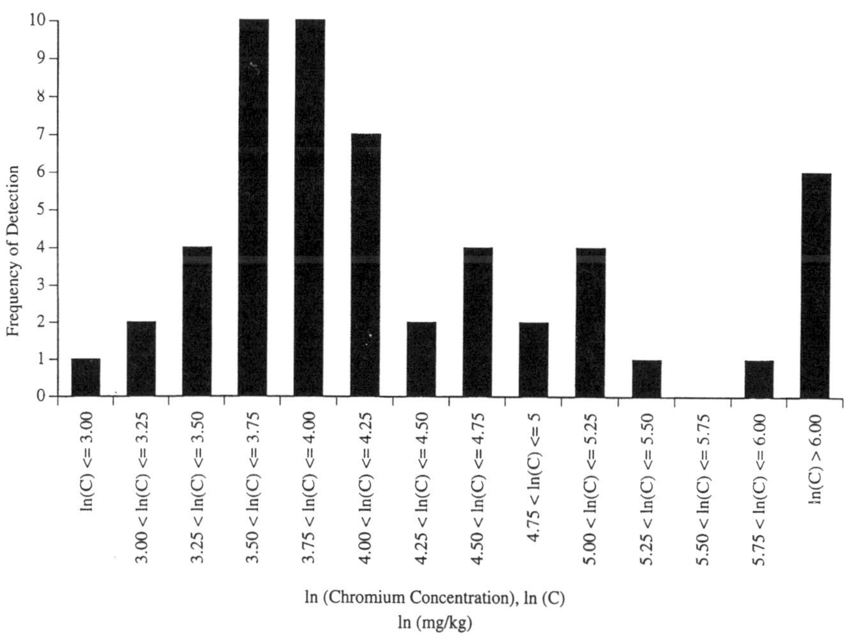

TABLE 2. **Numerical Sorting of Concentrations of Total Chromium in Soil Samples, Former Steel Specialty Products Manufacturing Facility, California**

mg/kg	ln(mg/kg)	
20	3.00	
22	3.09	
23	3.14	
28	3.33	
28	3.33	
28	3.33	
31	3.43	
35	3.56	
35	3.56	
35	3.56	
37	3.61	
37	3.61	
38	3.64	
39	3.66	
40	3.69	
40	3.69	
42	3.74	
43	3.76	
44	3.78	
45	3.81	
45	3.81	
47	3.85	
49	3.89	
52	3.95	
54	3.99	
54	3.99	
54	3.99	
55	4.01	
56	4.03	
61	4.11	
63	4.14	
64	4.16	
68	4.22	
69	4.23	
81	4.39	
85	4.44	
99	4.60	
100	4.61	
110	4.70	
110	4.70	Background
140	4.94	Cutoff Value
140	4.94	
150	5.01	
175	5.16	
188	5.24	
190	5.25	
210	5.35	
380	5.94	
710	6.57	
920	6.82	
928	6.83	
1100	7.00	
2850	7.96	
6700	8.81	
Count: 54	54	
Mean: 51.7	3.85	
Std. Dev.: ---	0.43	
+2 Norm. Dev.111.3	4.71	

- There must be an arguable case for using the method. If it is known that the risk-based cleanup level would be lower than the expected background concentration, or if the risk-based level is not calculated, or cannot be calculated, for various reasons, then establishing a background concentration would be appropriate.
- A sufficient number of data must be available for performing the analysis. If the background concentrations are expected to be very different from the contaminant concentrations, then a histogram can be generated from potentially as few as 10 or 12 pieces of background data in order to have confidence in the calculated statistics. However, as the contaminant concentrations get closer to the background concentrations, the number of background data required to maintain confidence in the calculated statistics increases correspondingly. When specific locations of contamination are not known initially, and data have been collected to locate and delineate contamination, then it is likely that a number of background measurements will have been made in the process of locating contamination. These measurements can be aggregated with any data from "background" or nonoperations areas to establish whether sufficient data are available for applying the method.
- If the background mean and the contaminant concentration mean are too close, it will not be possible to differentiate the "noise" (background variation) from the "signal" (chemical release). However, if the means are not too different, it might be argued that little "contamination" is actually being added to the environment, since it cannot be distinguished from naturally-occurring concentrations.
- The reconversion of transformed statistics into the original database (that is, estimating the mean and standard deviation of the log-transformed data, and then reconverting these results into nontransformed statistics) introduces a bias into the statistics. As long as the lognormal distribution is not highly-skewed, and as long as the number of measurements is sufficiently large (i.e., more than about 20 pieces of data), this bias should be less than 5%. Under other circumstances, unbiased estimators should be used (Gilbert, 1987).
- The percentage of background data that is chosen to match with the background cutoff value needs to be carefully considered. While this percentage can be chosen partly based on statistical factors such as a large difference between background measurements and obvious contaminant measurements, it will often be based on what the responsible parties and the oversight regulatory agencies can agree on as an acceptable error level in decision making.

REFERENCES

Cairney, T., Ed. 1993. *Contaminated Land: Problems and Solutions*, p. 2. Boca Raton, FL, Lewis Publishers.

(CALEPA) California Environmental Protection Agency. 1992. Department of Toxic Substances Control, *Supplemental Guidance for Human Health Multimedia Risk Assessments of Hazardous Waste Sites and Permitted Facilities*, July.

(CALEPA) California Environmental Protection Agency. 1994. Department of Toxic Substances Control, *Preliminary Endangerment Assessment Guidance Manual*, January.

Dragun, J. and Chiasson, A. 1991. *Elements in North American Soils*, Hazardous Materials Control Resources Institute, Greenbelt, MD.

Gilbert, R.O. 1987. *Statistical Methods for Environmental Pollution Monitoring*, p. 152-155, New York, Van Nostrand Reinhold.

Hahn, G.J. and Meeker, W.Q. 1991. *Statistical Intervals: A Guide for Practitioners*, New York, John Wiley & Sons.

Hwang, J.H. 1987. *Descriptions of Univariate Statistical Models for Use in Environmental Data Analyses and Means for Predicting Their Goodness-of-Fit*, New York, Technical Bulletin #530, National Council of the Paper Industry for Air and Stream Improvement, Inc., June.

Mackay, D., Paterson, S., Shiu, W.Y., and Cheung, B. 1983. Experimental Modeling Studies of Hazardous Substances in Ontario, In: *Proceedings of the Conference of the Canadian Ministry of Environment, General and Air Pollution Research Technology Transfer*, November 29-30, 1983, v. 1, p. 370.

Miller, M.L, Fix, J.J., and Bramson, P.E. 1978. Lognormal Analysis of Naturally Occurring Radionuclides in Soil and Vegetation of the Hanford Area, In: *Proceedings of the 3rd International Symposium, DOE/Univ. Texas Natural Radiation in the Environment,* Houston, April 23-28, 1978, v. 1, p. 826.

National Research Council Water Science and Technology Board. 1994. *Alternatives for Ground Water Cleanup*, September.

Ott, W.R. October 1990. A Physical Explanation of the Lognormality of Pollutant Concentrations, *J. Air & Waste Management Assn.*, **40**, 1378.

Pinder, J.E. and Smith, M.H. 1981. Frequency Distributions of Radiocesium Concentrations in Soil and Biota, *Proceedings of the Symposium on Mineral Cycling in Southeastern Ecosystems*, U.S. Dept. of Agriculture, January 23.

Shacklette, H.T. 1984. *Elemental Concentrations in Soils and Other Surficial Materials of the Coterminous United States*, USGS Professional Paper 1270.

(USEPA) U.S. Environmental Protection Agency. 1994. Office of Emergency and Remedial Response, *Soil Screening Guidance*, EPA/540/R-94/101, December.

(USEPA) U.S. Environmental Protection Agency. 1995. Region IX, Memorandum, Subject: *Region IX Preliminary Remediation Goals (PRGs) First Half 1995*, February 1.

Wilson, D.J. and Clark, A.N., Eds. 1994. *Hazardous Waste Site Soil Remediation: Theory and Application of Innovative Technologies*, New York, Marcel Decker.

Glossary of Acronyms

AAL	Allowable Ambient Limits
AAS	Aquifer Air Sparging
ACM	Asbestos Containing Material
AFCEE	Air Force Center for Environmental Excellence
ALCD	Advanced Landfill Cover Demonstration
AOC	Areas of Concern
ARAR	Applicable, Relevant and Appropriate Requirements
AS	Air Sparging
ASTM	American Society for Testing and Materials
ASW	Air Sparging Wells
ATSDR	Agency for Toxic Substances and Disease Registry
BACT	Best Available Control Technology
BCT	BRAC Cleanup Team
BEC	BRAC Environmental Coordinator
BMDL	Below Method Detection Level
BOD	Biological Oxygen Demand
BRAC	Base Realignment and Closure
BTEX	Benzene, Toluene, Ethylbenzene and Xylenes
CA	Carbon Adsorption
CAM	Continuous Air Monitors
CATOX	Thermal Catalytic Oxidation System
CERCLA	Comprehensive Environmental Response, Compensation, and Liability Act
CFB	Carbon Based Fluidized Bed Process
CFU	Colony Forming Units
CLP	Contract Laboratories Program
CO	Catalytic Oxidizer
DOD	Department of Defense
DOE	Department of Energy
DPC	Diphenyl Carbazide
DQP	Date Quality Objectives
DSR	Dose to Source Ration
DUSD(ES)	Deputy Under Secretary of Defense for Environmental Security
DWS	Drinking Water Standards
EBS	Environmental Baseline Survey
EFA	Eastern Fill Area
EIA	Enzyme Immunoassay
EIS	Environmental Impact Statement
EOD	Explosive Ordnance Division
EP	Extraction Procedure
EPA	Environmental Protection Agency
EPRI	Electric Power Research Institute
ERA	Ecological Risk Assessment
EZ	Exclusion Zones
FBR	Fluidized Bed Reactor

FFA	Federal Facilities Agreement
FISC	Fleet and Industrial Supply Center
FOSL	Finding of Suitability for Lease
FOST	Finding of Suitability for Transfer
FS	Feasibility Study
FTIR	Fourier Transform Infrared Spectroscopy
GAC	Granular Activated Carbon
GC	Gas Chromatograph
GC/MS	Gas Chromatograph/Mass Spectrometry
GCIRMS	Gas Chromatograph Isotope Ratio Mass Spectometry
GIS	Geographic Information System
GLIM	Generalized Linear Interactive Model
GPM	Gallons per Minute
GPR	Ground Penetrating Radar
GRI	Gas Research Institute
GWA	Ground Water Associates
GWPC	Groundwater Protection Criteria
HSP	Health and Safety Plan
HTGC	High Temperature Gas Chromatograph
HWCP	Hazardous Waste Cleanup Project
IDL	Interactive Data Language
IHTD	Indirect Heating Thermal Desorption
IR	Installation Restoration
IRF	Incineration Research Facility
IRIS	Integrated Risk Information System
IRM	Interim Remedial Measure
IROD	Interim Record of Decision
ITA	Innovative Treatment Areas
JCP&L	Jersey Central Power and Light Company
LAFB	Loring Air Force Base
LAMS	Landfill Assessment and Monitoring System
LBP	Lead Based Paint
LIBS	Laser Induced Breakdown Spectroscopy
LIF	Laser Induced Fluorescence
LLRW	Low Level Radioactive Waste
LNAPL	Light Non-aqueous Phase Liquids
LOG	Common Logarithm
MAWS	Minimum Additive Waste Stabilization
MB	Mixed Basins
MEDEP	Maine Department of Environmental Protection
MGP	Manufactured Gas Plant
MLSS	Mixed Liquor Suspended Solids
MPA	Microwave Plasma Analyzer
MR	Milliwave Radiometer
MSP	Mixed South Plants
MTBE	Methyl-t-butyl Ether

MTD	Maximum Tolerated Dose
NACIP	Navy Assessment and Control of Installation Pollutants
NAS	Naval Air Station
NCP	National Oil and Hazardous Substances Pollution Control Plan
NCP	National Contingency Plan
NETC	Naval Education and Training Center
NFESC	Naval Facilities Engineering Service Center
NGVD	National Geodetic Vertical Datum
NJNG	New Jersey Natural Gas
NMOC	Non-Methane Organic Carbon
NOEAL	No-Observable Adverse Effect Level
NOEC	No Observed Effect Concentration
NPDES	National Pollution Discharge Elimination System
NPL	National Priorities List
NRAD	National Command Control and Ocean Surveillance Center Research and Development Division
NTP	Notice to Proceed
NTS	National Test Site
NYSDEC	New York State Department of Environmental Conservation
OECD	Organization for Economic Cooperation and Development
OLF	Outlying Landing Field
OMB	Office of Management and Budget
ORP	Oxidation Reduction Potential
PADEP	Pennsylvania Department of Environmental Protection
PAH	Polycyclic Aromatic Hydrocarbons
PCB	Polychlorinated Biphenyls
PCDD	Polychlorinated Dibenzo-p-dioxins
PCDF	Polychlorinated Dibenzofurans
PCE	Tetrachloroethylene
PDF	Probability Density Function
PID	Photoionization Detector
PLFA	Phospholipid Fatty Acids
PMC	Pollutant Mobility Criteria
POL	Petroleum, Oils and Lubricants
POTW	Publicly Owned Treatment Works
PP	Pollution Prevention
PP&L	Pennsylvania Power and Light Co.
ppb	parts per billion
PPE	Personal Protective Equipment
PQL	Practical Quantitation Limit
PRG	Preliminary Remediation Goals
PRP	Potentially Responsible Parties
PSIG	Pounds per Inch Gauge
PVC	Polyvinyl Chloride
QC	Quality Control
QPP	Quality Project Plans
QSAR	Quantitative Structure Activity Relationships

QSC	Quaker State Corporation
RA	Removal Action
RAB	Restoration Advisory Board
RAGS	Risk Assessment Guidance for Superfund Sites
RAO	Remedial Action Objectives
RBCA	Risk Based Corrective Action
RBR	Risk Based Remediation
RBSL	Risk Based Screening Level
RCRA	Resource Conservation and Recovery Act
REV	Representative Elementary Volume
RfDs	Reference Doses
RFH	Radio Frequency Heating
RI	Remedial Investigation
RIC	Radio Isotopic Committee
RIDEM	Rhode Island Department of Environmental Management
RISC	Remediation of Industrial Soil Constituents
RKS	Rotary Kiln System
RMA	Rocky Mountain Arsenal
ROC	Receptors of Concern
ROD	Record of Decision
RP	Responsible Party
RPM	Remedial Project Managers
RSR	Remedial Standard Regulation
RTDM	Real Time Dust Monitor
RTECS	Registry of Toxic Effects of Chemical Substances
SAO	Sampling and Analysis Outline
SAP	Sampling and Analysis Plan
SARA	Superfund Amendments and Reauthorization Act
SCAPS	Site Characterization and Analysis Penetrometer system
SCFM	Standard Cubic Feet per Minute
SCG	Standards, Criteria, and Guidelines
SCWO	Super Critical Water Oxidation
SDIV	Naval Facilities Engineering Command Southern Division
SFB	Sand Based Fluidized Bed Process
SI	Site Inspection
SITE	Superfund Innovative Technologies Evaluation
SMP	Site Management Plan
SOCS	Subsurface Ordnance Characterization System
SOW	Statement of Work
SPLP	Synthetic Precipitation Leaching Procedure
SSTL	Site Specific Target Levels
STARS	Spill Technology and Remediation Series
STCG	Site Technology Coordination Groups
STOMP	Subsurface Transport Over Multiple Phases
STR	Stirred Tank Reactor
SVE	Soil Vapor Extraction
SWA	State Wide Average

TATS/PATS	Technology Application Teams/Process Action Teams
TCA	Trichloroethane
TCAAP	Twin Cities Army Ammunition Plant
TCE	Trichloroethylene
TCLP	Toxicity Characteristic Leaching Procedure
TEL	Threshold Effect Levels
TEQ	Toxicity Equivalency
TKN	Total Kjeldahl Nitrogen
TNT	2,4,6 Trinitrotoluene
TPH	Total Petroleum Hydrocarbons
TRU	Transuranic
TSCA	Toxic Substance Control Act
TSDF	Treatment, Storage and Disposal Facility
TVH	Total Volatile Hydrocarbons
UES	Ultimate Excavation Surface
UNS	Ultrasonic Navigation System
UST	Underground Storage Tanks
UXO	Unexploded Ordnance
VETEM	Very Early-Time Electromagnetic System
VEW	Vapor Extraction Wells
VMP	Vapor Monitoring Points
VOC	Volatile Organic Compounds
VSWR	Voltage Standing Wave Ratio
WGA	Western Governors Association
WMA	Western Manufacturing Area
WRI	Western Research Institute
WSA	Weapons Storage Area
XRF	X-Ray Fluorescence
ZLCF	Zero Order Leaching Correction Factor

Contributors

Mohamed S. Abdel-Rahman, New Jersey Medical School, Newark, NJ
Jane W. Adams, US Army Engineer Waterways Experiment Station, Vicksburg, MS
Carl Alderson, New York City Dept. of Parks and Restoration, Staten Island, NY
Dudley W. Alleman, HRP Associates, Inc., New London, CT
J. Allen, University of Oklahoma, Norman, OK
James N. Baker, Parsons Engineering Science, Inc., Liverpool, NY
Michael S. Battle, EA Engineering, Newburgh, NY
Hollis H. Bennett, US Army Engineer Waterways Experiment Station, Vicksburg, MS
Andrew Bergen, New York City Dept. of Parks and Recreation, Staten Island, NY
Mark Besmer, Rocky Mountain Arsenal, Commerce City, CO
Andrew Bouie, BDM Federal, Inc., Giathersburg, MD
Matthew K. Bowers, Wenck Associates, Inc., Maple Plain, MN
Gordon Brookman, Environmental Risk Limited, Bloomfield, CT
David A. Brown, Parsons Engineering Science, Inc., Liverpool, NY
James Brown, US DOE, Aiken, SC
James L. Brown, OHM Remediation Services Corp., Trenton, NJ
Stephen H. Brown, HRP Associates, Inc. New London, CT
John H. Carson,Jr., OHM Remediation Services, Corp., Findlay, OH
Mark D. Casey, Massachusetts DEP, Woburn, MA
D.B. Chan, Naval Facilities Engineering Service Center, Port Hueneme, CA
Daniel P. Claycomb, Environmental Standards, Inc., Valley Forge, PA
John P. Collins, Naval Facilities Engineering Service Center, Port Hueneme, CA
Matthew A. Cousino, ENSR Consulting and Engineering, Pittsburgh, PA
Mark E. Davidson, Naval Facilities Engineering Command, North Charleston, SC
A. Davis, Geomega, Boulder, CO
Michael J. Dette, US Army Environmental Center, Aberdeen Proving Ground, MD
Maureen A. Dooley, ABB Environmental Services, Inc., Wakefield, MA
Roy Eckart, University of Cincinnati, Cincinnati, OH
Stephen E. Eikenberry, Naval Facilities Engineering Service Center, Port Hueneme, CA
Robert A. Ellgas, Oakland, CA
Robert J. Fares, Environmental Standards, Inc., Valley Forge, PA
Steven Farrick, TRC Environmental Corporation, Windsor, CT
David L. Faust, KAI Technologies, Inc., Portsmouth, NH
John J. Fitzgerald, Massachusetts DEP, Woburn, MA
Paul E. Flathman, OHM Remediation Services Corp., Findlay, OH
Thomas H. Flor, Naval Facilities Engineering Service Center, Port Hueneme, CA
Keri J. Ford, Ogden Environmental & Energy Services Co. Inc., Somerset, NJ
Jason A. Gerrish, Remediation Technologies, Inc., Concord, MA
Jaime N. Goncalves, Environmental Standards, Inc., Valley Forge, PA
Thomas Gorrell, Pace University, New York, NY
M. Gowdy, Argonne National Laboratory, Argonne, IL

Paul J. Gretsky, SECOR, Chantilly, VA
Gary M. Grey, HydroQual, Inc., Mahwah, NJ
Gregory J. Gromicko, Groundwater Technology, Inc., Moon Township, PA
William Guarini, Envirogen, Princeton, NJ
John B. Gustafson, Shell Development Company, Houston, TX
Marilyn K. Guthrie, ARCO Products Company, Los Angeles, CA
Elaine Hammick, Environmental Risk Limited, Bloomfield, CT
Gregory P. Hatchett, Naval Facilities Engineering Command, Norfolk, VA
Stanley O. Hewins, USAF Center for Environmental Excellence, Brooks AFB, TX
K. Hoenke, Chevron Research and Technology, Richmond, CA
Bruce G. Hyde, Office of Development & Planning, New London, CT
Douglas E. Jerger, OHM Remediation Services Corp., Findlay, OH
Lyle A. Johnson, Western Research Institute, Laramie, WY
Paul C. Johnson, Arizona State University, Tempe, AZ
Abdelrazak M. Kadry, New Jersey Medical School, Newark, NJ
Raymond S. Kasevich, KAI Technologies, Inc., Portsmouth, NH
M. Keating, Geomega, Boulder, CO
Mark E. Kelley, Battelle Memorial Institute, Columbus, OH
William B. Kerfoot, K-V Associates, Inc., Falmouth, MA
Robert C. Klingensmith, Groundwater Technology, Inc., Moon Township, PA
M. Kombrink, University of Oklahoma, Norman, OK
Jennifer C. Kotanchik, Environmental Standards, Inc., Valley Forge, PA
Susan M. Lasdin, Department of Environmental Quality, Albany, NY
Kristin G. Lawrence, Environmental Standards, Inc., Valley Forge, PA
Paul R. Lear, OHM Remediation Services Corp., Findlay, OH
Michael Levandowsky, Pace University, New York, NY
Joshua E. Lieberman, ENSR Consulting and Engineering, Acton, MA
Andrew J. Lonergan, ABB Environmental Services, Inc., Tallahassee, FL
Ian D. MacFarlane, EA Engineering, Science, and Technology, Inc.,Sparks, MD
Joy A. Maiello, Envirogen, Princeton, NJ
Michael P. Mailloux, Chevron USA Products Company, Inc., Atlanta, GA
Bridgette A. Manderfeld, Federal Cartridge Company, Arden Hills, MN
David Mark, US DOE, Washington, DC
Martin R. McCleery, Department of the Army, Arden Hills, MN
Mark W. Moeller, Remediation Technologies, Inc., Concord, MA
Jaffer Mohiuddin, US DOE, Washington, DC
George R.Mussoline, Environmental Standards,Inc., Valley Forge,PA
David A. Nickerson, Parsons Engineering Science, Inc., Liverpool, NY
Nikolaos Nikolaidis, University of Connecticut, Storrs, CT
Vern Novstrup, Naval Facilities Engineering Service Center, Port Hueneme, CA
Timothy M. O'Connor, Department of Environmental Management, Providence, RI
Christopher M. O'Neill, Department of Environmental Conservation, Albany, NY
James A. O'Neill,II, Remediation Technologies, Inc., Fort Collins, CO

Michael T. Paczkowski, Geraghty & Miller, Inc., Plainview, NY
Timothy J. Peck, EA Engineering, Science, and Technology, Inc., Sparks, MD
Christopher J. Perry, Battelle Memorial Institute, Columbus, OH
R.P. Philp, University of Oklahoma, Norman, OK
David A. Pillard, ENSR Engineering and Consulting, Fort Collins, CO
Stephen L. Price, KAI Technologies, Inc.,Portsmouth, NH
Terence R. Regan, Geraghty & Miller, Inc., Plainview, NY
James Ridenour, New York State Department of Health, NY
Kelly A. Rinehimer, Environmental Standards, Inc., Valley Forge, PA
Peter Rissell, US Army Environmental Center, Aberdeen Proving Ground, MD
Stephen H. Rosansky, Battelle Memorial Institute, Columbus, OH
Dennis D. Rounds, South Dakota Petroleum Release Compensation Fund, Pierre, SD
Lisa M. Routhier, ABB Environmental Services, Inc., Orange Park, FL
O. Karl Scheible, HydroQual, Inc., Mahwah, NJ
Arnold Schiffman, Schiffman Associates, Ewing, NJ
George Schneider, US DOE, Idaho Falls, ID
Karen M. Scruton, Parsons Engineering Science, Inc., Liverpool, NY
Thomas Seery, Environmental Standards, Inc., Valley Forge, PA
Derek Sharp, Geraghty & Miller, Inc., Plainview, NY
Duane K. Simpson, Quaker State Corporation, Oil City, PA
Gloria A. Skowronski, New Jersey Medical School, Newark, NJ
Curtis C. Stanley, Shell Development Company, Houston, TX
Robert A. Stewart, HRP Associates, Inc., New London, CT
Sabina Strautman, Consolidated Rail Corporation, Philadelphia, PA
Judy A. Strickland, USAF Center for Environmental Excellence, Brooks AFB, TX
Paul Sutton, P.M. Sutton & Associates, Bethel, CT
Carl P. Swanstrom, Argonne National Laboratory, Argonne, IL
Scott C. Swanson, EA Engineering, Science, and Technology, Inc., Sparks, MD
Kenneth G. Symms, Environmental Standards, Inc., Valley Forge, PA
Brian D. Symons, Remediation Technologies, Inc., Fort Collins, CO
William D. Sweet, US Air Force Center for Environmental Excellence, Brooks AFB, TX
Roy Tan, Tetra Tech, Inc., Lafayette, CA
Jeffrey J.Tasca, Ogden Environmental & Energy Services Co., Inc., Somerset, NJ
Darryl L. Terho, Federal Cartridge Company, Arden Hills, MN
Rita M. Turkall, University of Medicine and Dentistry, Newark, NJ
Dennis G. Tuttle, ABB Environmental Services, Inc., Wakefield, MA
Edward P. Van Doren, ABB Environmental Services, Inc., Wakefield, MA
Curtis J. Varner, EA Engineering, Science, and Technology, Newburgh, NY
James F. Villaume, Pennsylvania Power & Light Company, Allentown, PA
Rock J. Vitale, Environmental Standards, Inc., Valley Forge, PA
Denice H. Wardrop, Environmental Standards, Inc., Valley Forge, PA
Jeffrey C. Weatherly, OHM Remediation Services Corp., Glen Allen, VA
L. Andrew White, HRP Associates, Inc., New London, CT

Mark D. White, Battell Memorial Institute, Columbus, OH
Patrick M. Woodhull, OHM Remediation Services Corp., Findlay, OH
James M. Wrona, OHM Remediation Services Corp., Findlay, OH
J. Wyatt, Chevron Chemical Company, San Ramon, CA
S. Laura Yeh, Naval Facilities Engineering Service Center, Port Hueneme, CA
George G. Yu, Battelle Memorial Institute, Columbus, OH
Paul Zielinski, US DOE, Washington, DC
Zhijiang Zhang, TRC Environmental Corporation, Windsor, CT

Index